COMBUSTION PROCESSES

I. Thermodynamics and Physics of Matter. Editor: F. D. Rossini
II. Combustion Processes. Editors: B. Lewis, R. N. Pease, H. S. Taylor
III. Fundamentals of Gas Dynamics. Editor: H. W. Emmons
IV. Laminar Flows and Transition to Turbulence. Editor: C. C. Lin
V. Turbulent Flows and Heat Transfer. Editor: C. C. Lin
VI. General Theory of High Speed Aerodynamics. Editor: W. R. Sears
VII. Aerodynamic Components of Aircraft at High Speeds. Editors: A. F. Donovan, H. R. Lawrence
VIII. High Speed Problems of Aircraft and Experimental Methods. Editors: A. F. Donovan, H. R. Lawrence, F. Goddard, R. R. Gilruth
IX. Physical Measurements in Gas Dynamics and Combustion. Editors: R. W. Ladenburg, B. Lewis, R. N. Pease, H. S. Taylor
X. Aerodynamics of Turbines and Compressors. Editor: W. R. Hawthorne
XI. Design and Performance of Gas Turbine Power Plants. Editors: W. R. Hawthorne, W. T. Olson
XII. Jet Propulsion Engines. Editor: O. E. Lancaster

VOLUME II

HIGH SPEED AERODYNAMICS

AND JET PROPULSION

COMBUSTION PROCESSES

EDITORS

B. LEWIS · R. N. PEASE

H. S. TAYLOR

PRINCETON, NEW JERSEY

PRINCETON UNIVERSITY PRESS

1956

London: GEOFFREY CUMBERLEGE, OXFORD UNIVERSITY PRESS

L. C. CARD 55-8070

PRINTED IN THE UNITED STATES OF AMERICA BY
THE MAPLE PRESS COMPANY, INC., YORK, PENNA.

FOREWORD

On behalf of the Editorial Board, I would like to make an acknowledgement to those branches of our military establishment whose interest and whose financial support were instrumental in the initiation of this publication program. It is noteworthy that this assistance has included all three branches of our Services. The Department of the Air Force through the Air Research and Development Command, the Department of the Army through the Office of the Chief of Ordnance, and the Department of the Navy through the Bureau of Aeronautics, Bureau of Ships, Bureau of Ordnance, and the Office of Naval Research made significant contributions. In particular, the Power Branch of the Office of Naval Research has carried the burden of responsibilities of the contractual administration and processing of all manuscripts from a security standpoint. The administration, operation, and editorial functions of the program have been centered at Princeton University. In addition, the University has contributed financially to the support of the undertaking. It is appropriate that special appreciation be expressed to Princeton University for its important over-all role in this effort.

The Editorial Board is confident that the present series which this support has made possible will have far-reaching beneficial effects on the further development of the aeronautical sciences.

Theodore von Kármán

PREFACE

Rapid advances made during the past decade on problems associated with high speed flight have brought into ever sharper focus the need for a comprehensive and competent treatment of the fundamental aspects of the aerodynamic and propulsion problems of high speed flight, together with a survey of those aspects of the underlying basic sciences cognate to such problems. The need for a treatment of this type has been long felt in research institutions, universities, and private industry and its potential reflected importance in the advanced training of nascent aeronautical scientists has also been an important motivation in this undertaking.

The entire program is the cumulative work of over one hundred scientists and engineers, representing many different branches of engineering and fields of science both in this country and abroad.

The work consists of twelve volumes treating in sequence elements of the properties of gases, liquids, and solids; combustion processes and chemical kinetics; fundamentals of gas dynamics; viscous phenomena; turbulence; heat transfer; theoretical methods in high speed aerodynamics; applications to wings, bodies and complete aircraft; nonsteady aerodynamics; principles of physical measurements; experimental methods in high speed aerodynamics and combustion; aerodynamic problems of turbo machines; the combination of aerodynamic and combustion principles in combustor design; and finally, problems of complete power plants. The intent has been to emphasize the fundamental aspects of jet propulsion and high speed aerodynamics, to develop the theoretical tools for attack on these problems, and to seek to highlight the directions in which research may be potentially most fruitful.

Preliminary discussions, which ultimately led to the foundation of the present program, were held in 1947 and 1948 and, in large measure, by virtue of the enthusiasm, inspiration, and encouragement of Dr. Theodore von Kármán and later the invaluable assistance of Dr. Hugh L. Dryden and Dean Hugh Taylor as members of the Editorial Board, these discussions ultimately saw their fruition in the formal establishment of the Aeronautics Publication Program at Princeton University in the fall of 1949.

The contributing authors and, in particular, the volume editors, have sacrificed generously of their spare time under present-day emergency conditions where continuing demands on their energies have been great. The program is also indebted to the work of Dr. Martin Summerfield who guided the planning work as General Editor from 1949–1952. The cooperation and assistance of the personnel of Princeton University Press and of the staff of this office has been noteworthy. In particular, Mr. H. S.

Bailey, Jr., the Director of the Press, and Mr. R. S. Snedeker, who has supervised the project at the Press and drawn all the figures, have been of great help. Special mention is also due Mrs. H. E. H. Lewis and Mrs. E. W. Wetterau of this office who have handled the bulk of the detailed editorial work for the program from its inception.

Joseph V. Charyk
General Editor

PREFACE TO VOLUME II

This volume is concerned with combustion processes in their various aspects, encompassing chemical kinetics, the kinetics of transport processes, fluid dynamics, and thermodynamics. It deals, therefore, with rate processes in chemical reactions, with the propagation of chemical reaction by the mechanisms of combustion waves and detonation waves, with the effect of turbulence on combustion waves, with processes of simultaneous mixing and combustion of fuels and oxidants, and with chemical equilibria. These subjects are basic for an understanding of the role of combustion in propulsion processes. After a survey of basic principles the presentation continues with oxidation and flame propagation in gaseous systems, and the combustion of liquid and solid fuels and propellants. Final sections of the book are devoted to detonation processes and the principles of energy production by nuclear reaction.

The volume editors express their appreciation of the cooperation received from the authors. The bulk of the first drafts of the several sections were received in 1951, and revised copy late in 1952 and early 1953. Delays unfortunately occurred in securing the necessary permission to publish certain manuscripts. This delayed the publication date and the position of this volume in the publication sequence. The volume editors believe that the difficulties arising from these delays have been minimized as far as is possible, but emphasize these aspects in defining the book as a presentation of current science.

The duties of General Editor of the present volume have largely been discharged by Professor Irvin Glassman of Princeton University. His assistance to us is deeply appreciated. Mr. R. S. Snedeker has continued to render his conspicuous service in the preparation of all the figures and in the supervision at Princeton University Press. The volume will reveal the same high quality of typography and arrangement that have been observed in the earlier volumes. To all, authors and collaborators, we extend our sincere thanks.

B. Lewis
R. N. Pease
H. S. Taylor

CONTENTS

PART 1. THERMODYNAMICS OF COMBUSTION

EDITOR: B. LEWIS

CONTENTS

PART 2. CHEMICAL KINETICS OF COMBUSTION

EDITORS: H. S. TAYLOR AND R. N. PEASE

Chapter 3. Combustion Waves in Closed Vessels

PART 4. COMBUSTION OF LIQUIDS AND SOLIDS

EDITOR: R. N. PEASE

PART ONE

THERMODYNAMICS OF COMBUSTION

SECTION A

HIGH TEMPERATURE EQUILIBRIUM

JAMES M. CARTER
DAVID ALTMAN

A,1. Problems in Combustion. There are two principal problems in determining the nature of high temperature combustion. The first is the determination of the nature and amount of the combustion products which may exist and also the thermodynamic properties of the mixture at one or more sets of combustion conditions.

These combustion conditions are always idealized for purposes of calculation. Thus in a rocket or jet motor, steady state conditions are assumed; combustion during starting or stopping is considered separately by approximate methods from a knowledge of steady state conditions.

In order for thermochemistry and thermodynamics to apply to combustion, either certain other simplified conditions must be assumed, or much more detailed information is required than is usually available. Thus, in effect, calculations on combustion in rocket and jet motors are made on the assumption that the combustion process occurs in a small mass element which is injected into a motor near the center of an infinite stream of identical elements and which then travels through the motor with negligible velocity and at constant pressure. The combustion chamber of the motor is assumed long enough to permit equilibrium to be attained, and the walls are assumed far enough removed to exert no effect since the element in question is surrounded by identical elements.

Similarly in explosions such as may occur in guns, or in intermittent devices, calculations are actually made for a mass element of given density, surrounded by identical elements of the same density, all in a container large enough to have no effect on the element being considered, except to keep its density constant.

It is obvious that in most cases these conditions are far from fulfilled, and that effects due to combustion speed, mixing, loss of heat to walls, and other phenomena must be considered. However, the simplified conditions assumed make calculation much simpler and do provide an upper limit for the efficiency of combustion processes. When these conditions are met, the reaction equilibria among the substances present, as discussed in I,A and C, do determine the combustion products and their thermodynamic properties.

Further simplification is obtained by assuming that combustion products are perfect gases, which is nearly true for most processes treated in rockets or jet motors. Deviations which may occur at high pressures, or in ultrafast processes, are considered in I,A and C and in other texts on thermodynamics [*1,2,3*].

The second problem is the determination of the heat which may be released by the combustion, the maximum combustion temperature which can be reached, or the work which may be extracted, subject to the possible compositions as determined above and to the conditions under which the combustion occurs. The usual conditions assumed are those for adiabatic combustion at constant pressure or at constant volume. Other conditions (those involving heat transfer or external work) are considerably more complicated. Discussion of these conditions has been given by Lewis and von Elbe [*4*] and by Hirschfelder [*5*].

The combustion process may be represented by

$$\text{combustibles } (p_1, V_1, T_1) = \text{combustion products } (p_2, V_2, T_2) + Q \text{ kilocalories} \quad (1\text{-}1)$$

with Q determined by the expression

$$E_{V_2,T_2} - E_{V_1,T_1} = -Q - \int_{p_1,V_1}^{p_2,V_2} p dV$$

In the above, the composition of the combustibles and the initial state are known; the composition of the products and their state, together with Q, are to be determined.

This problem may conveniently be divided into two parts: the determination of equilibrium composition and thermodynamic properties; and the determination of flame temperatures and heat released or work done for various final states of the combustion products.

A,2. Determination of Equilibrium Composition and Thermodynamic Properties. The composition and thermodynamic properties of the equilibrium products of combustion are uniquely determined by the atomic composition, the temperature, and the pressure (or volume of the system), as is shown in I,A and C. In particular, at a specified temperature and pressure they do not depend on the heats of formation of the combusting materials or on their heat of reaction. (However, these quantities will determine the temperature range over which the composition and thermodynamic properties are of interest, since they determine the initial energy of the system.)

For example, the equilibrium composition and thermodynamic properties of the combustion products of the combustible mixtures represented by $6C + 3H_2 + 4O_2$, $3C_2H_2 + 4O_2$, and $C_6H_6 + 4O_2$ are identical at any chosen temperature and pressure. Flame temperatures, heat released, or work available, differ widely because of different initial energies of the three systems. These energies are determined by the basic conservation-of-energy equations.

EQUILIBRIUM COMPOSITION. Equilibrium conditions in the combustion products can be obtained from thermodynamic data and from the total amounts of each atomic species involved. For many reactions, values of the equilibrium constants are available either from experimental measurements or as calculated values from spectroscopic data. Other equilibrium constants may be calculated from free energy data. The theory of equilibrium conditions is treated in I,A and sources of data are also included in I,A.

There is always an element of arbitrariness involved in choosing the equilibria regarded as significant. For instance, a choice frequently made is to neglect all equilibria involving species present in less than 0.01 or 0.1 per cent of the total. But this is not all-conclusive; obviously an equilibrium involving 0.01 per cent of the total moles present for a reaction with an energy change of 150 kilocalories per mole is more significant, in determining the state of the system, than an equilibrium involving 0.1 per cent of the moles for a reaction with an energy change of 10 kilocalories per mole.

A rough calculation which involves the heat given off in a combustion process resulting in products stable at ambient temperature, and the specific heats of these products, serves to set an upper limit on the combustion temperature. At this temperature, the equilibria which may exist among products involving all the atomic species present are examined. For example, the combustion of gasoline (taken as octane C_8H_{18}) with oxygen may be used. A typical reaction for rocket motors may be approximated by

$$C_8H_{18} + 6O_2 \rightarrow 8CO + 4H_2O + 5H_2 \tag{2-1}$$

(In combustion reactions involving carbon, hydrogen, and oxygen, a useful rule in estimating reaction products is, (1) to oxidize carbon to CO, (2) to use remaining oxygen to oxidize hydrogen to water, and (3) if any oxygen remains, to oxidize CO to CO_2.)

If this reaction occurred at 300°K, with the products remaining at 300°K, the heat released would be about 425 kilocalories. Cooling the products to 0°K would release an additional 35 kilocalories. Thus the total heat released can be expressed as

$$C_8H_{18}\ (l) + 6O_2\ (g, 300°K) = 8CO\ (g) + 4H_2O\ (g) + 5H_2\ (g, 0°K) + 460\ \text{kcal}$$

Enthalpy tables for the products establish a temperature near 3000°K, if the entire heat release is employed in raising the temperature.

Examination of the equilibrium constants for the various reactions which might be involved shows that the water-gas reaction is such that CO_2 is present, that steam will dissociate to OH, H_2, and O_2; H_2 and O_2 will dissociate to H and O respectively, but that because of the large excess of H_2, O_2 and especially O will be present in very small amounts. Therefore an initial estimate of the species present would include CO, CO_2, H_2, H_2O, OH, H.

In the calculation of equilibrium compositions, the starting point is a series of equations; these equations will equal in number, all of the unknowns to be determined.

These equations are of two kinds: the material balance equations,

$$n = \sum n_i \tag{2-2}$$

$$\begin{aligned}(\mathrm{M}) &= \sum (n_{i\mathrm{M}}) \cdot n_i \\ (\mathrm{N}) &= \sum (n_{i\mathrm{N}}) \cdot n_i \\ &\text{etc.,}\end{aligned} \tag{2-3}$$

and the equilibrium equations,

$$K_i f = \frac{f_{\mathrm{C}}^c \cdot f_{\mathrm{D}}^d \cdots}{f_{\mathrm{A}}^a \cdot f_{\mathrm{B}}^b \cdots} \tag{2-4}$$

etc.

The first equation states that the total number of moles is equal to the sum of the number of moles of the molecular components. The second set of equations expresses the fact that the total number of gram atoms of each atomic constituent, denoted by (M), (N), etc., is distributed among the molecular components n_i, with $n_{i\mathrm{M}}$ atoms of M in the component i. The third set of equations are the usual equilibrium expressions for the gas reactions

$$a\mathrm{A} + b\mathrm{B} + \cdots \leftrightarrows c\mathrm{C} + d\mathrm{D} + \cdots$$

In the equilibrium equations, f is the fugacity. Conversion to the usual forms for pressure, mole fraction, etc. is made by use of appropriate multiplying factors Q_p, Q_x, etc. involving fugacity as a function of pressure or density. These relations are given in I,A and C. The following equilibrium expressions are those most frequently involved in combustion involving carbonaceous fuels. The numbering used follows that given by Hirschfelder [*5*], and all subsequent references to equilibrium constants or reactions follow this numbering. It should be emphasized that while the expressions are given in terms of pressure, this is accurate only insofar as the perfect gas laws apply. When appreciable deviations occur, pressures must be replaced by fugacities.

$$K_1 = \frac{p_{CO} \cdot p_{H_2O}}{p_{CO_2} \cdot p_{H_2}} \qquad K_8 = \frac{p_N}{p_{N_2}^{\frac{1}{2}}}$$

$$K_5 = \frac{p_{CO}}{p_{O_2}^{\frac{1}{2}}} \qquad K_9 = \frac{p_H}{p_{H_2}^{\frac{1}{2}}}$$

$$K_6 = \frac{p_{O_2} \cdot p_{H_2}^2}{p_{H_2O}^2} \qquad K_{10} = \frac{p_{OH} \cdot p_{H_2}^{\frac{1}{2}}}{p_{H_2O}}$$

$$K_7 = \frac{p_O \cdot p_{H_2}}{p_{H_2O}}$$

In the simplest cases, when there are only one or two atomic species and two or three molecular species, these equations may be solved simply by algebraic means. In general, however, the number and complexity of the equations make this impossible.

For the reaction given in Eq. 2-1, the equilibrium equations would be set up as follows:

$$n = n_{CO} + n_{CO_2} + n_{H_2} + n_{H_2O} + n_H + n_{OH} \tag{2-2a}$$

$$\begin{aligned} (C) &= 8 = n_{CO} + n_{CO_2} \\ (H) &= 18 = 2n_{H_2} + 2n_{H_2O} + n_H + n_{OH} \qquad (2\text{-}3a) \\ (O) &= 12 = n_{CO} + 2n_{CO_2} + n_{H_2O} + n_{OH} \end{aligned}$$

and if the perfect gas laws hold, so that

$$pV = nRT, \qquad \frac{f}{p} = 1, \quad \text{and} \quad p_i = p\frac{n_i}{n}$$

$$\begin{aligned} K_{1p} &= \frac{n_{CO} \cdot n_{H_2O}}{n_{CO_2} \cdot n_{H_2}} \\ K_{9p} &= \left(\frac{p}{n}\right)^{\frac{1}{2}} \frac{n_H}{n_{H_2}^{\frac{1}{2}}} \qquad (2\text{-}4a) \\ K_{10p} &= \left(\frac{p}{n}\right)^{\frac{1}{2}} \frac{n_{OH} \cdot n_{H_2}^{\frac{1}{2}}}{n_{H_2O}} \end{aligned}$$

The above set of seven equations involving seven unknowns does not lend itself to easy algebraic solution. In this case, which is relatively simple, the algebraic solution can be obtained, but ordinarily, with more molecular species and more equations involved, such a method is virtually impossible.

Recourse must then be had to one of a number of approximate methods. These may be classified as (1) trial and error methods, (2) iterative methods, (3) graphical methods and use of published tables, and (4) punched-card or machine methods.

1. Trial and error methods. The straightforward method of solving for an equilibrium composition may be accomplished by the insertion into Eq. 2-4 of trial values consistent with the material balance Eq. 2-3 until all equations are satisfied. Simple rules, such as the one given previously in the example employing octane and oxygen, may be employed to assign initial trial values. This direct method is the least efficacious and is recommended only in those cases where the number of components is small and a reasonably good guess can be made of the composition.

The trial and error method, however, can effectively be applied to more complicated systems if the number of working equations are first reduced by algebraic substitutions. A general method for handling C, H, O, and N systems of up to ten components is demonstrated in the following treatment.

Let the number of moles of the components be represented by the following symbols:

$$a = n_{H_2} \qquad f = n_{N_2}$$
$$b = n_{H_2O} \qquad g = n_{NO}$$
$$c = n_{CO} \qquad h = n_{OH}$$
$$d = n_{CO_2} \qquad i = n_{H}$$
$$e = n_{O_2} \qquad j = n_{O}$$

These ten unknowns are related by means of Eq. 2-3 and 2-4 as follows:

$$\frac{bc}{ad} = K_1 \tag{2-5}$$

$$\frac{ga}{f^{\frac{1}{2}}b} = K_3\left(\frac{n}{p}\right)^{\frac{1}{2}} \tag{2-6}$$

$$\frac{a^2e}{b^2} = K_6\left(\frac{n}{p}\right) \tag{2-7}$$

$$\frac{aj}{b} = K_7\left(\frac{n}{p}\right) \tag{2-8}$$

$$\frac{i}{a^{\frac{1}{2}}} = K_9\left(\frac{n}{p}\right)^{\frac{1}{2}} \tag{2-9}$$

$$\frac{ha^{\frac{1}{2}}}{b} = K_{10}\left(\frac{n}{p}\right)^{\frac{1}{2}} \tag{2-10}$$

$$c + d = (\mathrm{C}) \tag{2-11}$$

$$b + c + 2d + 2e + g + h + j = (\mathrm{O}) \tag{2-12}$$

$$2a + 2b + h + i = (\mathrm{H}) \tag{2-13}$$

$$2f + g = (\mathrm{N}) \tag{2-14}$$

where (C), (O), (H), and (N) represent the total number of gram atoms of the elements, n is the total number of moles in the equilibrium mixture, and p is the total pressure. The subscripts on the K's denote the specific equilibria as given in Eq. 2-4. Although n is accurately given by the equation

$$n = a + b + c + d + e + f + g + h + i + j$$

the concentrations of the components are not sensitive to small variations in n. This fact permits the choice of an approximate n value in Eq. 2-6 to 2-10 with only very little loss in accuracy.

The above ten equations may now be simplified to yield the following three working equations:

$$b = \frac{(\mathrm{H}) - 2a - K_9(n/p)^{\frac{1}{2}}a^{\frac{1}{2}}}{2 + \dfrac{K_{10}(n/p)^{\frac{1}{2}}}{a^{\frac{1}{2}}}} \tag{2-15}$$

$$g = (\mathrm{O}) - (\mathrm{C})\,\frac{K_1 a + 2b}{K_1 a + b} - b\left[\frac{K_{10}(n/p)^{\frac{1}{2}}}{a^{\frac{1}{2}}} + \frac{K_7(n/p)^{\frac{1}{2}}}{a}\right] - 2K_6\left(\frac{n}{p}\right)\frac{b^2}{a^2} \tag{2-16}$$

$$\frac{2g^2a^2}{K_3^2(n/p)b^2} + g - (\mathrm{N}) = 0 \tag{2-17}$$

Equations 2-15, 2-16, and 2-17 are the three basic equations upon which the trial and error method is applied. A likely value of $a = n_{H_2}$ is chosen. Values of b and g are then obtained directly from Eq. 2-15 and 2-16. The values of a, b, and g so obtained are now substituted into Eq. 2-17. Nonzero solutions of Eq. 2-17 require new trial values of a until that equation is satisfied. Once Eq. 2-17 is satisfied, all other components are obtained simply from the individual Eq. 2-5 to 2-14.

The introduction of the component n_N, which occurs at very high temperatures, or components containing other elements such as sulfur and fluorine can easily be made by this scheme.

2. Iterative methods. Briefly, this method consists in determining (by inspection or otherwise) which species are present in largest amounts, and of solving the reduced number of equations involving these species only. The minor species are then determined from the major ones by use of the equilibrium equations, and the concentration of the major species is corrected for the presence of minor ones by using the mass balance equations. New relations among the major species are then determined, using the corrected values, and a second set of values for the minor species is calculated. This process is continued until no further change is produced by repeating the process. Obviously, the method can be continued to give as accurate values as desired. Usually the process is terminated at some predetermined limit of accuracy, such as 0.1 or 0.01 mole per cent.

In the example above, the first step would be taken by assuming that

H and OH are not present. Eq. 2-2, 2-3, and the water-gas equilibrium of Eq. 2-4 are then solved algebraically for the major components. This involves only a simple quadratic, since p and n do not enter into the equilibrium equation. H and OH are then calculated from the second and third equations in Eq. 2-4. Then second approximations to (H) and (O) are calculated from Eq. 2-3 by the relation $(H)' = (H) - n_H - n_{OH}$ and $(O)' = (O) - n_{OH}$. These second approximate values are used as in the first step to calculate new values for the major components, and these in turn are used again for H and OH.

As a final step, equilibria which may have been neglected in setting up the equations are examined. (In the present case the dissociation of steam to hydrogen and oxygen would be most probable.) If the amount of oxygen does prove negligible, based on the equilibrium values found for steam and hydrogen above, the procedure is justified. However, if the amount of oxygen is comparable to the amounts of other minor constituents, Eq. 2-2 to 2-4 must be rewritten to include this component, and the process must be gone through again.

The process above is comparatively easy when the combustion mixture is relatively far from stoichiometric proportions. Fortunately this is true for most rocket propellants, which are usually underoxidized, and for fuel-air mixtures in turbojets, etc., which are overoxidized. This means that several molecular species are present in large amounts, which condition will be changed little by the presence of minor species. The iteration process therefore converges very rapidly.

As stoichiometric proportions are approached in a combustion mixture, however, the iteration process becomes progressively less efficient, particularly at the higher temperatures involved. Thus in hydrogen-oxygen mixtures, if either excess hydrogen or excess oxygen is present, it can safely be assumed that H_2O and either H_2 or O_2 are the two major components, and the concentrations of OH, H, O, and either O_2 or H_2 can be fairly easily determined.

In stoichiometric proportions the concentrations of many or all species other than H_2O are roughly equal, and a change in one introduces changes in the others which are proportionately of the same magnitude. Under these conditions, the iterative process converges very slowly. For exact stoichiometric proportions as in steam, carbon dioxide, and some other important substances, the labor involved in calculation is not too serious, since only one final composition is involved. Most combustion mixtures are not in exact stoichiometric proportions and also involve more than one combustion reaction.

Brinkley and others [*6,7,8*] have devised a number of systematized methods of iteration. Some of the more complex methods will even converge fairly rapidly [*9*, p. 187] for highly dissociated systems near stoichiometric proportions.

One method (Sachsel, et al. [*10*, p. 620]), which is applicable when electrical computers are available, is to transform all of the equilibrium equations, Eq. 2-4, to logarithmic form. These then become simple linear equations and amenable to machine calculations.

3. Graphical methods and tables for use within definite composition ranges for several important systems have been developed by Winternitz, Huff, Calvert, and Kassner [*10*, p. 620; *11*; *12*]. These, together with (4) machine and punched-card methods of computation, are more fully described in Sec. C of this volume.

From the above, it is apparent that at the combustion temperatures usually encountered, determination of equilibrium compositions is a complicated and tedious process. Even if the only information required is the composition at the flame temperature, two or three such composition calculations must be made, since an initial estimate of the flame temperature (for which the composition is required) will usually be in error by an appreciable amount. If compositions are required not only at flame temperatures over a range of pressures or densities, but at lower temperatures and pressures in order to follow expansion processes, the number of calculations which must be made is greatly increased.

Interpolation of equilibrium compositions. Frequently it is desired to know the equilibrium compositions at temperatures or pressures other than those for which calculations have been made.[1] For different temperatures, the most accurate method is the use of the equilibrium constants for the temperature in question. These can be interpolated quite accurately, since in general a plot of ln K vs. T is nearly linear. When compositions are determined at other pressures, a new set of equilibrium calculations must be made since the pressure enters into Eq. 2-4 and these equations must, therefore, be altered for the new pressure.

There is, however, a simple graphical method of obtaining equilibrium compositions at temperatures or pressures intermediate to those for which two or more calculations have been made. It can also be used for a limited amount of extrapolation. When the amounts of the various components are plotted on the logarithmic scale of semilogarithmic paper against either temperature (at one pressure) or pressure (at one temperature), it is found that nearly straight lines are obtained. Therefore the procedure is to plot the amounts of the minimum number of components (starting with the least abundant) which will completely determine the composition, on semilogarithmic paper, and to read off the amounts at intermediate temperatures or pressures. By plotting the components least abundant, the maximum accuracy is obtained, since these components undergo the greatest percentage changes with temperature or pressure. The amounts

[1] There is a definite need for sets of tables or charts for the more important systems, at reasonably close intervals of composition, and all based on the *same data* for equilibrium constants, equations of state, etc. Such a basis might be the Bureau of Standards Tables of Chemical Thermodynamic Quantities [*13*].

of the other (major) components are then determined from Eq. 2-3, and the total number of moles of gas from Eq. 2-2. Even in cases where this method does not give the accuracy desired, it can be used to eliminate all but the final steps in an exact calculation by the iterative method.

Thermodynamic Properties. As in the case of the compositions, the thermodynamic properties of gases are point functions of atomic composition, temperature, and pressure or volume. However, for energy quantities there is the requirement of a base point. The choice of this point may be entirely arbitrary, but once the choice is made all thermodynamic quantities must be referred to it. Several such bases are in common use. Two bases frequently employed are the system stable at room temperature (25°C, 300°K, 70°F) and at a pressure of one atmosphere, and the system hypothetically stable at 0°K and one atmosphere, with combustion products ordinarily gaseous in the perfect gas state at this temperature. Since more and more tables of enthalpies, entropies, etc., based on 0°K are becoming available, this latter base has much to recommend it. (See Sec. C and I,A.)

For calculations involving shifts in chemical equilibrium during expansion, two other bases are more logical. The first is a base with the elements at ambient temperature, and the second, with the elements at 0°K. The latter has the advantage of corresponding more closely with theoretical considerations for absolute enthalpy, entropy, etc. It has the practical disadvantage at present that many heats of formation at 0°K are not known, and that in some cases enthalpy differences between ambient conditions and 0°K are not known. However, this situation is being remedied rapidly [*13*]. In the following discussion, the use of this base is assumed, as it results in the simpler formulas. If some other base is used, a constant is added to all thermodynamic quantities. Since behavior on combustion and expansion always involves differences, this constant cancels out in any computation.

The determination of enthalpy and internal energy for the equilibrium compositions at various temperatures and pressures (or densities) may be divided into two parts. The first is that due to the heat or energy of reaction at the base temperature, from the standard composition at base temperatures to the equilibrium composition at the temperature T. The second is the change in enthalpy or internal energy in heating this mixture from the base temperature to the temperature T, plus the effects in going from the perfect gas state at one atmosphere to the pressure or density under consideration. These relations are expressed in the equations below.

$$H = \sum n_i \cdot (\Delta H_f^0)_i + \sum n_i(H_T^0 - H_0^0)_i + n \int_{p_0}^{p} \left(\frac{\partial H}{\partial p}\right)_T dp \quad \text{(2-18)}$$

$$E = \sum n_i \cdot (\Delta E_f^0)_i + \sum n_i(E_T^0 - E_0^0)_i + n \int_{V_0}^{V} \left(\frac{\partial E}{\partial V}\right)_T dV \quad \text{(2-19)}$$

Here, n_i is the number of moles of component i present at temperature T and ΔH_f^0 or ΔE_f^0 is the heat or energy of formation of the component i at 0°K, from the elements at 0°K. $H_T^0 - H_0^0$ or $E_T^0 - E_0^0$ is the integral of the specific heat of component i from 0°K to T.

If the elements at ambient temperature (298°K) are taken as a base point, Eq. 2-18 becomes

$$H = \sum n_i \cdot [\Delta H^0_{f(298)}]_i + \sum n(H_T^0 - H_{298}^0)_i + \cdots \tag{2-18a}$$

and similarly for Eq. 2-19. If a base of stable composition at ambient temperature is used, the first term on the right of Eq. 2-18a is replaced by $\sum \Delta n_i \Delta H^0_{f(298)}$, where Δn_i is the change in component i in going from the composition stable at 298°K to the equilibrium composition at T. (It should be noted as an advantage of the base at 0°K that $\Delta H_f^0 = \Delta E_f^0$ and $H_0^0 = E_0^0$.) The last terms in the equations correct for the changes in enthalpy or internal energy with pressure or volume. For perfect gases these terms are zero. In the general case, as listed by Bridgman [*14*] for these and other thermodynamic formulas,

$$\left(\frac{\partial H}{\partial p}\right)_T = V - T\left(\frac{\partial V}{\partial T}\right)_p \tag{2-20}$$

$$\left(\frac{\partial E}{\partial V}\right)_T = T\left(\frac{\partial p}{\partial T}\right)_V - p \tag{2-21}$$

The numerical values of the expressions above are determined from the equation of state for the gas mixture.

Entropy values for combustible mixtures may be determined in a number of ways. For many substances entropy tables are available for the standard state at a pressure of one atmosphere [*13*]. The entropy of the mixture is then given by the expression

$$S^0 = \sum n_i S_i^0 - nR \sum X_i \ln X_i, \qquad X_i = \frac{n_i}{n} \tag{2-22}$$

For other materials, tables of the free energy function are available [*13*]. By use of the relation $F = H - TS$, the entropy may be determined as

$$S^0 = \frac{1}{T}\sum n_i(H_T^0 \cdot H_0^0)_i - \sum n_i\left(\frac{F^0 - H_0^0}{T}\right)_i - nR\sum X_i \ln X_i \tag{2-23}$$

In both the above equations, the term $nR\sum X_i \ln X_i$ is the entropy of mixing all the components, each originally present at one atmosphere, to give the mixture at one atmosphere.

To obtain the entropy at the desired pressure or volume, there must be added to the above values the change of entropy with pressure or volume. This is given by the equation

$$S - S^0 = -\int_{p_0}^{p} \left(\frac{\partial V}{\partial T}\right)_p dp = \int_{V_0}^{V} \left(\frac{\partial p}{\partial T}\right)_V dV \tag{2-24}$$

For most of the applications met with in jet and rocket propulsion, gases may be assumed to obey the perfect gas laws, so that the above equations reduce to

$$S - S^0 = -nR \ln \frac{p}{p_0} = nR \ln \frac{V}{V_0} \tag{2-25}$$

Specific heats for the combustion products are obtained from the expressions

$$C_p = \left(\frac{\partial H}{\partial T}\right)_p, \qquad C_V = \left(\frac{\partial E}{\partial T}\right)_V \tag{2-26}$$

In the above, it is necessary to keep in mind that the changes in enthalpy or internal energy include not only the amounts due to change in temperature for each species, but also those due to the change in equilibrium composition with temperature, according to the equations

$$\left(\frac{\partial H}{\partial T}\right)_p = \sum \left(\frac{\partial n_i}{\partial T}\right)_p (\Delta H_f^0)_i + \sum \left(\frac{\partial n_i}{\partial T}\right)_p (H_T^0 - H_0^0)_i + \sum n_i \left(\frac{\partial (H_T^0 - H_0^0)_i}{\partial T}\right)_p - T \int_{p_0}^{p} \left(\frac{\partial^2 V}{\partial T^2}\right)_p dp \tag{2-27}$$

$$\left(\frac{\partial E}{\partial T}\right)_V = \sum \left(\frac{\partial n_i}{\partial T}\right)_V (\Delta E_f^0)_i + \sum \left(\frac{\partial n_i}{\partial T}\right)_V (E_T^0 - E_0^0)_i + \sum n_i \left(\frac{\partial (E_T^0 - E_0^0)_i}{\partial T}\right)_V + T \int_{V_0}^{V} \left(\frac{\partial^2 p}{\partial T^2}\right)_V dV \tag{2-28}$$

The first two terms in the above give the heat involved in changes in composition, the third terms give the perfect gas value of the specific heat at fixed composition, and the last terms give the effect due to gas imperfections.

In view of the complex nature of these equations, a convenient method of obtaining average values for the specific heats over a range of temperatures is by differences from tables of enthalpy or internal energy, if these values are determined at fairly close intervals, both in temperature and in pressure (or volume).

For some purposes, $\gamma = C_p/C_V$ is required. This can be obtained from enthalpy and energy tables as above. For more accurate determinations, use is made of the thermodynamic relationship

$$C_p - C_V = -T \left(\frac{\partial V}{\partial T}\right)_p^2 \left(\frac{\partial p}{\partial V}\right)_T \tag{2-29}$$

This can be cast into the form

$$\frac{1}{\gamma} = \frac{C_V}{C_p} = 1 - \frac{T}{C_p}\left(\frac{\partial p}{\partial T}\right)_V \left(\frac{\partial V}{\partial T}\right)_p \tag{2-30}$$

where C_p is the heat capacity at constant pressure, T is the absolute temperature, V is the volume, and p is the pressure. All quantities must be expressed in consistent units. The partial derivatives are obtained from the compositions and the equation of state for the particular mixture under consideration. In particular for nonreacting mixtures obeying the perfect gas law the above expression reduces to

$$\frac{C_V}{C_p} = 1 - \frac{R}{C_p} \tag{2-31}$$

Another thermodynamic property which is useful in dealing with propellants is the velocity of sound. This is given by the expression

$$a^2 = \left(\frac{\partial p}{\partial \rho}\right)_S \tag{2-32}$$

It is shown in thermodynamics texts [*1*] that this reduces to

$$a^2 = \frac{C_p}{C_V}\left(\frac{\partial p}{\partial \rho}\right)_T \tag{2-33}$$

Here again all of the required quantities may be obtained from those previously calculated and from the equation of state of the gas mixture.

The extent of the calculations made for composition, equation of state, and thermodynamic properties will vary with the information desired. This may range from a fairly rough estimate of performance characteristics for one particular mixture at one pressure or density, to detailed calculations for a range of mixtures, at a number of pressures or densities, expanded over a series of expansion conditions. The latter in effect requires the preparation of a number of Mollier or equivalent charts, one for each mixture investigated. If anything more than a single performance value is required, it is surprising to one who has not worked with the calculations before to see how little additional effort is required to cover a wide range of combustion and expansion conditions.

A,3. Determination of Heat Release and Flame Temperature. As has been stressed above, the equilibrium molecular composition and thermodynamic properties of any particular atomic composition do not depend on the initial state of the mixture. In other words, the particular nature, heats of formation, etc., of a combustible mixture have no effect on the ultimate composition or thermodynamic properties of the mixture after combustion, when these properties are taken as functions of the temperature and pressure or density.

However, it is just these factors which determine the heat released in a combustion and the flame temperature which may be reached. To evaluate these effects it is necessary to return to Eq. 1-1 and to separate this into two parts. Eq. 1-1 reads:

$$\text{combustibles } (p_1, T_1, V_1) = \text{combustion products } (p_2, T_2, V_2) + Q \text{ kilocalories}$$

This may be separated into two steps:

$$\text{combustibles } (p_1, T_1, V_1) \rightarrow \text{products, standard } (p_0, T_0, V_0) + Q_0 \text{ kilocalories} \quad (3\text{-}1)$$

and

$$\text{products, standard } (p_0, T_0, V_0) \rightarrow \text{products } (p_2, T_2, V_2) + Q_2 \text{ kilocalories} \quad (3\text{-}2)$$

In the preceding articles, the thermodynamic quantities involved in the second of these processes have been discussed. Q_2 is simply the negative of the enthalpy or internal energy at p_2, T_2, V_2. All that is required to determine performance, flame temperatures, and other required data is to evaluate the thermodynamic quantities involved in the first process.

The heat or energy values involved in the first process above will of course depend on the thermochemical properties of the combustibles, on their initial state, and on the base taken for p_0, T_0, V_0. It is here that the advantage of the base taken at ambient conditions is most apparent. In such a case Q_0 is merely the heat or internal energy involved in the combustion at constant temperature. If the base at 0°K is taken it is necessary to add to the usual heat or energy of combustion the heat involved in cooling the combustion products from ambient temperature to 0°K.

Several specific processes may be mentioned in relation to the above. Maximum flame temperature is determined by the relation $Q_0 + Q_2 = 0$. At either constant pressure or volume, if all of the heat or energy released in going from the combustibles to the base state is absorbed in going from the base state to the combustion conditions, no heat or energy is lost in the process, and the maximum temperature is attained. Several examples are given in Sec. L and M.

Furthermore, the heat or energy converted to kinetic energy or work is directly obtainable from the above equations. If the final state of the combustion products has been determined after expansion from the combustion conditions as shown in Sec. B, the extent of energy conversion is determined by

$$W = Q_0 + Q_2$$

where Q_2 denotes the heat or energy involved in going from standard conditions to final condition. It should be noticed that this equation applies not only to expansion in which equilibrium composition of the combustion products is maintained during expansion, but also when the composition is assumed frozen, either at expansion conditions or at some intermediate temperature, since Q_2 is determined by composition as well as by temperature, according to Eq. 2-18 and 2-19.

A,4. Gas Imperfection. The effect of gas imperfection in modifying equilibrium compositions and thermodynamic properties has been indicated in some of the equations above. For actual calculations involving gas imperfection, it is necessary to employ an equation of state for the gases involved. By the use of such an equation, the relations between pressure, volume, and temperature of a gas are determined over the range of interest, and corrections can be applied to the equilibrium constants and to the thermodynamic properties.

Equations of state. There are various equations of state which have been proposed to correct for deviations from the perfect gas laws. Among these are the van der Waals equation [*15*], Berthelot's equation [*16*], the Beattie-Bridgeman equation [*17*], the Hirschfelder equation [*5*], particularly adapted to high temperature and pressure, and numerous more or less empirical equations. Graphical representations have also been employed. These are treated in I,C.

A general representation of an equation of state for a pure gas may be given by

$$pV = RT\left(1 + \frac{b}{V} + \frac{c}{V^2} + \frac{d}{V^3} + \cdots\right) \tag{4-1}$$

where b, c, d, etc. may be functions of temperature. The values of b, c, d, etc., are determined by the nature of the gas. They are usually determined from experimental measurements, although in some cases they may be calculated from theoretical considerations.

For a single pure gas, Eq. 4-1 shows a definite relationship involving temperature, volume, and pressure. Furthermore, the fugacity, on which corrections to the equilibrium constants depend, can be calculated for a pure gas.

The exact definition of fugacity and consideration of its application is given in I,A. It is sufficient to note here that fugacity is determined by the expression

$$\ln\frac{f_2}{f_1} = \frac{1}{RT}\int_{p_1}^{p_2} V dp \quad (T = \text{const}) \tag{4-2}$$

It can be seen that for a perfect gas $f_2/f_1 = p_2/p_1$, but that in general for imperfect gases this is not true.

$$\lim_{p_1 \to 0} p_1 = f_1$$

Real difficulty arises with gas mixtures, both in determining p, V, T relationships and to a greater extent in determining fugacity. Experimental data for the p, V, T behavior of gas mixtures is not abundant, and in any case the entire range of compositions cannot be covered. It is not immediately apparent how the constants in Eq. 4-1 for pure gases should be combined to give an equation for a gas mixture. Further, since fugacities are usually required for individual components of a mixture, there is difficulty in choosing the quantities and limits in the integral in Eq. 4-2.

Early attempts to compute the p, V, T relationships of gas mixtures were expressed in Dalton's and in Amagat's laws. Dalton's law states that the pressure exerted by a gas mixture is the sum of the individual pressures each gas would exert if it occupied the total volume alone

$$p = n_1 \cdot p_1(T, V) + n_2 \cdot p_2(T, V) + \cdots \qquad (4\text{-}3)$$

Amagat's law states that the volume occupied by a gas mixture is the sum of the individual volumes which would be occupied by each gas alone at the same total pressure:

$$V = n_1 \cdot V_1(T, p) + n_2 \cdot V_2(T, p) + \cdots \qquad (4\text{-}4)$$

For perfect gases, both laws are true. Amagat's law is in general somewhat more accurate, but neither is a good approximation at even moderate pressures. Closer approximations require methods for combining the constants in the equation of state Eq. 4-1 for pure gases to represent the behavior for gas mixtures. Consideration of these questions is given in I,C.

Evaluation of constants. Constants in the equations of state may be evaluated in various ways. A series of p, V, T measurements over a wide range furnishes data for evaluation of the constants; they can also be evaluated from critical data and from data on intermolecular forces.

A fairly extensive treatment of the calculation of the constants of the equation of state for combustion gases is given by Hirschfelder, et al [*5*]. The equation of state for gases at high temperatures adopted for this purpose is a somewhat simplified van der Waals equation:

$$\frac{pV}{RT} = 1 + \frac{b}{V} + \frac{0.625b^2}{V^2} + \frac{0.2869b^3}{V^3} + \frac{0.1928b^4}{V^4} \qquad (4\text{-}5)$$

which at high densities merges into the limiting form

$$\frac{pV}{RT} = \left[1 - 0.6962\left(\frac{b}{V}\right)^{\frac{1}{3}}\right]^{-1} \qquad (4\text{-}6)$$

In these equations attractive forces between molecules are disregarded (terms involving a in the van der Waals equation). The numerical coefficients are derived theoretically for the overlapping of rigid spheres. In the range of combustion temperatures (1500 to 5000°K) values of b are nearly independent of temperature and are assumed constant. Values of b for individual gases are evaluated from theoretical considerations on intermolecular forces. For gas mixtures, values of b may be calculated approximately at moderate pressures according to the relation

$$b = \sum X_i b_i$$

where X_i is the mole fraction of component i. Fowler [*18*] gives considerations for calculation of constants involving interaction between different molecular species. With such simple equations as those above, the p, V, T relations for gas mixtures are obtained rather simply. Also, the changes in enthalpy, internal energy, and entropy with pressure or density at constant temperature are readily obtained by the usual thermodynamic formulas:

$$\left(\frac{\partial H}{\partial p}\right)_T = V - T\left(\frac{\partial V}{\partial T}\right)_p \tag{2-20}$$

$$\left(\frac{\partial E}{\partial V}\right)_T = T\left(\frac{\partial p}{\partial T}\right)_V - p \tag{2-21}$$

$$\left(\frac{\partial S}{\partial p}\right)_T = -\left(\frac{\partial V}{\partial T}\right)_p \tag{4-7}$$

$$\left(\frac{\partial S}{\partial V}\right)_T = \left(\frac{\partial p}{\partial T}\right)_V \tag{4-8}$$

Changes in equilibrium constants with pressure or density are, however, more complicated since the fugacities of the individual components are involved. To obtain reasonably simple results, it is convenient to assume that a gas mixture is a perfect solution [*19*] or that

$$\frac{pV}{RT} - 1 = \sum_i \left[\frac{pV}{RT} - 1\right]_i X_i \tag{4-9}$$

By use of this assumption and the equation of state, activity coefficients and fugacities for the components in a mixture can be calculated. A detailed description of the method of calculation and tabular values of activity coefficients for the common gases are given in reference [*5*].

One difficulty in the application of the corrections above to the calculation of thermodynamic properties, or even to the calculation of p, V, T relationships, is that all of these equations apply only to gases of constant composition. That is, all of the partial derivations above should contain another symbol to show constant composition such as $(\partial H/\partial p)_{T,n}$ and all other quantities also should be understood to refer to a gas of fixed composition. When the deviations from the perfect gas laws are appreciable, this requires a large number of calculations. Thus in obtaining the enthalpy of an equilibrium mixture at one temperature for a series of pressures, it is not enough to integrate one equation of the form of Eq. 2-7 at a series of pressures. For each pressure a different expression must be used, since the equilibrium composition and hence the equation of state will be different for each pressure.

Effect of deviations from perfect gas behavior. It is fortunate that in most of the applications required in jet and rocket propulsion, gas mixtures may be considered to obey the perfect gas laws.

Deviations from perfect gas behavior affect both combustion conditions and thermodynamic properties. Combustion conditions are affected largely by fugacity changes, which may be appreciable when pressures are high. Thus the water-gas equilibrium is properly written

$$K_{1f} = \frac{f_{CO} \cdot f_{H_2O}}{f_{CO_2} \cdot f_{H_2}} \tag{4-10}$$

where f is fugacity and is equal to pressure only at low pressures.

This may be altered to the usual form involving partial pressures by introducing activity coefficients a, where $a = f/p$ (see Brinkley and Kirkwood, [*10*, p. 586]). Then the water-gas equilibrium expression becomes

$$K_{1p} \cdot G_1 = \frac{p_{CO} \cdot p_{H_2O}}{p_{CO_2} \cdot p_{H_2}}, \qquad G_1 = \frac{a_{CO_2} \cdot a_{H_2}}{a_{CO} \cdot a_{H_2O}} \tag{4-11}$$

If all activities are assumed unity then $G_1 = 1$, and the usual equilibrium expression is obtained. At high gas densities, deviations from the gas laws alter activity coefficients markedly, and the values of G may depart widely from unity. Values of G for a number of equilibria are given in [*5*] for a number of conditions. A few are given in the table below for illustration of the magnitudes involved.

Table A,4a. Corrections to equilibrium constants.

$\frac{V_0(273°K,\ 1\ atm)}{V(T,\ p)}$	G_1^*	$G_1^2G_6$	G_6	G_9
50	1.090	1.090	0.916	1.010
100	1.223	1.241	0 830	1.021
200	1.736	1.991	0.661	1.045
500	17.78	74.7	0.237	1.147
1000	11,670 (3000°K)	1.3×10^6 (3000°K)	0.00956	1.476

* G_1 is somewhat affected by temperature at the highest densities.

Water-gas equilibrium $$K_1G_1 = \frac{p_{CO} \cdot p_{H_2O}}{p_{CO_2} \cdot p_{H_2}}$$

Dissociation of carbon dioxide $$K_1^2K_6G_1^2G_6 = \frac{p_{CO}^2 \cdot p_{O_2}}{p_{CO_2}^2}$$

Dissociation of steam $$K_6G_6 = \frac{p_{H_2}^2 \cdot p_{O_2}}{p_{H_2O}^2}$$

Dissociation of hydrogen $$K_9G_9 = \frac{p_H}{p_{H_2}^{\frac{1}{2}}}$$

The figures above show that increase of density has a pronounced effect in throwing the water-gas equilibrium to the side of carbon monoxide and steam. If the gases were perfect this equilibrium would remain unaffected by density or pressure. The dissociation of carbon dioxide is increased manyfold over what would occur for perfect gases, so much so that at the highest densities, dissociation *increases* with pressure.

On the other hand, dissociation of water is suppressed below what would be expected, and dissociation of hydrogen is little affected.

The thermodynamic properties are much less affected by the deviations from the gas laws, but the effects are appreciable. The principal quantities to be considered are the internal energy, the enthalpy, and the entropy. It has been found both experimentally and from the equations of state, that internal energy varies nearly linearly with the *density*, and that $\Delta(pV)$, and hence enthalpy, and the entropy difference from perfect gas behavior also vary almost linearly with density up to moderate density values. This near-linear behavior does not hold if these deviations are plotted against pressure. An indication of the magnitude of the deviations involved is given below.

Table A,4b. Deviations from gas laws; $T = 3000°K$, $V = V(1\ \text{atm})/100$.

Gas	pV/RT	p, atm	$\Delta(pV)$, cal/mole	ΔE, cal/mole	ΔH, cal/mole	$\delta\Delta S$*, cal/mole°C
H_2	1.060	106	+35.8	+5.1	+40.9	−0.1003
N_2	1.155	116	+92.4	+3.5	+95.9	−0.2804
CO_2	1.311	131	+186.0	−18.2	+168.0	−0.6265
H_2O	1.043	104	+25.6	−20.0	+5.6	−0.1522

* $\delta\Delta S = \Delta S - R \ln (V/V_0)$

The relative effect of these deviations can be judged from the fact that C_V at this temperature is some 10 to 12 calories per mole per degree; flame temperatures are therefore altered only a few degrees by direct effects of deviations. Temperature alterations may be much greater because of composition changes induced by fugacity. Similarly, the ideal entropy change for a volume change by a ratio of 100 is about 9 entropy units; even the largest deviation is only 7 per cent of this.

A,5. Failure to Maintain Equilibrium in Combustion.

Effect of incomplete combustion. All of the above discussion has been predicated on reaching an equilibrium condition in the combustion process. This is not always so, and except in a few cases combustion is never complete. There are a number of factors which may limit the completion of combustion. In rocket motors, they may be placed into two general classes, the first involving factors limiting complete mixing, and the second involving the relation between reaction rates and the time that materials remain in the combustion chamber. Among the factors affecting mixing may be listed injector configuration, spray patterns, and the configuration of the combustion volume. Factors involving the time include L^*, and the absolute size of the combustion volume. It is apparent that if a rocket or jet motor is scaled up geometrically, the time from injection to reaching the expansion nozzle is increased. The effect of these factors is given in XI,B and H, and in XII,G and H.

In closed bombs, and to a first approximation in guns [*4*], the completeness of combustion may be affected by the presence of massive cold walls, by lack of turbulence, and by the extremely short time interval available for combustion.

The evaluation of the effects of incomplete combustion is quite difficult. In the first place, either extensive experimental data on the composition of the combustion products is required, or more or less ad hoc assumptions must be made. For these assumptions, several choices are possible: part of the fuel may be assumed to be completely unreacted and only heated, or it may be more or less decomposed without combustion; part of the oxidizing material may be similarly considered; or the combustion may be assumed to proceed to more or less arbitrary intermediate products. Consideration of possible reaction processes is given in Sec. D and E.

There is little or no difficulty involved in the determination of thermodynamic properties, heat release, or flame temperature when a composition arising from combustion has been chosen. The principal difficulty in this case is in deciding which composition most nearly approaches the actual course of combustion.

Equilibrium lag. Somewhat allied to incomplete combustion, but more amenable to theoretical treatment, is the question of lag in equilibrium. This may arise from several factors, among which may be mentioned the persistence of metastable states, failure of internal energy states to participate in the combustion process, and detonation phenomena. Examples of the first will be recalled by anyone who has seen a rocket motor operate with a nitric acid oxidizer below full efficiency. Equilibrium conditions in combustion require the presence of only minute amounts of the oxides of nitrogen, but the yellow exhaust is strong evidence for the presence of large amounts of NO or NO_2. Similarly, the odors arising from the use of organic fuels in rocket motors indicate the presence of intermediate products, since the equilibrium combustion products CO and CO_2 are odorless.

There is less evidence for the failure of internal energy states to participate in the combustion equilibrium, as shown in I,H. Some experimental evidence that rotational and vibrational states may not always be in equilibrium has been obtained by means of extremely minute orifices or venturi tubes used in obtaining gas samples. However, for most combustion processes in jet and rocket motors, the scale is many times larger than for sampling devices, so that the times involved are longer by several orders of magnitude. In addition, water vapor is present in most combustion products and has a pronounced catalytic effect in promoting vibrational and rotational equilibrium. Recent experimental work has shown that the relaxation time in pure CO_2 for vibrational energy is some 10^{-4} seconds or longer, while the presence of small amounts of water vapor reduced this to 10^{-8} seconds [*20*].

Detonation phenomena. In combustion processes having to do with propulsion, detonations are unwanted and indeed may lead to failure of equipment. The one exception to this is in the use of detonations for the production of high speed (ca. 20,000 ft/sec) metallic jets, as in the well-known bazooka charge.

Ordinarily in propulsion equipment, detonations occur only by accident. In jet motors and liquid fuel rocket motors, they may arise from ignition failure which permits the combustion chamber to be filled with a detonable mixture. In solid fuel rocket motors, breakup or fracture of the solid propellant will lead to rapid burning or deflagration which may under certain conditions transform into detonation.

Damage from detonations is due to two causes, (1) the extreme rapidity of combustion leading to high static pressures because of the large amount of combustion products present, and (2) the shock waves which are always present.

Combustion equilibrium as discussed above plays no part in detonation processes. The stability of the detonation is dependent upon a sharp discontinuity which travels through the mass at extreme speeds. Furthermore, in the region of the shock wave combustion products are moving at high velocities, of the order of half the velocity of the wave itself (see Brinkley and Kirkwood [*10*, p. 586]).

As a consequence of the great velocity of the detonation wave and the nearly as great "stream velocity" of the combustion products, there is little opportunity for any but translational degrees of freedom to be involved. Thus it can be shown (see Ubbelohde, [*10,* p. 566]) that with known detonation velocities and thicknesses of the wave region, a molecule can make only a few collisions (3 to 6) before the wave has passed. Since the combustion products then act as if their specific heats were those of a monatomic gas, the temperatures reached in a detonation wave are very high, even when allowance is made for the energy used in producing the "stream velocity" of the products. It is also highly probable that a large proportion of the combustion products are in excited states.

Behind the detonation wave, conditions are more calculable on an equilibrium basis, since the rotational and vibrational degrees of freedom have time to "catch up" with the rest of the process, and molecules lose their energy of excitation.

The conditions of detonation are based upon the Rankine-Hugoniot [*21*] equations for the continuity of mass, energy, and momentum across the detonation wave. These equations give the "dynamic adiabatic" of Hugoniot, but do not predict any one stable detonation velocity. This velocity is determined by application of the Chapman-Jouguet condition [*22,23*] which states that the stable velocity is that given by drawing a tangent to the Hugoniot curve in the p, V plane from the initial condition point p_0, V_0. There is recent work by Brinkley and others to indicate that the Chapman-Jouguet hypothesis concerning detonation velocity may be proved by making some reasonable assumptions (see Brinkley and Kirkwood, [*10,* p. 586]; also [*24*]). For further discussion of conditions of detonation, see III,D.

One of the principal difficulties in obtaining good numerical results from the above lies in lack of knowledge of the equation of state for combustion products. With initially gaseous materials no serious difficulties are encountered, but with solid explosives having densities equal to or greater than that of water, detonation pressures may be of the order of 10^5 atm, and the nature of the equation of state is uncertain. Unfortunately, as has been shown by Jones [*10,* p. 590], accurate measurements of detonation phenomena, used in conjunction with the Hugoniot relations, do not conversely serve to determine an equation of state. This is because temperatures are not included in the equations (only p and V). The means used to calculate temperature also influence the equation of state which is derived.

The phenomena of detonation are treated in greater detail in Sec. N, in III,D and III,G, and in the references given above, some of which contain fairly extensive bibliographies.

A,6. Cited References and Bibliography.

Cited References

1. Glasstone, S. *Textbook of Physical Chemistry.* Van Nostrand, 1946.
2. Epstein, P. S. *Textbook of Thermodynamics.* Wiley, 1937.

3. Steiner, L. *Introduction to Chemical Thermodynamics.* McGraw-Hill, 1941.
4. Lewis, B., and von Elbe, G. *Combustion, Flames and Explosion of Gases.* Macmillan, 1938.
5. Hirschfelder, J. O., et al. *Natl. Defense Research Comm. Rept. A-116*, 1942.
6. Brinkley, S. R., Jr., *J. Chem. Phys. 14*, 563 (1946); *15*, 107 (1947).
7. Krieger, F. J., and White, W. B. *J. Chem. Phys. 16*, 358 (1948).
8. Damköhler, G., and Edse, R. *Z. Elektrochem. 49*, 178 (1943).
9. Scarborough, J. B. *Numerical Mathematical Analysis.* Johns Hopkins Univ. Press, 1930.
10. *Third Symposium on Combustion, Flame, and Explosion Phenomena.* Williams & Wilkins, 1949.
11. Huff, V. N., and Calvert, C. S. *NACA Tech. Note 1653*, 1948.
12. Kassner, R. *Tech. Rept. GS-USAF (No. F-TR-2185 ND)*, 1948.
13. Rossini, F. D., Wagman, D. D., Evans, W. H., Levine, S., and Jaffe, I. Selected values of chemical thermodynamic properties. *Natl. Bur. Standards Circ. 500*, 1952.
14. Bridgman, P. W. *Phys. Rev. 3*, 273 (1914).
15. van der Waals, J. D. *Die Kontinuität des gasförmigen und flüssigen Zustandes, 1. Bd.*, 2. Folge. Leipzig, 1898.
16. Berthelot, P. *Compt. rend. 126*, 954 (1898).
17. Beattie, J. A., and Bridgeman, O. C. *Proc. Am. Acad. Arts Sci. 63*, 229 (1928).
18. Fowler, R. H. *Statistical Mechanics.* Cambridge Univ. Press, 1936.
19. Lewis, G. N., and Randall, M. *Thermodynamics.* McGraw-Hill, 1923.
20. Griffith, W. *J. Appl. Phys. 21*, 1319 (1950).
21. Rankine, W. J. M. *Trans. Roy. Soc. London A160*, 277 (1880).
22. Chapman, D. L. *Phil. Mag. 47 (5)*, 90 (1899).
23. Jouguet, E. *Compt. rend. 132*, 573 (1901).
24. Jones, H. *Proc. Roy. Soc. London A189*, 415 (1947).

Bibliography

Brason, F. W. *U.S. Bur. Mines Tech. Paper 632*, 1941.

Hottel, H. C., et al. *Trans. Am. Soc. Mech. Engrs. 70*, 667 (1948).

Huff, V. N., and Morrell, V. E. *NACA Tech. Note 2113*, 1950.

SECTION B

EXPANSION PROCESSES

DAVID ALTMAN
JAMES M. CARTER

B,1. Classification of Flow Processes. The statement of the conservation of energy for flow processes in gases where useful work other than expansion is neglected can be written in the following form:

$$dE = dq - pdV + dW_f \tag{1-1}$$

where dq is the increment of heat absorbed by the gas and dW_f the increment of frictional work done on the gas. For purposes of generality, another term may be included representing work other than expansion or friction. Since such work is not considered here, it is excluded in the following treatment in order to simplify subsequent equations. An alternate expression involving the enthalpy or heat content H can be derived through the relation

$$H = E + pV \tag{1-2}$$

or in differential form

$$dH = dE + pdV + Vdp \tag{1-3}$$

Combining Eq. 1-3 with 1-1 results in the alternate formulation

$$dH = dq + Vdp + dW_f \tag{1-4}$$

It is shown in the following treatment that either of these simple formulations of the energy equation can be utilized directly to interpret the effects of heat transfer, flow with friction, and heat release due to transitions among internal energy states and due to chemical reactions. The element of volume, however, is so chosen that uniform properties are assumed at any instant of flow within the differential element.[1] If the cross-sectional area of this element can be made as large as that of the duct (implying uniform properties throughout any cross section of flow), then simple relations will generally result relating the various thermodynamic quantities. Unless otherwise specified, it is assumed that this condition is realized for the integrated relations obtained in this section.

[1] For the flow of rarefied gases, it may not be possible to choose such an element because of the large mean free path. The discussion of such flows is contained in I,I and IV,E.

Corrections for nonuniform flow can be made with the relations developed from the transport properties of gases as developed in I,D.

For the complete representation of the thermodynamic parameters during a flow process, the conservation of energy equation must be combined with the conservation of momentum. For one-dimensional flow, the Bernoulli equation per unit mass of gas is

$$-\frac{dp}{\rho} = udu + dw_f \tag{1-5}$$

where w_f is the frictional work per unit mass. Noting that $\rho = m/V$, Eq. 1-5 may also be expressed as

$$-Vdp = mudu + dW_f \tag{1-5a}$$

for a gas of total mass m. Combination of Eq. 1-4 and 1-5a gives the relation between the enthalpy and the velocity,

$$dH = dq - mudu \tag{1-6}$$

Eq. 1-1 through 1-6 may be employed to treat the thermodynamics of any type of flow process. Although hydrodynamically it is convenient to classify fluid flow according to the state of motion of the fluid element, i.e. steady, irrotational, one-dimensional, etc., the present classification is made in a thermodynamic sense. Accordingly, the flow process is categorized as follows:

ADIABATIC. Any process occurring in a thermally insulated system so that $q = 0$ is an adiabatic process.

Isentropic. Any adiabatic and reversible process with $\Delta S = 0$, where S is the entropy, is an isentropic process. This category applies to systems with free flow or with work extracted, for which the internal energy states and chemical composition are either completely frozen or in equilibrium so that the composition is determined by the parameters of state.

Nonisentropic. Any adiabatic irreversible process with $\Delta S \neq 0$ is a nonisentropic process. The irreversibility may be due to many causes, some of which are nonequilibrium chemical reaction or internal energy transitions during flow, occurrence of friction, free expansion (isoenergetic, $\Delta E = 0$), and throttled flow (Joule-Thomson porous plug experiment which is isenthalpic, $\Delta H = 0$).

NONADIABATIC. Any process occurring in which heat is transferred to or from the surroundings so that $q \neq 0$ is a nonadiabatic process.

Heat transfer. Because this process must occur with a finite temperature gradient, it is necessarily irreversible.

Isothermal. This process represents the ideal case of heat transfer occurring so slowly that the temperature gradients are negligible and the temperature remains essentially constant. Although these requirements are difficult to realize in dynamic systems, the condition can be approached for low flow velocities. Such flow is usually reversible.

B,2. Thermodynamic Relations for Flow Processes.

ADIABATIC FLOW. Adiabatic flow is the most usual type of gas flow in expansion processes occurring in practice because the duration of flow is generally so short that heat flow to or from the surroundings can be considered negligible. If the process is reversible, the flow is isentropic and exact expressions can be obtained which relate the temperature, pressure, and velocity of the gas. If friction or combustion occurs during flow, the process is nonisentropic, and relations between T and p can be obtained only if the irreversible quantities can be related, either theoretically or empirically, to the temperature, pressure, or velocity.

Isentropic flow.

1. Free flow of nonreacting gases. For the case of the reversible free flow where no external work is done, Eq. 1-1 can be employed directly, setting $dq = dW_f = 0$. Therefore,

$$dE = -pdV \tag{2-1}$$

If the gases are perfect and if the heat capacity is independent of temperature, a simple integration is possible as follows (for n moles of gas):

$$C_V dT = -\frac{nRT}{V} dV \tag{2-2}$$

$$\frac{dT}{T} = -\frac{nR}{C_V}\frac{dV}{V} \tag{2-3}$$

and finally,

$$TV^{nR/C_V} = TV^{\gamma-1} = \text{const} \tag{2-4}$$

where $\gamma = C_p/C_V$ and $C_p - C_V = nR$. Alternative expressions to Eq. 2-4 can be obtained by means of the gas law $pV = nRT$. They are

$$pV^{\gamma} = \text{const} \tag{2-5}$$

and

$$Tp^{-nR/C_p} = Tp^{(1-\gamma)/\gamma} = \text{const} \tag{2-6}$$

The kinetic energy acquired by the gas is obtained by direct integration of Eq. 1-6 noting again that $dq = dW_f = 0$.

$$H_2 - H_1 = -\tfrac{1}{2}m(u_2^2 - u_1^2) \tag{2-7}$$

or

$$H + \tfrac{1}{2}mu^2 = \text{const} \tag{2-8}$$

for a gas which has a total mass m. Whereas Eq. 2-2 to 2-6 are based on perfect gases and constant specific heat, Eq. 2-7 and 2-8 are obtained directly from the conservation of energy and Bernoulli's equation, and hence are true for any real gas.

2. Isentropic flow with work extracted. If the gas does expansion work during flow and if its own kinetic energy is small relative to the work

done, then by means of Eq. 2-1 the change in internal energy is

$$E_2 - E_1 = -\int_{V_1}^{V_2} pdV \tag{2-9}$$

Noting the relation between p and V in Eq. 2-5, the change in E for perfect nonreacting gases is found to be

$$E_2 - E_1 = -\frac{p_2V_2^\gamma}{1-\gamma}\left[V_2^{1-\gamma} - V_1^{1-\gamma}\right] \tag{2-10}$$

where subscripts 1 and 2 refer to initial and final states respectively. Noting that $p_2V_2^\gamma = p_1V_1^\gamma = nRT_1V_1^{\gamma-1}$ and that $C_V = nR/(\gamma - 1)$, Eq. 2-10 can be shown to reduce to

$$\begin{aligned} E_2 - E_1 &= \frac{nRT_1}{1-\gamma}\left[1 - \left(\frac{V_2}{V_1}\right)^{1-\gamma}\right] \\ &= C_VT_1\left[\left(\frac{V_2}{V_1}\right)^{1-\gamma} - 1\right] \end{aligned} \tag{2-10a}$$

For the case where the kinetic energy of the gases cannot be neglected (as in gun barrels), a more detailed treatment is necessary which takes into account the pressure and velocity distribution of the gas in the duct, [*1*]. For such cases, Eq. 2-10a is usually taken as a first approximation. In evaluating various powder grains in a given gun where V_2/V_1 is fixed, the quantity nRT_1, known as the impetus, is found to be a very useful parameter.

3. **Effect of equilibrium chemical reaction and internal energy state transitions.** For calculations with gases for which the specific heat varies markedly with temperature (resulting from internal energy state changes), or in systems with shifting chemical equilibria, Eq. 2-4, 2-5, and 2-6 are approximate, subject to the choice of a proper average C_p or γ for the process. Accurate calculations in these cases are made by use of the statement for isentropic flow:

$$\Delta S = 0 \tag{2-11}$$

with

$$S = \sum n_i(S^0_{298})_i + \sum n_i \int_{298}^{T} C_{p_i}d\,(\ln T) - nR\ln p - \sum n_iR\ln\frac{n_i}{n} \tag{2-12}$$

For systems involving equilibrium chemical reaction during flow, the composition of the system, given by the n_i values, is determined at a trial temperature for a given pressure. The entropy S, at the temperature, may now be calculated from Eq. 2-12 and compared with the entropy in the initial state. Interpolation between several trial temperatures is then made to yield that value of S satisfying Eq. 2-11.

For shifting chemical equilibria, the enthalpy H is defined so that it contains the heat of chemical reaction resulting from composition changes. This definition is (for perfect gases)

$$H = \sum n_i[\Delta H_{f(298)}]_i + \sum n_i \int_{298}^{T} C_{p_i} dT \tag{2-13}$$

where $[\Delta H_{f(298)}]_i$ is the heat of formation per mole of species i from the elements at the base temperature 298°K. For constant composition, $(d/dT)\sum n_i[\Delta H_{f(298)}]_i = 0$ and so the enthalpy of each species is simply taken relative to its own value at the base temperature.[2] Fairly complete tabulations of the standard entropy

$$S = S^0_{298} + \int_{298}^{T} C_p d\,(\ln T)$$

the relative enthalpy

$$H = \int_{298}^{T} C_p dT$$

and the standard heat of formation $\Delta H_{f(298)}$ are given by the National Bureau of Standards [*2*].

In systems where the chemical composition is frozen but where internal energy state transitions occur, causing the specific heat to vary with the temperature, it is found that the Eq. 2-4 to 2-8 can be applied directly with good accuracy if an appropriate average $\overline{C}_p$ is employed. Several methods for choosing $\overline{C}_p$ have been considered [*3,4*], a simple and fairly accurate one being $\overline{C}_p = (H_1 - H_2)/(T_1 - T_2)$ where T_2 is a tentative final temperature. This value of $\overline{C}_p$ is very close to that calculated at the temperature $\frac{1}{2}(T_1 + T_2)$. These approximations are good because the enthalpy change is not highly sensitive to small variations in C_p or γ.

If analytic functions for the specific heat are available, direct integration of the energy equation is possible. Thus if the heat capacity of the gas can be expressed as

$$C_p = a + bT + cT^2 \tag{2-14}$$

the T, p relation is found to be

$$\frac{T_2}{T_1} = \left(\frac{p_2}{p_1}\right)^{\frac{nR}{a}} e^{-\left[\frac{b}{a}(T_1 - T_2) + \frac{c}{2a}(T_1^2 - T_2^2)\right]} \tag{2-15}$$

and the enthalpy change between T_2 and T_1 is

$$\Delta H_{T_2}^{T_1} = a(T_1 - T_2) + \frac{b}{2}(T_1^2 - T_2^2) + \frac{c}{3}(T_1^3 - T_2^3) \tag{2-16}$$

Nonisentropic flow: effects of friction and combustion. The most common causes for nonisentropic flow in adiabatic systems are the occurrence of frictional effects and combustion. Although the viscosity of the gas can generally be neglected, the frictional effects at the wall may become appreciable for very small ducts.

[2] The choice of the base temperature at 298°K, 0°K, or any other temperature is entirely arbitrary and would yield identical results.

1. Friction. It will be assumed that (1) the velocity is uniform across the section of the duct and (2) the viscosity is concentrated at the wall with the force determined empirically. Reference to Fig. B,2a shows that the pressure force acting on an elementary volume of length dx is

$$-Adp = A\rho udu + fldx \tag{2-17}$$

where f is the frictional force per unit area and l the periphery. Dividing through by $A\rho$ results in

$$\frac{-dp}{\rho} = udu + \frac{fldx}{A\rho} \tag{2-18}$$

Eq. 2-18 is essentially the same as Eq. 1-5 where the term dw_f is found to be equal to $fldx/A\rho$ which represents the frictional heat developed per

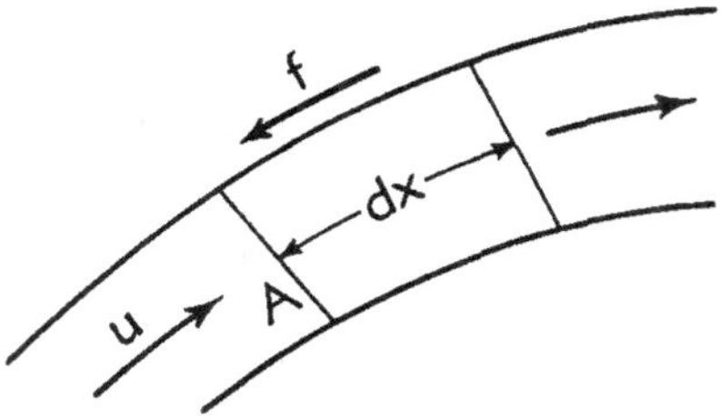

Fig. B,2a.

unit mass of gas as a result of the frictional work. The velocity can be determined directly from Eq. 1-6 where $dq = 0$ for adiabatic flow. The result is

$$dH + md\left(\frac{u^2}{2}\right) = 0 \tag{2-19}$$

Eq. 2-19 is significant in that it shows the velocity to be given directly by the enthalpy drop as in frictionless flow. The effect of friction, therefore, must be only to decrease the enthalpy drop between two pressure levels. Eq. 2-19 cannot yet be employed to determine the flow velocity until the T, p relation has been found which permits the enthalpy drop between two pressure levels to be calculated. This relation can be obtained from Eq. 1-4 setting $dq = 0$ and $dW_f = Vfldx/A$. The result is

$$dH = Vdp + \frac{Vfldx}{A} \tag{2-20}$$

Exact integration of this equation cannot be made unless the dependence of f on p and x is known. However, since the frictional work is completely manifested by heat, let $Vfldx/A = dq_i$. For a perfect gas with constant specific heat, Eq. 2-20 can now be rewritten as

$$C_p\frac{dT}{T} = nR\frac{dp}{p} + \frac{dq_i}{T} \tag{2-21}$$

in which dq_i/T can be identified with dS_i, the entropy gained through frictional heating. Integration from T_1 and p_1 to T and p yields

$$\frac{T}{T_1} = \left(\frac{p}{p_1}\right)^{\frac{nR}{C_p}} e^{\frac{\Delta S_i}{C_p}} \tag{2-22}$$

with $T > T_0$ for all frictional processes. Fig. B,2b and B,2c show the p, T and H, S diagrams for isentropic flow and flow with friction. The

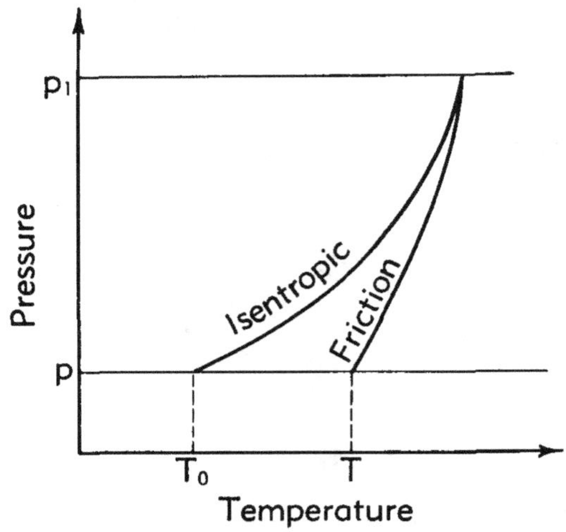

Fig. B,2b. *pT* diagram for isentropic flow and flow with friction.

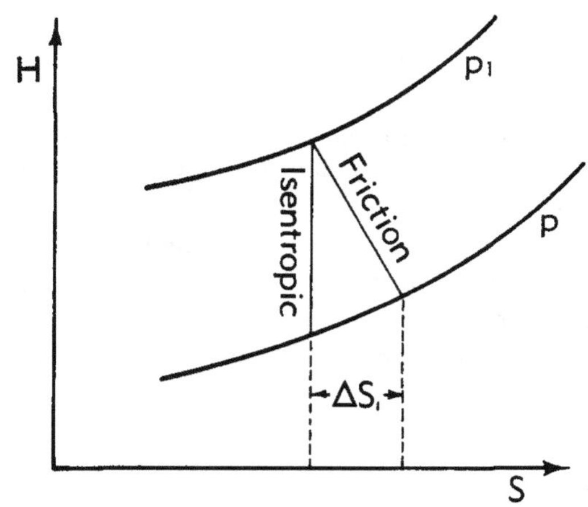

Fig. B,2c. *HS* diagram for isentropic flow and flow with friction.

irreversible entropy gained, ΔS_i, can be evaluated from Eq. 2-23 in terms of T and T_0 to yield

$$\Delta S_i = C_p \ln\left(\frac{T}{T_0}\right) \tag{2-23}$$

The adiabatic efficiency η, defined as the ratio of kinetic energy with friction to that without friction, is given by the expression

$$\eta = \frac{T_1 - T}{T_1 - T_0} \tag{2-24}$$

The expansion law may be obtained from Eq. 2-24 by multiplying both sides of the equation by $(T_1 - T_0)/T_1$ and identifying T_0/T_1 as $(p/p_1)^{nR/C_p}$. The result is[3]

$$\frac{T}{T_1} = \eta\left(\frac{p}{p_1}\right)^{\frac{nR}{C_p}} + (1 - \eta) \tag{2-25}$$

This law is sometimes represented with a polytropic expansion constant [5,6] which is an approximate form of Eq. 2-25. Since $nR/C_p < 1$, Eq. 2-25 may be expanded to yield

$$\begin{aligned}\frac{T}{T_1} &= 1 - \eta + \eta + \frac{\eta nR}{C_p}\ln\left(\frac{p_1}{p}\right) + \cdots \\ &\cong 1 + \frac{\eta nR}{C_p}\ln\frac{p}{p_1}\end{aligned} \tag{2-26}$$

Again, since $\eta nR/C_p < 1$, it will be observed that Eq. 2-26 gives the first two terms of the expansion for $(p/p_1)^{\eta nR/C_p}$ and so there finally results

$$\frac{T}{T_1} \cong \left(\frac{p}{p_1}\right)^{\frac{\eta nR}{C_p}} \tag{2-27}$$

which may also be expressed in the equivalent form (by use of the perfect gas law):

$$pV^k \cong \text{const} \tag{2-28}$$

where $k = \gamma/[\gamma - \eta(\gamma - 1)]$ is the polytropic constant.

These polytropic expansion laws have also been derived [5,6] by means of the equation $dq_i = C_p(dT - dT_0)$ and Eq. 2-21. However, it should be noted that this relation for dq_i is only approximate since the difference in the expansion work terms $\int Vdp$ for the viscous and inviscid gas between the fixed pressure levels has been neglected.

2. Combustion and nonequilibrium reaction during flow. The case of combustion during adiabatic flow, unlike that of equilibrium chemical reassociation, is nonisentropic because the chemical composition is not a unique function of the equation of state of the gas. The energy equation governing such flow is given by Eq. 1-4 with $dq = dW_f = 0$ for adiabatic frictionless flow.[4] The heat content H is given by Eq. 2-13 presented

[3] Strictly speaking η may not be constant during flow, but for approximate calculations the assumption of constancy is fairly good [5,6].

[4] Some authors prefer to treat combustion during flow as nonadiabatic, identifying the chemical heat with q [7]. This treatment will yield equivalent results if the reference state for H is defined as is done for a nonreactive gas.

earlier:

$$H = \sum n_i[\Delta H_{f(298)}]_i + \sum n_i \int_{298}^{T} C_{p_i} dT \tag{2-13}$$

where $[\Delta H_{f(298)}]_i$ is the heat of formation of the ith species at 298°K, the reference state being the elements at 298°K. For perfect gases, H is a function only of T and n_i, and taking the differential of H gives

$$\begin{aligned} dH &= \sum [\Delta H_{f(298)}]_i dn_i + \sum dn_i \int_{298}^{T} C_{p_i} dT + \sum n_i C_{p_i} dT \\ &= \sum [\Delta H_f]_i dn_i + \sum n_i C_{p_i} dT \end{aligned} \tag{2-29}$$

where $[\Delta H_f]_i$ is the heat of formation at the temperature of the gas. The energy equation may now be written as

$$\sum [\Delta H_f]_i \frac{dn_i}{dT} dT + \sum n_i C_{p_i} dT = \frac{nRT}{p} dp \tag{2-30}$$

where $n = \sum n_i$. Dividing through by nT and denoting $C_p = \sum n_i C_{p_i}$ as the average specific heat of gas results in

$$\sum \frac{[\Delta H_f]_i}{n} \frac{dn_i}{dT} \frac{dT}{T} + \frac{C_p}{n} \frac{dT}{T} = R \frac{dp}{p} \tag{2-31}$$

This equation is not integrable since n_i cannot be expressed as a unique function of the state of the gas because of its dependence on time. For a fixed geometry, however, the parameters T and p can be related to the velocity and hence to the time. If the reaction rate constants of the various combustion reactions are known, then expressing dn_i/dT as $(dn_i/dt)(dt/dT)$ will permit solution of Eq. 2-31 because these derivatives can be expressed analytically. This method of treatment has been applied to the dissociation of NO during flow through a nozzle [*8*].

Approximate solutions to Eq. 2-31 can be obtained if one knows the total chemical heat released and assumes the heat to be generated uniformly. The quantity $\sum([\Delta H_f]_i/n)(dn_i/dT)$ is replaced by $\Delta Q/\bar{n}(T_1 - T_2')$ where ΔQ is the total chemical heat of reaction, $\bar{n}$ is the average number of moles of gas considered, and T_2' is a tentative value of the final temperature. It will be observed that the quantity $\Delta Q/(T_1 - T_2')$ has the dimensions of heat capacity and can be denoted by C_p'. Eq. 2-31 now integrates to

$$T = \text{const} \times p^{\frac{\bar{n}R}{C_p + C_p'}} = \text{const} \times p^{\frac{\bar{\gamma}-1}{\bar{\gamma}}} \tag{2-32}$$

with an apparent

$$\bar{\gamma} = \frac{C_p + C_p'}{C_V + C_V'}$$

The low value of the apparent $\bar{\gamma}$ will lead to higher stream temperatures as shown in Fig. B,2d.

3. Throttled flow: isenthalpic. Steady state adiabatic flow from a high pressure region p_1 through a valve or porous plug to a low pressure region p_2 occurs at constant enthalpy. This result was first obtained in the classical experiments of Joule and Thomson in their famous porous plug experiments.

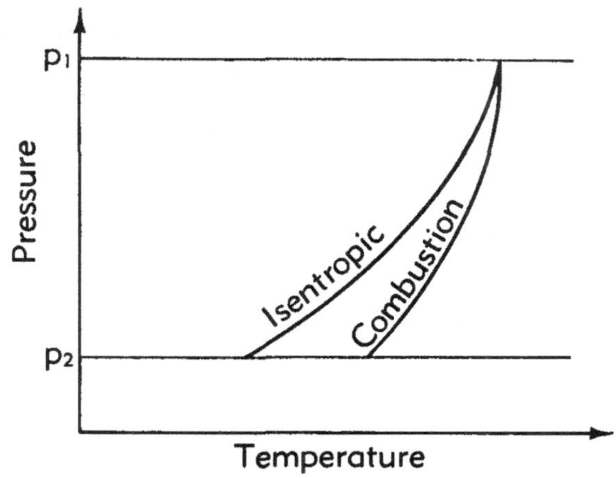

Fig. B,2d. pT diagram for isentropic flow and flow with combustion.

Employing the integrated form of Eq. 1-1 in which friction is neglected, the energy change per mole of gas in going from p_1 to p_2 is given by the equation

$$E_2 - E_1 = (p_1V_1 - p_2V_2) \tag{2-33}$$

where the net work produced by the gas is the work done by the gas at p_1, i.e. p_1V_1, minus that done on the gas at p_2, i.e. p_2V_2. Noting that the enthalpy H is given by the equation, $H = E + pV$, it is seen from Eq. 2-33 that the quantity

$$H_2 = E_2 + p_2V_2 = E_1 + p_1V_1 = H_1 \tag{2-34}$$

remains constant. If the gas is perfect so that H is a function only of T, the process is also isothermal. If the gas is imperfect, or if chemical equilibria exist which are pressure dependent, the process is still isenthalpic but not isothermal. This principle has found application in gas generation devices where the gas is formed at a high pressure and does work at a lower pressure.

4. Free expansion: isoenergetic. Free expansion of a gas occurs in the transfer of the gas from a higher to a lower pressure level without the appearance of external work. If the system is adiabatic and frictionless, it is obvious from Eq. 1-1 that the over-all process occurs at constant internal energy for the system, i.e. $\Delta E = 0$. For imperfect gases, E is a

function of both T and p and so the energy change associated with the change in p must be exactly compensated by a change in T. At low temperatures, this effect results in a cooling of the gas. This phenomenon is made use of in one method for the liquefaction of low boiling-point gases.

NONADIABATIC FLOW. In any real system, the assumption of adiabatic behavior can only be employed when the heat transferred through the walls is small with respect to heat effects produced in the gas. Nonadiabatic flow, on the other hand, is favored when (1) wall surface per unit mass of gas is large, (2) thermal conductivity of the gas and wall material is high, (3) wall temperatures are widely different from that of the gas, and (4) the flow velocity is low. The degree of irreversibility of the heat transfer process may be measured by the difference in temperature between the wall and the gas. In a thermodynamic sense, this irreversibility resulting from the transfer of heat is given by the entropy increase of the entire system:

$$\Delta S_1 + \Delta S_2 = \frac{q}{T_1} - \frac{q}{T_2} = q\,\frac{\Delta T}{T_1 T_2} > 0 \tag{2-35}$$

where T_1 and T_2 are the lower and higher temperatures, respectively, and ΔS_1 and ΔS_2 the corresponding entropy changes. As $\Delta T \rightarrow 0$ in Eq. 2-35, the process becomes isothermal and reversible. The conditions for this behavior are discussed under *Isothermal expansion* on page 39.

Heat transfer. It is assumed that the heat is transferred from the walls into the gas stream normal to the flow velocity and that the heat flow along the flow path is negligible. It is obvious that in any real case, a temperature gradient will necessarily become established normal to the flow because of the limitation of thermal conductivity.

The treatment given here will be limited to the thermodynamic behavior of a small element of fluid in which average properties can be chosen. The relations so obtained can be applied quite successfully to low velocity turbulent streams with large l/D ratios. Application may also be made to systems where the heat transferred, q, is much less than the enthalpy potential C_pT. It can be shown that subject to this limitation, specific consideration of the cross-sectional temperature contour will give rise only to second order corrections to the flow parameters. This fact is of practical significance since it permits the direct application of the following treatment to heat transfer in supersonic nozzles where $q \ll C_pT$ and where the cross-sectional contour cannot be uniform.

The energy equation (1-4) can be applied by setting $dW_f = 0$. Division by C_pT results in

$$\frac{dT}{T} = \frac{dq}{C_pT} + \frac{nRdp}{C_pp} \tag{2-36}$$

Direct integration from T_1 and p_1 to T and p yields the equation

$$\frac{T}{T_1} = \left(\frac{p}{p_1}\right)^{\frac{nR}{C_p}} e^{\frac{\Delta S}{C_p}} \tag{2-37}$$

which is identical to that which was given for flow with friction, Eq. 2-22, showing that thermodynamically the effect of friction on the T, p relation can be treated simply as a heat transfer process. One significant point of difference must be noted, however, namely, that the transfer of a definite quantity of heat between two fixed pressure levels will in general give a different T, p relation for the two processes. This arises from the fact that the temperature rate of transfer may be different for the two flows resulting in different values of ΔS.

The flow velocity and kinetic energy of the stream is obtained by integrating Eq. 1-6.

$$H_2 - H_1 + \tfrac{1}{2}m(u_2^2 - u_1^2) = q \tag{2-38}$$

where q represents the quantity of heat transferred to the gas between T_1 and T_2.

Eq. 2-37 and 2-38 cannot be employed for quantitative calculations unless the entropy change ΔS can be evaluated. Since q is not a function of state, such an evaluation can be made only when the geometry and heat transfer rates are known for the system. In certain applications with fixed geometry, however, simple approximate relations between q, S, and T can be obtained. One such case occurs in the heat transfer through rocket nozzles.

1. Approximate T, p relation for flow through rocket nozzles with heat transfer. Measurements of heat transfer rates through sectional nozzles [*5*] show that the maximum rate per unit area of nozzle occurs at the throat. However, per unit mass of gas, the maximum occurs somewhat upstream of the throat. Since the throat temperature given by $T_t = (2/(\gamma + 1))T_c$, is generally close to the chamber temperature T_c, it is obvious that an average temperature defined through the relation

$$\int_{T_e}^{T_c} \frac{dq}{T} = \frac{q}{T_{av}}$$

will be close to T_t. The fact that heat transferred downstream of the throat at lower temperatures will cause a greater change in entropy is partially compensated by the fact that the maximum heat transferred per unit mass lies upstream of the throat [*5*]. These facts appear to justify the approximate relation, where q_N is the heat transferred out of the gas during flow through the nozzle,

$$\Delta S \cong \frac{q_N}{T_t} = -\frac{(\gamma + 1)}{2T_c} q_N \tag{2-39}$$

for nozzles of conventional design (see XII,G).

Combination of Eq. 2-37 and 2-39 will now yield the following equation between T, p and q_N:

$$\frac{T}{T_c} = \left(\frac{p}{p_c}\right)^{\frac{nR}{C_p}} e^{-\frac{(\gamma+1)q_N}{2C_pT_c}} \tag{2-40}$$

or in terms of γ:

$$\frac{T}{T_c} = \left(\frac{p}{p_c}\right)^{\frac{\gamma-1}{\gamma}} e^{-\frac{(\gamma^2-1)}{2\gamma}\frac{q_N}{nRT_c}} \tag{2-40a}$$

Subject to the earlier assumption of $q_N \ll C_pT_c$, the exponential may be expanded to yield

$$\frac{T}{T_c} = \left(\frac{p}{p_c}\right)^{\frac{\gamma-1}{\gamma}} \left[1 - \left(\frac{\gamma+1}{2}\right)\frac{q_N}{C_pT_c}\right] \tag{2-41}$$

Eq. 2-38 and 2-40 establish the relations necessary for correcting exhaust velocities in rockets for nozzle heat transfer.

2. Exhaust velocity corrections resulting from nozzle heat transfer. The exhaust velocity in an adiabatic nozzle resulting from expansion of the gases at rest at T_c and p_c to T_e and p_e is given by the equation

$$\frac{1}{2}mu_e^2 = H_c - H_e = C_p(T_c - T_e) = C_pT_c\left(1 - \frac{T_e}{T_c}\right) \tag{2-42}$$

Denoting the quantity

$$1 - \frac{T_e}{T_c} = 1 - \left(\frac{p_e}{p_c}\right)^{\frac{\gamma-1}{\gamma}}$$

by η_i, the thermodynamic efficiency, Eq. 2-42 may be written simply as

$$\tfrac{1}{2}mu_e^2 = \eta_iC_pT_c \tag{2-42a}$$

It will be of interest now to compare the exhaust velocity in a water-cooled nozzle u_w with that in a regeneratively cooled nozzle u_r where the heat transferred, q_N, is fed back into the combustion zone at T_c and p_c. For the case where the heat transferred to the surroundings is lost (water cooling), the kinetic energy equation is given by Eq. 2-38. For expansion nozzles with small divergence angles or large expansion ratios, T_{av} should be chosen somewhat less than T_t.

$$\frac{1}{2}mu_w^2 = C_pT_c\left(1 - \frac{T_e}{T_c}\right) - q_N \tag{2-43}$$

Substituting for T_e/T_c from Eq. 2-41, there finally results

$$\frac{1}{2}mu_w^2 = \eta_iC_pT_c\left[1 + \left(\frac{\gamma+1}{2}\right)\frac{q_n}{C_pT_c}\left(\frac{1-\eta_i}{\eta_i}\right)\right] - q_N \tag{2-44}$$

When the heat transferred through the nozzle, q_N, is fed back into the combustion chamber, the chamber temperature is increased by the factor $(1 + q_N/C_pT_c)$. Aside from this change, the remainder of the process is the same as that for water cooling.[5] Hence, there results the equation

$$\frac{1}{2}mu_r^2 = \eta_i C_p T_c\left(1 + \frac{q_N}{C_pT_c}\right)\left[1 + \left(\frac{\gamma+1}{2}\right)\frac{q_N}{C_pT_c}\left(\frac{1-\eta_i}{\eta_i}\right)\right] - q_N \tag{2-45}$$

The exhaust velocity ratios for these three systems may be formulated as follows:

$$\frac{u_w}{u_e} = 1 - \frac{1}{2\eta_i}\left[1 - \tfrac{1}{2}(\gamma+1)(1-\eta_i)\right]\frac{q_N}{C_pT_c} \tag{2-46}$$

$$\frac{u_r}{u_e} = 1 + \frac{1}{4\eta_i}\left[(1-\eta_i)(\gamma-1)\right]\frac{q_N}{C_pT_c} \tag{2-47}$$

Eq. 2-47 yields an interesting result for it indicates a means of obtaining higher than adiabatic performance through regenerative heat transfer. This improvement in performance results from the fact that heat which is abstracted at a low pressure level is put back into the system at a higher pressure level. For very large pressure ratios, however, this effect is diminished since as $u_r/u_e \to 1$, $\eta_i \to 1$. The result in this limiting case is a direct consequence of the conservation of energy since at $\eta_i = 1$ the entire enthalpy is converted into kinetic energy in the adiabatic nozzle and obviously no scheme can cause an improvement.

Isothermal expansion. In practice, expansion processes which are isothermal are generally fairly difficult to realize because of limitations of thermal conductance; nevertheless, it is instructive to analyze the thermodynamic behavior of such flow for comparison with the more usual adiabatic case. These equations have found practical utility in systems with heat transfer to the gas from an external thermostat at the stagnation temperature of the stream (gas pressurization systems). Under these conditions, the isothermal expansion provides an upper limit for the efficiency of the expansion process. Isothermal expansion processes in rocket nozzles have been considered theoretically [*9*] but because of the short residence time, practical application does not seem feasible.

For isothermal expansion in a frictionless stream, the heat transferred to the gas between two pressure levels is uniquely determined and hence its transfer is reversible. This permits immediate identification of the entropy increment accordingly:

$$dq \equiv TdS \tag{2-48}$$

The energy equation may be applied to the process in the following form

[5] For the case where q_N is an appreciable fraction of C_pT_c, the value of q_N in the water-cooled and regeneratively cooled nozzles may not be the same because of the higher temperature in the regenerative case. Second order corrections of this sort are neglected in this treatment.

$(dW_f = 0)$:

$$\left(\frac{\partial E}{\partial p}\right)_T dp = dq - pdV = TdS - pdV \tag{2-49}$$

where the term $(\partial E/\partial p)_T$ contains the energy of imperfection. For perfect gases, this term is zero and direct integration of Eq. 2-49 yields for the heat absorbed by the gas

$$q = nRT \ln\left(\frac{V_2}{V_1}\right) = T(S_2 - S_1) \tag{2-50}$$

The equation for the velocity is obtained from Eq. 1-6 and 2-48 to give

$$dH - TdS + mudu = 0 \tag{2-51}$$

Since $F = H - TS$, the first two terms on the left may be identified with dF for an isothermal process. Integration finally yields

$$F + \tfrac{1}{2}mu^2 = \text{const} \tag{2-52}$$

which may be contrasted with Eq. 2-8 for adiabatic flow. The free energy change is given in terms of the pressure gradient of F, i.e. $(\partial F/\partial p)_T = V$ which integrates to

$$F_2 - F_1 = \int_{p_1}^{p_2} Vdp \tag{2-53}$$

The kinetic energy increment in terms of the pressure drop is obtained from Eq. 2-52 and 2-53 to yield

$$\tfrac{1}{2}m(u_2^2 - u_1^2) = \int_{p_2}^{p_1} Vdp \tag{2-54}$$

and for perfect gases, this reduces to

$$\frac{1}{2}m(u_2^2 - u_1^2) = nRT \ln\left(\frac{p_1}{p_2}\right) \tag{2-55}$$

B,3. Determination of Performance Parameters for Isentropic Flow.

Exhaust velocity and specific impulse. As given in Art. 1 and 2 above, the exhaust velocity or specific impulse for isentropic expansion is simply obtained from the initial and final states. For a gas with negligible initial velocity, the exhaust velocity is accurately given by the formula (cf. Eq. 2-42):

$$u_e = \sqrt{\frac{2\Delta H}{m}} \tag{3-1}$$

If the gases are perfect and chemical reactions during flow are negligible, Eq. 3-1 can be expressed in any one of the following forms where T_c is the chamber temperature and C_p is the heat capacity (assumed constant):

$$u_e = \sqrt{\frac{2C_pT_c}{m}\left[1 - \left(\frac{p_e}{p_c}\right)^{\frac{nR}{C_p}}\right]} \tag{3-1a}$$

or

$$u_e = \sqrt{\frac{2\gamma RT_c}{(\gamma - 1)\mathfrak{M}}\left[1 - \left(\frac{p_e}{p_c}\right)^{\frac{\gamma-1}{\gamma}}\right]} \tag{3-1b}$$

The enthalpy of the combustion gases at combustion conditions can be assumed accurately known from the composition and thermochemical data. The enthalpy at exhaust conditions may be calculated more or less accurately for various conditions of flow as outlined in Art. 2.

The computed exhaust velocity may be compared with the experimental thrust by the formula:[6]

$$F = \lambda u_e\dot{m} + (p_e - p_0)A_e \tag{3-2}$$

or

$$c = \lambda u_e = \frac{F - (p_e - p_0)A_e}{\dot{m}} \tag{3-3}$$

where $\dot{m} = dm/dt$ is the mass rate of flow and p_0 the ambient pressure.

In most cases the expansion nozzle is so proportioned that p_e is very nearly equal to p_0. For this condition the pressure term drops out, and the effective exhaust velocity is simply the ratio of thrust to mass rate of flow.

The parameter specific impulse, I_{sp}, is defined as the impulse per unit weight of propellant, Ft/w, or the thrust per unit weight rate of flow, $F/\dot{w}$. Both definitions are equivalent for steady state conditions. By means of Eq. 3-2, this definition results in the equation:

$$I_{sp} = \frac{c\dot{m} + (p_e - p_0)A_e}{\dot{w}} = \frac{c}{g} + \frac{(p_e - p_0)A_e}{\dot{w}} \tag{3-4}$$

For the nozzle so designed that $p_e = p_0$, I_{sp} is simply related to c as follows:

$$I_{sp} = \frac{c}{g} \tag{3-4a}$$

For most propellants for which combustion is known to be efficient, the experimental velocity as determined by Eq. 3-3 will be some 90 per cent or more of the value computed by Eq. 3-1. In some cases when heat loss to the combustion chamber and nozzle are known, and when other experimental conditions are ideal, the experimental velocity may be as high as 98 per cent of the computed value.

Thus the enthalpy drop converted to kinetic energy will usually range from 80 to 95 per cent of that computed by thermodynamic methods.

[6] For converging-diverging nozzles, only the axial component of the velocity will produce a net thrust. In comparing experimental and theoretical thrust values, therefore, one must apply a geometric correction λ to u_{eff} [*14*], which has the value $\lambda = \frac{1}{2}(1 + \cos\alpha)$ where α is the half angle in the divergence section. The "effective velocity" λu_{eff} is sometimes designated by c.

Among the factors which contribute to the loss are (1) loss of heat to the walls of combustion chamber and nozzle, (2) friction in the nozzle, and (3) angle of divergence of the nozzle. These are inherent in any rocket motor. Other less apparent factors are incomplete combustion, combustion continuing in the nozzle, uncertainty as to maintenance of equilibrium during expansion, and shock effects which may arise as soon as sonic velocity is reached.

Characteristic velocity. Although over-all performance as given by the exhaust velocity or specific impulse is easily correlated with theoretical data, the same is not true for some other performance parameters. In any experimental rocket motor test, the following quantities are ordinarily measured or known:

1. Thrust F
2. Combustion pressure p_c
3. Exit pressure p_e (usually made equal to p_0)[7]
4. Nozzle throat area A_t
5. Nozzle exit area A_e
6. Weight rate of propellant flow $\dot{w} = dw/dt$
7. Propellant composition

If exhaust and ambient pressures are equal, or if their difference is measured, the experimental exhaust velocity is determined from items (1) and (6). It may also be computed under ideal conditions from (2), (3), and (7).

In some test work, it is convenient to separate, as far as possible, the combustion and expansion processes. For a given propellant combination, the combustion process should determine the intrinsic capability to generate kinetic energy; the expansion process determines the effectiveness of use of the heat (or work) originally present.

For this reason, use is made of two experimental quantities, known as the "characteristic velocity" (c^*) and the "nozzle thrust coefficient" (C_F). These are defined by the following relations:

$$c^* = \frac{p_c A_t}{\dot{m}} \tag{3-5}$$

and

$$C_F = \frac{F}{p_c A_t} \tag{3-6}$$

It is obvious that the product of c^* and C_F is the effective exhaust velocity which equals c in Eq. 3-3 when $p_e = p_0$. The "characteristic velocity" is then the reciprocal of the mass rate of flow per unit nozzle throat area and per unit pressure in the combustion chamber. The "nozzle thrust coefficient" is the thrust per unit throat area and per unit pressure in the combustion chamber.

[7] For flight conditions with varying p_0, p_e is obtained from the area ratio A_e/A_t.

The theoretical expression for c^* is obtained from the definition, Eq. 3-5, and the equation of continuity

$$\rho u A = \dot{m} \tag{3-7}$$

where ρ is the mass density. These two equations yield (the subscript t applies to conditions at the throat)

$$c^* = \frac{p_c}{\rho_t u_t} \tag{3-8}$$

The throat is defined as that region in the nozzle with minimum cross section. From Eq. 3-7 it is seen that this condition occurs when the product ρu is a maximum. For systems where the gases are extensively dissociated and chemical equilibrium assumed, the accurate way of calculating c^*, therefore, is to determine the maximum value of ρu by calculating this quantity at several assumed pressure levels near the throat. Equations for calculating these quantities can be found in Art. 2. Once the maximum value of $\rho u = \rho_t u_t$ has been found, Eq. 3-8 can be applied directly.

The method for calculating c^* as outlined above can be very laborious because of the many equilibrium composition calculations that must be made at the various trial pressures (or temperatures) in the vicinity of the throat. For this reason, many approximate relations have been derived which have been used almost exclusively up to this time. These equations depend upon the fact that the relations for nondissociated gases with constant specific heats can be employed for dissociated systems if a proper average value of γ or C_p is chosen (see Art. 2).

Let γ_t be the average value of γ in the vicinity of the throat and γ' be the average value between the chamber and the throat. Under the assumption that the velocity in the nozzle throat is sonic under local conditions and that the perfect gas laws hold, the velocity at the throat is given by

$$u_t = \sqrt{\left(\frac{dp}{d\rho}\right)_s} = \sqrt{\frac{\gamma_t R T_t}{\mathfrak{M}_t}} \tag{3-9}$$

since $p = \text{const} \times \rho^\gamma$ according to Eq. 2-5. The equation of state at the throat is $p_t \mathfrak{M}_t = \rho_t R T_t$ which can be incorporated with Eq. 3-9 into Eq. 3-8 to give for c^* the following:

$$c^* = \frac{p_c}{p_t}\sqrt{\frac{RT_t}{\gamma_t \mathfrak{M}_t}} = \left(\frac{p_c}{p_t}\right)\left(\frac{T_t}{T_c}\right)^{\frac{1}{2}}\sqrt{\frac{RT_c}{\gamma_t \mathfrak{M}_t}} \tag{3-10}$$

The critical pressure and temperature ratios are further given by the expression [*15*]:

$$\frac{T_t}{T_c} = \frac{2}{\gamma' + 1} \quad \text{and} \quad \frac{p_t}{p_c} = \left(\frac{2}{\gamma' + 1}\right)^{\frac{\gamma'}{\gamma' - 1}} \tag{3-11}$$

Eq. 3-10 and 3-11 give finally

$$c^* = \sqrt{\frac{RT_c}{\gamma_t \mathfrak{M}_t}\left(\frac{\gamma' + 1}{2}\right)^{\frac{\gamma'+1}{\gamma'-1}}} \tag{3-12}$$

or

$$c^* = \frac{\sqrt{RT_c/\gamma_t \mathfrak{M}_t}}{\left(\dfrac{2}{\gamma' + 1}\right)^{\frac{\gamma'+1}{2(\gamma'-1)}}} \tag{3-12a}$$

For a nonreacting gas with constant specific heat, identifying marks on γ and $\mathfrak{M}$ may be removed and Eq. 3-12 is accurate.

Eq. 3-12 or 3-12a is still rather cumbersome for the chemically reacting gas because of the necessity of evaluating $\mathfrak{M}_t$, γ_t, and γ'. On the other hand, when performance calculations are normally made for rocket propellants, the quantities $\mathfrak{M}$ and γ are known in the chamber and in the exhaust. Let $\mathfrak{M}_c$ be the average molecular weight in the chamber and $\overline{\mathfrak{M}}$ be the average during the expansion with a corresponding definition for γ_c and $\bar{\gamma}$. Employing these definitions, various investigators have found the following equations quite reliable and useful. The quantity $\sqrt{\gamma}\,[2/(\gamma + 1)]^{(\gamma+1)/2(\gamma-1)}$ has been replaced by $\Gamma(\gamma)$, values for which have been tabulated [*15*].

$$c_c^* = \frac{\sqrt{RT_c/\mathfrak{M}_c}}{\Gamma(\gamma_c)} \tag{3-12b}$$

$$\overline{c^*} = \frac{\sqrt{RT_c/\overline{\mathfrak{M}}}}{\Gamma(\bar{\gamma})} \tag{3-12c}$$

The c^* equations (3-12a, 3-12b, and 3-12c) generally do not differ by more than 2 per cent for moderately dissociated systems, with c_c^* giving the maximum value in all cases.

From the nature of the formula for c^*, it is apparent that for non-dissociated gases and in the absence of such complicating effects as heat transfer, friction, shock effects, boundary layers, etc., the value of c^* is a function only of propellant composition and combustion efficiency since only these quantities can cause a variation in T_c, $\mathfrak{M}$, and γ. For this reason, it is frequently stated that c^* is independent of expansion conditions. For chemically dissociated systems, however, the chamber pressure will influence the state of chemical equilibrium and hence the values of T_c, $\mathfrak{M}$, and γ during expansion causing c^* to be indirectly a function of pressure. Because of the rate effect in adjusting to chemical equilibria during flow, c^* will also be a function of nozzle length in the contraction section as this length affects the residence time. Generally speaking, c^* can be considered to be affected by conditions up to the throat but independent of conditions downstream of that point for uniform flow.

Thrust coefficient. The thrust coefficient C_F is defined through the relation in Eq. 3-6, and in terms of thermodynamic quantities is given by the expression

$$C_F = \frac{\lambda u_e \dot{m} + (p_e - p_0)A_e}{p_c A_t} = \frac{c}{c^*} + \left(\frac{p_e - p_0}{p_c}\right)\frac{A_e}{A_t} \tag{3-13}$$

For $\lambda = 1$ and $p_e = p_0$, $C_F = C_F^0$ and is given by

$$C_F^0 = \frac{u_e}{c^*} \tag{3-14}$$

These equations show that unlike c^*, C_F is a function of nozzle geometry downstream of the throat, and of exit pressure.

For a perfect nonreacting gas with constant specific heat, C_F^0 is given by the expression:

$$C_F^0 = \sqrt{\frac{2\gamma^2}{\gamma - 1}\left(\frac{2}{\gamma + 1}\right)^{\frac{\gamma+1}{\gamma-1}}\left[1 - \left(\frac{p_e}{p_0}\right)^{\frac{\gamma-1}{\gamma}}\right]} \tag{3-15}$$

which was obtained from Eq. 3-1b, 3-12c, and 3-14. For a real gas with variable specific heat, an equation for C_F^0 similar to that given for c^* in Eq. 3-12 can be formulated with specified γ values. One such equation is

$$C_F^0 = \sqrt{\frac{2\bar{\gamma}\gamma_t}{\bar{\gamma} - 1}\left(\frac{2}{\gamma' + 1}\right)^{\frac{\gamma'+1}{\gamma'-1}}\left[1 - \left(\frac{p_e}{p_c}\right)^{\frac{\bar{\gamma}-1}{\bar{\gamma}}}\right]} \tag{3-16}$$

where γ', $\bar{\gamma}$, and γ_t are as defined previously (cf. Eq. 3-12, 3-12a, and 3-12c).

B,4. Nonequilibrium Effects.

Incomplete combustion. Jet engines are so designed that combustion reactions are essentially completed in the chamber prior to expansion through the nozzle. The size of the combustion chamber is determined not only by the mass rate of flow or thrust of the engine but also by the rate at which the combustion reaction proceeds. Under the assumption that the combustibles flow uniformly down the chamber without the occurrence of large eddy currents or stagnation regions (the flow, however, need not be laminar), the completeness of combustion may be measured in terms of the residence time t_c in the chamber. The values of the residence time in order to effect a certain completeness of reaction will not only be a function of the chemistry of the propellants but also of the hydrodynamic factors in injection which influence combustion rate, i.e. atomization, mixing, evaporation, etc.

The residence time is related simply to the rocket motor geometry and performance of the propellants as follows, restricted to the previous assumption of uniform flow:

$$t_c = \frac{m_c}{\dot{m}} \tag{4-1}$$

where m_c is the mass of the propellant in the combustion chamber at the steady state and $\dot{m}$ is the mass rate of flow.
Further,

$$m_c = \frac{p_c V_c \overline{\mathfrak{M}}_c}{R \overline{T}_c} \tag{4-2}$$

$$\frac{1}{\dot{m}} = \frac{c^*}{p_c A_t} \quad \text{(see Eq. 3-5)} \tag{4-3}$$

where $\overline{\mathfrak{M}}_c$ and $\overline{T}_c$ denote the molecular weight and temperature in the chamber averaged over the entire chamber contents. Eq. 4-1, 4-2, and 4-3 result in

$$t_c = \left(\frac{\overline{\mathfrak{M}}_c}{R\overline{T}_c}\right)\left(\frac{V_c}{A_t}\right) c^* \tag{4-4}$$

The quantity V_c/A_t has been termed L^* [*1*] and is directly proportional to residence time. Eq. 4-4 is sometimes further simplified [*1*] by means of a relation of the type:

$$c^* = \frac{\sqrt{RT_c/\mathfrak{M}_c}}{\Gamma(\gamma)}$$

where $\Gamma(\gamma)$ is a function only of γ (see preceding article). Identification of $\overline{T}_c$ with T_c and $\overline{\mathfrak{M}}_c$ with $\mathfrak{M}_c$ cannot be made in a strict sense, however, since the unbarred quantities refer to conditions at the entrance of the nozzle rather than averages in the chamber. Since in general T_c will be greater than $\overline{T}_c$ and $\mathfrak{M}_c$ less than $\overline{\mathfrak{M}}_c$, the following inequality can be derived:

$$t_c > \frac{L^*}{c^*[\Gamma(\gamma)]^2} \tag{4-4a}$$

The use of the parameter $L^* = V_c/A_t$ has been found very useful in the engineering development of rocket engines (see XII,G).

The result of an improperly designed combustion chamber is generally a low value of the chamber temperature T_c. The equations developed in Art. 3 for the performance parameters show that such a condition always leads to a loss in exhaust velocity. The continuation of combustion during expansion will generally permit a partial recovery of the lost combustion heat, but full recovery is theoretically impossible since the liberation of heat at lower pressure levels causes a lowering of the ideal thermodynamic efficiency. The effect of combustion during flow is to cause the rate of temperature drop to be diminished thereby decreasing the apparent value of γ. (The mechanism of this effect has been discussed more completely in the latter part of Art. 2 under the heading *Isothermal expansion.*) Incomplete combustion can generally be detected in rocket motors by the appearance of a large exhaust flame resulting from afterburning.

Reassociation reactions. The occurrence of reassociation reactions, as distinct from incomplete combustion, is not directly related to chamber design but results specifically from the changes in temperature and pressure during expansion. For high temperature propellant systems where the gases are extensively dissociated into "unstable" molecules[8] and atoms, reassociation of these components at the lower temperatures during expansion leads to a heat effect, similar to that of continuing combustion, which increases the exhaust velocity. The types of reactions which occur during expansion are listed below together with the heat release at 298°K [*2*].

$$\begin{aligned} H_2O + CO &= H_2 + CO_2 + 9.45\text{ kcal} \\ \tfrac{1}{2}H_2 + OH &= H_2O + 68.10\text{ kcal} \\ 2H &= H_2 + 103.80\text{ kcal} \\ 2N &= N_2 + 185.1\text{ kcal}^9 \\ 2O &= O_2 + 118.20\text{ kcal} \\ NO &= \tfrac{1}{2}N_2 + \tfrac{1}{2}O_2 + 21.60\text{ kcal} \\ CO + \tfrac{1}{2}O_2 &= CO_2 + 67.61\text{ kcal} \\ H_2 + \tfrac{1}{2}O_2 &= H_2O + 57.80\text{ kcal} \end{aligned}$$

The effect on the exhaust velocity of extensive reassociation may be as high as a 10 per cent improvement (e.g. the H_2—O_2 or H_2—F_2 propellant system) but normally does not exceed 5 per cent for carbonaceous propellants. The extent to which any one of the reactions listed above occurs is determined by the reaction rate for that equation and the residence time during flow. Since these rate laws are of the form

$$\frac{dn_i}{dt} = kf(n_1, n_2, \cdots)e^{-E/RT} \tag{4-5}$$

where $f(n_1, n_2, \cdots)$ is some function of the molecular concentrations, it is apparent that unless complete equilibrium is maintained, the major portion of the reaction occurs in the early stages of expansion where $e^{-E/RT}$ is large, the molecular concentrations are high, and residence times are long because of low flow velocities. Theoretically, it is possible to determine the extent to which each one of these reactions occurs during expansion, by means of the equations of Art. 2 if the rate law is known. Because of the tedious procedure of calculating quantitatively the exact concentration of minor components at each stage of expansion, it is

[8] By "unstable" molecules is meant constituents like OH, CN, CH, NH, etc., which do not normally exist at standard temperature and pressure.

[9] This value is presently in doubt, and the value 225 kcal has recently received some support.

usually convenient to employ a criterion to determine whether or not near-equilibrium is maintained with respect to any given reaction for which the rate constants are known.

Criteria for maintenance of equilibria. For high temperature systems in which the chemical species are extensively dissociated, reassociation reactions during the expansion process result in a significant modification of the expansion law. In a formal way, the expansion law including the effect of the individual rates of reaction may be expressed as follows (cf. Eq. 2-31):

$$\sum \frac{[\Delta H_f]_i}{n} \frac{dn_i}{dt} \frac{dt}{dT} \frac{dT}{T} + \frac{C_p}{n} \frac{dT}{T} = R \frac{dp}{p} \tag{4-6}$$

where dt/dT is the residence time of the gas per unit temperature interval. A fairly accurate evaluation of dt/dT may be made by applying an equation such as Eq. 2-32 to the flow process taking into account the geometry

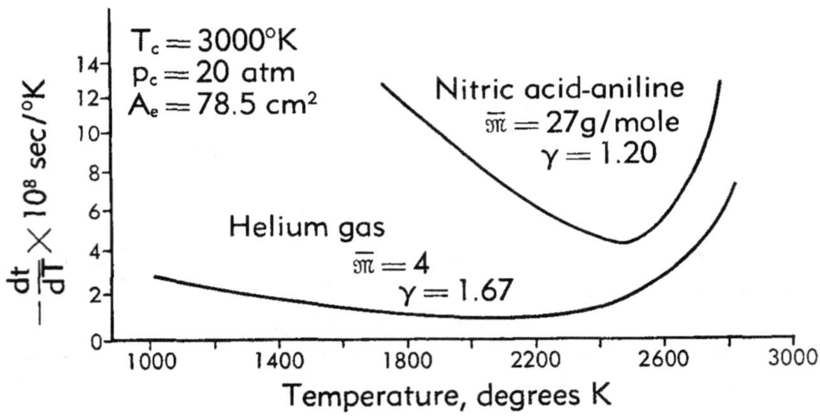

Fig. B,4. Typical curves of dt/dT during expansion through a rocket nozzle.

of the duct. For rocket motors of about 1000 pounds thrust, Altman and Penner have shown that such calculations yield residence times of the order of 10^{-8} to 10^{-7} sec/°K [*8*]. Fig. B,4 shows some typical curves of dt/dT during expansion through a rocket nozzle.

Because of the Arrhenius factor and the pressure effect in chemical rate expressions, reaction rates are fastest at the start of the expansion and monotonically decrease as the pressure falls. At each temperature, the residence time dt/dT can be used to determine the equilibrium values of dn_i/dt, and comparison with the kinetic values can be made to determine the approach to equilibrium. If such a comparison is made at the end state and equilibrium is followed, then it is certain that equilibrium was maintained throughout the entire expansion. Schaefer [*10*] and Penner [*11,12*] have utilized this concept to develop criteria for the maintenance of equilibria during flow through nozzles.

The treatment, essentially that of the above authors, is given here as applied to the reassociation reaction of atoms to form molecules. The method, however, can be generally applied to any reaction [*12*].

Consider the combination of atoms A and B with the aid of a third molecule M to form the molecule AB,[10]

$$\mathrm{A} + \mathrm{B} + \mathrm{M} \underset{k_b}{\overset{k_f}{\rightleftarrows}} \mathrm{AB} + \mathrm{M} \tag{4-7}$$

where k_f and k_b denote the forward and reverse (backward) reaction rates respectively. The equilibrium constant K_c for this reaction is

$$K_c = \frac{(\mathrm{AB})}{(\mathrm{A})(\mathrm{B})} = \frac{k_f}{k_b} \tag{4-8}$$

where the quantities in ( ) designate the concentrations of the species in moles per unit volume. Thermodynamic equilibrium during flow for this reaction can be maintained if

$$\frac{d \ln K_c}{dt} = \frac{d \ln (\mathrm{AB})}{dt} - \frac{d \ln (\mathrm{A})}{dt} - \frac{d \ln (\mathrm{B})}{dt} \tag{4-9}$$

The quantities $d \ln (\ )/dt$ refer to the total change of the concentrations of the species with time. Because the process is applied to an adiabatic expansion, this change results from two factors: (1) the temperature and pressure change cause a concentration change (change of state), and (2) chemical reaction causes a concentration change. The total change will therefore be the sum of these two changes.

Let $[d \ln (\ )/dt]_s$ be the fractional rate of concentration change resulting from a change in state and $[d \ln (\ )/dt]_{ch}$ that resulting from the chemical reaction. Then

$$\frac{d \ln (\ )}{dt} = \left[\frac{d \ln (\ )}{dt}\right]_s + \left[\frac{d \ln (\ )}{dt}\right]_{ch} \tag{4-10}$$

Assuming that a suitable average γ can be chosen for the expansion process, the concentration-temperature relation can be obtained directly from Eq. 2-4 noting that V is the reciprocal of the concentration. Differentiation of the logarithm of the concentration with respect to time can be accomplished by means of the temperature derivative as follows:

$$\ln (x) = \ln \text{const} + \frac{1}{\gamma - 1} \ln T \tag{4-11}$$

$$\left[\frac{d \ln (x)}{dt}\right]_s = \frac{d \ln (x)}{dT}\frac{dT}{dt} = \frac{1}{(\gamma - 1)T}\frac{dT}{dt} \tag{4-12}$$

[10] The combination of free atoms to molecules generally requires the presence of a third body to carry away the heat of reaction generated.

where x is any species in the gas. The term dT/dt is the rate of temperature change during flow, typical curves of which are given in Fig. B,4.

The concentration change resulting from chemical reaction alone is given by the rate law

$$\left[\frac{d(\mathrm{AB})}{dt}\right]_{\mathrm{ch}} = -\left[\frac{d(\mathrm{A})}{dt}\right]_{\mathrm{ch}} = -\left[\frac{d(\mathrm{B})}{dt}\right]_{\mathrm{ch}}$$

$$= k_f(\mathrm{A})(\mathrm{B})(\mathrm{M}) - k_b(\mathrm{AB})(\mathrm{M}) \tag{4-13}$$

$$= k_f(\mathrm{A})(\mathrm{B})(\mathrm{M})\left[1 - \frac{(\mathrm{AB})}{(\mathrm{A})(\mathrm{B})K_c}\right]$$

The ratio (AB)/(A)(B) at any temperature T may be replaced by the equilibrium constant K_c' at the temperature T'. For equilibrium flow, $T' = T$, whereas for near-equilibrium flow, $|T' - T|$ is a small number. Substituting K_c' for (AB)/(A)(B) in Eq. 4-13 there result the following equations for the logarithmic changes:

$$\left[\frac{d \ln (\mathrm{AB})}{dt}\right]_{\mathrm{ch}} = \frac{k_f(\mathrm{M})}{K_c'}\left[1 - \frac{K_c'}{K_c}\right] \tag{4-14}$$

$$\left[\frac{d \ln (\mathrm{A})}{dt}\right]_{\mathrm{ch}} = -k_f(\mathrm{B})(\mathrm{M})\left[1 - \frac{K_c'}{K_c}\right] \tag{4-14a}$$

$$\left[\frac{d \ln (\mathrm{B})}{dt}\right]_{\mathrm{ch}} = -k_f(\mathrm{A})(\mathrm{M})\left[1 - \frac{K_c'}{K_c}\right] \tag{4-14b}$$

Now, it may be noted that for near-equilibrium flow, $K_c' \cong K_c$ and therefore

$$1 - \frac{K_c'}{K_c} = -\frac{\Delta K_c}{K_c} \cong -\frac{d \ln K_c}{dT}(T' - T) \tag{4-15}$$

Substitution of Eq. 4-13, 4-14, 4-14a, and 4-14b into Eq. 4-9 by means of Eq. 4-10 yields

$$\frac{d \ln K_c}{dt} = -\frac{1}{(\gamma' - 1)T}\frac{dT}{dt}$$

$$- k_f(\mathrm{M})\left[(\mathrm{A}) + (\mathrm{B}) + \frac{1}{K_c'}\right]\left(\frac{d \ln K_c}{dT}\right)(T' - T) \tag{4-16}$$

which can be solved for $(T' - T)$ to yield

$$(T' - T)k_f(\mathrm{M})\left[(\mathrm{A}) + (\mathrm{B}) + \frac{1}{K_c'}\right]$$

$$= \left[1 + \frac{1}{(\gamma - 1)T(d \ln K_c/dT)}\right]\left(-\frac{dT}{dt}\right) \tag{4-17}$$

Eq. 4-17 may be simplified further by noting the thermodynamic relation $(d \ln K_c)/dT = \Delta E/RT^2$. Therefore $(\gamma - 1)T \cdot (d \ln K_c)/dT = (\gamma - 1)\Delta E/RT = \Delta E/C_V T$ where $-\Delta E$ is the heat evolved when the reaction occurs at constant volume of the gas. The temperature lag now has the specific expression

$$(T' - T) = \frac{[1 + (C_V T/\Delta E)](-dT/dt)}{k_f(\mathrm{M})[(\mathrm{A}) + (\mathrm{B}) + 1/K_c']} \tag{4-18}$$

Generally speaking, values of $-\Delta E$ for most atomic reactions vary from 50 to 150 kcal per mole, and in most cases, $-\Delta E > C_V T$. For the unique point where $-\Delta E = C_V T$, $T' = T$ for any value of k_f. This results from the fact that the concentration change resulting from the change in state is just sufficient to maintain equilibrium. For the case where $-\Delta E < C_V T$, the change of state causes concentration changes greater than those demanded for equilibrium and the reaction proceeds the other way, i.e. $T' < T$. The dominant terms in Eq. 4-18 are $-dT/dt$ and k_f. For rockets of about 1000 lbs thrust, $-dT/dt$ ranges from 1 to 2×10^7 °K/sec whereas for larger rockets like the V-2, the value ranges from about 1.5 to 3×10^6 °K/sec. (dT/dt is approximately inversely proportional to $\sqrt{\text{thrust}}$.) For atomic recombinations resulting from triple collisions, $k_f \sim 10^{11}$ (moles/liter)$^{-2}$ sec^{-1} [*13*]. Since $1 + C_V T/\Delta E < 1$, it appears that near-equilibrium can generally be maintained for atomic reactions unless the actual concentrations (A) and (B) are very small, in which case further recombination will have only negligible effect.

The equation for temperature lag of other types of chemical reactions can easily be formulated by the methods just given. The reaction types of greatest interest and their temperature lags are summarized below [*12*]:

1. Deactivation of excited components

$$\mathrm{A}^* = \mathrm{A} \tag{4-19}$$

$$(T' - T) = \frac{(-dT/dt)}{k_f(1 + 1/K_c')} \tag{4-20}$$

2. Bimolecular double decomposition

$$\mathrm{A} + \mathrm{B} = \mathrm{C} + \mathrm{D} \tag{4-21}$$

$$(T' - T) = \frac{(-dT/dt)}{k_f\{(\mathrm{A}) + (\mathrm{B}) + [(\mathrm{C}) + (\mathrm{D})]/K_c'\}} \tag{4-22}$$

3. Double decomposition reaction[11]

$$2\mathrm{AB} = \mathrm{A}_2 + \mathrm{B}_2 \tag{4-23}$$

$$(T' - T) = \frac{(-dT/dt)}{k_f\{4(\mathrm{AB}) + [(\mathrm{A}_2) + (\mathrm{B}_2)]/K_c'\}} \tag{4-24}$$

[11] Eq. 4-24 contains the factor 4 in the denominator which was omitted in the original reference.

B,5. Two-Phase Flow.

Systems of Application. The expansion process can become quite complicated if a condensed phase is present during all or part of the expansion. Such cases arise when metallic elements or salts are included in fuels or rocket propellants, in the detonation of TNT, and in the expansion of initially saturated vapors. The treatment in this section is confined to a semiquantitative discussion of the effects in open nozzles. This is considered in two parts, first for a condensed phase present in fixed amount during the entire expansion process, and second for condensation during expansion, when the amount of the condensed phase present is a function of the pressure, temperature, and rate of condensation.

General Theory for Permanent Condensed Phases. The problem of determining the course of the expansion process when a condensed phase is present may be divided somewhat arbitrarily into two parts: the hydrodynamic problem, dealing primarily with kinetic energy distribution between solid and gas, and the thermodynamic problem, dealing primarily with temperature distribution. Actually, the problems are interconnected and dependent on each other.

It is apparent that for particles of large size, a relatively long time will be required for the particles to attain the velocity and temperature of the gas. As the particle size becomes smaller, its behavior becomes gaslike and equilibration with the gas stream is very rapid. Although mathematically it is possible to represent the transition from solid (or liquid) to gas as particle size diminishes, it is known from nucleation theory that condensation requires nuclei and that the particles initially formed are of a minimum critical size dependent on the degree of supersaturation of the vapor [*16,17*]. In the treatment in a later part of this section, it is shown that both hydrodynamic and thermal equilibrium can be approached for particle sizes less than 10^{-4} cm in expansion processes and in regions of active combustion.

The energy equation and the "gaslike" solid. Under the assumptions that the flow process is adiabatic and the gases are perfect, the energy equation for frictionless flow is

$$C_{p_g}dT_g + C_{p_s}dT_s = V(dp_g + dp_s) \tag{5-1}$$

where the subscripts g and s refer to the gas and solid (or liquid) state respectively. The term p_s denotes the pressure exerted by the solid particles resulting from their thermal or Brownian motion. If n_s is the number of moles of solid for n_g moles of gas and if $\bar{n}_s$ is the average number of molecules per solid particle, then $n_s/\bar{n}_s$ is the number of moles of solid particles. These particles can be considered to behave like a gas of high molecular weight with the following equation of state:

$$p_s V = \left(\frac{n_s}{\bar{n}_s}\right) R T_g \tag{5-2}$$

whereas the equation of state of the gas is

$$p_g V = n_g R T_g \tag{5-3}$$

The equation of state of the particles is given in terms of T_g because the particles are assumed to be in translational equilibrium with the gas. Its internal energy, however, corresponds to another temperature T_s which approaches T_g for very high values of thermal conductivity or low time rates of temperature change. The heat capacities of n_s moles of solid are

$$C_{V_s} = C^0_{V_s} + \left(\frac{n_s}{\bar{n}_s}\right) 3R \tag{5-4}$$

$$C_{p_s} = C^0_{p_s} + \left(\frac{n_s}{\bar{n}_s}\right) 4R \tag{5-4a}$$

where the superscript ^{0} refers to the internal property of the solid (including the surface energy)[12] and the second term is the contribution resulting from the rotation and translation of the particles. For large values of $\bar{n}_s$, the two specific heats are very nearly equal, whereas for small values of $\bar{n}_s$, $C_{p_s} - C_{V_s}$ approaches R. Eq. 5-1, 5-2, and 5-3 may be combined to give

$$\left[C_{p_g} + \left(\frac{n_s}{\bar{n}_s}\right) 4R\right] \frac{dT_g}{T_g} + C^0_{p_s} \frac{dT_s}{T_g} = \left(n_g + \frac{n_s}{\bar{n}_s}\right) R \frac{dp}{p} \tag{5-5}$$

For large particles ($>10^{-7}$ cm so that $n_s/\bar{n}_s < 10^{-2} n_g$) Eq. 5-5 reduces to

$$C_{p_g} \frac{dT_g}{T_g} + C^0_{p_s} \frac{dT_s}{T_g} = n_g R \frac{dp}{p} \tag{5-6}$$

This equation cannot be integrated unless T_s is known as a function of T_g or p. Since T_s is determined by heat transfer conditions and particle size, it is obvious that there will be some lag between T_s and T_g. In the event that $T_s - T_g$ is negligible, Eq. 5-6 can be integrated directly to give ($T_s = T_g = T$):

$$\frac{T_e}{T_c} = \left(\frac{p_e}{p_c}\right)^{\frac{n_g R}{C_{p_g} + C^0_{p_s}}} \tag{5-7}$$

whereas for the other extreme where T_s remains at T_c, the initial temperature, $dT_s = 0$ and therefore

$$\frac{T_e}{T_c} = \left(\frac{p_e}{p_c}\right)^{\frac{n_g R}{C_{p_g}}} \tag{5-8}$$

represents the equation for the gas.

[12] For very small particles ($<10^{-7}$ cm), $C^0_{V_s}$ and $C^0_{p_s}$ may be expected to deviate from the values for the macroscopic solid.

The hydrodynamic problem. For larger aggregates of condensed phases, the problems may be approached from the opposite side, that of the usual laws of aerodynamics and heat transfer. These may be to a large extent separated. The force or drag acting on a spherical body with velocity relative to a gas may be represented by

$$F = C_D A \tfrac{1}{2}\rho_g(\Delta u)^2 \tag{5-9}$$

where

F = force acting on the particle
C_D = drag coefficient
ρ_g = gas density
A = frontal area of body = πr^2
Δu = relative velocity between gas and particle = $u_g - u_s$

The acceleration of the particle is determined by the force acting on it and its mass:

$$F = ma = \tfrac{4}{3}\pi\rho_s r^3 a \tag{5-10}$$

where

ρ_s = density of solid
r = radius of particle
a = acceleration of particle = du_s/dt

From Eq. 5-9 and 5-10 the particle acceleration is

$$\frac{du_s}{dt} = \frac{3\rho_g C_D}{8\rho_s r}(u_g - u_s)^2 \tag{5-11}$$

At low Reynolds numbers when the velocity lag is small ($Re < 2$), Stokes' law holds, and the acceleration is found to be

$$\frac{du_s}{dt} = \frac{9\mu}{2r^2}(u_g - u_s) \tag{5-12}$$

where μ is the gas viscosity.

In principle, Eq. 5-11 and 5-12 can be solved with the aid of the energy equation, $\frac{1}{2}m_g u_g^2 + \frac{1}{2}m_s u_s^2 = C_p(T_0 - T)$ and the equation for the rate of temperature change dT/dt (see Art. 2). The equations show that small values of u_g and dT/dt favor small velocity lags, $u_g - u_s$. In nozzles, both quantities increase continually during flow and only very small particles ($<10^{-5}$ cm diameter) can be further accelerated by an appreciable amount downstream of the throat. On the other hand, particles of sizes greater than 10^{-3} cm lag considerably behind the gas velocity even upstream of the throat.

The thermal problem. The problem of temperature equilibrium between the particles and gas is considerably more complex than the hydrodynamic problem because of the additional factors that are involved. There are three major influences which must be considered here:

1. Convective heat transfer to or from the particle.
2. Heat transfer within the particle.
3. Radiative heat transfer between the particle and the surroundings.

1. Convective heat transfer. The problem of convective heat transfer from the solid particle may be treated in a manner similar to that given in the hydrodynamic treatment. This loss of heat to the gas per unit time may be represented by

$$\frac{dq}{dt} = hA(T_s - T_g) \tag{5-13}$$

where h is the average coefficient of heat transfer and A the effective surface area. Assuming instantaneous heat transfer within the particle, the rate of heat loss dq/dt can be set equal to $-C_s(dT_s/dt)$ where C_s is the heat capacity of the particle. This substitution gives

$$-\frac{dT_s}{dt} = \frac{hA}{C_s}(T_s - T_g) \tag{5-14}$$

For spherical particles, $A = 4\pi r^2$ and $C_s = \frac{4}{3}\pi r^3 \rho_s c_s$ where c_s is the specific heat per unit weight and ρ_s the density. Eq. 5-14 now assumes the form

$$-\frac{dT_s}{dt} = \frac{3h}{c_s \rho_s r}(T_s - T_g) \tag{5-15}$$

which shows that high rates of heat transfer are favored by small particles.

A simple order-of-magnitude calculation may be instructive in determining the approximate particle size necessary to maintain thermal equilibrium in nozzles. For low Reynolds numbers ($Re < 2$) the Nusselt number $2hr/k = 2$, whereas for $Re > 2$ it has a larger value. A lower limit to the heat transfer can be obtained, therefore, using the value $hr/k = 1$ in Eq. 5-15. Setting $c_s \rho_s = 1$ and using an average value for k of 3×10^{-4} cal/sec°C cm for a typical combustion gas, Eq. 5-15 reduces to

$$-\frac{dT_s}{dt} = \frac{9 \times 10^{-4}}{r^2}(T_s - T_g) \tag{5-15a}$$

For thermal equilibrium, dT_s/dt should be equal to dT_g/dt which is of the order of 10^6 to 10^8 °K/sec for conventional supersonic nozzles [*8*]. Taking the average value of 10^7 °K/sec, one sees that if a temperature differential no greater than 100°K is to be maintained, r must be smaller than 10^{-4} cm.

2. Heat transfer within the particle. Assuming the heat transfer process at the surface of the particle to be fast, the limitation on thermal transfer will then result from conductivity within the particle itself. An indication of the time required to establish thermal equilibrium under these conditions can be obtained from an examination of the thermal conduction equation as applied to spheres with an appropriate boundary condition.

Assume the sphere at an initial temperature T_0 immersed in a medium of temperature zero. The solution for the temperature at the center of the sphere of radius r after a time t is found to be [*18*]

$$\frac{T}{T_0} = 1 - \frac{r}{\sqrt{\pi \alpha t}} \sum_{n=0}^{\infty} e^{-\frac{(2n+1)^2 r^2}{4\alpha t}} \tag{5-16}$$

Assuming $T_0 \cong 10°K$, thermal equilibrium (within 10°) can be maintained if the time required for T/T_0 to reach a value 0.5 is approximately equal to the time required for the gas to cool 10°. For rocket nozzles, this time may be chosen approximately as 10^{-6} sec [*8*]. Choosing a value of 10^{-3} cm^2/sec for α, the thermal diffusivity,[13] Eq. 5-16 shows that this criterion leads to the conclusion that thermal transfer within the particle is fast for particle sizes less than 10^{-5} cm and slow for sizes greater than 10^{-4} cm. For conductive particles with $\alpha \cong 10^{-1}$ cm^2/sec, the borderline for thermal transfer would lie between 10^{-3} and 10^{-4} cm, inasmuch as particle size varies as the square root of α as shown in Eq. 5-16.

3. Radiative heat transfer. Unless the gas has an absorptivity of unity, heat lost by radiation from the solid particles is largely not recovered by the system except for the case where the walls are regeneratively cooled. The limiting case of a particle radiating in a transparent gas is analyzed here.

For a particle of radius r, temperature T, and emissivity ϵ, the rate of energy loss through radiation alone is

$$-\frac{dE}{dt} = 4\pi r^2 \epsilon \sigma T^4 \tag{5-17}$$

where σ is the Stefan-Boltzmann radiation constant. For spherical particles, the rate of temperature fall of the particle is calculated to be

$$-\frac{dT}{dt} = \frac{3\epsilon\sigma T^4}{\rho_s C_s r} \tag{5-18}$$

Integrating between the initial temperature T_o and T after time t yields

$$T^{-3} = T_o^{-3} + \frac{\rho \epsilon \sigma t}{\rho_s C_s r} \tag{5-19}$$

Setting $\rho_s c_s$ and ϵ equal to unity, the equation reduces to

$$T^{-3} = T_o^{-3} + \frac{12.2 \times 10^{-12} t}{r} \tag{5-19a}$$

Total transit times in rocket nozzles are generally of the order of 10^{-4} to 10^{-3} seconds, depending on nozzle size and composition of gases [*8*].

[13] The particles are generally metallic oxides or halides with low conductivity.

Assuming an initial temperature of 2500°K, the following tabulation shows the final particle temperatures attained for various particle sizes.

Table B,5a. Temperature loss for radiating particles (initial T_0 = 2500°K; emissivity = 1; $c_s\rho_s$ = 1.)

Particle radius, cm	T°K ($t = 10^{-4}$ sec)	T°K ($t = 10^{-3}$ sec)
10^{-2}	≅2500	2475
10^{-3}	2475	2280
10^{-4}	2280	1555
10^{-5}	1555	(698)*
10^{-6}	(698)*	—

* These temperatures are probably low because of neglect of back radiation to particle.

The data in Table B,5a show that for particle sizes greater than 10^{-2} cm, very little energy is lost through radiation under these conditions. For particle sizes less than 10^{-5} cm, however, the rate of temperature loss can exceed that of the gas during adiabatic expansion. Under these conditions, the temperature of the particle will be determined by the radiation density in the gas and by convection.

Results in limiting cases. While it is difficult to calculate the actual distribution of kinetic and thermal energy between a gas and a solid or liquid during expansion, the effect on performance of limiting cases is easily determined. The treatment for a few of these cases (1 and 4) has been given previously [*19,20*].

Four ideal limiting cases may be distinguished:

1. Complete kinetic and thermal equilibrium, $u_s = u_g$, $T_s = T_g$. This corresponds to considering the solid particles as large gas molecules.
2. Kinetic equilibrium, thermal insulation, $u_s = u_g$, $T_s = T_0$ (constant).
3. Thermal equilibrium, kinetic nonequilibrium, $u_s = 0$, $T_s = T_g$.
4. Complete insulation, $u_s = 0$, $T_s = T_0$. This last case corresponds to assuming very large particles.

Let

x = weight fraction of solid
m = mass of system
C_{p_g}, C_{p_s} = heat capacities of gas and solid respectively
n_g = moles of gas in the mass $(1 - x)m$

The adiabatic flow parameters for the four cases listed above may now be derived, assuming the solid particles sufficiently large so that their thermal motion can be neglected.

Case 1. $u_s = u_g$, $T_s = T_g$
The energy equation for the expansion is

$$dH = (C_{p_g} + C_{p_s})dT = Vdp \tag{5-20}$$

The equation of state, assuming the gas to be perfect, is

$$pV = n_g RT \tag{5-21}$$

Eq. 5-20 and 5-21 may be combined and integrated, assuming constant specific heat, to give the following relation between T and p:

$$\frac{T_e}{T_c} = \left(\frac{p_e}{p_c}\right)^{\frac{n_g R}{C_{p_g}+C_{p_s}}} \tag{5-22}$$

The enthalpy change between T_c and T_e is calculated to be

$$\Delta H_{T_e}^{T_c} = (C_{p_g} + C_{p_s})(T_c - T_e) = \tfrac{1}{2}mu^2 \tag{5-23}$$

from which the specific impulse may be obtained:

$$I_{sp} = \frac{u}{g} = \frac{1}{g}\sqrt{\frac{2(C_{p_g} + C_{p_s})T_c[1 - (p_e/p_c)^{n_g R/(C_{p_g}+C_{p_s})}]}{m}} \tag{5-24}$$

Case 2. $u_s = u_g$, $T_s = T_c$
The equations for this case are similar to the ones just obtained except that $C_{p_s} = 0$.

Case 3. $u_s = 0$, $T_s = T_g$
Eq. 5-22 for the T, p relation is unaltered, but since all of the enthalpy change of both gas and solid goes into kinetic energy of the gas, Eq. 5-24 is replaced by

$$I_{sp} = \frac{(1-x)u_g}{g} = \frac{(1-x)}{g}\sqrt{\frac{2(C_{p_g} + C_{p_s})(T_c - T_e)}{(1-x)m}}$$

$$= \frac{1}{g}\sqrt{\frac{2(1-x)(C_{p_g} + C_{p_s})(T_c - T_e)}{m}} \tag{5-25}$$

This particular case is approached in those systems where the particles leave the nozzle at low velocities or remain in the combustion chamber.

Case 4. $u_s = 0$, $T_s = T_c$
Eq. 5-22 and 5-25 can be employed setting $C_{p_s} = 0$.

Wimpress [*20*] has obtained corresponding results for cases 1 and 4, and in addition has given formulas for the discharge coefficients in these systems.

It should be noted that any temperature decrease in the solid resulting from radiation does not contribute to the momentum, and such energy, therefore, is not to be included in the equations of the previous cases.

In any real system, the solid particles will emerge from the nozzle at some fraction of the gas velocity and at some temperature between T_c and T_e. Calculations for the specific impulse for such cases can be made by means of the relations:

$$\tfrac{1}{2}m[(1 - x)u_g^2 + xu_s^2] = C_{p_g}(T_c - T_{e_g}) + C_{p_s}(T_c - T_{e_s}) \quad (5\text{-}26)$$

and

$$I_{sp} = \frac{1}{g}[(1 - x)u_g + xu_s] \quad (5\text{-}27)$$

if the ratios u_s/u_g and T_{e_s}/T_{e_g} are known.

An assumed set of conditions may be chosen to illustrate the relative importance of the effects of nonequilibrium. Let:

T_c = 3000°K
p_e/p_c = 20.4
x = 0.20 (weight fraction of solid)
m = 1 gram (total mass)
$\mathfrak{M}_g$ = molecular weight of gas = 20 g/mole
n_g = 0.80/20 = 0.04 mole
C_{p_g} = (0.04)(15) = 0.60 cal/deg. The assumed value of 15 cal/deg/mole for the molar heat capacity of the gas is an apparent value (cf. Art. 2).
C_{p_s} = 0.2 × 0.5 = 0.1 cal/deg

The following tabulation shows the I_{sp} values for both complete thermal equilibrium and lack of thermal equilibrium for this system at various assumed particle velocities, expressed as fractions of the gas velocity.

Table B,5b.

u_s/u_g	$I_{sp}(T_s = T_g)$, sec	$I_{sp}(T_s = T_c)$, sec
0	206*	203
0.25	217	214
0.50	224	221
1.00	230	227
no solid	254	254

* Two limits are approached. For a very small velocity, where the solid moves out of chamber during time of gas efflux, I_{sp} = 206. If $u_s = 0$, I_{sp} = 203 since $T_s = T_c$ throughout flow.

From a practical point of view, the table shows that the question of thermal equilibrium is of far lesser importance than is particle velocity. For the assumed set of conditions, the extremes of possible particle temperatures result in at most a 1½ per cent change in I_{sp} while the effect of velocity exceeds 10 per cent. It is interesting to note that were no solid present, the last row in the table shows that the calculated I_{sp} for the gas at an initial temperature of 3000°K would be 254 seconds. This value may be compared with 230 seconds for the case where the gas contains a solid ($x = 0.20$) in thermal and kinetic equilibrium to show the penalty that must be paid by the appearance of a condensed phase. The introduction of metallic fuels into the combustion mixture of rocket engines to increase performance is based on the fact that the effect of increasing the initial chamber temperature will more than offset the effect of the solid particles. The net result is that a significant improvement can be obtained only when the original gas temperature is low.

Flow with Phase Change.

Effects of condensation during flow. The problem of two-phase flow and of condensation during adiabatic expansion has long been known to engineers of steam turbines [*6*] and also to aerodynamicists in wind tunnel operations (III,F and VIII,E,3,4). More recently, rocket engineers have become interested in these phenomena as a result of the consideration of the use of metallic constituents in the propellants which generally form oxides, hydroxides, or halides in the combustion chamber. Since the phase change from gas to liquid or solid is accompanied by the release of heat and a decrease in the number of moles of gas, it is obvious that flow conditions are altered considerably when condensation occurs.

Assuming thermal and kinetic equilibrium for the condensed phase, the energy equation characterizing such flow is (effects of chemical dissociation are neglected)

$$dH = C_{p_g}dT + C_{p_l}dT + \Delta H_v dn_g = Vdp \tag{5-28}$$

where the subscripts g and l refer to gas and liquid (or solid), respectively, and ΔH_v is the molar heat of vaporization. By applying the gas law $pV = n_g RT$ it can readily be shown that Eq. 5-28 is transformed to

$$\left[C_{p_g} + C_{p_l} + \Delta H_v \left(\frac{dn_g}{dt}\right)\frac{dt}{dT}\right]\frac{dT}{T} = n_g R \frac{dp}{p} \tag{5-29}$$

This equation can be solved only if the quantity dn_g/dt, giving the rate of condensation or nucleation, is known as a function of the parameters of state, T and p. A considerable amount of work has been done on rates of nucleation (Sec. E) [*16,17,21,22,23*], but the expressions obtained for nuclei formation are too difficult for simple solution of Eq. 5-29. Qualitatively, the theory shows that the rate of condensation increases with the ratio of the actual vapor pressure to the saturation vapor pressure. During a flow process, therefore, the rate of condensation will always lag behind that value required to maintain equilibrium, and in some supersonic nozzles, may even be completely inhibited [*6*].

The relation between T and p during adiabatic flow will be markedly affected by the occurrence of condensation which results in a smaller value of γ, or larger value of C_p as can be seen from Eq. 5-29. From a practical point of view in rocket applications, however, there is a fortunate cancellation of effects which results in very similar values for the exhaust velocity regardless of whether or not condensation occurs. Whereas a larger value of C_p tends to increase the exhaust velocity, condensation decreases the number of moles of gas thereby lowering the expansion work. The net effect is usually to increase performance slightly when condensation occurs. As an example, if pure KCl vapor at 2300°K and at its equilibrium vapor pressure of 20 atm expands isentropically to one atm, the final temperatures are 1675°K, for condensation occurring at equilibrium, and 1180°K if the gas does not condense. However, the enthalpy changes per mole of KCl are 10,690 and 10,040 calories respectively.

Equilibrium expansion of a pure condensible vapor. The determination of the flow parameters for a pure condensible vapor under equilibrium conditions can be calculated rather simply from the vapor pressure and entropy equations. It may be instructive to examine this case because it sets limiting values to the temperature, pressure, and velocity which can be obtained for flow with condensation. At each pressure level, both the temperature and velocity will be higher than for any other nonequilibrium process.

Consider the expansion of a pure condensible vapor from the initial temperature and pressure T_0 and p_0. At some point in the expansion, say T_1 and p_1, the gas may reach its saturation pressure, and beyond that point the temperature and pressure relation will be given simply by the vapor pressure equation. The thermodynamic condition that the vapor condense during expansion is simply that the term $d(\ln p)/d(\ln T)$ for the vapor-liquid equilibrium mixture be greater than for the expansion of the pure gas. It can readily be shown that this inequality leads to the condition that $n_g \Delta H_v > C_p T$ where ΔH_v is the enthalpy of vaporization and C_p the total heat capacity.[14] For every vapor, therefore, there is a temperature $T = n_g \Delta H_v / C_p$ above which the isentropic expansion of a two-phase liquid-vapor system would lead to vaporization rather than condensation. The thermodynamic treatment in this case is analogous to the one to be given on condensation.

Because most two-phase systems in chemical equilibrium encountered in combustion chambers obey the condition $n_g \Delta H_v > C_p T$, only the condensation phenomenon is treated here. Consider condensation starting at T_1 and p_1. Since the values of T and p at any point farther down the streamline are given simply by the vapor pressure equation, the only other unknown that need be determined is the composition. This can best be done by means of the isentropic condition $\Delta S = 0$. Let $X_g =$ the mole fraction of vapor at the temperature T, and S_g^T and S_l^T the corresponding

[14] This condition may also be applied when the condensible vapor is only a small fraction of the gas.

molar entropies of the gas and liquid. The isentropic condition then states that

$$S_g^{T_1} = X_g S_g^T + (1 - X_g) S_l^T \tag{5-30}$$

Noting that $S_g - S_l = \Delta S_v$ the molar entropy of vaporization, Eq. 5-30 can be rewritten as

$$S_g^{T_1} = S_g^T - (1 - X_g)\Delta S_v^T \tag{5-31}$$

and solving for the mole fraction of liquid, $1 - X_g$, there results

$$1 - X_g = \frac{S_g^T - S_g^{T_1}}{\Delta S_v^T} = \frac{R \ln (p_1/p) - \int_T^{T_1} C_{p_g} d \ln T}{\Delta S_v^T} \tag{5-32}$$

If C_{p_g}, the gaseous molar specific heat, can be taken as constant in the interval T_1 to T, Eq. 5-32 further simplifies to the following

$$X_l = 1 - X_g = \frac{R \ln (p_1/p) - C_{p_g} \ln (T_1/T)}{\Delta H_v/T} \tag{5-33}$$

where $\Delta S_v^T = \Delta H_v/T$ for condensation at equilibrium.

The enthalpy change per mole of mixture in this temperature interval is

$$H_1 - H = \int_T^{T_1} C_{p_g} dT + (1 - n_g)\Delta H_v^T \tag{5-34}$$

The kinetic energy increase in this region of condensation is

$$\tfrac{1}{2}m(u^2 - u_1^2) = H_1 - H \tag{5-35}$$

from which the specific impulse may be easily calculated.

B,6. Cited References.

1. Corner, J. *Theory of the Interior Ballistics of Guns.* Wiley, 1950.
2. Rossini, F. D., Wagman, D. D., Evans, W. H., Levine, S., and Jaffe, I. Selected values of chemical thermodynamic properties. *Natl. Bur. Standards Circ. 500,* 1952.
3. Stosick, A. J. Method used in calculating propellant properties. *Calif. Inst. Technol. Jet Propul. Lab. Progr. Rept. 1–25,* 1945.
4. Glassner, A., and Winternitz, P. F. A comparison of methods for the calculation of specific impulses in rocket motors. *Reaction Motors T.P. 1,* 1950.
5. *Jet Propulsion. Air Tech. Service Command, Guggenheim Aeronaut. Lab., Calif. Inst. Technol.,* 1946.
6. Stodola, A. *Steam and Gas Turbines,* Vol. 1. McGraw-Hill, 1927.
7. Hicks, B. L., Montgomery, D. J., and Wasserman, R. H. *J. Appl. Phys. 18,* 891 (1947).
8. Altman, D., and Penner, S. S. *J. Chem. Phys. 17,* 56 (1949).
9. Seifert, H. S., and Altman, D. A comparison of adiabatic and isothermal expansion processes in rocket nozzles. *Calif. Inst. Technol. Jet Propul. Lab. Mem. 9–15,* 1950.
10. Schaefer, K. On the thermodynamics of rocket propulsion, I. Available through Headquarters, Air Materiel Command, Wright Field, Dayton, Ohio. Feb. 1947.
11. Penner, S. S. *J. Am. Chem. Soc. 71,* 788 (1949).

12. Penner, S. S. *J. Franklin Inst. 249*, 441 (1950).
13. Amdur, I. *J. Am. Chem. Soc. 60*, 2347 (1935).
14. Malina, F. J. *J. Franklin Inst. 230*, 452 (1940).
15. Zucrow, M. J. *Jet Propulsion and Gas Turbines*. Wiley, 1948.
16. Frenkel, J. *Kinetic Theory of Liquids*. Oxford Univ. Press, 1946.
17. Volmer, M. *Kinetik der Phasenbildung*, 4. Bd. Bonhoeffer, Dresden and Leipzig, 1939.
18. Carslaw, H. S., and Jaeger, J. C. *Conduction of Heat in Solids*. Oxford Univ. Press, 1947.
19. Maxwell, W. R., Dickinson, W., and Caldin, E. F. Adiabatic expansion of a gas stream containing solid particles. *Aircraft Eng.*, 1946.
20. Wimpress, R. N. *Internal Ballistics of Solid Fuel Rockets*. McGraw-Hill, 1950.
21. Reiss, H. *J. Chem. Phys. 20*, 1216 (1952).
22. Fisher, J. C., Hollomon, J. H., and Turnbull, D. *J. Appl. Phys. 19*, 775 (1948).
23. Das Gupta, N. N., and Ghosh, S. K. *Revs. Mod. Phys. 18*, 225 (1946).

SECTION C

COMPUTATIONAL METHODS IN COMBUSTION CALCULATIONS

STUART R. BRINKLEY, JR.

C,1. Introduction. In the first two sections of this volume, methods of determining the thermodynamic properties of combustion gases have been described, and the application of these data to determine performance parameters of a working fluid composed of a combustion gas has been discussed. In this section it is proposed to consider in some detail computational methods for determining the thermodynamic properties of these gases and to discuss briefly typical computational methods for utilizing these results.

We shall designate by the word *program* the succession of elementary arithmetic operations leading to solution of a particular problem. The program must have certain characteristics when the problem is to be solved with automatic computing equipment. In particular, for problems to be solved by an iterative procedure, the method of proceeding from one approximation to the next must be well defined. Furthermore, it is evident that the program should consist of as small a number of steps as possible in order to hold computing time to a minimum. The methods described in this section are also well suited to manual computation using conventional desk calculators.

When a problem is to be solved for many different individual systems, it is desirable that the formulation of the explicit program for a particular system be reduced to routine not involving extensive analysis of the particular system.

We begin the formulation of a general computational procedure for determining the equilibrium composition and thermodynamic properties of a multiconstituent system by considering the stoichiometry of such systems and by deriving the conditions for equilibrium. The discussion serves to introduce the notation employed consistently in the remainder of the section and results in a formulation of the equilibrium conditions in a symmetrical form that is particularly well suited to provide the basis for a computational procedure. It is then possible to formulate in general terms a procedure applicable to any multiconstituent system. The explicit program for any given system can then be written down in a routine way.

C,2. Calculation of the Equilibrium Composition.

Formulation of the Problem. The composition and the thermodynamic properties of a mixture at equilibrium are independent of the path by which equilibrium is attained. From the point of view of thermodynamics, the molecular form in which the different elements are initially introduced to the system is therefore irrelevant, and it suffices to specify the gross composition of the system in terms of the elements contained by it. Since the composition of the system at equilibrium is most appropriately expressed in terms of intensive quantities (mole fractions), the results are independent of the total size of the system, which may be taken to be any convenient value. Therefore, if the number of different elements in the system is m, it is sufficient for a unique description of the system to specify the number of gram atoms of $m - 1$ elements relative to an arbitrary amount of the mth element, i.e. to specify values of the quantities $\{X^{(k)}/X^{(m)}\}$, $k = 1, 2, \cdots m - 1$, where

$$Q_k = \left\{\frac{X^{(k)}}{X^{(m)}}\right\} Q_m \tag{2-1}$$

and Q_k and Q_m are the number of gram atoms of the kth and mth elements, respectively, available to the system. Alternatively, one may specify $m - 1$ different linearly independent combinations of these ratios. We note that it is not always necessary to specify as many as $m - 1$ different quantities.

In a system containing many constituents, it is possible to select certain constituents which are sufficient for the complete thermodynamic description of the composition. By this it is meant that if the system is conceived to consist of the selected constituents only, its gross composition (in terms of the amounts of each chemical element present) is completely defined. The constituents thus sufficient to describe the composition are called the *components* of the system. An analytic criterion has been published [*1*] for the choice of the components. In terms of this criterion, the conditions for equilibrium may be written in a form which has a high degree of symmetry and is particularly well adapted for formulating a computational method for the calculation of the equilibrium composition.

The number of constituents of any system depends upon the accuracy with which it is desired to describe its composition. The constituents to be considered must be chosen a priori, and this choice usually will imply the neglect of certain equilibria that may be expected to exert a negligible effect on the composition of the system at equilibrium (see A,2).

Consider a closed system containing s different substances, which are assumed to be in chemical equilibrium. The molecular formula of the ith substance may be represented by

$$Y^{(i)} = X^{(l)}_{\alpha_{il}} \cdots X^{(k)}_{\alpha_{ik}} \cdots X^{(m)}_{\alpha_{im}} \tag{2-2}$$

$i = 1, 2, \cdots, s$, where $X^{(k)}$ is the symbol of the kth element, α_{ik} is the subscript (which may be zero) to this symbol in the formula of the ith substance, and m is the total number of elements represented in the system. For every i, the array of subscripts α_{ik}, $k = 1, 2, \cdots, m$, may be said to define a vector

$$y_i = (\alpha_{il}, \cdots, \alpha_{ik}, \cdots, \alpha_{im}) \tag{2-3}$$

which may be called the *formula vector* of substance i. If the rank of the matrix of the vector elements α_{ik} is c, it follows from a well-known theorem of algebra that there are c linearly independent vectors, and if $c \leqq s$, there are $s - c$ linearly dependent vectors which may be expressed as linear combinations of the independent vectors. It may be assumed that the independent vectors are designated by the values $1, 2, \cdots, c$ of their index. Then the dependent vectors may be expressed as linear combinations of the form

$$\sum_{j=1}^{c} \nu_{ij} y_j = y_i \tag{2-4}$$

$i = c + 1, c + 2, \cdots, s$. To Eq. 2-4 there correspond $s - c$ conceivable chemical reactions

$$\sum_{j=1}^{c} \nu_{ij} Y^{(j)} = Y^{(i)} \tag{2-5}$$

resulting in the formation, from the c substances with linearly independent formula vectors, of the $s - c$ substances with linearly dependent formula vectors. It follows that the specification of c substances such that their formula vectors are linearly independent is sufficient for a description of the composition of the system. Therefore, the number of components of the system equals the rank c of the matrix of the subscripts to the symbols of the elements in the formulas of the substances comprising the system. The number of components is usually, but not necessarily, equal to the number of different elements in the system. It may be noted that the choice of independent vectors is not, in general, unique and that, in consequence, the choice of c substances as components and the expression of the remaining $s - c$ substances as products of reactions involving only the chosen components is usually not unique. However, for the heterogeneous case consisting of a gas mixture in equilibrium with a solid phase composed of a single constituent, these criteria can be satisfied only by the selection of that constituent as a component.

This discussion has demonstrated the possibility of a choice of components which makes it possible to express each of the dependent constituents as products of reactions involving components only. Our computational procedure for the determination of the equilibrium composition is based upon the possibility of writing down for the case of interest the reactions that are expressed by Eq. 2-5. In most cases, it is possible to write these reactions immediately by intuition. In some cases, it may be necessary to formulate the reactions of Eq. 2-5 by applying to the system under consideration the steps indicated by Eq. 2-2 to 2-4.

If the number of moles of the jth component contained in the hypothetical system consisting of components only is denoted by q_j, it is evident that

$$\sum_{j=1}^{c} \alpha_{jk} q_j = Q_k \tag{2-6}$$

$k = 1, 2, \cdots, m$. We shall assume henceforth that the number of components c is equal to the number of different elements m. Introducing the ratios defined by Eq. 2-1, we obtain

$$\sum_{j=1}^{c} \alpha_{jk} q_j - \left\{\frac{X^{(k)}}{X^{(m)}}\right\} Q_m = 0 \tag{2-7}$$

It is usually convenient to assume that

$$\sum_{j}' q_j = 1 \tag{2-8}$$

where the operator $\sum_j'$ denotes a summation over the gaseous components only. Eq. 2-7 and 2-8 constitute $c + 1$ linear nonhomogeneous equations in the quantities Q_m and q_j, $j = 1, 2, \cdots, c$, and they may be employed to express the q_j in terms of the ratios $\{X^{(k)}/X^{(m)}\}$ and thus in terms of the parameters employed for the description of the system. Eq. 2-8 is equivalent to an assumption as to the total size of the system, the weight M_0 of the gaseous phase being given by

$$M_0 = \sum_{j}' q_j \mathfrak{M}_j \tag{2-9}$$

where $\mathfrak{M}_j$ is the molecular weight of the jth component. We shall call the constants q_j, determined by means of Eq. 2-7 and 2-8, the normalized stoichiometric constants.

The discussion of equilibrium in systems of many constituents assumes a particularly simple and symmetrical form in terms of the definitions that have been introduced. We consider the general case where the s substances are distributed among Π coexisting phases. The conservation of mass in the system as a whole requires that

$$\sum_{=1}^{\pi}\left\{\sum_{j=1}^{c} n_j^{(k)}\mathfrak{M}_j + \sum_{i=c+1}^{s} n_i^{(k)}\mathfrak{M}_i\right\} = \sum_{j=1}^{c} q_j\mathfrak{M}_j \tag{2-10}$$

where $n_i^{(k)}$ and $n_j^{(k)}$ are the number of moles in the kth phase of the ith and jth substances, respectively, and $\mathfrak{M}_i$ and $\mathfrak{M}_j$ are the corresponding weights of one mole. Since mass is conserved by each of the reactions of Eq. 2-5, one may substitute

$$\mathfrak{M}_i = \sum_{j=1}^{c} \nu_{ij}\mathfrak{M}_j \tag{2-11}$$

into Eq. 2-10. On equating the coefficients of $\mathfrak{M}_j$, there are obtained

$$\sum_{=1}^{\pi}\left\{ n_j^{(k)} + \sum_{i=c+1}^{s} \nu_{ij}n_i^{(k)}\right\} = q_j \tag{2-12}$$

$j = 1, 2, \cdots, c.$

The derivation by means of the method employed by Gibbs [*2*, pp. 63–82] subject to the constraints (Eq. 2-12) of the conditions for equilibrium is straightforward, and it is unnecessary to reproduce it here. The usual conditions for thermal and mechanical equilibrium are obtained, and the conditions for chemical equilibrium take the form,

$$\mu_j^{(k)} = \lambda_j \tag{2-13}$$

$$\mu_i^{(k)} = \sum_{j=1}^{c} \nu_{ij}\lambda_j \tag{2-14}$$

$k = 1, 2, \cdots, \Pi;\ j = 1, 2, \cdots, c;\ i = c+1, c+2, \cdots, s$, where $\mu_i^{(k)}$ and $\mu_j^{(k)}$ are the chemical potentials in the kth phase of the ith and jth substances respectively, the λ are constants which may be eliminated from Eq. 2-13 and 2-14, and where it has been assumed for simplicity that each substance of the system is included in every phase. If a particular component is excluded a priori from some phase, the corresponding member of Eq. 2-13 disappears as being without meaning, and similar considerations obtain if a particular dependent constituent is excluded a priori from some phase.

Eq. 2-12, together with Eq. 2-13 and 2-14 provide, by reason of their symmetry, the simplest basis for calculating the concentrations at equilibrium of the several constituents of the system. In particular, if the substances are limited to a single phase, the equations become

$$n_j + \sum_{i=c+1}^{s} \nu_{ij}n_i = q_j \tag{2-15}$$

$$\mu_i - \sum_{j=1}^{c} \nu_{ij}\mu_j = 0 \tag{2-16}$$

$j = 1, 2, \cdots, c; i = c + 1, \cdots, s$, for equilibrium at constant temperature and pressure. Eq. 2-15 and 2-16 are also applicable to systems composed of a gas phase in equilibrium with a single pure solid phase, since in this case it is not necessary to retain explicitly the index label of the phase.

The chemical potential of the kth constituent of a gas mixture can be expressed [*3*, p. 12] as

$$\mu_k = \mu_k^0(T) + RT \ln f_k \tag{2-17}$$

where f_k is the fugacity of the kth constituent in the mixture and $\mu_k^0(T)$, the chemical potential of the kth constituent in the standard state of unit fugacity, is a function of the temperature T only. The chemical potential of a constituent, say the mth, existing as a pure solid phase, can be written as a function of temperature only,

$$\mu_m = \mu_m^0(T) \tag{2-18}$$

for the range of pressures commonly considered. Eq. 2-16 can be written

$$K_i = f_i \prod_j{}' (f_j)^{-\nu_{ij}} \tag{2-19}$$

where

$$RT \ln K_i = \sum_{j=1}^{c} \nu_{ij}\mu_j^0 - \mu_i^0 \tag{2-20}$$

and where the operator Π_j' denotes that the repeated product is to be taken over gaseous components only. The thermodynamic equilibrium constant K_i of the reaction (2-5) for the formation of the ith dependent constituent from the components, is a function of the temperature only. Eq. 2-20 is valid both for homogeneous gas mixtures and for heterogeneous systems composed of a gas mixture in equilibrium with a single pure solid phase. The equilibrium constants K_i are most easily obtained by the relations,

$$K_i = K_{f_i} \prod_{j=1}^{c} (K_{f_j})^{-\nu_{ij}} \tag{2-21}$$

where K_{f_i} and K_{f_j} are the equilibrium constants for the formation of the ith dependent constituent and jth component, respectively, from the elements.

An activity coefficient γ_k of the kth gaseous species is defined by the relation,

$$f_k = X_k\gamma_k p$$
$$\lim_{p\to 0} \gamma_k = 1 \tag{2-22}$$

where X_k is the mole fraction of the kth species. Eq. 2-19 can be written in the form,

$$X_i = k_i \prod_j{}' X_j^{\nu_{ij}} \tag{2-23}$$

$i = c + 1, \cdots, s$, where the k_i are functions of temperature and pressure given by

$$k_i = K_i p^{\nu_i - 1} \Gamma_i$$

$$\Gamma_i = \gamma_i^{-1} \prod_j{}' \gamma_j^{\nu_{ij}} \tag{2-24}$$

with

$$\nu_i = \sum_j{}' \nu_{ij} \tag{2-25}$$

In general, the Γ_i are functions of the temperature and pressure and depend weakly on the composition. For an ideal gas mixture $\Gamma_i = 1$ for every i.

Computational Methods for the Ideal Gas. The composition of a system at chemical equilibrium is easily calculated where there is only a single reaction (one dependent constituent) to be considered. In this case, the concentration of each constituent can be related to a single variable, "the degree of reaction," and the solution of the mass action equation is straightforward. Difficulties are encountered if this method is extended to a consideration of two simultaneous equilibria, and when the number of such simultaneous equilibria becomes large, the ordinary methods become very laborious. A systematic general procedure for calculating the equilibrium composition of a system of many constituents has been described [*4*]. In that publication, the method was developed for systems of a very general nature. In subsequent communications [*5,6*] application of the method was restricted to the calculation of the equilibrium composition of mixtures consisting of a single homogeneous gas phase or of a gas phase in equilibrium with a single pure solid phase, and it was assumed that the gas phase is adequately described by the ideal-gas equation of state. By taking advantage of these restrictions, it was possible to formulate a computational method, applicable to these particular cases, which is substantially simpler and more systematic than the more general method. In addition, it was possible to formulate a special method, applicable in restricted cases, that has the advantage of converging more rapidly than the more systematic general method.

In this article we shall limit consideration to the cases of a homogeneous gaseous system or of a heterogeneous system composed of a gas mixture in equilibrium with a single pure solid phase. More general systems can be treated by methods described elsewhere [*4*].

A general method for computations at specified pressure. We consider first the formulation of a generally applicable method for calculating the equilibrium composition for specified pressure. The system is assumed to obey the ideal gas law and to consist either of a homogeneous gas phase or of a gas phase in equilibrium with a pure solid phase.

For each gaseous component, we employ Eq. 2-15 in the form,

$$X_j + \sum_{i=c+1}^{s} \nu_{ij} X_i = \frac{q_j}{n} \tag{2-26}$$

where X_i and X_j are the mole fractions in the equilibrium mixture of the ith dependent constituent and the jth gaseous component, respectively, and n is the total number of moles of gas in the equilibrium mixture corresponding to the normalized constants q_j. In view of Eq. 2-8

$$\sum' X_j + \sum_{i=c+1}^{s} \nu_i X_i = \frac{1}{n} \tag{2-27}$$

where

$$\nu_i = \sum_j' \nu_{ij}$$

The mole fractions of the gaseous constituents are subject to the identity relation,

$$\sum_j' X_j + \sum_{i=c+1}^{s} X_i = 1$$

Therefore, Eq. 2-27 becomes

$$\frac{1}{n} = 1 + \sum_{i=c+1}^{s} (\nu_i - 1) X_i$$

and Eq. 2-26 may be written

$$X_j = q_j - \sum_{i=c+1}^{s} \{\nu_{ij} - q_j(\nu_i - 1)\} X_i \tag{2-28}$$

there being a member of the set (2-28) for each gaseous component.

The computation of the equilibrium composition requires the simultaneous solution of Eq. 2-23 and 2-28. If $X_i \ll X_j$ for all i and j, the solution may be carried out by a simple iteration method [7, pp. 191–195]. An approximate set of values is chosen for the X_j. (In the absence of any criteria for the choice of the initial set, one may take $X_j = q_j$.) Eq. 2-23 are employed in the computation of corresponding values of the X_i. These in turn are employed with Eq. 2-28 for the determination of an improved set of values for the X_j. This iterative process is continued until the difference between successive approximations to the X_j is less than the desired precision of the computation.

The convergence of this simple iteration method is very slow for larger relative values of the X_i, and when the X_i and X_j are of the same order of magnitude, this method may not converge at all for any choice of components. A more powerful computational procedure is provided by the Newton-Raphson method [7, pp. 178, 187]. Eq. 2-28 may be written in the form,

$$F_j = 0 \tag{2-29}$$

for each gaseous component, where

$$F_j = q_j G - X_j - \sum_{i=c+1}^{s} \nu_{ij} X_i$$

$$G = 1 + \sum_{i=c+1}^{s} (\nu_i - 1) X_i$$

We seek the solution of Eq. 2-29 subject to Eq. 2-23. If the functions F_j are expanded in Taylor series about an approximate set of values of the variables X_j with neglect of terms involving derivatives of second and higher orders, there results a set of linear equations which can be compactly represented in the notation of matrices by

$$[A_{jk}^{(r)}][h_k^{(r)}] = [F_j^{(r)}] \tag{2-30}$$

where the rth and $(r + 1)$th approximations to the composition are related by

$$X_j^{(r+1)} = X_j^{(r)}\{1 + h_j^{(r)}\} \tag{2-31}$$

and where the elements of the matrix are given by

$$A_{jk} = U_{jk} - q_j V_k \tag{2-32}$$

where

$$U_{jk} = X_j \delta_{jk} + \sum_{i=c+1}^{s} \nu_{ij} \nu_{ik} X_i$$

$$V_k = \sum_{i=c+1}^{s} \nu_{ik} (\nu_i - 1) X_i$$

and where $U_{jk} = U_{kj}$. The superscript r indicates that the designated quantity is to be evaluated with the rth approximation to the composition of the system, and δ_{jk} is the Kronecker delta.

For computations employing automatic computers, the Newton-Raphson method is advisable. For manual computations employing desk calculations, a graphical method [*14*] is valuable in obtaining rapid convergence of successive iterations that may be preferable if the computers are unfamiliar with the Newton-Raphson method. The method consists in plotting the mole fraction of the ith dependent constituent against the mole fraction of the same constituent from the previous iteration. At equilibrium, the points lie on the 45° line $x = y$. Before equilibrium concentrations have been determined, two such successive points will determine a line that will intersect the 45° line at a concentration that is a better approximation than either of the points. Three such points establish a curve that will intersect the 45° line at an even better approximation. A set of values thus determined can be employed with Eq. 2-28 to determine values of the mole fractions of the components to be employed in the next iteration.

Criteria for the choice of components that results in the most rapid convergence of the iteration process can be developed from the remainders to the two-term Taylor series expansions of functions F_j. However, the resulting expressions are too cumbersome for practical utility, and in practice the convergence will be found to be satisfactory if the components are selected so as to minimize the quantities k_i, $i = c + 1, \cdots, s$.

An example: lean carbon, hydrogen, oxygen, nitrogen systems. We digress from the development of methods for the computation of equilibrium composition to show how the notation and methods that we have introduced are applied to a particular system. Fuel-oxidant mixtures containing an excess of oxygen over the amount required for stoichiometric conversion of the carbon to carbon dioxide and of the hydrogen to steam are called lean mixtures. The predominant constituents of the combustion products satisfying the condition of independence required for their choice as components are CO_2, H_2O, O_2, N_2, except at very high temperatures. If the oxidant is air, a fifth component, argon, must be considered.

The solution of Eq. 2-7 and 2-8 for this choice of components results in the equations

$$q(\mathrm{CO_2}) = \frac{4(\mathrm{C/O})}{2 + (\mathrm{H/O}) + 2(\mathrm{N/O})}$$

$$q(\mathrm{H_2O}) = \frac{2(\mathrm{H/O})}{2 + (\mathrm{H/O}) + 2(\mathrm{N/O})}$$

$$q(\mathrm{O_2}) = \frac{2 - 4(\mathrm{C/O}) - (\mathrm{H/O})}{2 + (\mathrm{H/O}) + 2(\mathrm{N/O})}$$

$$q(\mathrm{N_2}) = \frac{2(\mathrm{N/O})}{2 + (\mathrm{H/O}) + 2(\mathrm{N/O})}$$

relating the normalized stoichiometric constants to the atomic ratios.

In lean mixtures, we assume that methane and ammonia are absent from the equilibrium mixture, an assumption that is confirmed by test calculations for the temperatures of interest in combustion research. The chemical reactions in the form of Eq. 2-5 for the formation of the dependent constituents from the components are

$$
\begin{aligned}
CO_2 - \tfrac{1}{2}O_2 &= CO \\
H_2O - \tfrac{1}{2}O_2 &= H_2 \\
\tfrac{1}{2}O_2 &= O \\
\tfrac{1}{2}H_2O + \tfrac{1}{4}O_2 &= OH \\
\tfrac{1}{2}H_2O - \tfrac{1}{4}O_2 &= H \\
\tfrac{1}{2}N_2 + \tfrac{1}{2}O_2 &= NO \\
\tfrac{1}{2}N_2 &= N
\end{aligned}
$$

in which the symbols of all of the components are written on the left-hand side of the equations. The coefficients ν_{ij} of these equations can be summarized in matrix form,

$$
(\nu_{ij}) = \begin{array}{l} \\ CO \\ H_2 \\ O \\ OH \\ H \\ NO \\ N \end{array}
\begin{array}{c} \begin{array}{ccccc} CO_2 & H_2O & O_2 & N_2 & A \end{array} \\
\begin{bmatrix}
1 & 0 & -\tfrac{1}{2} & 0 & 0 \\
0 & 1 & -\tfrac{1}{2} & 0 & 0 \\
0 & 0 & \tfrac{1}{2} & 0 & 0 \\
0 & \tfrac{1}{2} & \tfrac{1}{4} & 0 & 0 \\
0 & \tfrac{1}{2} & -\tfrac{1}{4} & 0 & 0 \\
0 & 0 & 0 & \tfrac{1}{2} & 0 \\
0 & 0 & 0 & \tfrac{1}{2} & 0
\end{bmatrix} \end{array}
$$

The constants k_i of the mass action expressions, Eq. 2-24, are given for stated values of the temperature and pressure by

$$
\begin{aligned}
k(CO) &= p^{-\frac{1}{2}}K(CO) & k(H) &= p^{-\frac{3}{4}}K(H) \\
k(H_2) &= p^{-\frac{1}{2}}K(H_2) & k(NO) &= K(NO) \\
k(O) &= p^{-\frac{1}{2}}K(O) & k(N) &= p^{-\frac{1}{2}}K(N) \\
k(OH) &= p^{-\frac{1}{4}}K(OH)
\end{aligned}
$$

where the thermodynamic equilibrium constants can be calculated from the equilibrium constants of formation by means of the relations

$$
\begin{aligned}
K(CO) &= \frac{Kf(CO)}{Kf(CO_2)} & K(H) &= \frac{Kf(H)}{[Kf(H_2O)]^{\frac{1}{2}}} \\
K(H_2) &= \frac{1}{Kf(H_2O)} & K(NO) &= Kf(NO) \\
K(O) &= Kf(O) & K(N) &= Kf(N) \\
K(OH) &= \frac{Kf(OH)}{[Kf(H_2O)]^{\frac{1}{2}}}
\end{aligned}
$$

In writing down these relations, we have taken account of the fact that the equilibrium constant of formation of an element is unity.

The mole fractions of the dependent constituents are determined, for assumed values of the mole fractions of the components, by Eq. 2-23 in the form

$$
\begin{aligned}
X(\mathrm{CO}) &= \frac{k(\mathrm{CO})X(\mathrm{CO_2})}{[X(\mathrm{O_2})]^{\frac{1}{2}}} \\
X(\mathrm{H_2}) &= \frac{k(\mathrm{H_2})X(\mathrm{H_2O})}{[X(\mathrm{O_2})]^{\frac{1}{2}}} \\
X(\mathrm{O}) &= k(\mathrm{O})[X(\mathrm{O_2})]^{\frac{1}{2}} \\
X(\mathrm{OH}) &= k(\mathrm{OH})[X(\mathrm{H_2O})]^{\frac{1}{2}}[X(\mathrm{O_2})]^{\frac{1}{4}} \\
X(\mathrm{H}) &= k(\mathrm{H})[X(\mathrm{H_2O})]^{\frac{1}{2}}[X(\mathrm{O_2})]^{\frac{1}{4}} \\
X(\mathrm{NO}) &= k(\mathrm{NO})[X(\mathrm{O_2})]^{\frac{1}{2}}[X(\mathrm{N_2})]^{\frac{1}{2}} \\
X(\mathrm{N}) &= k(\mathrm{N})[X(\mathrm{N_2})]^{\frac{1}{2}}
\end{aligned}
$$

The values of the mole fractions of the components leading to an exact calculation of the equilibrium composition satisfy Eq. 2-29,

$$
F(\mathrm{CO_2}) = 0, \qquad F(\mathrm{H_2O}) = 0, \qquad F(\mathrm{O_2}) = 0, \qquad F(\mathrm{N_2}) = 0
$$

where the functions are given by

$$
\begin{aligned}
F(\mathrm{CO_2}) &= Gq(\mathrm{CO_2}) - X(\mathrm{CO_2}) - X(\mathrm{CO}) \\
F(\mathrm{H_2O}) &= Gq(\mathrm{H_2O}) - X(\mathrm{H_2O}) - X(\mathrm{H_2}) - \frac{X(\mathrm{OH})}{2} - \frac{X(\mathrm{H})}{2} \\
F(\mathrm{O_2}) &= Gq(\mathrm{O_2}) - X(\mathrm{O_2}) + \frac{X(\mathrm{CO})}{2} + \frac{X(\mathrm{H_2})}{2} - \frac{X(\mathrm{O})}{2} + \frac{X(\mathrm{OH})}{4} \\
&\qquad + \frac{X(\mathrm{H})}{4} - \frac{X(\mathrm{NO})}{2} \\
F(\mathrm{N_2}) &= Gq(\mathrm{N_2}) - X(\mathrm{N_2}) - \frac{X(\mathrm{NO})}{2} - \frac{X(\mathrm{N})}{2}
\end{aligned}
$$

and where

$$
G = 1 - \frac{X(\mathrm{CO})}{2} - \frac{X(\mathrm{H_2})}{2} - \frac{X(\mathrm{O})}{2} - \frac{X(\mathrm{OH})}{4} - \frac{3X(\mathrm{H})}{4} - \frac{X(\mathrm{N})}{2}
$$

The application of the Newton-Raphson method to the improvement of an approximate set of values of the mole fractions of the components requires the solution of Eq. 2-30 in the form

$$
\begin{bmatrix}
A_{11} & A_{12} & A_{13} & A_{14} \\
A_{21} & A_{22} & A_{23} & A_{24} \\
A_{31} & A_{32} & A_{33} & A_{34} \\
A_{41} & A_{42} & A_{43} & A_{44}
\end{bmatrix}
\begin{bmatrix}
h(\mathrm{CO_2}) \\ h(\mathrm{H_2O}) \\ h(\mathrm{O_2}) \\ h(\mathrm{N_2})
\end{bmatrix}
=
\begin{bmatrix}
F(\mathrm{CO_2}) \\ F(\mathrm{H_2O}) \\ F(\mathrm{O_2}) \\ F(\mathrm{N_2})
\end{bmatrix}
$$

where $h(\mathrm{CO_2})$, $h(\mathrm{H_2O})$, $h(\mathrm{O_2})$, and $h(\mathrm{N_2})$ are estimates of the fractional errors of a particular approximation to the mole fractions of the respective components, to be employed as indicated by Eq. 2-31 in obtaining the next approximation, and where the coefficients, defined by Eq. 2-32, are given by

$$A_{11} = U_{11} - V_1 q(CO_2) \qquad A_{31} = U_{13} - V_1 q(O_2)$$
$$A_{12} = U_{12} - V_2 q(CO_2) \qquad A_{32} = U_{23} - V_2 q(O_2)$$
$$A_{13} = U_{13} - V_3 q(CO_2) \qquad A_{33} = U_{33} - V_3 q(O_2)$$
$$A_{14} = U_{14} - V_4 q(CO_2) \qquad A_{34} = U_{34} - V_4 q(O_2)$$
$$A_{21} = U_{12} - V_1 q(H_2O) \qquad A_{41} = U_{14} - V_1 q(N_2)$$
$$A_{22} = U_{22} - V_2 q(H_2O) \qquad A_{42} = U_{24} - V_2 q(N_2)$$
$$A_{23} = U_{23} - V_3 q(H_2O) \qquad A_{43} = U_{34} - V_3 q(N_2)$$
$$A_{24} = U_{24} - V_4 q(H_2O) \qquad A_{44} = U_{44} - V_4 q(N_2)$$

with

$$U_{11} = X(CO_2) + X(CO)$$
$$U_{12} = 0$$
$$U_{13} = -\frac{X(CO)}{2}$$
$$U_{14} = 0$$
$$U_{22} = X(H_2O) + X(H_2) + \frac{X(OH)}{4} + \frac{X(H)}{4}$$
$$U_{23} = \frac{X(H_2)}{2} + \frac{X(OH)}{8} - \frac{X(H)}{8}$$
$$U_{24} = 0$$
$$U_{33} = X(O_2) + \frac{X(CO)}{4} + \frac{X(H_2)}{4} + \frac{X(O)}{4} + \frac{X(OH)}{16} + \frac{X(H)}{16} + \frac{X(NO)}{4}$$
$$U_{34} = \frac{X(NO)}{4}$$
$$U_{44} = X(N_2) + \frac{X(NO)}{4} + \frac{X(N)}{4}$$

and

$$V_1 = -\frac{X(CO)}{2}$$
$$V_2 = -\frac{X(H_2)}{2} - \frac{X(OH)}{8} - \frac{3X(H)}{8}$$
$$V_3 = \frac{X(CO)}{4} + \frac{X(H_2)}{4} - \frac{X(O)}{4} - \frac{X(OH)}{16} + \frac{3X(H)}{16}$$
$$V_4 = -\frac{X(N)}{4}$$

The total number of moles n contained in M_0 grams of mixture is equal to G^{-1}, where

$$M_0 = 44.010q(CO_2) + 18.016q(H_2O) + 32.000q(O_2) + 28.016q(N_2) + 39.944q(A)$$

Special methods. It is sometimes possible to take advantage of accidental regularities in the coefficients ν_{ij} to formulate special computational methods applicable to particular systems (cf. A,2). Such methods will sometimes converge in a smaller number of iterations than the general method described above. However, the special method will usually require a longer and more complicated program for each iteration than the general method and only rarely does there result any reduction in computing time.

Mention may be made of two special methods of interest because of the frequent occurrence of systems to which they are applicable. In lean carbon, hydrogen, oxygen, and nitrogen systems at temperatures less than ca. 4000°K, it is appropriate to select CO_2, H_2O, N_2, and O_2 as the components. We employ Eq. 2-15,

$$n_j + \sum_{i=c+1}^{s} \nu_{ij} n_i = q_j \tag{2-15}$$

$j = CO_2, H_2O, N_2, O_2$ and Eq. 2-23 in the form,

$$n_i = k_i n^{1-\nu_i} \prod_j{}' n_j^{\nu_{ij}} \tag{2-33}$$

We define a parameter μ,

$$\mu = \sqrt{nO_2} \tag{2-34}$$

Then an algebraic solution of Eq. 2-33, 2-34, and those of Eq. 2-15 for which $j \neq O_2$ can be found in the form,

$$n_i = n_i(n, \mu) \tag{2-35}$$
$$n_j = n_j(n, \mu)$$

where n is the total number of moles of gas, requiring the solution of equations of degree not higher than quadratic. The solution must satisfy

$$n = 1 + \sum_{i=c+1}^{s} (1 - \nu_i) n_i \tag{2-36}$$

obtained from Eq. 2-8 and 2-15, together with that member of Eq. 2-15 for which $j = O_2$. These two relations can be employed as the basis of a Newton-Raphson procedure for the improvement of an approximate pair of values of n and μ. The explicit program for the method outlined above has been published elsewhere [*8*]. It may be mentioned that this method converges with satisfactory rapidity for the exactly stoichiometric mixture. In this case, in order to avoid divisions by zero, Eq. 2-34 is replaced by

$$\mu = \sqrt{nO_2} + \epsilon$$

where ϵ is a number slightly less than the precision to which nO_2 is to be determined.

In rich homogeneous carbon, hydrogen, oxygen, and nitrogen systems at temperatures less than ca. 4000°K it is appropriate to select CO, H_2, H_2O, and N_2 as components. We define a parameter μ,

$$\mu = \frac{n_{H_2O}}{n_{H_2}} \tag{2-37}$$

If the temperature is high enough so that species such as CH_4, NH_3, etc. can be neglected, an algebraic solution of Eq. 2-33 and 2-37 with those of Eq. 2-15 for which j = CO, N_2 together with the relation,

$$n_{H_2} + n_{H_2O} + \sum_{i=c+1}^{s} (\nu_{iH_2} + \nu_{iH_2O})^{n_i} = q_{H_2} + q_{H_2O} \tag{2-38}$$

can be found in the form (2-35), requiring the solution of equations of degree not higher than quadratic. The solution must satisfy Eq. 2-36 and that member of Eq. 2-15 for which $j = H_2O$. These two relations can be employed as the basis of a Newton-Raphson procedure for the improvement of an approximate pair of values of n and μ. The explicit program for this method has been published elsewhere [*5*]. We note that μ defined by Eq. 2-37 is determinate even when the system contains no hydrogen.

Sufficient conditions for the existence of special methods of the type described have been formulated [*8*].

Modifications for computation at a specified density. Many of the applications of thermodynamics to combustion processes are such that it is natural to specify the state by means of the variables temperature and pressure. The methods for the calculation of the equilibrium composition that have been described assume that the temperature and pressure have been explicitly specified.

In certain applications, it may be more natural to specify the state by means of the variables temperature and density (or specific volume). The methods for calculating the equilibrium composition at a specified density are analogous to those at a specified pressure, but they differ somewhat in detail.

The conditions for equilibrium, Eq. 2-23, can be advantageously written in the form

$$n_i = k_i' \prod{}' n_j^{\nu_{ij}} \tag{2-39}$$

$i = c + 1, \cdots, s$, where the k_i are functions of temperature and density which for the ideal gas, are given by

$$k_i' = K_i\left(\frac{RT\rho}{M_0}\right)^{\nu_i-1} \tag{2-40}$$

Eq. 2-40 determines the concentrations of the dependent constituents for assumed values of the concentrations of the components. In order that these calculations lead to an exact calculation of the equilibrium composition, it is necessary that the results satisfy the stoichiometric conditions, Eq. 2-15, which we write in the form,

$$F_j' = 0 \tag{2-41}$$

where

$$F_j' = q_j - n_j - \sum_{i=c+1}^{s} \nu_{ij}n_i$$

The application of the Newton-Raphson method to the improvement of an approximate set of values of the concentrations of the components requires the solution of equations analogous to Eq. 2-30,

$$[A_{jk'}'^{(r)}][h_k'^{(r)}] = [F_j'^{(r)}] \tag{2-42}$$

where the rth and $(r + l)$th approximations to the composition are related by

$$n_j^{(r+1)} = n_j^{(r)}\{1 + h'^{(r)}\} \tag{2-43}$$

and where the elements of the matrix are given by

$$A_{jk'}' = n_j\delta_{jk'} + \sum_{i=c+1}^{s} \nu_{ij}\nu_{ik'}n_i \tag{2-44}$$

The total number of moles n contained in M_0 grams of gas mixture can be calculated with Eq. 2-36.

The special methods that have been described are easily applied without substantial modification to calculations at specified density. We again employ the stoichiometric conditions as expressed by Eq. 2-15 and the equilibrium conditions as expressed by Eq. 2-29. The constants k_i' appearing in the equilibrium conditions are independent of the total number of moles n for specified temperature and density. Consequently, it is possible to seek a solution of the form

$$n_j = n_j(\mu)$$
$$n_i = n_i(\mu)$$

$j = 1, 2, \cdots, c; i = c + 1, \cdots, s$, for those cases where the introduction of a variable μ results in a separation of variables.

C · COMPUTATIONAL METHODS

Computational Methods for the Imperfect Gas. At the temperatures of interest in the consideration of combustion gases, the assumption of the validity of the ideal gas law is adequate for most purposes. However, for combustion processes occurring at high pressures, as in closed bombs or in gun barrels, the ideal gas equation of state is no longer adequate and it is necessary to consider the effects of gas imperfection on the calculation of the equilibrium composition. We here indicate the modifications to the previously described computation methods that are necessary to take account of gas imperfections, and we assemble the relations required for this purpose. A comprehensive discussion of the evaluation of the thermodynamic properties of mixtures of real gases, with derivations of the necessary relations, has been given by Beattie [9].

For calculations that are carried out at a specified pressure, the mass action constant k_i is related to the thermodynamic equilibrium constant K_i by Eq. 2-24,

$$k_i = K_i p^{\nu_i - 1} \Gamma_i$$

where

$$\Gamma_i = \gamma_i^{-1} \prod_j{}' \gamma_j^{\nu_{ij}} \tag{2-45}$$

and where the activity coefficient of the kth constituent is the ratio of the fugacity to the partial pressure of the kth constituent in the gas mixture, tending to unity as the pressure tends to zero. In the general case of non-ideal gas, the Γ_i are functions of temperature, pressure, and composition.

If the equation of state is explicit in the pressure and of the form,

$$\frac{p}{\rho} = \frac{nRT}{M_0} \Phi(p, T) \tag{2-46}$$

where $\Phi\ (p, T)$ is generally a function of the composition as well as of the temperature and pressure, the activity coefficient of the kth constituent of the gas mixture is given by

$$\ln \gamma_k = \int_0^p \left(\Phi - 1 + n \frac{\partial \Phi}{\partial n_k}\right) \frac{dp}{p} \tag{2-47}$$

This expression can be evaluated explicitly for a particular equation of state and employed to determine the activity coefficients of the constituents of the mixture as functions of temperature, pressure, and, in general, composition. Eq. 2-45 can then be employed to evaluate the functions Γ_i that determine the effect of gas imperfection on the mass action constants k_i.

If the activity coefficients are functions of the composition of the system, the Γ_i must be evaluated with each successive approximation to the composition. In order to calculate the Γ_i to be employed in the $(r + l)$th iteration, one may employ values of the mole fractions of the components given by Eq. 2-31,

$$X_j^{(r+1)} = X_j^{(r)}\{1 + h_j^{(r)}\}$$

and estimates of the corresponding mole fractions of the dependent constituents

$$X_i^{(r+1)} = X_i^{(r)}\left\{1 + \sum_j{}' \nu_{ij}h_j^{(r)}\right\} \tag{2-48}$$

which are correct to the first power in the $h_j^{(r)}$. The Γ_i are usually relatively insensitive to the composition, and, consequently, it is frequently sufficiently accurate to employ the rth approximation to the composition in calculating the Γ_i to be employed in the $(r + l)$th iteration.

Lewis and Randall [10] have proposed as a basis for an approximate thermodynamic treatment of imperfect gas mixtures the rule,

$$f_k = X_k f_k^0(\mathrm{p}) \tag{2-49}$$

where $f_k^0(\mathrm{p})$ is the fugacity of the kth constituent as a pure gas at the temperature and total pressure of the mixture. If the kth constituent obeys the gas law,

$$pV = n_k RT\Phi_k(p, T) \tag{2-50}$$

where Φ_k is a function of pressure and temperature only and V is the volume occupied by n_k moles at temperature T and pressure p, the Lewis and Randall rule results in the expression,

$$\ln \gamma_k = \int_0^p [\Phi_k - 1]\frac{dp}{p} \tag{2-51}$$

for the activity coefficient of the kth constituent. Where applicable, Eq. 2-51 is more convenient than the exact form, Eq. 2-47, since it is independent of the composition of the mixture. The range of validity of the approximate expression for a particular application may be determined by comparison between results obtained from Eq. 2-47 and 2-51 with an explicit form of the equation of state.

For calculations at a specified density, the equation of state may be assumed to be of the form,

$$\frac{p}{\rho} = \frac{nRT}{M_0}\Psi(\rho, T) \tag{2-52}$$

where $\Psi(\rho, T)$ is generally a function of composition as well as of temperature and density. The mass action constants k_i' for this case can be related to the corresponding thermodynamic equilibrium constants by the expressions,

$$k_i' = K_i\left(\frac{RT\rho}{M_0}\Psi\right)^{\nu_i-1} \tag{2-53}$$

The activity coefficient γ_k of the kth constituent of the mixture is given by the general thermodynamic relation,

$$\ln \gamma_k = \int_0^\rho \left(\Psi - 1 + n\frac{\partial \Psi}{\partial n_k}\right)\frac{d\rho}{\rho} - \ln \Psi \tag{2-54}$$

This expression can be evaluated explicitly for a particular equation of state and employed to determine the activity coefficients of the constituents of the mixture as functions of temperature, density, and, in general, composition. Eq. 2-45 can then be employed to evaluate the functions Γ_i that determine the effect of gas imperfection on the mass action constants k_i'.

If the activity coefficients are functions of the composition of the system, the Γ_i must be evaluated with each successive approximation to the composition. In order to calculate these quantities for the $(r + l)$th iteration, one may employ an estimate to the composition given by Eq. 2-43,

$$n_j^{(r+1)} = n_j^{(r)}\{1 + h_j^{(r)}\}$$

and by the analogues of Eq. 2-48,

$$n_i^{(r+1)} = n_i^{(r)}\left\{1 + \sum_j{}' \nu_{ij}h_j^{(r)}\right\} \tag{2-55}$$

As a basis for the approximate thermodynamic treatment of gas mixtures, Gibbs [*2*, pp. 155–158] has proposed that the pressure of a mixture of gases be assumed to be equal to the sum of the pressures of the different gases as existing each by itself at the same temperature and with the same value of its chemical potential. If the kth constituent obeys the gas law,

$$pV_k = RT\Psi_k(T, V_k) \tag{2-56}$$

where V_k is the molar volume of the kth constituent, then the assumption of Gibbs has been shown by Beattie [*9*] to lead to the approximate relation,

$$\ln \gamma_k = \int_0^\rho \left(\Psi_k - 1 + n_k\frac{\partial \Psi_k}{\partial n_k}\right)\frac{d\rho}{\rho} - \ln \Psi_k \tag{2-57}$$

for the activity coefficient of the kth constituent. This expression is not in general independent of the composition of the mixture. It may, however, lead to simpler expressions than those derived from the general relation, Eq. 2-54. The range of validity of the approximate expression for a particular application may be determined by comparison between results obtained from Eq. 2-54 and 2-57 with an explicit form of the equation of state.

Computational Methods for Heterogeneous Systems. The computational methods described above are applicable to homogeneous gas systems and to the gas phase of heterogeneous systems in which the gas phase is in equilibrium with a single pure solid phase. Methods for systems of more general nature have been described in [4].

The gross composition of the system can be described (Art. 2) by the specification of independent atomic ratios, say R_k, $k = 1, 2, \cdots, c - 1$, or by an equal number of linearly independent combinations of the atomic ratios, where we assume that the number of elements contained by the system is equal to the number of components c. If the system exists in a single phase, it is necessary according to the phase rule to specify values of each of the $c - 1$ ratios R_k in order that the equilibrium conditions be determinate for a specified thermodynamic state. If the system exists in two phases, it possesses one less degree of freedom and it is necessary to specify values of only $c - 2$ ratios R_k in order that the equilibrium conditions be determinate. This implies the existence of an implicitly defined relation,

$$F(R_1, \cdots, R_{c-1}; p, T) = 0 \tag{2-58}$$

For any specified values of the state variables, Eq. 2-58 represents a surface in the space of $c - 1$ dimensions with coordinates $R_1, \cdots, R_{c-1}$, which is the limiting surface for the coexistence of the solid and gaseous phases. Eq. 2-58 can be transformed, in practice by numerical methods from a tabular presentation of the relations, to a relation defining a critical value of one of the ratios, R_1 say, as a function of the remaining ratios and of the thermodynamic state.

$$R_1^* = R_1^*(R_2, \cdots, R_{c-1}; p, T) \tag{2-59}$$

If the gross composition of a system is defined by a set $(R_1, \cdots, R_{c-1})$ and if the representative point for this system lies in the volume representative of the heterogeneous system, this volume being bounded by the surface, Eq. 2-58 or Eq. 2-59, then the appropriately formulated computational program for the heterogeneous system is employed to determine the composition of the gaseous phase of the mixture. If the solid component is designated by the index $j = j^*$, then the number of moles n_{j^*} of this component in equilibrium with a weight M_0 (Eq. 2-9) of the gaseous phase is given by the relation

$$n_{j^*} = q_{j^*} - \frac{1}{G} \sum_{i=c+1}^{s} \nu_{ij^*} X_i \tag{2-60}$$

or by

$$n_{j^*} = q_{j^*} \sum_{i=c+1}^{s} \nu_{ij^*} n_i \tag{2-61}$$

where q_{j^*} is determined, together with the other stoichiometric constants for the gaseous components, by Eq. 2-7 and 2-8 for specified values of the parameters descriptive of the gross composition of the system.

THE USE OF AUTOMATIC COMPUTING EQUIPMENT. The calculation of the equilibrium composition is essentially sequential in nature. The program consists of a succession of elementary arithmetic operations, and it requires the retention in some kind of memory of various intermediate results. The calculations can thus be carried out by automatic sequential calculators of adequate memory capacity. If the program is carefully formulated in its details, it turns out that the memory requirements are quite modest. A number of calculations for four component systems have been successfully carried out on the ENIAC calculator [*11*], which presently provides facilities for the long term storage of sixteen ten-digit numbers. The new automatic calculators are provided with much larger memory capacity, and the formulation of programs for these calculators will occasion no difficulty. The IBM Card Programmed Electronic Calculator is commercially available. It can be operated as a sequence calculator of moderate speed and with ample internal memory capacity for the problem at hand. When sequence equipment is employed for a systematic series of computations, it is possible to employ results already obtained in obtaining the first approximation for a new calculation.

With the exception of the IBM Card Programmed Electronic Calculator, existing punched-card computing equipment must usually be operated in parallel fashion. By this, we mean that all of the systems of interest are taken as a group through the succession of steps specified by the program. Intermediate results are stored in punched cards. Although considerably less efficient than sequential operation, calculation in this manner can effect substantial savings in cost and time over manual methods. It is usually necessary to wire a control panel for each of the steps in the program. The number of basically different panels required by the methods described is held to a minimum [*5*].

When parallel type equipment is employed, it is convenient to linearize Eq. 2-23 or Eq. 2-39 by taking the natural logarithm of both sides of the equations. It is then necessary, during each iteration, to determine the logarithms of each of the X_i (or n_i) and to determine the exponential function of $\ln X_i$ (or $\ln n_i$). These operations can be efficiently carried out by standard "table look-up" methods employing punched-card tables of the logarithm and exponential functions. When a large number of systems is to be considered, the first approximations to the composition are conveniently obtained by subtabulation of the results for a small number of initial calculations, which may be carried out manually or which may involve a preliminary run on the computing machines.

C,3. Calculation of the Thermodynamic Properties at Equilibrium. The thermodynamic properties of a gas mixture of known composition are easily determined by means of well-known relations that are discussed in detail in standard treatises on thermodynamics (see also I,C). When these relations are written within the framework of the notational structure that has been developed, it is possible by rearrangement to phrase them in forms that are particularly well adapted to economical computation.

The gross composition has been quantitatively described for the calculation of the equilibrium composition by the parameters q_j, $j = 1, 2, \cdots, c$. The normalization of the q_j, expressed by Eq. 2-8, implies that *q_j is the number of gram moles of the jth component in the hypothetical mixture consisting of components only whose gaseous phase is of weight M_0*, where

$$M_0 = \sum_j{}' q_j \mathfrak{M}_j \tag{3-1}$$

and where $\mathfrak{M}_j$ is the molecular weight of the jth component. The total number of moles of gas contained in a gas phase of weight M_0 has been shown to be

$$n = \frac{1}{G} \tag{3-2}$$

where G can be computed by the relation

$$G = 1 + \sum_{i=c+1}^{s} (\nu_i - 1)X_i \tag{3-3}$$

We have also shown that

$$n = 1 + \sum_{i=c+1}^{s} (1 - \nu_i)n_i \tag{3-4}$$

for a computation conducted in terms of the number of moles n_k per M_0 grams of the kth constituent, $k = 1, 2, \cdots, s$, instead of the corresponding mole fraction X_k. The mean molecular weight $\mathfrak{M}$ of the gas phase, defined by

$$\mathfrak{M} = \sum_{k=1}^{s} X_k \mathfrak{M}_k \tag{3-5}$$

may evidently be calculated by

$$\mathfrak{M} = M_0 G = \frac{M_0}{n} \tag{3-6}$$

We have assumed that the gas phase is described by a law in one of the alternative forms

$$\frac{p}{\rho} = \frac{RT}{\mathfrak{M}} \Phi(p, T) \tag{3-7}$$

$$\frac{p}{\rho} = \frac{nRT}{M_0} \Psi(\rho, T) \tag{3-8}$$

where $\Phi = \Psi = 1$ for an ideal gas. The first expression determines the density as a function of pressure, temperature, and composition. The second expression determines the pressure as a function of density, temperature, and composition. The molar volume V and the density are related by

$$V = \frac{\mathfrak{M}}{\rho} = \frac{M_0}{n\rho} \tag{3-9}$$

The enthalpy and energy of a system are undefined to the extent of an arbitrary additive constant. This fact is of no consequence in applications, since one always wishes to know the difference of enthalpy (or energy) between two different states, and this difference is unambiguously defined. The enthalpy or energy of an equilibrium mixture of combustion gases must therefore be evaluated relative to some convenient state taken as standard. Evidently, in calculating the difference of enthalpy or energy between two states, the functions for each state must be evaluated relative to the same standard state.

As standard state for the equilibrium mixture of combustion gases, it is usually convenient to employ the stoichiometrically equivalent mixture of elements, each in its customary standard state. Alternative standard states may be preferable in some cases and can be employed without substantial change to the computational procedure (A,2).

Computations for the ideal gas at specified pressure. We define the relative specific enthalpy of a gas mixture as the difference between the enthalpy of unit weight of the mixture and the enthalpy of the stoichiometrically equivalent mixture of elements in their standard states. For an ideal gas mixture, we write

$$h^*(p, T) = \frac{1}{\mathfrak{M}} \left\{ \sum_{k=1}^{s} X_k (H^0 - E_0^0)_k + \sum_{k=1}^{s} X_k [\Delta E_{f(0)}^0]_k \right\} \tag{3-10}$$

where the asterisk is employed to specify that the quantity is evaluated for an ideal gas mixture. In this expression, $(H^0 - E_0^0)_k$ is the change in enthalpy when one mole of the kth constituent is taken from its standard state at 0°K to a state of temperature T and sufficiently low pressure such that the ideal gas law and the Gibbs-Dalton law are applicable, and $[\Delta E_{f(0)}^0]_k$ is the change of energy (equal at 0°K to the change of enthalpy) accompanying the formation of one mole of the kth constituent in its standard state from the elements in their standard states. The quantity $(H^0 - E_0^0)_k$ is a function of the temperature only and can be calculated by statistical methods from the spectroscopically observed energy levels of the kth constituent (I,A); $[\Delta E^0_{f(0)}]_k$ is a constant for a given k that can be deduced from appropriate calorimetric experiments. The relative specific enthalpy h^* is a function of both temperature and pressure since the mole fractions X_k are functions of both temperature and pressure.

Eq. 2-28 can be written in the form,

$$X_j = q_j G - \sum_{i=c+1}^{s} \nu_{ij} X_i \tag{3-11}$$

and used to eliminate the mole fractions of the components from Eq. 3-10. Then, after rearrangement, one obtains

$$h^*(p, T) = \eta_J(T) + \eta_I(p, T) \tag{3-12}$$

where

$$\eta_J(T) = \frac{1}{M_0} \sum_{j=1}^{c} q_j \{ (H^0 - E^0_0)_j + (\Delta E^0_{f(0)})_j \}$$

$$\eta_I(p, T) = \frac{1}{\mathfrak{M}} \sum_{i=c+1}^{s} X_i (\Delta H^0_f)_i$$

and where $(\Delta H^0_f)_i$ is the change in enthalpy accompanying the formation of one mole of the ith dependent constituent from the components at temperature T.

$$(\Delta H^0_f)_i = (H^0 - E^0_0)_i + [\Delta E^0_{f(0)}]_i - \sum_{j=1}^{c} \nu_{ij} \{ (H^0 - E^0_0)_j + [\Delta E^0_{f(0)}]_j \}$$

The function η_J is a function of temperature requiring only a knowledge of the gross composition. It is independent of pressure and can be evaluated without reference to the equilibrium composition of the system. The summation for the function η_I extends only over the dependent constituents of the mixture. The function η_I is independent of the choice of standard state for the system. Eq. 3-12 can be employed more economically for computation than Eq. 3-10.

The specific entropy of an ideal gas mixture is given by

$$s^*(p, T) = \frac{1}{\mathfrak{M}} \left\{ \sum_{k=1}^{s} X_k S^0_k - R \sum_{k=1}^{s} X_k \ln X_k - R \ln p \right\} \tag{3-13}$$

where S^0_k is the entropy of the kth constituent as an ideal gas at unit pressure and temperature T. This quantity can be calculated by statistical methods from the spectroscopically determined energy levels of the kth constituent.

Employing Eq. 3-11 and rearranging, the first summation can be written in the form

$$\sum_{k=1}^{s} X_k S_k^0 = G \sum_{j=1}^{c} q_j S_j^0 + \sum_{i=c+1}^{s} X_i(\Delta S_i^0) \tag{3-14}$$

where

$$(\Delta S_i^0) = S_i^0 - \sum_{j=1}^{c} \nu_{ij} S_j^0$$

Employing Eq. 2-23 and Eq. 3-11, the second summation can be written in the form

$$\sum_{k=1}^{s} X_k \ln X_k = (G - 1) \ln p + G \sum_{j=1}^{c} q_j \ln X_j + \sum_{i=c+1}^{s} X_i \ln K_i \tag{3-15}$$

Furthermore, for gaseous equilibria,

$$R \ln K_i = (\Delta S_i^0) - \frac{1}{T} (\Delta H_f^0)_i \tag{3-16}$$

Combining Eq. 3-13 to 3-16, one finally obtains

$$s^*(p, T) = \sigma_J(T) + \sigma_I(p, T) - \frac{R}{M_0} \ln p \tag{3-17}$$

where

$$\sigma_J(T) = \frac{1}{M_0} \sum_{j=1}^{c} q_j S_j^0 - \frac{R}{M_0} \sum_{j=1}^{c} q_j \ln q_j$$

$$\sigma_I(p, T) = -\frac{R}{M_0} \sum_{j=1}^{c} q_j \ln \frac{X_j}{q_j} + \frac{1}{T} \eta_I(p, T)$$

The function σ_J is a function of temperature requiring only a knowledge of the gross composition. It is independent of pressure and can be evaluated without reference to the equilibrium composition of the system. In evaluating σ_I, it is only necessary to look up logarithms of quantities equal in number to the number of components, and with arguments that are of order of magnitude of unity, since the second term in the expression for σ_I involves η_I, a function that is employed in determining the specific enthalpy. Eq. 3-17 is thus in a form that is substantially better suited for computation than is Eq. 3-13.

An expression for the specific heat at constant pressure is obtained by differentiating Eq. 3-12.

$$c_p^*(p, T) = \eta_J'(T) + \frac{\partial \eta_I(p, T)}{\partial T} \tag{3-18}$$

where

$$\eta_J'(T) = \frac{1}{M_0}\sum_{j=1}^{c} q_j(C_p^0)_j$$

$$\frac{\partial \eta_I(p, T)}{\partial T} = \frac{1}{\mathfrak{M}}\sum_{i=c+1}^{s} X_i(\Delta C_p^0)_i - \eta_I \frac{1}{G}\left(\frac{\partial G}{\partial T}\right)_p + \frac{1}{\mathfrak{M}}\sum_{i=c+1}^{s}\left(\frac{\partial X_i}{\partial T}\right)_p (\Delta_i H^0) \tag{3-19}$$

and where

$$(\Delta C_p^0)_i = (C_p^0)_i - \sum_{j=1}^{c} \nu_{ij}(C_p^0)_j$$

The first terms of Eq. 3-18 and 3-19 give the frozen composition contribution to the heat capacity. The remaining terms account for the effect on the heat capacity of changes in composition due to a change in temperature at constant pressure.

Several methods are available for evaluating the second term of Eq. 3-18. The simplest and least accurate is to carry out the indicated differentiation of the function η_I by numerical methods. It is to be noted that this procedure is considerably more accurate than a numerical differentiation of the complete enthalpy function h^*, since η_I is generally a small part of h^*.

A more accurate numerical method is also available. We define a function φ of temperature and pressure by

$$\varphi(p, T) = \eta_I(p, T) - \frac{1}{\mathfrak{M}}(p)\int_{T_0}^{T}\sum_{i=c+1}^{s} X_i(T')(\Delta C_p^0)_i\, dT' \tag{3-20}$$

where the integral is to be evaluated by numerical integration along a path of constant p and where T_0 is some temperature for which the integrand vanishes. Then

$$\frac{\partial \eta_I(p, T)}{\partial T} = \frac{1}{\mathfrak{M}}\sum_{i=c+1}^{s} X_i(\Delta C_p^0)_i + \frac{\partial \varphi(p, T)}{\partial T} \tag{3-21}$$

where the function φ is to be differentiated by numerical methods.

The most accurate procedure for evaluating the heat capacity is to carry out an exact calculation of the derivatives appearing in Eq. 3-19. If Eq. 2-23 and 2-28 are differentiated with respect to temperature at constant pressure, and if terms involving the dependent constituents are eliminated from the resulting expressions, there results a set of simultaneous linear equations which can be compactly represented in the notation of matrices by

$$[A_{jk}]\left[\left(\frac{\partial \ln X_k}{\partial \ln T}\right)_p\right] = [f_j] \tag{3-22}$$

$j = k = 1, 2, \cdots, c$, where

$$f_j = \frac{1}{RT}\sum_{i=c+1}^{s}[q_j(\nu_i - 1) - \nu_{ij}]X_i(\Delta H_f^0)_i$$

and where the matrix elements A_{jk} are given by Eq. 2-32. The coefficients of Eq. 3-22 are thus identical with those of Eq. 2-30 for the Newton-Raphson method of determining the equilibrium composition. Upon solution of Eq. 3-22, the derivatives appearing in Eq. 3-19 can be evaluated by

$$\frac{\partial X_i}{\partial T} = \frac{X_i}{T}\left\{\frac{(\Delta H_f^0)_i}{RT} + \sum_{j=1}^{c}\nu_{ij}\frac{\partial \ln X_j}{\partial \ln T}\right\} \tag{3-23}$$

$$\frac{\partial G}{\partial T} = \sum_{i=c+1}^{s}(\nu_i - 1)\frac{\partial X_i}{\partial T} \tag{3-24}$$

The polytropic form of the adiabatic expansion law is derived on the assumption that the heat capacity is independent of temperature (B,1). If the heat capacity is evaluated in connection with applications based upon the polytropic expansion law, it is usually sufficiently accurate to employ the frozen composition approximation,

$$\frac{\partial \eta_f(p, T)}{\partial T} = \frac{1}{\mathfrak{M}}\sum_{i=c+1}^{s}X_i(\Delta C_p^0)_i \tag{3-25}$$

in place of the exact expression, Eq. 3-19.

An expression for the heat capacity at constant volume can be obtained from the thermodynamic relation,

$$c_p - c_V = -T\left(\frac{\partial V}{\partial T}\right)_p^2\left(\frac{\partial p}{\partial V}\right)_T$$

One obtains

$$c_V^* = c_p^* - \frac{R}{\mathfrak{M}}\left(1 - \frac{T}{G}\frac{\partial G}{\partial T}\right)^2\left(1 + \frac{p}{G}\frac{\partial G}{\partial T}\right)^{-1} \tag{3-26}$$

The derivatives appearing in this expression can be evaluated by numerical differentiation of the function G or by exact methods. The temperature derivative is given by Eq. 3-24. If Eq. 2-23 and 2-28 are differentiated with respect to pressure at constant temperature, and if terms involving the dependent constituents are eliminated from the resulting expressions, there results a set of linear simultaneous equations,

$$[A_{jk}]\left[\frac{\partial \ln X_k}{\partial \ln p}\right] = [g_j] \tag{3-27}$$

$j = k = 1, 2, \cdots, c$, where

$$g_j = \sum_{i=c+1}^{s} [g_j(\nu_i - 1) - \nu_{ij}](\nu_i - 1)X_i$$

and where the matrix elements A_{jk} are given by Eq. 2-32 and are identical with those of Eq. 3-22. Upon solution of Eq. 3-27, the pressure derivative of Eq. 3-26 can be evaluated by

$$\frac{\partial G}{\partial p} = \sum_{i=c+1}^{s} (\nu_i - 1)\frac{\partial X_i}{\partial p} \tag{3-28}$$

with

$$\frac{\partial X_i}{\partial p} = \frac{X_i}{p}\left\{(\nu_i - 1) + \sum_{j=1}^{c} \nu_{ij}\frac{\partial \ln X_j}{\partial \ln p}\right\} \tag{3-29}$$

If the calculations are made in connection with applications based upon the polytropic expansion law, it is usually sufficiently accurate to employ the frozen composition approximation,

$$c_p^* - c_V^* = \frac{R}{\mathfrak{M}} \tag{3-30}$$

instead of the exact expression, Eq. 3-25.

The acoustic velocity a is given by

$$a^2 = \left(\frac{\partial p}{\partial \rho}\right)_S = \gamma\left(\frac{\partial p}{\partial \rho}\right)_T \tag{3-31}$$

$$\gamma = \frac{c_p}{c_V}$$

An exact expression involving previously discussed quantities can be obtained from the ideal gas equation of state and Eq. 3-26.

$$a^2 = \frac{p}{\rho}\left\{1 + \frac{p}{G}\frac{\partial G}{\partial p} - \frac{R}{\mathfrak{M}c_p^*}\left(1 + \frac{T}{G}\frac{\partial G}{\partial T}\right)^2\right\}^{-1} \tag{3-32}$$

For approximate calculations, the frozen composition approximation is

$$a^2 = \frac{p}{\rho}\left(\frac{c_p^*}{c_p^* - R/\mathfrak{M}}\right) \tag{3-33}$$

A final comment should be made relative to the evaluation of the heat capacity, adiabatic exponent γ, or acoustic velocity. If the application of these quantities is such as to warrant high precision in their calculation, it is more convenient to employ the exact methods of computation than the less accurate methods based upon numerical differentiation. The elements of the matrix inverse to $[A_{jk}]$ are determined in connection with the calculation of the equilibrium composition (Eq. 2-30). They can be retained for use in the solution of Eq. 3-22 and 3-27. The resulting program is essentially sequential in nature and well adapted to automatic computing machines. On the other hand, the computations involving numerical differentiation are nonsequential in that they require, for an evaluation at a given temperature and pressure, the results for a number of adjacent values of these variables. This may result in difficulties with respect to the input and storage of the necessary quantities, and it will result in a cumbersome and nonsystematic computational program.

Computations for the ideal gas at specified density. The relations that are employed for calculating the thermodynamic properties of an ideal gas mixture at specified density are entirely analogous to those for specified pressures and can be obtained from them by the use of well-known thermodynamic relations.

We define the relative specific energy content of a gas mixture as the difference between the energy content of unit weight of the mixture and the energy content of the stoichiometrically equivalent mixture of elements in their standard states. For an ideal gas mixture,

$$e^*(\rho, T) = \frac{1}{M_0}\left\{\sum_{k=1}^{s} n_k(E^0 - E_0^0)_k + \sum_{k=1}^{s} n_k[\Delta E_{f(0)}^0]_k\right\} \tag{3-34}$$

where $(E^0 - E_0^0)_k$ is the change in energy content when one mole of the kth constituent is taken from its standard state at 0°K to a state of temperature T and sufficiently low pressure such that the ideal gas law is applicable. Employing Eq. 2-15, this expression can be written in a form analogous to Eq. 3-12,

$$e^*(\rho, T) = \epsilon_J(T) + \epsilon_I(\rho, T) \tag{3-35}$$

where

$$\epsilon_J(T) = \frac{1}{M_0}\sum_{j=1}^{c} q_j[(E^0 - E_0^0)_j + [\Delta E_{f(0)}^0]_j$$

$$\epsilon_I(\rho, T) = \frac{1}{M_0}\sum_{i=c+1}^{s} n_i(\Delta E_i^0)$$

and where

$$(\Delta E_i^0) = (E^0 - E_0^0)_i + [\Delta E_{f(0)}^0]_i - \sum_{j=1}^{c} \nu_{ij}[(E^0 - E_0^0)_j + [\Delta E_{f(0)}^0]_j$$

The function ϵ_I is independent of the choice of standard state for the system.

Eq. 3-13 for the specific entropy can be transformed to a form explicit in the density,

$$s^*(\rho, T) = \frac{1}{M_0}\left\{\sum_{k=1}^{s} n_k S_k^0 - R\sum_{k=1}^{s} n_k \ln n_k - nR \ln \frac{nRT\rho}{M_0}\right\} \tag{3-36}$$

where S_k^0 is the molar ideal gas entropy of the kth constituent at unit pressure. Proceeding in a manner that is analogous to the derivation of Eq. 3-17, and employing for this purpose Eq. 2-15 and 2-39, one obtains

$$s^*(\rho, T) = \sigma_J(T) + \sigma_I(\rho, T) - \frac{R}{M_0} \ln \frac{RT\rho}{M_0} \tag{3-37}$$

where σ_J has been defined in connection with Eq. 3-17 and where

$$\sigma_I(\rho, T) = -\frac{R}{M_0}\sum_{j=1}^{c} q_j \ln \frac{n_j}{q_j} + \frac{R}{M_0}(1 - n) + \frac{1}{T}\epsilon_I(\rho, T)$$

An expression for the specific heat at constant volume is obtained by differentiating Eq. 3-35,

$$c_V^*(\rho, T) = \epsilon_J'(T) + \frac{\partial \epsilon_I(\rho, T)}{\partial T} \tag{3-38}$$

where

$$\epsilon_J'(T) = \frac{1}{M_0}\sum_{j=1}^{c} (q_j C_V^0)_j$$

$$\frac{\partial \epsilon_I(\rho, T)}{\partial T} = \frac{1}{M_0}\left\{\sum_{i=c+1}^{s} n_i(\Delta C_V^0)_i + \sum_{i=c+1}^{s} \frac{\partial n_i}{\partial T}(\Delta E_i^0)\right\} \tag{3-39}$$

and where

$$(\Delta C_V^0)_i = (C_V^0)_i - \sum_{j=1}^{c} \nu_{ij}(C_V^0)_j$$

The frozen composition approximation to Eq. 3-39 is

$$\frac{\partial \epsilon_I(\rho, T)}{\partial T} = \frac{1}{M_0}\sum_{i=c+1}^{s} n_i(\Delta C_V^0)_i$$

Numerical methods for a more precise evaluation of this derivative can be formulated that are completely analogous to those described for the case at constant pressure. It is also possible to carry out an exact calculation of the derivatives appearing in Eq. 3-39. If Eq. 2-15 and 2-39 are differentiated with respect to temperature at constant density, and if terms involving the dependent constituents are eliminated from the resulting expressions, there results a set of simultaneous linear equations,

$$[A'_{jk}]\left[\left(\frac{\ln n_k}{\ln T}\right)\right] = [f'_j] \tag{3-40}$$

$j = k = 1, 2, \cdots, c,$ where

$$f'_j = -\frac{1}{RT}\sum_{i=c+1}^{s} \nu_{ij} n_i (\Delta E_i^0)$$

and where the matrix elements A'_{jk} are given by Eq. 2-44. The coefficients of Eq. 3-40 are thus identical with those of Eq. 2-42 for the Newton-Raphson method of determining the equilibrium composition. Upon solution of Eq. 3-40, the derivatives appearing in Eq. 3-39 can be evaluated by

$$\frac{n_i}{T} = \frac{n_i}{T}\left\{\frac{(\Delta E_i^0)}{RT} + \sum_{j=1}^{c} \nu_{ij}\frac{\partial \ln n_j}{\partial \ln T}\right\} \tag{3-41}$$

An expression for the heat capacity at constant pressure can be obtained from the thermodynamic relation

$$c_p - c_V = -T\left(\frac{\partial p}{\partial T}\right)_V^2 \left(\frac{\partial v}{\partial p}\right)_T$$

One obtains

$$c_p^* = c_V^* + \frac{nR}{M_0}\left(1 + \frac{T}{n}\frac{\partial n}{\partial T}\right)^2\left(1 + \frac{\rho}{n}\frac{\partial n}{\partial \rho}\right)^{-1} \tag{3-42}$$

The derivatives appearing in this expression can be evaluated by numerical differentiation of the function n or by exact methods. In view of Eq. 3-4,

$$\frac{\partial n}{\partial T} = \sum_{i=c+1}^{s} (1 - \nu_i)\frac{\partial n_i}{\partial T} \tag{3-43}$$

and

$$\frac{\partial n}{\partial \rho} = \sum_{i=c+1}^{s} (1 - \nu_i)\frac{\partial n_i}{\partial \rho} \tag{3-44}$$

The temperature derivatives of Eq. 3-43 are evaluated by means of Eq. 3-40 and 3-41. If Eq. 2-15 and 2-40 are differentiated with respect to density at constant temperature, and if terms involving the dependent constituents are eliminated from the resulting expressions, there results a set of linear simultaneous equations,

$$[A'_{jk}]\left[\frac{\ln n_k}{\ln}\right] = [g'_j] \tag{3-45}$$

$j = k = 1, 2, \cdots, c$, where

$$g'_j = -\sum_{i=c+1}^{s} \nu_{ij}(\nu_i - 1)n_i$$

and where the matrix elements A'_{jk} are given by Eq. 2-42 and are identical with those of Eq. 3-40. Upon solution of Eq. 3-45, the density derivative of Eq. 3-42 can be evaluated by Eq. 3-44 with

$$\frac{\partial n_i}{\partial \rho} = \frac{n_i}{\rho}\left\{(\nu_i - 1) + \sum_{j=1}^{c} \nu_{ij}\frac{\partial \ln n_j}{\partial \ln \rho}\right\} \tag{3-46}$$

As in the case of calculations at specified pressure, if the application is such as to warrant high precision in the calculation, it is more convenient to employ the exact methods embodied in Eq. 3-40 to 3-46 instead of less accurate methods based upon numerical differentiation. The elements of the matrix inverse to $[A'_{jk}]$ are determined in connection with the calculation of equilibrium composition and can be retained for use in the solution of Eq. 3-40 and 3-45.

The acoustic velocity a is given by Eq. 3-31. An exact expression involving previously discussed quantities can be obtained from the equation of state and Eq. 3-42,

$$a^2 = \frac{p}{\rho}\left\{1 + \frac{\rho}{n}\frac{\partial n}{\partial \rho} + \frac{nR}{M_0 c_V^*}\left(1 + \frac{T}{n}\frac{\partial n}{\partial T}\right)^2\right\} \tag{3-47}$$

The thermodynamic properties of imperfect gas mixtures. Beattie [*9*] has given a comprehensive discussion of the thermodynamics of imperfect gas mixtures with a systematic derivation of the general relations required to take account of the effect of gas imperfection on the calculation of the thermodynamic properties. In the present part, we assemble those relations of interest in the consideration of combustion gases.

If the equation of state is explicit in the pressure and of the form

$$\frac{p}{\rho} = \frac{nRT}{M_0}\Phi(p, T)$$

one obtains the following relations

$$h(p, T) = h^*(p, T) - \frac{RT^2}{\mathfrak{M}} \int_0^p \frac{\partial \Phi}{\partial T} \frac{dp}{p}$$

$$s(p, T) = s^*(p, T) - \frac{R}{\mathfrak{M}} \int_0^p \left[\Phi - 1 + T \frac{\partial \Phi}{\partial T} \right] \frac{dp}{p} \tag{3-48}$$

for the relative specific enthalpy h, and specific entropy s of the real gas mixture at temperature T and pressure p. The quantities designated by an asterisk are the ideal gas contributions and are evaluated as described in an earlier part. The integrals can be evaluated explicitly for particular forms of the equation of state.

If the equation of state is explicit in the density and of the form,

$$\frac{p}{\rho} = \frac{RT}{\mathfrak{M}} \Psi(\rho, T)$$

one obtains the following relations,

$$e(\rho, T) = e^*(\rho, T) - \frac{nRT^2}{M_0} \int_0^\rho \frac{\partial \Psi}{\partial T} \frac{d\rho}{\rho}$$

$$s(\rho, T) = s^*(\rho, T) - \frac{nR}{M_0} \int_0^\rho \left[\Psi - 1 + T \frac{\partial \Psi}{\partial T} \right] \frac{d\rho}{\rho} \tag{3-49}$$

for the relative specific energy content e, and specific entropy s, at temperature T and density ρ. The quantities designated by an asterisk are the ideal gas contributions and are evaluated as described previously. The integrals can be evaluated explicitly for particular forms of the equation of state.

Because the exact expressions for the specific heat at constant pressure or at constant volume become very cumbersome, these quantities are most satisfactorily evaluated by numerical methods. If the computation is at specified pressure, the specific heat capacity may be written in the form

$$c_p(p, T) = c_p^*(p, T) + \frac{\partial}{\partial T} \chi(p, T) \tag{3-50}$$

where c_p^* is the ideal gas contribution to the specific heat, evaluated as described above and where the second term contains the effect of gas imperfection. The function χ can be evaluated by

$$\chi(p, T) = h(p, T) - h(p, T_0) - \int_{T_0}^T c_p^* dt \tag{3-51}$$

where T_0 is any convenient reference temperature, the integration being carried out numerically along a path at constant pressure. The contribution of gas imperfection to the specific heat can then be evaluated by carrying out the indicated numerical differentiation of the function χ. An analogous method can be employed for calculation of the specific heat at constant volume at specified density, and similar methods can be devised to evaluate the acoustic velocity and specific heat ratio.

The use of automatic computing equipment. The methods described for calculating the thermodynamic properties are essentially sequential in nature. They can be applied to automatic sequential calculators in a straightforward way. Parallel type punched-card equipment can also be profitably employed in these calculations in cases where the work volume is comparatively large.

C,4. Evaluation of Performance Factors of Fuel-Oxidant Systems. The utilization of thermodynamic data in the construction of expansion curves and the analysis of working cycles has been described in Sec. B. Calculations based upon the polytropic expansion law are straightforward and require no special consideration. The results of an evaluation of the thermodynamic properties over a range of states consist of a table that is in every way analogous to the familiar Mollier chart and from which such charts can readily be constructed. The table can be directly exploited in the analysis of working cycles in exactly the same way that such charts are employed [*12*], utilizing the numerical processes of direct and inverse interpolation that are analogous to the usual graphical procedures. Methods of employing punched-card equipment have been described [*13*] for carrying out these processes efficiently and rapidly with a punched-card edition of the tabular form of the Mollier chart. (In some cases, the table for the combustion gas must be supplemented by similar auxiliary tables, such as tables for air.) It is worth noting that analyses of this type are free of the approximations limiting the accuracy of the polytropic expansion law, since they employ a tabular presentation of the exact caloric equation of state for the combustion gas.

C,5. Cited References.

1. Brinkley, S. R., Jr. *J. Chem. Phys. 14*, 563–564; 686 (1946).
2. Gibbs, J. W. *Collected Works*, Vol. 1. Longmans, Green & Co, 1928.
3. Guggenheim, E. A. *Modern Thermodynamics by the Method of Willard Gibbs.* Methuen, London, 1933.
4. Brinkley, S. R., Jr. *J. Chem. Phys. 15*, 113–114 (1947).
5. Brinkley, S. R., Jr., and Smith, R. W., Jr. *Proc. Seminar Sci. Computations*, 1949; *Intern. Business Machines Corp.*, New York, 1950.
6. Kandiner, H. J., and Brinkley, S. R., Jr. *Ind. Eng. Chem. 42*, 850–855 (1950).
7. Scarborough, J. B. *Numerical Mathematical Analysis.* Johns Hopkins Univ. Press, 1930.

8. Brinkley, S. R., Jr., and Lewis, B. *U.S. Bur. Mines Rept. Invest. 4806*, 1952.
9. Beattie, J. A. *Chem. Revs. 44*, 141–192 (1949).
10. Lewis, G. N., and Randall, M. *Thermodynamics*. McGraw-Hill, 1923.
11. Brinkley, S. R., Jr., and Lewis, B. *Chem. Eng. News 27*, 2540–2541 (1949).
12. Hottel, H. C., Williams, G. C., and Satterfield, C. N. *Thermodynamic Charts for Combustion Processes*. Wiley, 1949.
13. Brinkley, S. R. Jr. *Ind. Eng. Chem. 43*, 2471–2475 (1951).
14. Carter, J. M. Private communication.

PART TWO

CHEMICAL KINETICS OF COMBUSTION

SECTION D

FUNDAMENTALS OF CHEMICAL KINETICS

HUGH S. TAYLOR

D,1. Introduction. It is pertinent, at the outset, to inquire into the conditions in which the science of chemical kinetics is involved in the problem of high speed propulsion. The majority of propulsive devices depend for their energy and motive power on the occurrence of chemical reactions of one or more substances. The steam engine and steam turbine utilize the heat energy of a burning fuel such as coal, oil, or gas. The internal combustion engine and Diesel engine involve the interactions of hot hydrocarbon fuel-air mixtures. The energy so generated can be transformed by suitable machinery into traction on land, on or under the sea, and, by means of propellers, into the motion of conventional airplanes. Electrically driven vehicles depend directly or indirectly on some process of chemical change. The motions of shells and rockets are acquired by still more energetic chemical processes, involving the rapid reaction or the explosion of suitably chosen chemicals. The advent of the jet-propelled vehicle was made possible by the development of methods for transforming chemical reactants into the thrust of gaseous reaction products at high temperatures and pressures at high velocities. Normally, the reactions involved are combustion or oxidation processes. The combustible material is termed the fuel, and air or oxygen is the normal oxidizing agent. Recently, in the search for the solution of the problem of high speed propulsion, the range of inquiry and of test has passed beyond these more usual ingredients, and it is now possible to consider any energetically rich reaction as a potential propulsive device. In this latter category must now be included the nuclear reactions which a number of atomic nuclei are known to undergo. In the realm of ordinary chemical reactions, however, a variety of processes must also be considered. Thus, the interactions between aniline or hydrocarbons and concentrated nitric acid, so-called self-igniting reactions, are illustrative of energy-rich chemical reactions between two liquids. The exothermic decomposition of nitromethane or of compounds such as acetylene and its derivatives must be included in the list of monopropellants capable of furnishing motive power.

The speed with which a chemical reaction occurs determines, in part, its utility as a source of propulsion energy. The slow combustion of gasoline in, for example, a spirit lamp contrasts markedly in this respect with the same chemical reactions occurring in the internal combustion, Diesel, or jet engines. The slow burning of a stick of dynamite does not yield the same propulsive power that is developed by the ordinary techniques of utilizing dynamite in explosions. The speed of chemical processes is affected by the prevailing conditions of temperature, pressure, and composition of the reacting mixture, and by the presence or absence of other substances including surfaces, as well as on sources of energy introduced into the reacting system from without. Normally, increase of temperature accelerates the rate of chemical reactions. The influence of pressure can be practically illustrated by the free burning of fuels at normal atmospheric pressure and by their failure to burn continuously at very high altitudes. Mixtures of carbon monoxide and air or oxygen when completely dry and free from hydrogen-containing materials burn or explode only with difficulty; in the presence of even small quantities of water vapor, rapid and explosive combustion of carbon monoxide can be achieved. This points to a definite intervention of the water into a sequence of chemical reactions whereby, finally, the carbon monoxide is oxidized to carbon dioxide, but by a different, more rapid reaction path from that which occurs in the absence of the water vapor. The water may be called a catalyst for the combustion process. Surfaces also behave catalytically, accelerating a given over-all process by providing a path of greater speed. Impinging streams of hydrogen and oxygen preheated to 500°C fail to ignite away from a surface, but do so readily when, for example, a rod of silica is introduced into the impinging streams at their junction. The slow process in the absence of a surface is converted into an explosively rapid process initiated at the silica surface.

Chemical kinetics is that branch of physical chemistry which is concerned with the factor of speed or reaction velocity in chemical processes. It seeks to define the rate of chemical reactions in terms of the several possible variables, temperature, pressure, composition, state of aggregation, external energy sources, and presence or absence of catalysts. It seeks a more precise definition of the detailed mechanism of any chemical reaction by a study of the factors that influence the rate at which the over-all process or its component stages occur.

D,2. General Considerations. The simplest reaction system is that in which a single reaction is proceeding without any complicating secondary reactions occurring simultaneously. We may term such a reaction an *elementary reaction*. In principle, and in all cases under suitable conditions, every reaction is reversible so that a reaction in one direction may be complicated by the reverse process occurring simultaneously. The products of the first reaction may also undergo further reactions to produce new products. Normally, therefore, the study of reaction kinetics is concerned with the velocities of a group of *simultaneous reactions*. It is the objective of the science of reaction kinetics to determine the velocity with which each elementary reaction in a group of simultaneous or successive reactions proceeds.

An elementary reaction is characterized by its *order of reaction*, sometimes termed its molecularity. Reactions of the first order or unimolecular reactions involve changes of a single molecule into products. Reactions of the second order or bimolecular reactions involve the reaction between two molecules or an atom and a molecule to yield the products. Third order reactions involve interactions between three species simultaneously. There are few third order reactions between molecules. Reactions between two atoms require the intervention of a third body. Reaction between two radicals or between an atom and a radical may require a third body for its completion. Whether the third body is required in such cases must be ascertained in each specific case. Orders of reaction higher than the third order may be regarded as nonexistent. The low probability of a simultaneous encounter of four or more molecular species is a sufficient reason for this observation. It is obvious that a succession of bimolecular reactions may be the mechanism of reaction of third or higher order reactions (see Art. 11).

The order of an elementary reaction is revealed by the mathematical equation relating the observed velocity of reaction to the concentration of the reactants. Thus, in a first order reaction of the generalized form

$$\mathrm{A} \rightarrow \text{products}$$

we can express the velocity of reaction by the differential equation

$$\frac{dx}{dt} = k_1(a - x) \tag{2-1}$$

where a is the initial concentration of the reactant A and x is the amount of A which has changed after time t. The velocity of reaction is proportional at any moment to the first power of the concentration of the reactant. The constant of proportionality is k_1. This equation may be converted into two equivalent integrated forms:

$$k_1t = \ln\frac{a}{a - x}, \qquad a - x = ae^{-k_1t} \tag{2-2}$$

A plot of the logarithm of the concentration against time will yield a straight line from the slope of which the constant k_1 can be computed. The constant k_1 is evidently the specific reaction rate (Arrhenius, 1887).

A second order reaction of the form

$$2\mathrm{A} \rightarrow \text{products}$$

yields an equation for the velocity of reaction which is proportional at any moment to the square of the reactant concentration:

$$\frac{dx}{dt} = k_2(a - x)^2 \tag{2-3}$$

of which the integrated form is

$$k_2t = \frac{x}{a(a - x)} \tag{2-4}$$

The same equation holds when there are two reactants A and B present in equal initial concentrations. In the more general case, with initial concentrations respectively a and b, the differential equation becomes

$$\frac{dx}{dt} = k_2(a - x)(b - x) \tag{2-5}$$

which on integration gives

$$k_2t = \frac{1}{a - b} \ln \frac{b(a - x)}{a(b - x)} \tag{2-6}$$

The corresponding equations for third order reactions are

$$3A \rightarrow \text{products}$$

$$\frac{dx}{dt} = k_3(a - x)^3, \qquad k_3t = \frac{1}{2}\left[\frac{1}{(a - x)^2} - \frac{1}{a^2}\right] \tag{2-7}$$

and

$$A + B + C \rightarrow \text{products}$$

$$\frac{dx}{dt} = k_3(a - x)(b - x)(c - x) \tag{2-8}$$

$$k_3t = \frac{(b - c)\ln\frac{(a - x)}{a} + (c - a)\ln\frac{(b - x)}{b} + (a - b)\ln\frac{(c - x)}{c}}{(a - b)(b - c)(c - a)} \tag{2-9}$$

For the special case indicated by the equation

$$2A + B \rightarrow \text{products}$$

the corresponding third order equations are

$$\frac{dx}{dt} = k_3(a - 2x)^2(b - x) \tag{2-10}$$

and

$$k_3t = \frac{1}{(a - 2b)^2}\left[\frac{2x(2b - a)}{a(a - 2x)} + \ln\frac{b(a - 2x)}{a(b - x)}\right] \tag{2-11}$$

When one of the reactants in a second or a third order reaction is in large excess, it is evident that its concentration will remain substantially unchanged when the other constituents of the reaction are already consumed. The order of reaction will appear, consequently, to be one unit lower than the actual order. A second order process becomes a pseudo first order reaction. A third order process becomes pseudo second order.

D,3. The Half Life of a Reaction. When a given constituent in a reaction has been reduced in concentration to half its initial value it is convenient to designate the time interval required as the half life $t_{0.5}$. From the equations just developed it follows that, in a first order reaction,

$$t_{0.5} = \frac{1}{k_1} \ln 2 \tag{3-1}$$

and is hence independent of the initial concentration. For the second order reaction, 2A → products, the corresponding equation is

$$t_{0.5} = \frac{1}{k_2 a} \tag{3-2}$$

and for the third order reaction, 3A → products, the equation becomes

$$t_{0.5} = \frac{3}{2k_3 a^2} \tag{3-3}$$

Hence, quite generally for an nth order reaction, nA → products,

$$t_{0.5} \sim \frac{1}{a^{n-1}} \tag{3-4}$$

so that the half life becomes, in its dependence on the initial concentration, an index of the order of reaction. Any other fraction of the life, $t_{0.2}$, etc., can equally well be employed as an index of reaction order.

D,4. The Order of Simultaneous Reactions. The reversible reaction

$$A + B \rightleftarrows C + D$$

represents a simple form of such simultaneous reactions. The differential equation for equal concentrations of A and B is obviously

$$\frac{dx}{dt} = k_2(a - x)^2 - k_2' x^2 \tag{4-1}$$

where k_2 and k_2' are the rate constants of the forward and backward second order processes. The integrated form of this expression becomes

$$k_2 = \frac{x_e}{2ta(a - x_e)} \ln \frac{x(a - 2x_e) + ax_e}{a(x_e - x)} \tag{4-2}$$

In this equation x_e is the value of x when the system reaches equilibrium.

It is evident that, in this particular case, the half life is inversely proportional to the reactant concentration. Two opposing second order reactions behave as a simple second order process in this respect. Bodenstein [1] investigated this case in detail in the hydrogen-iodine-hydrogen iodide reaction determining k_2, k_2', the equilibrium constant $K = k_2'/k_2$, and the influence of temperature on the velocity of forward and backward reactions and on the equilibrium constant. The same reaction was studied by Kistiakowsky [2] over a pressure range of 1–150 atm, determining the influence of the deviations from ideality of the reactant gases. Bodenstein's data were reexamined by Taylor and Crist [3] and extended to deuterium-iodine reactions; they also analyzed the data from the standpoint of statistics (see Art. 8). These kinetic measurements constitute the classic data of reaction kinetics.

The generalized reaction for recombination of two radicals

$$2A \rightleftarrows A_2$$

yields a differential equation of the form,

$$\frac{dx}{dt} = k_2(a - x)^2 - k_1 x \tag{4-3}$$

the integrated form of which is

$$k_2 = \frac{x_e}{t(a^2 - x_e^2)} \ln \frac{x_e(a^2 - x x_e)}{a_2(x_e - x)} \tag{4-4}$$

It is again to be noted that the half life is inversely proportional to the radical concentration.

The variations in pattern of complex and successive reactions are so numerous that it is not possible to give an exhaustive treatment of all the possibilities in this outline. One well-investigated case of successive reactions is set forth in some detail since it provides a general method of treatment of wide applicability in many complex reaction processes. The example selected is the formation of hydrogen bromide from its elements, exhaustively studied by Bodenstein and his coworkers. It finally emerged (Christiansen [4], Polanyi [5], and Herzfeld [6]) that the process occurs in a sequence of reactions

(i) $$Br_2 = 2Br$$

(ii) $$Br + H_2 = HBr + H$$

(iii) $$H + Br_2 = HBr + Br$$

(iv) $$H + HBr = H_2 + Br$$

(v) $$2Br = Br_2$$

The velocity constants of the five successive reactions are designated by

k_1 to k_5. It will be noted that (iv) is the reverse of (ii) and (v) that of (i). The solution to a complex kinetic process of this type is based on a consideration advanced by Bodenstein [7] which assumes that in a brief interval after the start of the reaction there is a stationary state set up in which all intermediate species are present in small but constant concentration. These stationary state concentrations follow from a consideration of the successive steps. Thus, the stationary state concentration of bromine atoms is given by the expression

$$\frac{d[\mathrm{Br}]}{dt} = 2k_1[\mathrm{Br_2}] - k_2[\mathrm{Br}][\mathrm{H_2}] + k_3[\mathrm{H}][\mathrm{Br_2}] + k_4[\mathrm{H}][\mathrm{HBr}] - 2k_5[\mathrm{Br}]^2 = 0 \quad (4\text{-}5)$$

The corresponding expression for hydrogen atoms is

$$\frac{d[\mathrm{H}]}{dt} = k_2[\mathrm{Br}][\mathrm{H_2}] - k_3[\mathrm{H}][\mathrm{Br_2}] - k_4[\mathrm{H}][\mathrm{HBr}] = 0 \quad (4\text{-}6)$$

Subtraction of these two equations gives at once

$$k_1[\mathrm{Br_2}] = k_5[\mathrm{Br}]^2$$

whence

$$[\mathrm{Br}] = \left(\frac{k_1[\mathrm{Br_2}]}{k_5}\right)^{\frac{1}{2}} \quad (4\text{-}7)$$

which is evidently the equilibrium concentration of bromine atoms in the system $Br_2 = 2Br$. Insertion of Eq. 4-7 in Eq. 4-6 permits the solution for the stationary state concentration of hydrogen atoms

$$[\mathrm{H}] = \frac{k_2\left(\frac{k_1}{k_5}\right)^{\frac{1}{2}}[\mathrm{H_2}][\mathrm{Br_2}]^{\frac{1}{2}}}{k_3[\mathrm{Br_2}] + k_4[\mathrm{HBr}]} \quad (4\text{-}8)$$

This is quite evidently not an equilibrium concentration of hydrogen atoms. The rate of formation of hydrogen bromide is given by the expression

$$\frac{d[\mathrm{HBr}]}{dt} = k_2[\mathrm{Br}][\mathrm{H_2}] + k_3[\mathrm{H}][\mathrm{Br_2}] - k_4[\mathrm{H}][\mathrm{HBr}]$$

Introducing the values for [H] and [Br] and carrying through the simplification possible, one obtains the expression

$$\frac{d[\mathrm{HBr}]}{dt} = \frac{2\left(\frac{k_2k_3}{k_4}\right)\left(\frac{k_1}{k_5}\right)^{\frac{1}{2}}[\mathrm{H_2}][\mathrm{Br_2}]^{\frac{1}{2}}}{\frac{k_3}{k_4} + \frac{[\mathrm{HBr}]}{[\mathrm{Br_2}]}} \quad (4\text{-}9)$$

This agrees with the empirical expression originally found by Bodenstein and Lind [8],

$$\frac{d[\mathrm{HBr}]}{dt} = \frac{k[\mathrm{H_2}][\mathrm{Br_2}]^{\frac{1}{2}}}{m + \frac{[\mathrm{HBr}]}{[\mathrm{Br_2}]}} \tag{4-10}$$

The five reaction rate constants determine the values of the empirical constants k and m. Further analysis resulted in a detailed evaluation of all the rate constants. It emerged that the reactions (iii) and (iv) of hydrogen atoms with bromine and hydrogen bromide were very fast reactions (see Art. 5). The reaction (ii) of bromine atoms was a slow process. It is an endothermic reaction to the extent of 16.4 kcal, and this is the minimum energy which must be brought to a collision process to result in reaction. Since the reactions (i) and (v) maintain the equilibrium concentration of bromine atoms at a given temperature, it is evident that reaction (ii) will be the slowest step in the whole sequence, its rate determining the over-all rate. It is the "bottleneck" reaction in the sequence.

At a given temperature the stationary state concentration of bromine atoms, and therefore the velocity of the over-all reaction, can be increased by absorption by bromine of light energy coming to the system from without. With appropriate wavelengths of light, bromine molecules are dissociated into bromine atoms. Bodenstein and his coworkers showed that illumination gave rise to the same sequence of reactions as in the thermal reaction, with the same rate constants. Taking into account the photodissociation of molecular bromine by the absorbed light I_{abs}, measured in quanta of magnitude $h\nu$,

$$\mathrm{Br_2} + I_{\mathrm{abs}} = 2\mathrm{Br}$$

the kinetic expression for the photochemical formation of hydrogen bromide becomes, since [Br] now becomes $(2I_{\mathrm{abs}}/k_5)^{\frac{1}{2}}$,

$$\frac{d[\mathrm{HBr}]}{dt} = \frac{2\left(\frac{k_2k_3}{k_4}\right)[\mathrm{H_2}]\left(\frac{2I_{\mathrm{abs}}}{k_5}\right)^{\frac{1}{2}}}{\frac{k_3}{k_4} + \frac{[\mathrm{HBr}]}{[\mathrm{Br_2}]}} \tag{4-11}$$

This example should serve to emphasize that chemical kinetics is independent of the nature of the process whereby the reactive species are produced, whether thermally, by light, X rays, or fast particles including electrons, neutrons, protons, or alpha particles. Once a given set of intermediates exists in steady state in the reaction system, the velocities with which the various reactions occur are independent of the mode of origin of the intermediates. A word of caution must be inserted, however, since X rays and swift particles can give rise to species, notably ions, which are present in negligible amounts with normal thermal excitation and photoexcitation.

One further variant in the kinetics of the photoreaction between hydrogen and bromine occurs at low light intensity, small diameters of the reaction vessels, or at low pressures. Here, the recombination of bromine atoms in the gas phase is replaced by a recombination on the walls of the container, a first order process in bromine atoms:

$$Br \xrightarrow{\text{wall}} \tfrac{1}{2}Br_2$$

In this case the kinetic expression contains the absorbed light energy to the first power since $[Br] = 2I_{abs}/k_5$, and the rate equation becomes

$$\frac{d[HBr]}{dt} = \frac{2k_2[H_2]\dfrac{2I_{abs}}{k_5}}{1 + \dfrac{k_4}{k_3}\dfrac{[HBr]}{[Br_2]}} \tag{4-12}$$

In all reactions of this type involving atoms (and radicals) as intermediate species, the intrusion of the wall reaction into the recombination processes may be expected either at low reaction pressures or in systems of small cross section. The well-known use of "explosion arresters" in the form of plugs placed in gas lines carrying explosive gas mixtures is one practical application of this "wall effect."

The sequence of reactions included in the over-all process of hydrogen bromide formation constitutes a reaction chain. The chain is initiated by the dissociation of bromine into atoms. It is propagated by reactions (ii) and (iii), each of which results in hydrogen bromide formation and the generation of a species capable of continuing the chain propagation. These two reactions together comprise one unit (or link) in the chain since reaction (iii) reproduces the active species in reaction (ii). Reaction (iv) inhibits hydrogen bromide formation since the hydrogen atoms destroy the desired product. However, as the process involves the simultaneous production of the reactive species [Br] which can carry on the chain, we term the process one of chain inhibition. Reaction (v), which removes reacting species from the system, is a process of chain termination or chain breaking. The number of times the reactions (ii), (iii), and (iv) occur between initiation in (i) and termination in (v) may be termed the chain length, normally expressed in terms of the *net* product formed, related to the number of chains started. In the present case, since reaction (ii) is known to be the slow process and (iii) and (iv) very fast, the chain length obviously depends on the relative rates of (ii) and (v). Since (ii) is very slow at room temperatures, the formation of hydrogen bromide is negligible and the chain length is vanishingly small. As the temperature is raised, the competition of (ii) with (v) changes in favor of the former so that chain lengths of several thousand may be recorded around 300°C.

The reaction of hydrogen and chlorine occurs by a similar chain mechanism. At room temperature, with chain initiation by light energy,

$$Cl_2 + I_{abs} = 2Cl$$

the chain propagation occurs in two stages

$$Cl + H_2 = HCl + H$$

$$H + Cl_2 = HCl + Cl$$

Chain inhibition by hydrogen chloride

$$H + HCl = H_2 + Cl$$

is not important. Hence, since the chain propagation steps are both very fast, it follows, as was first shown by Bodenstein [7], that as many as 10^5–10^6 molecules of hydrogen chloride may result from the initial absorption of one light quantum producing two chlorine atoms. Chain inhibition may occur not only by interaction with a reaction product as in the case last discussed but also with other molecular species present in the system. In the hydrogen-chlorine reaction, nitrogen-containing species (albumens) are extremely efficient traps for chlorine atoms. Such materials inhibit the reaction so severely that negligible amounts of hydrogen chloride are produced until such species are completely chlorinated. The chain reaction displays an induction period. Oxygen molecules are effective traps for hydrogen atoms interacting with them and thus inhibiting hydrogen chloride formation. Such inhibitors enter the kinetic expression, as did hydrogen bromide, in the denominator. Thus, the original Bodenstein expression for photoreaction of hydrogen and chlorine in presence of oxygen is

$$\frac{d[HCl]}{dt} = k\frac{I[Cl_2]}{m + n[O_2]} \tag{4-13}$$

The efficiency of an inhibitor is a function not only of its reactivity with the chain carrier which it traps but also of the chain length in its absence. It is thus understandable why minimal amounts of oxygen exercise an enormous inhibition on the hydrogen-chlorine chain which, in pure gases, reaches a chain length of 10^5–10^6. In this reaction, also, the action of walls in terminating chains is well known. Mixtures of hydrogen and chlorine which react explosively in vessels of large diameter react negligibly when illuminated with similar light intensities in capillary glass tubes. Here, evidently, the chains are efficiently terminated on the glass walls.

D,5. Temperature and the Velocity of Chemical Reaction. The influence of temperature on the velocity of chemical reactions, applicable whether the reactions are homogeneous, induced by radiation, or occurring at surfaces, was formulated empirically by Arrhenius in 1889. The specific velocity constant k varies with temperature according to an equation with two empirical constants A and B

$$k = Ae^{-B/T} = Ae^{-E/RT} \tag{5-1}$$

The plot of $\ln k$ vs. $1/T$ should yield a straight line from the slope of which the activation energy $E = RB$ can be calculated, the intercept on the $\ln k$ ordinate yielding A. Because of the similarity of this temperature variation with that found in the variation of equilibrium with temperature, Arrhenius postulated that, in any reaction system, there was an equilibrium between normal species S_n and activated species S_a, the latter in small concentration and determining the rate of product formation.

$$S_n \rightleftarrows S_a \rightarrow \text{products}$$

Trautz [*9*] in 1914 and later, more specifically, W. C. McC. Lewis [*10*] in 1918 in the case of elementary bimolecular processes, defined the empirical Arrhenius constant A in terms of the collision number Z_{12} between the molecules 1 and 2 of the reactant system. The kinetic theory of gases expresses the collision number by the equation

$$Z_{12} = 2\sqrt{2\pi}\, n_1 n_2 \sigma_{12}^2 \sqrt{\frac{kT}{\mu}} \tag{5-2}$$

where n_1 and n_2 are the number of each species per cm^3, σ_{12} the distance between nuclei at the moment of collision, and μ is the reduced mass, $m_1 m_2/(m_1 + m_2)$.

When the colliding species are identical, the corresponding collision number Z_{11} has the form,

$$Z_{11} = 2n^2\sigma^2 \sqrt{\frac{\pi kT}{m}} \tag{5-3}$$

which can be derived from Z_{12} by setting $n_1 = n_2$ and noting that the number of collisions is halved, since the method of derivation in the former case counts collisions of 1 with 2 and of 2 with 1 whereas these are identical when identical molecules are colliding.

Lewis found that the data of Bodenstein on the rate of decomposition of hydrogen iodide could be closely calculated from the kinetic theory value of Z_{11} for hydrogen iodide and the observed activation energy E_{obs} derived from the slope of the $\ln k$ vs. $1/T$ plot, with the equation

$$k_2 = Z_{11} e^{-E_{obs}/RT} \tag{5-4}$$

On this view the expression $e^{-E/RT}$ becomes actually the ratio of effective collisions to total collisions, and is therefore a measure of the efficiency of the collision process. In the case of the hydrogen iodide reaction with

$E_{obs} = 43.7$ kcal at about 730°K this efficiency is of the order of $e^{-30} \cong 10^{-12}$.

The data of the hydrogen-bromine chain sequence, as analyzed by Christiansen, Polanyi, and Herzfeld, indicated that the exothermic reactions (iii) and (iv), Art. 4, between atoms and molecules occurred on the average with a collision efficiency of the order of 0.1 to 1, i.e. $k \sim Z_{12}$. For endothermic reactions, the collisions would require at least the heat absorbed in the reaction; so that, in the case of the reaction (ii), Art. 4, $Br + H_2 = HBr + H$ with $\Delta H = 16.4$ kcal, the rate of reaction cannot exceed

$$k_2 = Z_{12}e^{-16.4/RT}$$

Actually the experimental work of Jost [*11*] showed that the best value for this rate constant is given by the following expression:

$$\ln k_2 = -\frac{17600}{4.57T} + 0.5 \ln T + 12.308$$

The second term on the right takes account of the influence of temperature on the collision number Z, namely $\sim T^{0.5}$. On this basis the activation energy for reaction (ii) becomes $E_2 = 17.6$ kcal; since the reaction is endothermic by 16.4 kcal, the activation energy of the reverse process, reaction (iv) of Art. 4, $H + HBr = H_2 + Br$ must be $E_4 = 17.6 - 16.4 = 1.2$ kcal. Since it can be shown experimentally that the ratio k_3/k_4 is independent of temperature, over a range from room temperature to 200°C in the photoreaction, it is evident that the activation energies of the two reactions must be very closely alike. Hence E_3 is also $\cong 1.2$ kcal.

If now the experimental value for the rate of reaction (ii) be compared with the collision number Z_{12} and a collision efficiency given by an activation energy of 17.6 kcal, it is found that there is a divergence between the two of the order of 0.1. To account for the discrepancy it was assumed that a steric factor might be involved since a particular line of approach of the two reacting species might be more favorable than some other geometry of collision. The collisional expression for reaction rate would then take the form

$$k = PZ_{12}e^{-E/RT} \tag{5-5}$$

where P is the appropriate steric factor. In the hydrogen-bromine reaction, P had values of 0.1, 1, and 0.1 for the respective reactions (ii), (iii), and (iv). Such values for P are not improbable on the basis of the geometry of collision. When, however, a whole spectrum of "fast" and "slow" reactions with values of P ranging from 10^6 to 10^{-8} was found, it became evident that other interpretations of the P factor in the collision equation were necessary. Indeed, all the terms in the equation require further consideration. The collision number Z can only have a limited significance, since its evaluation is based on the kinetic theory concepts of collision between elastic spheres with diameters deduced from, for example, viscosity data. The activation energy E was an experimental datum requiring measurements of velocity at not less than two temperatures. The nature of the energy involved was not defined other than to say that it represented the minimum energy of the colliding partners for reaction, the fraction of the collisions with an energy in excess of E having the form $e^{-E/RT}$, for collisions involving only two degrees of freedom. For complex molecules where n degrees of freedom can contribute to the required activation energy E the expression for the fraction of the collisions having an excess of energy E takes the form (see Hinshelwood [*11a*, p. 53])

$$\frac{e^{-\frac{E}{RT}}\left(\frac{E}{RT}\right)^{\frac{n}{2}-1}}{\left(\frac{n}{2}-1\right)!} \tag{5-6}$$

which reduces to the simpler form when the number of degrees of freedom $n = 2$.

D,6. The Potential Energies of 2-, 3-, and 4-Atom Systems. The theoretical approach to the concept of activation energy came with the development of ideas concerning the potential energy of systems of two or more atoms.

The potential energy of a system of two atoms forming a diatomic molecule as a function of the distance of separation of the two nuclei can be represented by Fig. D,6a. The zero of potential energy on the ordinate may be taken as the potential energy of the two atoms at infinite distance from each other. As this distance diminishes, attractive forces come into play resulting in an increased kinetic energy of approach and a corresponding decrease in potential energy. At short distances between atoms repulsive forces become increasingly important and finally predominate so that the potential energy passes through a minimum followed by a rise to very large and positive values.

At a distance corresponding to the zero-point energy ($= \frac{1}{2}h\nu$) above the potential energy minimum the lowest vibrational level ($v = 0$) of the diatomic molecule is located. Higher vibrational levels are indicated corresponding to $v = 1, 2, 3, 4$, etc. With each vibrational level are associated rotational energy levels characterized by their own quantum number J. The motion of the vibrating molecule in the lowest vibrational state is described by the potential energy curve around the minimum. The motion is closely harmonic. The molecule in the lowest vibrational state has zero kinetic energy at the distances represented by the extremities of $v = 0$ on the potential energy curve. At the minimum the molecule has the largest kinetic energy, and therefore the molecule spends a minimum of time in such a configuration defined by the distance r_e. At the higher vibrational levels the motion becomes progressively more and more anharmonic till finally the anharmonicity passes over into a complete separation of the two atoms.

Such potential energy curves can be constructed from spectroscopic data when these are available. Morse [12], however, developed a useful empirical equation to describe the potential energy curves of such molecules having the form

$$V = D_e[e^{-2a(r-r_e)} - 2e^{-a(r-r_e)}] \tag{6-1}$$

D_e is the value of V at the minimum in the potential energy curve and is obviously equal in magnitude to the heat of dissociation ΔH of the molecule plus the zero-point energy

$$D_e = \Delta H + \tfrac{1}{2}h\nu_0 = \Delta H + \tfrac{1}{2}hc\omega_e \tag{6-2}$$

where ν_0 and ω_e are the frequencies of the vibrator expressed in sec^{-1} and cm^{-1} respectively. The constant a is related also to the frequency ω and to D_e by the expression $a = 0.1227\omega\sqrt{\mu/D_e}$ where μ is the reduced mass of the molecule $(= m_1m_2/(m_1 + m_2))$. If ω is expressed in wave numbers (cm^{-1}), D_e, which is ordinarily expressed in calories or electron volts, should be expressed in the same units of reciprocal length by use of the appropriate conversion factors (352 wave numbers = 1 kcal; 8100 wave numbers = 1 ev). With ω and D_e the constant a is received in angstrom units (10^{-8} cm) which is the usual form in which r is expressed. Inspection of the Morse equation shows immediately that it satisfies two boundary conditions, namely $V = 0$ for large r and $V = -D_e$ for $r = r_e$.

Two atoms approaching each other may enter a state which is continuously repulsive, indicated by the dashed curve in Fig. D,6a. The interpretation of these two curves in the quantum mechanics arises from the representation of the binding energy P of a system of two atoms as the sum or difference of two quantities, the classical coulombic interactions of the electric charges in the system, A, and a quantum mechanical interchange or exchange energy, α.

$$P = A \pm \alpha \tag{6-3}$$

In the potential energy diagram, the state leading to a stable molecule is given by $P = A + \alpha$; the repulsive state arises from $P = A - \alpha$. It is evident that the exchange forces α are numerically superior to the coulombic forces A.

A system of three univalent atoms X, Y, Z can form only atom-molecule systems X + YZ; Y + ZX; Z + XY. Chemical reaction in such a system involves a transfer of a valence bond from one pair of atoms to another:

$$X + YZ = XY + Z$$

The potential energy of a system of three atoms in any configuration which will be linear or triangular was expressed by London [*12a*] in terms of three magnitudes $A \pm \alpha$, $B \pm \beta$, and $C \pm \gamma$ representing the energies between the three possible pairs of atoms. In the triangular configuration

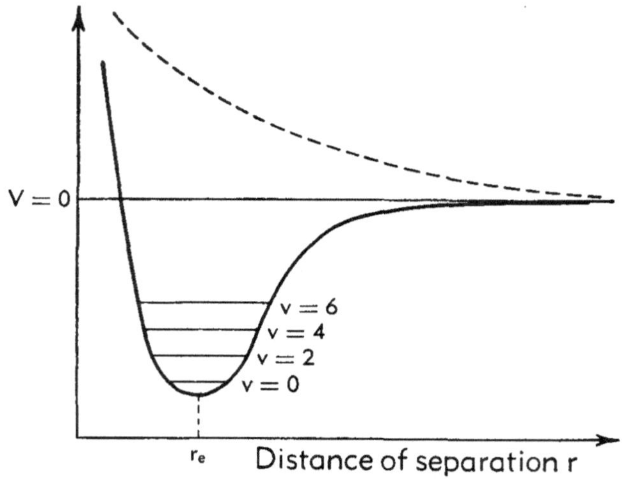

Fig. D,6a. Potential energies of 2-atom system as a function of nuclear separation; attractive and repulsive states.

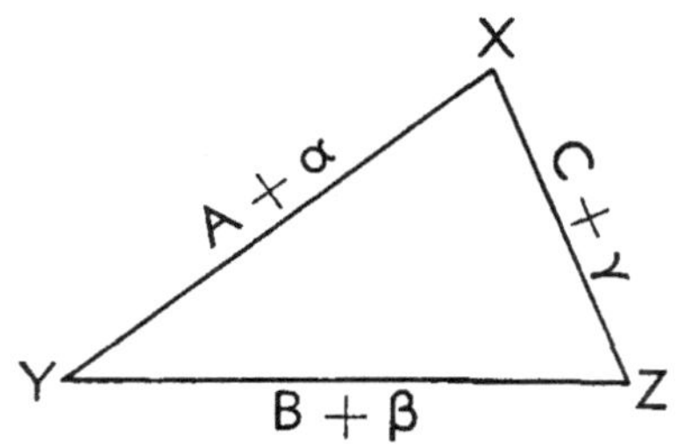

Fig. D,6b. Binding energies of a triangular array of three atoms XYZ.

of Fig. D,6b, the binding energy of the 3-atom system is given by the expression

$$P_{XYZ} = A + B + C + \sqrt{\alpha^2 + \beta^2 + \gamma^2 - \alpha\beta - \beta\gamma - \gamma\alpha} \quad (6\text{-}4)$$

in which the first three terms on the right are the additive coulombic terms, the exchange terms being given by the quantity under the square root sign. This quantity can be derived graphically by representing α, β, and γ by lines of length proportional to the energies in question, set off successively at angles of 60° to one another. The line joining the two extremities of the three-line diagram is, by the properties of 60° triangles, equal to the exchange terms in Eq. 6-4 [*13*].

For 4-atom systems WXYZ involving six interatomic distances, the corresponding binding energy is given by an equation

$$P_{WXYZ} = A_1 + A_2 + B_1 + B_2 + C_1 + C_2 + [(\alpha_1 + \alpha_2)^2 + (\beta_1 + \beta_2)^2 + (\gamma_1 + \gamma_2)^2 - (\alpha_1 + \alpha_2)(\beta_1 + \beta_2) - (\beta_1 + \beta_2)(\gamma_1 + \gamma_2) - (\gamma_1 + \gamma_2)(\alpha_1 + \alpha_2)]^{\frac{1}{2}} \quad (6\text{-}5)$$

where A_1, A_2, α_1, α_2, etc., are the quantities indicated in Fig. D,6c.

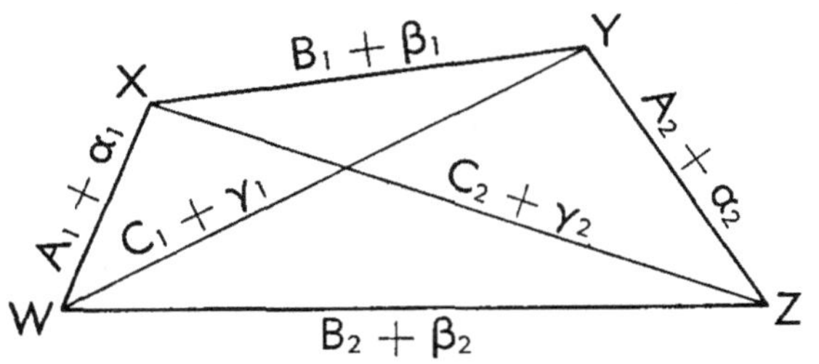

Fig. D, 6c. Binding energies of a planar array of four atoms WXYZ.

D,7. Potential Energy and Activation Energy. The concept of an activation energy follows directly from these ideas concerning the potential energies of 2-, 3-, and 4-atom systems and can conveniently be illustrated by the system of three hydrogen atoms in which the reaction

$$H + H_2 = H_2 + H$$

can occur and can be experimentally studied in the conversion of ortho- to para-hydrogen either thermally at about 500°C or photochemically at room temperatures with atomic hydrogen produced by photosensitization with mercury vapor excited by the resonance line $\lambda = 2536.7A =$ 112 kcal.

Eyring and Polanyi [*14*], using the London formula, showed that the potential energy of a system of three hydrogen atoms was lowest when there was a linear configuration of approach of the atom to the molecule. The potential energies of linear arrays of three hydrogen atoms H_I—H_{II}—H_{III} for varying distances r_1 and r_2 between the three can be represented by an energy contour map of the form shown in Fig. D,7a. The process of interaction between an atom and a molecule can be followed on the map with an initial state in the right-hand bottom valley with $r_2 = 0.75A$ for the molecule H_{II}—H_{III} and r_1 a relatively large distance. The final state is the similar position at the top left-hand corner where r_1 is now 0.75A and r_2 is large. As the atom approaches the molecule, or as r_1 decreases, the energy of the 3-atom system is representable by a point on the map passing from the initial to the final state. The easiest method of transit is along the valley floor over the saddle located at $r_1 = r_2 = 0.9$A in the left angle of the map and then along the bed of the second energy valley to the final state. This progression from initial

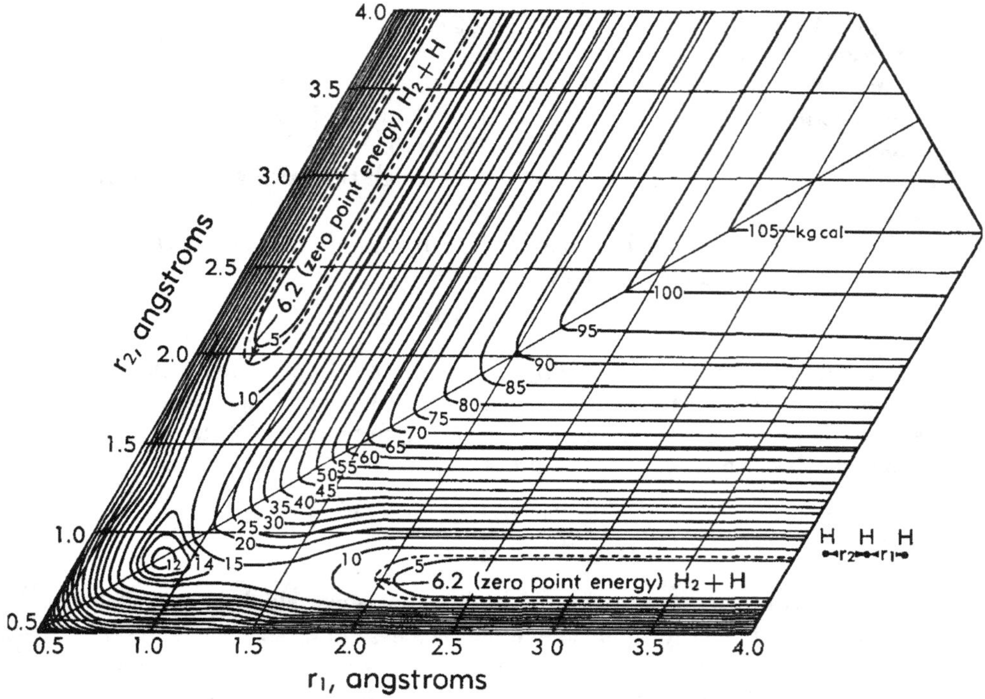

Fig. D,7a. Potential energy contour map of a linear array of three hydrogen atoms.

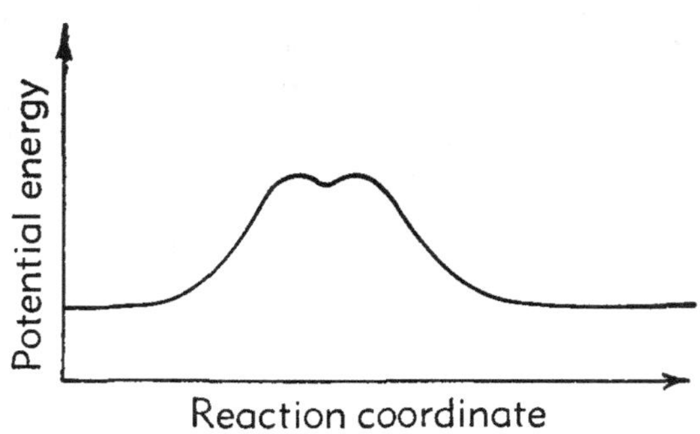

Fig. D,7b. Potential energy as a function of the reaction coordinate.

to final state can be represented by a diagram (Fig. D,7b) plotting energy as a function of the progression, represented as a "reaction coordinate." The slight depression at the saddle point is indicative of a slight potential energy minimum of the 3-atom system. The potential energy maximum is that of the "activated complex" in the process

$$\underset{\text{initial state}}{H_I + H_{II}H_{III}} \rightarrow \underset{\text{activated complex}}{H_I—H_{II}—H_{III}} \rightarrow \underset{\text{final state}}{H_IH_{II} + H_{III}}$$

The potential energy difference between the initial state and the activated complex represents the activation energy of the reaction in the forward direction. Eyring and Polanyi and, in later work, Eyring and his pupils *calculated* thus a best value for this reaction of three hydrogen atoms. The calculation yielded a value of 7 kcal in good agreement with the experimental value of 7.25 kcal. The theoretical calculation of activation energies was thus, in principle, solved, although the calculations involved a great deal of empirical approximation.

By similar methods, in the system of four hydrogen atoms it can be shown that the bimolecular collision process

$$H_IH_{II} + H_{III}H_{IV} = H_IH_{III} + H_{II}H_{IV}$$

has a calculated activation energy in excess of 90 kcal. Given this high value for the activation energy of the bimolecular collision it is at once apparent why the interaction of two hydrogen molecules actually proceeds through the two-stage process

(i) $$H_2 + H_2 = H_2 + H + H$$

(ii) $$H + H_2 = H_2 + H$$

The equilibrium concentration of hydrogen atoms which is established accounts for an energy equal to one-half the heat of dissociation, 103/2 = 51.5 kcal and, for stage (ii), the activation energy, found and calculated, is 7.25 kcal. The summation of these energies indicates them to be energetically easier, by at least 30 kcal, than the simple bimolecular collision.

In the reactions of hydrogen with the halogens, Eyring [*15*] calculated that the bimolecular collision of hydrogen and iodine molecules was energetically easier than the atomic path, while with bromine, chlorine, and fluorine the path via halogen atoms was energetically more favorable, in agreement with experiment. Eyring could calculate that the bimolecular reaction between hydrogen and fluorine was slow, contrary to experimental data then available. It is now known that impurities in fluorine were probably responsible for the earlier experimental results in this system.

D,8. Statistical Treatment of the Activated Complex. Consider the generalized reaction between two reactants A and B proceeding through an activated complex M′ to the formation of reaction products.

$$A + B \rightarrow M' \rightarrow \text{reaction products}$$

The rate of formation of products must obviously be determined by the concentration of activated complex, M′, at the top of the potential barrier separating reactants and products, and by the frequency with which the complex passes the barrier.

$$\text{rate of reaction} = \frac{\bar{c}^*\bar{u}}{\delta} \tag{8-1}$$

where $\bar{c}^*$ is the number of complexes per unit volume in a length δ representing the activated state at the top of the barrier and $\bar{u}$ is the mean velocity of crossing.

Eyring pointed out [*16*] that the activated complex could be regarded as an ordinary molecule, with all the usual thermodynamic and statistical properties, save in one respect. One degree of freedom which, in a normal molecule, would be a vibrational degree of freedom consists, in the activated complex, of a translational motion in the direction of the reaction coordinate leading to a decomposition of the complex to yield the reaction products. Eyring emphasized that such activated complexes could be treated statistically as normal molecules by replacing one degree of freedom of vibration by translation along the reaction coordinate. Thus

$$\bar{c}^* = c^* \frac{(2\pi m^* kT)^{\frac{1}{2}}\delta}{h} \tag{8-2}$$

in which we recognize the second term as the partition function for one degree of freedom of translation, m^* being the mass of the activated complex, and c^* the concentration derivable from the partition functions of all the remaining degrees of freedom of the complex. Since the mean velocity $\bar{u}$ of crossing the barrier is $(kT/2\pi m^*)^{\frac{1}{2}}$, the rate equation becomes

$$\begin{aligned}\text{rate of reaction} &= c^* \frac{(2\pi m^* kT)^{\frac{1}{2}}\delta}{h}\left(\frac{kT}{2\pi m^*}\right)^{\frac{1}{2}} \cdot \frac{1}{\delta} \\ &= c^* \frac{kT}{h}\end{aligned} \tag{8-3}$$

This equation indicates that the effective velocity of crossing the energy barrier is the universal frequency kT/h multiplied by a concentration term c^* whose nature has been already indicated. Comparing this equation with the rate in terms of the specific reaction rate k' and the concentrations of reactants

$$\text{rate of reaction} = k' c_A c_B$$

it follows that

$$k' = \frac{kT}{h}\frac{c^*}{c_A c_B} = \frac{kT}{h} K^* \tag{8-4}$$

where K^* is the equilibrium constant in the process

$$\mathrm{A + B \rightleftarrows M'}$$

with the reservation as to the stated nature of the concentration term c^*, and its consequence in the definition of K^*. Since, thermodynamically, we can represent an equilibrium constant by the expression $\Delta F^* = -RT \ln K^*$ we can therefore write

$$k' = \frac{kT}{h} e^{-\Delta F^*/RT} \tag{8-5}$$

or, using the familiar relation,

$$\Delta F = \Delta H - T\Delta S$$

we may write

$$k' = \frac{kT}{h} e^{-\Delta H^*/RT} e^{\Delta S^*/R} \tag{8-6}$$

where the thermodynamic quantities ΔF^*, ΔH^*, and ΔS^* refer to the free energies, heat contents or enthalpies, and entropy changes involved in K^* as defined. We may term these quantities the changes of free energy, heat content or enthalpy, and of entropy *of activation* of the reaction process.

Further, since the assumed equilibrium process $A + B \rightleftarrows M'$ can also be treated by the methods of statistical mechanics we can also write the specific reaction rate k' in terms of the respective partition functions Q so that,

$$k' = \frac{kT}{h} K^* = \frac{kT}{h} \frac{Q^*}{Q_A Q_B} e^{-E_0/RT} \tag{8-7}$$

where E_0 is the difference between the zero-point energy of the complex and that of the sum of the reactants. That the zero-point energies of these quantities enter the rate equations accounts for the various isotopic separation processes that stem from differences of rate among "light" and "heavy" isotopic atomic or molecular species. Hirschfelder, Eyring, and Topley [*17*] calculated the reaction rates of hydrogen atom isotopes with hydrogen isotope molecules at 1000°K and compared their data with experiments of Farkas and Farkas [*18*]. A. H. Taylor and Crist [*3*] studied the decomposition and synthesis of hydrogen and deuterium iodides, and found that at 700°K the hydrogen iodide rate constant was 1.60 times that of the deuterium iodide. The ratio of the synthesis rates was 1.93. Wheeler, Topley, and Eyring calculated higher values of the two ratios, viz. 1.92 and 2.34 respectively.

These several equations for the specific reaction rate k' are in principle derivable entirely from the thermodynamic and statistical properties of the molecules involved. They are the equations basic to what Eyring has termed the *absolute rate theory* in chemical kinetics.

D,9. Absolute Rate Theory and Collision Theory. A Comparison. The two theories, if exact, would give the same rate constant in every rate process. It is interesting to record that the two theories give the same expression in one case only, that of the combination of two atoms A and B. In this case the activated complex would resemble a diatomic molecule AB save that the vibration is replaced by a translational degree of freedom. The partition function Q^* will therefore contain translational terms and a rotational term

$$Q^* = \frac{[2\pi(m_A + m_B)kT]^{\frac{3}{2}}}{h^3} \cdot \frac{8\pi^2 IkT}{h^2} \tag{9-1}$$

The moment of inertia I of the complex is given by definition as

$$I = \sigma_{AB}^2 \frac{m_A m_B}{m_A + m_B}$$

The partition functions of the two atoms A and B are exclusively translational of the form $Q_A = (2\pi m_A kT)^{\frac{3}{2}}/h^3$. Inserting these values in Eq. 8-7 the expression for k' becomes

$$k' = \sigma_{AB}^2 \left[8\pi kT \left(\frac{m_A + m_B}{m_A m_B} \right) \right]^{\frac{1}{2}} e^{-\frac{E_0}{RT}} \tag{9-2}$$

which is at once recognizable as $Z_{AB}e^{-E_0/RT}$. The absolute rate theory gives a clearer definition of the quantity σ_{AB}. This is strictly defined as the internuclear distance in the complex at the top of the potential energy barrier. This distance is in general smaller than the distances used in collision theory, drawn from kinetic theory of gases, generally from viscosity data.

Applied to molecules rather than to atoms it is at once apparent that $Q^*/Q_A Q_B$ becomes considerably more complicated and the collision number no longer emerges. This is equivalent to saying that the unmodified collision theory applies only to reaction between hard spherical atoms with translational degrees of freedom. With more complex systems, with rotational partition functions of the complex and the many vibrational terms, the simple collision theory becomes the grossest kind of approximation which, by introduction of the probability or steric factor P, it is attempted empirically to refine. The absolute rate theory expressed in Eq. 8-6 when compared with the collisional equation $k' = PZe^{-E/RT}$ reveals that the existence of "fast" or "slow" rates must be dependent on the term $e^{\Delta S^*/R}$. A negative value of ΔS^* will give a slow rate, a positive value a fast process. Thus, in the dimerization of ethylene where the activated complex must closely resemble a butene molecule one expects a negative entropy of activation. Actually, the data indicate an entropy change of -30.1 kcal/deg and a rate of reaction only 10^{-3} times that expected from the collisions and the observed activation energy. The scission of a cycloparaffin to form a diradical will be accompanied by an entropy increase so that a fast reaction is anticipated. The closing of a polymethylene chain to form a cycloparaffin would be accompanied by an entropy decrease so that a slow rate process would be expected. Experiments confirm these conclusions. Soper pointed out [*19*] that P factors parallel the *over-all* entropy changes in several reactions. Thus, in the reaction between dimethylaniline and methyl iodide $P = 0.5 \times 10^{-7}$ while $e^{\Delta S/R} = 0.9 \times 10^{-8}$. In the isomerization of ammonium cyanate solutions $P = 1$ while $e^{\Delta S/R} = 0.1$.

In the most complex rate process involving the formation of an activated complex by the collision of two polyatomic molecules A and B the partition functions, q, involved are: for A, 3 translational, 3 rotational, and $3n_A - 6$ vibrational partition functions; for B the same number of translational and rotational partition functions and $3n_B - 6$ vibrational functions. For the activated complex there will be 3 translational, 3 rotational, and $3(n_A + n_B) - 7$ vibrational partition functions since one vibration is now the translational degree of freedom along the reaction coordinate. If we assume the approximate equality of the several partition functions of translation, rotation, and vibration, respectively, and factor out of the quotient the value of Z derived from the 2-atom case already discussed, i.e. q_r^2/q_t^3, there remains to account for the P term a quotient of the product of 5 vibrational terms divided by 5 rotational terms

$$P = \frac{q_v^5}{q_r^5}$$

Since the q_v terms differ little from the unity at room temperatures and q_r terms may be of the order of 10 to 100, it follows that P may have values of the order of 10^{-5} to 10^{-10}. The more complex the colliding molecules the greater the values of q_r and consequently the smaller the P factor. These considerations indicate the reason for the success of the collision theory in reactions among atoms and diatomic molecules and for its failure in more complex systems which belie the simple assumptions involved in simple collision theory.

Finally the equation of absolute rate theory in the form given by Eq. 8-5

$$k' = \frac{kT}{h} e^{-\Delta F^*/RT}$$

indicates that since kT/h is a frequency factor, constant in magnitude, the rate of any reaction process is determined by the free energy of activation at a given temperature, independent of the nature of the reaction. It is the especial merit of this approach to rate processes that it is not limited to chemical reactions but is equally applicable to a wide variety of physical rate processes, including viscosity and diffusion, electrical conductivity, overvoltage in electrode processes, creep in metals,

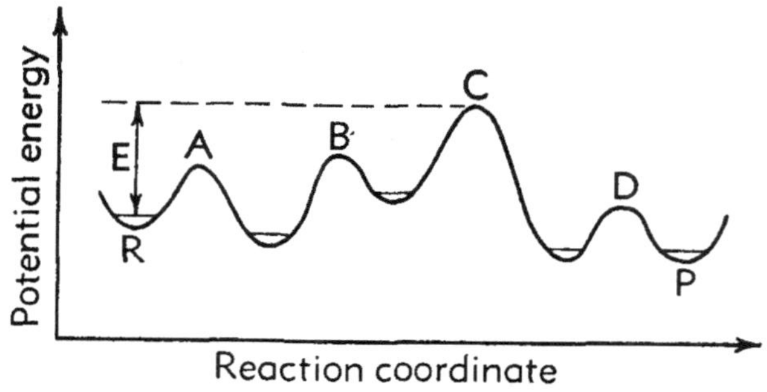

Fig. D,9a. Potential energy curve for successive reactions from reactants R to products P.

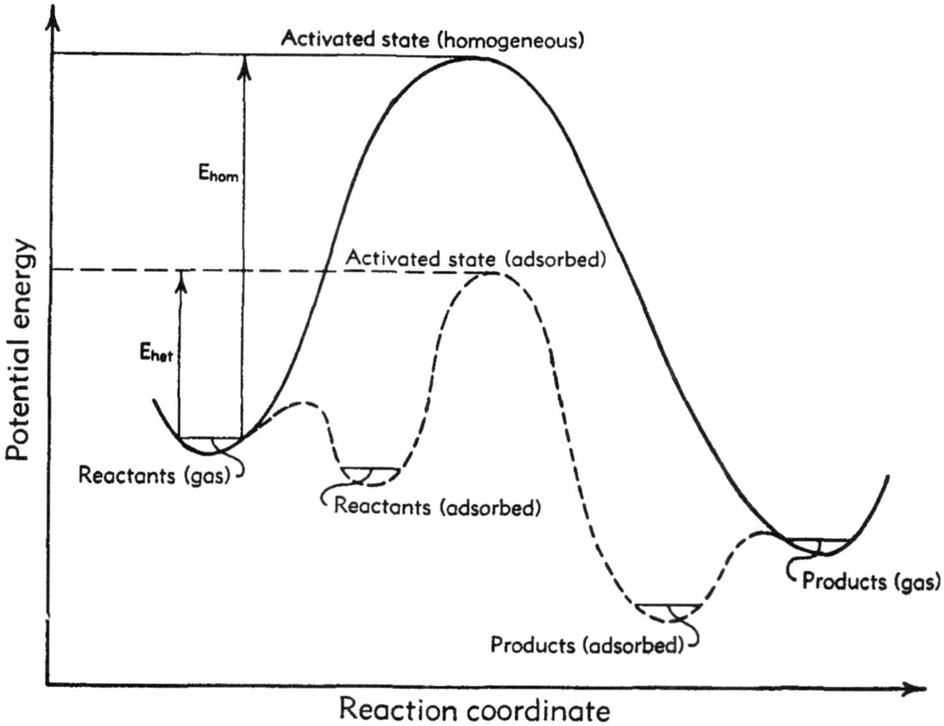

Fig. D,9b. Potential energy curves for a homogeneous reaction and the same occurring catalytically via adsorbed species.

the motion of glaciers, the upthrust of mountains, and the influence of pressure and temperature in biological processes, notably the reactions in presence of enzymes and those capable of yielding cold light.

In a sequence of reactions involving a succession of intermediate states between reactants and products, it is the free energy of activation of the complex which has the highest value of potential energy which determines the over-all rate of reaction. This fact leads as a consequence to the conclusion that the rate of the over-all reaction can tell little with respect to the detailed sequence of reaction steps. They may be many or few and without kinetic significance so long as one activated complex lies high relative to all the others in the potential energy-reaction coordinate diagram as in Fig. D,9a.

On the absolute rate theory the role of a catalyst is to provide a reaction path, in one or many steps, of which the potential energy of the complex which is rate determining in the new sequence of reactions is *lower* than the potential energy of the activated complex in the uncatalyzed reaction. Here again the rate of the catalyzed reaction tells little or nothing concerning the detailed mechanism of the catalytic reaction. Such details can be obtained only by measurements of the rate of the assumed steps or by inductive processes as exemplified in Art. 4 for the hydrogen-bromine reaction. The role of a surface as a catalyst in a given reaction is illustrated by the potential energy diagram, Fig. D,9b.

D,10. The Theory of First Order Processes. The theory of first order processes was first sought in an interaction between the reactant and the radiation present in the system, principally infrared radiation. This false start was remedied when Lindemann [*20*] pointed out that a first order reaction equation could result from a bimolecular collision process, yielding an activated complex which required a definite time interval before separation into reaction products. During this time interval the complex might suffer deactivating collisions. Such a process could be represented schematically thus:

$$\begin{array}{c} \mathrm{A} + \mathrm{A} \underset{k_2}{\overset{k_1}{\rightleftarrows}} \mathrm{A} + \mathrm{A'} \\ \qquad\qquad \downarrow k_3 \\ \qquad\qquad\qquad \text{reaction products} \end{array}$$

The rates of formation and disappearance of the activated complex A′ in the steady state are given by the expression

$$\frac{d[\mathrm{A'}]}{dt} = k_1[\mathrm{A}]^2 - k_2[\mathrm{A'}][\mathrm{A}] - k_3[\mathrm{A'}] = 0$$

whence

$$[\mathrm{A'}] = \frac{k_1[\mathrm{A}]^2}{k_2[\mathrm{A}] + k_3}$$

The rate of formation of products is given by $k_3[\mathrm{A'}]$, whence

$$\frac{d[\text{products}]}{dt} = \frac{k_3k_1[\mathrm{A}]^2}{k_2[\mathrm{A}] + k_3}$$

The condition for a first order process is evidently given by $k_2[\mathrm{A}] \gg k_3$

in which case

$$\frac{d[\text{products}]}{dt} \cong \frac{k_1k_3}{k_2}[\text{A}]$$

In the reverse condition $k_3 \gg k_2[\text{A}]$ the process is a straightforward bimolecular process

$$\frac{d[\text{products}]}{dt} \cong k_1[\text{A}]^2$$

Since this latter condition is favored by a large value of k_3 or at low concentrations of [A] it follows that a first order equation may be expected in complex molecules where a longer time interval is required for the activation energy to assemble in the bond which is to be broken and that the first order rate may turn into a second order rate law when the concentration of [A] is low enough.

The decomposition of nitrogen pentoxide first studied by Daniels and Johnston [*21*] is an example of a reaction found to be kinetically first order even down to pressures of the order of 0.02 mm as well as in liquid systems with the reactant in solution. In some other first order decompositions the transition to second order kinetics occurs at pressures of the order of 100 mm or higher. A foreign gas can serve in place of the reactant to maintain by collisions with A the concentration of activated complexes. The efficiency of such collisions is a function of the foreign gas added. This evidently depends on the efficiency of energy transfer by collision, a branch of the science of chemical kinetics which is still in its infancy with rudimentary ideas as to the factors which operate importantly.

The absolute rate theory gives a simple formulation for the rate of such first order reactions. Assuming the equilibrium between reactants and activated complex already indicated, the rate equation becomes

$$k' = \frac{kT}{h}\frac{Q'}{Q}e^{-E_0/RT}$$

If the ratio Q'/Q is practically unity, or if $\Delta S^* = 0$ which is the alternative mode of statement, then

$$k_1 = \frac{kT}{h}e^{-E_0/RT}$$

The value of kT/h is of the order of 10^{13} sec^{-1}. Polanyi and Wigner [*22*] showed that in a large number of first order processes a majority have frequency factors within one order of magnitude of this value. Large deviations from this frequency factor can be rationalized on the basis that the decompositions are "nonadiabatic," that is they require transitions from one potential energy surface to another for decomposition. It is to this factor that the slow decomposition of N_2O is ascribed, its rate of decomposition being low by a factor of the order of 10^{-3}.

D,11. The Theory of Third Order Processes. Consideration of the potential energy curve of the 2-atom system indicates that two atoms colliding with any excess kinetic energy cannot produce a stable diatomic molecule. The system will travel up the potential energy curve on the repulsive sector until above $V = 0$ and then separate to infinity, the atoms carrying with them the kinetic energy with which they approached each other. For formation of a stabilized molecule, a third body is necessary to carry off the excess energy of recombination and the kinetic energy of approach. The 3-body process between atoms A and B and the third body M thus becomes

$$A + B + M \rightarrow ABM \rightarrow AB + M$$

The reaction is third order because the activated complex contains the three reactants. When A = B we have the recombination of atoms to form homonuclear molecules. This has been investigated experimentally by Rabinowitch and Wood [*23*] for the recombination of iodine and of bromine atoms. That the third body M can have varying efficiency is indicated by the table below.

Table D,11. Values of k_3 in atom recombination; $-dX/dt = k_3[X]^2[M]$; $k_3(\text{cm}^3/\text{mole})^2\text{sec}^{-1} \times 10^{32}$

M =	He	A	H_2	N_2	O_2	CH_4	CO_2	C_6H_6
X = Br	0.76	1.3	2.2	2.5	3.2	3.6	5.4	—
X = I	1.8	3.8	4.0	6.6	10.5	12	18	100

Theory is inadequate to account satisfactorily for the varying efficiencies of M in all these 3-body processes. The concept of a ternary collision is indefinite. It is ordinarily assumed that the ratio of binary collisions Z_{11} to ternary collisions Z_{112} is equal to the ratio of the mean free path to the diameter of the molecule. That is to say, a ternary collision can occur within the time interval when two colliding species are within a molecular diameter of each other. Since the diameters are of the order of 10^{-8} cm and the mean free path is of the order of 10^{-5} cm at 1 atm pressure, the ratio of binary to ternary collisions should be about 1000 at 1 atm total pressure.

The reactions of nitric oxide with oxygen, chlorine, bromine, hydrogen, and deuterium are molecular processes which follow the third order rate expression. In agreement with the observation concerning 3-body collisions they are relatively slow processes. It should be observed, however, that these processes can all be formulated in terms of two successive bimolecular processes,

(i) $$2NO \rightleftarrows (NO)_2$$

(ii) $$(NO)_2 + O_2 \xrightarrow{k_2} 2NO_2$$

If equilibrium is established in the first reaction, $[(NO)_2] = K[NO]^2$. Then, from the second reaction,

$$\frac{d[NO_2]}{dt} = k_2[(NO)_2][O_2] = k_2K[NO]^2[O_2]$$

It is in agreement with this interpretation that the velocity of this third order process decreases with increasing temperature. Bodenstein [*24*] interprets this as due to a heat of formation of $(NO)_2$, equal to 4 kcal, and the bimolecular process (ii) with an activation energy of 2 kcal. The same reaction (i) with the same K can be used to interpret the rates of reaction of nitric oxide with chlorine and bromine which, in contrast to oxygen, increase with increasing temperature. On this interpretation the activation energies of reaction (ii) with chlorine and bromine, in contrast to the case of oxygen must be somewhat greater than 4 kcal. Hydrogen only reacts at elevated temperatures. The activation energy is 47 kcal. Deuterium reacts at about 0.7 the rate with hydrogen [*25*].

D,12. The Reaction of Molecules with Atoms or Radicals. Both theory and experiment confirm that atom-molecule and radical-molecule reactions normally require much smaller activation energies than reactions between two molecular species. Polanyi [*26*] systematically studied the influence of the nature of the molecule in such processes by studying elementary reactions which can be generalized by the equation

$$Na + RX = NaX + R$$

Polanyi showed that the collision efficiency progressively diminishes in the reaction of sodium atoms with CH_3X as X is varied in the sequence I, Br, Cl, F. Whereas with CH_3I every collision is efficient, CH_3F requires an activation energy of 25 kcal. Polanyi systematized his results on the nature of R in the following summary in which the numbers refer to the collisions necessary for one reaction:

Increase in reaction rate is produced by

1. Lengthening of the carbon chain

$$\underset{10000}{CH_3Cl} \rightarrow \underset{7000}{CH_3CH_2Cl} \rightarrow \underset{4400}{CH_3(CH_2)_2Cl} \rightarrow \underset{3300}{CH_3(CH_3)_3Cl}$$

2. Passage from primary to secondary to tertiary carbon-halogen linkage

$$\underset{7000}{CH_3CH_2Cl} \rightarrow \underset{3300}{CH_3CHClCH_3} \rightarrow \underset{1500}{(CH_3)_3CCl}$$

3. Presence of carbon double bond in a neighboring position

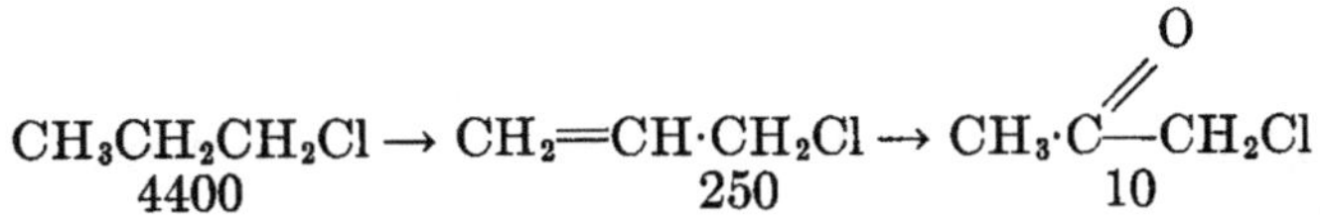

4. Multiple substitution

$$\underset{4400}{CH_3CH_2CH_2Cl} \rightarrow \underset{300}{CH_3CH_2CHCl_2} \rightarrow \underset{180}{CH_3CCl_2CH_3}$$

Decrease in reaction rate is produced by an adjacent double bond

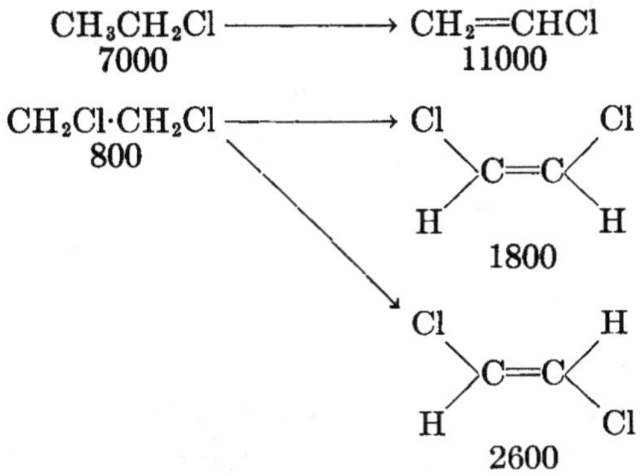

No effect on the reaction rate is produced by branching of the hydrocarbon chain

$$\underset{3300}{CH_3CH_2CH_2CH_2Cl} \rightarrow \underset{3500}{(CH_3)_2CH\cdot CH_2Cl}$$

$$\underset{2200}{CH_3CH_2CH_2CH_2\cdot CH_2Cl} \rightarrow \underset{2000}{(CH_3)_2CH\cdot CH_2CH_2Cl}$$

As Polanyi pointed out, "the laws prevailing are very explicit and simple." The reactions of radicals with hydrocarbons are equally explicit. Thus, the data of Smith and Taylor [*27*] indicate that the activation energy for the reaction of methyl radicals with hydrogen in paraffinic hydrocarbons decreases in the order, E primary CH $>$ E secondary CH $>$ E tertiary CH. An adjacent double bond markedly increases the activation energy. The presence of a carbon double bond in a neighboring position decreases the activation energy. Thus,

$$CH_3\cdot + CH_3CH_2CH_3 \rightarrow CH_4 + R\cdot \qquad E = 6 \text{ kcal}$$

$$CH_3\cdot + CH_2{=}CH\cdot CH_3 \rightarrow CH_4 + R\cdot \qquad E = 3 \text{ kcal}$$

$$CH_3\cdot + C_6H_6 = CH_4 + R\cdot \qquad E > 20 \text{ kcal}$$

$$CH_3\cdot + C_6H_5CH_3 = CH_4 + R\cdot \qquad E = 5.6 \text{ kcal}$$

More refined measurements made recently [*28*] change the values of the activation energies but maintain the generalizations. Such results point to variations in bond energies C—X or to interpretations based on resonance. In the case of toluene just cited the resonance of the benzyl radical may be assessed at 24.5 kcal (Szwarc [*29*]) or the energy of removal of a hydrogen atom from the CH_3 group in toluene as 78 kcal in comparison with values > 90 kcal in paraffinic CH_3 groups.

The results of these simple type systems can be extended to more complex systems. Thus:

1. The reactivity of α-methylenic hydrogen atoms in unsaturated substances is high

$$—\underset{\alpha}{CH_2}{\cdot}CH{=}CH{\cdot}\underset{\alpha}{CH_2}— + R \rightarrow RH + —\underset{\cdot}{CH}{\cdot}CH{=}CH{\cdot}CH_2—$$

2. Monomers with α-methylene hydrogens do not polymerize readily by a free radical mechanism. Allyl acetate, 1-octene, propylene, isobutylene, and isopropyl acetate polymerize with difficulty to give low molecular weight materials.
3. The reactivity of free radicals will be influenced by resonance stabilization. Stabilized radicals will be less reactive than radicals not stabilized by resonance.
4. Such stabilization plays an important part in chain transfer mechanisms and in the role of the solvent in reactions in solutions.
5. The data bear on the mechanisms of pyrolysis. Thus, in the pyrolysis of toluene which might yield (i) $C_6H_5{\cdot} + CH_3{\cdot}$ or (ii) $CH_3{\cdot}C_6H_4{\cdot} + H{\cdot}$ or (iii) $C_6H_5CH_2{\cdot} + H{\cdot}$, it is the third which is important. The products are hydrogen and dibenzyl.
6. The same rules are equally applicable and explicit with respect to self-disproportionation of radicals on pyrolysis, as outlined by Rice and Rice [*30*]. A vinyl radical,·

$$(CH_2CHX)_m \vdots CH_2CHX \cdot CH_2{\cdot}\dot{C}X(CH_2CHX)_n—$$

cleaves at the position shown by the vertical dotted line to yield

$$(CH_2CHX)_m \vdots CH_2CHX{\cdot} + CH_2{=}CX(CH_2CHX)_n$$

the radical undergoing decomposition at the position shown by the vertical dotted line to yield

$$(CH_2CHX)_{m-1} \vdots CH_2{\cdot}CHX{\cdot} + CH_2{=}CHX$$

which, proceeding as before, again breaks down yielding another molecule of monomer, the depolymerization process being the exact reverse of the chain addition of monomer to a radical in the process of polymerization. Both polymerization and depolymerization are chain processes in virtue of the low activation energies involved in the molecule-radical reactions. These views on polymerization-depolymerization were reached by Taylor and Tobolsky [*31*] and established experimentally by Grassie and Melville [*32*].

Similar regularities obtain in the interaction of molecules with surfaces, the unsaturation of which surfaces, as Polanyi [*33*] pointed out, makes them equivalent to free atoms or radicals. Thus, the dehydrogenation-cyclization of *n*-octane leads to the formation of *o*-xylene by removal of secondary hydrogens from the 2,7 carbon atoms with ring formation, rather than to the formation of ethyl benzene by hydrogen removal involving one secondary and one primary CH bond in the compound. In the same way on a given catalyst the dehydration or dehydrogenation of a secondary alcohol is more easily achieved than that of a primary alcohol.

While many atom-molecule reactions occur with activation energies which are small compared with molecule-molecule reactions and thus permit such reaction processes at low temperatures, there are atom- or radical-molecule reactions which have appreciable activation energies. Attention has already been drawn to the reaction $Na + CH_3F$ where $E = 25$ kcal. Of importance in combustion phenomena is the low reactivity of oxygen atoms. As Kistiakowsky and Harteck and Kopsch [*34,34a*] showed, the oxygen atom reacts slowly with both hydrogen and carbon monoxide, the activation energies required reaching 7 ± 2 and >8 kcal, respectively.

D,13. Chain Mechanisms. A variety of chemical processes proceed by mechanisms which involve both free radicals or atoms and a sequence of reactions forming a chain. These processes include thermal and photochemical decomposition processes, polymerization and depolymerization processes under the influence of heat and light, as well as a variety of oxidations and halogenations which can give rise to the production of flames and explosions.

Atomic hydrogen produced by mercury photosensitization with the resonance line of mercury, $\lambda = 2536.7$A, can initiate such chains. Thus:

$$H_2 + Hg(^1S_0) + h\nu(2536.7\text{A}) \rightarrow 2H + Hg$$

In presence of ethylene, polymerization of the latter or the formation of higher saturated hydrocarbons may occur [*35*], e.g.

$$H + C_2H_4 \rightarrow C_2H_5$$

$$C_2H_5 + C_2H_4 \rightarrow C_4H_9 \text{ etc.}$$

$$C_2H_5 + H_2 \rightarrow C_2H_6 + H$$

$$C_2H_5 + C_2H_5 \rightarrow C_4H_{10}$$

Metallic alkyls on pyrolysis or photolysis produce free radicals which can initiate similar sequences of reactions [*36*,*37*].

The pyrolysis of complex organic molecules may also proceed by formation of radicals and a chain sequence. Thus the pyrolysis of acetaldehyde may proceed by the following sequence of reactions

$$\text{(i)} \qquad CH_3CHO \rightarrow CH_3 + CHO$$

$$\text{(ii)} \qquad CH_3CHO + CH_3 \rightarrow CH_4 + CH_3CO$$

$$\text{(iii)} \qquad CH_3CO \rightarrow CH_3 + CO$$

$$\text{(iv)} \qquad 2CH_3 \rightarrow C_2H_6$$

It is evident that reaction (i) is a chain initiation, (ii) and (iii) are chain steps, and (iv) is a chain-ending process. Rice and Herzfeld [*38*] showed that such chain processes might give first, second, or $\frac{3}{2}$ order processes depending on the nature of the chain and the chain-breaking steps. In the example cited, a primary rupture by (i) would be insignificant if (ii) and (iii) were ready reactions producing long chains. The kinetics of the suggested mechanism follows from the steady state treatment of the CH_3 radical concentration

$$\frac{d[CH_3]}{dt} = 0 = k_1[CH_3CHO] - k_4[CH_3]^2$$

since the CH_3 removed in (ii) is reproduced in (iii) or

$$[CH_3] = \left(\frac{k_1}{k_4}\right)^{\frac{1}{2}} [CH_3CHO]$$

The reaction rate of methane formation becomes

$$\frac{d[CH_4]}{dt} = k_2[CH_3][CH_3CHO] = k_2\left(\frac{k_1}{k_4}\right)^{\frac{1}{2}} [CH_3CHO]^{\frac{3}{2}}$$

The experimental data are indecisive as to whether the reaction is $\frac{3}{2}$ order or a first order reaction falling to second order at lower pressures. The chain length can be approximated in two ways. Leermakers found [*39*] that the chain length in the photodecomposition of acetaldehyde was as high as 200 molecules of acetaldehyde decomposed per light quantum absorbed at 300°C. Allen and Sickman [*40*] induced the decomposition of acetaldehyde below the temperature at which it will decompose of itself thermally by decomposing azomethane in the presence of acetaldehyde at 300°C. The ratio of aldehyde molecules decomposed to azomethane molecules decomposed increased with increase in the ratio $P_{ald}/P_{azom}^{\frac{1}{2}}$. Chain lengths between 22 and 500 were thus found. With such chain lengths it is evident that the main products of decomposition will be $CH_4 + CO$ with only traces of C_2H_6. An alternative chain-ending process could be

$$CH_3 + CHO = CH_4 + CO$$

Recent mass spectrometric analyses of the photodecomposition of acetaldehyde indicate [*41*] the presence of some ethane in the hydrocarbons formed at 125°C.

The effect of inhibitors can be studied to gain evidence of radical chains in some of these reactions. Nitric oxide is a useful agent for such purposes. The presence of 2 mm of NO reduces the velocity of decomposition of diethyl ether to less than 30 per cent of the uninhibited rate. The proportion of chain reaction as measured by the use of nitric oxide decreases as the number of carbon atoms in the ether increases. The action of nitric oxide in inhibiting the thermal decomposition of paraffin hydrocarbons has been studied by Echols and Pease [*42*], Steacie and Folkins [*43*], and more recently by Stubbs and Hinshelwood [*44*]. Small amounts of nitric oxide cause a marked reduction in the initial rate of decomposition [*42,44*]. Increasing amounts reduce the initial rate to a constant limiting value, independent, over a wide range, of the nitric oxide concentration. Steacie and Folkins suggested that the limiting low rate pertains to the primary reaction which normally starts the chain process. Echols and Pease suggested that a stationary state occurred in which nitric oxide, having complexed with the radical, regenerates from the complex as many radicals as are removed by the inhibitor. Stubbs and Hinshelwood ascribe the limiting rate to a simultaneous intramolecular reaction, nonchain in nature, in which a hydrogen atom migrates, yielding a lower paraffin and an olefin. They reject as a result of their experiments the interpretations of the earlier workers.

Other inhibitors are effective. Propylene inhibits the decomposition of paraffins, less effectively than nitric oxide. One mm of the latter is equal in inhibiting action to 12.5 mm propylene with pentane at 100 mm and 530°C. In these ratios, however, the two inhibitors produce the same limiting value, approximately 14 per cent of the initial rate in absence of inhibitor in the case of pentane just cited. The effect is certainly complex since it is affected in magnitude by treatment of the glass surfaces of the container, for example with potassium chloride. In a series of paraffins from propane to *n*-decane the apparent chain length at 530° and 100 mm pressure varies from 10 to 3. This serves to measure the proportion of chain reaction to intramolecular process on the assumption of Stubbs and Hinshelwood with respect to mechanism. The nature of the wall surface is, in the view of Echols and Pease, critical and must be included in any final interpretation of these undoubtedly complex phenomena.

D,14. Branching Chains and Explosions. In the examples of chain steps cited in the preceding articles the disappearance of one chain carrier has always been accompanied by the formation of one chain carrier

$$Br + H_2 = HBr + H$$
$$H + Br_2 = HBr + Br$$
$$CH_3 + CH_3CHO = CH_4 + CH_3CO$$
$$CH_3CO = CO + CH_3$$

We may term such chain steps "simple chains." However, it is apparent that, in some chain steps, the disappearance of one atom or radical may result in the formation of two or more chain carriers. A simple example involves the reaction of oxygen atoms with hydrogen molecules where

$$O + H_2 = OH + H$$

the oxygen gives rise to two carriers, the hydroxyl radical and a hydrogen atom, each of which can continue chain processes. The reaction of H with O_2 may, when it occurs, also produce two chain carriers

$$H + O_2 = OH + O$$

This process of multiplication of chain carriers is known as chain-branching and, in the special cases just considered, where the multiplication is by a factor two it is evident that an original oxygen atom can give two carriers of which one can give two more, and so on. Such multiplication in ten generations yields 1024 carriers from the original chain carrier; in 20 generations the number exceeds 10^6 and in 30 generations it can exceed 10^9.

The influence of chain-branching on the kinetics can be illustrated in the general case. Let a reactant A give rise to a chain carrier C

$$A \xrightarrow{k_1} C$$

Let the chain step give rise to branching in which αC are produced from one carrier together with a product P

$$C + A \xrightarrow{k_2} P + \alpha C$$

The carriers are destroyed in any manner with a velocity k_3

$$C \xrightarrow{k_3} \text{chain-ending}$$

In the steady state,

$$\frac{d[C]}{dt} = 0 = k_1[A]^n - k_2[C][A] + \alpha k_2[C][A] - k_3[C]$$

The steady state concentration of the chain carrier is then

$$[C] = \frac{k_1[A]^n}{k_2[A](1 - \alpha) + k_3} \tag{14-1}$$

The destruction process k_3 may occur as a gas-phase reaction (k_g) or at the walls of the containing vessel (k_w). We may therefore write

$$[C] = \frac{k_1[A]^n}{k_2[A](1-\alpha) + k_g + k_w} \tag{14-2}$$

With simple or nonbranching chains where α never exceeds unity, the concentration [C] will always be finite, dependent on the rates of formation and destruction. When, however, α can exceed unity, it may result that the first term in the denominator becomes so large and negative as

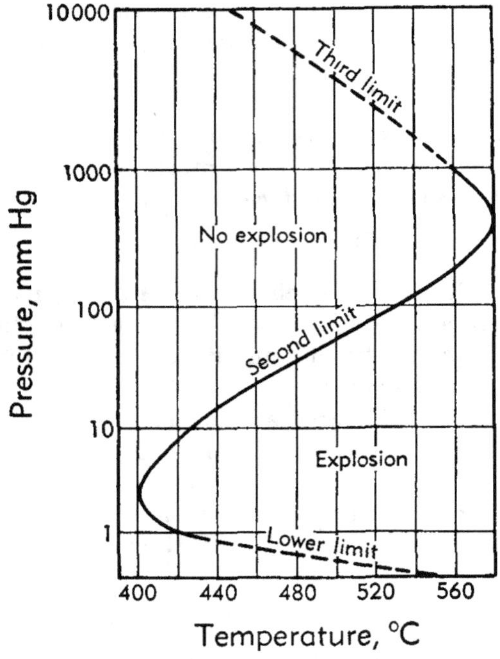

Fig. D,14. The explosion limits in the hydrogen-oxygen system.

to equal the combined magnitude $k_g + k_w$; at this moment the denominator becomes zero and the concentration of carrier rises rapidly to infinity. A measurable rate at finite [C] passes rapidly into an infinitely rapid rate. The result is described as a branching-chain explosion.

Such explosions are to be sharply differentiated from thermal explosions where, owing to the exothermicity of the reaction and the difficulties attending heat removal, the temperature of the system rises so rapidly that extremely rapid reaction and explosion can occur. This type of thermal explosion can take place even though α never reaches unity and in the case of simple chain steps. With branched chains, explosion can occur even though isothermal conditions are maintained, when $k_2[A](\alpha - 1) = k_g + k_w$.

Whenever chain destruction can occur in the gas phase and at the wall in a branched chain reaction there will, at any temperature, be two limits of pressure within which explosion occurs, outside of which a non-explosive rate process will obtain. These two limits depend on the magnitudes k_g and k_w. At low pressures the influence of the walls and therefore k_w will be dominant. At high pressures the destruction of chain carriers will occur as 2- or 3-body processes in the gas phase. This upper limit is usually sharply temperature-sensitive increasing with increase of temperature. The lower explosion limit is much less sensitive to temperature. The dependence of explosion on pressure and temperature in the hydrogen-oxygen system is shown in Fig. D,14. The detailed discussion of such limits is given in Sec. E.

D,15. Wall Reactions. The importance of the walls of the containing vessel in all rate processes needs to be determined. Even in molecular reactions the effect may be large. Ethylene and bromine vapor do not react readily at room temperature in the gas phase. The rapid process observed occurs on the glass walls of the container. If this be coated with paraffin wax the velocity of reaction is reduced to negligible values, the process occurring probably only at imperfections in the coating. With varying surfaces, varying velocities are obtained. Stearic acid surfaces are more active than glass which is more active than cetyl alcohol, which, in its turn, is more active than paraffin wax. The importance of surface or wall reactions can sometimes be assessed by variation of the ratio of surface to gas volume in the kinetic studies.

In chain processes and in reactions involving the recombination of atoms and radicals the walls may play a controlling part in the rate process. Hydrogen atoms are strongly held by glass surfaces which cause rapid recombination. Phosphoric acid covering the glass surface markedly reduces the recombination process. Chlorine and bromine atoms react rapidly on glass surfaces. For this reason it is impossible to explode even sensitive hydrogen-chlorine mixtures in glass capillary tubes. Explosion traps in the feedlines containing explosive gas mixtures owe their effectiveness, in part at least, to the influence of walls on the recombination of atoms and radicals thus destroying chain mechanisms of explosion. Their effect in cooling the gases may also be significant.

Walls may initiate chain processes as well as end chain reactions. Passage of hydrogen containing traces of oxygen over platinized asbestos can be made to yield hydrogen atoms to the downstream gas flow. This can be shown by the reaction of the hydrogen atoms with sulfur some distance beyond the platinum [*45*]. Glass at high temperatures can chemisorb hydrogen [*46*] probably as atoms and these may be specially reactive in combustible gas mixtures. Alyea and Haber [*47*] demonstrated the influence of surface on such chain processes by showing that impinging streams of preheated hydrogen and oxygen would not ignite until a quartz rod was lowered into the stream at the junction point, explosion then resulting immediately. The influence of coating vessels as Pease and Cheseboro showed with potassium chloride [*48*] in hydrocarbon oxidation is further evidence of the role of walls in chain processes.

Atoms or radicals generated in the gas volume must reach the walls by a process of diffusion. This depends on the distance to be traversed, and on the mean free path λ. Smoluchowski showed that the number of collisions n suffered in traveling a distance l is given by

$$n = \frac{3\pi l^2}{4\lambda^2} \tag{15-1}$$

For parallel walls a distance d apart, the average number of collisions suffered by a particle in reaching the wall can be shown to be

$$\bar{n} = \frac{\pi d^2}{16\lambda^2} \tag{15-2}$$

If, as in the hydrogen-chlorine chain, each stage regenerates a chain carrier, H or Cl, and if, on the average, g collisions are necessary for the atom to react, then the chain length ν is

$$\nu = \frac{\pi d^2}{16g\lambda^2} \tag{15-3}$$

It is for this reason that long chains are absent in capillary tubes.

Inert gases present in the reacting system will affect the speed of chain processes by their influence on the mean free path. Semenov [*49*] showed that the chain oxidation of phosphorus was dependent on the square of the diameter of the vessel, on the square of the mean free path, and also on the pressure p_i of the inert gas. With a cylindrical vessel the chain length is given by

$$\nu = \frac{6\pi d^2}{4\lambda^2} p_{P_4} p_{O_2} \left(1 + \frac{p_i}{p_{P_4} + p_{O_2}}\right) \tag{15-4}$$

More detailed study of the influence of the dimensions of the reaction vessel on the velocity of chain reactions can yield additional information as to the nature of the chain-starting and chain-breaking processes. Lewis and von Elbe [*49a*, pp. 16, 17] summarize this aspect of the matter as follows:

"If a chain reaction is found to be independent of diameter two cases are possible: chain initiation and chain breaking at the wall, and the case of chain initiation and breaking in the gas phase. In the latter case the reaction rate will remain independent of vessel size for all diameters, whereas in the former case branching causes the rate eventually to increase with increasing diameter. Conversely, if a reaction rate is found to be dependent on vessel size a decrease of diameter should lead to a decision between chain starting and breaking at the wall and the two other alternatives, namely, chain initiation in the gas phase and breaking at the wall and vice versa.

"For an *un*branched chain reaction in which both chain initiation and breaking occur at the wall, the diameter criterion fails to distinguish the mechanism from the case of a chain reaction branched or unbranched in which initiation and breaking occur in the gas phase. There are, however, other criteria, particularly the fact that the rate per unit area and the chain breaking efficiency of the wall depend on the nature of the surface. Other cases of unbranched chain reactions are covered by the diameter criterion.

"There does not seem to be any equally well-defined criterion other than the diameter dependence for distinguishing the different cases of branched and unbranched chain reactions."

D,16. Gas Reactions at Surfaces. General Principles. The example of the ethylene-bromine reaction, already cited, indicates that surfaces are specific in producing chemical reaction. By varying the surface, a molecule may be caused to decompose in different ways. Thus, formic acid on metallic surfaces decomposes into hydrogen and carbon dioxide,

$$HCOOH = H_2 + CO_2$$

On some oxides, for example alumina, the decomposition yields water and carbon monoxide:

$$HCOOH = H_2O + CO$$

On glass surfaces and with certain oxides, both reactions occur in ratios dependent on the nature of the glass and the temperature. Similar processes of dehydrogenation and dehydration of alcohols occur on various solid surfaces. The interaction between the reactant and the surface is specific. Chemical forces are therefore involved.

With active surfaces, reaction may occur at the surface when no reaction is measurable in the surrounding gas phase. In such cases, the volume of the gas phase plays no significant role in the chemical process which is confined to a volume element, to be defined, in the neighborhood of the surface. Since chemical forces are involved and since these are extremely short range forces, Langmuir [*50*] concluded that the volume element is measured by the area of the surface and the dimension of the reacting molecule or, alternatively, reaction occurs in a monolayer of the reactant on the active surface. The monolayer is held at the surface by forces of adsorption.

D,17. Adsorption. Two types of adsorption can now be distinguished: (1) physical or van der Waals adsorption in which the molecules are held to the surface, at distances of the order of 3A, by van der Waals forces of the same nature as the forces operative between gas molecules; (2) chemisorption in which the molecules or their dissociation products are held to the surface, at distances of the order of 1A, by chemical forces of the same nature as the valency forces operative between atoms in molecules. It is the latter type of adsorption which is normally responsible for chemical reaction at surfaces. The van der Waals adsorption should only produce reaction in those few cases where reaction is facilitated by compression or liquefaction of the reactants. It also serves to define surface areas [*61*, pp. 1-28].

Chemisorption may be rapid, as is true of the chemisorption of hydrogen on a number of metals, notably nickel, iron, platinum, and tungsten. Hydrogen is rapidly chemisorbed by evaporated nickel films even at 20°K [*51*]. It may also be slow, as is true of the chemisorption of nitrogen on evaporated iron films or of methane on evaporated nickel films. In contrast with hydrogen, methane is measurably chemisorbed on evaporated nickel at temperatures above 100°C. On technical catalyst surfaces the rates of chemisorption are markedly slower than on evaporated metal films. Thus, the rate of chemisorption of nitrogen on reduced synthetic ammonia iron catalysts is conveniently measurable in the temperature interval of 200–400°C [*52*]. On a variety of oxide surfaces, including ZnO, Cr_2O_3, ZnO-Cr_2O_3, ZnO-MoO_x, Al_2O_3, and SiO_2 the rate of chemisorption of hydrogen has been measured at temperatures in the range of 0–600°C. Over a range of temperatures, the velocity of adsorption can be expressed [*53*] in terms of an Arrhenius-type equation: rate $= ke^{-E/RT}$ where E represents an activation energy obtained from the slope of the line obtained by plotting the logarithm of the observed rate as a function of $1/T$. It is now known that the observed rates in slow adsorptions are, by many orders of magnitude, slower than the rate at which molecules strike the area of surface with an energy equal to or greater than the observed activation energy. The most recent interpretations of this result [*54,55,56*] suggest that the observed activation energy is composite of two quantities $E_{\text{obs}} = E + u$ where E is the minimum energy of the colliding gas molecules and u is an energy required to create an adsorbing site in the solid surface, the number of such sites increasing with temperature in proportion to $e^{-u/RT}$; with hydrogen, E is small as is undoubtedly true in the case of metals and u is negligible. With oxides, u may be considerable. The magnitude of E will be dependent on the gas, small with hydrogen on certain metals and oxides, but measurable with some other surfaces such, for example, as carbon. With nitrogen, methane, and the saturated hydrocarbons the activation energy required by the gas molecule on metal films and technical catalysts may be many kilocalories in amount.

The rate of adsorption is dependent on the pressure and on the fraction of the surface covered. An equation of the form

$$\frac{d\theta}{dt} = k_a e^{-g\theta}$$

where g is a constant and θ the fraction of surface covered, appears to satisfy a variety of experimental measurements [*57,58,59*]. It obviously must no longer be valid at high coverage since it does not reduce to zero when $\theta = 1$.

D,18. Adsorption Equilibrium. The Adsorption Isotherm. Langmuir [*50*] developed the concept of adsorption as due to a time lag between condensation of a gas on a surface and evaporation from the surface. At equilibrium the two processes are equal in rate so that

$$a_1(1 - \theta)p = b_1\theta e^{-\Delta H/RT} \tag{18-1}$$

where θ is the fraction of the surface covered at pressure p, a_1 and b_1 are the proportionality constants for condensation and evaporation respectively. The heat of adsorption per mole is ΔH. The condensation occurs by gas molecules striking the bare surface $(1 - \theta)$ with an accommodation coefficient α which Langmuir assumed to be unity. Eq. 18-1 yields, for the fraction of the surface covered, the expression

$$\theta = \frac{\frac{a_1}{b_1} e^{\Delta H/RT} p}{1 + \frac{a_1}{b_1} e^{\Delta H/RT} p} \tag{18-2}$$

For constant ΔH and constant temperature this yields the expression

$$\theta = \frac{bp}{1 + bp} \tag{18-3}$$

where $b = (a_1/b_1)e^{\Delta H/RT}$. It is readily seen that at low pressures $(1 \gg bp)$ the surface covered is proportional to the pressure, $\theta = bp$. At high pressures $(bp \gg 1,\ \theta = 1)$ the surface is completely covered, according to Langmuir, with a single monolayer, owing to the short range of chemical forces, in the case of chemisorption. Langmuir's treatment assumes a uniform surface, one adsorbed molecule to a vacant site, and no interaction between adsorbed molecules. The modifications introduced by nonuniformity of surface sites and those due to interactions and change in accommodation coefficient with coverage require special treatment. See [*50*; *60*; *61*, pp. 1–28].

D,19. Desorption Phenomena. Since adsorption is always an exothermic process, the desorption of a gas will always be slower than the adsorption process at equal areas of bare and covered surface. If the rate of adsorption is given by $v_a = k'e^{-E/RT}$ the corresponding velocity for desorption will be $v_d = k''e^{-(E+\Delta H/RT)}$ where ΔH is the heat of adsorption. As a function of the surface covered, if $+d\theta/dt = k_a e^{-g\theta}$, the rate of desorption would be expressed by an equation $-d\theta/dt = k_d e^{-h\theta}$. This again is not applicable over the whole range of θ since the rate of desorption does not become zero for $\theta = 0$. This proposal of Langmuir [*62*] does satisfy a number of experimental measurements. It also must be modified for nonuniform surfaces and interactions between the adsorbed species.

D,20. Kinetics of Chemical Reactions at Surfaces. It is apparent that any reaction at a solid surface can be broken down into the following five elements:

1. Transport of reactants to the surface
2. Chemisorption of reactants on the surface
3. Chemical reaction between reactants adsorbed on the surface or between an adsorbed reactant and an impinging molecule
4. Desorption of products from the surface
5. Transport of products away from the surface

It is obvious also that if any one of these five stages is very slow compared with the others it will control the rate of the over-all process.

Langmuir [*63*] developed the kinetics of reactions at surfaces in those cases when stage (3) the chemical reaction on the surface is rate-controlling, (1), (2), (4), and (5) being relatively much faster. Langmuir postulated that the rate of reaction was proportional to the fraction of the surface θ covered by reactant. This postulate leads in the case of a single reactant to the equations

$$v = \frac{dp}{dt} = k\theta = k_1 p \quad \text{at low coverages}$$

$$v = \frac{dp}{dt} = k\theta = k \quad \text{at high coverages}$$

At intermediate coverages, the velocity would be proportional to a fractional power of p. The first case at low coverage is known in many examples, the classical example being that of van't Hoff and Kooij, the decomposition of phosphine on glass. The decomposition of hydrogen iodide on gold is an example of the case of high coverage, the rate being independent of pressure over a marked range of pressures. Ammonia decomposition behaves similarly on tungsten filaments. A fractional dependence on pressure was found by Stock and Bodenstein for the decomposition of antimony hydride on antimony films,

$$v = kp_{SbH_3}^{0.6}$$

With two reactants A and B, the fractions of the surface occupied by the reactants are given by Langmuir expressions of the form

$$\theta_A = \frac{bp_A}{1 + bp_A + b'p_B}, \quad \theta_B = \frac{b'p_B}{1 + bp_A + b'p_B} \tag{20-1}$$

Many variants of the kinetics will occur dependent on the magnitudes of b and b'. Using the reaction $C_2H_4 + H_2 = C_2H_6$ as an example, Pease found, in agreement with the Langmuir postulates, that with an inactive copper catalyst at 200°C with ethylene and hydrogen weakly adsorbed

$$v = kp_{H_2} \cdot p_{C_2H_4} \tag{20-2}$$

With an active copper catalyst operating at 0°C the surface is largely covered with adsorbed ethylene. Hydrogen can only enter the surface $(1 - \theta)$ left bare of ethylene. At constant temperature and θ nearly equal to 1, the equilibrium equation for condensation and evaporation

$$k_1p(1 - \theta) = k_2\theta$$

becomes approximately

$$k_1p(1 - \theta) = k_2$$

or

$$1 - \theta = \frac{k_2}{k_1p}$$

The bare surface is inversely proportional to the pressure of a strongly adsorbed gas. In this case Pease found, at 0°C

$$v = \frac{kp_{H_2}}{p_{C_2H_4}} \tag{20-3}$$

With comparable values of b and b' for the two gases, the Langmuir expressions for θ_A and θ_B must be used (see Jungers [*64*]). Catalyst poisons which are strongly adsorbed reactants or impurities thus appear in the denominator of the kinetic expression. Thus, in the oxidation of sulfur dioxide and oxygen on platinum surfaces, the kinetic expression becomes

$$v = k\frac{p_{SO_2}}{p_{SO_3}^{0.5}} \tag{20-4}$$

pointing to a strong adsorption of sulfur trioxide product, proportional to the square root of the pressure.

The absence of ethane product in the kinetic expression for the hydrogenation of ethylene arises from the low adsorbability of this gas, which, as fast as formed, leaves the copper surface so that its concentration, governed by the magnitude $b''p_{C_2H_6}$, is negligible on the surface. Oxygen plays a minor role in the kinetics of sulfur dioxide oxidation presumably because the platinum surface is constantly covered by oxygen atoms, the surface being substantially PtO bonds.

D,21. Adsorption and Desorption as Rate-Determining Processes. The synthesis and decomposition of ammonia on iron synthetic ammonia catalysts, composed of iron with small admixtures of K_2O and Al_2O_3, are the classical examples of chemical reactions of gases at surfaces governed as to rate by adsorption and desorption in synthesis and decomposition respectively. The total process occurs in a series of reversible steps which may be represented thus in terms of the surface S and adsorbed gases and radical intermediates

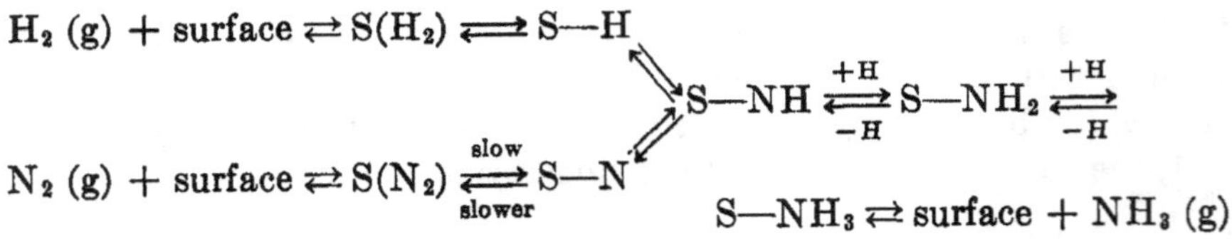

It is now known that all the steps in the above sequence are fast compared with the slow adsorption of nitrogen in the synthesis and its slower desorption in the decomposition process. The rate of chemisorption of nitrogen has been measured and shown to be of the right magnitude to account for the rate of synthesis [*65*]. This has been confirmed by measurements of the rates of the isotopic reactions

$$H_2 + D_2 = 2HD$$

$$NH_3 + ND_3 = NH_2D + ND_2H$$

$$N_2^{14} + N_2^{15} = 2N^{14}N^{15}$$

The first two, involving the making and breaking of H—H and N—H bonds, are fast even at ordinary temperatures on iron catalysts [*66,67*]. The third reaction occurs at synthesis temperatures only at rates comparable with those demanded by synthesis data [*65*].

The kinetics of the decomposition process yields the following interesting rate expression on $Fe\text{-}K_2O\text{-}Al_2O_3$ catalysts.

$$-\frac{d[NH_3]}{dt} = k\frac{p_{NH_3}^{0.6}}{p_{H_2}^{0.9}} \tag{21-1}$$

As shown by Temkin and Pyzhew [*68*] and by Brunauer, Love, and Keenan [*69*] this expression is fully explained if the desorption of nitrogen is the rate-determining process and if the surface reactions yielding S—N, S—H, S—NH, and S—NH_2 obey at all times the equilibrium

$$K = \frac{p_{H_2}^3 \cdot p_{N_2}}{p_{NH_3}^2}$$

The rate expression then becomes approximately

$$-\frac{d[NH_3]}{dt} = K[N_s]^2$$

where $[N_s]$ becomes the concentration of adsorbed atomic nitrogen, S—N. Emmett and Brunauer showed that the concentration of chemisorbed nitrogen atoms varied as the $\frac{1}{6}$ power of the molecular concentration of nitrogen in the gas phase. The rate of ammonia decomposition should therefore be proportional to $[N_2\ (g)]^{0.33}$ or, with the above equilibrium holding, it should be proportional to $[NH_3]^{0.66}/[H_2]^{0.9}$ as found experimentally.

It must be emphasized that in the reactions in which adsorption and desorption are the rate-determining processes the Langmuir kinetical interpretation no longer applies. On the Langmuir basis, Eq. 21-1 would suggest that ammonia decomposition occurred on the small area of surface left bare by hydrogen which was strongly adsorbed. Experiment shows this to be contrary to the facts. Hydrogen is not strongly adsorbed by iron synthetic ammonia catalysts at 400–500°C, the reaction temperature. On the contrary, nitrogen which does not appear in the kinetic expression is strongly adsorbed in this temperature range. The Langmuir interpretation does not apply to this case because stage (3), chemical reaction on the surface, is not rate-determining. That synthesis and decomposition are governed by the rates of adsorption and desorption respectively, stages (2) and (4) of the preceding article, are the reasons for the kinetic expression actually observed.

D,22. Transport to and from the Surface as Rate-Determining Steps. In pure gas reactions, transport to and from the surface becomes important in one special case. When the rate of reaction at the surface is excessively fast, for example, at high temperature and in a flow system, it may well happen that the total reaction occurring is governed by the rate at which the reactants are supplied to the surface of reaction. A reaction which is governed by Langmuir kinetics at low temperatures may turn over into a rate governed by transport to or from the surface in a much higher temperature range. In many combustion processes accelerated by incandescent surfaces this condition often prevails.

In another class of surface reactions the transport of reactants or products may also be rate-determining. When access of the reactants to the surface involves slow processes of diffusion through a condensed medium surrounding the reaction surface, transport will be rate-determining. The exchange reaction $H_2\ (g) + HDO\ (g) \rightleftarrows HD\ (g) + H_2O$, for example, is a very fast reaction at a variety of metal surfaces when all the reactants are in the gaseous state. When, however, the process is operated below the dew point of water or when the catalyst is immersed in liquid water, the rate of exchange becomes slower by several orders of magnitude. Evidently, the rate of reaction is then determined by the rate at which the gas, hydrogen, can diffuse through the water surrounding the catalyst to reach the surface. This achieved, the reaction must occur at the surface as rapidly as it does in the gaseous state.

Even in gaseous systems the rate of reaction will often be dependent on the nature of the flow in the experimental arrangement. Laminar flow past a surface is accompanied by a stationary layer of gas next to the surface through which reactants must diffuse to reach the surface. Again, the rate-determining step will be governed by the relative rates of diffusion or reaction on attaining the surface. Turbulent flow past a surface decreases the thickness of the stationary layer and so decreases the time requirement in the diffusion step. Similar factors will influence rates of reaction in homogeneous gaseous systems where interdiffusion of two gas streams will be dependent on whether the motion of the two streams is laminar or turbulent.

D,23. Kinetics of Reactions in Solid and Solid-Gas Systems. A variety of decomposition processes in solids may be employed for propulsive processes. The use of double-base propellants and of mixtures of oxidizers and fuels such as perchlorate-polymer mixtures are illustrative of such reactions in solids.

The kinetics of chemical reactions involving breakdown of solids or of reaction between gases and solids is less developed than with other chemical rate processes. In one general area it is possible to identify the factors that govern the rates of the processes occurring. This area includes those processes in which a solid phase decomposes to yield a second solid phase and a gas and also those in which a gas reacts with a solid to yield another gas and a separate solid phase. Typical of these two cases are the following reactions:

(i) $$Ag_2O\ (s) \rightarrow Ag\ (s) + \tfrac{1}{2}O_2\ (g)$$

(ii) $$CaCO_3\ (s) \rightarrow CaO\ (s) + CO_2\ (g)$$

(iii) $$CuSO_4{\cdot}5H_2O\ (s) \rightarrow CuSO_4{\cdot}3H_2O\ (s) + 2H_2O\ (g)$$

(iv) $$CuO\ (s) + H_2\ (g) \rightarrow Cu\ (s) + H_2O\ (g)$$

It is characteristic of all these examples that, in accordance with the phase rule, the equilibrium is determined solely by the gas pressures: (i) $K = p_{O_2}$, (ii) $K = p_{CO_2}$, (iii) $K = p_{H_2O}$, and (iv) $K = p_{H_2O}/p_{H_2}$. Kinetically, all these reactions are interface reactions, the reaction being confined to the interface between the two solid phases, as was first observed by Faraday and reemphasized by Langmuir [*70*]. The reactions are initiated at nuclei where the two phases start to coexist. The interface grows in area as the nuclei enlarge, passes through a maximum and then diminishes as the reactant is progressively converted to product. Since the rate is proportional to the interfacial area the rate curve is first sigmoidal and after passing through its maximum at maximum area of interface it then falls progressively to zero (see Fig. D,23). Lewis [*71*] first demonstrated this in the case of silver oxide, and numerous examples of this behavior are now known. Pease and Taylor [*72*] examined the interface reaction (iv) in great detail. In the field of combustion this behavior is characteristic of the low temperature decomposition of potassium perchlorate investigated by Fry and Otto [*73*]

$$KClO_4\ (s) = KCl\ (s) + 2O_2\ (g)$$

It should be emphasized that the interface as the rate-determining factor may well apply only to a particular set of conditions, probably the slow rates of such reactions. For fast decompositions and reactions,

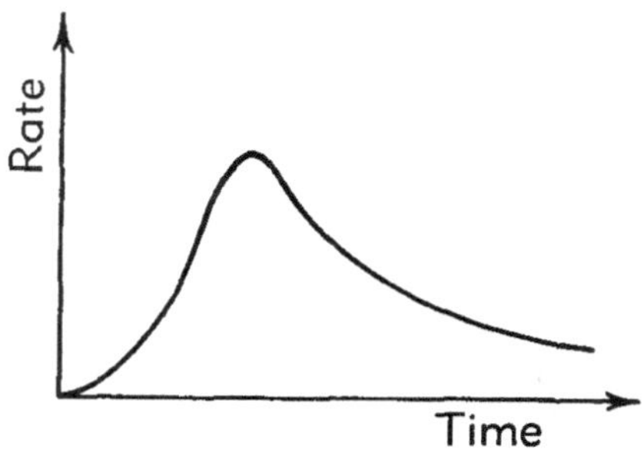

Fig. D,23. The velocity of reaction as a function of time at an interface between two solid phases.

transport of products or of energy input can readily become the rate-determining steps.

The decompositions of solid and also of liquid explosives belong to the class of fast reactions leading to gaseous decomposition products. A number of the cases investigated have been shown to be characterized by both high activation energies and high frequency factors. Activation energies ranging from 48 to 60 kcal and frequency factors from 10^{18} to 10^{27} (cf. values for unimolecular processes, normally 10^{13}) have been recorded with such materials as trinitrotoluene, picric acid, tetryl, nitrocellulose, and nitroglycerine. The high activation energy implies a high degree of stability at room temperatures but, with the high frequency factor, a rapid development of instability with increase of temperature. Semenov [*74*, p. 423–443] analyzed these cases in detail and came to the conclusion that they all involved the development of branching chains. He traced out these concepts in reference to data of Garner and his coworkers with lead styphnate, barium azide, and mercury fulminate.

More recent work by Garner and Maggs [*75*] and by Wischin [*76*] on the photochemical and thermal decomposition of barium and strontium azides has led to a theory of the decomposition of metallic azides by Mott [*77*] which can have general significance for the interpretation of the kinetics of interface reactions involving solids. These authors doubt the applicability of chain reactions in the interpretation, especially since the endothermic decomposition of copper sulfate pentahydrate and of silver carbonate share with the explosives the characteristic high frequency factors and high energies of activation.

The photochemical and thermal decomposition of barium azide reveal the formation of nuclei whose rate of growth and increase in number with time can be measured. The nuclei increase in radius at a linear rate and increase in number as the third power of the time. The activation energy for growth is 23.5 kcal and for nuclei formation 74 kcal. These observations on growth and number lead to the conclusion that decomposition should result in a pressure increase proportional to the sixth power of the time. Values of n in the expression $p = \text{const} \times t^n$ have yielded values of n from 6 to 8, which might indicate an increase in number as the fourth power of the time.

Mott regards barium azide as an ionic conductor at room temperature with metal ions in interstitial positions in the lattice in thermal equilibrium and mobile. The N_3^- ions are assumed immobile. Under ultraviolet irradiation, electrons are ejected from the azide ions into the conduction band of the crystal. The positive holes left in the full band, corresponding to the N_3^- ions, are mobile and diffuse to the surface. When the "holes" reach the surface the N_3 radicals can escape, forming molecular nitrogen. The electron in the conduction band wanders until it is trapped by a speck of metallic barium. The resulting electric field will attract the interstitial metal ions; the speck grows by a process of electrolysis. With heat, nitrogen is lost from the surface, excess metal atoms going into solid solution, and on dissociation increase the concentration of interstitial ions and add to the number of free electrons. The thermal energy causes azide ions to lose electrons to the metal, thus forming azide radicals. These being in the interior of the salt cannot break up, but acceptance of an electron from an adjacent ion can yield a diffusion of the "positive hole" away from the nucleus to the surface where it can decompose to nitrogen. Mott shows that the process of growth of nuclei parallels that occurring in a halide grain of a photographic emulsion [*78*]. It can be treated quantitatively, yielding with respect to rate of growth proportionality to the third power of the time, if it is assumed that 2 electrons are necessary to form a stable nucleus. On this basis a nucleus with 2 electrons and thus one interstitial barium ion is stable or can catalyze the thermal decomposition. Somewhat different interpretations of these data have recently been discussed by Thomas and Tompkins [*79*].

This approach to the problem deemphasizes the role of the solid-solid interface as the seat of enhanced reaction. It assigns to the properties of defect lattice structures the important role in these solid-solid-gas reactions and their kinetics.

D,24. Fast Reactions: Introduction. The present and preceding articles set forth the basic principles and discuss some typical examples of what may be described as classical reaction kinetics. In the main these principles and their exemplification are concerned with and have been based upon reactions proceeding leisurely toward an equilibrium state. For example, in the hydrogen-bromine reaction for which a detailed analysis of the kinetic data of Bodenstein and Lind [*8*] has been attained it is pertinent to observe that the fastest reaction velocity studied by these authors in their classical investigations occupied a period of 25 minutes reaction time. It is obvious that such slow processes are of real interest and importance to those who would produce reactions for high speed aerodynamics and jet propulsion, where reaction times are measured in units of milliseconds, only when the conclusions from the classical slow processes are directly transferable to fast chemical kinetics.[1] It is a matter for study and investigation how far toward fast rates the data accumulated in slow kinetic studies can be extrapolated without modification or adjustment. In brief, can the Arrhenius or alternative expressions for classical kinetics be extrapolated to reaction times of the order of fractions of seconds rather than minutes or hours? Are there factors, which are insignificant in the leisurely process, that become dominant in the faster areas? It is to such matters that the following articles of this section are addressed.

D,25. Rates of Fast Reactions. We can focus attention on the elements of this problem by consideration of a bimolecular process, $A + B \rightarrow$ products, with reactants present in equal concentrations, the kinetic expression being

$$-\frac{dA}{dt} = k[A][B]$$

For such a reaction the half life $t_{0.5} = k^{-1}A_0^{-1}$ where A_0 is the initial concentration of A. Table D,25 gives the half lives in relation to rate constants, assuming a maximum rate of $k = 10^{14}$ cm³ mole⁻¹ sec⁻¹, with the corresponding activation energies E, with steric factors P varying from unity to 10^{-4} and for two initial pressures of A, namely $A_0 = 2.5 \times$

[1] This summary on fast reactions embodies ideas formulated by a group engaged at the James Forrestal Research Center, Princeton University, on a study of the basic science of the kinetics of fast reactions. Acknowledgement is made to the contributions of R. N. Pease, F. J. Long, D. Garvin, and M. J. Boudart.

10^{-5} moles/cm³ and 2.5×10^{-8} moles/cm³ corresponding to initial pressures for A_0 of 0.5 and 0.5×10^{-3} atm at standard temperature and pressure.

It is apparent that a bimolecular reaction with little or no activation energy proceeds with a speed characteristic even of a detonation throughout any convenient concentration range. A broad range of reaction speeds exists between that limit and those amenable to classical chemical kinetics and methods. The quenching of excited atoms must normally occur within a time of 10^{-8} sec. Ion recombinations occur within an interval of 10^{-6} sec. In a detonation wave traveling at 2000 m/sec, if the reaction zone were 1 cm thick the reaction time would be of the order of 10^{-5} to 10^{-6} sec. With ordinary Bunsen flames, with an effective flame front a few tenths of a millimeter thick, and with a linear flow velocity

Table D,25. Half lives of fast reactions.

k	P	$\frac{E}{RT}$	$t_{0.5}$, sec (0.5 atm)	$t_{0.5}$, sec (0.5×10^{-3} atm)
10^{14}	1	0	4×10^{-10}	4×10^{-7}
10^{11}	10^{-3}	0	4×10^{-7}	4×10^{-4}
10^{11}	1	7	4×10^{-7}	4×10^{-4}
10^{8}	10^{-4}	4.6	4×10^{-4}	0.4
10^{8}	1	13.8	4×10^{-4}	0 4
10^{5}	10^{-4}	11.5	0.4	400
10^{5}	1	20.7	0.4	400
10^{3}	10^{-4}	16.1	400	—
10^{3}	1	25.3	400	

of feed gases of 100 cm/sec the reaction time is of the order of 10^{-3} to 10^{-4} sec. In conventional flow systems the velocity of sound gives the limiting flow velocity, some 3.5×10^{4} cm/sec for air at 25°C. Passage over 1 cm of length would require 3×10^{-5} sec, suggesting that studies between say 1 and 10^{-3} sec reaction time might be accessible to experimental study in conventional flow systems.

D,26. Experimental Methods.

General considerations. In rapid reactions, experimental techniques assume especial importance in at least three directions: (1) the heating of reactants to reaction temperature, (2) the mixing of the reactants, and (3) the freezing of the composition and analysis of the products.

The conventional flow method. This has, in general, advantages of simplicity over other methods which require specialized apparatus and techniques, so that, at least for exploratory purposes, it can be used for reactions in the range of 1 to 10^{-2} sec reaction time. Mixing of reactants to secure completely random mixing is necessary. The normal molecular diffusion processes are too slow even in gaseous systems and must be assisted by methods which secure a high degree of small scale turbulence in which rapid eddy diffusion can occur. Turbulence cannot be secured readily in normal scale laboratory apparatus. An air stream with a velocity of 1 liter/min in a 2-mm capillary at 850°C is still remote from the turbulent region. It follows therefore that premixing of reactants is preferable prior to introduction into the reaction zone. Techniques of mixing include (1) passage of the reactants through packed beds or sintered microsphere plates (which may also serve as heating elements), which are easy of construction and can be readily tested for efficiency, (2) jet impingement with gas flows of large Reynolds numbers, or axial jets discharging into a cylinder and mixing with a second component entering through the annulus between jet and cylinder walls, which yields complete mixing 6 to 7 cylinder diameters from the jet (Alexander, Baron, and Comings [*80*]), and (3) cyclone mixers, which have been investigated experimentally by Shepherd and Lapple [*81*]. A cyclone may be profitably employed to sweep out the total volume of a cylindrical reactor making it more available for homogeneous reaction. A disk mixer (4) devised by Johnston and Yost [*82*] has been employed for mixing and tested for efficiency in actual fast kinetic experiments. It consists of a flat disk 0.2 cm^3 in volume, 18 mm in diameter, and 1 mm thick with four tangential leads distributed round the circumference. The gas exit is a 2-mm capillary along the axis of the disk. Mixing was found to be 99.5 per cent complete in 0.01 sec. A propeller or mechanical agitator (5) in the reaction stream can be employed when premixing is precluded because of reaction considerations.

When radiant transfer predominates (above approximately 750°C) heat transfer is relatively rapid and can be accomplished in capillaries where the average distance of molecules from the radiating or absorbing walls is relatively small. These considerations are applicable alike to the preheating and the final cooling process. When the temperatures of reaction are lower, with convective and conductive processes operative, the heat transfer problem becomes analogous to the mixing problem just discussed.

Analytic techniques involve (1) optical absorption methods, (2) conductometric methods, (3) mass spectrographic methods, and (4) gravimetric or volumetric methods. In the last case, gravimetric methods involving the absorption of a product and its weighing have elements of simplicity and easily attainable accuracy which recommend them at least in exploratory phases of such work.

Oscillographic method. Johnston and Yost [*82*] have exploited a method for studying the rapid reactions of nitrogen oxides, in which a sample of freshly mixed reagents is isolated and the change in concentration of one component is followed using light absorption. The metered gases are mixed and are passed through a small long reaction cell at a high rate of flow so that the average age of the mixture in the cell is 0.01 to 0.05 sec. When a steady state is attained, the cell is isolated from the flow system by closing a stopgate, and the change in concentration of one component is followed by measuring the absorption of light of a given frequency as a function of time. Chopped light is passed through the cell and falls on a phototube which in turn feeds to a cathode ray oscillograph, and the transient pattern with a suitable time scale is photographed and then analyzed to provide data on the chemical conversion.

Fast scanning infrared spectrographic method. The technique just described may be extended with the use of an infrared spectrograph under development by Crawford and Cowan [*83*] and by various industrial firms, having a detector the time constant of which is about 0.002 sec. With such instruments it should be possible to follow simultaneously the change of concentration of several components.

The mass spectrographic technique. This technique employs the mass spectrometer to withdraw samples rapidly from steady state reaction systems and may permit an identification of the intermediates present in such a system. An application to hydrocarbon decomposition was attained by Eltenton [*84*]. The technique is extending with further refinements of the mass spectrometer [*85*].

The diffusion flame technique. Rapid simple reactions that occur within 10^5 collisions may be studied by the diffusion flame method introduced by Polanyi [*26,86*] or by one of its variants. A reagent B, diluted with an inert carrier gas is fed into a cylindrical reactor at low pressure and constant composition of B. Reagent A, also diluted with an inert carrier, is fed at a lower molar flow rate but at a higher linear velocity into B through a small nozzle axially placed in the reactor. The reaction occurs at a rate controlled by the local concentrations of A and B. A steady state is reached in which diffusion and reaction rates balance, and in which A is confined to a roughly spherical zone with high concentration at the nozzle center and a vanishingly small concentration at the sphere's edge. Kinetic analysis is carried out by studying some property of the steady state pattern. The reaction between sodium atoms and various halides could be followed by the Polanyi school with measurement of the Na atom concentration by use of resonance light emitted by the atomic sodium on illumination by the sodium D line.

In a variant of this method employed by Polanyi, two reactants were chosen to interdiffuse at low pressures, the steady state process being revealed by the deposition along the reaction tube of a solid product. Garvin and Kistiakowsky [*87*] utilized the first type of technique studying the steady state "temperature pattern" within the reaction zone.

The best experimental half life range is from 5×10^{-5} to 10^{-2} sec.

The flash photolysis method. Intermediates capable of rapid reaction produced by thermal processes may also be produced at lower temperatures and in controlled quantities by photolysis. Porter [*88*] and Norrish have utilized this technique in their development of flash photolysis. High concentrations of intermediates may be produced within microseconds or fractions thereof by spark discharges of high intensity and brief duration using suitable condensers and discharge tube containing inert gas for the flash. The kinetics of the reactions succeeding the flash may be followed spectrometrically by photographing the absorption spectra of the constituents of the reaction tube utilizing a second smaller flash discharge and its continuous emission as the light source. In this case each photograph involves a separate filling of the reaction tube with reactants and a separate flash photolysis. More recently by the use of phototubes it has been found possible to follow a single photolysis in time and thus arrive at the kinetics of the rapid reactions in which the intermediates produced participate [*89*].

Rapid compression methods. These depend on the mixing of reagents at low temperatures at which they do not react, and on rapid compression and consequent adiabatic heating to the desired experimental temperatures. Internal combustion engines driven at a high rate of revolution [*90,91*] or dead stop piston compression machines [*92*] have been used. These methods have yielded few data for kinetic analysis. Zeldovich [*93*] has used a static method for ignition limit studies of carbon disulfide-oxygen mixtures which, initially at pressures less than atmospheric, are suddenly compressed by removal of a stopper from a 1.5-meter long, 20-mm inside diameter tube, the inrushing air producing a compression wave, in the limit a shock compression wave, which heats the reagents to a temperature dependent on the characteristics of the gas. For observation of the progress of reaction with time, this compression must be coupled with a rapid quench, for example, by rapid evacuation into an empty vessel, thus quickly cooling the reaction products. A flow analogue of such a static compression method can also be devised for kinetic studies. Davidson and his coworkers have utilized shock waves for the study of fast reactions [*94*].

Liquid systems. Roughton and his coworkers have followed reaction velocities in liquid systems by following the time course of heat effects over a period of 0 to 0.01 sec even when the total temperature change was no greater than 0.01° [*95,96,97*]. Applied to the kinetics and rapid thermochemistry of carbonic acid, Roughton measured the temperature rise due to the chemical reaction between hydrochloric acid and sodium bicarbonate solutions at distances of 2, 5, 8, and 11 cm from the orifice of the mixing chamber where the reactants met. He was able in this way to separate the two stages of the reaction

(i) $$H^+ + HCO_3^- = H_2CO_3$$

and

(ii) $$H_2CO_3 = CO_2 + H_2O$$

and to determine the velocity constant of (ii) over the range 0 to 37°C. This reaction is of importance in controlling the speed of elimination of carbon dioxide by mammals in the expired air. The first reaction is practically instantaneous, being complete in times less than 0.001 sec, reaction (ii) being followed over the time range to 0.05 sec.

Detonation velocities. Kistiakowsky and his coworkers have recently examined experimentally the velocity of detonation in explosive mixtures in varying concentrations, in presence and absence of added gases [*98*]. Stationary waves in hydrogen-oxygen mixtures with and without addition of argon and helium have been determined in tubes of two diameters over a wide range of concentrations. The detonation waves were measured electronically in cylindrical pipes, 1.2 and 10 cm diameter, by the use of miniature piezoelectric gauges. Evidence is obtained to show that deviations from theory may be due to the inability of the heat capacity of polyatomic gases to be fully excited in the prevailing time intervals.

The impact of plane shock waves on detonable gas mixtures follows a pattern of phenomena with increasing strength of shocks. Weak shocks are transmitted by such mixtures with slow decay. Stronger shocks induce a slowly rising velocity. The strongest shocks produce immediate detonation. Measurements of detonation velocities of cyanogen and oxygen at several pressures and in tubes of several diameters have been made by Kistiakowsky, Knight, and Malina [*99*] and used by them to compute heats of dissociation of nitrogen and carbon monoxide. For these, the authors find 9.76 ev and 11.11 ev respectively, actually the upper values in a spectrum of values for these quantities suggested by spectroscopic, electron impact, and other methods. The authors reject objections based on the applicability of the theory of detonation waves and on systematic errors with simultaneous nonequilibrium conditions in the reaction. This assumption of equilibrium must, according to Bauer [*100*, pp. 95–123], be "seriously questioned." Kistiakowsky and his colleagues used added argon to test the hypothesis that the entire reaction energy is not released to the wave or that some internal degrees of freedom are not excited within the available time. The addition of argon gave the expected change in detonation velocity, dismissing these hypotheses in their view.

Sound-absorption methods. The use of sound velocity measurements to determine the rate of rapid gas reactions was suggested by Nernst and Keutel in 1910 [*101*]. Richards in 1939 reviewed [*102*] a number of experimental efforts in this direction. Deviations of the gases from ideal behavior complicate the computations, and measurements at low pressures are difficult.

Measurements of relaxation times. Kantrowitz [*103*] pointed out that heat capacity lags shown by some polyatomic gases from sound-absorption methods could also be followed by studies of flow of gases past an impact tube. The method involves the measurement of a small pressure difference between a properly prepared gas sample and that flowing past an impact tube. Bauer [*100*, pp. 95–123] has shown that this method is applicable to the case of heat capacity lags due to general chemical reaction rather than to slow attainment of equipartition between the various degrees of freedom and has examined the problem theoretically in detail. With Gustavson he has explored its applicability to the reaction $2NO_2 = N_2O_4$. The techniques involved have been discussed by Kantrowitz [*103*] and by Griffith [*104*].

D,27. Nonequilibrium and Nonequipartition Systems in Fast Reactions. The Arrhenius-Eyring equation for the velocity of a chemical reaction

$$v = Ae^{-B/T} = PZe^{-E/RT} = \kappa \frac{kT}{h} \frac{Q^*}{Q_A Q_B} e^{-E/RT}$$

involves several assumptions for its validity. It is assumed that (1) the reaction is adiabatic, i.e. it takes place on a single potential energy surface, (2) on this surface there exists a region defining an activated complex, (3) there is equilibrium between reactants and activated complexes, and (4) the transmission coefficient κ is constant and in classical adiabatic reactions is not sensibly different from unity.

One or more of these assumptions are probably invalid in many fast reactions. The first assumption is probably never valid in hot flames where as Laidler and Shuler point out [*105*] atom and radical reactions are often nonadiabatic. The second assumption is not satisfactory for radical association reactions which must play a role in high temperature kinetics. The third assumption is invalid in flames where intermediates with anomalous temperatures have been observed spectroscopically. The failure of the assumptions is usually corrected for by introducing an all-purpose transmission coefficient κ into the rate expressions. In the absence of evidence to the contrary this coefficient is normally given a value of unity. Attempts have been made to determine actual values of κ [*106*] without significant result. For fast reactions at high temperatures and small values of E/RT, recent calculations by de Vogelaare and Boudart [*107*] indicate that in a model reaction $X + X_2 = X_2 + X$ values of κ differing from unity may set in at critical energy values $E_{cr} \cong 1.5E_s$ where E_s is the saddle-point energy. The calculations predict sticking collisions at high energy values of collision with low values of κ. Boudart finds some evidence in support of his view in the recent studies by Schwarz, Williams, and Hamill [*108*] of abnormal kinetics of "hot" atoms produced in the photolysis of hydrogen iodide which vary significantly from normal kinetics unless the "hot" atoms are moderated to thermal velocities by added inert gases.

Shuler has recently [*109*] discussed in detail the origin of nonequilibrium distributions in flames. He recalls that Eyring, Walter, and Kimball [*110*] emphasized that, in a chemical reaction involving transfer of energy among various degrees of freedom, there will be a different reaction rate for each energy level of the final state. The distribution of products among the energy levels, obtained by summing the rates for each level will then, in general, not be an equilibrium distribution. Golden and Peiser [*111*] making such calculations for the Br + H_2 reaction appear to confirm this prediction. The equilibrium distribution of products is established through collisions after reaction has taken place.

Shuler summarizes the following specific processes in flames which may give rise to nonequilibrium distributions.

1. Electronic nonequilibrium distributions arise if an adiabatic chemical reaction gives rise to products in an electronically excited state rather than in the ground state. As an example he cites the carbon monoxide-oxygen flame where CO_2 is produced in the electronically excited $CO_2^*(^3\pi)$ state,

$$O(^3P) + CO(^1\Sigma^+) + M \rightarrow CO_2^*(^3\pi) + M$$

The reaction leading to the ground state CO_2 is "forbidden" by the Wigner spin conservation principle and occurs at a lower rate. The excess of CO_2^* over that expected thermally is borne out by experimental measurements [*112*].

Electronic excitation of atoms, radicals, or molecules can involve transfer of vibrational energy to electronic energy in inelastic collisions. Quenching of electronically excited species yields quenching molecules in vibrationally excited states. The reverse process is therefore possible. The excitation of sodium by vibrationally excited sodium chloride in the Polanyi diffusion flame experiments

$$NaCl' + Na(^2S) = Na^*(^2P) + NaCl$$

has recently been discussed by Magee [*113*].

Electronic excitation also results by transfer of the energy of a reaction to a third body

$$A + B + C = AB + C^*$$

The excitation of mercury by recombination of hydrogen atoms, the reverse of the process of quenching the excited mercury by hydrogen molecules, is a familiar example. The electronic excitation of metals such as Fe, Pb, and Tl in flames as observed by Gaydon and Wolfhard

[*114*] has been attributed by Thomas [*115*] to this process. Iron lines with energy equivalent to 170 kcal have been observed requiring a very energetic association process for which $C_2 + O_2 = 2CO$ or $2C + O_2 = 2CO$ have been postulated.

2. Vibrational nonequilibrium distribution can arise by transfer of energy, from electronically excited reaction products in excess of their equilibrium concentrations, and by inelastic collisions. Shuler cites, as a possible example of this type, the $OH(^2\Sigma^+)$ radicals in the interconal zone of the oxyacetylene flame which appears to have a vibrational "temperature" of 3750°K compared to rotational and adiabatic flame temperatures of 3000°K. Redistribution of the initial excess vibrational energy of the quencher by collisions may give rise to a type of equilibrium nonequipartition distribution of the kind thus found in oxyacetylene flames.

   Golden and Peiser's work [*111*] already cited suggests that intermediates and products of chemical reactions are formed *initially* in vibrational nonequilibrium distributions. Predissociations, radiationless transitions, and perturbations in general may also affect distribution among vibrational (and rotational) energy levels. They do not *bring about* nonequilibrium distributions, but they may serve as a very sensitive *test* for the presence of nonequilibrium conditions. The work of Durie [*116*] showed that the effects of the predissociation in the $^2\Sigma^-$-$^2\Pi$ system of CH are suppressed as the conditions in the flame approach thermal equilibrium. The predissociation of $OH(^2\Sigma^+)$ and $OH(^2\Sigma^-)$ in conjunction with the excess concentration of hydrogen or oxygen atoms manifests itself as a vibrational nonequilibrium distribution of $OH(^2\Sigma^+)$ in the inner cones of both oxyhydrogen and oxyacetylene flames.

3. Rotational nonequilibrium distribution can occur by direct production in chemical reactions and by physical quenching of electronically excited species. Further, the breakup of polyatomic molecules under impact of energetic particles may produce fragments in nonthermal rotational excitation or distribution. For this to show up in spectra, it is necessary that the fission be accompanied by simultaneous electronic excitation. It is to this mechanism that abnormal rotational excitation in flames is often attributed.

4. Translational nonequilibrium distributions are doubtless also produced in fast reactions, but the experimental difficulties of determining translational distributions means that demonstrable proof of such is extremely slight [*117*].

It is perhaps of interest to recall in connection with the problem of nonequilibrium distribution of products of reaction the work carried out by Audubert prior to World War II in which the presence of nonequilibrium products was sought with the aid of photon counters, especially a CuI photoelectric counter. Audubert admits [*118*, p. 263] that his counters behaved very capriciously and that the researches are "extremely thankless" so that few physical chemists have as yet had recourse to such studies. Nevertheless, Audubert believes that a large number of chemical reactions are accompanied by the emission of ultraviolet radiation with energies of 150–160 kcal not easily interpreted by the data of classical chemical kinetics. The reactions studied include the oxidation of carbon and the dissociation of hydrazoic acid and its salts. In the latter, excited nitrogen molecules with a mean life of 2×10^{-3} sec at ordinary pressure rising to 5×10^{-2} sec at 2 cm Hg have been found. This work suggests that, with the developments that have taken place in recent years in respect to photon counters, a reexamination of chemical reactions and especially of fast reactions would be particularly welcome.

D,28. The Persistence of Nonequilibrium Conditions. This persistence will depend, according to Shuler, on (1) the extent and type of departure from equilibrium distribution arising from the chemical reactions and the subsequent *specific* energy transfer processes, (2) the number and effectiveness of the equilibrating collisions, and (3) the interplay of these two factors.

As to the effect of number of collisions, the pressure effect in some flames indicates that, in the oxyacetylene flame, as examined by Gaydon and Wolfhard [*119*] and Shuler and Broida [*120*], the rotational "temperature" of $OH(^2\Sigma^+)$ falls steadily from about 9000°K at 1.5 mm to temperatures near the adiabatic temperature of 3320°K at pressures of 760 mm. As regards the other factors, available data are meagre.

Shuler [*109*] has attempted a theoretical analysis of the nonequilibrium steady state in flames, where the system is considered schematically as a homocontinuous open reaction system. This model, which assumes both invariance of composition with respect to time and with respect to spatial coordinates, can be treated more easily. In the Bodenstein steady state treatment it is assumed that the lifetimes of the radicals and atoms are short compared with the duration of reaction so that $d[\mathrm{R}]/dt = 0$, where [R] is the steady state of the intermediate. In the case of open reaction systems the steady state is truly time-invariant for all species, and the analysis proceeds by setting up mass balance equations for all the species entering reaction. Shuler sets forth a simple example of such treatment for the determination of change of nonequilibrium electronic distribution as a function of pressure.

D,29. Cited References.

1. Bodenstein, M. *Z. physik. Chem. 13,* 56 (1894); *22,* 1 (1897); *29,* 295 (1898).
2. Kistiakowsky, G. B. *J. Am. Chem. Soc. 50,* 2315 (1928).
3. Taylor, A. H., and Crist, R. H. *J. Am. Chem. Soc. 63,* 1377 (1941).

4. Christiansen, J. A. *Kgl. Danske. Videnskab. Selskab. Mat.-fys. Medd. 1*, 14 (1919).
5. Polanyi, M. *Z. Elektrochem. 26*, 50 (1920).
6. Herzfeld, K. F. *Ann. Physik 57*, 635 (1919); *Z. Elektrochem. 25*, 301 (1919).
7. Bodenstein, M. *Z. physik. Chem. 85*, 329 (1913).
8. Bodenstein, M., and Lind, S. C. *Z. physik. Chem. 57*, 168 (1907).
9. Trautz, M. *Z. anorg. u. allgem. Chem. 102*, 81 (1919).
10. Lewis, W. C. McC. *J. Chem. Soc. London 113*, 471 (1918).
11. Jost, W. *Z. physik. Chem. B3*, 95 (1929).
11a. Hinshelwood, C. N. *Kinetics of Chemical Change*. Oxford Univ. Press, 1940.
12. Morse, P. M. *Phys. Rev. 34*, 57 (1929).
12a. London, F. *Z. Elektrochem. 35*, 552 (1929).
13. Altar, W., and Eyring, H. *J. Chem. Phys. 4*, 661 (1936).
14. Eyring, H., and Polanyi, M. *Z. physik. Chem. B12*, 279 (1931).
15. Eyring, H. *J. Am. Chem. Soc. 53*, 2537 (1931); *54*, 3191 (1932); *Chem. Revs. 10*, 103 (1932).
16. Eyring, H. *J. Chem. Phys. 3*, 107 (1935).
17. Hirschfelder, J. O., Eyring, H., and Topley, B. *J. Chem. Phys. 4*, 170 (1936).
18. Farkas, A., and Farkas, L. *Proc. Roy. Soc. London A152*, 124 (1935).
19. Soper, F. G. *J. Chem. Soc. 138*, 1393 (1935).
20. Lindemann, F. A. *Trans. Faraday Soc. 17*, 598 (1922).
21. Daniels, F., and Johnston, E. H. *J. Am. Chem. Soc. 43*, 53 (1921).
22. Polanyi, M., and Wigner, E. *Z. physik. Chem. 139*, 439 (1928).
23. Rabinowitch, E., and Wood, W. C. *Trans. Faraday Soc. 32*, 907 (1936); *J. Chem. Phys. 4*, 497 (1936).
24. Bodenstein, M. *Helv. Chim. Acta 18*, 743 (1935).
25. Hinshelwood, C. N., and Mitchell, J. W. *J. Chem. Soc. 139*, 378 (1936).
26. Polanyi, M. *Atomic Reactions*. Williams and Norgate, London, 1932.
27. Smith, J. O., and Taylor, H. S. *J. Chem. Phys. 7*, 390 (1939); *8*, 543 (1940).
28. Trotman-Dickenson, A. F., Burchard, J. R., and Steacie, E. W. R. *J. Chem. Phys. 19*, 163–169 (1951).
29. Szwarc, M. *J. Chem. Phys. 16*, 128 (1948).
30. Rice, F. O., and Rice, K. K. *The Aliphatic Free Radicals*. Johns Hopkins Univ. Press, 1935.
31. Taylor, H. S., and Tobolsky, A. V. *J. Am. Chem. Soc. 67*, 2063 (1945).
32. Grassie, N., and Melville, H. W. *Proc. Roy. Soc. London A199*, 1–14 (1949).
33. Polanyi, M. *Z. Elektrochem. 35*, 561 (1929).
34. Kistiakowsky, G. B. *J. Am. Chem. Soc. 52*, 1868 (1930).
34a. Harteck, P., and Kopsch, U. *Z. physik. Chem. B12*, 327 (1931).
35. Jungers, J. C., and Taylor, H. S. *J. Chem. Phys. 6*, 325 (1938).
36. Taylor, H. S., and Jones, W. H. *J. Am. Chem. Soc. 52*, 111 (1930).
37. Taylor, H. S., and Jungers, J. C. *Trans. Faraday Soc. 33*, 1353 (1937).
38. Rice, F. O., and Herzfeld, K. F. *J. Am. Chem. Soc. 56*, 284 (1934).
39. Leermakers, J. A. *J. Am. Chem. Soc. 56*, 1899 (1934).
40. Allen, A. O., and Sickman, D. V. *J. Am. Chem. Soc. 56*, 2031 (1934).
41. Dodd, R. E. *Trans. Faraday Soc. 47*, 56 (1951).
42. Echols, L. S., and Pease, R. N. *J. Am. Chem. Soc. 61*, 208, 1024 (1939).
43. Steacie, E. W. R., and Folkins, H. O. *Can. J. Research B17*, 105 (1939); *B18*, 1 (1940).
44. Stubbs, F. J., and Hinshelwood, C. N. *Proc. Roy. Soc. London A200*, 458 (1950); *A201*, 18 (1950); with Ingold, K. U. *A203*, 486 (1950).
45. Mitchell, A. E., and Marshall, A. L. *J. Chem. Soc. London 123*, 2448 (1923).
46. Alyea, H. N. *J. Am. Chem. Soc. 53*, 1324 (1931).
47. Alyea, H. N., and Haber, F. *Z. physik. Chem. B10*, 193 (1930).
48. Pease, R. N., and Cheseboro, P. *Proc. Natl. Acad. Sci. U.S. 14*, 472 (1928).
49. Semenov, N. *Z. physik. Chem. 46*, 109 (1927).
49a. Lewis, B., and von Elbe, G. *Combustion, Flames and Explosions of Gases*. Academic Press, 1951.
50. Langmuir, I. *J. Am. Chem. Soc. 40*, 1361 (1918).

51. Beeck, O. *Advances in Catalysis*, Vol II. Academic Press, 1950; *Discussions Faraday Soc. 8*, 118 (1950).
52. Emmett, P. H., and Brunauer, S. *J. Am. Chem. Soc. 62*, 1732 (1940).
53. Taylor, H. S. *Trans. Faraday Soc. 38*, 137 (1932).
54. Volkenstein, F. F. *J. Phys. Chem. U.S.S.R. 21*, 163 (1947); *23*, 917 (1949).
55. Volkenstein, F. F. *J. Phys. Chem. U.S.S.R. 22*, 311 (1948).
56. Boudart, M. *Thesis*. Princeton Univ., 1950.
57. Zeldovich, J. *Acta Physicochim. U.R.S.S. 1*, 449 (1934).
58. Roginskiĭ, S., and Zeldovich, J. *Acta Physicochim. U.R.S.S. 1*, 554–575 (1934).
59. Taylor, H. A., and Thon, N. *J. Am. Chem. Soc. 74*, 4169 (1952).
60. Halsey, G., and Taylor, H. S. *J. Chem. Phys. 15*, 624 (1947).
61. Taylor, H. S. *Frontiers in Chemistry*, Vol. 8. Interscience, 1949.
62. Langmuir, I. *J. Am. Chem. Soc. 54*, 2798 (1932).
63. Langmuir, I. *Trans. Faraday Soc. 17*, 607–621 (1922).
64. Jungers, J. C., et al. *Bull. soc. chim. belge 54*, 303 (1945); *55*, 160 (1946); *57*, 618 (1948); *56*, 210, 246 (1949); *J. chim. phys. 47*, 139 (1950).
65. Kummer, J. T., and Emmett, P. H. *J. Chem. Phys. 19*, 289 (1951).
66. Taylor, H. S., and Jungers, J. C. *J. Am. Chem. Soc. 57*, 660 (1935).
67. Kummer, J. T., and Emmett, P. H. *Brookhaven Conf. Rept. 2*, 1 (1949).
68. Temkin, M., and Pyzhew, V. *Acta Physicochim. U.R.S.S. 12*, 327 (1940).
69. Brunauer, S., Love, K. S., and Keenan, R. G. *J. Am. Chem. Soc. 64*, 751 (1942).
70. Langmuir, I. *J. Am. Chem. Soc. 38*, 2263 (1916).
71. Lewis, G. N. *Z. physik. Chem. 52*, 310 (1905).
72. Pease, R. N., and Taylor, H. S. *J. Am. Chem. Soc. 43*, 2179 (1921).
73. Fry, H. S., and Otto, C. E. *J. Am. Chem. Soc. 45*, 1138 (1923).
74. Semenov, N. *Chemical Kinetics and Chain Reactions*. Oxford Univ. Press, 1935.
75. Garner, W. E., and Maggs, J. *Proc. Roy. Soc. London A172*, 299 (1939).
76. Wischin, A. *Proc. Roy. Soc. London A172*, 314 (1939).
77. Mott, N. F. *Proc. Roy. Soc. London A172*, 325 (1939).
78. Gurney, R. W., and Mott, N. F. *Proc. Roy. Soc. London A164*, 151 (1938).
79. Thomas, J. G. N., and Tompkins, F. C. *Proc. Roy. Soc. London A209*, 550 (1951).
80. Alexander, L. G., Baron, T., and Comings, E. W. Transport of momentum, mass and heat in turbulent jets. *Contract N6ori-71 Tech. Rept. 8*, Urbana, 1950.
81. Shepherd, C. B., and Lapple, C. E. *Ind. Eng. Chem. 31*, 972–984 (1939); *32*, 1246 (1940).
82. Johnston, H. S., and Yost, D. C. *J. Chem. Phys. 17*, 386 (1949).
83. Crawford, B. R., and Cowan, G. R. *Nav. Ord. Contract 10346 Progress Rept.* Univ. Minn., Jan. 1952.
84. Eltenton, G. C. *J. Chem. Phys. 15*, 455, 465, 474 (1947).
85. Lossing, F. P., and Tickner, A. W. *J. Chem. Phys. 20*, 907 (1952).
Ingold, K. U., and Lossing, F. P. *Can. J. Chem. 31*, 30 (1953).
86. Bawn, C. H. E. *Ann. Repts. on Progress Chem., Chem. Soc. London 39*, 36 (1942).
87. Garvin, D., and Kistiakowsky, G. B. *J. Chem. Phys. 20*, 105 (1952).
88. Porter, G. *Proc. Roy. Soc. London A200*, 284 (1950); *Discussions Faraday Soc. 9*, 60 (1950).
89. Porter, G., and Wright, F. J. *Z. Elektrochem. 56*, 782 (1952).
90. Retailliau, E., Ricardo, H. A., and Jones, M. C. K. *Am. Scientist 39*, 656 (1951).
91. Walcutt, C., and Rifkin, E. B. *Ind. Eng. Chem. 43*, 2844 (1951).
92. Taylor, C. F., et al. *S. A. E. Quart. Trans. 4*, 231 (1950).
93. Gershanik, T., Zeldovich, Ya. B., and Rozlovskiĭ, A. I. *J. Phys. Chem. U.S.S.R. 24*, 85 (1950).
94. Carrington, T., and Davidson, N. *J. Chem. Phys. 19*, 1313 (1951).
95. Hartridge, H., and Roughton, F. J. W. *Proc. Cambridge Phil. Soc. 22*, 426 (1924).
96. Bateman, J. B., and Roughton, F. J. W. *Biochem. J. London 29*, 2622 (1935).
97. Roughton, F. J. W. *J. Am. Chem. Soc. 63*, 2930 (1941).
98. Beretz, D. J., Greene, E. F., and Kistiakowsky, G. B. *J. Am. Chem. Soc. 72*, 1080–1086 (1950).

99. Kistiakowsky, G. B., Knight, H. T., and Malina, M. E. *J. Chem. Phys.* *20*, 876 (1952).
100. Bauer, S. H. *L. Farkas Memorial Volume. Research Council Israel Special Publication 1.* Jerusalem, 1952.
101. Keutel, F. *Inaug. Dissertation.* Univ. Berlin, 1910.
102. Richards, W. T. *Revs. Mod. Phys.* *11*, 36 (1939).
103. Kantrowitz, A. *J. Chem. Phys.* *14*, 150 (1946).
Huber, P. W., and Kantrowitz, A. *J. Chem. Phys.* *15*, 275 (1947).
104. Griffith, W. C. *Ph.D. Dissertation.* Harvard Univ., 1949; *J. Appl. Phys.* *21*, 1319 (1950).
105. Laidler, K. J., and Shuler, K. E. *Chem. Revs.* *48*, 153 (1951).
106. Glasstone, S., Laidler, K. J., and Eyring, H. *The Theory of Rate Processes.* McGraw-Hill, 1941.
107. De Vogelaare, R., and Boudart, M. Unpublished researches. *Forrestal Research Center*, Princeton, 1952.
108. Schwarz, H. A., Williams, R. R., and Hamill, W. H. *J. Am. Chem. Soc.* *74*, 6007 (1952).
109. Shuler, K. E. *J. Phys. Chem.* *57*, 396 (1953).
110. Eyring, H., Walter, J., and Kimball, G. E. *Quantum Chemistry*, Chap. XVI. Wiley, 1944.
111. Golden, S., and Peiser, A. M. *J. Chem. Phys.* *17*, 630 (1949).
112. Kondratjewa, E., and Kondratjewa, V. *Acta Physicochim. U.R.S.S.* *4*, 547 (1936); *6*, 625, 748 (1937).
113. Magee, J. L. *J. Chem. Phys.* *8*, 687 (1940).
114. Gaydon, A. G., and Wolfhard, H. G. *Proc. Roy. Soc. London* *A205*, 118 (1951).
115. Thomas, N. *J. Chem. Phys.* *20*, 899 (1952).
116. Durie, R. A. *Proc. Phys. Soc. London* *A65*, 125 (1952).
117. Prigogine, I., and Xhronet, E. *Physica* *15*, 913 (1949).
Prigogine, I., and Mahieu, M. *Physica* *16*, 51 (1950).
Takayanagi, K. *Progress Theoret. Phys. Japan* *6*, 486 (1951).
118. Audubert, R. *Contribution to the Study of Molecular Structure.* Desoer, Liège, 1947.
119. Gaydon, A. G., and Wolfhard, H. G. *Proc. Roy. Soc. London* *A194*, 169 (1948).
120. Shuler, K. E., and Broida, H. P. *J. Chem. Phys.* *20*, 1383 (1952).
Broida, H. P., and Lalos, G. T. *J. Chem. Phys.* *20*, 1466 (1952).
Kane, W. R., and Broida, H. P. *J. Chem. Phys.* *21*, 347 (1953).

SECTION E

KINETICS OF SEVERAL OXIDATION REACTIONS

ROBERT N. PEASE

Having laid down the general characteristics of rate processes and the means by which the experimental data are utilized, it will be of interest to describe in a little detail the behavior of individual systems.[1]

A word of caution is appropriate at this point. Although the interest of the student of combustion is in conditions which ultimately lead to high temperatures and rates, almost all of the physicochemical studies have been made at relatively low temperatures where the rates are slow enough to permit observation. This has the advantage of spreading out the course of the reaction over a period of time. Its disadvantage is that the prevailing elementary reactions may differ in importance from those governing high temperature processes, and there is the further question of the emphasis to be placed on chemical kinetics in flames and explosions as compared to microscopic and macroscopic transport processes (diffusion, conduction, convection, turbulence, and the like). Over and above this is the fact that the details of even the simplest reactions are not yet fully disentangled. Thus the material which follows should be regarded more as the contemporary expression of a point of view about the course of combustion processes than as a set of final statements.

E,1. The Hydrogen-Oxygen Reaction. The first reaction to be described is the deceptively simple one between the two elementary gases, hydrogen and oxygen. Since water is the final product, it would be easy to write [*12*]

$$2H_2 + O_2 \rightarrow 2H_2O\ (g), \qquad \Delta H^0_{298} = -115.6 \text{ kg cal}$$

In these terms the reaction would be trimolecular (or of third order), and it was in fact originally so characterized. However, the process has proven to be far more complex than this, as will be seen from what follows. As many as a dozen individual steps (elementary reactions) have been assumed to contribute to the over-all process [*2,3,4*].

[1] The monograph of Lewis and von Elbe [*1*] should be consulted for a more complete discussion of these reactions than is possible here.

The reaction proves to be of the branched chain type in which the chain carriers are believed to be hydrogen and oxygen atoms, the hydroxyl radical, and the HO_2 quasi-molecule. Its outstanding characteristic is the existence of a limited region of explodibility terminated abruptly at an upper pressure, as was first discovered by Hinshelwood [*4*]. Further investigation has revealed the existence of a corresponding low pressure limit, and finally of still a third limit at higher temperatures and pressures.

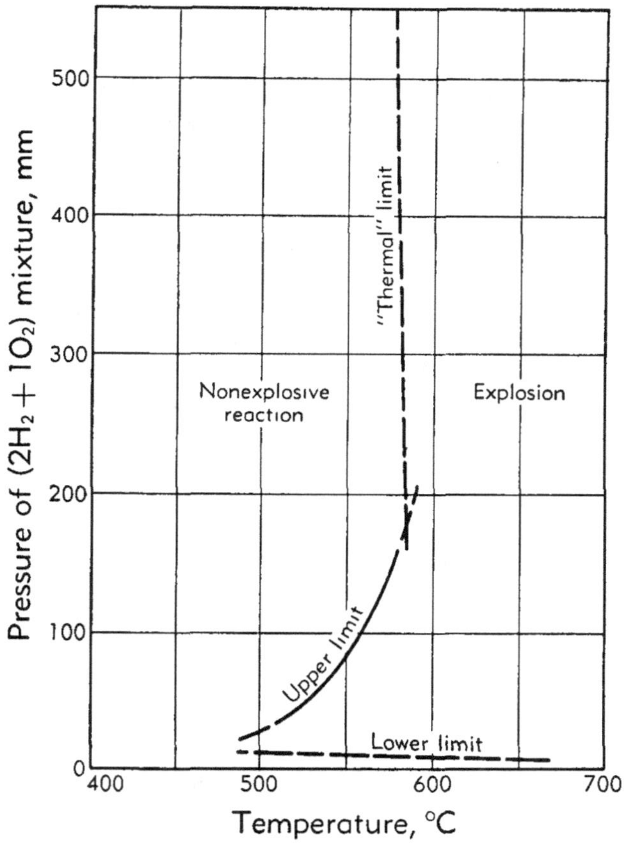

Fig. E,1a. Explosion limits for hydrogen-oxygen mixtures ($2H_2 + O_2$).

The general behavior may be better understood from a study of Fig. E,1a, which refers to a $2H_2{:}1O_2$ mixture in a silica reaction bulb. The three limits are here designated as the lower, upper, and thermal limits—alternatively the first, second, and third limits. The *upper* or *second* limit may be reached from above by first running hydrogen into the reaction bulb to a pressure substantially above the limit, followed by the requisite oxygen. When the bulb is then slowly evacuated, explosion occurs when the limit is reached. In a similar way the *lower* or *first* limit may be reached by slowly running the reaction mixture into an evacuated bulb until explosion is signaled by a pressure pulsation and a feeble flash.

The possibility of chain-branching in general was first raised by Christiansen and Kramers [*5*]. The concept was applied to the low pressure oxidation of phosphorus by Semenov [*6*], and to the low pressure hydrogen-oxygen explosions by Hinshelwood [*4,7*]. For the latter, the accepted mechanism involves the following reactions [*1*]

(i)	$OH + H_2 \rightarrow H_2O + H,$	$\Delta H^0_{298} = -15.8$ kg cal
(ii)	$H + O_2 \rightarrow OH + O,$	$\Delta H^0_{298} = +17.2$ kg cal
(iii)	$O + H_2 \rightarrow OH + H,$	$\Delta H^0_{298} = +3.0$ kg cal
(iv)	$H + O_2 + M \rightarrow HO_2 + M,$	($\Delta H^0_{298} = -70.$ kg cal)[2]
(iv*a*)	H, O, OH $\rightarrow$ recombination at wall	

Branching occurs via reactions (ii) and (iii) supplemented by reaction (i). By this means the number of atoms, and hence the reaction rate, tends to increase in geometrical ratio. Opposing these processes are two others, (iv) and (iv*a*), which consume atoms and are held to be responsible for the second and first limits respectively. It will be noted that no initial atom-yielding reaction is specified. Actually any chance process would suffice, for example, ionization by residual radioactivity or by a cosmic ray. In principle, just one atom need be produced.

The second limit. With respect to the second, or upper, pressure limit it will be noted that reaction (iv) represents an alternative to (ii) for the reaction of $H + O_2$. In the presence of a third body M, which may be any other molecule present, a trimolecular collision is assumed to stabilize the HO_2 complex. This quasi-molecule is believed to possess sufficient stability to permit it to reach the wall and react with others of its kind to produce water or hydrogen peroxide,[3] e.g.

$$2HO_2 \rightarrow H_2O_2 + O_2$$

Since reaction (iv) is of higher order, its rate increases with pressure relative to (ii) and ultimately puts a stop to branching.

Analysis of this mechanism by the conventional steady state assumption proceeds as follows. In the nonexplosive region just above the second limit, the rates of chain-branching (ii) and (iii) and of chain-breaking (iv) are substantially equal,

$$k_2[H][O_2] + k_3[O][H_2] = k_4[H][O_2][M]$$

Further, the rates of formation and consumption of oxygen atoms in (ii)

[2] Based on $2H + O_2 \rightarrow H_2O_2$; $\Delta H^0_{298} = -138.0$ kg cal

[3] Hydrogen peroxide is found as a product of the nonexplosive reaction [*8,9*].

and (iii) may be set approximately equal,

$$k_2[H][O_2] = k_3[O][H_2]$$

so that

$$2k_2[H][O_2] = k_4[H][O_2][M]$$

whence

$$2k_2 = k_4[M]$$

which is taken as the condition at the second limit. Explosion will then occur if $k_4[M]$ is reduced below $2k_2$. This is what happens when the total pressure, and hence [M], is decreased below the limit.[4]

This conclusion may be put to experimental test in the following way. Each kind of molecule M will have its own value of k_4, depending on its collision cross section, average velocity, and specific activity in stabilizing HO_2. If only hydrogen and oxygen are present, $k_4[M]$ may be broken up into two terms, $k_{H_2}[H_2]$ and $k_{O_2}[O_2]$. Thus

$$2k_2 = k_{H_2}[H_2] + k_{O_2}[O_2]$$

whence

$$[H_2] = \frac{2k_2}{k_{H_2}} - \frac{k_{O_2}}{k_{H_2}}[O_2]$$

A plot of $[H_2]$ vs. $[O_2]$ for various limit mixtures should be linear, the slope giving the ratio of the relative effectiveness of the two kinds of stabilizing molecules, as shown in Fig. E,1b taken from the work of Frost and Alyea [*10*]. These data give

$$p_{H_2} = 28.9 - 0.304p_{O_2}$$

The value of $k_{O_2}/k_{H_2} = 0.304$ may be compared with a calculated value of 0.42 based on kinetic theory collision calculations (Table E,1a).

The method may be extended to other types of molecules M, as third bodies, by writing

$$2k_2 = k_{H_2}[H_2] + k_{O_2}[O_2] + k_M[M]$$

$$[H_2] + \frac{k_{O_2}}{k_{H_2}}[O_2] = \frac{2k_2}{k_{H_2}} - \frac{k_M}{k_{H_2}}[M]$$

Knowing the value of k_{O_2}/k_{H_2}, a plot of $[H_2] + k_{O_2}/k_{H_2}[O_2]$ against [M] will give k_M/k_{H_2}, which is the efficiency of the added gas M relative to hydrogen. Proceeding in this way, von Elbe and Lewis [*2*], among others, have compared the efficiencies of a number of foreign gases with hydrogen. Results are given in Table E,1a. "Calculated" ratios are obtained by comparing products of the squares of the average molecular diameters and reduced masses for HO_2 and the gas in question.

[4] In the above the k's are rate constants for the reactions corresponding to the subscripts. Square brackets stand for concentrations.

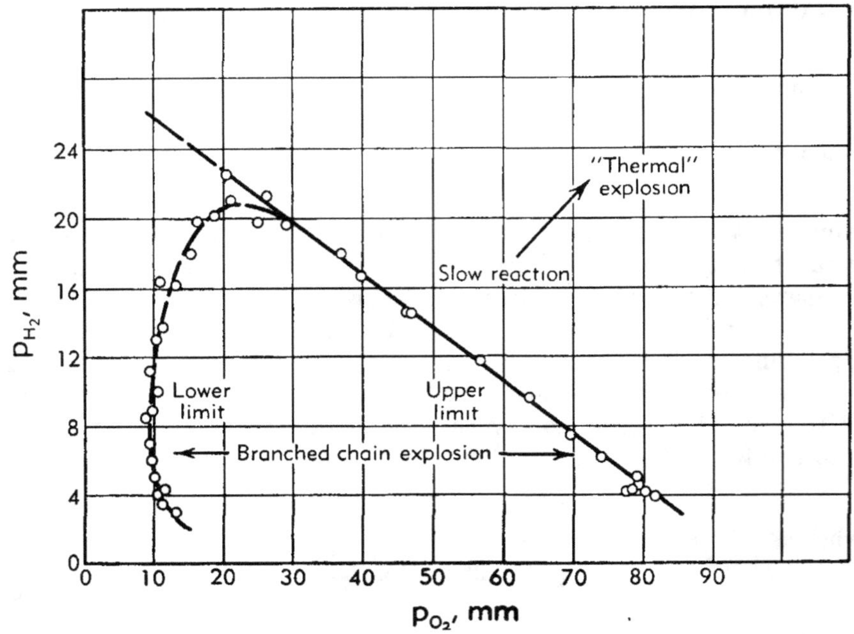

Fig. E,1b. Explosion limits of hydrogen-oxygen mixtures at 500°C (Frost and Alyea).

It will be noted that CO_2 and H_2O are more effective than would be expected on a purely collisional basis—that is, these gases suppress the explosion at lower than calculated partial pressures. Such a result might be ascribed to the higher heat capacities of CO_2 and H_2O, or to a specific chemical interaction of some type. The high efficiency of water vapor, especially, is of importance since water is a reaction product.

Table E,1a. Comparison of gases as third bodies in suppressing the hydrogen-oxygen explosion. Second limit.

Gas mixture	*Experimental*	*Calculated*
O_2/H_2	0.35	0.42
N_2/H_2	0.43	0.46
CO_2/H_2	1.47	0.43
H_2O/H_2	14.3	0.6–0.9
He/A	1.80	1.60

The above formulation implies that there should be no effect of surface on the second limit since reactions (i)–(iv) are all gas-phase processes; reaction (iv*a*), which represents the loss of atoms at a surface, is reserved for the first limit effect at much lower pressures. Such independence of surface was indicated by the original experiments of Thompson and Hinshelwood [*4*], who found substantially the same limits in an empty silica bulb, in the same bulb filled with lengths of silica tubing, and in a porcelain bulb—for a $2H_2:1O_2$ mixture at 500°C. On the other hand, the data of Frost and Alyea (Fig. E,1b) indicate that such 2:1 mixtures lie just outside the explosion region at 500°C, though they come close to the transition zone between the first and second limits. The essential difference in the two sets of experiments lies in the fact that Frost and Alyea employed a reaction bulb which was coated with potassium chloride, a procedure which has been found to suppress the nonexplosive reaction above the second limit [*11*]. The variability of the data in the first-second limit transition zone is in fact strikingly shown by the results of von Elbe and Lewis [*2*] on the effect of vessel size and of coating material (Fig. E,1c). Judging from the data of Frost and Alyea (Fig. E,1b) it would be desirable to make such comparisons of surface effects in oxygen-rich mixtures, or at least to confirm that conditions correspond in fact to the second limit.

Further information about the second limit may be obtained from the temperature dependence. The equation

$$[H_2] = \frac{2k_2}{k_{H_2}} - \frac{k_{O_2}}{k_{H_2}}[O_2]$$

permits determination of the ratios $2k_2/k_{H_2}$ and k_{O_2}/k_{H_2}. The data of Frost and Alyea indicate that the latter has the nearly constant value 0.30, between 480° and 540°C. As already pointed out, this is close to the estimate 0.42, based on a collision mechanism for reaction (iv); and since only the square root of T is involved no appreciable temperature dependence would be expected. Graphical determination of the intercepts, $2k_2/k_{H_2}$, gives the values 20 at 480°, 30 at 500°, 44 at 520°, and 62 at 540°C, the units being mm. A plot of $\ln 2k_2/k_{H_2}$ vs. $1/T$ based on the Arrhenius equation

$$\ln k = -\frac{E_A}{2.3RT} + \text{const}$$

gives 23,000 calories for the net activation energy. This would be the difference ($E_2 - E_{H_2}$) between the activation energies of reactions (ii) and (iv), the latter for hydrogen as the stabilizing molecule. If now the activation energy for reaction (iv) is close to zero, as might be supposed in view of the fact that no bonds are broken in the process and the probability that the reaction is exothermic, this activation energy applies largely to reaction (ii).

Thermochemical data for reaction (ii) yields [*12*]

$$\text{(ii)} \qquad H + O_2 \rightarrow OH + O, \qquad \Delta H^0_{298} = +17{,}200 \text{ cal}$$

The activation energy for this endothermal process must be equal to or in excess of this value, a condition which is fulfilled if $E_2 = 23{,}000$ cal. Thus on all counts the proposed mechanism is self-consistent.

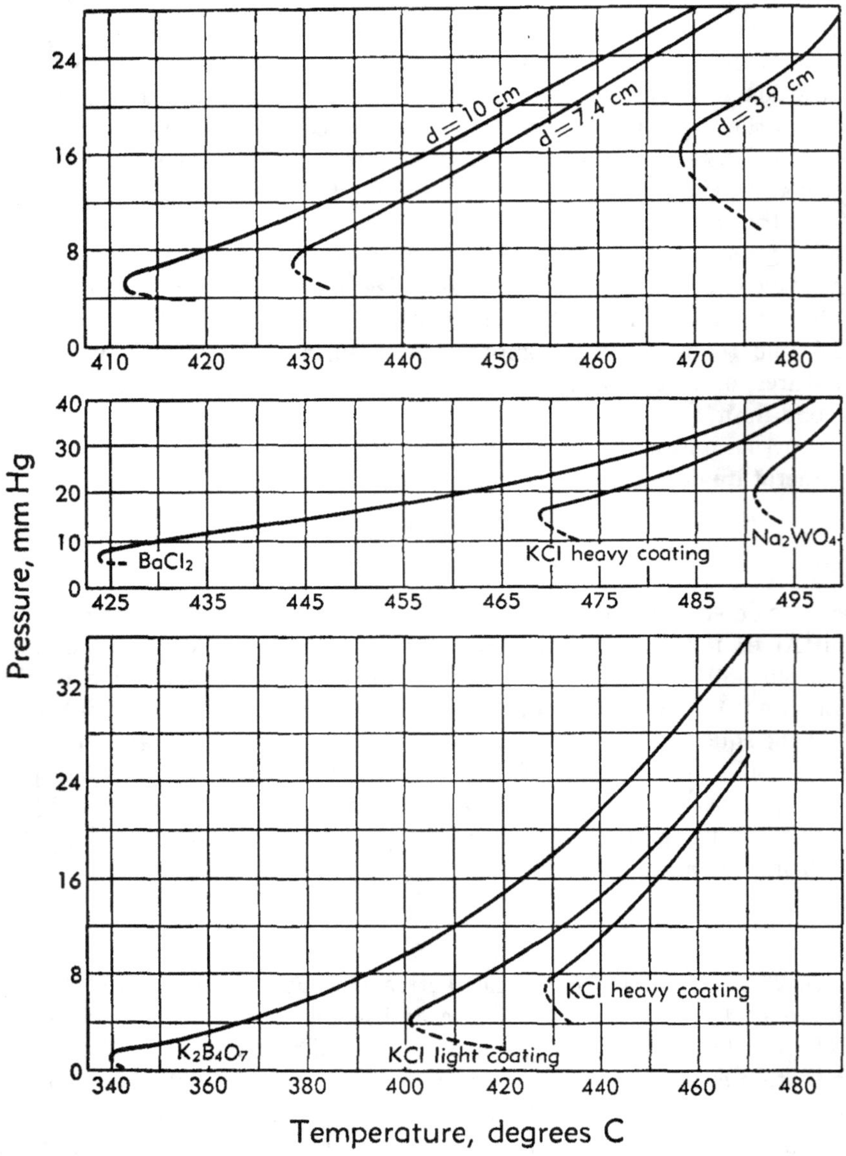

Fig. E,1c. Explosion of hydrogen-oxygen mixtures. Junctions of first and second explosion limits as affected by vessel diameter and surface coating (von Elbe and Lewis). Top: Variation of vessel diameter *d*. Heavy KCl coating. Middle: Variation of coating material. Vessel diameter 3.9 cm. Bottom: Variation of coating, potassium salts only. Vessel diameter 7.4 cm.

The first limit. With respect to the first or lower limit it is assumed that chains are broken by adsorption of chain carriers (H, O, OH) on the walls, followed by their mutual destruction by recombination (iv*a*). This process should obviously be facilitated by reducing the pressure, which increases the diffusion coefficients of the chain carriers, or by decreasing the vessel diameter, but it will also depend on the efficiency of the wall itself. For the latter reason it has proven difficult to obtain reproducible results. Not only does a given surface require conditioning, but also different wall surfaces yield quite different limits. This has already been illustrated by the observations of von Elbe and Lewis (Fig. E,1c). Further evidence is obtained from experiments by Semenova, Biron, and Nalbandjan [*13*], who found that with freshly fused Pyrex the lower limit (50 per cent mixture, 515°C) was of the order of 0.2 mm, whereas in a quartz vessel it was nearer 10 mm.

An efficient surface is one which neutralizes most atoms or radicals in a single contact—one, that is, for which the reflectivity is small. Under these circumstances a concentration gradient is set up, and relatively high pressures of reactants, or of added inerts, are required to effect explosion. This is evidently the case for quartz surfaces, or for KCl-coated Pyrex (Frost and Alyea). On the other hand, if many or most chain carriers are reflected, their concentration gradients tend to flatten out, and the concentrations critical for explosion may be reached at much lower reactant pressures (fused Pyrex). Hence in quartz an "inert gas effect" may be expected ($p_{H_2} + p_{O_2}$ is lowered as inert is added), while in fused Pyrex the effect should be less pronounced. This is borne out by data of Biron and Nalbandjan. At 515°C the fused Pyrex vessel gave a lower limit of 0.15 mm, which was unaffected by addition of 0.15 mm argon (total pressure 0.30 mm); whereas in quartz a lower limit of 11.2 mm was decreased to 7.3 mm by addition of 14.5 mm argon. A similar inert gas effect in quartz had previously been observed by Hinshelwood and Moelwyn-Hughes [*14*]. Thus in a qualitative way, the dependence of the lower limit on surface has been confirmed.

Further evidence of the importance of surface has been obtained by the study of the effect of vessel diameter on lower limit pressure. Other things being equal, it would be expected that explosion would be easier to achieve (the limit would be lower) in a vessel of larger diameter since this would have lower surface-volume ratio. This may be put on a somewhat more quantitative basis in terms of the path by which a chain carrier ultimately reaches the wall. Evidently, if only a small fraction of collisions between chain carrier and reactant molecule is chemically effective—that is, if the reaction has an appreciable activation energy—the chain carrier may reach the wall and be neutralized before a branched chain is developed. According to the Einstein-Smoluchowski relation, the average displacement $\bar{x}$ of a molecule or atom in a time t is given by [15]

$$\bar{x}^2 = \frac{4}{3\pi}\bar{u}\lambda t$$

where $\bar{u}$ is the average velocity and λ is the mean free path. Since the number of collisions Z' which the molecule (or atom) undergoes in time t is

$$Z' = \frac{\bar{u}t}{\lambda}$$

then

$$Z' = \frac{3\pi}{4}\frac{\bar{x}^2}{\lambda^2}$$

This is the average number of collisions suffered by the molecule during a net displacement $\bar{x}$.

Suppose now that $\bar{x}$ is identified with the radius r of the reaction vessel, and that Z' is taken to be the average number of collisions required to ensure adequate chain-branching for explosion. Then, since λ varies inversely as the total pressure p (constant composition), the condition for explosion may be expressed as

$$p^2 \cdot r^2 = \text{const}$$

or alternatively

$$p \cdot d = \text{const}$$

where d is the vessel diameter.

This relation has been found to hold roughly. For example, the data of Semenova [13] and of Hinshelwood and Moelwyn-Hughes [14] show fair constancy (Table E,1b). It is obvious, however, that individual differences among the surfaces of reaction bulbs have a very considerable effect.

Table E,1b. Effect of vessel diameter on lower limit.

	p, mm Hg	*d, mm*	*pd*
Fused Pyrex vessels (Semenova)			
420°C	1.06	10.2	10.8
	0.76	15.7	11.9
	0.6	20.0	12.0
	0.4	30.0	12.0
564°C	0.32	10.2	3.3
	0.3	15.7	4.7
	0.2	20.0	4.0
	0.14	30.0	4.2
Quartz vessels (Hinshelwood)			
550°C	3.64	18	65.0
	2.85	32	91.0
	0.91	49	45.0
	0.76	54	73.0
	1.36	54	41.0

This is about as much as may be discussed profitably with respect to the lower limit. Experimental difficulties have so far prevented more quantitative and detailed treatment. Nevertheless, it may be said that the hypothesis of chain-breaking at the walls (reaction iv*a*) is qualitatively confirmed.

Other specific effects on low pressure explosions. Before leaving the subject of upper and lower limits, mention should be made of some further specific effects of added gases. It is found, for example, that iodine (introduced as isopropyl iodide) completely suppresses the low pressure explosion [*16*], perhaps via the reactions

$$H_2 + I_2 \rightarrow 2HI$$
$$H + HI \rightarrow H_2 + I$$
$$2I \rightarrow I_2$$

On the other hand, nitrogen peroxide in small amounts has the effect of lowering considerably the temperature at which explosion can occur [*17,18*] (from about 470 to 370°C) with amounts of nitrogen peroxide as small as a fraction of a millimeter pressure. Further, there is an upper as well as a lower limit of nitrogen peroxide pressure, as if this gas could assist in breaking chains as well as in branching.

Equally striking are experiments of Haber and Alyea [*19*] on the effects of surface in initiating explosion. Hot streams of hydrogen and oxygen were made to impinge on each other in a compartment through which nitrogen was flowing, and the point of crossing was observed through a window. No inflammation was observed under the conditions of the low pressure explosion region until a rod of glass, silica, or porcelain was introduced. Inflammation then occurred, but ceased as soon as the rod was withdrawn. Iron and copper, but not aluminum, gave the same effect. This might make it appear that a surface is essential for setting up the branching condition. It is possible that a surface reaction is somehow responsible for the primary formation of atoms and radicals, and that this is the reaction which is facilitated by the glass or silica rod. In this connection it may be mentioned that Garstang and Hinshelwood were unable to produce low pressure explosions in an aluminum vessel [*20*], although a surface reaction was observed. Further investigation of these effects would appear to be desirable.

The nonexplosive reaction and the third limit. Reaction may be quite slow just above the second limit (see Fig. E,1a); but the rate rises as temperature and pressure are increased until explosive conditions are once again encountered at the third limit.

The system is quite as temperamental between the second and third limits as it is near the first, the rate being very sensitive to the nature of the surface and the surface-volume ratio of the reaction bulb. At lower temperatures and pressures an increase in surface (i.e. a packing of wall material in the form of beads or small tubes) may increase the rate, as if a surface-catalyzed reaction predominated [*21*]. At higher temperatures and pressures the positive "catalytic" effect of surface gives way to inhibition, and the rate is less in a packed reaction bulb than in an empty one. This inhibitory effect of surface is very pronounced. For example, a 50 per cent mixture at 1 atm may explode (third limit) in a carefully cleaned Pyrex reaction tube at 550°C; whereas by simply rinsing out the tube with a potassium chloride solution the rate at this temperature may be reduced to negligible amount and the explosion temperature raised to 625°C. If now the tube is packed with broken glass and also treated with potassium chloride, explosions are not observed short of the softening point of Pyrex (about 700°). Comparable rates are reached at 540°, 610°, and 650°, respectively [*23,24*]. Reaction kinetics in this range is complex. The rate may vary as the third or fourth power of the pressure, while apparent activation energies run from 95 kg cal (KCl-Pyrex) to 150 kg cal (quartz) [*25,26*].

Further evidence of surface inhibition is afforded by the acceleration observed on addition of inert gases such as helium, argon, and nitrogen [*21,22*]. These substances apparently impede the passage of active particles to the walls where they would normally be suppressed. Even the effect of oxygen on the reaction rate may be of this nature—it is approximately the same as that of an equal amount of nitrogen. Water vapor, the final product, also increases reaction rates, giving the process an autocatalytic aspect. The accelerating effects of such gases stand roughly in the ratio

$$H_2:N_2(O_2):A:H_2O = 1:3:4:5$$

as might be expected if a diffusion process is involved.

To account for the reemergence of this surface-impeded reaction above the second limit, it is obviously necessary to elaborate further the reaction scheme already discussed (reactions i to iv). The most fruitful hypothesis [*2,3*] concerns the fate of the HO_2 molecule. Up to this point it has been implied that every such molecule survives until it reaches the wall, where it is destroyed. It is now assumed that toward higher temperatures a new gas-phase reaction

$$HO_2 + H_2 \rightarrow H_2O + OH \text{ (or} \rightarrow H_2O_2 + H)$$

becomes possible, regenerating an active particle. This reaction will now be in competition with the diffusion-limited reaction

$$HO_2 + \text{wall} \rightarrow \text{destruction } (\rightarrow H_2O \text{ or } H_2O_2)$$

The steady state relations characterizing the second limit then no longer hold, since the new source of OH or H permits again the reaction sequence (i) to (iv). The way is thus prepared for a third explosion limit.

Hinshelwood [3] expresses the third limit condition in the following way. The elementary reactions[5] are

initiating reaction → H or OH; rate $= f_1$

(i) $OH + H_2 \rightarrow H_2O + H$

(ii) $H + O_2 \rightarrow OH + O$

(iii) $O + H_2 \rightarrow OH + H$

(iv) $H + O_2 + M \rightarrow HO_2 + M$

(v) HO_2 (+ wall) $\rightarrow H_2O_2, H_2O$

(vi) $HO_2 + H_2 \rightarrow H_2O + OH$

Application of the usual steady state assumption gives for the rate of formation of water[6]

$$\frac{d[H_2O]}{dt} = \frac{f_1 \dfrac{1.5k_5 + 2k_6[H_2]}{k_5 + k_6[H_2]}}{1 - \dfrac{2k_2}{\sum k_4[M]} - \dfrac{k_6[H_2]}{k_5 + k_6[H_2]}}$$

The explosion condition (infinite rate) is then provided by setting the denominator equal to zero:

$$1 - \frac{2k_2}{\sum(k_4[M])} - \frac{k_6[H_2]}{k_5 + k_6[H_2]} = 0$$

whence

$$\frac{2k_2}{\sum(k_4[M])} = \frac{k_5}{k_5 + k_6[H_2]}$$

It will be noted that when $k_6[H_2] \ll k_5$, this equation reduces to the second limit condition:

$$2k_2 = \sum(k_4[M])$$

thus suggesting that there is a range over which both second and third limits (and their junction) may be observed. This is found to be the case, and permits experimental evaluation of the constants in the last equation in the form (hydrogen and oxygen only)

$$2k_2 = k_{H_2}[H_2] + k_{O_2}[O_2]$$

$$[H_2] = \frac{2k_2}{k_{H_2}} - k'_{O_2}[O_2]$$

[5] The initiating reaction is not further specified for the moment since it does not affect the third limit condition.

[6] This equation is exact if OH radicals are assumed in the initiating reaction. If H atoms are assumed, there is an added term $(-f_1)$ which is negligible if chains are long.

where $k'_{O_2} = k_{O_2}/k_{H_2}$. These can then be introduced into the third limit condition.

An elaboration of the constant k_5 is also required. The corresponding reaction (v), whose rate has been set equal to $k_5[HO_2]$, provides for the diffusion of HO_2 molecules to the surface and for their destruction there. If it is assumed that all such molecules are in fact destroyed on reaching the walls, diffusion becomes the rate-limiting process. For a given reaction vessel, k_5 can then be assumed proportional to the diffusion coefficient D:

$$k_5 = K_5 \cdot D$$

with the understanding that the diffusion coefficient itself depends on the gas composition.

The mixture law for diffusion is arrived at as follows: For the mutual diffusion of two gases, the Stefan-Maxwell equation yields [*27*, p. 228]

$$D_{12} = \frac{(C_1^2 + C_2^2)^{\frac{1}{2}}}{3\pi\sigma_{12}^2(N_1 + N_2)} \text{ cm}^2 \text{ sec}^{-1}$$

where

$$C = \left(\frac{8kT}{\pi m}\right)^{\frac{1}{2}} \quad \text{(the average velocity)}$$

$$\sigma_{12} = \tfrac{1}{2}(\sigma_1 + \sigma_2) \quad \text{(the mean molecular diameter)}$$

$$N_1, N_2 = \text{concentrations in molecules per cm}^3$$

When one substance is present in small amount (e.g. HO_2)

$$N_1 + N_2 \cong N_2$$

The equation may then be written

$$D_{12} = \frac{D'_{12}}{N_2}$$

It is now assumed that for diffusion through a mixture of gases [*28*]

$$\frac{1}{D} = \frac{1}{D_{12}} + \frac{1}{D_{13}} + \cdots$$

$$= \frac{N_2}{D'_{12}} + \frac{N_3}{D'_{13}} + \cdots$$

Finally,

$$k_5 = K_5 \cdot D = \frac{K_5}{\dfrac{N_2}{D'_{12}} + \dfrac{N_3}{D'_{13}} + \cdots}$$

or more specifically for hydrogen and oxygen only

$$k_5 = \frac{K_5}{\dfrac{[H_2]}{D_{H_2}} + \dfrac{[O_2]}{D_{O_2}}}$$

where D_{H_2} and D_{O_2} represent D'_{12} and D'_{13} with appropriate corrections for transforming concentration in molecules per cm³ into pressures in mm Hg. Alternatively,

$$k_5 = \frac{K_5 \cdot D_{H_2}}{[H_2] + D'_{O_2}[O_2]}$$

where $D'_{O_2} = D_{H_2}/D_{O_2}$. The constants D_{H_2} and D'_{O_2} may then be calculated and substituted in the equation.

The third limit equation may be rewritten

$$\frac{\sum(k_4[M])}{2k_2} = 1 + \frac{k_6[H_2]}{k_5}$$

or for hydrogen and oxygen only

$$\frac{[H_2] + k'_{O_2}[O_2]}{2k_2/k_{H_2}} = 1 + \frac{k_6[H_2]}{K_5 D_{H_2}}([H_2] + D'_{O_2}[O_2])$$

$$[H_2] + k'_{O_2}[O_2] - \frac{2k_2}{k_{H_2}} = \frac{2k_2 k_6[H_2]}{k_{H_2} K_5 D_{H_2}}([H_2] + D'_{O_2}[O_2])$$

$$\frac{[H_2] + D'_{O_2}[O_2]}{1 + k'_{O_2}\frac{[O_2]}{[H_2]} - \frac{2k_2}{k_{H_2}[H_2]}} = \frac{k_{H_2} K_5 D_{H_2}}{2k_2 k_6} = \text{const}$$

This equation was tested by determining limits in a KCl-coated silica vessel at 586°C, by plotting the data (Fig. E,1d, curve 1), and by comparing with the theoretical curve (curve 2). For the latter the constants were given the values

$$D'_{O_2} = 3.0 \quad \text{(calculated 3.8)}$$

$$k'_{O_2} = 0.38 \quad \text{(0.33 from second limit)}$$

$$\frac{2k_2}{k_4} = 144 \quad \text{(150 from second limit)}$$

$$\frac{k_{H_2} K_5 D_{H_2}}{2k_2 k_6} = 1100 \quad \text{(not otherwise determinable)}$$

The experimental values involve some uncertainty because of the occurrence of an induction period. Nevertheless the agreement is reasonably good, justifying to this extent the approximate analysis. Additional runs with 100 mm of nitrogen added were also made (curve 3) and calculated (curves 4 and 5) with reasonable agreement. It may therefore be granted that the mechanism suggested for the third limit is essentially correct.

It would now be possible in principle to investigate the original equation for the slow reaction between the second and third limits. However, this requires a knowledge of the initiating reaction (rate = f_1, above). About this, there have been two suggestions. Hinshelwood [*3*] has sug-

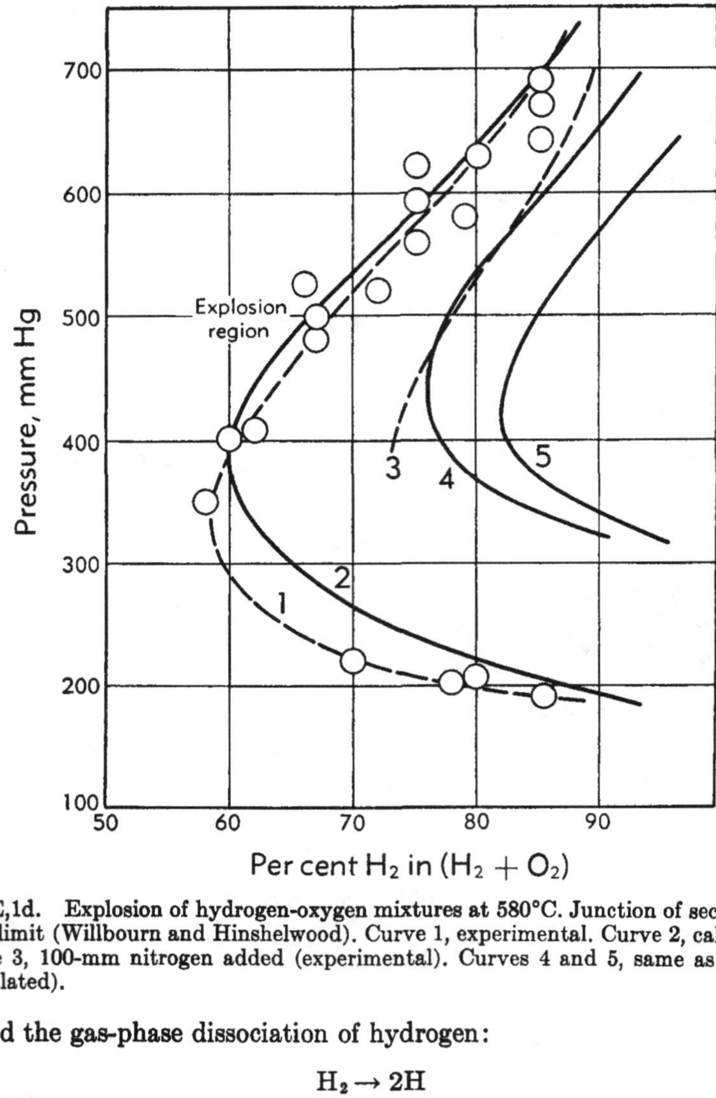

Fig. E,1d. Explosion of hydrogen-oxygen mixtures at 580°C. Junction of second and third limit (Willbourn and Hinshelwood). Curve 1, experimental. Curve 2, calculated. Curve 3, 100-mm nitrogen added (experimental). Curves 4 and 5, same as curve 3 (calculated).

gested the gas-phase dissociation of hydrogen:

$$H_2 \rightarrow 2H$$

Von Elbe and Lewis [*2,29,30*] contend that this reaction would not be sufficiently fast, and assume that hydrogen peroxide (from HO_2 formed by the branching-chain mechanism) is the initiator:

$$H_2O_2 \rightarrow 2OH$$

In any event the extreme variability in the rate of the slow reaction due to surface effects makes this problem somewhat difficult to handle. But perhaps enough has been said to indicate how elaborately varied the phenomena associated with a branched chain reaction can be.

E,2. The Carbon Monoxide-Oxygen Reaction. Since carbon monoxide is often a reactant or an intermediate product of combustion, its mode of oxidation is a matter of special interest. The process bears a formal likeness to that for hydrogen, but there are additional complications due to surface action and to sensitization by hydrogen and by water vapor (common impurities).

It may be stated with confidence that a low pressure explosion range exists with fairly definite lower and upper limits, pointing (as is the case for hydrogen) to a basic branched chain mechanism. Clear evidence for such behavior was first presented by Semenov, et al [*31*]. A carbon monoxide-oxygen mixture, stored over water, was passed slowly and at constant (low) pressure through a silica tube, the temperature of which was raised until inflammation was observed. In this way an "explosion peninsula"

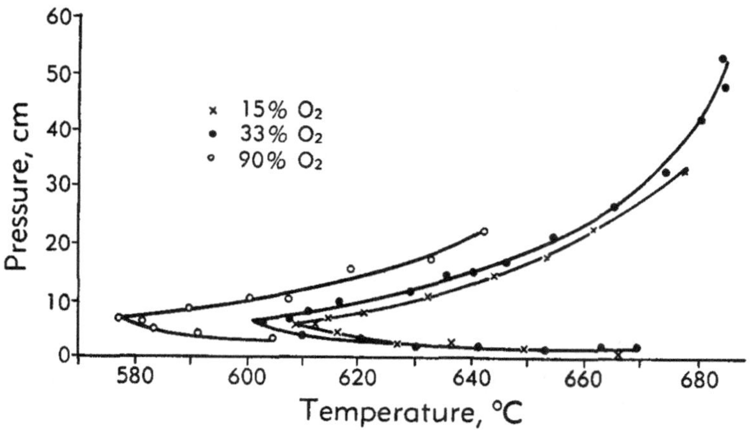

Fig. E,2a. Explosion limits for carbon monoxide-oxygen mixtures (Semenov).

was mapped out for mixtures containing 15, 30, and 90 per cent oxygen, as shown in Fig. E,2a. The field to the right of a curve corresponds to explosion, while that to the left corresponds to negligible reaction, if temperatures are not too high. Points on the upper limit branch give good straight lines when plotted as ln p against $1/T$, the slope corresponding to an activation energy of 35,000 cal as compared to 22,000 cal for the hydrogen-oxygen system (which was also studied). Minimum explosion temperatures are about 150° higher than those for hydrogen-oxygen.

Later work by Hinshelwood, et al [*32,33*] brought out a number of additional features. Dixon and others [*34,35*] had shown that spark ignition of very dry carbon monoxide-oxygen mixtures was extremely difficult. Following Semenov, Hinshelwood introduced carefully dried carbon monoxide-oxygen mixtures into a silica tube, and gradually heated until inflammation occurred. Upper limits were substantially as reported by Semenov. However, a particularly puzzling result was observed when the explosion region was approached in a different way. In the hydrogen-oxygen system it had been possible to bypass the upper limit by first introducing hydrogen to the reaction tube at a sufficiently high pressure, this being followed by oxygen in the amount desired. The upper limit could then be located by slowly evacuating the system ("withdrawal" method). When this procedure was applied to the carbon monoxide-oxygen system, no reaction whatever was observed. Yet, if the mixing were initiated just below the upper limit, a faint glow would appear which would burst into incandescence on evacuating the system. Hinshelwood speaks of this as "metastability," and suggests an analogy to change-of-state phenomena where both nucleation (creation of active centers) and growth must be considered. This should correspond to an induction period, and just this has been reported by other investigators. Topley [*36*], for example, comments that dry mixtures occasionally exploded after as much as 2 minutes. Again, Dixon [*37*], who studied the effects of mixing the gases in a heated tube, reported induction periods ranging from 0.5 to 10 seconds even with moisture present. In the case of the hydrogen-oxygen system it will be recalled that special emphasis was given to the unimportance of the chain-initiating step in the branching-chain mechanism. In principle just one chain carrier was required if branching conditions were otherwise favorable. For carbon monoxide-oxygen this is apparently no longer true. Chain initiation as well as branching must be taken into account.

Bypassing this difficulty and centering attention on the branching process, the question of possible chain carriers arises. In the dried gases, these must be derived from carbon monoxide and oxygen. The most promising suggestion is due to von Elbe and Lewis [*38*], who assume that the effective carriers are atomic oxygen and ozone, with a possible assist from the hypothetical higher oxide, CO_3. Their reaction scheme runs as follows:

$$
\begin{array}{lll}
\text{(i)} & O + O_2 + M \rightarrow O_3 + M, & \Delta H^0_{298} = -25.2 \text{ kg cal} \\
\text{(ii)} & O + CO + M \rightarrow CO_2 + M, & \Delta H^0_{298} = -126.9 \text{ kg cal} \\
\text{(iii)} & O_3 + CO \rightarrow CO_2 + 2O, & \Delta H^0_{298} = +16.7 \text{ kg cal} \\
\text{(iv)} & O_3 + CO \rightarrow CO_3 + O, & \Delta H^0_{298} = \ ? \\
\text{(v)} & CO_3 + CO \rightarrow 2CO_2, & \Delta H^0_{298} = \ ? \\
\text{(vi)} & O_3 + CO + M \rightarrow CO_2 + O_2 + M, & \Delta H^0_{298} = -101.7 \text{ kg cal} \\
\text{(vii)} & O_3 + CO + CO \rightarrow CO_3 + CO_2, & \Delta H^0_{298} = \ ?
\end{array}
$$

Branching occurs in the bimolecular reaction (iii) which produces two oxygen atoms, and is limited on the high pressure side by the trimolecular reactions (ii), (vi), and (vii). The essential feature of the hydrogen-oxygen mechanism is thus preserved.

If all reactions but (i), (iii), and (vi) are ignored, the upper limit condition for the low pressure explosion region may be expressed by the equality of rates of (iii) and (vi):

$$k_3[O_3][CO] = k_6[O_3][CO][M]$$

This implies that up to the limit the extra ozone molecule produced via (iii) and (i) is always destroyed in reaction (vi). The condition then becomes

$$k_6[M] = k_3$$

which corresponds to the upper limit condition in the hydrogen-oxygen system. For the complete scheme, reactions (i) to (vii), von Elbe and Lewis derived the limit relation as

$$k_6[M] = \frac{k_3 - (k_3 + k_4)k_2[CO]/k_1[O_2]}{1 + k_2[CO]/k_1[O_2]}$$

This equation goes over into the above form when $k_1[O_2]$ is large. For lower values, the term on the right may increase with oxygen concentration, so that the limit pressure represented by [M] increases as reported by Semenov.

Reaction (i), involving the formation of ozone from oxygen atoms and molecules, is a reasonable one. It is the basis for photochemical ozone formation at short wavelengths.[7] Evidence regarding reaction (ii) between oxygen atoms and carbon monoxide indicates that it is about $\frac{1}{100}$ as rapid as (i), but some carbon dioxide is formed in the photochemical reaction [*39,40*].

Reaction (iii) is somewhat endothermic ($\Delta H^0_{298} = +16.7$ kg cal) indicating the need for activation energy. The value 33 kg cal, calculated by Semenov from the temperature coefficient of the upper explosion pressure limit would be consistent with this. Further, the equilibrium is probably favorable. Direct observation of the ozone-carbon monoxide reaction at 150° to 250°C indicates that small amounts of carbon dioxide are produced (0.2 to 1 per cent) and that the reaction is chemiluminescent, whereas luminescence is not observed in purified air—thus it is not due to ozone decomposition [*41,42*]. Further, the chemiluminescent spectrum is like the carbon monoxide flame spectrum, which in turn has recently been identified with emission by neutral oxygen molecules in high vibrational quantum states [*43*]. This suggests that reaction (iii) should probably be broken down into two elementary processes:

(iii*a*) $$O_3 + CO \rightarrow CO_2 + O_2^*$$

(iii*b*) $$O_2^* \rightarrow 2O$$

with the understanding that some excited oxygen molecules O_2^* radiate. Lines are observed indicating an excited state corresponding to 6.09 electron volts or 141 kg cal. Since reaction (iii*a*) involves 102 kg cal, an

[7] The initial step is photochemical dissociation of oxygen at 1750A or less.

activation energy of about 39 kg cal would be needed. This is perhaps still not too far from Semenov's value of 33 kg cal. The excitation is obviously sufficient to account for oxygen dissociation (141 vs. 118 kg cal). Thus, the von Elbe and Lewis reaction scheme seems not unreasonable. though the postulated higher oxide, CO_3, remains unauthenticated.

The influence of small amounts of hydrogen on the low pressure explosion limits of carbon monoxide and oxygen presents additional features of interest. The effect is quite drastic according to Buckler and Norrish [*44*], as little as 0.2 mm hydrogen being sufficient to lower the

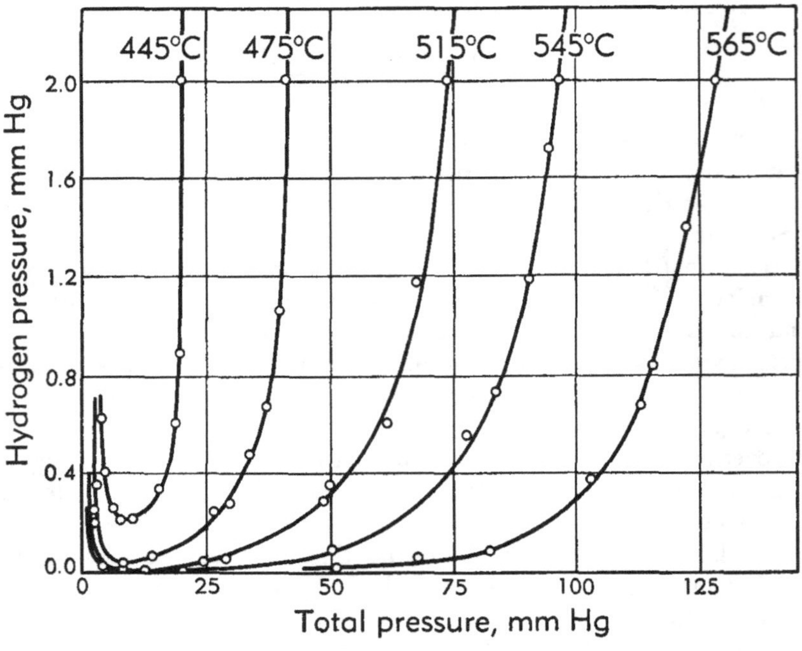

Fig. E,2b. **Variation of explosion limits of dry ($2CO + O_2$) mixtures with hydrogen pressure (Buckler and Norrish). Pyrex reaction vessel 2.7 cm diameter.**

explosion temperature by as much as 150°C (Fig. E,2b). Since conditions now correspond more nearly to the hydrogen-oxygen explosion regime, it would be plausible to assume that the process being studied is the latter as affected by the presence of carbon monoxide both as third body and as reactant. There are not enough data to explore this situation in detail, though it is possible that the reaction

$$OH + CO \rightarrow CO_2 + H$$

serves to tie the two systems together.

The nonexplosive reaction between carbon monoxide and oxygen also reveals distinctive features. The sensitivity to water vapor of spark-induced explosions and flames is again met with in the slow reaction. Thus, Hinshelwood, et al [*32*] found that with carefully dried gases the rate was only measurable at a little under 700°C, and was increased by a packing of silica spheres in a silica bulb, in rough proportion to the increase in surface. The reaction order in the unpacked bulb was somewhat less than the first with respect to each reactant. With water vapor present, the reaction could be conveniently studied at temperatures 150° lower, i.e. water vapor accelerates the process. At 560°C, rates were found to be proportional to water vapor and to carbon monoxide concentrations. On the other hand, the rate was inversely proportional to the oxygen concentration—oxygen is an inhibitor. Further, the rate was less in smaller reaction vessels and was further decreased by packing, that is, surface inhibition occurs. Hinshelwood inclines to the view that chain carriers, perhaps hydrogen atoms, are destroyed by oxygen at the surface. In an attempt to determine if hydrogen produced by the water-gas shift,

$$H_2O + CO \rightarrow CO_2 + H_2$$

could be the actual initiator of chains, mixtures of carbon monoxide and water vapor were heated in a silica vessel. Results were variable, but hydrogen (along with carbon dioxide) was formed, though only at 700 to 800°C at an appreciable rate. The over-all picture, therefore, may include chain initiation by hydrogen, and termination at the walls by reaction of hydrogen atoms with oxygen.

With respect to the slow combustion of hydrogen-carbon monoxide-oxygen mixtures at 550°C, Prettre [*45*] concludes that the initial rate is proportional to carbon monoxide, and independent of oxygen concentration. With respect to hydrogen, the effect may be represented by a Langmuir adsorption-isotherm type of term:

$$\frac{[H_2]}{1 + 0.48[H_2]}$$

as if adsorbed hydrogen were involved. Presumably as a matter of coincidence, rates are not very different from those of Hinshelwood with comparable amounts of water vapor in place of hydrogen.

It will be clear from the above that further study of the carbon monoxide-oxygen system will be required before quantitative conclusions about mechanism can be reached.

E,3. The Oxidation of Paraffin Hydrocarbons. The oxidation of a hydrocarbon is inevitably a more complex process than that of hydrogen or carbon monoxide, and quantitative information is understandably less complete. There is little doubt that chains are involved, with radicals like CH_3, CH_3O, and CH_3O_2 participating in addition to H, O, OH, and HO_2, but the number of subsequent possibilities for reaction becomes so great that it is never entirely clear just what elementary processes are critically important. Consequently it is convenient to deal mainly with general qualitative behavior. This procedure is the more attractive since the families of hydrocarbons possess many common properties.

The general characteristics of the oxidation of saturated aliphatic (paraffin) hydrocarbons above ethane are conveniently presented by reference to a pressure-temperature diagram such as Fig. E,3a [*46*]. The solid curves running from upper left to lower right give thermal explosion limits for lean mixtures of *n*-octane, *n*-heptane, and 2,2,4-trimethylpentane (iso-octane) in air. Points to the right of these curves are in the explosion field. To the left of the curves is the region of nonexplosive reaction; but this proves to contain a restricted zone (shaded region), in which so-called cool flames are observed. In this zone, reaction is characterized by an induction period following which the rate suddenly accelerates and a pale bluish-green luminescence spreads through the reaction mixture, sometimes repeating itself as many as 2 to 5 times. Outside of this zone (extreme left) reaction becomes slow and subject to measurement. This region of slow reaction has its own striking characteristics. There may be observed an induction period lasting for several hours, after which reaction is completed in a matter of minutes. Further, at pressures just below the lower cool flame limit the reaction rate may pass through a maximum and a minimum with rising temperature (region of negative temperature coefficient).

From the above description it will be evident that experimental observations often exhibit the most striking vagaries. For example, referring again to Fig. E,3a, suppose that an iso-octane-air mixture is studied at 5 to 6 atm, with rising temperature. Below about 280°C, reaction would be slow with long induction periods; between 280° and 300°, cool flames would be observed; from 300° to 330°, the reaction would be explosive; from 330° to 380°, again cool flames; from 380° to 410°, a region of rapid, but nonexplosive reaction; and finally from 410° up, only explosions would be encountered.

Observations of this general nature originally led Edgar and his colleagues to an appreciation of the considerable differences between iso-octane and *n*-heptane (or *n*-octane) as fuels [*47,48*]. Comparisons were made in a flow system at 1 atm with increasing temperature. In air, *n*-octane (or *n*-heptane) began to react at about 210°C, the amount of oxygen consumed rising to about one half of stoichiometric at 400°. Hardly any further change was then observed until 650° was attained, when explosion set in. On the other hand, iso-octane was almost entirely unaffected until the temperature reached 500°, when rates started to mount, and explosion followed at about 550°. Reference to Fig. E,3a will reveal that for *n*-octane the cool flame region was entered at about

250°; whereas for iso-octane this region would have been entirely bypassed, and only the zone of slow reaction (with its induction period) encountered until the explosion field was reached at about 550°. The consequences of such observations are well known in their bearing on the improvement of fuels for Otto-cycle engines.

Behavior such as that outlined above is observed generally for saturated hydrocarbons except methane, and for related series of compounds—olefins, ethers, alcohols, aldehydes, and the like. In all cases

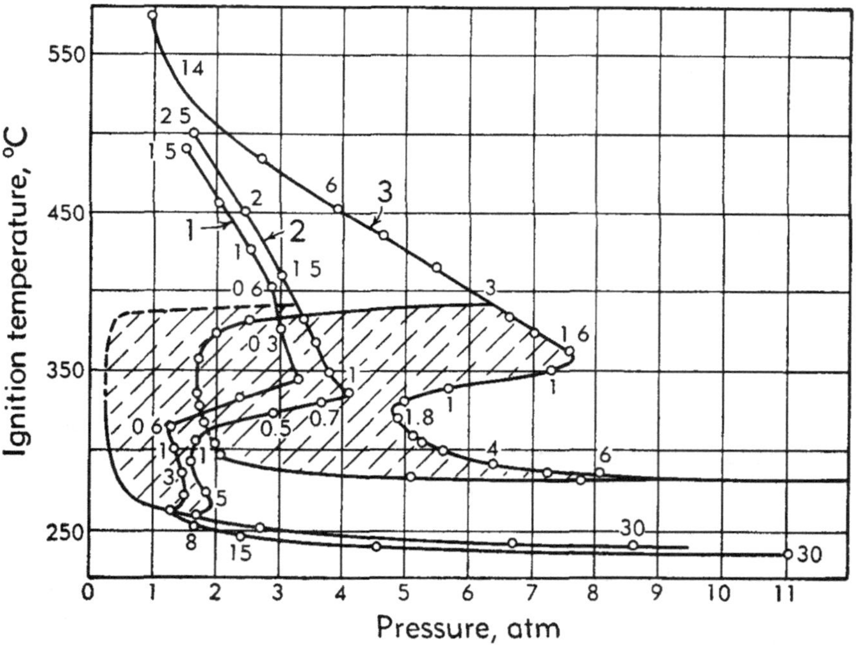

Fig. E,3a. Explosion limits for hydrocarbon-air mixtures (MacCormac and Townend). Curve 1, *n*-octane. Curve 2, *n*-heptane. Curve 3, iso-octane. Hydrocarbon equivalent to 0.65 of that corresponding to stoichiometric reaction with the oxygen of the air. Times noted are induction periods in seconds.

there is a more or less well-defined explosion "peninsula" pointing toward lower pressures[8] and bounded by a cool flame zone, with a surrounding region of slow self-induced reaction. Chemical analysis has revealed a great array of products including not only carbon oxides and water but also peroxides, aldehydes, alcohols, and products of thermal "cracking" such as olefins and lower members of the paraffin series. There is then a considerable amount of experimental material about such oxidation reactions from which it should be possible to draw conclusions as to the types of elementary processes involved. Without attempting a

[8] It should be noted that the range of limited explosion in hydrogen-oxygen mixtures points to lower *temperatures*—not lower *pressures*.

general quantitative solution, it will be of interest to explore some of the possibilities.

A tentative approach to the mechanism of such processes may be conveniently formulated in terms of Rice's free radical theory of the "cracking" of hydrocarbons and related compounds [*49,50*]. The gist of this proposal may be summarized in the statement that the primary cracking reaction consists of dissociation into free radicals rather than into stable product molecules. Ethane, for example, does *not* yield primarily ethylene and hydrogen:

$$CH_3 - CH_3 \rightarrow \begin{array}{c} CH_2{=}CH_2 \\ + \\ H{-}H \end{array}$$

but rather two methyl radicals:[9]

$$CH_3 - CH_3 \rightarrow CH_3\cdot + CH_3\cdot$$

The latter are then assumed to set up a reaction sequence involving intermediate formation of ethyl radicals and hydrogen atoms:

$$\begin{aligned} CH_3\cdot + C_2H_6 &\rightarrow CH_4 + C_2H_5\cdot \\ C_2H_5\cdot &\rightarrow C_2H_4 + H\cdot \\ H\cdot + C_2H_6 &\rightarrow H_2 + C_2H_5\cdot \end{aligned}$$

Ethylene and molecular hydrogen are thus represented as indirect products of a radical reaction chain.[10] Such chains are ultimately broken by radical recombination, for example:

$$H\cdot + C_2H_5\cdot \rightarrow C_2H_6$$

Mechanisms of this type are substantiated by detection of free radicals using Paneth's metallic mirror removal technique [*51,52*], by the effects of inhibitors and accelerators, and in some cases by observation of large photochemical quantum yields.

In considering how such free radical mechanisms might be elaborated to include oxidation, attention may be directed to the fact that oxygen is itself an accelerator of cracking reactions. Cracking in absence of oxygen is observable at temperatures well within the oxidation range,[11] but the amount of cracking is much more extensive when small amounts of oxygen are added [*54*]. For example, Appleby and Avery [*55*] find that at about 500°C and 1 atm in a KCl-treated reaction vessel, the cracking rate of *n*-butane is increased 100-fold by as little as 0.5 per cent oxygen. It is evident that a reaction chain is set up, since 15 moles of butane are found to react per mole of oxygen disappearing. As the oxygen content is

[9] The dots represent unpaired electrons.

[10] Methane, which is always observed, should also be formed in relative amounts depending on the chain length.

[11] For example, cracking of iso-pentane has been studied from 300 to 575°C [*53*].

increased up to 10 per cent, the chain length decreases, though the total amount of reaction increases. Under these conditions, water becomes an important product, and is in fact the *only* important oxygen-containing product of the reaction. At the same time the proportion of butene among the olefins increases. Analytic data then give for the initial product distribution (10 per cent O_2 at 500°) the relations:

$$C_4H_{10} + \tfrac{1}{2}O_2 \rightarrow C_4H_8 + H_2O, \qquad 84 \text{ per cent}$$
$$C_4H_{10} \rightarrow C_3H_6 + CH_4, \qquad 7 \text{ per cent}$$
$$C_4H_{10} \rightarrow C_2H_4 + C_2H_6, \qquad 9 \text{ per cent}$$

These figures may be contrasted with those obtained for cracking at 500° in the absence of oxygen, namely [*56,57,58*],

$$C_4H_{10} \rightarrow C_4H_8 + H_2, \qquad 10 \text{ per cent}$$
$$C_4H_{10} \rightarrow C_3H_6 + CH_4, \qquad 60 \text{ per cent}$$
$$C_4H_{10} \rightarrow C_2H_4 + C_2H_6, \qquad 30 \text{ per cent}$$

Here then is the beginning of the oxidation process. Such results indicate that the initial attack of oxygen in this "high temperature" region involves selective dehydrogenation of the hydrocarbon molecule. The normal primary dissociation of the hydrocarbon into alkyl radicals,

$$\left.\begin{aligned} C_4H_{10} &\rightarrow 2C_2H_5\cdot \\ &\rightarrow CH_3\cdot + C_3H_7\cdot \end{aligned}\right\} \Delta H^0 = +80 \text{ kg cal } [59]$$

seems to be overshadowed by what is probably an energetically more favorable process:[12]

$$\text{(i)} \qquad C_4H_{10} + O_2 \rightarrow C_4H_9\cdot + HO_2, \qquad \Delta H^0 = +30 \text{ kg cal}$$

Subsequent reaction then depends on the fate of the radicals ($C_4H_9\cdot$ and $HO_2\cdot$) produced in reaction (i). The $HO_2\cdot$ radical is presumed to reach the wall where it is converted into water or hydrogen peroxide, depending on the nature of the surface

$$HO_2\cdot \xrightarrow{\text{wall}} H_2O,\ H_2O_2$$

The butyl radical either dissociates or reacts with oxygen. Dissociation is preferred at high temperatures and low oxygen concentration, and leads to the normal cracking reactions,

$$\text{(ii)} \qquad \left.\begin{aligned} C_4H_9\cdot &\rightarrow CH_3\cdot + C_3H_6 \\ &\rightarrow C_2H_5\cdot + C_2H_4 \end{aligned}\right\} \Delta H^0 = +25 \text{ kg cal } [60\text{, p. } 346]$$

$$\text{(iii)} \qquad CH_3\cdot + C_4H_{10} \rightarrow CH_4 + C_4H_9\cdot, \qquad \Delta H^0 = 0$$

[12] ΔH based on 100 kg cal for loss of a hydrogen atom from butane and half the value for the reaction:

$$2H + O_2 \rightarrow H_2O_2, \qquad \Delta H^0 = -136 \text{ kg cal } [12]$$

Reaction with oxygen suggests two possibilities. A bimolecular process not unlike reaction (i) yields,

$$\text{(iv)} \qquad C_4H_9\cdot + O_2 \rightarrow C_4H_8 + HO_2\cdot, \qquad \Delta H^0 = -35 \text{ kg cal}^{13}$$

This reaction is essentially chain-ending if $HO_2\cdot$ is stabilized at the wall.[14] Alternatively a peroxyalkyl radical may be formed by a three-body process:

$$\text{(v)} \qquad C_4H_9\cdot + O_2 + M \rightarrow C_4H_9O_2\cdot + M, \quad \Delta H^0 = -70 \text{ kg cal}^{15}$$

with the understanding that the collision complex formed in (v) would be more stable at lower temperatures.

This scheme indicates the manner in which Rice's free radical mechanism might be extended to cover the preliminary stages of hydrocarbon oxidation. Ignoring reaction (v) for the moment, the remaining reactions (i) to (iv) give a particularly simple scheme which seems to account reasonably well for "high temperature" experiments like those of Appleby and Avery (10 per cent oxygen in butane at 500°C and 1 atm). A steady state solution gives

$$-\frac{d[C_4H_{10}]}{dt} = k_1[C_4H_{10}][O_2] + \frac{k_1k_2}{k_4}[C_4H_{10}]$$

This differs from their experimental relation

$$-\frac{d[C_4H_{10}]}{dt} = K[C_4H_{10}]^{\frac{3}{2}}[O_2]^{\frac{1}{2}}$$

but in view of the limited range of conditions neither equation is ruled out. (Both stress butane as compared to oxygen.) However, the experimental activation energy (20 kg cal) seems rather low for a reaction occurring with only moderate speed at 500°C. A value not less than ΔH^0 for reaction (i) (30 kg cal) might be expected. Allied experiments on propane oxidation [*62,63*] in which a rough second order dependence was indicated,

$$-\frac{d[C_3H_8]}{dt} = K[C_3H_8][O_2]$$

[13] Based on:

$$\begin{aligned} C_4H_9\cdot + H &\rightarrow C_4H_{10}, & \Delta H^0 &= -100 \text{ kg cal } [61] \\ C_4H_{10} &\rightarrow C_4H_8 + 2H, & \Delta H^0 &= +134 \text{ kg cal } [62] \\ H\cdot + O_2 &\rightarrow HO_2\cdot, & \Delta H^0 &= -70 \text{ kg cal} \end{aligned}$$

[14] The reaction

$$C_4H_{10} + HO_2 \rightarrow C_4H_9\cdot + H_2O_2, \qquad \Delta H^0 = +30 \text{ kg cal}$$

might also be considered here [*55*].

[15] Based on value for $HO_2\cdot$, and the similar dissociation energies for RH and H_2 ($\Delta H^0 = +100$ kg cal).

give a somewhat larger activation energy (38 kg cal), which is rather more consistent with the tentative value of ΔH^0 assigned to reaction (i). However, more experimental work is needed. Both rate and activation energy will vary with the hydrocarbon being studied.

It is likely that the mechanism for the initial process as developed above is valid up to considerably higher temperatures. The relative amount of induced cracking via reactions (ii) and (iii) may be expected to increase with temperature [*64*]; and since cracking is an endothermic process, the temperature rise due to oxidation will be moderated thereby in fuel-rich mixtures. However, as the oxygen-fuel ratio is increased, exothermic reactions must begin to make a contribution. The process:

$$CH_3\cdot + O_2 \rightarrow CH_3O_2\cdot \rightarrow \cdot CH_2OOH \rightarrow CH_2O + OH\cdot, \qquad \Delta H^0 = -50 \text{ kg cal } [12]$$

may be expected to compete with the cracking reaction (iii) for methyl radicals; while further oxidation of the formaldehyde (CH_2O) and of other intermediates leads ultimately (after an induction period) to local temperature rise, explosion, and the establishment of thermodynamic equilibrium at the flame temperature. Pressure-temperature limits for such explosions are represented by the curves in the upper left-hand corner of Fig. E,3a.

At lower temperatures a somewhat different state of affairs exists. This region is of importance because it is here that the influence of structure of the fuel molecule on oxidation characteristics is most pronounced (Fig. E,3a); and because low temperature oxidation rates correlate in a general way with fuel knock properties, as Edgar [*47,48*] and many subsequent investigators have emphasized [*1*, p. 188–202].

It might have been expected that oxidation rates would decrease continuously as temperature is lowered, but this turns out to be only partly true. Reaction is normally preceded by an induction period, whose length increases as expected (i.e. rate decreases). However, following this the reaction autoaccelerates as if an active but relatively stable intermediate were accumulating. A plot of fraction reacted vs. time then produces an S-shaped curve which may be broken from time to time by the incidence of "cool flames,"[16] as shown in Fig. E,3b [*62,63*] for 50 per cent propane-oxygen mixtures at 360 mm and various temperatures. If now the *maximum* rates are estimated from the inflection points of the curves, it is found that such rates first *rise* as temperature is increased (curves F and E) and then *fall* (curves B and A). Finally at still higher temperatures the maximum rate reverses itself once more and increases to the point of explosion. Thus there are four characteristic regions for reaction:

[16] The faint bluish glow characterizing cool flames has been identified with the formaldehyde spectrum [*65*]. Only about one molecule of formaldehyde in a million actually radiates [*66*].

(i) Slow reaction at lower temperatures and pressures
(ii) Cool flame reaction
(iii) Slow reaction at intermediate temperatures and pressures
(iv) Explosion

These regions may be identified in Fig. E,3c [*63*], which refers again to the combustion of 50 per cent propane-oxygen mixtures. Further details of the minimum in rate characterizing region (iii) are shown in Fig. E,3d [*62,63*]. Similar results are obtained with other paraffin hydrocarbons.

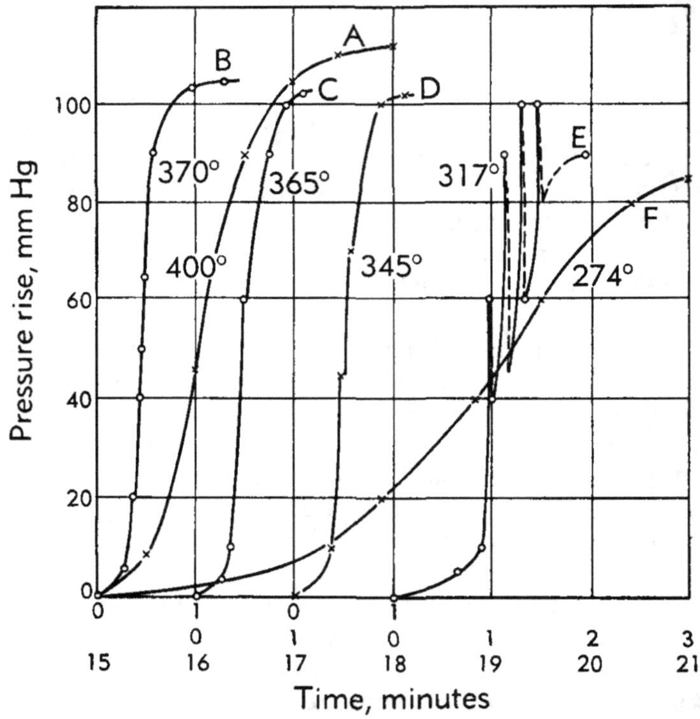

Fig. E, 3b. Pressure-time curves for 50 per cent propane-oxygen mixtures at 360 mm and various temperatures (Newitt and Thornes).

The problem of accounting for the transition from high to low temperatures in terms of a Rice-type free radical mechanism is a particularly puzzling one. Reference to Fig. E,3d will show that the transition is continuous at sufficiently low pressures. It would therefore be reasonable to assume that the initial reactions already presented are still in effect. However some change in emphasis is required.

It is generally believed that reaction (v) producing a peroxyalkyl radical is the process which is coming into prominence as temperature is

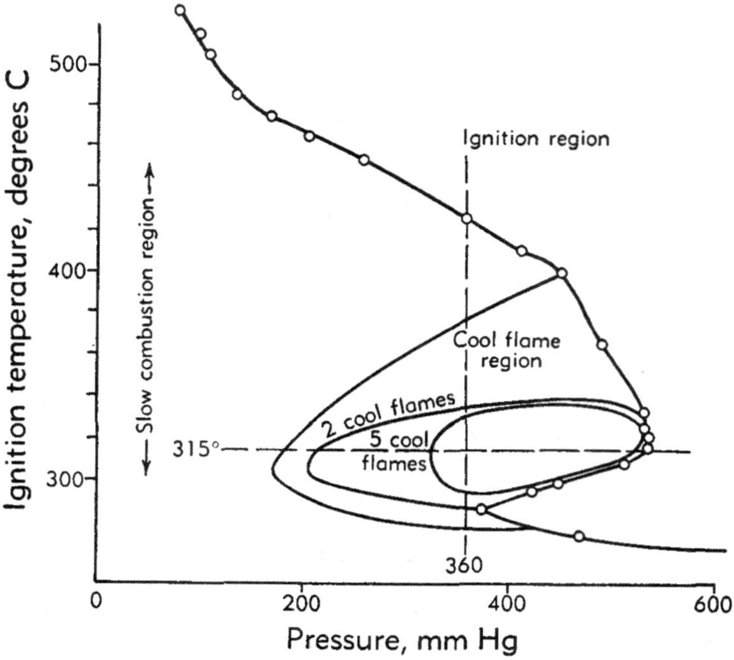

Fig. E,3c. Combustion regions of 50 per cent propane-oxygen mixtures (Newitt and Thornes).

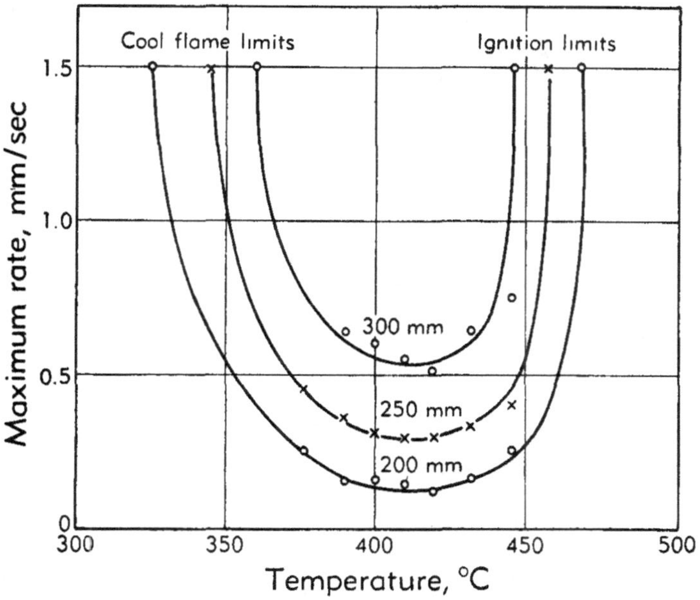

Fig. E,3d. Rate of reaction of 50 per cent propane-oxygen mixtures in the region between cool-flame limits and explosion limits (Newitt and Thornes).

lowered. This may be written

(v) $$R\cdot + O_2 + M \rightarrow RO_2\cdot + M$$

It competes for alkyl radicals R· with reaction (iv),

(iv) $$R\cdot + O_2 \rightarrow \text{olefin} + HO_2\cdot$$

and both are in competition with the cracking reactions (ii) and (iii),

(ii) $$R\cdot \rightarrow \text{olefin} + R'\cdot$$
(iii) $$R'\cdot + RH \rightarrow R'H + R\cdot$$

Higher pressures should promote reaction (v). In addition it should be the preferred path at lower temperatures since stabilization of the peroxyalkyl radical $RO_2\cdot$ is involved, and activation energies for (ii) and (iv) are presumed to be larger than for (v). However, this in itself is not sufficient to account for the observed autoacceleration and net increase in rate as temperature is decreased. In fact, comparison with the analogous process

$$H\cdot + O_2 + M \rightarrow HO_2\cdot + M$$

met with in the hydrogen-oxygen reaction might have led to the conclusion that reaction (v) would be a source of chain-breaking rather than branching. Yet branching of some kind is essential to autoacceleration.

A great deal of effort has been directed to the solution of this problem. The branching process cannot be like that in low pressure hydrogen-oxygen explosions since there are no sharp limits and acceleration is gradual. The best guess is that a moderately long-lived intermediate accumulates slowly, and eventually gives rise to new chains.[17] For example, it might be assumed that the peroxyalkyl radical dissociates into such an intermediate I and a new radical R″·:

(vi) $$RO_2\cdot \rightarrow I + R''\cdot$$

The radical then continues the main chain

(vii) $$R''\cdot + RH \rightarrow R''H + R\cdot$$

while the intermediate accumulates but eventually starts a new chain

(viii) $$I \rightarrow \text{stable products} + R'\cdot$$

In this way a delayed type of branching would be achieved leading to autoacceleration. Finally, a reaction

(ix) $$R\cdot + O_2 \xrightarrow{\text{wall}} \text{stable products}$$

may be added to take account of suppression of the low temperature

[17] The over-all process has been characterized by Semenov as "degenerate branching" [*67,68*].

reaction in a packed reaction tube [*64*], though it is rather to be expected that *all* radicals will be subject to suppression by the walls.

From the chemist's standpoint, the real difficulty lies in attempting to identify the intermediate I in the above mechanism. Literal interpretation of reaction (vi) in the case of *n*-butane yields

(vi) 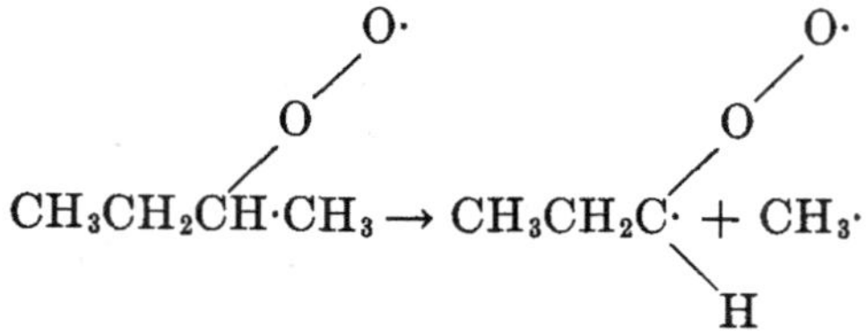

This intermediate would correspond in composition to propionic acid

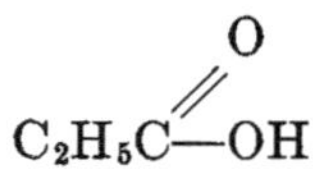

but there is no evidence that acids are involved, or are present in any quantity. Alternatively, carbon monoxide (CO) might be split out leaving ethyl alcohol (C_2H_5OH). Carbon monoxide is certainly formed in quantity, and alcohols have been identified (though usually as methyl alcohol), but there is no indication that alcohols are generally responsible for auto-oxidation. Again, a series of changes might produce olefins:

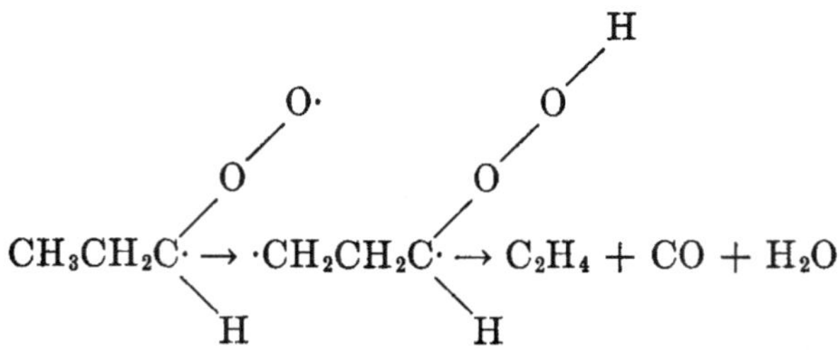

However, olefinic products appear to be even less subject to oxidation than the saturated hydrocarbons.

Prevailing opinion seems to favor the alkyl peroxides (ROOH or ROOR) as the effective agents [*69,70,71*], yet there are difficulties here also. It is supposed that peroxide is formed by the reactions

$$\begin{aligned} ROO\cdot + RH &\rightarrow ROOH + R\cdot \\ &\rightarrow ROOR + H\cdot \end{aligned}$$

Branching then results from dissociation:

$$ROOH \rightarrow RO\cdot + \cdot OH$$

If this mechanism is to be effective, the peroxide must have sufficient stability to accumulate, so that it should be detectable by analysis. It is true that *hydrogen* peroxide is commonly found if a clean Pyrex or silica reaction vessel is employed, and is usually accompanied by its addition product with aldehyde, for example

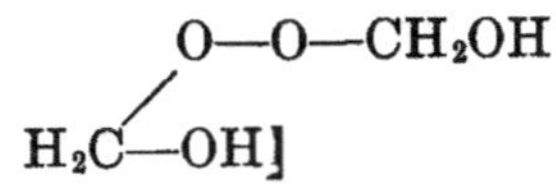

from formaldehyde [*72,73,74,75,76*]. Nevertheless, these substances are no longer obtained in salt-coated vessels although reaction still occurs. Evidence for the presence of true alkyl peroxides is less conclusive, though they have been reported among the oxidation products [*77*]. Direct observation has shown that such peroxides are rather unstable. For example, di-*tert*-butyl peroxide, $(CH_3)_3COOC(CH_3)_3$, decomposes readily at 150°C, and would hardly survive at 300° and above [*78*]. Ethyl hydroperoxide, C_2H_5OOH, decomposes at 140 to 200°C, largely on the walls; while diethyl peroxide, $C_2H_5OOC_2H_5$, becomes explosive above 200°C at pressures of 10 mm or more. Significantly enough, such explosions can occur in the presence of a propane-oxygen mixture without immediately affecting the latter [*79,80*]. Thus the evidence for alkyl peroxides themselves as intermediates leaves something to be desired. Possible decomposition products—in particular, aldehydes—have also enjoyed a vogue. For example, the oxyalkyl radical, RO·, presumed to be formed from the peroxide, might dissociate to yield formaldehyde:

$$R'CH_2O\cdot \rightarrow R'\cdot + HCHO$$

This is an interesting possibility because formaldehyde is a common oxidation product; and the faint bluish glow associated with cool flames has been identified with the formaldehyde spectrum [*70*]. Further, formaldehyde is stable up to 500°, and is only slowly oxidized at 300°C [*81,82*]. However, this substance *inhibits* the low temperature oxidation of hexane [*83*], though it does accelerate methane oxidation at 400°C [*84*].

The most recent candidate for the role of intermediate in low temperature oxidation is a diketone. Observations of the ultraviolet absorption spectra of reacting hydrocarbon-oxygen systems reveals the development of characteristic bands between 2000A and 4000A [*85,86,87*]. Absorption at shorter wavelengths (2000A) is attributed to peroxides, acids, higher aldehydes, and perhaps diolefins and benzene, while at 3000–4000A, formaldehyde bands appear against a background attributed to higher aldehydes and ketones. It is the intermediate region which is of special interest. With *n*-butane and higher hydrocarbons, but not with propane or isobutane, a strong maximum develops at about 2600, which decays rapidly when a cool flame passes through the mixture. Comparison spectra indicate rather conclusively that with *n*-pentane the product is 2,4-pentanedione [*88*], i.e. acetylacetone,

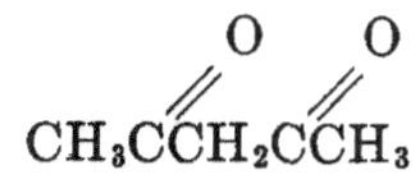

A rough estimate indicates that at least 10 per cent of the *n*-pentane reacting was converted to this substance. The over-all reaction could be written:

but doubtless both formation and subsequent reaction are actually chain processes. Precisely how this recent discovery is to be fitted into the general mechanism remains to be seen. Reactions like

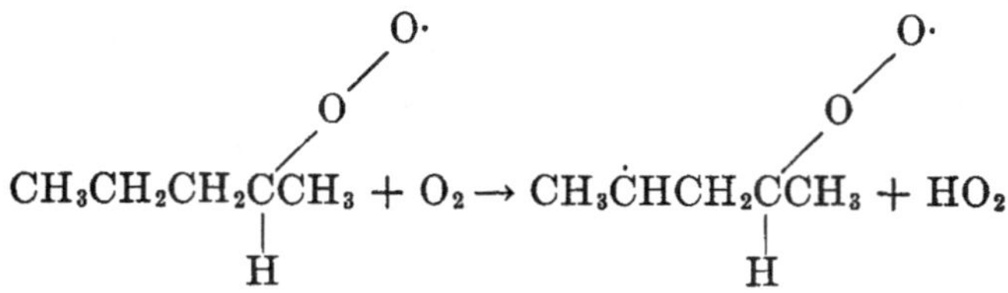

or

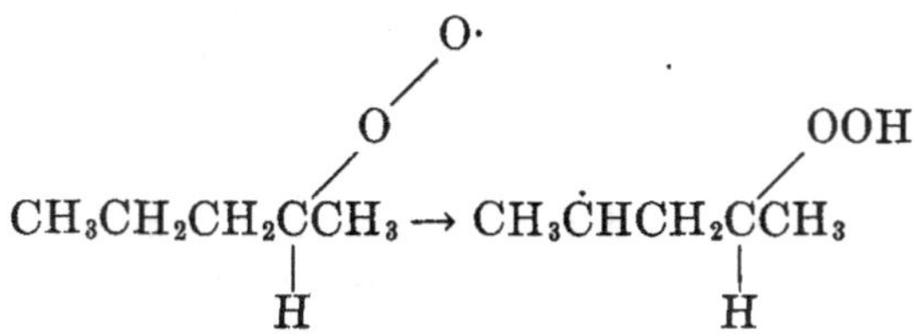

followed by addition of a second oxygen molecule are a possibility. However, it will be clear that final conclusions about the nature of the low temperature intermediate and its fate must be deferred.

Some further comment on explosion limits may be desirable. A glance at Fig. E,3a and E,3c will reveal that the "high temperature" limit curve breaks rather sharply to lower pressures when the cool flame zone is encountered. As might be expected, true explosion is preceded by rather heavy cool flame reaction to give so-called "2-stage" ignition. It is to be presumed that the development of explosion is strongly influenced by the accumulation of intermediates during the first or cool flame stage. It is further evident that this sequence of events will be closely allied to development of knock in the internal combustion engine (compression of the end gas). Consideration of this matter is somewhat beyond the scope of the present discussion, however. For further information, the reader is referred to the literature [*1*].

E,4. Some Other Exothermic Reactions. The chemical kinetics of many other exothermic reactions having potential interest in the propulsion field have been investigated in at least a preliminary way. Some of these are discussed briefly in the following paragraphs.

Oxidation of the boron hydrides. Compounds of boron with hydrogen form an interesting group whose structures and properties are only beginning to be clarified. All are thermodynamically unstable with respect to the elements [*12*] but at least three can be stored at room temperature for moderate periods. Of these, diborane (B_2H_6) is a gas condensing at -92.5°C, pentaborane (B_5H_9) is a liquid (boiling point 58°C) and decaborane ($B_{10}H_{14}$) is a solid (melting point 99.5°C). The compound BH_3 which would satisfy classical valence requirements is not isolable, though its analogues BCl_3, BF_3, and $B(CH_3)_3$ are stable vapors at room temperature and pressure.[18]

Although diborane is reasonably stable at low temperatures, it begins to decompose at an appreciable rate above about 80°C, yielding higher hydrides and hydrogen in a series of quasi-reversible reactions [*91–94*]. The process probably involves initial dissociation of diborane,

$$B_2H_6 \rightleftarrows 2BH_3$$

followed by successive additions of diborane to the BH_3 radical with loss of hydrogen, yielding such intermediates as pentaborane and decaborane. The latter possess somewhat greater stability, but ultimately at 700 to 900°C all break down into elementary boron and hydrogen. Evidently these processes can be expected to occur in the course of combustion.

Mixtures of diborane and dry air or oxygen do not react to a measurable extent at room temperature; though, if the gases are moist, a fog of boric acid forms at once [*95*].[19] There does not seem to be any zone of slow reaction; but, above about 100°C, dry mixtures may become explosive within a range of composition and pressure, though only after an induction period of varying length [*96,97*]. Sensitized decomposition of diborane does not appear to play a role in this reaction as it does in hydrocarbon oxidation. For example, in 2 hours at 103°C the fraction of diborane

[18] The occurrence of only the higher hydrides beginning with diborane presents something of a problem structurewise, since all these compounds are short of bonding electrons [*89*]. In diborane, for example, there are only enough electrons to form covalent bonds between hydrogen and boron atoms with no obvious source of a B—B bond. This difficulty is usually resolved by assuming that two of the hydrogen atoms assist in some way in binding the two halves of the molecule together. One such representation depends upon a so-called "protonated double bond" in which two protons are associated with a double bond between the boron atoms [*90*]

$$\begin{array}{c} H^+ \\ H_2B{=\!=}BH_2 \\ H^+ \end{array}$$

Analogous structures can be written for the higher hydrides. Alternatives, some involving resonance among several forms, seem rather less satisfactory but the structure problem is still under investigation.

[19] Occasionally, liquid diborane will take fire if the container is broken.

decomposed (14 per cent at 80 mm pressure) is unaffected by small amounts of air (3 to 15 mm), though larger additions (30 mm) lead to explosion (1 minute). Again, with 25 mm each of diborane and oxygen, initial normal decomposition of diborane may be observed by pressure rise though explosion follows (5 to 15 minutes).

Observations such as these suggest that neither diborane nor by inference the intermediate, BH_3, (believed to take part in the decomposition) are directly attacked by oxygen. This leaves as possible initiators of explosion the products of decomposition formed during the induction period—in particular pentaborane, which is known to be spontaneously inflammable at room temperature. This would also explain the fact that hydrogen may suppress explosion just as it does decomposition. For example, a mixture of 25 mm each of diborane and oxygen explodes within 2 minutes at 120°C (150 cm^3 reaction bulb coated with products); but if 50 mm hydrogen is also added there is no explosion at least for a 15-minute period. It may be added that ethylene is also a suppressor, whereas equal amounts of helium are without effect. Even oxygen in excess may at least prolong induction periods. For example, with 10 mm diborane at 145°C, explosions occur within a few seconds for oxygen pressures between 2 and 25 mm; larger additions of oxygen suppress explosion at least beyond 1000 seconds. This may lead to the semblance of a 2- or 3-limit system as in the hydrogen-oxygen system. It would therefore be logical to infer a branched chain mechanism, but the details are unknown. In particular, the consequences of formation in the gas phase of solid nuclei of boric acid (or boron, with excess of diborane) are not at all clear.

Pentaborane is less readily decomposed or hydrolyzed than diborane, but it is more susceptible to oxygen [*91*]. At room temperature small additions of oxygen to pentaborane vapor at 1 to 4 mm pressure produce a bluish luminescence which fades in the space of 1 to 2 minutes. Larger amounts of oxygen lead to explosion. For example, in a reaction bulb of 4.95-cm radius coated with reaction products, explosion occurs when pentaborane and oxygen are each at about 1 mm pressure [*98*]. Limiting pressures obey the relation:

$$p_{O_2} = \frac{0.370}{p_{B_5H_9}} + 0.74$$

the numerical constants being about inversely proportional to the bulb radius. It is therefore evident that pentaborane could readily trigger the diborane explosion as indicated above. Further information about the details of the pentaborane-oxygen interaction would be of special interest.

The hydrogen-fluorine reaction. A somewhat different type of exothermic process is the combination of hydrogen and fluorine [*12*]:

$$\tfrac{1}{2}H_2 + \tfrac{1}{2}F_2 \rightarrow HF\ (g), \qquad \Delta H^0_{298} = -64.2\ \text{kg cal}$$

This reaction is still in a technical sense an oxidation, with fluorine taking the place of oxygen as oxidant. Its advantage over the reaction with oxygen,

$$H_2 + \tfrac{1}{2}O_2 \rightarrow H_2O\ (g), \qquad \Delta H^0_{298} = -57.8 \text{ kg cal}$$

lies in the stability of the product at the high temperatures encountered in flames [*99,100*]. For example, at 3000°K, the equilibrium constant for formation of hydrogen fluoride from the elements (as above) is calculated to be 7.0×10^4 [*101,102*], while that for water vapor is only 22.3 [*61*].

In terms of reaction kinetics, the hydrogen-fluorine reaction is believed to follow the classical pattern of the hydrogen-chlorine reaction, though experimental details are less completely understood [*103–105*]. The dissociation energy of the fluorine molecule proves to be much less than that of chlorine ($\Delta H^0_{298} = +36.6$ and $+58.0$ kg cal/mole respectively) [*12*] with the result that the chain

$$\begin{aligned} F_2 &\rightarrow 2F, & \Delta H^0_{298} &= +36.6 \text{ kg cal} \\ F + H_2 &\rightarrow HF + H, & \Delta H^0_{298} &= -30.4 \text{ kg cal} \\ H + F_2 &\rightarrow HF + F, & \Delta H^0_{298} &= -98.0 \text{ kg cal} \end{aligned}$$

should proceed much more readily than the corresponding hydrogen-chlorine chain. Direct experiments turn out to be seriously complicated by reactions between fluorine or hydrogen fluoride and the walls of the containing vessel. For example, Bodenstein [*105*] found that in a glass vessel which had been used for some time, the reaction was slow at room temperature; yet if the vessel were first heated and evacuated, admission of hydrogen followed by fluorine resulted in ignition. This was true even at liquid air temperature. The inertness in the used vessel was presumed to be due to inhibition by silicon fluoride (SiF_4) which might have accumulated. Again, ignition would occur in a silver vessel at room temperature, but reaction was slow in platinum or magnesium. Differences in surface activity for recombination of hydrogen and fluorine atoms are presumed to account for these effects, but details are lacking. However, it is clear enough that hydrogen-fluorine mixtures can be explosive at room temperature and below; and this conclusion is further substantiated by the experiments of Priest and von Grosse [*100*] on the hydrogen-fluorine welding torch. By way of contrast, the thermal (nonphotochemical) reaction of hydrogen-chlorine mixtures is slow even at 200°C [*106,107*]. The experimental facts about the hydrogen-fluorine combination are thus qualitatively consistent with expectations based on an atom-chain mechanism; but information comparable to that which is available for the hydrogen-chlorine reaction (photochemical quantum yields, influence of inhibitors, and the like) is not available.

Decomposition of nitromethane. Ordinary explosives incorporate combustible and oxidant in a single package. One of the simplest molecules of this sort is nitromethane (CH_3NO_2). This substance is a liquid at room temperature (boiling point 100.8°C). Its vapor is subject to slow decomposition in the neighborhood of 400°C, yielding roughly 1 mole NO, 0.6 mole H_2O, 0.3 mole CO, 0.3 mole CH_4, and smaller amounts of a number of other substances [*108–111*]. As will be noted, there is insufficient oxygen for complete combustion; and even so only about half the available oxygen is utilized in the above process, the remainder appearing as nitric oxide. Nevertheless the reaction should be exothermic—for example [*12*],

$$CH_3NO_2\ (g) \rightarrow NO + \tfrac{1}{2}CO + \tfrac{1}{2}H_2O\ (g) + \tfrac{1}{2}CH_4, \qquad \Delta H^0_{298} = -17.3 \text{ kg cal}$$

If all the oxygen were involved, the reaction might be represented by the somewhat more exothermic process

$$CH_3NO_2\ (g) \rightarrow \tfrac{1}{2}N_2 + CO + H_2O\ (g) + \tfrac{1}{2}H_2, \qquad \Delta H^0_{298} = -72.1 \text{ kg cal}$$

In any event, liquid nitromethane can be made to detonate violently by shock or by sudden heating [*112,113*]. There is no information available on thermal explosion limits of the vapor.

Studies of the rate of slow decomposition of the vapor at about 400°C and at pressures less than 1 atm indicate that the process is approximately first order and homogeneous [*108–111*]. Added gases such as helium, nitrogen, hydrogen, carbon dioxide, and even nitric oxide (a product) do not affect the rate materially, but oxygen seems to alter the course of the reaction.

The reaction scheme is in some doubt. A bimolecular process such as

$$2CH_3NO_2 \rightarrow 2CH_3NO + O_2$$

is ruled out by the first order requirement; while the corresponding unimolecular reaction

$$CH_3NO_2 \rightarrow CH_3NO + O$$

is perhaps energetically unfavorable. The alternative [*12*],

$$CH_3NO_2 \rightarrow CH_3 + NO_2, \qquad \Delta H^0_{298} = -52.2 \text{ kg cal}$$

is attractive since the energy requirement corresponds rather closely to the experimental activation energy—given variously as 61.0, 52.7, and 50.6 kg cal. Another possibility is the isomerization reaction to methyl nitrite

$$CH_3NO_2 \rightarrow CH_3ONO$$

followed by dissociation

$$CH_3ONO \rightarrow CH_3O\cdot + NO$$

Such a scheme would be consistent with the observation that methyl nitrite decomposes readily at substantially lower temperatures (near 200°C) [*114,115*]. The further steps by which intermediates are converted into final products are evidently contingent on the nature of the initial reaction, and cannot be specified with any certainty.

The recombination of hydrogen atoms. In concluding this section it is tempting to comment briefly on hydrogen atom recombination as an example of a process involving free atoms, and characterized by high specific rate and high energy output.

The reaction is usually studied [*116*, p. 328] by generating hydrogen atoms in a Wood discharge tube [*117*] through which flows a stream of molecular hydrogen at a pressure of 1 mm or less. The partially dissociated exit gas passes directly to a reaction tube, along which the decay in atom concentration is determined by microcalorimetric methods and pressure measurements. A time scale is provided by the known flow rate and the reaction tube dimensions.

There is no doubt that recombination takes place very rapidly, but the kinetics turn out to be rather complex. Direct union of isolated atom pairs by a collision mechanism is an impossibility, because the heat liberated in molecule formation [*12*]

$$2\mathrm{H} \rightarrow \mathrm{H_2}, \qquad \Delta H^0_{298} = -104.2 \text{ kg cal}$$

is just that required for immediate redissociation. For stabilization purposes, the collision process must include either the wall of the container or another molecule—a "third body"—so that the reaction energy may be partly dissipated at the moment of liberation.[20] Such processes as

$$\mathrm{H} + \mathrm{H} + \mathrm{H} \rightarrow \mathrm{H_2} + \mathrm{H}$$

and

$$\mathrm{H} + \mathrm{H} + \mathrm{H_2} \rightarrow \mathrm{H_2} + \mathrm{H_2}$$

are therefore anticipated, as well as wall reactions. Allowing that the latter might be either first or second order, the rate equation becomes

$$-\frac{d[\mathrm{H}]}{dt} = k_1[\mathrm{H}] + k_2[\mathrm{H}]^2 + k_3[\mathrm{H}]^3 + k_4[\mathrm{H}]^2[\mathrm{H_2}]$$

Early work [*118–121*] sufficed to fix the order of magnitude of the rate, the general dependence on concentration, and the activation energy (no more than 1 kg cal). In a later exhaustive analysis of new experimental data, Amdur [*122*] concluded that no one of the four terms in the above equation could be completely ruled out, although it was not possible to distinguish between a number of arbitrary combinations. If all four terms

[20] The alternative process of conversion of two normal hydrogen atoms to an electronically excited molecule (followed by radiation of the energy) turns out to be a forbidden transition; and in any case the energy requirement is too high (96,690 cm^{-1} or 262 kg cal per mole).

were included, the best values of the constants at 30°C were:

$$k_1 = 0.69 \text{ sec}^{-1}$$
$$k_2 = 2.33 \times 10^8 \text{ (cm}^3\text{/mole) sec}^{-1}$$
$$k_3 = 1.70 \times 10^{16} \text{ (cm}^3\text{/mole)}^2 \text{ sec}^{-1}$$
$$k_4 = 1.00 \times 10^{16} \text{ (cm}^3\text{/mole)}^2 \text{ sec}^{-1}$$

Converting to mm Hg as a somewhat more realistic concentration unit, the values become:

$$k_1 = 0.69 \text{ sec}^{-1}$$
$$k_2 = 12.3 \text{ mm}^{-1} \text{ sec}^{-1}$$
$$k_3 = 47.6 \text{ mm}^{-2} \text{ sec}^{-1}$$
$$k_4 = 28.0 \text{ mm}^{-2} \text{ sec}^{-1}$$

These figures, which give rates in mm per second at pressures of 1 mm indicate that in the particular reaction tube employed the gas-phase reactions, corresponding to k_3 and k_4, predominate over wall recombination $(k_1 + k_2)$. The half life characteristic of the 3-atom reaction (k_3) would be about 0.03 second at 1 mm, or 5×10^{-8} seconds at 760 mm initial pressure of hydrogen atoms. These conclusions are consistent with those of earlier workers [*118–121*].

Hydrogen atom recombination thus emerges as a process with high rate and high heat release. All that is needed to put it on a practical basis is a satisfactory means of packaging hydrogen atoms.

E,5. Cited References and Bibliography.

Cited References

1. Lewis, B. and von Elbe, G. *Combustion, Flames, and Explosions of Gases.* Academic Press, 1951.
2. von Elbe, G., and Lewis, B. *J. Chem. Phys. 10,* 366 (1942).
3. Hinshelwood, C. N. *Proc. Roy. Soc. London A185,* 353; *A188,* 1 (1946).
4. Thompson, H. W., and Hinshelwood, C. N. *Proc. Roy. Soc. London A122,* 610 (1929).
5. Christiansen, J. A., and Kramers, H. A. *Z. physik. Chem. 104,* 451 (1923).
6. Semenov, N. *Z. Physik 46,* 109 (1927).
7. Grant, G. H., and Hinshelwood, C. N. *Proc. Roy. Soc. London A141,* 29 (1933).
8. Pease, R. N. *J. Am. Chem. Soc. 52,* 5106 (1930); *53,* 382 (1931).
9. Holt, R. B., and Oldenberg, O. *Phys. Rev. 71,* 479 (1947).
10. Frost, A. A., and Alyea, H. N. *J. Am. Chem. Soc. 55,* 3227 (1933).
11. Pease, R. N., and Chesebro, P. R. *Proc. Natl. Acad. Sci. 14,* 472 (1928).
12. Rossini, F. D., et al. Selected values of chemical thermodynamic properties. *Natl. Bur. Standards Circ. 500,* 1952.
13. Semenova, N., Biron, A., and Nalbandjan, A. *Acta Physicochim. U.R.S.S. 6,* 25, 43 (1937).
14. Hinshelwood, C. N., and Moelwyn-Hughes, E. A. *Proc. Roy. Soc. London A138,* 311 (1932).
15. Hinshelwood, C. N. *Kinetics of Chemical Change in Gaseous Systems,* 3rd ed. Oxford Univ. Press, 1933.

16. Garstang, W. L., and Hinshelwood, C. N. *Proc. Roy. Soc. London A130*, 640 (1931).
17. Gibson, C. H., and Hinshelwood, C. N. *Trans. Faraday Soc. 24*, 559 (1928).
18. Thompson, H. W., and Hinshelwood, C. N. *Proc. Roy. Soc. London A124*, 219 (1929).
19. Haber, F., and Alyea, H. N. *Z. physik. Chem. B10*, 193 (1930).
20. Garstang, W. L., and Hinshelwood, C. N. *Proc. Roy. Soc. London A134*, 6 (1931).
21. Hinshelwood, C. N., and Thompson, H. W. *Proc. Roy. Soc. London A118*, 170 (1928).
22. Gibson, C. H., and Hinshelwood, C. N. *Proc. Roy. Soc. London A119*, 591 (1928).
23. Pease, R. N., and Chesebro, P. R. *Proc. Natl. Acad. Sci. 14*, 472 (1928).
24. Pease, R. N. *J. Am. Chem. Soc. 52*, 5106 (1930).
25. Prettre, M. *J. chim. phys. 33*, 189 (1936).
26. Oldenberg, O., and Sommers, H. S. *J. Chem. Phys. 9*, 432 (1941).
27. Tolman, R. C. *Statistical Mechanics*. Chemical Catalog Co., 1927.
28. Melville, H. M. *Trans. Faraday Soc. 28*, 308 (1932).
29. von Elbe, G., and Lewis, B. *Rev. Inst. franç. pétrole et Ann combustibles liquides 4*, 363 (1949).
30. Oldenberg, O. *J. Chem. Phys. 19*, 196 (1951).
31. Kopp, D., Kowalsky, A., Sagulin, A., and Semenov, N. *Z. physik. Chem. B6*, 307 (1930).
32. Hadman, G., Thompson, H. W., and Hinshelwood, C. N. *Proc. Roy. Soc. London A137*, 87; *A138*, 297 (1932).
33. Garner, W. E. *Trans. Faraday Soc. 24*, 470 (1928); *26*, 190 (1930); *27*, 176 (1931).
34. Dixon, H. B. *Trans. Roy. Soc. London 175*, 617 (1884).
35. Bone, W. A., and Weston, F. R. *Proc. Roy. Soc. London A110*, 615 (1926).
36. Topley, B. *Nature 125*, 560 (1930).
37. Dixon, H. B. See Coward, H. F. *J. Chem. Soc. London*, 1382 (1934).
38. von Elbe, G., and Lewis, B. *J. Am. Chem. Soc. 59*, 2025 (1937).
39. Jackson, W., and Kistiakowsky, G. B. *J. Am. Chem. Soc. 52*, 3471 (1930); *56*, 2631 (1934).
40. Groth, W. *Z. physik. Chem. B37*, 315 (1937).
41. Trautz, M., and Seidel, W. *Ann. Physik 67*, 527 (1922).
42. Kondratjew, V., Solnyshkova, S., and Zaciorsky, M. *Acta Physicochim. U.R.S.S. 14*, 663 (1941).
43. Hornbeck, G. A., and Hopfield, H. S. *J. Chem. Phys. 17*, 982 (1949).
44. Buckler, E. E., and Norrish, R. G. W. *Proc. Roy. Soc. London A167*, 292, 318 (1938).
45. Prettre, M. *Compt. rend. 212*, 1090; *213*, 29 (1941).
46. MacCormac, M., and Townend, D. T. A. *J. Chem. Soc. London 141*, 240 (1938).
47. Edgar, G. *Ind. Eng. Chem. 19*, 145 (1927).
48. Edgar, G., et al. *J. Am. Chem. Soc. 51*, 1875, 2203, 2213 (1929).
49. Rice, F. O. *J. Am. Chem. Soc. 53*, 1959 (1931).
50. Rice, F. O., and Rice, K. K. *The Aliphatic Free Radicals* (summary of developments up to 1934). Johns Hopkins Univ. Press, 1934.
51. Paneth, F., and Hofeditz, W. *Ber. deut. chem. Ges. 62*, 1335 (1929).
52. Rice, F. O., Johnston, W. R., and Evering, B. L. *J. Am. Chem. Soc. 54*, 3529 (1932).
53. Frey, F. E., and Hepp, H. J. *Ind. Eng. Chem. 25*, 441 (1933).
54. Rice, F. O., and Polly, O. L. *Trans. Faraday Soc. 35*, 850 (1939).
55. Appleby, W. G., and Avery, W. H. *J. Am. Chem. Soc. 75*, 1809 (1953).
56. Neuhaus, M., and Marek, L. F. *Ind. Eng. Chem. 24*, 400 (1932).
57. Frey, F. E., and Hepp, H. J. *Ind. Eng. Chem. 25*, 441 (1933).
58. Echols, L. S., and Pease, R. N. *J. Am. Chem. Soc. 61*, 208 (1939).
59. Stevenson, D. P. *J. Chem. Phys. 10*, 291 (1942).
60. Steacie, E. W. R. *Atomic and Free Radical Reactions*. Reinhold, 1946.

61. Rossini, F. D., et al. *Selected Values of Physical and Thermodynamic Properties of Hydrocarbons and Related Compounds.* Carnegie Inst. Technol. Press, Pittsburgh, 1953.
62. Newitt, D. M., and Thornes, L. S. *J. Chem. Soc. London,* 1669 (1937).
63. Pease, R. N. *J. Am. Chem. Soc. 60,* 2244 (1938).
64. Pease, R. N. *J. Am. Chem. Soc. 51,* 1839 (1929).
65. Ubbelohde, A. R. *Proc. Roy. Soc. London A152,* 354 (1935).
66. Topps, J. E. C., and Townend, D. T. A. *Trans. Faraday Soc. 42,* 345 (1946).
67. Semenov, N. *Z. physik. Chem. B11,* 464 (1930).
68. Semenov, N. *Chemical Kinetics and Chain Reactions.* Oxford Univ. Press, 1935.
69. Ubbelohde, A. R., and Egerton, A. *Nature 135,* 67 (1935).
70. Ubbelohde, A. R. *Proc. Roy. Soc. London A152,* 354 (1935).
71. Walsh, A. D. *Trans. Faraday Soc. 42,* 269 (1946).
72. Lenher, S. *J. Am. Chem. Soc. 53,* 3752 (1931).
73. Pease, R. N., and Munro, W. P. *J. Am. Chem. Soc. 56,* 2034 (1934).
74. Harris, E. J., and Egerton, A. *Chem. Revs. 21,* 287 (1937).
75. Harris, E. J. *Trans. Faraday Soc. 44,* 764 (1948).
76. Kooijman, P. L. *Rec. trav. chim. 66,* 5, 205 (1947).
77. Mondain-Monval, P., and Quanquin, B. *Ann. chim. Paris 15,* 391 (1931).
78. Raley, J. H., Rust, F. F., and Vaughan, W. E. *J. Am. Chem. Soc. 70,* 88 (1948); *73,* 15 (1951).
79. Harris, E. J., and Egerton, A. *Nature 142,* 830 (1938).
80. Harris, E. J., and Egerton, A. *Proc. Roy. Soc. London A168,* 1 (1938).
81. Fletcher, C. J. M. *Proc. Roy. Soc. London A146,* 357 (1934).
82. Fort, R., and Hinshelwood, C. N. *Proc. Roy. Soc. London A129,* 284 (1930).
83. Cullis, C. F., and Hinshelwood, C. N. *Discussions Faraday Soc. 2,* 117 (1947).
84. Bone, W. A., and Gardner, J. B. *Proc. Roy. Soc. London A154,* 297 (1936).
85. Egerton, A. and Pidgeon, L. M. *Proc. Roy. Soc. London A142,* 26 (1933).
86. Ubbelohde, A. R. *Proc. Roy. Soc. London A152,* 378 (1935).
87. Egerton, A., and Young, G. H. S. *Trans. Faraday Soc. 44,* 750 (1948).
88. Thomas, J. R., et al. *Ind. Eng. Chem. 43,* 2761 (1951).
89. Bell, R. P., and Emeleus, H. J. *Quart. Revs. London 2,* 132 (1948).
90. Pitzer, K. S. *J. Am. Chem. Soc. 67,* 1126 (1945).
91. Stock, A. *Hydrides of Boron and Silicon.* Cornell Univ. Press, 1933.
92. Burg, A., and Schlesinger, H. I. *Chem. Revs. 31,* 1 (1942).
93. Bragg, J. K., McCarty, L. V., and Norton, F. J. *J. Am. Chem. Soc. 73,* 2134 (1951).
94. Clarke, R. P., and Pease, R. N. *J. Am. Chem. Soc. 73,* 2132 (1951).
95. Stock, A., et al. *Ber. deut. chem. Ges. 59,* 2210 (1926); *47,* 3115 (1914).
96. Price, F. P. *J. Am. Chem. Soc. 72,* 5361 (1950).
97. Whatley, A. T. *Ph.D. Thesis.* Princeton Univ., 1952.
98. Price, F. P. *J. Am. Chem. Soc. 73,* 2141 (1951).
99. Leonard, A. S. *J. Am. Rocket Soc. 72,* 10 (1947).
100. Priest, H. F., and von Grosse, A. *Ind. Eng. Chem. 39,* 431 (1947).
101. Cole, L. G., Farber, M., and Elverum, G. W. J. *J. Chem. Phys. 20,* 586 (1952).
102. Doescher, R. N. *J. Chem. Phys. 20,* 330 (1952).
103. Eyring, H. *J. Am. Chem. Soc. 53,* 2537 (1931).
104. Bodenstein, M., and Jockusch, H. *Sitzber. preuss. Akad. Wiss. Physik.-math. Kl. 2,* 27 (1934).
105. Bodenstein, M., and Jockusch, H. *Z. anorg. u. allgem. Chem. 231,* 24 (1937).
106. Christiansen, J. A. *Z. physik. Chem. B2,* 405 (1929).
107. Morris, J. C., and Pease, R. N. *J. Am. Soc. 61,* 391 (1939).
108. Taylor, H. A., and Vesselovsky, V. V. *J. Phys. Chem. 39,* 1095 (1935).
109. Cottrell, T. L., Graham, T. E., and Reed, T. J. *J. Chem. Phys. 18,* 1306 (1950).
110. Cottrell, T. L., Graham, T. E., and Reed, T. J. *Trans. Faraday Soc. 47,* 584 (1951).
111. Hillenbrand, L. J., and Kilpatrick, M. L. *J. Chem. Phys. 19,* 381 (1951).
112. McKittrick, D. S., et al. *Anal. Chem. 10,* 630 (1938).
113. Bellinger, F., et al. *Ind. Eng. Chem. 40,* 1320 (1948).

114. Steacie, E. W. R. *Proc. Roy. Soc. London A146*, 388 (1934).
115. Steacie, E. W. R. *J. Chem. Phys. 4*, 96, 223 (1936).
116. Schumacher, H. J. *Chemische Gas-reaktionen.* Steinkopff, Dresden, 1938.
117. Wood, R. W. *Proc. Roy. Soc. London A102*, 1 (1922).
118. Smallwood, H. M. *J. Am. Chem. Soc. 51*, 1985 (1929); *56*, 1542 (1934).
119. Amdur, I., and Robinson, A. L. *J. Am. Chem. Soc. 55*, 1395, 2615 (1933); *57*, 856 (1935).
120. Steiner, W. *Trans. Faraday Soc. 31*, 623 (1935).
121. Senftleben, H., and Reichemeier, O. *Ann. Physik 6*, 105 (1930).
122. Amdur, I. *J. Am. Chem. Soc. 60*, 2347 (1938).

Bibliography

Hinshelwood, C. N. *The Kinetics of Chemical Change.* Oxford Univ. Press, 1940.
Jost, W. *Explosions-und Verbrennungsvorgänge in Gasen.* Springer, Berlin, 1939.
Schumacher, H. J. *Chemische Gasreaktionen.* Steinkopff, Dresden, 1938.
Semenov, N. *Chemical Kinetics and Chain Reactions.* Oxford Univ. Press, 1935.
Steacie, E. W. R. *Atomic and Free Radical Reactions.* Reinhold, 1946.

PART THREE

FLAME PROPAGATION IN GASES

SECTION F

MECHANICS OF REACTING CONTINUA

JOHN M. RICHARDSON
STUART R. BRINKLEY, JR.

F,1. Fundamental Equations.

State variables. In macroscopic descriptions of physical systems, one's information always falls far short of the maximum possible information. The information concerning the state of a given system at a given instant of time may be considered adequate if one can obtain from it a reasonable estimate of the information to be yielded by this system at some later time. Perhaps even more pertinent is the possibility of performing the process serialwise (making each estimate from the previous estimate) over subintervals of a given time interval with the final result fairly insensitive to the lengths of the subintervals. One expects this procedure to fail for a given type of information if the subintervals become much smaller than a critical time interval which will be called the "time resolution."

This section is confined to the consideration of systems and of time resolutions for which adequate information consists of a knowledge of the spatial distribution of the following densities:

1. Concentrations $C_r, r = 1, \cdots, n$, of the relevant chemical species expressed in moles per unit volume.
2. Momentum density (or mass current) $\mathbf{m} = \rho\mathbf{q}$, where ρ is the mass density (= reciprocal of the specific volume v) and $\mathbf{q}$ is the particle velocity.
3. Total energy density $\rho(e + \frac{1}{2}q^2)$, where e is the specific internal energy as measured by an observer moving with the fluid, and where $\frac{1}{2}q^2$ ($q = |\mathbf{q}|$) is the specific kinetic energy of bulk motion as measured by a fixed observer.

If the above set of densities is inadequate, the set must be modified by the addition of new densities and/or by the decomposition of the former densities into finer categories. As an example of the latter, the momentum density might be broken down into the momentum densities for the chemical species considered individually. Another example might be the decomposition of the energy density into translational, rotational, and vibrational parts.

A kinematical theorem. Consider any property $G = G(\mathbf{r}, t)$ and a volume V whose surface S moves with a velocity[1] $\mathbf{q}$, which may or may not equal the particle velocity. Now it follows that[2]

$$\frac{d}{dt}\int_V dV\, G = \int_V dV\, \frac{\partial G}{\partial t} + \int_S d\mathbf{S} \cdot \mathbf{q}G \tag{1-1}$$

This relation may be proved by considering the infinitesimal change in $\int_V dV\, G$ occurring during the time interval dt to be a linear combination of the changes $(\partial G/\partial t)dtdV$ occurring in the contribution of each stationary volume element dV and the changes $(d\mathbf{S} \cdot \mathbf{q}dt)G$ associated with the surface shift.

If the velocity vector $\mathbf{q}$ is defined everywhere, not merely on the surface, Eq. 1-1 may be rewritten with the aid of the Gauss integral theorem in the form

$$\frac{d}{dt}\int_V dV\, G = \int_V dV \left[\frac{\partial G}{\partial t} + \nabla \cdot (\mathbf{q}G)\right] \tag{1-2}$$

Transport of chemical species. Consider an arbitrary mobile volume V whose surface S moves with the particle velocity $\mathbf{q}$. The time rate of change of the number of moles of species r in V is

$$\frac{d}{dt}\int_V dV\, C_r = -\int_S d\mathbf{S} \cdot \mathbf{J}_r + \int_V dV\, R_r, \qquad r = 1, \cdots, n \tag{1-3}$$

where $\mathbf{J}_r$ is the current of species r (moles/unit area/unit time) with respect to mobile axes (moving with the velocity $\mathbf{q}$), and where R_r is the rate of production (moles/unit volume/unit time) of r by chemical reaction.

The mass density is evidently

$$\rho = \sum_{r=1}^{n} \mathfrak{M}_r C_r \tag{1-4}$$

where $\mathfrak{M}_r$ is the molecular weight of r. Since the mass current with respect to mobile axes vanishes by definition (since mass current $= \mathbf{m} = \rho\mathbf{q}$ with respect to fixed axes) it follows that

$$\sum_{r=1}^{n} \mathfrak{M}_r \mathbf{J}_r = 0 \tag{1-5}$$

[1] This phrase is intended to mean that a point initially in the surface remains in the surface if it moves with a velocity $\mathbf{q}$.

[2] In the surface integral in Eq. 1-1, $d\mathbf{S} = \mathbf{n}dS$ where $\mathbf{n}$ is the outward pointing normal and dS is a differential surface area.

Since mass is conserved in chemical reactions,

$$\sum_{r=1}^{n} \mathfrak{M}_r R_r = 0 \tag{1-6}$$

Multiplying Eq. 1-3 by $\mathfrak{M}_r$, summing over r, and using Eq. 1-4, 1-5, and 1-6 one obtains the result

$$\frac{d}{dt}\int_V dV\rho = 0 \tag{1-7}$$

expressing the fact that the total mass contained in a mobile volume is a constant.

It is convenient to introduce a set of basic components i, $i = 1, \cdots, p$, with concentrations $C_i^{(b)}$ defined by an independent set of linear combinations of C_r,

$$C_i^{(b)} = \sum_{r=1}^{n} \nu_{ir} C_r, \qquad i = 1, \cdots, p \tag{1-8}$$

which are conserved in chemical reaction. Obviously, p must be less than n. The last requirement may be expressed in the form

$$\sum_{r=1}^{n} \nu_{ir} R_r = 0, \qquad i = 1, \cdots, p \tag{1-9}$$

Thus the basic components may be regarded as immutable building blocks from which molecules of a given species can be constructed (by addition and/or by subtraction). For example, a molecule of species r, say, can be regarded as composed of ν_{1r} particles of basic component 1, ν_{2r} particles of basic component 2, etc. As a possible realization of the above definitions, it usually suffices to regard the basic components as the largest chemical units unchanged by reaction.

The basic components have their own currents (referred to mobile axes) defined by

$$\mathbf{J}_i^{(b)} = \sum_{r=1}^{n} \nu_{ir} \mathbf{J}_r, \qquad i = 1, \cdots, p \tag{1-10}$$

Multiplying Eq. 1-3 by ν_{ir}, summing over r, and using Eq. 1-8 and 1-9 one obtains

$$\frac{d}{dt}\int_V dV\, C_i^{(b)} = -\int_S d\mathbf{S}\cdot \mathbf{J}_i^{(b)}, \qquad i = 1, \cdots, p \tag{1-11}$$

These relations are the form that the stoichiometric restrictions take in the nonuniform case.

Before proceeding further, it is desirable to bring the integral relations above into the more familiar differential forms. Using the fundamental kinematical theorem Eq. 1-2 and the Gauss integral theorem, Eq. 1-3 may be written

$$\int_V dV \left[\frac{\partial C_r}{\partial t} + \nabla \cdot (\mathbf{q}C_r + \mathbf{J}_r) - R_r \right] = 0, \qquad r = 1, \cdots, n$$

Since the integration extends over an *arbitrary* volume, the integrand must vanish, giving finally

$$\frac{\partial C_r}{\partial t} + \nabla \cdot (\mathbf{q}C_r) = -\nabla \cdot \mathbf{J}_r + R_r, \qquad r = 1, \cdots, n \tag{1-3a}$$

In a similar fashion, Eq. 1-7 is converted into

$$\frac{\partial \rho}{\partial t} + \nabla \cdot (\mathbf{q}\rho) = 0 \tag{1-7a}$$

the well-known equation of continuity. Eq. 1-11 is converted into

$$\frac{\partial C_i^{(b)}}{\partial t} + \nabla \cdot (\mathbf{q}C_i^{(b)}) = -\nabla \cdot \mathbf{J}_i^{(b)}, \qquad i = 1, \cdots, p \tag{1-11a}$$

Letting X_r be the chemical symbol for species r, one may express chemical reactions, $1, \cdots, s, \cdots, l$, in the following way

$$\sum_{r=1}^{n} X_r \kappa_{rs} = 0, \qquad s = 1, \cdots, l \tag{1-12}$$

in which the coefficients may be positive or negative. Let the rate of advancement λ_s of reaction s be defined by

$$R_{rs} = \kappa_{rs}\lambda_s \tag{1-13}$$

where R_{rs} is the contribution of reaction s to the rate R_r. Evidently,

$$R_r = \sum_{s=1}^{l} R_{rs} = \sum_{s=1}^{l} \kappa_{rs}\lambda_s, \qquad r = 1, \cdots, n \tag{1-14}$$

Since Eq. 1-6 must hold independently of the values of λ_s, one obtains

$$\sum_{r=1}^{n} \mathfrak{M}_r \kappa_{rs} = 0, \qquad s = 1, \cdots, l \tag{1-15}$$

expressing the conservation of mass for each reaction individually. Since these equations are of the same form as Eq. 1-12 they may be used alternatively to symbolize chemical reactions. Since Eq. 1-9 must also hold independently of the values of λ_s, one obtains the following identities connecting ν_{ir} and κ_{rs}:

$$\sum_{r=1}^{n} \nu_{ir}\kappa_{rs} = 0, \qquad i = 1, \cdots, p; s = 1, \cdots, l \tag{1-16}$$

Transport of momentum. Considering again the mobile volume V defined previously, one can obtain the equation for the transport of momentum by a straightforward application of Newton's second law of motion. The total momentum inside V is $\int_V dV \rho \mathbf{q}$. The total force acting upon the material in V may be written

$$\int_S d\mathbf{S} \cdot \sigma + \sum_{r=1}^{n} \int_V dV\, C_r \mathbf{F}_r$$

where σ is the stress tensor (a symmetric tensor) and $dV\, C_r \mathbf{F}_r$ is the force acting upon the $dV\, C_r$ moles of species r contained in the volume element dV. Application of Newton's second law yields

$$\frac{d}{dt} \int_V dV \rho \mathbf{q} = \int_S d\mathbf{S} \cdot \sigma + \sum_{r=1}^{n} \int_V dV\, C_r \mathbf{F}_r \tag{1-17}$$

The right-hand side of Eq. 1-17 requires further discussion. The first term is the surface contribution to the total force and is the sum of surface tractions $d\mathbf{S} \cdot \sigma$ acting on each element of surface $d\mathbf{S}$. The term $d\mathbf{S} \cdot \sigma$ represents the net momentum transfer through $d\mathbf{S}$ (as measured by an observer moving with the surface) added to the short range forces acting through $d\mathbf{S}$. The second term, the volume contribution, represents the sum of the long range forces (including external forces). There is some ambiguity in the split into volume and surface contributions. However, it will be found convenient to express as much as possible of the total force in the form of a surface integral.

The stress tensor σ may be regarded as being defined by Eq. 1-17. It is not necessarily equal to the stress defined thermodynamically, using the local equilibrium assumption. The detailed discussion of the form that σ takes for a viscous fluid and its relation to the "thermodynamic stress" will be taken up later. Nevertheless, it may be appropriate to consider the relation between σ and more commonly used quantities. First of all, the hydrostatic pressure p is defined as the negative of the average of the diagonal elements:

$$p = -\frac{1}{3} \sum_{\alpha=1}^{3} \sigma_{\alpha\alpha} \tag{1-18}$$

in which α labels the coordinates x, y, z. The shear stress tensor τ is defined by

$$\tau = \sigma + \mathbf{1}p \tag{1-19}$$

where $\mathbf{1}$ is the unit tensor.

The differential form of Eq. 1-17 is

$$\frac{\partial}{\partial t}\rho\mathbf{q} + \nabla\cdot(\mathbf{q}\rho\mathbf{q}) = \nabla\cdot\sigma + \sum_{r=1}^{n} C_r\mathbf{F}_r \tag{1-17a}$$

Transport of energy. The total energy contained in the mobile volume V is $\int_V dV\rho(e+\frac{1}{2}q^2)$. The time rate of change of this energy is equal to the rate at which heat[3] flows into V plus the rate at which work is performed on the material in V. Thus one obtains

$$\frac{d}{dt}\int_V dV\rho(e+\tfrac{1}{2}q^2) = -\int_S d\mathbf{S}\cdot\mathbf{J}_Q + \int_S d\mathbf{S}\cdot\sigma\cdot\mathbf{q} + \sum_{r=1}^{n}\int dV(C_r\mathbf{q}+\mathbf{J}_r)\cdot\mathbf{F}_r \tag{1-20}$$

where $\mathbf{J}_Q$ is the heat flow vector.

In the last term of Eq. 1-20, the action of external (and long range internal) forces has been treated separately for each component. It is necessary to do this since the deviation $\sum_r\int dV\,\mathbf{J}_r\mathbf{F}_r$ from the term $\sum_r\int dV\,C_r\mathbf{F}_r\cdot\mathbf{q}$, which corresponds to a grosser viewpoint, cannot be incorporated in $\int_S dS\cdot\mathbf{J}_Q$.

The differential form of Eq. 1-20 is

$$\frac{\partial}{\partial t}[\rho(e+\tfrac{1}{2}q^2)] + \nabla\cdot[\mathbf{q}\rho(e+\tfrac{1}{2}q^2)] = -\nabla\cdot\mathbf{J}_Q + \nabla\cdot\sigma\cdot\mathbf{q} + \sum_{r=1}^{n}(C_r\mathbf{q}+\mathbf{J}_r)\cdot\mathbf{F}_r \tag{1-20a}$$

Summary and alternative forms. Within the restrictions implied by the choice of state variables, a reacting continuum is described by Eq. 1-3a, 1-17a, and 1-20a whose form is independent of the states of particular systems. This description, of course, must be amended by other relations embracing the material properties of the particular system under consideration. For instance, the reversible parts of p and e are given by the thermal and caloric equations of state. The irreversible relations will be dealt with in the next article.

In the theory of combustion and detonation waves, many interesting phenomena are adequately described by these equations in one dimension in the absence of external forces (or long range interval forces) for inviscid liquids and gases. For this case the equations are summarized below:

[3] Here, as is always the case, heat and heat flow are defined so as to complete the energy balance. A modification in the definition of work must be accompanied by corresponding modification in the definition of heat flow.

$$\frac{\partial C_r}{\partial t} + \frac{\partial qC_r}{\partial x} = -\frac{\partial J_r}{\partial x} + R_r \qquad r = 1, \cdots, n \tag{1-3a$'$}$$

$$\frac{\partial \rho q}{\partial t} + \frac{\partial \rho q^2}{\partial x} = -\frac{\partial p}{\partial x} \tag{1-17a$'$}$$

$$\frac{\partial}{\partial t}[\rho(e + \tfrac{1}{2}q^2)] + \frac{\partial}{\partial x}[\rho q(e + \tfrac{1}{2}q^2)] = -\frac{\partial}{\partial x}(J_Q + pq) \tag{1-20a$'$}$$

In the above equations q, J_r, and J_Q are the x components of the corresponding vector quantities.

Before concluding this article, alternative forms of the fundamental transport equations will be considered. The first alternative form, actually more familiar than the differential or integral forms already derived, is a differential form in which the rates of change are those measured by a mobile observer; that is, one introduces the mobile time derivative $d/dt = \partial/\partial t + \mathbf{q} \cdot \nabla$ in place of the stationary time derivative $\partial/\partial t$ and, furthermore, one emphasizes quantities per unit mass measured by mobile observers. First, one notes that for any function g,

$$\frac{\partial(\rho g)}{\partial t} + \nabla \cdot (\mathbf{q}\rho g) = \rho \frac{dg}{dt}$$

using the equation of continuity (1-7a). Eq. 1-3a and 1-17a can then be transformed into

$$\rho \frac{dc_r}{dt} = -\nabla \cdot \mathbf{J}_r + R_r, \qquad r = 1, \cdots, n \tag{1-3b}$$

$$\rho \frac{d\mathbf{q}}{dt} = \nabla \cdot \sigma + \rho \sum_{r=1}^{n} c_r \mathbf{F}_r \tag{1-17b}$$

respectively, where $c_r = C_r/\rho$ is the number of moles of species r per unit mass. Eq. 1-3b are not independent since $\sum \mathfrak{M}_r c_r = 1$. This loss must be made up by the explicit introduction of the equation of continuity, which can be written

$$\frac{d \ln \rho}{dt} = -\nabla \cdot \mathbf{q} \tag{1-7b}$$

If the momentum transport equation multiplied by $\mathbf{q}$ is subtracted from the energy transport equation (1-20a), the result is

$$\rho \frac{de}{dt} = -\nabla \cdot \mathbf{J}_Q + \sigma : (\nabla \mathbf{q}) + \sum_{r=1}^{n} \mathbf{J}_r \cdot \mathbf{F}_r \tag{1-20b}$$

where

$$\sigma : (\nabla \mathbf{q}) = \sum_{\alpha,\beta=1}^{3} \sigma_{\alpha\beta} \frac{\partial q_\alpha}{\partial x_\beta}$$

in which q_1, q_2, q_3 are the Cartesian components of $\mathbf{q}$ and $x_1 = x$, $x_2 = y$, and $x_3 = z$. Since the stress tensor σ is symmetric, the energy equation depends only upon the symmetric part of $\nabla\mathbf{q}$, which is the rate of strain:

$$\dot{\epsilon}_{\alpha\beta} = \frac{1}{2}\left(\frac{\partial q_\alpha}{\partial x_\beta} + \frac{\partial q_\beta}{\partial x_\alpha}\right) \tag{1-21}$$

There is still another mode of expression for the basic equations, called the Lagrangian form. Here, positions are defined by the labels of a set of points, each of which moves with a velocity $\mathbf{q}$. Let the set of labels be a three-dimensional continuum of numbers ξ_a, $a = 1, 2, 3$. Now it is understood the independent variables are ξ_1, ξ_2, ξ_3, t, so that now the symbol $\partial/\partial t$ connotes time differentiation with constant ξ_a. Since constant values of ξ_1, ξ_2, and ξ_3 correspond to a point moving with the velocity $\mathbf{q}$, $\partial/\partial t$ in the Lagrangian form is identical to d/dt in the old (Eulerian) form.

The density (mass) ρ in real space is related to the density ρ_ξ in label space ξ by the equation

$$\rho = J^{-1}\rho_\xi \tag{1-22}$$

where

$$J = \left|\det\left(\frac{\partial x_\alpha}{\partial \xi_a}\right)\right| \tag{1-23}$$

Since ρ_ξ must be time independent (because of the definition of $\mathbf{q}$), Eq. 1-22 is the integrated form of the equation of continuity (1-76).

Eq. 1-3b, 1-17b, and 1-20b can now be recast in the forms

$$\rho\frac{\partial c_r}{\partial t} = -\sum_{a=1}^{3}\sum_{\alpha=1}^{3}\frac{\partial \xi_a}{\partial x_\alpha}\frac{\partial j_{r,\alpha}}{\partial \xi_a} + R_r \qquad r = 1, \cdots n \tag{1-3c}$$

$$\rho\frac{\partial q_\alpha}{\partial t} = \sum_{a=1}^{3}\sum_{\beta=1}^{3}\frac{\partial \xi_a}{\partial x_\beta}\frac{\partial \sigma_{\beta\alpha}}{\partial \xi_a} + \rho\sum_{r=1}^{n} c_r F_{r,\alpha} \tag{1-17c}$$

and

$$\rho\frac{\partial e}{\partial t} = -\sum_{a=1}^{3}\sum_{\alpha=1}^{3}\frac{\partial \xi_a}{\partial x_\alpha}\frac{\partial j_{Q,\alpha}}{\partial \xi_a} + \sigma:\dot{\epsilon} + \sum_{r=1}^{n}\mathbf{j}_r\cdot\mathbf{F}_r \tag{1-20c}$$

respectively. If one desires, further modifications can be effected through the use of Eq. 1-22 and the relation

$$\sum_{a=1}^{3}\frac{\partial}{\partial \xi_a}\left(J\frac{\partial \xi_a}{\partial x_\alpha}\right) = 0$$

The equations, discussed above and in preceding paragraphs, are incomplete in that they contain the quantities $\mathbf{J}_r$, R_r, σ, and $\mathbf{J}_Q$, which are as yet undefined. The relations connecting these quantities with the fundamental quantities, C_r, $\mathbf{u}$, and e, depend upon the particular system under consideration. However, the phenomenological theory of irreversible processes, considered briefly in the following article, allows one to make certain general statements about the form of these relations.

F,2. Application of the Phenomenological Theory of Irreversible Processes.

General remarks. Since the theory of irreversible processes is considered in I,J, only a brief discussion will be presented here, with emphasis upon only those features which are special to reacting continua. Furthermore, this subject is considered in detail in a recent book by de Groot [*1*], which the reader is urged to consult.

The equations discussed in the preceding article were applicable to any type of material system: gas, liquid, or solid (amorphous or crystalline). The discussion henceforth will be limited to isotropic materials exhibiting no elasticity in shear (that is, gases and ordinary liquids of low molecular weight).

The first important assumption underlying the theory of irreversible processes is that in a nonuniform system not in equilibrium one can define distributions of temperature T and chemical potentials μ_r related to the distributions of the fundamental densities C_r, $\rho\mathbf{q}$, and $\rho(e + \frac{1}{2}q^2)$. The next assumption, particularly relevant to the present discussion, is that T and μ_r can to a good approximation [*2*] be treated as point functions of the fundamental densities. This localizability assumption is ordinarily replaced by the local equilibrium assumption in the treatment of systems free of reactions and bulk viscosity, since in the latter case disequilibrium is associated only with gradients; thus one may regard a given volume element in equilibrium as long as the influence of neighboring elements is to be neglected. However, when reactions and bulk viscosity are present, one can no longer do this, because these effects produce disequilibrium in a given volume regardless of the states of neighboring elements.

These considerations indicate that one must, in considering a single volume element, use a quasi-thermodynamics, the justification of which lies beyond the scope of this article. However, the results [*3*] are that one can define a specific entropy s such that for a fixed volume element

$$Td(\rho s) = d(\rho e) - \sum_{r=1}^{n} \mu_r dC_r \tag{2-1}$$

in which T and μ_r are calculated as though the element were in thermodynamic equilibrium with the reactions inhibited.

For convenience in treating volume elements of fixed mass, Eq. 2-1 is transformed to

$$Tds = de + p'd\left(\frac{1}{\rho}\right) - \sum_{r=1}^{n} \mu_r dc_r \tag{2-2}$$

where

$$p' = \rho\left(Ts - e + \sum_{r=1}^{n} \mu_r c_r\right) \tag{2-3}$$

The quantity p' is evidently the thermodynamically defined pressure, which in general differs from the dynamically defined pressure p (see Eq. 1-18) because of bulk viscosity and chemical reaction.

Entropy transport. The rate of change of the specific entropy s as measured by a mobile observer is given by Eq. 2-2 if the differential operator d is replaced by the mobile time derivative $d/dt = \partial/\partial t + \mathbf{q} \cdot \nabla$. Using Eq. 1-20b, 1-7b, and 1-3b for the evaluation of de/dt, $d(1/\rho)/dt$, and dc_r/dt, one obtains

$$\rho \frac{ds}{dt} = -\nabla \cdot \left(\frac{\mathbf{J}_Q - \sum_{r=1}^{n} \mu_r \mathbf{J}_r}{T}\right) + \mathbf{J}_Q \cdot \nabla \frac{1}{T} + (\sigma + \mathbf{1}p'):\left(\frac{\nabla \mathbf{q}}{T}\right) + \sum_{r=1}^{n} \mathbf{J}_r \cdot \left(\frac{\mathbf{F}_r}{T} - \nabla \frac{\mu_r}{T}\right) - \sum_{r=1}^{n} \frac{R_r \mu_r}{T} \tag{2-4}$$

The first term on the right-hand side of Eq. 2-4 may be regarded as the negative divergence of the entropy current as measured by a mobile observer. Then the remaining terms must be regarded as making up the entropy source density.

The last term in Eq. 2-4 may be modified by the introduction of the rates of advancement defined by Eq. 1-13 so that

$$-\sum_{r=1}^{n} \frac{R_r \mu_r}{T} = \sum_{s=1}^{l} \frac{\lambda_s A_s}{T} \tag{2-5}$$

where

$$A_s = -\sum_{r=1}^{n} \mu_r \kappa_{rs} \tag{2-6}$$

is called the chemical affinity [4] and λ_s the rate of advancement, respectively, of reaction s.

It is evident that since $\sigma + \mathbf{1}p'$ is symmetric, $\nabla \mathbf{q}$ may be replaced by $\dot{\varepsilon} = \text{sym}\,(\nabla \mathbf{q})$, the rate of strain given in Cartesian form by Eq. 1-21.

Currents, forces, and kinetic coefficients. In a generalized sense, the quantities $\mathbf{J}_r$, $-(\sigma + \mathbf{1}p')$, $\mathbf{J}_Q$, and λ_s are called the generalized currents even though they are not all vector quantities. When multiplied by T the quantities multiplying the generalized currents in the entropy source density are called the generalized forces. These statements may be summarized in the table:

Table F,2.

Currents	*Forces*
$\mathbf{J}_r$	$\mathbf{X}_r = \mathbf{F}_r - T\nabla(\mu_r/T)$
$\boldsymbol{\pi} = -(\boldsymbol{\sigma} + \mathbf{1}p')$	$-\dot{\boldsymbol{\varepsilon}} = -\text{sym}(\nabla\mathbf{q})$
$\mathbf{J}_Q$	$\mathbf{X}_e = T\nabla(1/T) = -(\nabla T)/T$
λ_s	A_s

It follows from general considerations that the currents can be expressed as linear combinations of the forces with the coefficients depending only upon the state of the medium at the point in question. This is obviously true if the deviations from local equilibrium are small. However, the truth of the first statement actually depends only upon the validity of the localizability assumption and upon the validity of the quasi-thermodynamics discussed earlier. Taking into account the tensorial properties of an isotropic medium, one may write

$$\left.\begin{aligned} \mathbf{J}_r &= \sum_{r'=1}^{n} L_{rr'}\mathbf{X}_{r'} + L_{re}\mathbf{X}_e \qquad r = 1, \cdots, n \\ \mathbf{J}_q &= \sum_{r'=1}^{n} L_{er'}\mathbf{X}_{r'} + L_{ee}\mathbf{X}_e \end{aligned}\right\} \tag{2-7a}$$

$$\left.\begin{aligned} \boldsymbol{\pi} &= -L'_{qq}\mathbf{1}\dot{\epsilon} - L''_{qq}\dot{\boldsymbol{\varepsilon}} + \sum_{s'=1}^{l} L_{qs'}\mathbf{1}A_{s'} \\ \lambda_s &= -L_{sq}\dot{\epsilon} + \sum_{s'=1}^{l} L_{ss'}A_{s'} \end{aligned}\right\} \tag{2-7b}$$

where

$$\dot{\epsilon} = \sum_{\alpha=1}^{3} \epsilon_{\alpha\alpha} = \nabla\cdot\mathbf{q} = -\frac{d\ln\rho}{dt}$$

The L coefficients are called the kinetic coefficients.

Since $\sum_{r=1}^{n}\mathfrak{M}_r\mathbf{J}_r = 0$ (Eq. 1-5), it follows that

$$\sum_{r=1}^{n} \mathfrak{M}_r L_{rr'} = 0 \tag{2-8a}$$

It can be shown also that

$$\sum_{r=1}^{n} L_{rr'}\mathfrak{M}_{r'} = 0 \tag{2-8b}$$

There are also symmetry conditions on the matrix of L coefficients. These are the Onsager [5] reciprocal relations arising from the principle of microscopic reversibility when the deviations from local equilibrium are small. To apply this principle one must label the currents and forces odd or even depending on whether the corresponding densities are odd or even with respect to time reversal. One sees that all of the currents and forces are even except for $\boldsymbol{\pi}$ and $-\dot{\boldsymbol{\varepsilon}}$ which are odd since the corresponding density $\rho\mathbf{q}$ is odd with respect to time reversal. In an analogous way the indices of the L coefficients may be called odd or even. The Onsager reciprocal relations simply state that the L matrix is symmetrical with respect to the interchange of indices that are both odd or both even, and antisymmetric with respect to the interchange of an odd and an even index. Thus

$$\left.\begin{aligned} L_{rr'} &= L_{r'r} \\ L_{er'} &= L_{r'e} \\ L_{ss'} &= L_{s's} \\ L_{qs} &= -L_{sq} \end{aligned}\right\} \tag{2-9}$$

Finally it must be stated that the symmetric part of the L matrix is positive indefinite. This property is equivalent to the second law of thermodynamics.

Eq. 2-8a, 2-8b, and 2-9, and the positive indefinite property express all of the general conditions on the kinetic coefficients that are known at the present time. In the case of large deviations from equilibrium but not large gradients, the Onsager reciprocal relations, Eq. 2-9, break down but the other conditions still hold. In the case of large gradients, nonlocal connections between currents and forces must be introduced.

The writing of the kinetic coefficients in terms of the conventional transport properties such as shear viscosity, bulk viscosity, thermal conductivity, etc., is easily achieved and is not treated here.

If one transforms the currents and forces in such a way that the entropy source density remains a sum of products of currents and forces, the Onsager reciprocal relations and the positive indefinite property still hold. Eq. 2-8a and 2-8b transform to new equations of the same type. The possibility of performing these transformations accounts for the wide variety of equivalent forms encountered in the literature.

F,3. Cited References and Bibliography.

Cited References

1. de Groot, S. R. *Thermodynamics of Irreversible Processes*. Interscience, 1951.
2. Prigogine, I., and Defay, R. *Thermodynamique chimique conformément aux méthodes de Gibbs et de De Donder*, Vol. I and II. Desoer, Liège, 1944 and 1946.
3. De Donder, Th. *Bull. Acad. roy. Belg., Cl. Sc.* *23*, 685, 770, 936 (1937); *24*, 15 (1938).

Duhem, P. *Energétique.* Gauthier-Villars, Paris, 1911.
van Rysselberghe, P. *Chem. Revs. 16,* 29, 37 (1935).
4. De Donder, Th., and van Rysselberghe, P. *Affinity.* Stanford Univ. Press, 1936.
Guggenheim, E. A. *Thermodynamics; An Advanced Treatment for Chemists and Physicists.* North-Holland Publishing Co., Amsterdam, 1949.
5. Onsager, L. *Phys. Rev. 37,* 405 (1931); *38,* 2265 (1931).
Casimir, H. B. G. *Revs. Mod. Phys. 17,* 343 (1945).

Bibliography

Note: The works cited below are intended to supplement the above references in a general way. The descriptive phrases are supplied by the authors.

Bateman, H. *Proc. Natl. Acad. Sci. 25,* 388 (1939).
Macroscopic approach to reacting continua.
Born, M., and Green, H. S. *Proc. Roy. Soc. London, A190,* 455 (1947).
Statistical mechanical treatment of generalized hydrodynamical equations without explicit consideration of irreversibility.
Cox, R. T. *Revs. Mod. Phys. 22,* 238 (1950).
Statistical mechanical treatment of irreversible processes based upon equations for probability transfer among "coarse-grained" states.
Curtiss, C. F., and Hirschfelder, J. O. *J. Chem. Phys. 18,* 171 (1950).
Entropy transport in reacting continua.
Eckart, C. *Phys. Rev. 58,* 267, 269, 924 (1940).
Quasi-thermodynamic consideration of reacting continua.
a. Irving, J. H., and Kirkwood, J. G. *J. Chem. Phys. 18,* 817 (1950).
b. Irving, J. H., and Zwanzig, R. W. *J. Chem. Phys. 19,* 1173 (1951).
Statistical mechanical derivation of the ordinary hydrodynamical equations of a one-component nonreacting system in both classical (a) and quantum mechanical (b) cases. Irreversibility is not considered.
Prigogine, I. *Étude thermodynamique des phénomènes irréversibles.* Dunod, Paris; Desoer, Liege, 1947.
Tolman, R. C., and Fine, P. C. *Revs. Mod. Phys. 20,* 51 (1948).
Examples of entropy transport.

SECTION G

COMBUSTION WAVES IN NONTURBULENT EXPLOSIVE GASES

BERNARD LEWIS
GUENTHER VON ELBE

CHAPTER 1. THEORY OF COMBUSTION WAVES

G,1. Description and General Equations.[1] Explosive gases comprise chemical compounds such as ozone, acetylene, hydrazine, or azomethane as well as mixtures of two reactants such as fuel gas and oxygen within the limits of inflammability. In an explosive gas an ignition source such as an electric spark produces a local zone of chemical reaction which propagates from layer to layer by the transport of heat and chain carriers. This zone of rapid chemical reaction and steep gradients of temperature and concentrations constitutes a wave in the broad definition of the term and will be called a *combustion wave*. Its propagation is controlled by heat conduction from the burned to the unburned gas and diffusion of various molecular species comprising reactants, product molecules, and intermediates which may act as chain carriers and whose transport rate may therefore be of first order importance in determining the rate of propagation of the wave. Combustion waves are distinguished from detonation waves (Sec. N; III,D; III,G) which are shock waves sustained by the energy of the chemical reaction.

Fig. G,1 is a schematic representation of one-dimensional propagation of a combustion wave. The wave is thought to have stationary profiles of temperature and molecular concentrations and the reference point of the x axis is chosen at a fixed position with respect to the wave, so that the wave remains stationary and the gas flows through the wave in the x direction. In the upper half of the figure a stream tube is shown which traverses the wave from the boundary u of the unburned gas to the boundary b of the burned gas. Between these boundaries chemical reaction occurs and mass is flowing at the velocity S. Since the gas is expanding in the x direction the velocity increases from the smallest value S_u at the boundary u to a maximum S_b at the boundary b. Because the mass cross-

[1] The material in this section is a condensed version of Chap. VII of the authors' book, *Combustion, Flames and Explosions of Gases*, Academic Press, New York, 1951.

ing any layer in the wave is constant, the density ρ changes from ρ_u to ρ_b in such manner that

$$\rho_u S_u = \rho_b S_b \tag{1-1}$$

Molecules are transported not only by mass flow but also by diffusion which takes place in the direction of decreasing concentration; thus, molecules of reactants diffuse in the direction u to b, and molecules of final products diffuse in the direction b to u. The diffusion velocities must be

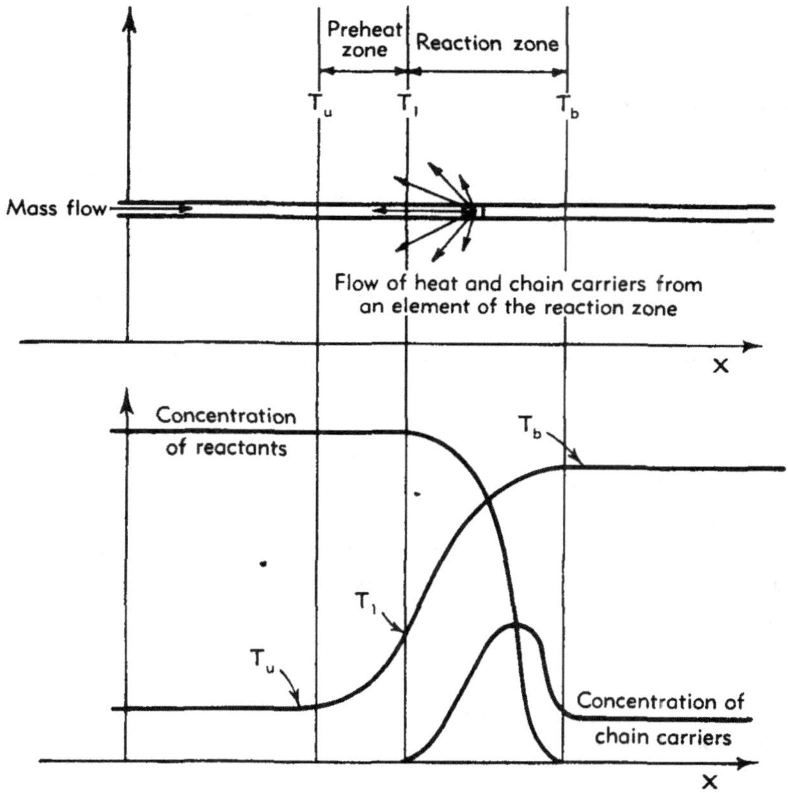

Fig. G,1. Scheme of the propagation and structure of a combustion wave.

added to or subtracted from the mass velocity, so that the reactant molecules are seen to move faster than the product molecules and mix with the latter to some extent. Molecules of intermediate products may attain a peak of concentration somewhere between boundaries u and b and thus diffuse in both directions. In the steady state all these transport processes occur in such a way that the total mass passing in a given time interval through the cross-sectional area of the stream tube is the same everywhere along the tube. If one follows an element of mass through successive stages between the boundaries u and b it is found that at first its temperature rises from the temperature T_u of the unburned gas to a temperature T_1 which marks the inflection point of the T, x curve. This curve is shown to be convex toward the x axis in the range T_u to T_1 corresponding to a positive value of the second derivative, d^2T/dx^2. In the theory of heat conduction the rate of rise of temperature for one-dimensional heat flow is given by $\kappa(\partial^2T/\partial x^2)$, where κ is the coefficient of thermal diffusivity and the positive sign thus indicates that the mass element receives by conduction more heat from the hotter elements downstream than it loses to the cooler elements upstream. On passing the inflection point the element changes from a heat sink to a heat source in the sense that it now loses by conduction more heat to the upstream elements than it receives from the downstream elements. Its temperature continues to increase owing to chemical heat evolution but the sign of $\partial^2T/\partial x^2$ is now negative and the T, x curve is concave toward the x axis. Analogous considerations apply to the changes of concentration by diffusion processes, notably the diffusion of intermediate products such as atoms and free radicals which may serve as chain carriers in the reaction; however, while the temperature curve rises uniformly until the temperature T_b of the burned gas is established, the chain carrier concentration may pass through a maximum and decrease to a value corresponding to the thermal equilibrium between atoms, free radicals, and neutral molecules at the boundary b. The two zones in front of and behind the inflection point temperature T_1 may be distinguished as the preheat and reaction zones. The transport of heat and chain carriers from an element in the reaction zone may be visualized by a vector construction as shown in the upper part of Fig. G,1. In a plane wave the vector components parallel to the wave layers cancel each other between adjacent stream tubes. It should be borne in mind that, though no mass can be transported across a stream tube boundary, the transport of heat by conduction and exchange of molecules by diffusion can occur across these boundaries, and ignition may thus propagate from one stream tube to the other. In the stream of explosive gases the combustion wave always tends to spread itself over the entire stream cross section.

The boundaries u and b of the wave are not exactly defined as the temperature curve approaches the limiting values T_u and T_b asymptotically; however, the asymptotic convergence is very rapid, so that the boundaries can be defined within narrow limits. Behind the wave the medium is either in thermodynamic equilibrium or in a metastable state in which further chemical and physical changes are so slow that under adiabatic conditions the gradients of temperature and concentrations become substantially zero.

The theoretical treatment of combustion waves requires the application of the hydrodynamic equations of change in conjunction with equations for diffusion and the rate of the chemical reaction. A formulation of the basic equations for the steady state combustion wave has been given by Damköhler [*1*; *2*, pp. 115–117]. A more complete formulation which includes also the nonsteady state equations has been given by Hirschfelder and Curtiss [*3*, p. 121; *4*]. It comprises (1) the equation of state, usually the perfect gas law; (2) the equation of motion; (3) the equation of conservation of energy; and (4) one equation of continuity, including terms arising from diffusion processes and chemical kinetics, for each chemical component. We shall first consider the derivation of the equations for the simplest case, namely, steady state one-dimensional wave propagation.

Mass movements induced by thermal expansion in combustion waves are in the subsonic range and have such low Mach numbers that the pressure drop caused by momentum changes across the wave is very small. This pressure drop can influence the gas flow pattern and therefore affect the wave shape; however, it has no significant effect on the temperature and concentration profiles in the wave and so can be ignored in all problems dealing with the inner structure of the wave.

The gas in any wave layer is composed of various molecular species (molecules of reactants and final reaction products, intermediate products such as atoms and free radicals) which may be labeled $1, 2, \cdots i \cdots,$ respectively. Each component i is flowing with respect to the wave layer at a velocity which is the sum of the average velocity S and the diffusion velocity V_i of component i. The diffusion velocities are obtained from Fick's law; these equations are given below. We note here that V_i may be positive or negative depending upon the sign of the concentration gradient of component i, and that the mass flow across a plane x is $\sum_i n_i \mathfrak{M}_i(S + V_i) = S\sum_i n_i \mathfrak{M}_i = \rho S$, where n_i is the number of moles per unit volume of component i, and $\mathfrak{M}_i$ the molecular weight, so that $\sum_i n_i \mathfrak{M}_i V_i = 0$. In a volume dx enclosed by two planes x and $x + dx$ of unit area, the number of moles of the ith component is $n_i dx$; in the steady state this number does not change with time so that any formation or disappearance of such molecules is compensated by a difference between the numbers $n_i(S + V_i)_x$ and $n_i(S + V_i)_{x+dx}$ of moles i entering and leaving the volume. The only process that causes formation or disappearance of molecules is chemical reaction, so that we may write

$$\frac{d[n_i(S + V_i)]}{dx} = \left(\frac{\partial n_i}{\partial t}\right)_{\text{chem}} = K_i \tag{1-2}$$

The term K_i is the rate at which n_i would change in a closed vessel due to chemical causes, under the same conditions of temperature, density, and composition as in the volume dx, but without any contributions from wall reactions. It is in this term that the mechanism of the chemical reaction is taken into account. One such continuity equation (1-2) and a corresponding reaction kinetic equation for K_i apply to each component. It is noted that, since mass is neither created nor destroyed by the chemical reaction, the K_i must satisfy the relation $\sum_i \mathfrak{M}_i K_i = 0$. Multiplying each Eq. 1-2 by $\mathfrak{M}_i$, taking the sum over all components, and remembering that

$$\sum_i n_i \mathfrak{M}_i (S + V_i) = \rho S \tag{1-3}$$

we find $d(\rho S)/dx = 0$. This integrates to $\rho S = \rho_u S_u = \dot{m}$, the symbol $\dot{m}$ being frequently used in the literature to denote the constant mass rate of flow measured in g/cm^2 sec.

For the equation of conservation of energy we consider a rectangular volume enclosed between a unit-area plane x inside the wave and a corresponding plane b outside the wave in the burned gas where all gradients of temperature and concentration have vanished. We denote by H_i the enthalpy per mole of the ith component, as defined by

$$H_i = (E_0^0)_i + \int_0^T \mathfrak{M}_i (c_p)_i dT$$

where $(E_0^0)_i$ is the energy of one mole of the ith component at absolute zero with respect to a standard state (it represents latent chemical energy), and $(c_p)_i$ is the specific heat at constant pressure per gram of the component. The enthalpy flux across the plane x is $\sum_i n_i(S + V_i)H_i$, and across the plane b it is $(\sum_i n_i S H_i)_b$ since all V_i vanish in plane b. The volume enclosed by x and b thus receives per unit time the enthalpy $\sum_i n_i(S + V_i)H_i - (\sum_i n_i S H_i)_b$, representing the total energy gain due to mass flow and diffusion processes. Since in the steady state the energy in the volume does not change with time, there is an energy loss of equal magnitude; this is represented by the heat flux $\lambda(dT/dx)$ across plane x, where λ is the coefficient of heat conductivity, so that the equation of conservation of energy becomes

$$\lambda \frac{dT}{dx} = \sum_i n_i(S + V_i)H_i - \left(\sum_i n_i S H_i\right)_b \tag{1-4}$$

Here the heat loss by radiation has been neglected. This is justifiable as combustion waves are of small width and the gases are highly transparent even in spectral ranges such as the absorption bands of CO_2 and H_2O, so that, according to Kirchhoff's law, the emissivity is small. In corroboration, attention is called to the fact that a flame in a closed vessel not in contact with the wall attains the theoretical adiabatic flame temperature even though it "sees" a cold nonreflecting wall.

Eq. 1-2 and 1-4 together with diffusion and reaction kinetic equations determine the steady state after introducing proper boundary conditions. At the "hot" boundary b the conditions are defined by the vanishing of the reaction rate and the gradients of temperature and concentration.

We may note that for adiabatic combustion the enthalpies of the burned and unburned gases are equal, so that $(\sum_i n_i S H_i)_b = S_u(\sum_i n_i H_i)_u$. The "cold" boundary u is more difficult to define, according to Hirschfelder and Curtiss, because the usual chemical rate expressions do not vanish and the integration from the "hot" to the "cold" boundary must be somewhat arbitrarily terminated. Hirschfelder and Curtiss have defined the "cold" boundary condition by means of a somewhat abstract "flame holder" which does not appear to have immediate physical significance in the case of a free adiabatic combustion wave but has the merit of resembling conditions near a burner rim or some other flame-stabilizing object in a gas stream. After specification of the boundary conditions a solution of the differential equations is obtained for only one value of the mass flow $\dot{m}$, and the wave structure is in detail determined.

Concerning the diffusion velocities V_i, reference is made to a paper by Curtiss and Hirschfelder [*5*] in which, on the basis of Fick's law, the general equation for V_i is found to be

$$V_i = \frac{n^2}{\rho n_i} \sum_j \mathfrak{M}_j D_{ij} \frac{d}{dx}\left(\frac{n_j}{n}\right) \tag{1-5}$$

where $n = \sum_i n_i$ is the total number of moles per cm³. The terms D_{ij} are not the usual diffusion constants $\mathfrak{D}_{ij}$ for binary mixtures, but they can be written in terms of the $\mathfrak{D}_{ij}$ and the composition. Eq. 1-5 is too complicated for practical use in connection with the foregoing equations and may be replaced by the approximate relation

$$V_i = -\frac{n}{n_i} D_i \frac{d}{dx}\left(\frac{n_i}{n}\right) \tag{1-6}$$

where the terms D_i are given by

$$D_i = \frac{1 - n_j \mathfrak{M}_i/\rho}{\sum_{j \neq i} n_j / n\mathfrak{D}_{ij}} \tag{1-7}$$

Eq. 1-6 is exact in the case of binary mixtures and gives the correct diffusion velocity of any component of a general mixture which is present in trace. In other cases it is still a good approximation.

We shall next set down the equations of change in their most general form. Introducing the vectors $\mathbf{U}$ and $\mathbf{V}_i$ of the average mass velocity (also called particle velocity and in this text occasionally referred to as gas velocity) and the diffusion velocity, the equation of continuity for the ith component becomes

$$\frac{\partial n_i}{\partial t} + \nabla \cdot [n_i(\mathbf{U} + \mathbf{V}_i)] = K_i \tag{1-8}$$

Here ∇ is the vector differential operator and the equation applies to nonsteady states as well as to mass flow and diffusion in all dimensions. In the one-dimensional steady state problem the time derivative vanishes and the operator ∇ becomes d/dx so that Eq. 1-2 results. In this case $\mathbf{U}$ is equal to S, but this is not so in the general case.

The equation of conservation of energy becomes

$$\frac{\partial}{\partial t}\sum_i n_i E_i + \nabla \cdot \left[\sum_i n_i H_i(\mathbf{U} + \mathbf{V}_i) - \lambda \nabla T\right] = \mathbf{U} \cdot \nabla p - \frac{2}{3}\mu(\nabla \cdot \mathbf{U})^2 + 2\mu\phi : \nabla \mathbf{U} \quad (1\text{-}9)$$

Here E_i is the internal energy per mole of component i and μ is the coefficient of viscosity. The equation carries all possible terms entering into the total rate of change of energy per unit volume of gas or fluid anywhere in the system at any time, even changes associated with a change of pressure p and with viscosity effects. It may be noted that pressure changes are generally negligible in the energy balance except in the case of transition to detonation, and viscosity effects are also generally negligible insofar as the energy balance is concerned, so that the right-hand side of Eq. 1-9 may usually be taken as zero. In explanation of the symbols we add that Φ is the symmetric tensor

$$[\Phi_{ij}] = \frac{1}{2}\frac{\partial V_i}{\partial x_j}\frac{1}{2}\frac{\partial V_j}{\partial x_i}$$

and the symbol : indicates the products of two tensors defined by $w:y = \sum_{ij} W_{ij} Y_{ij}$. If the right-hand side is negligible, the equation simply states that the rate of change of internal energy is equal to the rate of change of enthalpy due to mass flow minus the rate of change of heat due to thermal conduction. In the one-dimensional steady state problem, the time derivative vanishes, and Eq. 1-9 can be integrated once with respect to x, so that Eq. 1-4 results.

In burner flames the combustion wave is in a steady state and can generally be considered to be approximately plane because the radius of curvature is large compared to the wave width. Although the wave is inclined to the direction of flow, it can be shown that Eq. 1-2 and 1-4 for one-dimensional propagation are applicable.

For this purpose we write the second term of Eq. 1-8 in the form, consistent with the definitions of vector analysis,

$$\nabla \cdot (n_i \mathbf{U}) + \nabla \cdot (n_i \mathbf{V}_i) = (\nabla n_i) \cdot \mathbf{U} + n_i \nabla \cdot \mathbf{U} + (\nabla n_i) \cdot \mathbf{V}_i + n_i \nabla \cdot \mathbf{V}_i$$

where

$$(\nabla n_i) \cdot \mathbf{U} = \frac{\partial n_i}{\partial x} U_x + \frac{\partial n_i}{\partial y} U_y + \frac{\partial n_i}{\partial z} U_z$$

$$n_i \nabla \cdot \mathbf{U} = n_i \left(\frac{\partial}{\partial x} U_x + \frac{\partial}{\partial y} U_y + \frac{\partial}{\partial z} U_z\right)$$

and similarly for the terms $(\nabla n_i) \cdot \mathbf{V}_i$ and $n_i \nabla \cdot \mathbf{V}_i$. Eq. 1-9 can be expanded in an analogous manner. In the case of a plane steady state wave that is inclined against the gas flow and whose normal is taken to be the x direction, all derivatives with respect to y and z are zero, and since $U_x = S$, Eq. 1-2 and 1-4 are again obtained.

At the boundary u of the inclined wave the velocity component U_x has the value $U_x \cos \alpha$, where U_u is the velocity of the unburned gas entering the wave and α is the angle between the direction of the gas flow and the normal to the wave. At the boundary b the velocity component U_x has the value $U_b \cos \beta$ where U_b is the velocity of the burned gas leaving the wave and β is the angle between the direction of gas flow leaving the wave and the normal to the wave. Since

$$U_u \cos \alpha = S_u \tag{1-10}$$

and

$$U_b \cos \beta = S_b \tag{1-11}$$

it follows from Eq. 1-1 that

$$\rho_u U_u \cos \alpha = \rho_b U_b \cos \beta \tag{1-12}$$

Since there is no change of velocity in the plane of the wave, i.e. the tangential component of the gas velocity remains constant, one may write

$$U_u \sin \alpha = U_b \sin \beta \tag{1-13}$$

Combining Eq. 1-12 and 1-13 one obtains

$$\frac{\rho_u}{\rho_b} = \frac{\cot \beta}{\cot \alpha} \tag{1-14}$$

which relates the refraction of the stream tube to the change of density.

G,2. Special Solutions for One-Dimensional Steady State Propagation. Attempts at quantitative solution of the combustion wave equations started quite naturally from the simple case of one-dimensional steady state propagation as illustrated in Fig. G,1. The earliest such attempt was made by Mallard and Le Chatelier [*6*]. These authors visualized the temperature distribution in the wave essentially as shown in Fig. G,1 and proposed that the inflection point temperature T_1 constitutes an "ignition" temperature below which chemical reaction does not occur. Diffusion processes were naturally not considered in this early work and thus it was possible to set down the following equation describing the conservation of energy at plane x_1 corresponding to T_1:

$$\lambda \left(\frac{dT}{dx}\right)_1 = \rho_u S_u c_p (T_1 - T_u) \tag{2-1}$$

where c_p is the specific heat at constant pressure, assumed to be constant. The temperature rise from T_1 to T_b was assumed to be approximately linear along the x coordinate so that one could write

$$\lambda\left(\frac{dT}{dx}\right)_1 = \frac{(T_b - T_1)}{\Delta x} \tag{2-2}$$

where Δx represents a somewhat ill-defined length comparable in magnitude to the width of the reaction zone. The equation for the burning velocity is then written

$$S_u = \frac{\lambda(T_b - T_1)}{(T_1 - T_u)\rho_u c_p \Delta x} \tag{2-3}$$

This in substance is Mallard and Le Chatelier's equation. It constitutes essentially a dimensional analysis of the combustion wave problem. Other authors have since attempted to enlarge on this treatment by introducing expressions for the rate of the chemical reaction. In this way the unknown length Δx can be eliminated, but an equally unknown or oversimplified rate expression is introduced instead; that is, Δx is replaced by the product of burning velocity S_u and the time required for completing the chemical reaction after a mass element has passed plane x_1. Treatments of this kind, with variations as to detail, have been published by Jouguet and Crussard [*7,8,9*], Nusselt [*10*], Eucken [*11*, p. 859ff.], and Daniell [*12*].

One may hazard a guess of the value of the "ignition" temperature, perhaps estimating a lower bound from available kinetic evidence; an upper bound would be the flame temperature T_b. One may then calculate the lower part of the temperature distribution curve by integrating the equation

$$\lambda\left(\frac{dT}{dx}\right) = \rho_u S_u c_p (T - T_u) \tag{2-4}$$

between the limits T_1 and T_u. This in substance was originally proposed by Michelson (see [*13*]), and numerical calculations have been made by Mache [*13*] for a mixture of hydrogen and oxygen with $S_u = 130$ cm/sec. Such calculations lead to a rough estimate of the width of the combustion wave; the temperature rises exponentially along the x axis so that one may select a point x_u at which the temperature rise becomes just noticeable (a few per cent of T_u) and determine the distance from x_u to x_1 where the temperature is T_1. One finds for Mache's example that $x_1 - x_u \cong 10^{-2}$ cm, which gives the order of magnitude of the wave width.

In the earlier literature some steady state solutions have been worked out on the basis of assumed reaction mechanisms. However, the lack of chemical kinetic knowledge prevented inclusion of realistic intrawave reaction kinetics into the fabric of necessary assumptions; and the latter were tailored to fit only the problem of solving the differential equation of heat transport, since diffusion processes were not taken into consideration. Characteristically, the division of the wave into preheat and reaction zones was thought to correspond to an ignition temperature of more or less independent physical significance.

The more modern treatments are still handicapped by insufficient chemical kinetic knowledge. They introduce general expressions for unimolecular or bimolecular reactions if the treatment is intended to be general, or more or less realistic reaction mechanisms conforming to available chemical information for specific gas mixtures. Usually such treatments acknowledge and in some fashion deal with the problem of simultaneous solution of diffusion and heat conduction equations. The mathematical fiction of a timeless steady state makes it necessary to "prohibit" in some fashion the occurrence of even the slightest chemical change in the fresh gas that by assumption flows eternally from infinite distance toward the reaction zone. Hence, supplementary conditions are imposed which in effect state that the gas remains unaltered during its postulated infinite time of existence prior to combustion in the wave. This is done in various ways: (1) by outright assumption of an ignition temperature below which no reaction whatsoever occurs, as had been done in the older literature (Zeldovich, Frank-Kamenetsky, and Semenov [*14,15,16,17*], Bartholomé, et al. [*18,19,20,21*], Bechert [*22,23*]); (2) by boundary conditions stating that upstream at infinite distance the gas is in its original state and that all gradients of chemical change vanish (Lewis and von Elbe [*24*], Boys and Corner [*25,26*]); or (3) by simply assuming that upstream beyond some boundary the gas is in its original state (Hirschfelder, Curtiss, et al. [*3*, pp. 121–140; *27*; *28*; *29*] "flame holder;" Tanford, Pease, et al. [*30*; *31*; *32*; *3*, pp. 140–146] "finite width of reaction zone"). Since, in reality, an explosive gas is not in existence for such long periods of time that its composition is altered appreciably by slow chemical reaction prior to the reaction in the combustion wave, such devices are harmless mathematical fiction albeit without deeper physical significance. The problem of selecting an ignition temperature has been met by modern authors [*14,15,16,17,18,19,20,21,22,23*] with the remarkably simple assumption that the temperature throughout the reaction zone is nearly constant and equal to the flame temperature and that therefore for practical purposes the latter can be identified with the ignition temperature.

The problem of simultaneous solution of the differential equations of heat conduction and diffusion was avoided by Lewis and von Elbe [*24*] by means of the assumption that throughout the wave the sum of thermal and chemical energy per unit mass is constant. In this way the temperature is determined by the chemical composition, and heat conduction does not explicitly enter into the system of equations. In later work, Zeldovich, Frank-Kamenetsky, and Semenov [*14,15,16,17*] have adopted the same assumption though cast into somewhat different mathematical language: They explicitly equate the diffusional transport of chemical enthalpy from the unburned to the burned side to the conductive transport of heat from the burned to the unburned side. This alternative way of stating the assumption of constancy of thermal and chemical energy per unit mass serves the purpose of simplification equally well. Both groups of authors have avoided the problem of multiple diffusion of species, particularly the diffusion of chain carriers, by means of suitable assumptions fixing the relationship of the local chain carrier concentrations to the concentration of reactants. In the work of Hirschfelder, Curtiss, et al. [*3*, pp. 121–140; *27*; *28*; *29*] such compromises are not made and computing machines are resorted to in order to overcome the mathematical complexities.

Some of the modern authors have taken the fundamental approach of deriving numerical steady state solutions entirely from basic physical and chemical parameters. The first such paper was published by Lewis and von Elbe [*24*] and dealt with combustion waves in ozone-oxygen mixtures. Later papers of Zeldovich, Frank-Kamenetsky, and Semenov [*14,15,16,17*] deal with moist mixtures of carbon monoxide and oxygen. Taking the burning velocity as criterion for comparing experiment and theory, these treatments have yielded order-of-magnitude agreements. The papers of Boys and Corner [*25,26*] deal largely with mathematical formulism and more or less hypothetical examples of quantitative studies. In the papers of Hirschfelder, Curtiss, et al. [*3*, pp. 121–140; *27*; *28*; *29*], the problem of obtaining numerical solutions is attacked with least compromise, though precision solutions for specific explosive gases are not yet in sight due to insufficiency of information on chemical and physical parameters.

Other authors do not take this fundamental approach but incorporate into their equations coefficients derived from experimentally determined properties of combustion waves. Such coefficients may be chosen to give a best fit to experimental series such as burning velocity measurements at various mixture ratios, temperatures, and pressures. Examples worked out by Tanford and Pease [*3*, pp. 140–146; *30*; *31*; *32*] for moist carbon monoxide-oxygen and for hydrogen-oxygen mixtures, on the basis that the controlling transport process is the diffusion of hydrogen atoms, and other examples by Bartholomé, et al. [*18,19,20,21*] and Bechert [*22,23*] on the basis of a thermal theory (i.e. neglecting diffusion processes) have shown remarkably good agreement between measured and predicted trends of burning velocity. However, it appears [*33,34*] that such agreement can be obtained by various alternative treatments based on different sets of assumptions; that is, in view of the many parameters that enter into the final expression of burning velocity, such expressions have considerable flexibility for adjusting theoretical curves to experimental series. A detailed review of the various treatments is found elsewhere [*35*].

Boys and Corner [*25,26*] have stated that in their view "there are no real subtleties in the quantitative theory of flames, but there are very great mathematical complexities." This appears to be quite true when the theory of flames is narrowly conceived to mean only the computation of velocity and inner structure of steady state combustion waves. In such work one formulates a problem that is known to be solvable, the quality of the results being known beforehand. Disagreement between experiment and theory is properly attributed to insufficient knowledge of the reaction mechanism, or questionable simplifications of various kinds, or questionable reliability of data, and so on; but the fundamentals of the theory of combustion wave propagation are not in doubt. Indeed since Mallard and Le Chatelier's first analysis there exists a qualitative knowledge which should satisfy much of our intellectual curiosity concerning combustion waves in the steady state; whereas ability actually to compute burning velocities appears to be a refinement that is not necessarily essential to further exploration of a more broadly conceived theory of flames. This comprises the conditions for initiation of combustion waves, quenching by solids, and limits of inflammability. The steady state treatments referred to have provided no insight into these subjects. Yet it seems that these subjects can be approached with rather simple means, as shown by further developments in this section.

G,3. Wave Near a Heat Sink. Dead Space and Quenching Distance. Combustion waves are not generally plane and the combustion process is not generally adiabatic. To assist in the formulation of manageable boundary conditions and suitable approximation of the differential equations, it is necessary to introduce appropriate models of combustion waves for various conditions. One such model corresponding to conditions at the base of a Bunsen burner cone is shown in Fig. G,3a. The wave and the burner rim are shown in sectional view. The burner tube is assumed to be in a horizontal position. Explosive gas flows from the tube and maintains a stationary conically shaped combustion wave in the free jet beyond the tube rim. A series of isotherms and flow lines illustrates the distribution of temperature and the flow pattern. Thermal expansion causes a deflection of the gas flow from its course parallel to the tube toward the open atmosphere. The solid rim, which is a sink for heat (and probably also for chain carriers due to adsorption of atoms and free radicals), absorbs much of the chemically evolved heat from adjacent stream tubes, but its effectiveness as a sink decreases with increasing distance, so that for stream tubes in the interior of the stream the heat evolution in the combustion wave is substantially adiabatic. We differentiate adiabatic combustion from nonadiabatic combustion by affixing the superscript ^{0} to the symbols T_1 and T_b; thus, T_1^0 and T_b^0 denote the inflection point temperature and the final combustion temperature, respectively, in the wave at large distance from the rim. Two surfaces marked T_1 surface and T_b surface are shown which at large distance merge with the isothermal surfaces T_1^0 and T_b^0 but on approaching the rim cut across isotherms between T_1^0 and T_b^0 (not shown) and finally merge with each other. The temperatures T_1 and T_b along these nonisothermal surfaces are theoretically defined as the temperatures T_1 and T_b in the previously discussed case of one-dimensional wave propagation, the only difference being that the present case involves heat flow in three dimensions and we must write $\partial^2T/\partial x^2 + \partial^2T/\partial y^2 + \partial^2T/\partial z^2$, or the equivalent mathematical symbol ∇^2T, instead of $\partial^2T/\partial x^2$. Thus, within the shaded zone enclosed by T_1 surface and T_b surface the term ∇^2T is negative and outside the zone it is positive or zero (the latter case corresponding to constant

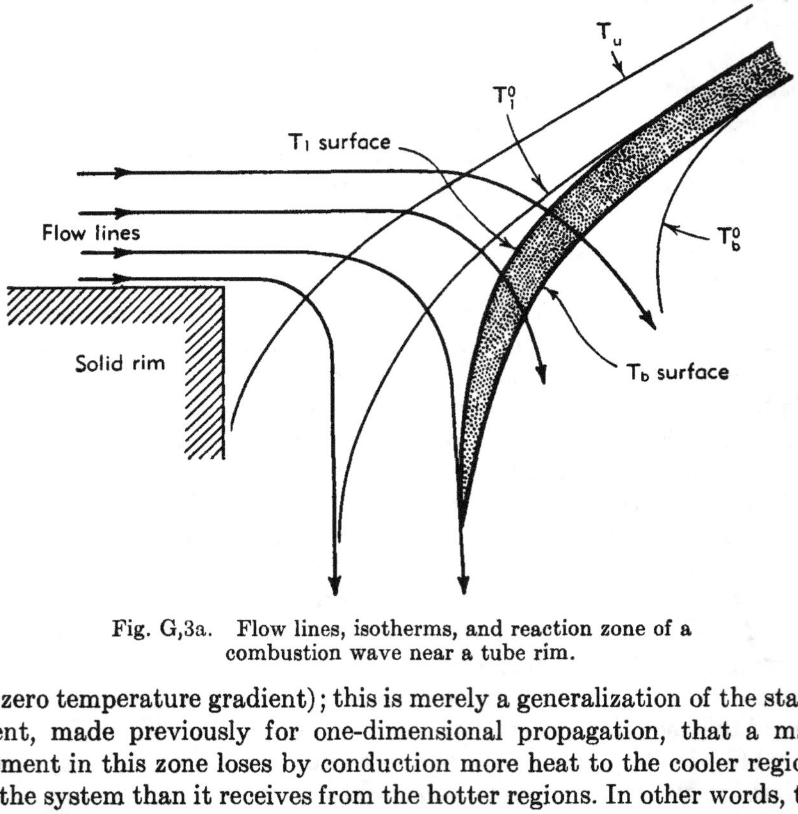

Fig. G,3a. Flow lines, isotherms, and reaction zone of a combustion wave near a tube rim.

or zero temperature gradient); this is merely a generalization of the statement, made previously for one-dimensional propagation, that a mass element in this zone loses by conduction more heat to the cooler regions of the system than it receives from the hotter regions. In other words, the heat flow from the shaded zone by conduction across the T_1 surface is larger than the heat flow into the zone by conduction across the T_b surface, and the resulting net loss of heat is balanced by the chemical heat evolution. The latter also raises the temperature of the gas stream from T_1 to T_b, and in the absence of a heat sink the temperature along the T_b surface would be constant and equal to T_b^0; however, the presence of the sink causes the temperature to decrease along the T_b surface. The change of temperature along the T_1 surface is speculative since experimental observation is difficult and a reliable theoretical treatment of the problem cannot be given at present. The T_1 surface marks the transition of a mass element from a state of virtually no chemical reaction to chemical reaction with heat evolution at a significant rate. In the usual case of a chain reaction, the rate of heat evolution at any point of the system is governed not only by the temperature and concentration of reactants, but also by the chain carrier concentration. It may be surmised that, in a combustion wave, chain carriers are formed at high temperature levels and that their concentration at the low temperature levels is dependent on diffusion. On this basis it would seem that the region of chain carrier formation becomes more and more remote from the T_1 surface, as the rim is approached; and in order to maintain a significant reaction rate at the T_1 surface the decreasing concentration of chain carriers must be compensated by an increase of T_1. For a flame with a nonchain reaction mechanism, should it exist, a theoretical analysis would probably show that the T_1 surface coincides substantially with the T_1^0 isotherm.

The zone enclosed by the T_1 and T_b surfaces constitutes a heat source in the gas stream. The heat is evolved by combustion of the gas that flows across the T_1 surface. We may assume that any mass element crossing the T_1 surface burns to completion while any mass element that bypasses the T_1 surface remains unburned. This is a somewhat simplified view since some chemical reaction takes place outside the T_1 surface and some interchange of combustion products and reactants occurs by diffusion. However, the simplification affords a working model of a combustion wave which is not likely to be very different from the actual phenomenon. On this basis, the T_1 surface, or, as we may call it, the front of the reaction zone, becomes the reference surface whose movement relative to the gas stream determines the direction and velocity of the combustion wave. Thus, the direction of the wave coincides everywhere with the normal to the reaction zone front, and the wave velocity is the normal component of the mass velocity relative to the front. As before, we use the symbols ρ and S to denote, respectively, the density and the normal component of the relative mass velocity in any wave surface. Using the subscript $_1$ to denote quantities relating to the T_1 surface or the front of the reaction zone, the normal component of mass flow across unit area of the flame front becomes $\rho_1 S_1$. For steady state conditions such as exist in a burner flame, this quantity may be correlated to the flow of unburned gas with the aid of the diagram in Fig. G,3b. A stream tube is shown which encloses the element dl_1 of the reaction zone front. In the unburned gas, at a distance from the wave where the temperature is T_u, the density and velocity in the stream tube are ρ_u and U_u, respectively, and the cross section is dy, so that the mass transported per unit time across the area dy is $\rho_u U_u dy$. In the steady state this must be equal to the mass transport across any area enclosed by the stream tube, so that $\rho_u U_u dy$ is equal to $\rho_1 S_1 dl_1$. We also consider a geometric surface in the shape of a straight tube of constant cross section, which encloses the element dl_1 and is normal to its plane. To define the wave velocity we shall assume that the axial component of mass flow along this tube is constant, so that $\rho_1 S_1 dl_1$ is equal to $\rho_u S_u dl_1$. The velocity S_u is the velocity of the reaction zone front with respect to the unburned gas and is called the *burning velocity*. Since $\rho_u U_u dy = \rho_1 S_1 dl_1 = \rho_u S_u dl_1$, S_u is related to U_u by the equation

$$S_u = U_u \frac{dy}{dl_1} \tag{3-1}$$

If the flow lines remained parallel in the wave the subtended area would

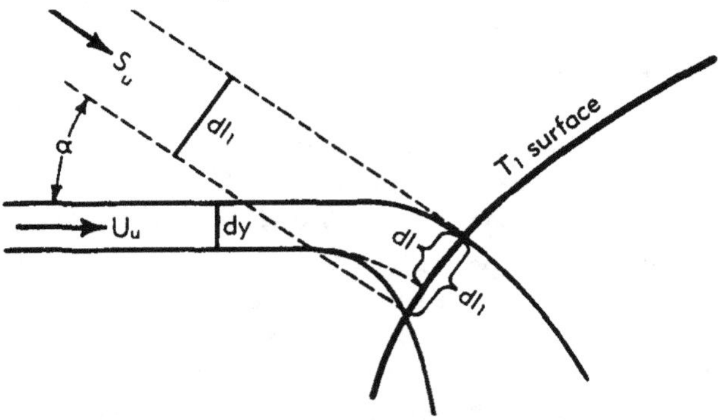

Fig. G,3b. Relation between mass velocity and burning velocity.

be $dl \leqq dl_1$. Denoting the angle enclosed by the velocity vectors of magnitude U_u and S_u by α, and since $dy/dl = \cos\alpha$,

$$S_u = U_u \left(\frac{dl}{dl_1}\right) \cos\alpha \tag{3-2}$$

which for the case of the plane wave, where $dl/dl_1 = 1$, reduces to Eq. 1-10. Assuming the surface element dl_1 to be of unit area, and extending the straight tube beyond dl_1 to the T_b surface, the tube encloses a volume of the reaction zone of the magnitude unit area × zone width. If $\bar{q}$ represents the average rate of heat evolution per unit of this volume, the total rate of heat evolution in the volume is represented by the product $\bar{q}$ × zone width. This quantity must be equal to the rate of supply of heat of combustion, or $Q\rho_u S_u$, where Q is the heat of combustion per unit mass of fresh gas. On approach of the reaction zone to a heat sink the rate $\bar{q}$ decreases because the temperatures and chain carrier concentrations are lowered and the rate of the chemical reaction therefore decreases. The zone width may also be expected to decrease because the T_1 and T_b surfaces must ultimately merge. Since Q and ρ_u are constant, it follows that S_u decreases and ultimately vanishes. We may speak of the region of vanishing burning velocity as the fringe of the combustion wave. Here the mass flow of explosive gas becomes parallel to the front of the reaction zone and bypasses the zone (Fig. G,3a), though it may react farther downstream by contact with hot burned gas. The region of the wave fringe that extends from the T_1 surface to the heat sink may be called the "dead space." It marks the closest possible approach of the reaction zone to the sink. Evidently, when two heat sinks, such as two solid plates, approach each other, the dead spaces must ultimately merge, and the combustion wave cannot propagate past the "zone of quenching" that is thus formed. On this principle it is possible to design flame traps which are grids or porous partitions whose channel width does not exceed a critical distance called the *quenching distance,* which is determined by the nature of the explosive gas. A well-known example is the grid of the Davy safety lamp. Also, it is general practice in the design of gas appliances such as range burners to make the flame ports so small that the combustion wave cannot penetrate into the burner duct and ignite the jet of fuel gas at the gas orifice.

It is occasionally a useful simplification to consider the wave to be a single surface across which transition from no reaction to complete reaction takes place, and which is inclined against the flow of unburned gas at the angle α defined by Eq. 3-2.

G,4. Curved Waves. As an example of a strongly curved wave we consider two plane combustion waves which are symmetrically inclined against the gas flow, as shown in Fig. G,4. A stream tube passing through the intersection of the waves is exposed to the converging heat flow from both waves; and if the gas mixture in the curved part of the wave is not seriously attenuated by diffusion effects as described below, a mass element traveling along the stream tube is activated to chemical reaction before it arrives at the point where the T_1 surfaces would intersect if the wave surfaces remained plane. Consequently the T_1 surface and with it all other wave surfaces move forward against the gas stream and acquire curvatures approximately as shown. The axial flow line passes at right angle through the T_1 surface so that in the center of the stream the burning velocity attains a maximum value and decreases gradually on both sides corresponding to the increase of the angle α between the direction of flow and the normal to the T_1 surface. The example illustrates that, on the basis of heat flow alone, a concave curvature increases the burning velocity, and it may be analogously shown that a convex curvature decreases the burning velocity. However, it is evident that the effect of curvature becomes significant only when the radius of curvature of the T_1 surface is of the order of magnitude of the width of the combustion wave. The wave width can often be estimated from shadow or schlieren photographs of flames; and a measure of the order of magnitude is provided by the readily measurable quenching distance. For flames burning at atmospheric pressure, the wave width ranges from the order of 1 mm for weak fuel-air mixtures to the order of 0.01 mm for fast-burning mixtures such as hydrogen and oxygen. These dimensions are so small that curvature ordinarily is significant only at very small flame dimensions such as exist at the early stage of flame development from a spark where the surface is strongly convex, or at the tip of the inner cone of a burner flame. In spherical flames which have grown to some size, and over most of the surface of a burner flame, effect of the curvature on the burning velocity is negligible and the wave layers may be considered to be substantially plane-parallel.

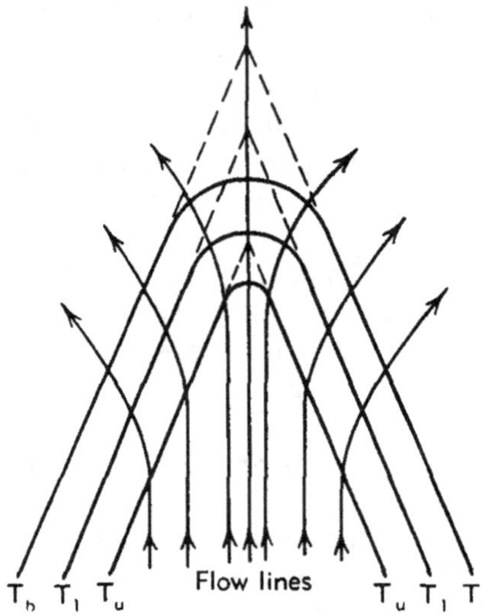

Fig. G,4. Curvature of wave layers at the intersection of two combustion waves.

A complication arises from the fact that in a curved wave the concentration gradients of the various molecular species are not parallel to the lines of flow but cut across the boundaries of the stream tubes in the sense that they converge toward, or diverge from, the central stream tube that passes through the apex of the curvature. Since the diffusion flow of each molecular species follows the course of the concentration gradients "downhill" from the zone of highest concentration to the zone of lowest concentration, it follows that the concentration of some molecules decreases in the central stream tube, and increases in adjacent tubes, while the concentration of other molecules increases in the central stream tube and decreases in adjacent tubes. If the composition of the explosive gas mixture is critically adjusted so that loss of some molecular species causes a sharp decrease of the burning velocity—as is the case for mixtures near the limits of inflammability—it may occur that a concave curvature does not increase but actually decreases the burning velocity. In that case, for example, the two waves of Fig. G,4 would not merge but would remain separated by a central stream of noninflammable gas against which the wave surfaces are inclined at a steep angle. Experimental observations of this effect, as well as of others attributable to mixture change by diffusion, are described in Art. 13 and 14.

G,5. Principles of Stabilization of Combustion Waves in Gas Streams. The most common example of a combustion wave in a gas stream is the inner cone of a Bunsen burner flame. We shall examine the mechanism by which the wave maintains a fixed position with respect to the burner rim [*36*; *37*, pp. 241–247].

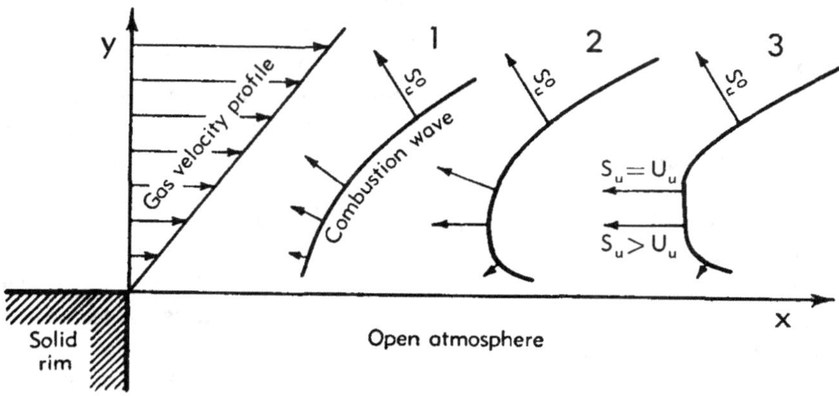

Fig. G,5a. Scheme of stabilization of a combustion wave in a laminar jet. The wave profile near the jet boundary is shown in three positions. The burning velocity is represented by vectors.

1. Combustion wave too close to solid rim. The gas velocity exceeds the burning velocity everywhere. The wave is driven back.
2. Equilibrium position. The increased distance from the solid rim causes the burning velocity to increase except at the outermost fringe where the mixture is diluted by the atmosphere. At one point, gas velocity and burning velocity are equal.
3. Wave too far from solid rim. The burning velocity partly exceeds the gas velocity. The wave moves forward toward the equilibrium position.

Consider a laminar jet of explosive gas issuing from a tube into the open atmosphere. Assuming Poiseuille flow, the gas velocity is zero at the stream boundary and increases to a maximum in the center of the stream. We shall confine ourselves to an examination of the conditions near the fringe of the combustion wave in the immediate vicinity of the burner rim, as illustrated in Fig. G,3a. The linear dimensions of the region of interest are usually very small. In slow-burning mixtures such as methane and air they are of the order of 1 mm; in fast-burning mixtures such as hydrogen and oxygen they may attain values of the order of 0.01 mm. Since usually the tube diameter is large in comparison, the gas velocity is represented by an approximately linear vector profile. This is shown in the schematic drawing of Fig. G,5a. It is further assumed that the flow lines of the free jet are parallel to the tube axis. Let us suppose that a combustion wave is formed in the stream, and that the fringe of the wave approaches the burner rim closely. Along the wave profile the burning velocity decreases as indicated by the magnitude of the vectors that have been drawn normal to the wave. At some distance from the fringe the burning velocity attains its maximum value S_u^0. Toward the fringe the burning velocity decreases. If the wave fringe is very close to the rim, as indicated by position 1, the burning velocity in any flow line is smaller than the gas velocity, and the wave is driven back by the gas flow. As the distance from the rim increases, the loss of heat and chain carriers decreases and the burning velocity becomes larger. Eventually, a position is reached in which at some point of the wave profile the burning velocity is equal to the gas velocity. This is indicated by position 2. The wave is now in equilibrium with respect to the solid rim. If the wave is displaced to a larger distance, such as position 3, the burning velocity at the indicated point becomes larger than the gas velocity and the wave moves back to the equilibrium position.

In this way a Bunsen flame attains a stable position at the rim of the burner tube. It is noted that equality of gas and burning velocity exists at only one point of the wave profile and that everywhere else the gas velocity exceeds the burning velocity. The wave profile is therefore inclined against the direction of the gas flow at an angle which is determined from Eq. 3-2. The shape of the combustion wave and the length of the burning velocity vectors in Fig. G,5a are drawn to conform with this equation, taking $dl/dl_1 = 1$. However, the profile of an actual wave fringe differs in appearance considerably, though not in principle, from the present idealized model in which the wave is represented by a single surface and the flow lines are assumed to retain their direction parallel to the x axis. In actual flames, the wave layers are widely spaced and the flow lines are bent outward as a result of thermal expansion.

When the tube diameter is sufficiently large so that the gas velocity profile at the wave fringe may be considered linear, one may write for the gas velocity the equation

$$U_u = gy \tag{5-1}$$

where y is the distance from the stream boundary and g is a constant which will be referred to as the *boundary velocity gradient*. If the gas flow is reduced, g decreases and the equilibrium position 2 shifts closer to the solid rim. On further decrease of the gas flow, a condition is reached at which the gas velocity at some point becomes smaller than the burning velocity and the combustion wave propagates against the gas stream into the tube. This is referred to as *flashback*, and the critical value of g at which this condition is first realized is denoted by g_F. Another critical value of the boundary velocity gradient refers to the condition of *blowoff* and is denoted by g_B. When the gas flow is increased the equilibrium position shifts away from the burner rim. It is noted that with increasing distance from the rim the explosive gas becomes progressively diluted by interdiffusion with the surrounding atmosphere, and the burning velocity in the outermost stream lines decreases correspondingly. This is indicated in Fig. G,5a by the increasing retraction of the wave fringe in positions 1 to 3. Thus, an ultimate equilibrium position of the wave exists beyond which the effect on the burning velocity of increased distance from the burner rim is overbalanced by the effect of dilution. If the boundary velocity gradient is so large that the combustion wave is driven beyond this position, the gas velocity exceeds the burning velocity in every streamline and the combustion wave blows off.[2]

Flashback and blowoff conditions may be illustrated by plotting the burning velocity and stream velocity as functions of distance from the

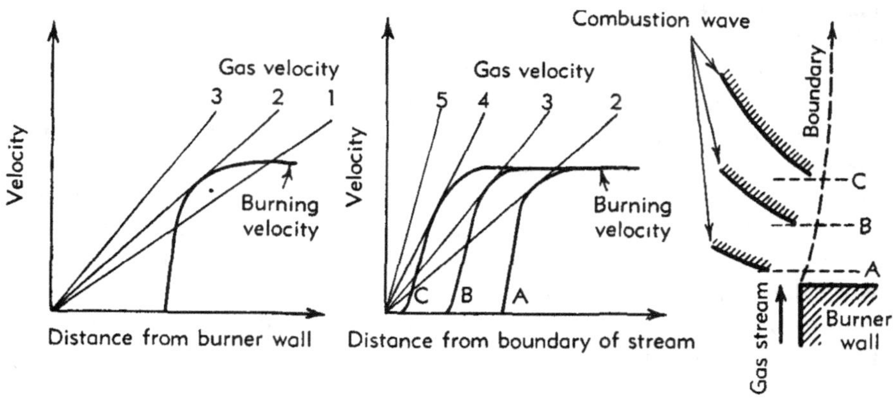

Fig. G,5b. Schematic illustration of flashback and blowoff. Left, burning velocity and gas velocity inside the burner tube. Curve 1, flashback; curve 2, flashback limit; curve 3, no flashback. Right, burning velocity and gas velocity above the tube rim. Curves *A*, *B*, *C*, burning velocity at heights *A*, *B*, *C*; curve 2, flashback limit; curve 3, stable flame; curve 4, blowoff limit; curve 5, blowoff.

stream boundary. This is shown in Fig. G,5b. In the diagram on the left, the heavy line represents the burning velocity for the case of the combustion wave inside the tube. Three light lines are shown which represent the stream velocity for three values of the gas flow. Curve 1 cuts across the burning velocity curve so that the burning velocity exceeds the stream velocity in some range and the combustion wave moves upstream against the flow, that is, it flashes back. Curve 2 is tangent to the burning velocity curve and therefore defines the upper limit of gas flow at which flashback occurs. Curve 3 lies outside the burning velocity curve so that the stream velocity exceeds the burning velocity everywhere and the combustion wave is driven out of the tube. The conditions in the free jet outside the tube are shown in the diagram on the right. To any gas velocity curve

[2] In this and the following articles the term "blowoff" refers to the mechanism just described. Blowoff due to lateral gas motion has not been studied extensively and is not considered here.

above curve 2 there corresponds a stable position of the flame above the orifice indicated by positions A, B, and C and a curve of burning velocity. As the gas velocity increases, the burning velocity curve progressively moves toward the boundary until a critical gas velocity (curve 4) is reached which is the blowoff limit. On further increase (curve 5) the gas velocity exceeds the burning velocity everywhere, and the flame blows off. The critical boundary velocity gradients g_F and g_B are given by the slopes of curves 2 and 4, respectively.

It is also possible to stabilize a combustion wave by inserting a solid object into the interior of the stream of explosive gas. Two examples are shown in Plate G,5. On the left a wire ring is mounted concentrically in a stream of natural gas and air at a considerable distance from the rim of a glass tube. The diameter of the ring is smaller than the inner diameter of the tube, and thus the ring is well within the undiluted stream. Since, by frictional drag, the gas velocity at the ring is very low, the combustion wave readily finds a position of equilibrium and forms a cone in the same way as it does at the rim of the tube. The wave is unable to travel upstream from the ring because the gas velocity below the ring is larger than the burning velocity over the entire cross section of the stream. On the right the obstruction consists of an axially mounted wire. Here the combustion wave finds a position of equilibrium at some distance above the wire where the gas velocity is retarded by frictional drag and the burning velocity is not reduced to zero by the quenching effect of the wire. Frictional drag occurs along the whole length of the wire, and consequently the gas velocity drops to zero at the wire surface. Under the conditions depicted, the wave cannot propagate upstream in this retarded zone because of the quenching effect of the wire. It can be made to travel upstream if the wire is heated, thus reducing the quenching effect. At sufficiently small gas flow the heat received by the wire from the flame is enough to bring this about. In the case of the ring obstruction, it is obviously impossible for the wave to travel upstream even if the ring becomes incandescent.

The stabilization mechanism can be visualized in the case of the wire obstruction by means of the diagram in Fig. G,5c. On the left is shown schematically the closing up of flow lines downstream from the wire. On the right, curves of burning velocities and gas velocities across the stream at three planes A, B, and C are shown. At A the burning velocity is zero over a distance from the wire corresponding to the quenching distance; the gas velocity is zero at the wire surface and increases along a gradient so that it remains always larger than the burning velocity. On moving downstream from plane A both the burning velocity and the gas velocity increase in the disturbed region; the burning velocity in this region increases more rapidly than the gas velocity. At plane B the two profiles meet each other tangentially, and the wave is stabilized. If the gas flow is below the critical value for blowoff, then above the plane *B* the burning velocity profile would intersect the gas velocity profile so that the wave would travel back to plane *B* if it were displaced from it. If the gas flow is increased to the blowoff limit, then above the plane *B* the gas velocity is everywhere larger than the burning velocity as illustrated for plane *C*; therefore, the wave cannot return to plane *B*.

As long as the heat received by the wire is conducted away efficiently, the region of zero burning velocity from the wire surface outward, i.e. the quenching distance, is independent of the wire diameter. If the wire diameter is only a small fraction of twice the quenching distance, the gas and burning velocity profiles at blowoff will be virtually independent of wire diameter; so that the critical boundary velocity gradient for blowoff is independent of wire diameter. However, for wire or rod diameters that

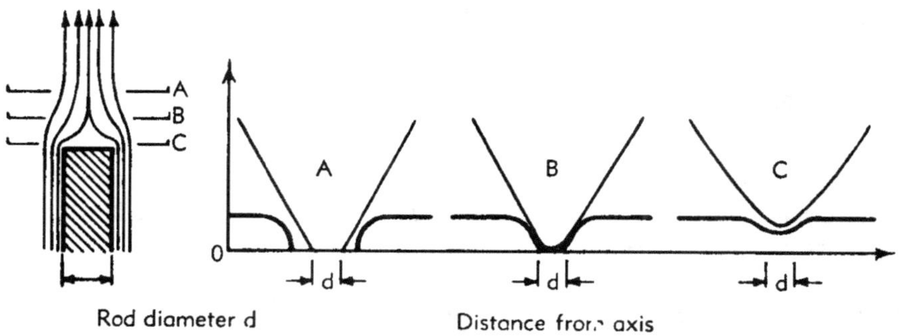

Fig. G,5c. Stabilization of combustion wave by a rod. Left, flow lines around the rod; right, burning velocity (heavy line) and gas velocity (light line) in planes *A*, *B*, and *C*.

are comparable to or larger than twice the quenching distance, the velocity profiles at blowoff change, as shown in Fig. G,5d. In the case of the larger rod the critical region is more effectively sheltered and the boundary velocity gradient for blowoff becomes larger as indicated by the steeper slope of the gas velocity profile.

A practical example of flame stabilization by an object in the stream is the Méker burner which utilizes a grid. This burner may be operated with fuel-lean mixtures at stream velocities exceeding the blowoff limit at its outer rim.

A discussion of the mechanism of flashback has also been given by Forsythe, Davies, and Townend [*38*; *3*, p. 99]. These authors have explained the phenomenon in a manner similar to that given above. In particular they recognize the visible dead space between the burner wall and the luminous reaction zone as the region of zero burning velocity. However, their theory does not include consideration of the region of decreased burning velocity in the fringe of the combustion wave. Instead their attention is focused upon the effect of various experimental parameters on the dead space and upon the development of correlations between change of dead space and the limiting gas flow for flashback. The significance of these correlations for an understanding of the general problem of flame stability or even the more limited problem of flashback is not immediately apparent. From what has been said above it is clear that the

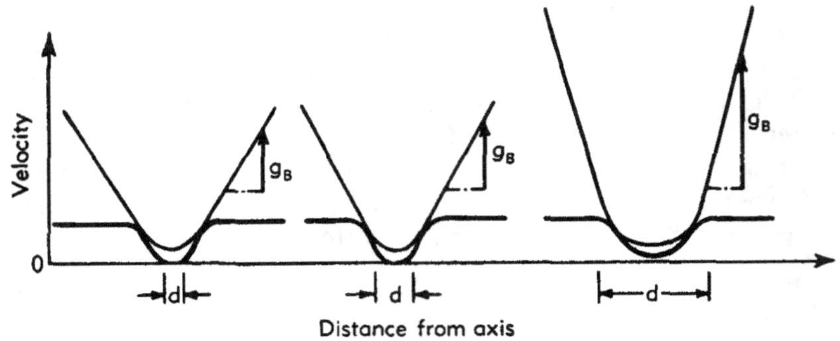

Fig. G,5d. Increase of critical boundary velocity gradient g_B at stabilizing rods with increasing rod diameter d.

dead space is only a part of the disturbed region around the fringe of the combustion wave and that the quantities most directly related to flame stability are the critical boundary velocity gradients.

G,6. Calculation of Wave Shape and Gas Flow Pattern.

Wave stabilized in cylindrical stream at axis or periphery. The problem to be treated concerns the shape of the ordinary Bunsen flame cone and the gas flow pattern. To simplify the analysis the following assumptions are made: (1) the combustion wave is represented by a single surface; (2) the burning velocity is constant over the whole wave surface; and (3) the flow lines retain their direction and velocity from the burner rim up to the wave surface. We shall consider a laminar gas stream in which the flow lines are parallel to the x coordinate. According to the Poiseuille equation, the gas velocity in any flow line is

$$U_u = n(R^2 - r^2), \tag{6-1}$$

where R is the distance from the center to the boundary of the stream and r is the distance of any flow line from the stream center. The constant n is equal to $-a/4\mu$, a being the constant hydrodynamic pressure gradient $\Delta p/\Delta x$ along the tube and μ the viscosity of the gas. The equation is equally applicable to the flow in a cylindrical tube and between two plane-parallel plates. The angle α between the velocity vectors U_u and S_u is given by the equation

$$\frac{dx}{dr} = \tan \alpha, \tag{6-2}$$

as shown in Fig. G,6a. Since

$$\tan \alpha = \frac{(U_u^2 - S_u^2)^{\frac{1}{2}}}{S_u} \tag{6-3}$$

it follows that

$$\frac{dx}{dr} = \pm \frac{(U_u^2 - S_u^2)^{\frac{1}{2}}}{S_u} \tag{6-4}$$

where U_u, as a function of r, is given by Eq. 6-1 and S_u is constant by

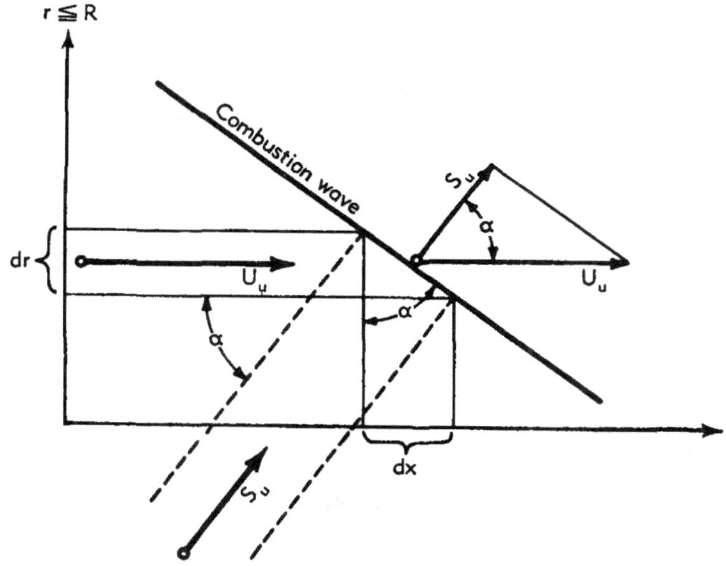

Fig. G,6a. Illustrating the relation $dx/dr = \tan \alpha = (U_u^2 - S_u^2)^{\frac{1}{2}}S_u$.

assumption. Integration [13] of Eq. 6-4 yields

$$x_0 - x = \pm \frac{\dot{m}}{3}\left[\sin \varphi \cos \varphi \Delta\varphi + \left(1 - \frac{1}{k^2}\right) F(k, \varphi) + \left(1 + \frac{1}{k^2}\right) E(k, \varphi)\right] \tag{6-5}$$

where $\dot{m} = \left(\frac{nR^2}{S_u - 1}\right)\left(\frac{nR^2 + S_u}{n}\right)^{\frac{1}{2}}$

$$k^2 = \frac{nR^2 - S_u}{nR^2 + S_u}$$

$$\sin^2 \varphi = \frac{nr^2}{nR^2 - S_u}$$

$$\Delta\varphi = (1 - k^2 \sin^2 \varphi)^{\frac{1}{2}}$$

and $F(k, \varphi)$ and $E(k, \varphi)$ are elliptical integrals of the first and second kind, $\int_0^\varphi d\varphi/\Delta\varphi$ and $\int_0^\varphi \Delta\varphi d\varphi$, respectively.

Eq. 6-5 yields the outline of the combustion wave in the gas stream. It corresponds to an approximately conical shape of the height $x_0 - x$. Depending upon the $\pm$ sign the cone points either downstream or upstream.

Additional considerations show that the cone is always on the downstream side of the object to which it is attached [*13*], and therefore a cone attached at the base will be upright in the stream and a cone attached at

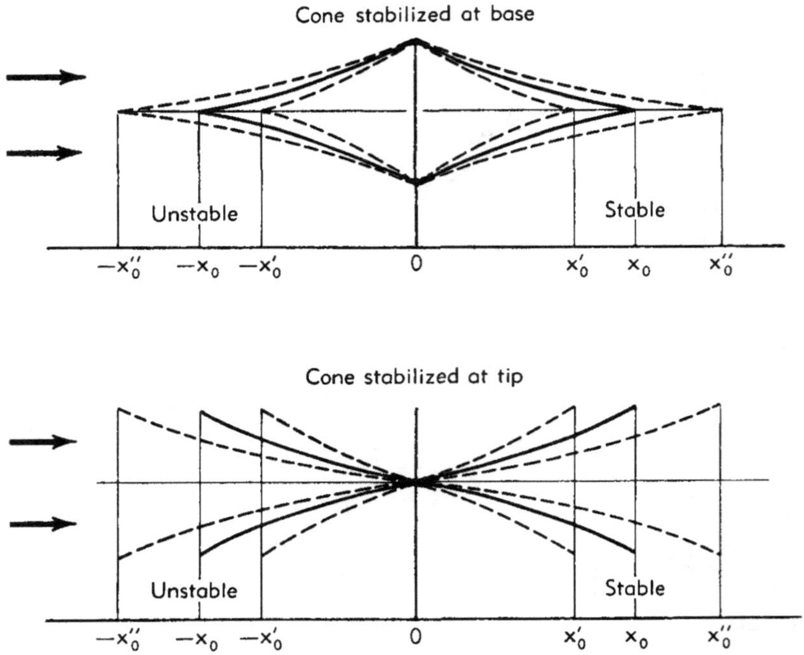

Fig. G,6b. Orientation of a conical combustion wave in the gas stream.

the tip will be inverted in the stream. This is illustrated in Fig. G,6b, which shows the two possible cones corresponding to Eq. 6-5. Suppose that the cone is stabilized at the base and that by some fortuitous fluctuation the tip of the downstream cone at the right is shifted from x_0 to x_0'. The angle of the gas flow against the normal of the wave surface then becomes everywhere smaller than α and the normal component of the gas velocity exceeds the burning velocity. The cone is thus driven back into its original position. When the cone tip is shifted to x_0'', the burning velocity exceeds the normal component of the gas velocity and the cone shifts back to its original position. The reverse situation applies to the upstream cone at the left. When the cone tip is displaced from $-x_0$ to $-x_0'$ or $-x_0''$, the combustion wave cannot regain its original position because in the former case the burning velocity is smaller and in the latter case it is larger than the normal component of the gas velocity. The upstream cone is therefore unstable. The same reasoning applies to stabilization at the tip of the cone as illustrated in the lower part of Fig. G,6b.

Another part of the analysis concerns the velocity and direction of the gas flow through the combustion wave. Values of U_u and α at any distance r from the axis are obtained from Eq. 6-1 and 6-3. If the density ratio ρ_u/ρ_b is known, the angle β between the flow of burned gas and the normal to the wave plane can be calculated from Eq. 1-14, and the velocity U_b behind the wave from Eq. 1-13.

The assumptions (1, 2, and 3) above introduce some oversimplifications. The calculated cone outline ends in a sharp tip. Such discontinuity of slope is evidently not physically possible. Actually, the burning velocity increases and a well-rounded tip is formed as discussed in Art. 4. At the base of the cone the burning velocity progressively decreases to zero and the actual wave profile becomes very different from the calculated shape. The calculation cannot be extended beyond a distance from the axis at which $U_u = S_u$ because beyond this distance Eq. 6-3 becomes imaginary. This is no physical problem because in a real stabilized flame the burning velocity vanishes more rapidly than the gas velocity toward the rim.

Development of combustion wave from a point ignition source at axis of laminar cylindrical stream. The preceding subarticles have dealt with the stabilization and structure of flames in streams of explosive gas. It is of interest to examine the various stages of flame development in a moving stream from the instant of application of an ignition source to the establishment of a stable flame.

The theoretical aspects of the problem have been treated by Dery [*3*, p. 235]. The ignition source is considered to be at a point on the axis of a stream in Poiseuille flow. The flame origin thus moves with the gas stream at the maximum velocity U_{max} which according to Eq. 6-1 equals nR^2. In order to simplify the problem we shall assume that no thermal expansion occurs so that the flow lines remain undisturbed. At any distance r from the axis the gas velocity relative to the origin is $U - U_{max}$, which from Eq. 6-1 is found to equal $-U_{max}r^2/R^2$. The ratio r/R is denoted by ξ. A coordinate system is adopted whose zero point moves with the origin along the stream axis at the velocity U_{max}. A point on the surface of the combustion wave moves with respect to this coordinate system at a velocity represented by the vector $\mathbf{u}$ which is equal to the vector sum $S_u\mathbf{n} - \xi^2\mathbf{U}_{max}$, where S_u is the magnitude of the burning velocity and $\mathbf{n}$ is the normal directed toward the unburned gas. One axis of the coordinate system is r, and the other coincides with the stream axis. We replace these coordinates by the reduced coordinates ξ and η, the latter being defined as the ratio of the distance from the origin along the stream axis to the stream radius R. We may then write the equation in the form

$$\frac{\mathbf{u}}{U_{\max}} = \left(\frac{S_u}{U_{\max}}\right)\mathbf{n} - \xi^2\mathbf{j} \tag{6-6}$$

where $\mathbf{j}$ is the unit vector, so that the configuration of the flame, in reduced coordinates, at any instant depends only on the initial conditions and the value of $S_u/U_{\max}$.

For the initial condition it is assumed that ignition occurs at the surface of a sphere whose radius in reduced coordinates is $(\xi_0^2 + \eta_0^2)^{\frac{1}{2}}$. The progress of the surface with time is represented by a series of lines, each representing the position of the combustion wave at some "reduced" time $\tau = U_{\max}t/R$. These lines may in principle be found by a simple graphical method in which Eq. 6-1 is approximated by

$$\Delta(\xi, \eta) = \left(\frac{S_u}{U_{\max}}\right)\Delta\tau\mathbf{n} - \xi^2\Delta\tau\mathbf{j} \tag{6-7}$$

The indicated vectorial addition is performed, and the next isochronous line is drawn through the terminal points of the vectors. This approximation can be further improved upon by a method given by Dery [*3*, p. 235]. The results of a graphical integration are shown in Fig. G,6c for the case $(\xi^2 + \eta^2)^{\frac{1}{2}} = 0.025$ and $S_u/U_{\max} = 0.25$. The wave surface is seen to develop gradually from a sphere to an oblong shape which is convex to the unburned gas facing downstream and becomes increasingly concave to the unburned gas facing upstream. The other family of curves gives the direction of the flame corresponding to the local gas velocity and burning velocity. The development is, of course, caused by the reduced gas velocity in the outer flow lines, and since in the present idealized case the burning velocity remains finite at the boundary while the gas velocity decreases to zero, the concave surface ultimately extends to the burner rim and develops into a stable flame cone or flashes back through the tube, depending upon the flow. The downstream surface grows indefinitely and eventually vanishes at a distance from the rim where the stream is broken up by eddies and dissolves into the surrounding atmosphere. Since in an actual laminar jet the burning velocity does not remain finite at the boundary but decreases to zero, it is also possible that the combustion wave never penetrates into a region where the burning velocity exceeds the gas velocity, in which case the flame blows off. This will be the case, even though the flow is below the blowoff limit, when the ignition source is located beyond some critical distance from the rim because the dilution of the stream boundary progressively increases with distance.

The general correspondence of theory and experiment is illustrated by a schlieren snapshot shown in Plate G,6. In taking the photograph the camera shutter remained open while the following series of events took

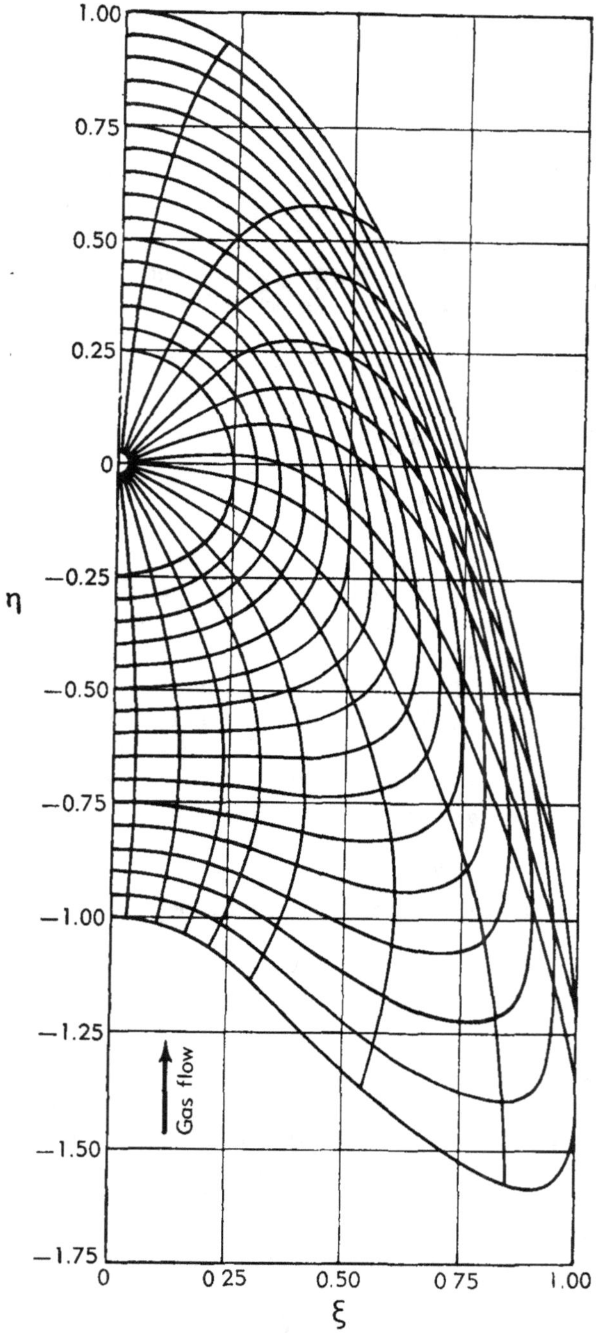

Fig. G,6c. Theoretical construction of a combustion wave with null lateral expansion of stream tubes in a gas in Poiseuille flow. Initiation is assumed on a spherical surface the ratio of whose radius to that of the gas stream is 0.025. The ratio of the burning velocity to the gas velocity is 0.25. A value of burning velocity of 50 cm/sec and a burner radius of 4 mm corresponds to an interval between isochronous lines of 4 milliseconds (Dery).

place: (1) passage of an electric spark across a jet of a somewhat rich mixture of natural gas and air; (2) a few milliseconds later, a strong light flash of very short duration, producing the schlieren image; and (3) development of a steady luminous flame cone, somewhat distorted by the spark electrodes dipping into the stream, which registered on the photographic film because of its comparatively long exposure time. The heart-shaped schlieren outline, representing the boundary between the inner burned and the outer unburned gas, is readily recognizable and is very similar in appearance to the theoretical outline; the concave surface facing upstream is not so regular as in theory, but this can hardly be otherwise considering the irregularities of spark ignition and the disturbance of the flow by the electrodes.

G,7. Momentum Change and Thrust Pressure. The momentum change induced by thermal expansion in the wave generally modifies the gas flow pattern very considerably. The pressure in the unburned gas exceeds somewhat the pressure in the burned gas. For the case of a plane steady state wave the calculation of this pressure difference, $p_u - p_b$, can be performed very simply. In this case only the normal component S of the mass velocity changes. Taking the normal direction as the direction of the x coordinate, the change of momentum of the gas stream per unit length of x becomes $\rho S(dS/dx)$. Since viscosity effects are negligible, this is equal to the pressure gradient $-dp/dx$. One thus obtains the equation $-dp = \rho S dS = \rho_u S_u^0 dS$, which by integration over the wave width yields

$$p_u - p_b = \rho_u S_u^0(S_b^0 - S_u^0) \tag{7-1}$$

Since $\rho_u S_u^0 = \rho_b S_b^0$, Eq. 7-1 may also be written

$$p_u - p_b = \rho_u (S_u^0)^2 \left(\frac{\rho_u}{\rho_b} - 1\right) \tag{7-2}$$

As mentioned earlier, the pressure difference $p_u - p_b$ across a combustion wave is very small compared to the total pressure. For example, in methane-air mixtures at atmospheric pressure $p_u - p_b$ is of the order of 0.01 mm Hg; and in mixtures of acetylene and oxygen it is of the order of 1 mm Hg. Such pressures have no significant effect on the inner wave structure, but, as will be seen later, they may give rise to gas motions which affect the flow pattern.

In an ordinary Bunsen burner flame the unburned gas is confined by the burner tube and by the combustion wave which rests on the tube rim in the shape of a cone. Except for regions at the base and the tip of the cone the wave may be considered plane, and the pressure p_u is therefore approximately constant in the space bounded by the wave and the plane of the tube rim. From the tube rim upstream the pressure increases due to the friction of the gas at the tube wall, which is calculable from

Poiseuille's equation. The burned gas flows into the open atmosphere and the pressure p_b is therefore equal to the ambient pressure p_0. In a spherical flame expanding against the atmosphere the unburned gas extends from the wave front to the open atmosphere, and the pressure p_u decreases from a maximum at the wave front to p_0 at infinite distance from the wave. Assuming the gas to behave as an incompressible fluid, that is, for burning velocities that are small compared to sound velocity, Silsbee [*39*] has developed an equation for the pressure p_u as a function of the distance r from the flame origin, viz.

$$p_u = p_0 + \rho_u S_b^2 \left(1 - \frac{\rho_b}{\rho_u}\right)\left[\left(\frac{2r_b}{r}\right) - \left(1 - \frac{\rho_b}{\rho_u}\right)\frac{r_b^4}{2r^4}\right] \qquad (7\text{-}3)$$

where $r_b \leqq r$ denotes the flame radius. The pressure p_b which is constant inside the sphere is obtained by combining Eq. 1-1, 7-2, and 7-3 for $r = r_b$, viz.

$$p_b = p_0 + \rho_u S_b^2 \left(1 - \frac{\rho_b}{\rho_u}\right)\left(1.5 - 0.5\frac{\rho_b}{\rho_u}\right) \qquad (7\text{-}4)$$

G,8. Propagation in Channels.

Diameter large compared to quenching distance. When the wave propagates in a tube, the flow induced by thermal expansion is restricted by the tube wall and is subject to viscous drag at the wall. The ensuing modification of flow pattern and shape and area of the combustion wave is a problem of considerable importance but has not as yet been explored sufficiently to permit more than a qualitative description and limited theoretical discussion of the phenomena.

Comparatively simple conditions exist in the case that a tube containing explosive gas is closed at one end and the gas is ignited at the open end. During the propagation of the combustion wave, the column of unburned gas flows from the wave toward the open end. Owing to viscous drag the flow of burned gas is retarded at the wall and accelerated at the tube axis. It is therefore forced into a flow pattern which, at some distance from the wave, should not differ greatly from Poiseuille flow if the Reynolds number is sufficiently low. In order to attain this state, the stream tubes near the wall must expand and the stream tubes near the axis must contract; that is, the flow lines that originate in the combustion wave must incline toward the tube axis approximately as shown schematically in Fig. G,8a. If the wave surface were plane perpendicular, the angle α would be zero while the angle β would be finite. This is inconsistent with the continuity relation as formulated in Eq. 1-14. Therefore, the wave must incline in such a way that an angle $\alpha > \beta$ is formed. This leads to a wave surface which is convex toward the closed end. Additional considerations lead to the conclusion that the apex of the wave in the stream axis must be blunt rather than sharp-tipped. If a sharp tip existed, the combustion wave would be inclined to the flow lines near the axis at a sharp angle, which means that the velocity of the unburned gas with respect to the wave would be much larger than the burning velocity. Since the axial flow line would enter the tip at a right angle, the burning velocity at the tip would be much larger than elsewhere. Although this is possible in an ordinary Bunsen cone on a burner where the wave surface is concave toward the unburned gas, it is not possible in the present instance where the wave surface is convex toward the unburned gas. The wave, therefore, must conform to a shape similar to that shown in Fig. G,8a. The formation of curved waves of this type is readily confirmed by casual visual observation of luminous combustion waves in tubes and has been studied photographically (see for e.g. [*40*]).

It is necessary to distinguish the curvature of the wave that is caused by the above-described mechanism from curvature caused by the decrease of burning velocity near the wall. The latter effect extends only over the depth of penetration of quenching, and we may minimize this contribution

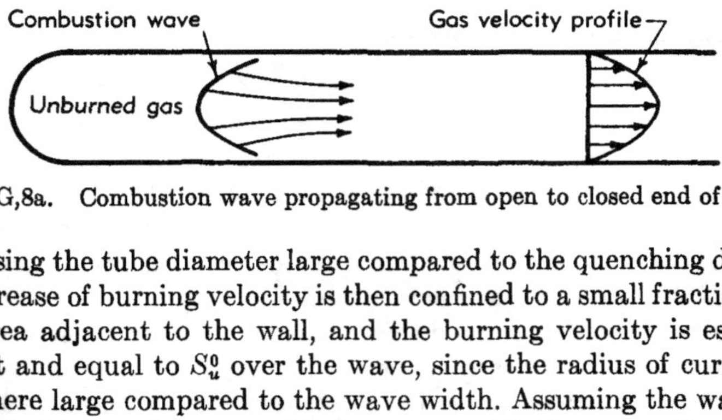

Fig. G,8a. Combustion wave propagating from open to closed end of tube.

by choosing the tube diameter large compared to the quenching diameter. The decrease of burning velocity is then confined to a small fraction of the wave area adjacent to the wall, and the burning velocity is essentially constant and equal to S_u^0 over the wave, since the radius of curvature is everywhere large compared to the wave width. Assuming the wave to be at rest, that is, moving a coordinate system with the wave, the tube and the enclosed unburned gas move against the wave with a velocity that may be denoted by $\bar{U}$, and the mass flow through the wave is $\rho_u \bar{U} \times$ tube cross section. Since the mass flow is approximately equal to $\rho_u S_u^0 \times$ wave area, and the wave area is larger than the tube cross section, it follows that $\bar{U} > S_u^0$. Thus the wave propagates with respect to a fixed point on the tube at a velocity exceeding the burning velocity. Since the normal component of the velocity of the unburned gas with respect to the wave surface is S_u^0, it follows that at the apex of the wave where the wave surface is normal to the gas flow, the flow of unburned gas with respect to the wave is slowed down from U_u to S_u^0. This means that in the region where the wave is normal to the tube axis the unburned gas is driven toward the closed end of the tube; but since the density of the column of gas between the wave and the closed end remains virtually constant, there must be a compensating mass flow in the opposite direction in other regions of the wave. In other words, the unburned gas near the wave surface is in circulatory motion somewhat as shown in Fig. G,8b(1). In this portion of the figure the tube is considered to be at rest, and a particle of unburned gas is displaced on approach of the wave along a path indicated by arrows until overtaken by the wave. If instead, we consider the wave to be at rest, the flow pattern may be visualized as shown in Fig. G,8b(2). Here the gas motion appears only as retarded and accelerated flow. The stream tubes near the axis increase in width on approaching the

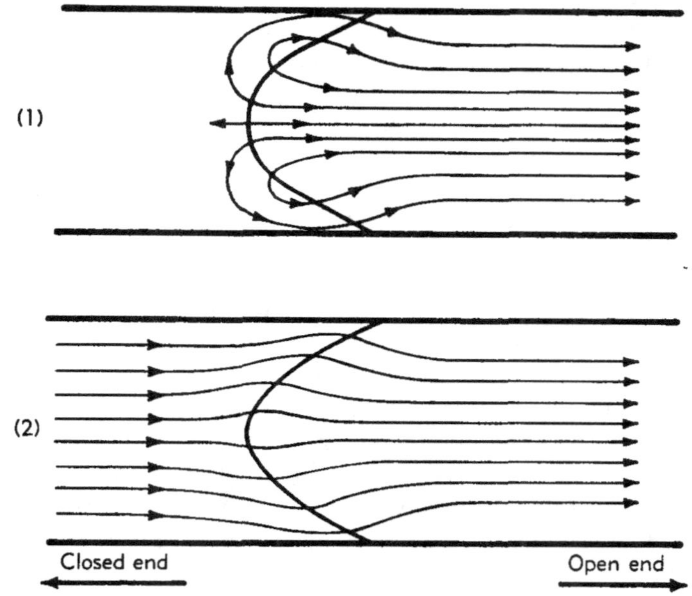

Fig. G,8b. Gas flow pattern for combustion wave propagation from open to closed end of tube. (1) Tube at rest. (2) Wave at rest.

combustion wave, corresponding to a decrease of the velocity of the unburned gas, and behind the wave they contract to smaller diameters. Stream tubes farther away from the axis contract before entering the wave, corresponding to increased velocity, and expand behind the wave to larger diameters.

From hydrodynamics (Bernoulli's principle) the loss of momentum in the axis stream tube is compensated by an increase of pressure. It is clear that this increase of pressure and corresponding loss of momentum is caused by the mass acceleration that occurs in the axial stream tube behind the wave. It is therefore this mass acceleration which causes the flow across the tube ahead of the wave. The latter flow causes movement of unburned gas at the wall against the combustion wave. The wall exerts a viscous drag on this flow and therefore acts as a brake on unlimited acceleration of mass flow across the tube. It may be proposed that this drag is responsible for establishment of the steady state. Otherwise it is difficult to see why the wave should not elongate indefinitely and the velocity of the wave increase without limit. Indeed, it is known that in sufficiently large tubes or ducts a steady state is not obtained; the flame becomes increasingly turbulent and very high speeds result.

Diameter comparable to quenching distance. The conditions at the fringe of a combustion wave near a solid surface have already been discussed in Art. 3. Here the front of the reaction zone (T_1 surface) becomes parallel to the gas flow, while at increasing distance from the solid surface the angle α of the flow lines with the normal to the T_1 surface decreases; the burning velocity correspondingly increases from zero to S_u^0. The quenching effect of the wall thus extends along the wave area to some fairly well-defined distance. When the tube is sufficiently narrow so that the depth of penetration of quenching extends to the tube axis, i.e when the tube diameter approaches the quenching limit, the burning velocity increases over the wave area from a value of zero at the fringe to a maximum value $(S_u)_{max} \leqq S_u^0$ in the center. The flow of burned gas is subject to viscous drag as before and would tend to a Poiseuille distribution were it not for the marked temperature gradient from the wall to the center and the corresponding decrease of density. The outer stream tubes are now of much smaller cross section and the inner stream tubes of much larger cross section than corresponds to isothermal Poiseuille flow. Consequently, it may be suggested that there is not a marked change of cross section in the stream tubes on passing through the wave, that is, the flow lines remain substantially parallel to the tube wall. Since there exists no appreciable crosswise flow, the wave travels along the tube at the velocity $\bar{U} = (S_u)_{max}$. The wave surface intersects the tube axis at right angles and elsewhere the normal to the surface is inclined against the gas flow at angles α corresponding to the equation $\cos \alpha = S_u/(S_u)_{max}$. We may visualize the flow pattern and the characteristic surfaces of the wave essentially as shown in Fig. G,8c. The profile of the reaction zone is shown as a stippled area, and its crescent shape corresponds to the fact that toward the wall the rate of the chemical reaction decreases. Thus, the closer a flow line is to the solid surface, the longer will be the time required for complete chemical reaction, and the longer becomes the path of a gas element traveling through the reaction zone. These paths are indicated by horizontal lines in the shaded area. The temperature distribution is represented by a series of isotherms. The solid surface and the gas at some distance in front of the wave are at the temperature T_u. For reasons that have already been discussed in Art. 3, the temperatures T_1 along the front of the reaction zone are thought to increase with decreasing distance from the wall. To the burning velocity $(S_u)_{max} \leqq S_u^0$ in the center, there corresponds the inflection point temperature $(T_1)_{min} \geqq T_1^0$

and the final temperature $(T_b)_{max} \leqq T_b^0$. From the center toward the fringe, T_1 increases and T_b decreases, and at the fringe T_b and T_1 become equal. The fringe temperature is denoted by T_0. The dotted lines indicate the paths of steepest temperature drop and thus represent lines of heat flow. It is noted that much of the heat received by the unburned gas before it enters the reaction zone, and in particular by the gas that flows in the "dead" space y_0, is acquired from the completely burned gas behind the wave rather than from the chemically reacting gas in the wave. These

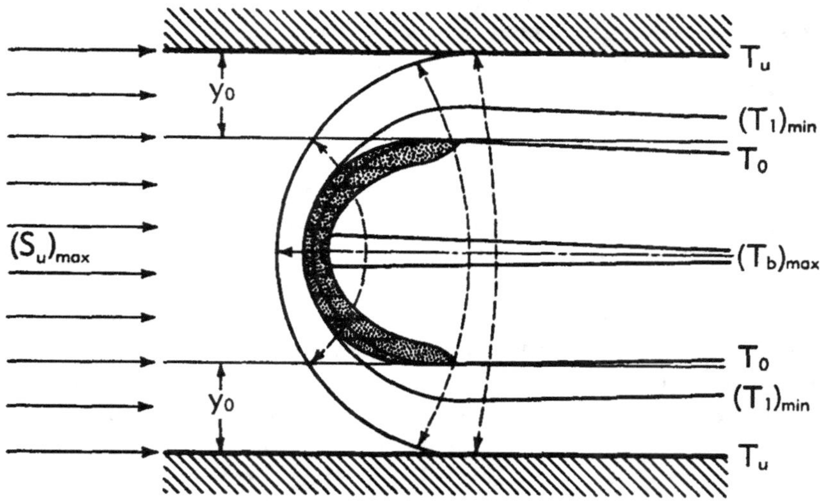

Fig. G,8c. Scheme of combustion wave between cooling surfaces.

considerations apply equally to a cylindrical tube and to a channel formed by two plane-parallel plates and are the basis of the theory of flame quenching by solid surfaces that is described further in Art. 15.

G,9. Thermal Model of Combustion Wave. Excess Enthalpy. We continue with the development of a model of the combustion wave which ultimately should permit quantitative comparisons with experimental data. For this purpose we shall further examine Eq. 1-4 which expresses the principle of conservation of energy for the plane adiabatic combustion wave. The equation may be rewritten in the form

$$\frac{\lambda(dT/dx) - \sum_i n_i V_i H_i}{\dot{m}} = \frac{\sum_i n_i H_i}{\rho} - \left(\frac{\sum_i n_i H_i}{\rho}\right)_b \tag{9-1}$$

which is obtained from the relation $S = \dot{m}/\rho$ and transposing. The term $\sum_i n_i H_i/\rho$ is the enthalpy per unit mass of the gas mixture, and the right-hand side of Eq. 9-1 denotes the difference of enthalpy per unit mass between some plane x inside the wave and a plane b where combustion is complete. If the diffusion terms $\sum_i n_i V_i H_i$ are smaller on the average than

$\lambda(dT/dx)$ over the wave width, the enthalpy per unit mass of the gas within the wave is larger than at the boundary b and also at the boundary u, since for adiabatic combustion these are all equal. In this case, integration of Eq. 9-1 over the wave width yields the excess enthalpy h per unit wave area which constitutes a wave of thermal energy traveling with the combustion wave through the explosive gas. In other words, the unburned gas in the preheat zone attains a temperature exceeding T_u without much change of chemical composition, so that the enthalpy increases. If the diffusion terms $\sum_i n_i V_i H_i$ are equal to or larger than $\lambda\ (dT/dx)$, an enthalpy excess would not exist. This would signify that the diffusion process so dilutes the unburned gas with combustion products, whose chemical energy is spent, that the enthalpy may be equal to or lower than the enthalpy of the fresh gas even though the temperature has risen due to heat conduction.

Proof or disproof of the existence of an energy wave associated with the combustion wave on purely theoretical grounds does not appear possible at present. If the excess enthalpy is real, it must be associated with the ignition process in the sense that it is identified with the minimum energy requirement of an ignition source. The authors are of the opinion that the excess enthalpy is never zero and that it constitutes in fact the energy barrier which prevents ignition of explosive gases by infinitesimal sources. Supporting evidence will be presented later which shows that minimum ignition energies are largely calculable from the assumption that the diffusion terms $\sum_i n_i V_i H_i$ in Eq. 9-1 are zero. Correctly, perhaps, these diffusion terms are not very much smaller than, but of comparable magnitude to, $\lambda(dT/dx)$. This is suggested by a discussion of Jost [*2*, pp. 113–115]. This author writes the approximate equation, derived from elementary kinetic theory,

$$\lambda \cong D\bar{c}$$

where D is the diffusion coefficient and $\bar{c} = C_p n$, where C_p is the molar specific heat and $\bar{c}$ the heat capacity per unit volume. In further gross approximation, the temperature and the concentration of the reactants are assumed to change linearly over the wave width Δx, so that the heat flow from the burned to the unburned gas becomes $\lambda(T_b - T_u)/\Delta x = D\bar{c}(T_b - T_u)/\Delta x$, and the diffusion flow of moles of reactants in the opposite direction becomes $Dn_0/\Delta x$, n_0 being the original molar concentrations. The corresponding enthalpy flux is obtained by multiplying the flow of moles of reactants by the heat of reaction; the latter is, per mole, equal to $C_p(T_b - T_u)$, so that one obtains, roughly, $D\bar{c}(T_b - T_u)/\Delta x$; that is, the heat flux of thermal conduction and the enthalpy flux of diffusion appear to be, in the first approximation, of equal magnitude. As Jost himself emphasizes, the conclusion is not compelling in view of the gross approximations introduced. Should the thermal conduction term exceed the diffusion terms by a factor of 2 or more, then Jost's estimate could still be sufficiently valid, whereas the assumption of zero diffusion terms would yield the correct order of magnitude of the excess enthalpy.

In neglecting the enthalpy transport by diffusion we are introducing an approximation which we believe is not seriously in error for problems involving calculation of the rate of heat transport across the wave. Thus, the temperature profile in the wave, calculated with this approximation, would not be excessively in error. On the other hand, concerning the problem of calculating burning velocities, there is good reason to believe that the determining process is not so much the heat transport but the diffusion of chain carriers, atoms, and free radicals into the unburned gas. In order to obtain manageable equations for estimating burning velocities it seems, on the contrary, admissible to take license with the temperature distribution in the wave and to approximate the equations in such a way that the diffusion terms remain preserved as closely as possible. Such procedures have had some success [*24*; *31*; *32*; *3*, pp. 140–146].

The thermal model that we propose to use for calculations of heat transfer problems is of no practical value in the treatment of the adiabatic steady state combustion wave, but it may be used to investigate nonsteady states produced by heat sources or fluctuations of the temperature distribution and the disturbance of the steady state by heat sinks. It is, therefore, of value for visualizing the mechanism of ignition, combustion wave stabilization, and quenching, and for deriving approximate quantitative relations between burning velocity, ignition energy, and quenching distance, which can be tested experimentally.

G,10. Principles of Ignition.

Transition from nonsteady to steady state for one-dimensional propagation. The steady state corresponds to a unique distribution of temperature and concentrations in the combustion wave. If small fluctuations occur that disturb the wave structure, some mechanism must operate by which the wave returns to the steady state, or else the wave would not be truly stable—it would continuously accelerate, or decay. Likewise, when a combustion wave is formed by an ignition source, the initial gradients of temperature and concentrations at the source are generally very different from those in the steady state wave, so that the steady state must develop from a series of transient, or nonsteady, states. The two problems, wave stability and ignition, thus appear basically identical. In this article, we shall attempt to describe qualitatively the mechanism by which the steady state develops from nonsteady states, and specify the critical conditions that must be satisfied if a stable combustion wave is to be formed. We may then deduce criteria for the effectiveness of an ignition source and also for the effect of dilution of an explosive medium by inert constituents. In this way we obtain a qualitative theory of ignition and limits of inflammability. Development of approximate quantitative relations which, however, do not include treatment of the limits of inflammability, is reserved for the next article.

The principle of energy wave associated with the combustion wave has been discussed above. We replace Eq. 9-1 by

$$\lambda \frac{dT}{dx} = \frac{\dot{m}}{\rho} \frac{dh}{dx} \tag{10-1}$$

where dh/dx denotes the enthalpy excess per unit volume in plane x. The superscript ^{0} is used, as before, to denote an adiabatic, plane steady state

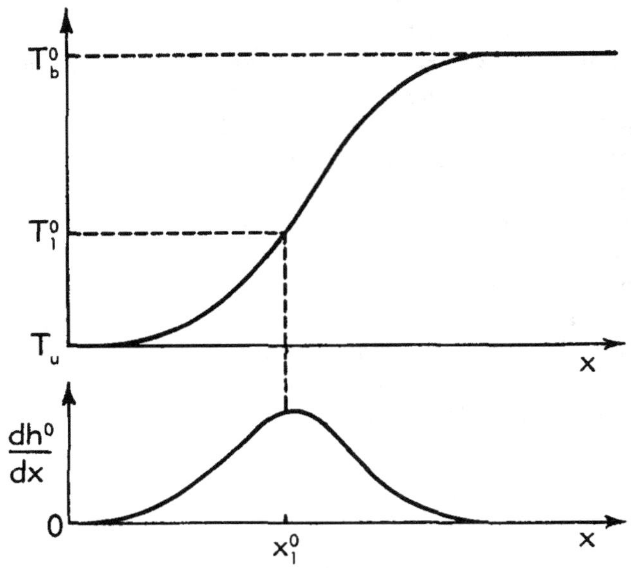

Fig. G,10a. Temperature and excess enthalpy in adiabatic steady state combustion wave.

combustion wave. The curve of dh^0/dx versus x is shown schematically in Fig. G,10a, in juxtaposition with the temperature curve. The curve has a maximum at the maximum of the temperature gradient in the plane x_1^0 of the inflection point on the temperature curve. The area h^0 below the curve represents the excess enthalpy over the whole wave width per unit area of wave surface.

Other equations describing the steady state in the thermal model are readily derived from the general Eq. 1-2 and 1-4. By differentiating Eq. 1-4 and substituting Eq. 1-2 one obtains

$$\lambda \frac{d^2T}{dx^2} = \sum_i H_i K_i + \sum_i n_i(S + V_i) \frac{dH_i}{dx} \tag{10-2}$$

The first term on the right-hand side represents the rate of change of enthalpy per unit volume, due to chemical reaction. Because components of higher enthalpy disappear in the process, the term is negative. Its positive value will be denoted by the symbol q. The derivative dH_i/dx may be written as $\mathfrak{M}_i(c_p)_i dT/dx$ so that the second term becomes the product of dT/dx and $\sum_i \mathfrak{M}_i n_i(S + V_i)(c_p)_i$. We introduce the average specific heat $c_p = \sum_i \mathfrak{M}_i n_i(S + V_i)(c_p)_i/\dot{m}$, which is not likely to be very different from the ordinary specific heat of the gas mixture, and rewrite Eq. 10-2 in the form

$$\lambda \frac{d^2T}{dx^2} + q = \dot{m}c_p \frac{dT}{dx} \tag{10-3}$$

In accordance with previously used symbols for the adiabatic plane combustion wave, the superscript ^{0} is affixed to $\dot{m}$, q, and dT/dx. According to the model, q^0 is very small between planes u and x_1^0, and as an approximation it appears permissible to neglect this term between these limits; furthermore, in approximate calculations c_p may be considered constant, so that between the limits u and x_1^0, Eq. 10-3 integrates to

$$\left(\lambda \frac{dT}{dx}\right)_1^0 = \dot{m}^0 c_p(T_1^0 - T_u) \tag{10-4}$$

where the subscript 1 refers to the plane x_1^0. Since $(dT/dx)^0$ vanishes at the limits u and b, integration of Eq. 10-3 between these limits yields

$$\int_{x^0}^{b} q^0 dx = \dot{m}^0 c_p(T_b^0 - T_u) \tag{10-5}$$

and by eliminating $\dot{m}^0 c_p$ from Eq. 10-4 and 10-5 one obtains

$$\left(\lambda \frac{dT}{dx}\right)_1^0 = \frac{T_1^0 - T_u}{T_b^0 - T_u} \int_{x_1^0}^{b} q^0 dx \tag{10-6}$$

We denote the left-hand side of Eq. 10-6 by u_1^0 and the right-hand side by W^0 so that

$$u_1^0 = W^0 \tag{10-7}$$

This equation expresses the steady state condition in simple terms that are useful for further considerations. The term u_1^0 denotes the total heat received per unit time and unit area of the wave by the unburned gas in front of the plane x_1^0; the term W^0 denotes that part of the heat produced in the reaction zone per unit time and unit area which must be abstracted from the zone in order to prevent the temperature from rising above T_b^0.

When an explosive medium is heated locally, nonsteady states result. For the present considerations we assume that by some means heat is imparted to a plane layer of the medium so that two plane temperature waves, which may develop into combustion waves, propagate symmetrically in each direction from the central plane of origin. The heat input per unit area of the plane of origin is $2h$ so that h represents the initial excess enthalpy of each wave. For such waves an equation analogous to Eq. 10-7 can be obtained from Eq. 1-8 and 1-9. These equations are adapted to the present one-dimensional problem by substituting $\partial/\partial x$ for the differential operator ∇. We consider that under the conditions of interest the pressure is substantially constant throughout the wave and that viscosity effects are negligible; thus, the equation of motion can be ignored, and the right-hand side of Eq. 1-9 becomes zero. Concerning the continuity equation (1-8), we remember the earlier discussion of its steady state counterpart, Eq. 1-2. Since, as before, $\sum_i \mathfrak{M}_i K_i = 0$, we may write

$$\sum_i \frac{\partial \mathfrak{M}_i n_i}{\partial t} + \sum_i \frac{\partial[\mathfrak{M}_i n_i (U + V_i)]}{\partial x} = 0 \tag{10-8}$$

Since $\sum_i \mathfrak{M}_i n_i (U + V_i)$ is the mass flow $\dot{m}$ and $\sum_i \mathfrak{M}_i n_i$ is the density ρ, we may write

$$\frac{\partial \rho}{\partial t} + \frac{\partial \dot{m}}{\partial x} = 0 \tag{10-9}$$

Since the pressure is substantially constant, the density ρ can change only by thermal expansion or contraction, or by a change of the number of molecules per unit volume due to chemical reaction. These two processes seem to have little bearing on the principles of combustion wave stabilization. We shall therefore simplify the model by considering $\partial \rho/\partial t$ to be substantially zero so that $\partial \dot{m}/\partial x = 0$. The mass flow is thus considered constant with respect to x although it may change with time. We carry out the indicated differential operations in Eq. 1-9 and obtain, after rearranging,

$$\lambda \frac{\partial^2 T}{\partial x^2} - \left\{ \sum_i E_i \frac{\partial n_i}{\partial t} + \sum_i H_i \frac{\partial[n_i(U + V_i)]}{\partial x} \right\} = \sum_i n_i (U + V_i) \frac{\partial H_i}{\partial x} + \sum_i n_i \frac{\partial E_i}{\partial t} \tag{10-10}$$

This is the nonsteady state analogue to the steady state equation (10-2). The expression in braces on the left-hand side is the rate of heat evolution due to chemical reaction and reduces to $\sum_i H_i K_i$ when the time derivative is zero. The expression in braces is, as before, denoted by q and we may rewrite Eq. 10-10, in analogy with Eq. 10-3, as

$$\lambda \frac{\partial^2 T}{\partial x^2} + q = \dot{m} c_p \frac{\partial T}{\partial x} + \rho c_v \frac{\partial T}{\partial t} \tag{10-11}$$

Here c_v denotes the specific heat at constant volume; this distinction arises from the first term in Eq. 1-9 which represents the change not of enthalpy H but of internal energy E with time, either due to temperature change or due to a change of the molecular concentrations.

The temperature distribution curve on either side of the plane of origin may be expected to be S-shaped analogous to the curve in Fig. G,10a, so that an inflection point x_1 exists which moves away from the origin as the wave progresses. We shall select the plane x_1 rather than the origin as the reference coordinate. The origin thus recedes from the fixed plane x_1 and the mass flow $\dot{m}$ has the same definition that it had in the steady state case. We assume, as before, that in front of the plane x_1 chemical heat evolution is negligible, and integrate each of the remaining three terms of Eq. 10-11 between the limit x_1 and a limit u in front of the wave beyond which all changes become negligible. This leads to

$$\left(\lambda \frac{\partial T}{\partial x}\right)_1 = \dot{m}c_p(T_1 - T_u) + \int_u^{x_1} \rho c_v \left(\frac{\partial T}{\partial t}\right) dx \tag{10-12}$$

Unlike the steady state case, the temperature in any plane x in front of the plane x_1 changes with time at the rate $\partial T/\partial t$. The area (T, x) under the curve of T versus x thus shrinks or expands at a rate represented by the integral expression on the right-hand side of Eq. 10-12. Since no chemical process operates in front of the plane x_1 by which thermal energy can either be consumed or produced, the T, x area can shrink only when energy is transported back into the reaction zone across the plane x_1. We may visualize this process as involving a decrease of the distance between the planes u and x_1; that is, the plane x_1 moves closer to u and "overruns" a part of the T, x area in its front; in this way the T, x area shrinks and simultaneously the temperature gradient between the planes u and x_1 becomes steeper. The net energy flow across the plane x_1 in the direction of the unburned gas is thus not $[\lambda(\partial T/\partial x)]_1$ but

$$u_1 = \left(\lambda \frac{\partial T}{\partial x}\right)_1 - \int_u^{x_1} \rho c_v \left(\frac{\partial T}{\partial t}\right) dx \tag{10-13}$$

This equation signifies that for positive values of the $\partial T/\partial t$ integral, viz. for rising temperatures, u_1 is smaller than $[\lambda(\partial T/\partial x)]_1$ because the T, x area shrinks. In other words, for positive values of the $\partial T/\partial t$ integral the distance between u and x_1 decreases, and the temperature gradient increases. The reverse argument is equally demonstrable: For negative values of the integral the distance between u and x_1 increases, and the temperature gradient decreases.

In order to investigate the change of temperature distribution with time over the entire wave width we must integrate every term in Eq. 10-11 between the limits u and b. The limit b might be taken as the plane of origin where the temperature gradient is zero. However, we prefer instead to place the limit b at a distance from x_1 beyond which chemical changes become negligible, viz. beyond which the function q becomes negligibly small. The boundary b is not sharply defined, but it is defined within narrow limits because the function q rapidly vanishes to zero. The boundary b thus chosen generally does not coincide with the plane of origin, and between b and the origin there may exist a layer of burned gas within which, under nonsteady state conditions, the temperature is not uniform. The temperature gradient $(\partial T/\partial x)_b$ is thus not generally zero, and some heat exchange takes place between the gas in the wave and the burned gas behind the wave. However, we are probably justified in considering that this heat exchange is of minor importance in the present problem of development of the steady state from the nonsteady state, and we shall therefore assume that the gradient $(\partial T/\partial x)_b$, though not zero, may be neglected. The gradient $(\partial T/\partial x)_u$ is zero, and q is zero below x_1 so that we obtain from Eq. 10-11

$$\int_{x_1}^{b} q dx = \dot{m}c_p(T_b - T_u) + \int_{u}^{b} \rho c_v \left(\frac{\partial T}{\partial t}\right) dx \tag{10-14}$$

For positive values of the $\partial T/\partial t$ integral in this equation, the q integral is seen to be larger than the mass flow term. This means that the molecules of reactants are being consumed more rapidly than they are supplied by mass flow across the boundary x_1, and therefore the total number of molecules of reactants between x_1 and b decreases with time. On a percentage basis the decrease is largest in a layer dx near the boundary b, and in the time dt the reactants become exhausted in a layer in front of the boundary b, so that a new boundary b is established somewhat closer to x_1. For negative values of the $\partial T/\partial t$ integral, the reverse argument is demonstrable. In conjunction with the previous argument for the boundaries u and x_1 we conclude that for positive values of the $\partial T/\partial t$ integral the wave width decreases, and the temperature gradient correspondingly increases everywhere; whereas for negative values of the integral the wave width increases, and the temperature gradient decreases everywhere.

Eliminating $\dot{m}c_p$ from Eq. 10-12 and 10-14, combining with 10-13 and introducing

$$W = \frac{T_1 - T_u}{T_b - T_u} \int_{x_1}^{b} q dx \tag{10-15}$$

we obtain

$$W - u_1 = \frac{T_1 - T_u}{T_b - T_u} \int_{u}^{b} \rho c_v \left(\frac{\partial T}{\partial t}\right) dx \tag{10-16}$$

When W is larger than u_1 the $\partial T/\partial t$ integral is positive, and vice versa. Eq. 10-16 therefore determines the direction which the development of the combustion wave is taking. When u_1 is larger than W, the wave expands and the gradients become flatter, and when W is larger than u_1 the wave contracts and the gradients become steeper. The steady state equation (10-7) is seen to represent a special case for which the wave width and the gradients do not change with time.

When W equals u_1 the $\partial T/\partial t$ integral is zero, but it does not follow that $\partial T/\partial t$ vanishes everywhere. The integral may be composed of positive and negative parts of equal magnitude and in time these parts may become unbalanced so that the wave either expands or contracts. In order to use Eq. 10-16 as a criterion for the steady state we must examine all possible states in which W equals u_1 and determine the singular state which represents the true equilibrium.

This state must satisfy Eq. 10-1 which on integration yields a value h^0 which represents the excess enthalpy of the steady state for unit area of the combustion wave. For the present we shall consider therefore only such states that have the excess enthalpy h^0 in common. We segregate these states into groups so that within each group the distance between the boundaries x_1 and b is constant. Each such group is further subdivided into smaller groups by specifying that the enthalpy excess between the planes x_1 and b, viz. the enthalpy excess in the reaction zone, shall be a certain percentage of h^0. Within any such subgroup an infinity of states is imaginable, each corresponding to some arbitrary distribution of the enthalpy excess, i.e. temperature distribution. We select a state for which the $\partial T/\partial t$ integral, that is, the difference $W - u_1$, is a minimum, and this state represents the closest possible approach to the steady state under the specified restrictions.

We consider first the case that the width of the reaction zone, or the distance between planes x_1 and b, is very small compared to its magnitude in the steady state. If an appreciable percentage of the enthalpy h^0 is imparted to the volume enclosed by the planes x_1 and b, the temperatures in the reaction zone become very large. Now, the q integral in the function W increases with temperature chiefly because q contains Arrhenius factors of the form $e^{-E/RT}$; however, these factors cannot increase beyond the finite value of 1. The value of the integral at small widths of the reaction zone is therefore chiefly determined by the difference between the integral limits, $b - x_1$, and at sufficiently small width is smaller than the steady state value. It remains dependent upon temperature, increasing when the percentage of h^0 in the reaction zone is increased and decreasing when that percentage is reduced. When the entire h^0 is imparted to the reaction zone, the temperature in front of the plane x_1 drops abruptly to T_u and the function u_1 is infinitely large. On distributing an increasing percentage of h^0 in front of the plane x_1 the value of u_1 decreases. W also decreases, and we may visualize the relation between the curves of W and u_1 versus the percentage of the excess enthalpy h^0 in the reaction zone somewhat as shown in Fig. G,10b(1). The curves do not intersect; that is, u_1 is always larger than W, and the wave must expand. As the wave expands,

u_1 decreases and W increases, because $b - x_1$ increases, so that for progressively increasing width of the zone the corresponding curves of W and u_1 approach each other until they meet tangentially as shown in Fig. G,10b(2). At the point of tangency W equals u_1. This point corresponds to a critical width and percentage of h^0 in the reaction zone. A fluctuation in the percentage, keeping the zone width constant, causes u_1

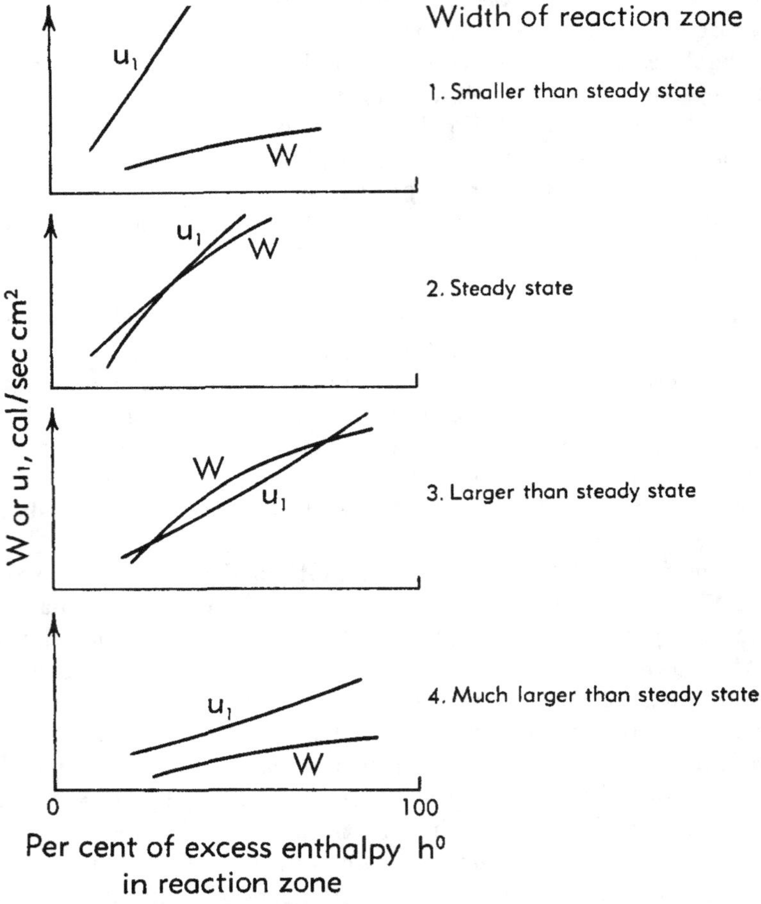

Fig. G,10b. Curves of W and u_1 versus per cent of excess enthalpy in the reaction zone at various widths of the reaction zone. Schematic illustration.

to be larger than W so that the zone must expand. If the medium is capable of supporting a combustion wave this brings about an overlapping of the curves as shown in Fig. G,10b(3) so that W becomes larger than u_1 and the zone must contract again. The wave is thus stable, and, depending upon the amount of overlapping of the curves W and u_1, the wave will be more or less resistant to fluctuations. The point of tangency must therefore be the steady state. As the zone width is increased indefinitely, the temperatures in the reaction zone become quite low and both u_1 and W decrease. The Arrhenius factors assume control so that W decreases more rapidly than u_1. The two curves draw apart, and beyond some limiting width the u_1 function is always larger than W, as shown in Fig. G,10b(4). From such states the wave cannot recover, and it decays to an ordinary temperature wave.

In Fig. G,10b(2) the ratio of W to u_1 is unity at the point of tangency, while elsewhere it is smaller than unity. Among all possible states in the group of states whose reaction zone width equals the steady state width, the steady state is therefore characterized by a maximum of the ratio W/u_1. If we plot the maximum values of W/u_1 in all groups of states against the wave widths[3] we obtain a curve somewhat as shown by the curve h^0 in Fig. G,10c. One point on this curve must correspond to the steady state, and this point must fall on an intersection of the curve with

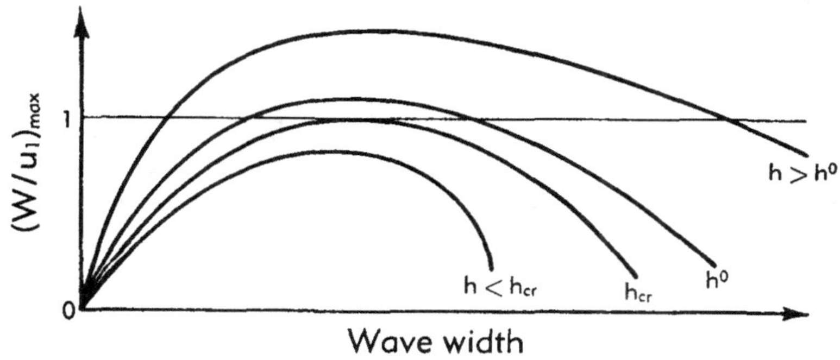

Fig. G,10c. Curves of maximum ratios W/u_1 versus wave width for various values of h. Schematic illustration.

the line of the height 1 above the abscissas. There are two such intersections. The first point, belonging to the smaller wave width, evidently corresponds to the steady state, while the second point represents the maximum width from which the wave can return to the steady state. At larger widths the ratio $(W/u_1)_{max}$ is smaller than 1. The wave continually expands and decays to a temperature wave once its limits are extended beyond the width defined by the second point.

To some value of h other than h^0 there corresponds another set of curves of the type shown in Fig. G,10b and a single curve of $(W/u_1)_{max}$ in Fig. G,10c. The previous argument is applicable. From any point on a curve h in Fig. G,10c between zero and the first point of intersection with the 1 level the wave expands, and from any point between the first and second point of intersection it contracts. The wave thus tends toward the first point of intersection. However, this point does not correspond to a

[3] Wave width may be substituted for zone width because the states under consideration have a uniquely defined enthalpy and temperature distribution.

steady state because Eq. 10-1 is not fulfilled, that is, $dh/dx \neq dh^0/dx$. In case h is larger than h^0, the temperature levels are larger for comparable wave widths, and T_b will exceed T_b^0. The development of the wave is determined by the fact that there is not enough chemical enthalpy in the unburned gas to maintain the combustion temperature at $T_b > T_b^0$. As the wave progresses, T_b gradually decreases, and behind the wave a flat temperature gradient is formed which rises from the boundary b to the plane of origin. The excess enthalpy between the boundaries u and b of the wave gradually decreases to h^0, and ultimately the steady state is established. In case the heat input h is smaller than h^0, the temperature levels are smaller for comparable wave widths. Between h^0 and some critical enthalpy $h_{cr} < h^0$, for which the curve $(W/u_1)_{max}$ meets the 1 level tangentially, the value of T_b at the first point of intersection is smaller than T_b^0, and there exists an enthalpy deficiency at the boundary b. However, as the wave progresses, as it must when $W \geqq u_1$, h increases gradually to h^0 and T_b to T_b^0 at the expense of chemical energy. Thus, a flat temperature gradient is formed behind the wave whose sign is opposite to that for $h > h^0$, the highest temperature being at the boundary b and the lowest at the plane of origin. When the excess enthalpy is less than h_{cr} there exists no state in which W exceeds or equals u_1, and the wave always decays even though some chemical reaction may occur.

Additional considerations suggest that the critical enthalpy h_{cr} can hardly be much smaller than the steady state enthalpy h^0. If a fraction of the sensible heat h^0 were removed from the steady state wave without changing the wave width, and the remaining enthalpy were distributed so that the ratio W/u_1 remained a maximum, a flatter temperature distribution would result. The function W is more sensitive to temperature than u_1 and would therefore decrease more than u_1 so that the ratio W/u_1 would become smaller than unity. As shown in Fig. G,10c, the critical wave width corresponding to h_{cr} and $(W/u_1)_{max} = 1$ is larger than the steady state width. If we now expand the wave from the steady state to the critical width, we tend to offset the decrease of W by widening the limits of the q integral (Eq. 10-15), and we decrease the value of u_1 because the temperature gradient becomes flatter, so that the ratio $(W/u_1)_{max}$ tends to increase. On the other hand, since the expansion of the wave also causes a decrease of the temperature levels in the reaction zone, the gain in the value of the function W due to widening of the integration limits is nullified to some extent. In view of the temperature sensitivity of the function W it seems that any considerable decrease of the excess enthalpy would make it impossible to restore W/u_1 to unity by increasing the wave width. We may therefore assume that h^0 and h_{cr} are of comparable magnitude, and in general probably so little different that for purposes of experimental investigations they may be considered to be identical.

We may summarize the most important conclusions from the foregoing discussion as follows: There exists a critical source energy h_{cr}, per unit area of the combustion wave, below which ignition is not possible even though the explosive medium adjacent to the source may be induced to react chemically. Furthermore, for any source energy h, there exists a limiting width of the combustion wave above which the wave decays to a temperature wave; this width is smallest for the critical source energy h_{cr} and increases with increasing values of h (see Fig. G,10c).

Principles of limits of inflammability. We now apply the principles of transition from nonsteady states to the steady state as illustrated in Fig. G,10b to the problem of the limits of inflammability. It is not immediately obvious that composition changes or reduced pressures should introduce true limits of wave stability. The general steady state equation for combustion waves is applicable to any heat-evolving medium and predicts only a widening of the wave and decrease of burning velocity with change in parameters in the direction of decreasing reaction rate. It might be supposed therefore that limits of inflammability are not true limits but are only practical limits imposed by finite vessel size and convectional disturbances in very slow moving combustion waves. This is improbable, because well-defined limits are observed in liquid and solid explosive media and observations on gas mixtures suggest that the burning velocity does not asymptotically decrease to zero as the limit is approached. It was shown in connection with Fig. G,10b that the stability of the steady state depends upon a region of overlapping of the curves W and u_1. This suggests that a theoretical criterion for limits of inflammability may be obtained by investigating the conditions required for an overlapping of the curves, that is, the relative changes of the functions W and u_1 when the wave is expanded beyond the steady state width. In mathematical terms, the derivatives of the steady state values of W and u_1 must be determined with respect to the width of the reaction zone. If the derivative of W is larger than that of u_1, W is also larger than u_1, and the wave returns to the steady state; if the derivative of W is smaller, the steady state is not stable inasmuch as a small change of wave width will cause the wave to degenerate to a temperature wave.

In the following we shall not use the superscript ^{0}, it being understood that all quantities refer to the steady state or to differential displacements from that state. We integrate Eq. 10-1 between the limits u and b, introduce an average density $\bar{\rho}$, and obtain

$$h = \frac{\lambda\bar{\rho}(T_b - T_u)}{\dot{m}} \tag{10-17}$$

Combining with Eq. 10-4,

$$u_1 = \frac{\lambda\bar{\rho}c_p(T_b - T_u)(T_1 - T_u)}{h} \tag{10-18}$$

since $\lambda(dT/dx)_1 = u_1$.

When the width $b - x_1$ of the reaction zone is changed, the total excess enthalpy h and the final temperature T_b of the wave remain constant, and the only term in Eq. 10-18 that responds to a displacement from the steady state is T_1. We write $b - x_1 = s$ and obtain

$$\frac{du_1}{ds} = \frac{u_1(dT_1/ds)}{T_1 - T_u} = \frac{W(dT_1/ds)}{T_1 - T_u} \tag{10-19}$$

since $u_1 = W$. Introducing an average value $\bar{q}$ of the rate of heat evolution between the limits x_1 and b, W may be written as $\bar{q}s(T_1 - T_u)/(T_b - T_u)$, and its derivative becomes

$$\frac{dW}{ds} = \frac{[\bar{q} + s(d\bar{q}/ds)](T_1 - T_u)}{T_b - T_u} + \frac{W(dT_1/ds)}{T_1 - T_u} \tag{10-20}$$

It is seen that dW/ds is larger than du_1/ds when the first term in Eq. 10-20 is positive. We are therefore concerned only with the relative magnitudes of the terms $\bar{q}$ and $s(d\bar{q}/ds)$. The latter is a negative quantity since an increase of the reaction zone width s causes the enthalpy h to be spread over a larger volume so that the average temperature in the zone drops and the average reaction rate decreases. The stability condition for the steady state becomes therefore

$$\bar{q} > -s\left(\frac{d\bar{q}}{ds}\right) \tag{10-21}$$

and the limit of inflammability is determined by the equality of these terms. If T_{av} is the temperature corresponding to the average reaction rate, the right-hand term may be expanded to the form $s(d\bar{q}/dT_{av})(-dT_{av}/ds)$ which is seen to be positive because $d\bar{q}/dT_{av}$ is positive and dT_{av}/ds is negative. The term $\bar{q}$ is proportional to $e^{-E/RT_{av}}$ and $d\bar{q}/dT_{av}$ thus becomes proportional to $(E/RT_{av}^2)e^{-E/RT_{av}}$. The condition for the limit of inflammability then becomes

$$\left(\frac{E}{RT_{av}}\right)\left(-\frac{dT_{av}}{ds}\right)s = 1 \tag{10-22}$$

The temperature T_{av} is obtained by averaging the values of q in the reaction zone so that $\bar{q}$ is obtained as a kinetic function approximately of the form $\bar{q} = c \times e^{-E/RT_{av}}$. In this manner the upper temperature levels are heavily weighted against the lower levels, and T_{av} has a value which is probably not much smaller than T_b. When the reaction zone width is increased, the difference $T_b - T_{av}$ increases corresponding to the decrease of the average value of the enthalpy excess per unit volume, dh/dx. As a first approximation, this suggests that $T_b - T_{av}$ is proportional to s, so that $(T_b - T_{av})/s = \text{const}$; and by differentiation one obtains

$$-\left(\frac{dT_{av}}{ds}\right)s = T_b - T_{av} = T_b\left(1 - \frac{T_{av}}{T_b}\right) \tag{10-23}$$

T_{av}/T_b is a number not much smaller than 1. With this equation, and assuming T_{av} to be substantially equal to T_b, the limit condition (10-22) becomes

$$\frac{E}{RT_b}\left(1 - \frac{T_{av}}{T_b}\right) = 1 \tag{10-24}$$

Although this relation can have only approximate validity and the factor $1 - T_{av}/T_b$ remains at present undetermined, it shows the type of correlation that should exist at the limit of inflammability between the flame temperature and characteristic kinetic quantities of the mixture.

Minimum ignition energy; critical source temperature; critical heating period. The previous considerations may be extended to two- and three-dimensional propagation of a combustion wave. For the one-dimensional case, it was assumed that a very thin slablike volume of the explosive medium is heated and that two waves propagate symmetrically in each direction from the central plane of origin. When the heated volume is a cylinder the wave propagates as an expanding cylindrical shell, and when it is a sphere the wave propagates as an expanding spherical shell. The functions W and u_1 are now dependent on the radius of curvature of the wave layers. It is not necessary to write the equations for the new functions as the previous criterion for the change of wave width (Eq. 10-16) does not change. However, in adapting arguments illustrated by Fig. G,10b and G,10c to the cylindrical and spherical cases, we must now specify that the curvature of the T_1 surface remains constant during the various hypothetical changes of wave width and excess enthalpy. A steady state is now not represented in the diagrams of W/u_1 versus wave width because as the wave propagates the radius of curvature increases. As before, there will be a value h_{cr}, referred to unit area of the T_1 surface, which is just large enough to permit the equality $W = u_1$ and thus to prevent the decay of the wave. For this condition the temperature T_b behind the wave will, as before, be lower than T_b^0. As the wave propagates the cylindrical or spherical T_1 surface increases in area and additional enthalpy must be gained by the wave layers if the critical value h_{cr} per unit area is to be maintained. An increase of radius from r to $r + dr$ requires an increase of enthalpy per unit area by an amount $dh_{cr} = h_{cr}dr/r$ for the cylindrical case, and $dh_{cr} = 2h_{cr}dr/r$ for the spherical case. Since the enthalpy of the burned gas in the volume $dr \times$ unit area behind the wave is $\rho_b c_p T_b \times dr$, and the enthalpy of adiabatic combustion is $\rho_b c_p T_b^0 \times dr$, the enthalpy gained by unit area of the wave for both cylinder and sphere is $\rho_b c_p(T_b^0 - T_b)dr$. If this amount is equal to or larger than dh_{cr}, the wave continues to propagate; if it is smaller than dh_{cr}, the wave decays. Since the minimum conditions for wave propagation, namely, $h = h_{cr}$ and $W = u_1$, determine a minimum value of T_b, one obtains by equating the enthalpy increase with the enthalpy requirement for the cylindrical case,

$$\rho_b c_p[T_b^0 - (T_b)_{\min}] = \frac{h_{cr}}{r_{\min}} \tag{10-25}$$

and a similar equation for the spherical case.

Eq. 10-25 shows that a minimum value of the "flame radius" r is associated with the critical enthalpy excess h_{cr}. Since the surface of a cylinder of unit length is $2\pi r$, it follows that for cylindrical propagation the energy supplied by the source must not be smaller than $2\pi r_{\min} \times h_{cr}$ per unit length. For spherical propagation the critical source energy, which we shall denote by the symbol H_{cr}, becomes

$$H_{cr} = 4\pi r_{\min}^2 h_{cr} \tag{10-26}$$

We shall make use of this equation later. It is important to note that the quantity H_{cr} represents the absolute minimum ignition energy. No matter how "hot" an electric spark may be, if its energy content is less than the minimum it cannot produce a self-propagating combustion wave, and the zone of chemical reaction penetrates into the explosive medium to only a small depth.

Eq. 10-25 shows that with increasing flame radius r the temperature T_b approaches T_b^0 so that behind the wave in the direction from the flame origin to the wave surface the temperature of the burned gas increases. For reasons discussed earlier, one may in general expect the difference $T_b^0 - (T_b)_{\min}$ to be very small so that the temperature gradient is probably not readily detectable by experiment. As T_b approaches T_b^0, the enthalpy excess h increases from h_{cr} to h^0, corresponding to the wave structure approaching asymptotically the ideal structure of the plane adiabatic combustion wave. When cylindrical or spherical ignition has been produced by a source energy exceeding the critical, the temperature at the origin is always larger than $(T_b)_{\min}$; and as the flame radius increases, the temperature T_b immediately behind the wave at first decreases as described for one-dimensional propagation. Since the wave requires an increasing amount of excess enthalpy as the flame radius increases, the demand for excess enthalpy ultimately exceeds the source energy and must be satisfied at the expense of chemical energy, as discussed before. Thus, at some flame radius, which depends on the magnitude of the source energy, the temperature T_b attains a minimum value $< T_b^0$, and as the flame radius increases further, T_b approaches T_b^0 asymptotically.

The foregoing illustrates that there are two phases to the ignition process. In the first phase, source energy is imparted to the explosive medium; in the second, the wave develops to a state in which propagation is independent of the source parameters. During the first phase, which we may call the "heating period," the source energy is distributed in the explosive medium in a manner dependent on source geometry, the amount of source energy, and the rate at which the energy is imparted to the medium. During the second phase, the wave passes through a series of nonsteady states which lead to either extinction or propagation. The energy distribution at the end of the first phase determines the events of the second phase. In the one-dimensional case, extinction is assured when the wave develops to a width exceeding the limiting width for the source energy h. In the two- and three-dimensional cases there exists a similar limit which we may call the limiting flame radius, and which for cylindrical propagation refers to the source energy per unit length and for spherical propagation refers to absolute source energy. The limiting wave width, or flame radius, represents the extreme condition at which the equality $W = u_1$ can be fulfilled and is thus dependent on a critical energy distribution at the end of the heating period and hence on a set of critical source parameters.

In developing a simple terminology for critical source parameters we shall use the term "source volume" to denote, not the volume occupied by some material source such as a hot wire, etc., but the volume of explosive medium over which the energy distribution extends at the end of the heating period. The term "source temperature" will be used to denote the temperature in the source volume at the point of origin of the source energy. To the limiting wave width or flame radius there corresponds a critical source temperature, a critical source volume, and a critical heating period. These quantities are interdependent but individually significant inasmuch as a given source energy cannot cause ignition of an explosive medium if either the initial temperature of the heat source is below a critical temperature, or the heated volume exceeds a critical limit, or if the heating period is too long.

Concerning the relation of these three quantities to one another, it has been noted above that in one-dimensional propagation the limiting wave width and temperature distribution for the source energy h_{cr} are nearly identical with the steady state. Therefore, the half width of the critical source volume is of the order of, or perhaps even considerably smaller than, the width of the steady state reaction zone between x_1 and b. The critical heating period is very short, comparable to or perhaps even considerably shorter than the time in which the steady state wave could travel a distance equal to the width of the reaction zone. The critical source temperature is very high, corresponding to the very rapid chemical reaction in the source volume; it may be comparable to or even considerably higher than the adiabatic combustion temperature T_b^0. As the source energy is increased the critical source volume increases, the critical heating period increases, and the critical source temperature decreases corresponding to the lower rates of the heat transport and chemical reaction for the larger source volume. The dependence of the reaction rate on the source temperature is governed by Arrhenius factors. Such factors not only become very small at low temperatures but also change by much larger fractions for equal temperature intervals. Thus, equal changes of temperature have a much larger effect on the rate at low than at high temperatures. It follows that for successive increments of source energy the decrements of the source temperature become progressively smaller until a virtual lower limit of the source temperature is reached. At this limit a large increase of source energy produces a small decrease of critical source temperature so that the medium may seem to possess a "constant" ignition temperature.

The latter possibility may be examined by making a rough calculation using a numerical example. We consider a hypothetical explosive in which the reaction rate follows an Arrhenius equation and estimate the change of the critical source temperature with source energy, when the energy is large and the Arrhenius factor is correspondingly small in the limit state. As a simplifying approximation we assume the source energy to be distributed instantly over a very long and wide but thin rectangular volume of the explosive medium, i.e. a semi-infinite slab, which is bounded on both sides by a large volume of the same medium. The initial temperature in the explosive mass outside the slab is taken as 0°C. The equation of rate of change of temperature at the origin is given by Eq. 10-11 for $\dot{m} = 0$. We write the equation in the form

$$\frac{\partial T}{\partial t} = \kappa \frac{\partial^2 T}{\partial x^2} + \frac{q}{\rho c_v} \tag{10-27}$$

where κ, the coefficient of thermal diffusivity, is $\lambda/\rho c_v$. The term $q/\rho c_v$ which represents the rate of temperature rise due to chemical reaction is assumed to be equal to const $\times\, e^{-E/RT}$, where E is the activation energy. The equation cannot be solved analytically. Instead, we shall consider separately two extreme ranges of source temperature, one corresponding to $q = 0$, and the other corresponding to the equation

$$\frac{\partial T}{\partial t} = \frac{q}{\rho c_v} = \text{const} \times e^{-E/RT} \tag{10-28}$$

In the former range the temperature is so low that chemical heat evolution is insignificant and ignition certainly does not occur, and in the latter range the temperature is so high that chemical heat evolution far exceeds heat loss by conduction and ignition occurs with certainty. The critical temperature T_{cr} lies somewhere between two temperatures defining an upper and lower bound, respectively, of these ranges. The solution of Eq. 10-27 [*41*] for the center of the heated volume and $q = 0$ is

$$\frac{T}{T_i} = \psi\left(\sqrt{\frac{1}{\vartheta}}\right), \qquad \vartheta = \frac{4\kappa t}{d^2} \tag{10-29}$$

where T_i is the initial temperature of the slab, d the half-width of the slab, t the time, and ψ is the probability function of the argument $\sqrt{1/\vartheta}$ (see for e.g. [42]).

Differentiation of Eq. 10-29 yields

$$\frac{d(T/T_i)}{d\vartheta} = -\left(\frac{1}{\sqrt{\pi}}\right)\vartheta^{-\frac{3}{2}}e^{-1/\vartheta} \tag{10-30}$$

which is a measure of the rate of decrease of temperature due to conduction and attains a maximum of 0.23 at $\vartheta = \frac{2}{3}$. The rate of temperature rise, assuming adiabatic heat evolution, is given by Eq. 10-28 which is rewritten

$$\frac{d(T/T_i)}{d\vartheta} = ke^{-E/RT} \tag{10-31}$$

and near the ignition threshold must yield values comparable to the range of values obtained from Eq. 10-30. We choose $ke^{-E/RT_i} = 0.01$, and e^{-E/RT_i}

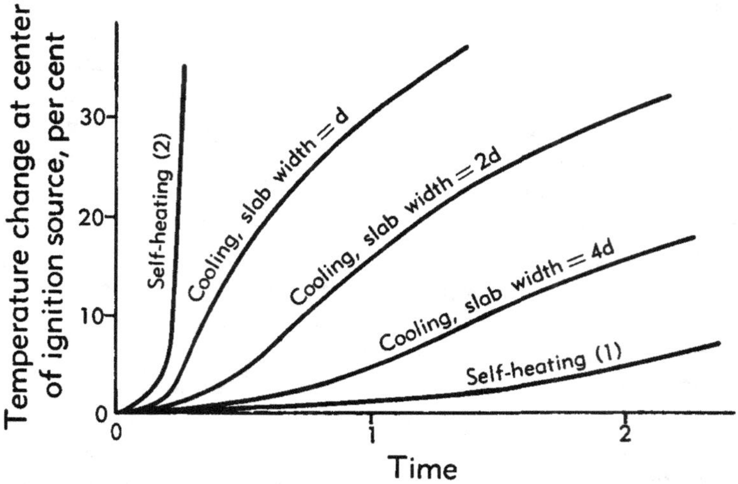

Fig. G,10d. Change of temperature with time at center of ignition source in an explosive medium. The source is a slablike volume within a larger volume of the medium and the source energy is assumed to be initially distributed uniformly over the source volume. Width of slab source, d, $2d$ or $4d$. Time in dimensionless units.

$= 10^{-20}$, the latter corresponding to an activation energy of 62,000 cal/mole for $T_i = 400°C$, so that $k = 10^{18}$ and

$$ke^{-E/RT} = 10^{(18-20T_i/T)} \tag{10-32}$$

If the temperature T_i in degrees Kelvin is increased by 5 per cent, Eq. 10-31 becomes

$$ke^{-E/RT} = 10^{(18-19T_i/T)} \tag{10-33}$$

Fig. G,10d is a plot showing the percentage change of temperature at the center of the source with time. The intermediate curves marked "cooling" are obtained from Eq. 10-29 and represent decrease of temperature with time if no heat is evolved in the medium. The lower and upper curves marked "self-heating" are obtained by numerical integration from Eq. 10-32 and 10-33, respectively, representing a difference of 5 per cent in T_i, in combination with Eq. 10-31. They represent increase of temperature with time, assuming adiabatic heat evolution in the medium. Under conditions assumed for curve 1 (Eq. 10-32), heat evolution is negligible by comparison with the cooling rate and the temperature at the source center decreases with time substantially as shown by the cooling curves for the various slab widths. On increasing the initial absolute temperature of the source by 5 per cent, curve 2 is obtained, and it is seen that the temperature rises so rapidly that cooling is negligible. Thus, at the lower temperature T_i the center of the source becomes cooler and at the higher temperature T_i it becomes hotter, even though the source energy is changed by a factor of four corresponding to a fourfold change of slab width; i.e. the lower and upper bounds of the regions of nonignition and ignition are insensitive to large changes of source energy. For this example, the difference between the upper and lower bound is only 34° and the critical temperature T_{cr} remains somewhere in this range. The result is not greatly different if a lower value is chosen for the activation energy, say, 40,000 cal/mole. Values for E of this order apply to solid and liquid explosives and such explosives are ignitible by contact with sources of quite low temperature, so that the argument appears applicable to such substances.

We now consider some aspects of actual ignition sources. In order to obtain ignition with an energy comparable to the theoretical minimum energy, a high temperature must be produced in a volume whose dimensions must not exceed the order of magnitude of the width of the combustion wave. This requires a very rapid and compact source of energy. In gases, electric capacitance sparks are well suited for the purpose. They may not always be small enough in size but they generally meet the requirement of high rate of energy release since the discharge time is of the order of 10^{-7} second. A problem common to localized ignition sources is the introduction of various heat-absorbing bodies into the medium. With sparks, the problem is minimized by separating the electrodes beyond the quenching distance. With sources such as electrically heated wires, a part of the source energy is absorbed by the wire. Let us assume that the wire is so fine that only a negligible part of the source energy is retained by it. Assuming a fixed electric resistance, the source energy is determined by the current strength and heating period, i.e. the time of current flow. For constant source energy, the lower the current and the longer therefore the heating period, the greater will be the volume over which the source energy is distributed in the medium; and above some critical heating period the source volume exceeds the critical limit so that ignition does not occur. At some very high current strength the critical heating period reduces to a minimum consistent with the theoretical minimum of the source energy. As the current strength is decreased, the critical heating period becomes larger, the source energy also becomes larger, and the critical source temperature decreases. Because of the exponential relations between reaction rate and temperature introduced by Arrhenius factors, the source energy as a function of critical heating period rises sharply until the source temperature has decreased to a

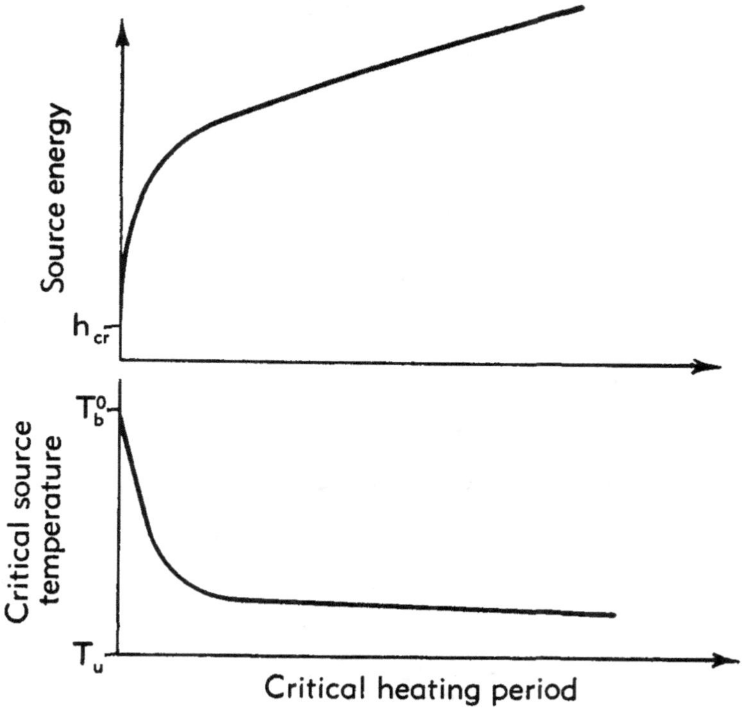

Fig. G,10e. Source energies and critical source temperatures versus critical heating period. Schematic illustration.

quasi-constant value, and then increases more or less linearly at a much slower rate. A schematic illustration of the interdependence of the critical source energy, temperature, and heating period is shown in Fig. G,10e.

Ignition by bodies of large heat capacity; catalytic and noncatalytic surfaces. The heat capacity of the material of an ignition source is usually not negligible. Let us consider that the explosive medium makes contact with a large body of high heat conductivity at a temperature somewhat lower than the adiabatic combustion temperature T_b^0, as is generally the case, for example, when a hot solid of sizable dimensions is introduced into an explosive gas. The sequence of events may be visualized somewhat as shown in Fig. G,10f(a) for a body with a noncatalytic surface. Heat flows into the medium and chemical reaction occurs which evolves additional heat. Stage 1 shows the temperature distribution at the instant that the temperature gradient at the source boundary becomes zero and heat flow into the medium ceases. At the boundary the temperature remains more or less constant because of the large heat capacity and heat conductivity of the body, but in the medium it continues to increase owing to reaction so that a temperature gradient is established along which some of the heat originally imparted to the medium is drained back into the body. When the chemical enthalpy at the surface is exhausted

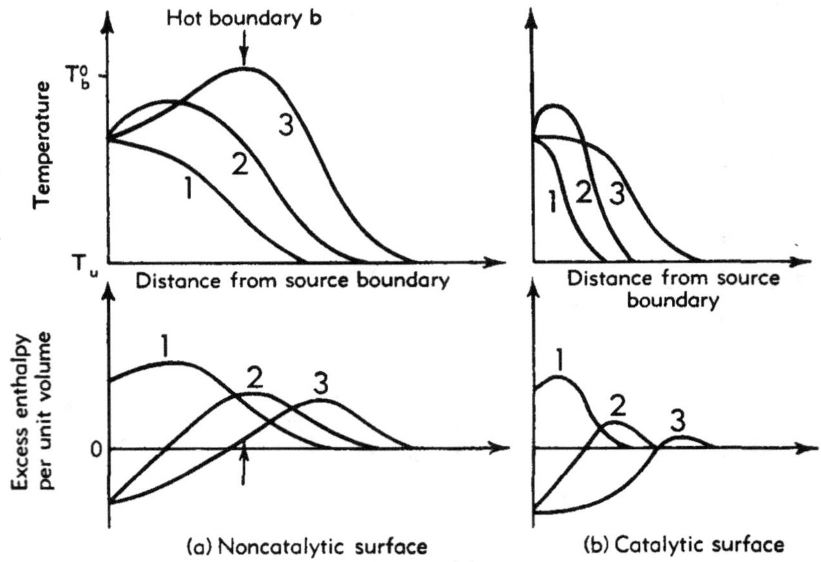

Fig. G,10f. Successive stages of temperature and excess enthalpy distribution in explosive medium near surface of a large heated body. Schematic illustration. Stage 1, zero temperature gradient at source boundary. Stage 2, hot boundary established at source boundary. Stage 3, hot boundary has moved into the medium.

(stage 2) a layer of medium is formed which is deficient in enthalpy content, while farther away an enthalpy excess persists. This excess enthalpy in the medium continues to drain into the body by heat flow along the reversed temperature gradient, but as the zone of zero chemical enthalpy, viz. the hot boundary b, moves away from the surface the temperature gradient becomes flatter and the heat drain slows down. When the boundary b has moved to the crest of the temperature wave (stage 3), the heat drain may have become quite small, and if sufficient excess enthalpy persists the wave will develop into a steady state combustion wave. This appears to be a certainty when the temperature of the body is only slightly below the adiabatic combustion temperature T_b^0. When the temperature of the body is decreased the curves of temperature and excess enthalpy become lower and flatter; finally, the residual excess enthalpy (stage 3) will be distributed over a wave width exceeding the limiting width, and the wave decays to a temperature wave. In the latter case, the temperature of the medium will ultimately decrease below that of the body and heat flow from the body to the medium will resume. Conceivably, this will cause the temperature in the unreacted medium to rise gradually and, as the temperature wave progresses, the medium may acquire sufficient excess enthalpy for ignition. This process can be arrested by heat dissipation at boundaries such as the wall of the vessel containing the explosive medium.

Some surfaces, such as platinum, are catalytically active in contact with a number of gas mixtures, i.e. a reaction of fuel and oxygen takes place with liberation of heat at the surface. Let us consider two large bodies, one catalytic and the other noncatalytic but otherwise identical, in contact with the medium and at a temperature corresponding to the ignition threshold of the noncatalytic body. The various stages of temperature and excess enthalpy distribution are compared schematically in Fig. G,10f(a) and (b). The reaction rate at the catalytic surface is very large, so that stage 1 is completed more rapidly, and less heat flows into the medium. In stage 2 the catalytic reaction causes a short and steep reversed temperature gradient which effectively drains the medium of its enthalpy content. The residual enthalpy excess becomes therefore much smaller compared to the noncatalytic surface, and instead of promoting ignition, the catalyst actually inhibits ignition. This does not apply if the mass of the catalytic body is so small that its temperature greatly increases as the result of the heat received from the medium; thus, a little platinum black in a stream of hydrogen and oxygen may accumulate enough heat from the catalysis of the medium to become very hot and ignite the mixture.

G,11. Simplified Equations for Calculations Involving Heat Transport Across the Combustion Wave.

Approximate determination of the temperature profile of the wave from minimum ignition energy and burning velocity. We shall simplify the equations for the thermal model of the plane adiabatic combustion wave by introducing the average rate of heat evolution $\bar{q}^0$. For reasons of systematics which will become apparent later, we replace the symbol b of the hot boundary by the symbol x_3^0 and write in analogy to Eq. 10-5

$$\bar{q}^0(x_3 - x_1)^0 = \rho_u S_u^0 c_p (T_b^0 - T_u) \tag{11-1}$$

where $\dot{m}^0$ is expressed as the product of density and burning velocity. When a mass element passes through the reaction zone where its chemical enthalpy is transformed into heat, it utilizes a part of its chemical enthalpy for a temperature increase from T_1^0 to T_b^0, and returns the remainder to the unburned gas by the process of heat conduction. From the steady state condition and the principle of conservation of energy it follows that the latter part is equal to the heat received by the element before it entered the reaction zone across the plane x_1^0. If we now assume that heat evolution proceeds at the constant rate $\bar{q}^0$ in all layers dx of the reaction zone, we must simulate the actual wave by allowing the mass element to retain its chemically evolved heat until it has attained the temperature T_b^0, and transfer the remainder to the unburned gas. Thus, we divide the zone into two parts: in the first part, extending from x_1^0 to a plane x_2^0, the heat evolution is adiabatic and its rate is therefore

$$\bar{q}^0(x_2 - x_1)^0 = \rho_u S_u^0 c_p (T_b^0 - T_1^0) \tag{11-2}$$

In the second part, extending from x_2^0 to x_3^0, the heat evolution is isothermal and its rate is

$$\bar{q}^0(x_3 - x_2)^0 = \rho_u S_u^0 c_p (T_1^0 - T_u) \tag{11-3}$$

as may be seen by subtracting Eq. 11-2 from 11-1, or immediately on the basis that the right-hand side represents the heat received by the mass flow in front of the plane x_1^0; the conservation principle is thus satisfied. This quantity is related to the temperature gradient at x_1^0 by Eq. 10-4 which we write in the form

$$\lambda \left(\frac{dT}{dx}\right)_1^0 = \rho_u S_u^0 c_p (T_1^0 - T_u) \tag{11-4}$$

Assuming that the quantities ρ_u, c_p, λ, T_b^0, T_u, S_u^0, and the reaction zone width $(x_3 - x_1)^0$ are known or measurable, there remain several quantities which must be calculated from the equations since they occur only in the wave model and not in the actual wave, and are therefore not independently determinable. These quantities are $\bar{q}^0$, T_1^0, $(dT/dx)_1^0$, and either $(x_3 - x_2)^0$ or $(x_2 - x_1)^0$. Thus far, only three independent equations have been formulated, Eq. 11-3 having been obtained by subtracting Eq. 11-1 and 11-2; a fourth equation is required to determine the four unknown quantities.

Since there shall be no loss or gain of heat by conduction in any layer dx between x_1^0 and x_2^0, $\lambda(d^2T/dx^2)$ must be zero between these limits. It follows that dT/dx is constant and equal to $(T_b^0 - T_1^0)/(x_2 - x_1)^0$ between x_1^0 and x_2^0. In order to obtain continuity of the temperature curve at the boundary of the preheat and reaction zones, the slope of the curve must not change in the plane x_1^0, and we obtain the fourth equation

$$\left(\frac{dT}{dx}\right)_1 = \frac{T_b^0 - T_1^0}{(x_2 - x_1)^0} \tag{11-5}$$

The model is now completely determined in terms of known or measurable quantities. The temperature distribution in the preheat zone is obtained from Eq. 10-3 by setting $q = 0$. Integrating once between the limit x_1^0 and a variable limit $x < x_1^0$, we obtain

$$\frac{dT}{dx} = \left(\frac{dT}{dx}\right)_1 e^{-\rho_u S_u^0 c_p (x_1^0 - x)/\lambda} \qquad x \leqq x_1^0 \tag{11-6}$$

A second integration yields

$$T - T_u = (T_1 - T_u)e^{-\rho_u S_u^0 c_p (x_1^0 - x)/\lambda} \tag{11-7}$$

The temperature is thus seen to increase exponentially from T_u to T_1^0.

The relation between the temperature distribution in the wave model and in the actual wave may be visualized somewhat as shown in Fig. G,11a. The model curve may be expected to circumscribe the actual curve

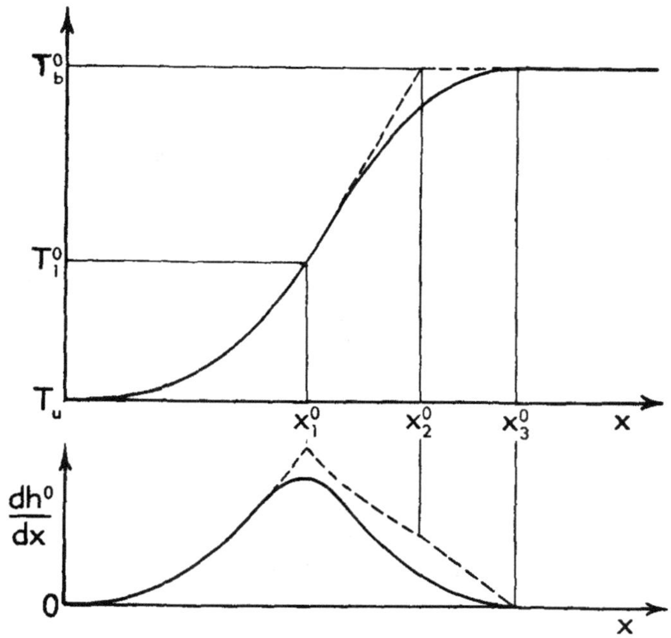

Fig. G,11a. Temperature and excess enthalpy in adiabatic steady state combustion wave. Dotted curve corresponds to approximate wave model.

rather closely and may well be substituted for the unknown actual distribution. The corresponding excess enthalpy curves are shown in juxtaposition. In the preheat zone, where the temperature distribution in the wave model and in the actual wave are nearly identical, the two curves are nearly identical. In the reaction zone of the wave model, a mass element retains the excess enthalpy which it possessed at the time of crossing the plane x_1^0 until it arrives at x_2^0 because no heat is lost or gained by conduction. The enthalpy excess per unit volume, dh^0/dx, is therefore simply proportional to the density. The latter is an inverse function of temperature so that the curve should slope downward between x_1^0 and x_2^0 somewhat as shown. Between x_2^0 and x_3^0 the enthalpy excess is equal to the residual chemical enthalpy in any layer because the temperature does not increase further. The enthalpy decreases at the constant rate $\bar{q}^0$, and the velocity of the mass flow has the constant value $\dot{m}/\rho_b = S_b$, so that dh^0/dx decreases linearly to zero, as shown.

If it is desired to calculate the total excess enthalpy h^0 of the adiabatic plane wave, the "averaged" wave structure defined by the foregoing equations is not required because fortunately Eq. 10-1 is of a simple form that lends itself to integration if an average density $\bar{\rho}$ is introduced. This integration has already been carried out in Eq. 10-17. For calculations of order-of-magnitude accuracy, ρ_u may be substituted for $\bar{\rho}$, and since $\dot{m}^0/\rho_u = S_u^0$,

$$h^0 \cong \frac{\lambda_u}{S_u^0}(T_b^0 - T_u) \tag{11-8}$$

This equation will be used later in discussing experimental data of ignition energies.

It may be questioned whether the reaction zone width $x_3 - x_1$ can actually be considered a measurable quantity because in an actual wave the positions of the boundaries x_3 and x_1 are not fixed by readily recognizable criteria. Consequently, it is desired to introduce some other experimental quantity in order to determine the quantities $\bar{q}$, T_1^0, $(dT/dx)_1^0$, and $(x_2 - x_1)_1^0$ of the wave model. For this quantity we choose the minimum ignition energy H of a point source. The system of equations by which this quantity is correlated with other quantities, either measurable or belonging to the wave model, is rather complex. It has been fully set down elsewhere [*37*]. Here it will suffice merely to indicate the principles on which this correlation is based.

As shown by Eq. 10-26 the quantity H together with h_{cr}, which can be identified with h^0 and may be regarded as a known quantity from Eq. 11-8, determines the radius r_1 of the T_1 surface for the minimum size of a spherical flame. We introduce the additional assumption that the temperature profile through the reaction zone is not greatly different from that in a plane wave, so that $x_2 - x_1$ can be taken as equal to $r_1 - r_2$, where, in the model, the temperature T_b is established over a sphere of radius r_2. Furthermore the temperatures T_b and T_1 are taken to be T_b^0 and T_1^0 of the plane wave. A boundary condition is required relating the rate of heat evolution in the volume of radius r_2 with the rate of heat flow across the T_1 surface. This condition correlates the magnitudes of r_1 and r_2 and since r_1 is known from above, quantity $r_1 - r_2$ is known. Since $r_1 - r_2$ is identified with $x_2 - x_1$ it now becomes possible to calculate T_1^0 and the remaining parameters of the model from Eq. 11-1 to 11-5.

Calculation of quenching distance and flashback limit. Fig. G,8c illustrates schematically the profile of a wave traveling between plates or in a tube of dimensions comparable to the quenching distance. The wave is in a steady state, and the mass flow is considered to be everywhere parallel to the wall. The problem is complicated by the fact that heat flow by conduction is not one-dimensional. We shall first consider the region in front of the reaction zone. From the general equation (1-9) of conservation of energy, the steady state condition becomes

$$\nabla \cdot \dot{m} c_p (T - T_u) - \lambda \nabla^2 T = 0 \tag{11-9}$$

where ∇ represents the vector differential operator, and $\dot{m}$ is the mass flow vector. In the case of a channel formed by two semi-infinite parallel

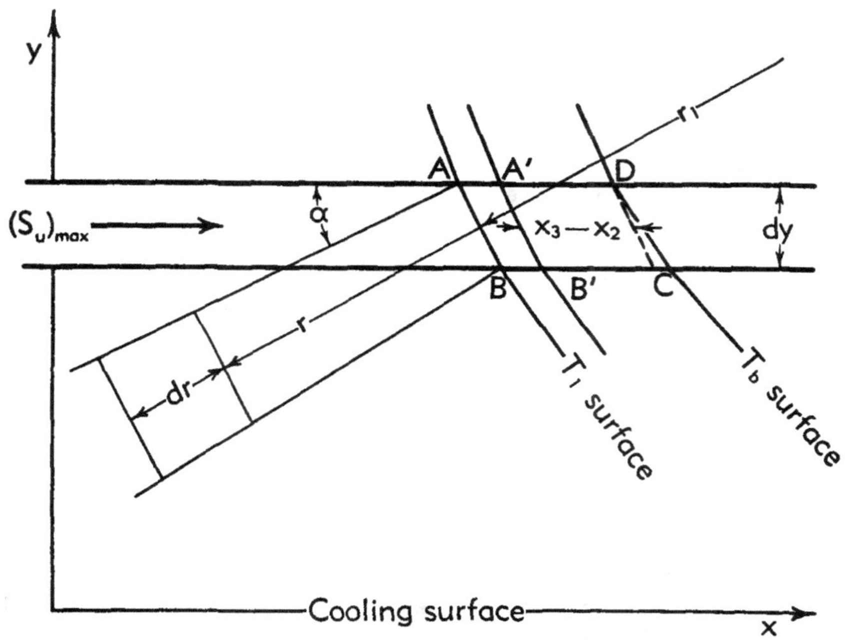

Fig. G,11b. Segment of a combustion wave near a cooling surface.

plates, the reaction zone and the isothermal surfaces in front of the zone assume curvatures somewhat similar to semicylindrical surfaces. A segment of the wave is shown in Fig. G,11b. The element AB of the reaction zone front has a cylindrical curvature of radius r_1 and its normal is inclined to the mass flow at an angle α. The element is subtended by the angle $d\alpha$, and since the tangent of a small angle is equal to the angle, $d\alpha = AB/r_1$. For simplicity, we shall assume that the isotherms are parallel to the reaction zone, as is the case at the center and the fringe of the wave; thus, within the angle $d\alpha$ the isothermal surfaces are cylindrically concentric to AB. Eq. 11-9 may therefore be expressed in cylindrical coordinates comprising the angle α and the radius r of any isothermal surface element. The vector $\dot{m}$ is $-\rho_u(S_u)_{max} \cos \alpha$, and $\nabla^2 T$ is

$(1/r)[d(rdT/dr)/dr]$. After introducing $a_m = \rho_u(S_u)_{max}c_p/\lambda$, the equation becomes

$$\frac{1}{r}\frac{d}{dr}\left(r\frac{dT}{dr}\right) + a_m \frac{d[(T - T_u)\cos\alpha]}{dr} = 0 \qquad (11\text{-}10)$$

or

$$\frac{d^2T}{dr^2} + \frac{1}{r}\frac{dT}{dr} = -a_m\left[\left(\frac{dT}{dr}\right)\cos\alpha - (T - T_u)\sin\alpha\frac{d\alpha}{dr}\right] \qquad (11\text{-}11)$$

At the fringe of the wave, the isothermal surfaces become parallel to the wall, so that $\alpha = 90°$, $\cos\alpha = 0$, $r = \infty$, and $d\alpha/dr$ vanishes. Eq. 11-11 reduces to $d^2T/dr^2 = 0$, and introducing coordinates x and y, where x is parallel and y normal to the wall, this becomes

$$\frac{d^2T}{dy^2} = 0 \text{ for } \alpha = 90° \qquad (11\text{-}12)$$

since for $\alpha = 90°$ the coordinate y coincides with the radius r.

At the center of the wave, α is zero so that $\sin\alpha = 0$ and $\cos\alpha = 1$. Integration of the remaining terms of Eq. 11-11 between r_1 and infinity yields

$$\lambda(T_1 - T_u) = u_1 r_1 e^{a_m r_1}[-Ei(-a_m r_1)]; \qquad \alpha = 0 \qquad (11\text{-}13)$$

The exponential integral can be expressed by the series

$$-Ei(-z) = e^{-z}\left(\frac{1}{z} - \frac{1}{z^2} + \frac{2!}{z^3} - \cdots\right)$$

so that

$$\lambda(T_1 - T_u) = \frac{u_1}{a_m}\left(1 - \frac{1}{z} + \frac{2!}{z^2} - \cdots\right); \qquad z = a_m r_1 \qquad (11\text{-}14)$$

As shown by calculations that have been given elsewhere [*37*, p. 418], $a_m r_1$ is generally of the order of 10, so that the sum inside the bracket of Eq. 11-14 exceeds unity by only about 10 per cent. We may therefore write, with good approximation, $\lambda(T_1 - T_u) = u_1/a_m$, or

$$u_1 = \rho_u(S_u)_{max}c_p(T_1 - T_u); \qquad \alpha = 0 \qquad (11\text{-}15)$$

where u_1/λ is the positive value of the temperature gradient $(dT/dr)_1$. This equation is obtained directly from Eq. 11-11 if the divergence term $(1/r)(dT/dr)$ is neglected.

Behind element AB, combustion occurs. Corresponding to the previous treatment of the combustion wave, the temperature curve is continued linearly with the slope u_1/λ until the temperature T_b is reached in a plane $A'B'$. Between AB and $A'B'$ the heat evolution is adiabatic, and between $A'B'$ and a plane CD it is isothermal. The distance $A'D$ is denoted by $x_3 - x_2$. The volume $ABCD$ is enclosed by a stream tube of unit cross section. Behind DC the temperature is assumed to remain substantially constant along the stream tube; that is, the cooling of gas is considered to be a relatively slow process so that the temperature change along the x coordinate can be neglected. Since the specific heat is taken constant, the sensible heat $c_p(T_b - T_u)$ of the gas flowing across plane $A'B'$ into the volume $A'B'CD$ is the same as that of the gas flowing across plane DC out of the volume, and the net enthalpy increase in the volume $A'B'CD$ per unit time is equal to the rate of heat evolution, which for unit cross section of the stream tube is $q(x_3 - x_2)$. Since the temperature does not change, this must be equal to the sum of the rates of heat flow by thermal conduction across the four planes $A'B'$, $A'D$, $B'C$, and CD. The planes $A'D$ and $B'C$ together contribute the rate of change of heat $d[(x_3 - x_2)\lambda dT_b/dy]/dy$. The heat flow across CD is $\lambda(dT_b/dy)[\tan\alpha - d(x_3 - x_2)/dy]$. The heat flow across $A'B'$ is the same as across AB, namely, $u_1/\cos\alpha$. Equating $q(x_3 - x_2)$ to these three terms, differentiating and rearranging, one obtains

$$(x_3 - x_2)\left[q + \lambda\left(\frac{d^2T_b}{dy^2}\right)\right] = \frac{u_1}{\cos\alpha} - \lambda\frac{dT_b}{dy}\tan\alpha \qquad (11\text{-}16)$$

since terms containing the factor $d(x_3 - x_2)/dy$ cancel each other.

The heat flow $u_1/\cos\alpha$ is partly used in raising the gas temperature from T_u to T_1 and is partly dissipated by the cooling surfaces upstream from the wave. The heat flow originates from the completely burned gas and from that part of the heat released within the reaction zone which is transferred to the unburned gas. The former contribution is $\lambda(dT_b/dr)/\cos\alpha$, and if we again neglect divergence terms as in Eq. 11-15, the latter contribution is $\rho_u S_u c_p(T_1 - T_u)$. Since

$$\frac{dT_b}{dr} = \frac{dT_b}{dy}\sin\alpha \qquad (11\text{-}17)$$

one obtains

$$\frac{u_1}{\cos\alpha} = \lambda\frac{dT_b}{dy}\tan\alpha + \rho_u S_u c_p(T_1 - T_u) \qquad (11\text{-}18)$$

Finally, Eq. 11-16 reduces to

$$(x_3 - x_2)\left[q + \lambda\left(\frac{d^2T_b}{dy^2}\right)\right] = \rho_u S_u c_p(T_1 - T_u) \qquad (11\text{-}19)$$

Eq. 11-19 is equivalent to a reaction zone front facing the gas stream everywhere perpendicularly. This model, which is simpler than that described previously in Fig. G,8c and G,11b, is shown in Fig. G,11c as a series of stream tubes and wave segments. The x_1 surface is a plane normal to the x axis. The gas velocity at any point is the burning velocity S_u. From the x_2 surface heat flows in the x direction only, in an amount necessary to raise the gas temperature from T_u to T_1. The x_2 surface, is therefore, also a plane normal to the x axis, and for this reason it is everywhere equidistant from plane x_1. Behind x_2 the temperature changes only in the y direction, and heat is lost to the wall. Since between x_1 and x_2 the gas temperature increases adiabatically to T_b by heat evolution,

$$(x_2 - x_1)q = \rho_u S_u c_p (T_b - T_1) \tag{11-20}$$

The heat flow from plane x_2 is the same as that from plane x_1 and is $u_1 = \rho_u S_u c_p (T_1 - T_u)$ or

$$\lambda \frac{T_b - T_1}{x_2 - x_1} = \rho_u S_u c_p (T_1 - T_u) \tag{11-21}$$

and since in the absence of heat loss to the wall the temperature T_b would

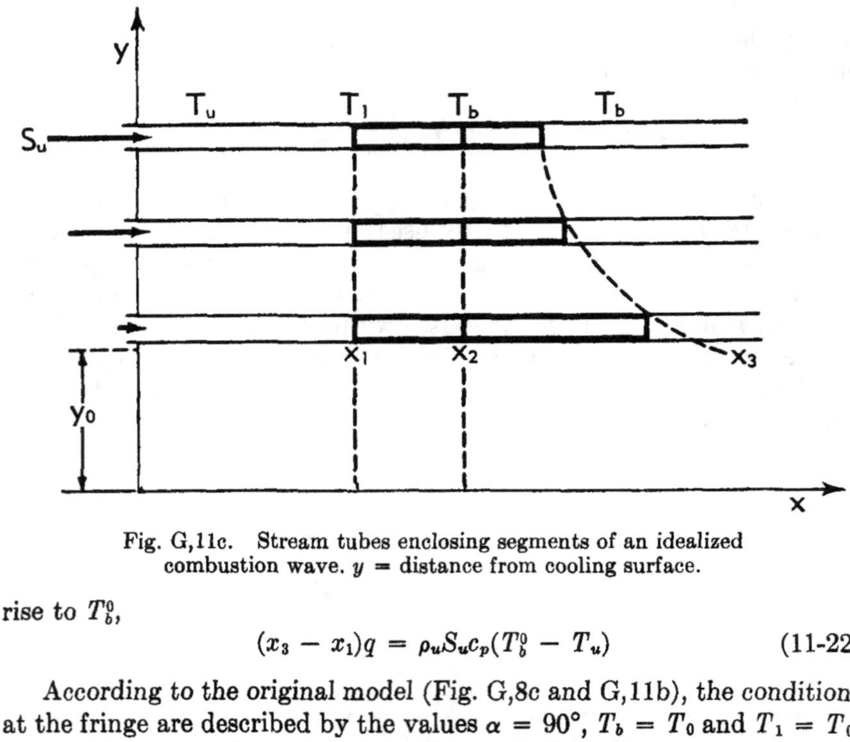

Fig. G,11c. Stream tubes enclosing segments of an idealized combustion wave. y = distance from cooling surface.

rise to T_b^0,

$$(x_3 - x_1)q = \rho_u S_u c_p (T_b^0 - T_u) \tag{11-22}$$

According to the original model (Fig. G,8c and G,11b), the conditions at the fringe are described by the values $\alpha = 90°$, $T_b = T_0$ and $T_1 = T_0$; hence $dT_b/dy = u_1/\lambda$, and $(x_3 - x_2)[q + \lambda(d^2T_b/dy^2)]$ becomes zero since the right-hand side of Eq. 11-16 becomes $u_1(1/\cos\alpha - \tan\alpha)$ or, as shown by trigonometry, equal to $u_1[\cos\alpha/(1 + \sin\alpha)]$, and therefore vanishes. From Eq. 11-19 and 11-22 it follows that S_u and q become zero; hence, at the fringe

$$\frac{d^2T_b}{dy^2} = 0 \tag{11-23}$$

Eq. 11-19 and the two Eq. 11-12 and 11-23, which have been formulated for the case of two surfaces separated by a distance comparable to the quenching limit, may also be applied to the case of widely separated surfaces, which is equivalent to a single surface bordering on a quasi-infinite depth of explosive medium. In this case $(S_u)_{max}$ becomes S_u^0, $(T_b)_{max}$ becomes T_b^0, and $(T_1)_{min}$ becomes T_1^0. If the combustion wave propagates not in a plane-parallel channel but in a cylindrical tube, it will be necessary to add only the divergence term $\lambda(dT_b/dy)/y$ to the terms within the bracket on the left-hand side of Eq. 11-19, and a corresponding term to Eq. 11-12 and 11-23.

Solving Eq. 11-20 and 11-21 for q, one obtains

$$\frac{q}{\lambda} = a^2(T_1 - T_u) \tag{11-24}$$

where $a = \rho_u S_u c_p/\lambda$, as before. Eq. 11-24 together with Eq. 11-20 and 11-22 yields

$$x_3 - x_2 = \frac{T_b^0 - T_b + T_1 - T_u}{a(T_1 - T_u)} \tag{11-25}$$

Substituting q/λ and $(x_3 - x_2)$ in Eq. 11-19 and transposing, one obtains

$$-\frac{d^2T_b}{dy^2} = \frac{a^2(T_1 - T_u)(T_b^0 - T_b)}{T_b^0 - T_b + T_1 - T_u} \tag{11-26}$$

Here T_b^0 and T_u are constant, and in order to make the equation integrable T_1 and a must be formulated as functions of T_b. The largest possible value of T_b is T_b^0 corresponding to infinite distance along the wave from the wall, and the smallest value is T_0 corresponding to the wave fringe. The corresponding values of T_1 are T_1^0 and T_0; and one may propose that the intermediate values are reasonably well represented by a linear function, viz.

$$T_1 - T_1^0 = T_b^0 - T_b \tag{11-27}$$

so that in narrow channels $(T_1)_{min} - T_1^0 = T_b^0 - (T_b)_{max}$.

The relation between a and T_b is determined by Eq. 11-21 which is rewritten in the form

$$a(x_2 - x_1) = \frac{T_b - T_1}{T_1 - T_u} \tag{11-28}$$

Since the distance $x_2 - x_1$ is a constant throughout the wave, one obtains

$$\frac{a}{a^0} = \frac{(T_b - T_1)(T_1^0 - T_u)}{(T_b^0 - T_1^0)(T_1 - T_u)} \tag{11-29}$$

Eq. 11-26 is now integrable. We introduce the dimensionless variable

$$\theta = \frac{T - T_0}{T_b^0 - T_0} \tag{11-30}$$

where T may be any temperature between T_b^0 and T_u. The temperature

$T = (T_b)_{\max} \leqq T_b^0$ corresponds to $\theta_{\max} \leqq 1$; $T = T_0$ corresponds to $\theta = 0$; and $T = T_u$ corresponds to $-\theta_u = (T_u - T_0)/(T_b^0 - T_0)$. In place of the distance y, we introduce the dimensionless variable

$$v = a^0 y \tag{11-31}$$

The width of the "dead" space, that is, the distance of the fringe of the wave from the wall, is y_0. In terms of the dimensionless variables v and θ, the fringe is defined by $v = v_0$ and $\theta = 0$. Setting

$$\tau = \frac{T_1^0 - T_u}{T_b^0 - T_1^0} \tag{11-32}$$

and since, according to Eq. 11-27, $T_0 = (T_b^0 + T_1^0)/2$, one may write Eq. 11-27 in the form

$$\frac{T_1 - T_u}{T_b^0 - T_0} = 1 - \theta + 2\tau \tag{11-33}$$

as can be readily verified. Further, $(T_b^0 - T_b)/(T_b^0 - T_0) = 1 - \theta$, and from Eq. 11-27 and 11-29,

$$\frac{a}{a^0} = \frac{2\theta\tau}{1 - \theta + 2\tau} \tag{11-34}$$

Since $(d^2T_b/dy^2)/[a^{02}(T_b^0 - T_0)] = d^2\theta/dv^2$, Eq. 11-26 may be written in the form

$$-\frac{d^2\theta}{dv^2} = \frac{2\theta^2(1 - \theta)\tau^2}{(1 - \theta + 2\tau)(1 - \theta + \tau)} \tag{11-35}$$

If, instead of the assumption of Eq. 11-27 concerning the change of T_1 along the wave, the extreme assumption is made that

$$T_1 = \text{const} = T_1^0 \tag{11-27a}$$

Eq. 11-34 becomes

$$\frac{a}{a^0} = \theta \tag{11-34a}$$

and Eq. 11-35 becomes

$$-\frac{d^2\theta}{dv^2} = \frac{\theta^2(1 - \theta)\tau}{1 - \theta + \tau} \tag{11-35a}$$

The integration of Eq. 11-35 and 11-35a in analytic form has not been attempted, but the equations can be integrated numerically for any given value of τ and assuming upper limits of integration, $\theta \leqq 1$. The required additional boundary condition is

$$\left(\frac{d\theta}{dv}\right)_{\theta_{\max}} = 0 \tag{11-36}$$

which applies to two cases, that of a single surface bordering on a quasi-infinite depth of explosive medium and that of an explosive medium contained by two relatively close parallel surfaces. In the former case, the temperature gradient $dT_b/dy \sim d\theta/dv$ approaches zero asymptotically with increasing distance from the surface, i.e. as T_b approaches T_b^0, corresponding to $\theta_{max} = 1$. In the latter case the temperature gradient passes through zero at $(T_b)_{max}$ halfway between the plates, and the temperature is always $< T_b^0$, i.e. $\theta_{max} < 1$. The integrations yield curves of θ versus $v = v_0$, and the complete solution requires the determination of the integration constant v_0.

The Eq. 11-12 for the dead space integrates to

$$-\frac{dT}{dy} = -\left(\frac{dT}{dy}\right)_{v_0} \tag{11-37}$$

Introducing variables θ and v, and integrating between limits $\theta = 0$ and θ_u, one obtains

$$v_0 = \theta_u\left(\frac{dv}{d\theta}\right)_0 \tag{11-38}$$

where $\theta_u = 1 + 2\tau$.

For a cylindrical tube, Eq. 11-35 and 11-35a are modified by adding the divergence term $(1/v)d\theta/dv$ to the second derivative. The definition of v changes to $v = a^0r$, where r is the distance from the tube axis. For $\theta = 0$, $v = v_0$; and for θ_u, $v = v_R = a^0R$, where R is the radius of the tube. The equation for the dead space then becomes

$$v\frac{d\theta}{dv} = v_0\left(\frac{d\theta}{dv}\right)_0 \tag{11-39}$$

and integration between limits v_0 and v_R yields

$$v_R = v_0e^{(\theta_u/v_0)/(d\theta/dv)_0} \tag{11-40}$$

The foregoing theory of the change of burning velocity and temperature with distance from a cooling surface is applicable to three measurable quantities: (1) the critical boundary velocity gradient for flashback, g_F; (2) the quenching distance between plane-parallel plates, $d_\parallel$; and (3) the quenching diameter of tubes, d_0.

We replace the critical boundary velocity gradient g_F by an equivalent quantity called the depth of penetration, d_P, defined by the equation

$$d_P = \frac{S_u^0}{g_F} \tag{11-41}$$

With reference to Fig. G,5b(left) it is seen that in order to determine the depth of penetration d_P it is necessary to determine the curve of S_u versus distance y from the boundary and draw a tangent to this curve from the origin. d_P is then the distance from the wall at which the tangent line attains the value of the maximum burning velocity. This is illustrated in the lower part of Fig. G,11d in dimensionless parameters for $\tau = 0.2$,

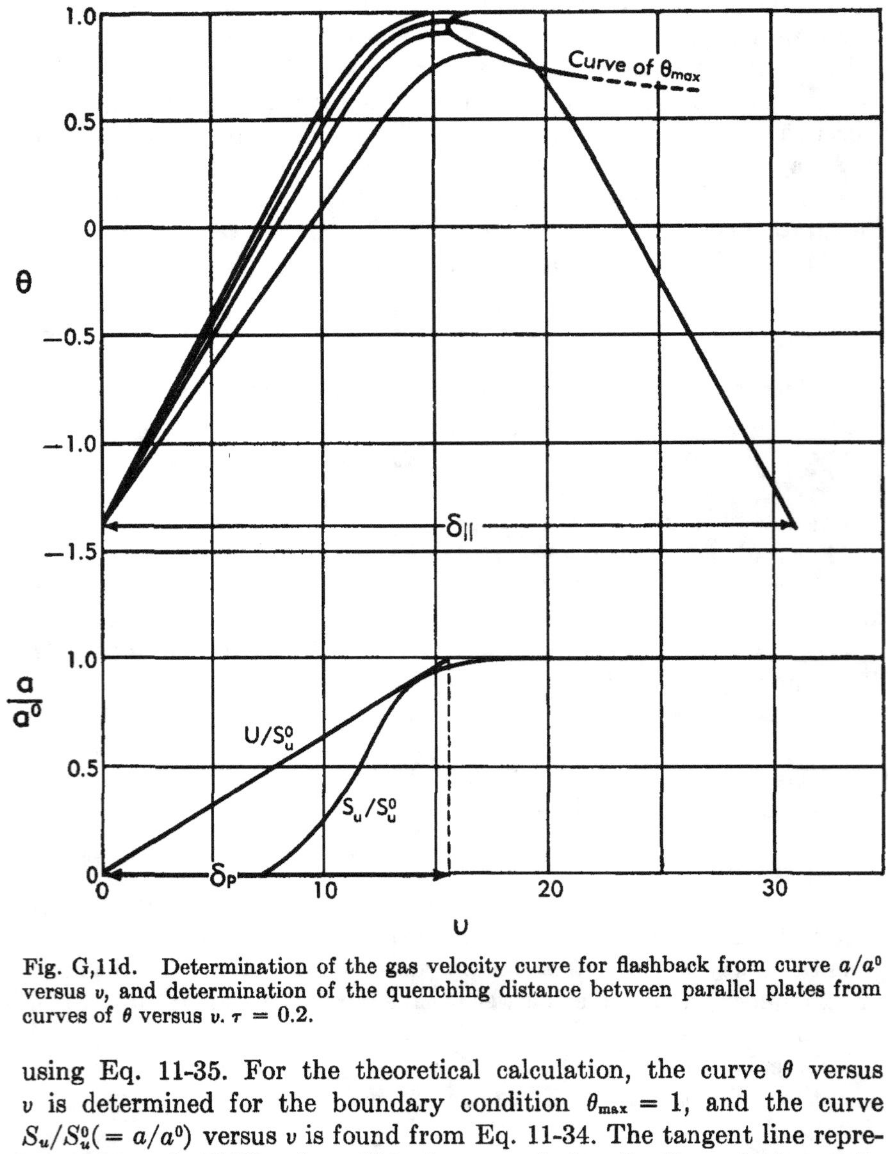

Fig. G,11d. Determination of the gas velocity curve for flashback from curve a/a^0 versus υ, and determination of the quenching distance between parallel plates from curves of θ versus υ. $\tau = 0.2$.

using Eq. 11-35. For the theoretical calculation, the curve θ versus υ is determined for the boundary condition $\theta_{max} = 1$, and the curve $S_u/S_u^0 (= a/a^0)$ versus υ is found from Eq. 11-34. The tangent line represents the ratio U/S_u^0, where U is the gas velocity. In dimensionless units the depth of penetration is

$$\delta_P = a^0 d_P \tag{11-42}$$

For the calculation of quenching distances and quenching diameters, a series of curves θ, υ is determined for $\theta_{max} < 1$. These curves constitute parabolalike arcs, and as the difference $1 - \theta_{max}$ is increased, it is found that the arcs at first contract and then expand again. This is shown in the upper part of Fig. G,11d for $\tau = 0.2$, using Eq. 11-35 as before. Several such arcs are plotted. It is evident that the arc with the smallest span represents the quenching distance and that lower arcs have no physical significance. The value of the abscissa of the smallest span is

$$\delta_{\parallel} = a^0 d_{\parallel} \tag{11-43}$$

or

$$\delta_0 = a^0 d_0 \tag{11-44}$$

depending on whether the calculation has been made for plane-parallel plates or cylindrical tubes. The curve drawn through the maxima is typical in illustrating that the smallest span occurs at values of θ_{max} somewhat below 1.

Since Eq. 11-35 or 11-35a can be integrated for any value of τ, one may calculate curves of δ_P, $\delta_{\parallel}$, and δ_0 as functions of τ. Such curves are shown in Fig. G,11e, using Eq. 11-35. The parameter τ is composed of the adiabatic combustion temperature T_b^0, which is readily calculated for any explosive gas, and the temperature T_1^0, which may be calculated from experimental minimum ignition energies by the procedure that has been indicated above. If the value of τ is known for an actual gas, values of δ_P, $\delta_{\parallel}$, and δ_0 may be read from the curves. The actual values of quenching distances and flashback limits, expressed by the parameters d_P, $d_{\parallel}$, and d_0 in centimeters, are obtained from Eq. 11-42, 11-43, and 11-44, a^0 being composed of measurable quantities.

Calculation of blowoff limits. The mechanism by which a burner flame maintains itself at the rim of the burner tube has been illustrated in Fig. G,5a, which shows a combustion wave in three positions with respect to the burner rim, one position corresponding to equilibrium and the other two corresponding to displacements from equilibrium either too close to or too far away from the rim. Fig. G,5a will be used as a model to formulate the equations for the equilibrium position.

The gas flow is assumed to be parallel to the x axis. x is taken as zero in the plane of the rim, and y is zero at the jet boundary. Divergence of the heat flow due to wave curvature is again neglected, so that heat flow is assumed to occur in the x direction only. For any point x, y the mass flow $\dot{m}$ equals $\rho_u U_u$, so that in front of the reaction zone where $q = 0$ the general steady state equation (10-2) becomes

$$\rho_u U_u c_p \frac{dT}{dx} = \lambda \frac{d^2T}{dx^2} \tag{11-45}$$

which, by integration between the limits T_u at $x = 0$ and T at x, becomes

$$\rho_u U_u c_p (T - T_u) = \lambda \frac{dT}{dx} - u_0 \tag{11-46}$$

The integration constant u_0 is the heat flow to the rim and is equal to λ

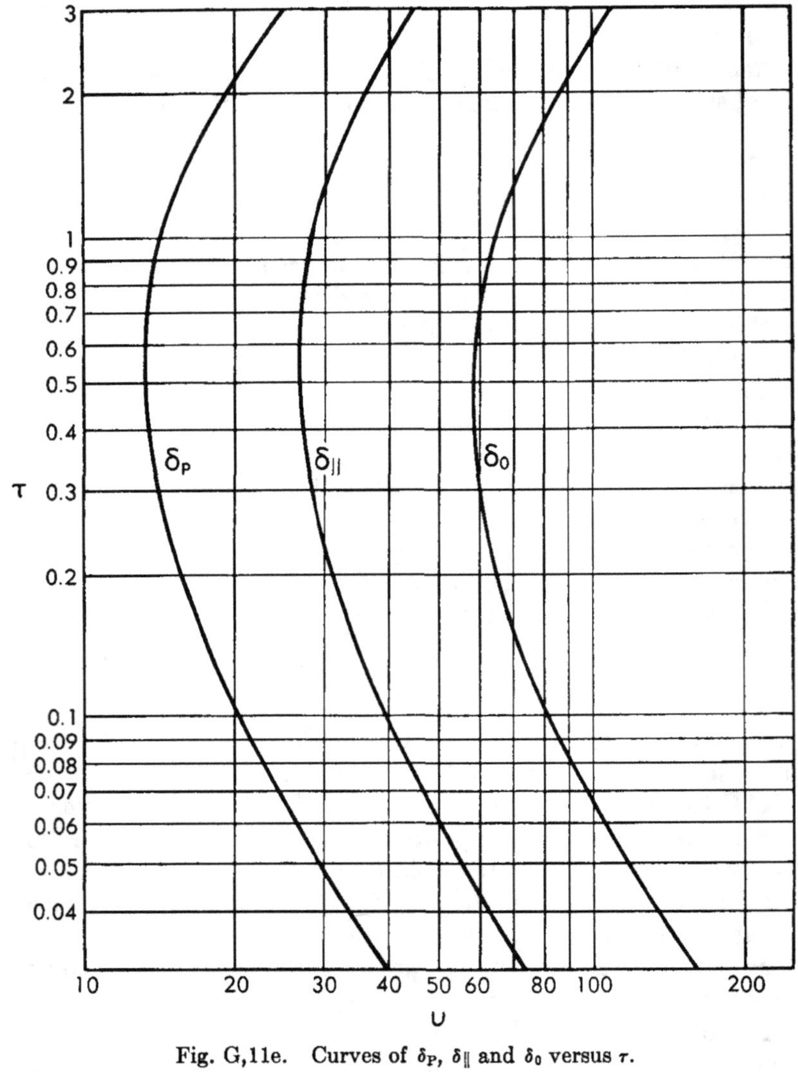

Fig. G,11e. Curves of δ_P, $\delta_\parallel$ and δ_0 versus τ.

times the temperature gradient at $x = 0$. Considering specifically a flow line in which the gas velocity equals the burning velocity S_u, Eq. 11-46 becomes

$$\rho_u S_u c_p (T - T_u) = \lambda \frac{dT}{dx} - u_0 \tag{11-47}$$

Integration between $x = 0$ and $x = x_1$ at the front of the reaction zone yields, with $a = \rho_u S_u c_p / \lambda$,

$$x_1 = \frac{1}{a} \ln \left[1 + \frac{a\lambda(T_1 - T_u)}{u_0} \right] \tag{11-48}$$

x_1 is the distance from the plane of the burner rim to the front of the reaction zone along any flow line in which the gas velocity equals the burning velocity. A flow line of equal gas velocity and burning velocity exists for the wave in unstable positions (position 3), as well as for the wave in the equilibrium position (position 2). For the latter position x_1 becomes a minimum.

Denoting by t_1 the time in which a gas element travels from $x = 0$ to x_1,

$$x_1 = S_u t_1 \tag{11-49}$$

and since $a/a^0 = S_u/S_u^0$, Eq. 11-48 is rewritten

$$a^0 S_u^0 t_1 = \left(\frac{a^0}{a}\right)^2 \ln\left[1 + \frac{a\lambda(T_1 - T_u)}{u_0}\right] \tag{11-50}$$

As before, the reaction zone is divided into an adiabatic and an isothermal zone. The adiabatic zone is defined by $(dT/dx)_1 = (T_b - T_1)/(x_2 - x_1)$, and hence from Eq. 11-47

$$\rho_u S_u c_p (T_1 - T_u) = \frac{\lambda(T_b - T_1)}{x_2 - x_1} - u_0 \tag{11-51}$$

The total heat evolved per unit time is $(x_3 - x_1)q$. The heat gained by the gas stream per unit time is $\rho_u S_u c_p(T_b - T_u)$, and the rate of heat loss to the surface is u_0. Therefore, the heat balance is

$$(x_3 - x_1)q = \rho_u S_u c_p (T_b - T_u) + u_0 \tag{11-52}$$

If no heat loss occurred, T_b would rise to T_b^0, and this equation would reduce to Eq. 11-22. Subtracting Eq. 11-52 from Eq. 11-22 and dividing by λ,

$$\frac{u_0}{\lambda} = a(T_b^0 - T_b) \tag{11-53}$$

Eq. 11-27, 11-29, and 11-32 remain applicable, as before. Combining with Eq. 11-52 one obtains by a series of algebraic operations

$$\frac{a\lambda(T_1 - T_u)}{u_0} = \frac{1 + 2\tau}{1 - a/a^0} \tag{11-54}$$

On substituting in Eq. 11-50

$$a^0 S_u^0 t_1 = \left(\frac{a^0}{a}\right)^2 \ln\left[\frac{2(1 + \tau) - a/a^0}{1 - a/a^0}\right] \tag{11-55}$$

At equilibrium, t_1 becomes a minimum since x_1 is a minimum. By minimizing Eq. 11-55, one obtains

$$2 \ln\left[\frac{2(1 + \tau) - a/a^0}{1 - a/a^0}\right] = \frac{a}{a^0}\,\frac{1 + 2\tau}{2(1 + \tau) - a/a^0}\,\frac{1}{1 - a/a^0} \tag{11-56}$$

with τ, a^0, and S_u^0 functions of mixture composition only. The mixture composition is determined by the dilution of the original mixture with gas from the atmosphere surrounding the jet. If the original mixture consists of fuel and air and if the surrounding atmosphere is air, the composition of any mixture formed by interdiffusion is defined by a single variable, namely, the mole fraction X_f of the fuel. To any value of X_f there corresponds a value of τ, a^0, and S_u^0. τ determines a/a^0 and $a^0S_u^0t_1$ from Eq. 11-55 and 11-56. Since $S_u = S_u^0a/a^0$, the value of S_u is also determined. Likewise, the value of t_1 is determined from S_u^0, a^0, and $a^0S_u^0t_1$. It is thus possible to construct curves of S_u and t_1 versus X_f, provided that data on S_u^0 and τ are available as functions of X_f.

An equation must now be developed for the diffusion process. Consider a layer of the original mixture of composition X_{f_0} emerging from the tube. The instant at which the layer enters the free jet is zero time for the diffusion process. The total diffusion time is t_1. To facilitate the treatment, it is assumed that the layer moves uniformly at the velocity S_u and that diffusion occurs in the y direction only. The diffusion equation then becomes [*41*, p. 537]

$$\frac{2X_f}{X_{f_0}} = 1 \pm \psi\left(\frac{y}{2\sqrt{Dt_1}}\right) \tag{11-57}$$

where ψ denotes the probability integral and D is the diffusion coefficient. Since, according to Eq. 5-1,

$$y = \frac{S_u}{g} \tag{11-58}$$

the argument of the probability function is $S_u/2g\sqrt{Dt_1}$. This quantity can be determined from Eq. 11-57 for any value of X_f/X_{f_0}, or, since X_{f_0} is known, for any value of X_f.

It is now possible to determine a series of values for g. For this purpose, a value $X_f < X_{f_0}$ is chosen. This determines $S_u/2g\sqrt{Dt_1}$ from Eq. 11-57 and 11-58. g is then found from data of the diffusion coefficient D and values of S_u and t_1 are obtained from the curves S_u and t_1 versus X_f. On choosing progressively smaller values of X_f, g increases at first and then decreases. The maximum value of g is obviously g_B, the critical boundary velocity gradient for blowoff.

This concludes the theoretical investigation of problems associated with the inner structure of the combustion wave. In the following article, the validity of the concept of associating minimum ignition energy and excess enthalpy is tested on the basis of experiment. In Art. 14, a comparison is made of calculated and experimental quenching, flashback and blowoff limits by means of values of T_1^0 or τ obtained from ignition experiments.

CHAPTER 2. EXPERIMENTAL PHENOMENA OF COMBUSTION WAVES

G,12. Ignition: Experiments and Comparison with Theory.

Ignition by electric sparks. The closest approach to an instantaneous point source of ignition is a capacitance spark. In determining the threshold energy of ignition it is necessary to take into account the quenching effect of the spark electrodes. Such experiments have been performed by Blanc, Guest, von Elbe, and Lewis [*43*; *3*, p. 363]. Since the details have been reported elsewhere [*37*, pp. 390–418], it will suffice to summarize the important results of these studies.

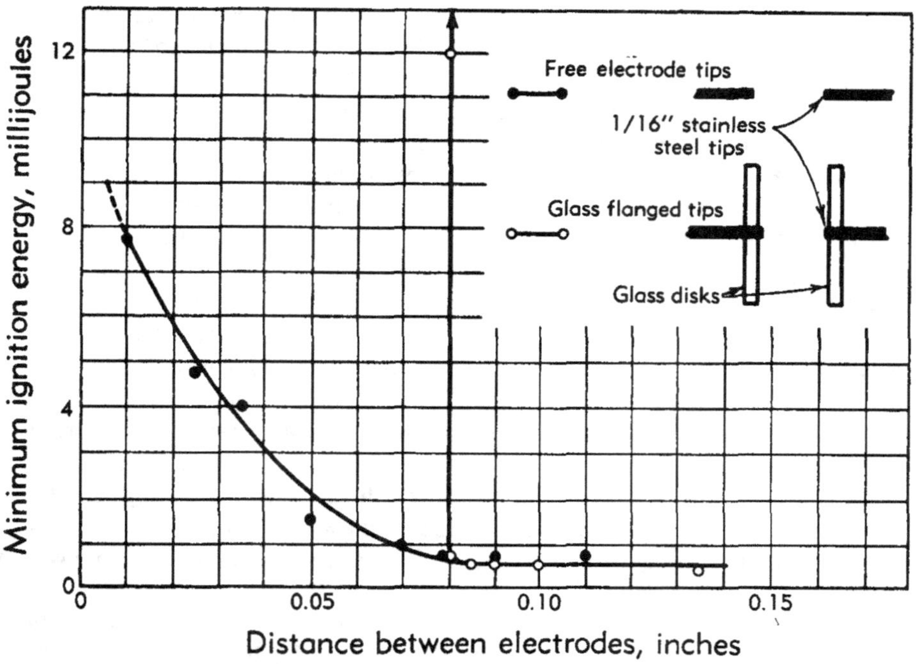

Fig. G,12a. Minimum ignition energies for free and glass-flanged electrode tips as function of electrode distance. Stoichiometric mixture of natural gas (about 83 per cent CH_4 + 17 per cent C_2H_6) and air at one atmosphere pressure.

Fig. G,12a shows minimum ignition energies as a function of distance of separation of the electrodes. The minimum ignition energies were determined for a given electrode setting by trial and error tests in which the capacitance of the discharge circuit was varied until the ignition threshold was reached. The value of the energy was computed as the product $\frac{1}{2}CV^2$, where C is the capacitance at the ignition threshold and

V, the voltage. As shown by Fig. G,12a, when the electrode tips are flanged by glass disks, a well-defined distance, the quenching distance, is found below which no ignition can be obtained even with very heavy sparks. If the flanges are removed, the gas can be ignited below this distance but the energy increases as the electrodes approach each other. This increase in the minimum energy occurs only below the quenching distance, and above this distance the value of the minimum ignition energy is the same for point or flanged electrodes.

Calorimetric measurements of the heat generated by sparks [*44*] have shown that immediately after a spark discharge over 95 per cent of the electric energy, $\frac{1}{2}CV^2$, is available in the gas in the form of heat. This heat distributes itself in time over the gas volume and the electrode material, the latter ultimately absorbing the major share. The measurements were carried out in a manner permitting observation of the rate of heat loss to the electrodes. It was thus possible to make an order-of-magnitude comparison of the time in which a significant fraction of the spark energy is lost to the electrodes with the time during which the spark develops to a combustion wave. Though there is no experimental criterion for the latter time, an upper bound is set by the ratio of wave width to burning velocity as may be inferred from the discussion in Art. 10. The comparison shows that the spark develops into a combustion wave long before any significant fraction of the spark energy is absorbed by the electrodes. This applies to electrode distances not only above but also somewhat below the quenching distance. Whether or not ignition occurs, i.e. the wave continues to propagate, depends on the shape and separation of the electrodes. If the latter constitute flat surfaces, propagation is possible only when the separation exceeds the quenching distance. Beyond the quenching distance, the rapidity of combustion wave development from the spark insures that heat loss to the electrodes is no factor in the threshold conditions of ignition, a fact which is confirmed by Fig. G,12a. With point electrodes below the quenching distance it is possible to compensate the initial heat loss of the spherically growing flame by increased spark energy.

The dependence of the values of minimum ignition energy on electrode distance and pressure is further illustrated by the family of curves in Fig. G,12b. Beyond the quenching distance, the minimum ignition energy is seen to be remarkably independent of electrode distance over a considerable range of distances and pressures. If the spark energy were distributed uniformly over the length of the spark gap, this fact would be difficult to understand; on the contrary, inasmuch as the ignition threshold should in this case correspond to a critical energy per unit length of the spark, one would expect the total critical energy to increase linearly with gap length. It appears therefore that some process operates which concentrates the energy into a more compact volume between the elec-

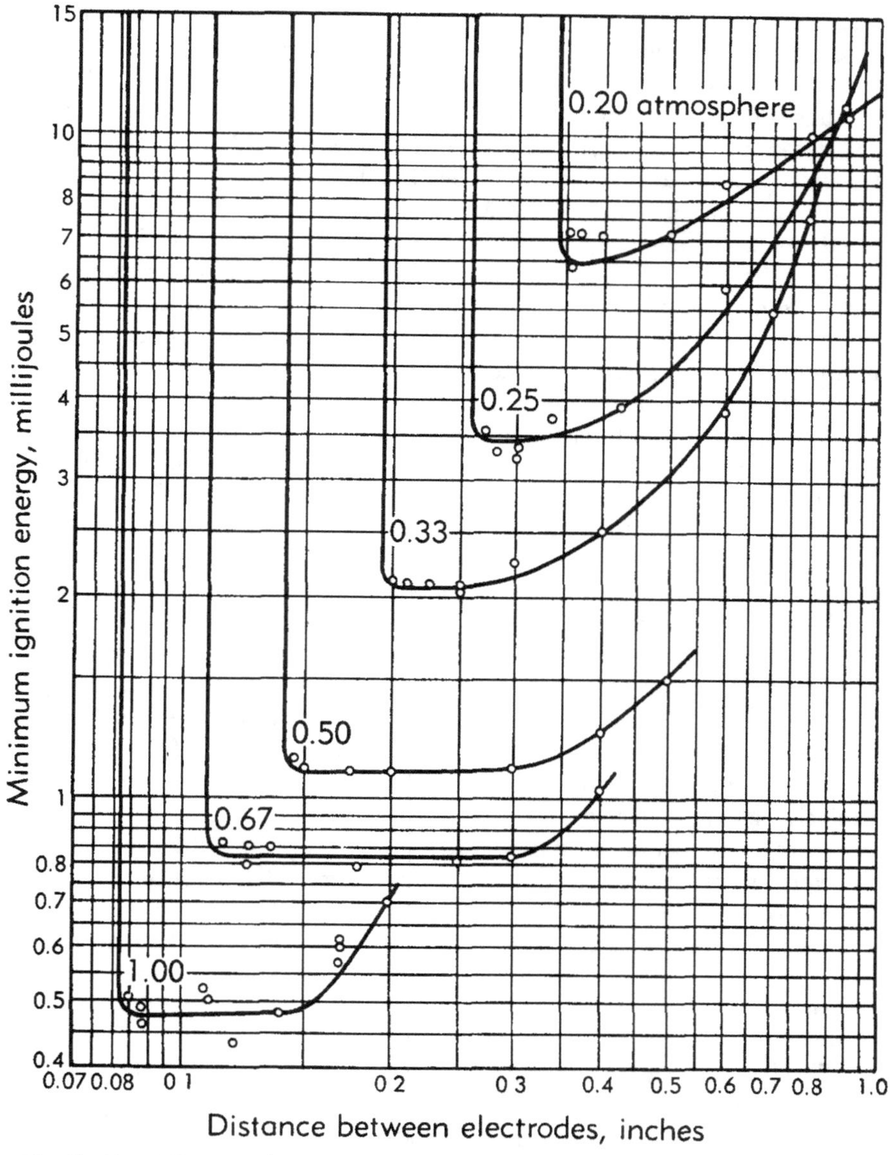

Fig. G,12b. Minimum ignition energies for glass-flanged electrode tips as functions of electrode distance and pressure. Stoichiometric mixture of methane and air.

trodes. Blanc, Guest, von Elbe, and Lewis have suggested that this is caused by nonuniform dissipation of electric energy along the gap length. They called attention to present theories of the sparking mechanism [*45*], according to which considerable space charges are formed before breakdown due to formation of positive ions in the gap, so that the potential distribution is not linear between the electrodes and, at least in the initial phase of the breakdown, the energy dissipation is not uniform. Alternatively, one may consider the possibility that processes which occur after the electric discharge are responsible for the effect. In particular, it may be suggested that within the spark channel a considerable mass flow is generated by movement of ions in the electric field and momentum transfer between ions and other molecules. This mass flow must continue after the discharge is completed and may conceivably contract the energized gas in the channel to a more compact volume by a particular flow configuration. Experimental evidence for such a process seems to be provided by high speed moving-picture records. Three records are shown in Plate G,12 which records the events after the passage of sparks near the minimum ignition energy threshold through jets of explosive methane-air mixtures, the jet orifice being outside of the picture below the electrodes. In Plate G,12(1) the flame is seen to approach extinction in subsequent frames. The sequences of frames show the development of a spark channel between the electrodes and the subsequent contraction of the channel to a much more compact volume. The photographs strongly suggest that this comes about by mass flow which, curiously, is directed away from the negative electrode and consolidates the energized gas toward the center of the gap but somewhat closer to the positive electrode. A satisfactory explanation of this phenomenon in terms of the mechanism of spark discharge and the aftereffects of the spark does not seem possible at present. The occurrence of mass flow in electric discharges has long been known as "electric wind," but thus far no systematic study of such effects seems to have been undertaken which has a bearing on the problem of ignition of explosive gases by sparks.

In this connection, mention is made of the striking observation that for very large spark energies the quenching distance between glass-flanged plate electrodes is found to be not smaller but larger than for low energies. For the explanation of this effect it is suggested that so much turbulence is produced by very large sparks that the heat loss to the plates outweighs the additional energy, and the plates must be separated farther in order to obtain ignition.

We shall restrict the meaning of the term "minimum ignition energy" to denote only the smallest values of energy that are found along curves of ignition energy versus electrode distance as exemplified by Fig. G,12b. Such minimum ignition energies are essentially only functions of mixture composition and pressure. Their values have been determined for numerous mixtures of hydrocarbons, or hydrogen with oxygen and inert gas at various pressures; quenching distances have been simultaneously determined. Typical families of curves are shown in Fig. G,12c for mixtures of ethane, oxygen, and nitrogen.

An interesting and very remarkable relation is found when the logarithms of minimum ignition energies are plotted against the logarithms

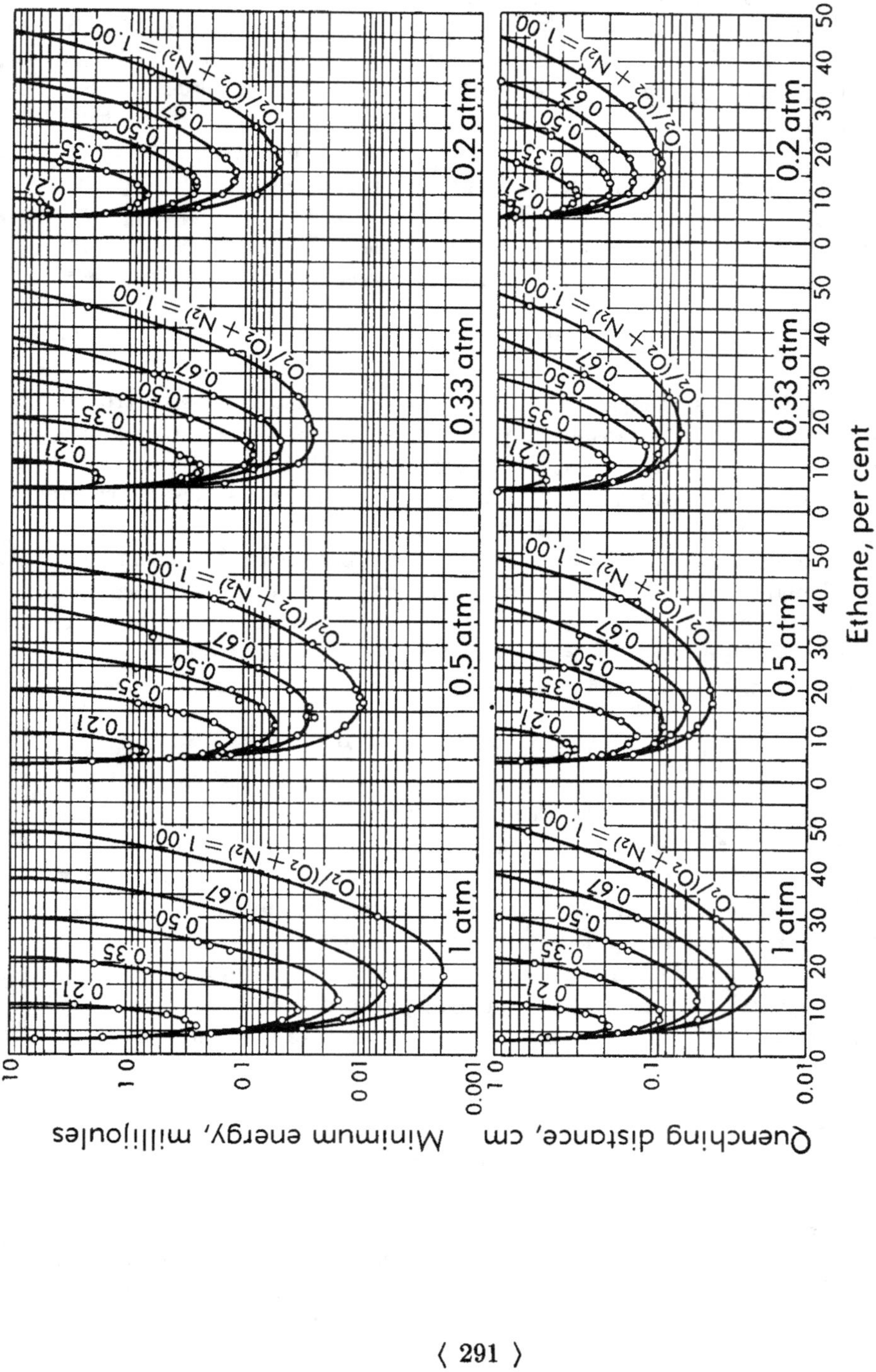

Fig. G,12c. Minimum spark ignition energies in millijoules of mixtures of ethane, oxygen, and nitrogen at one atmosphere pressure and lower, and quenching distances between flanged electrodes for the same mixtures. Curves correspond to constant ratios of oxygen and nitrogen.

of quenching distances. It is found that for a wide variety of fuel-oxygen-nitrogen mixtures over a large range of pressures the points are grouped within a narrow band whose slope has a value of 2, corresponding to the approximate relation $H_{cr} = Cd^2$ where H_{cr} is the minimum ignition energy defined earlier, d the quenching distance, and C a quasi-constant term. This equation may be compared with the theoretical Eq. 10-26 which shows that the minimum ignition energy is proportional to the critical value h_{cr} of the excess enthalpy per unit area of the combustion wave and to the square of the minimum flame radius. The latter is of the order of magnitude of $d/2$ so that the right-hand side of Eq. 10-26 is approximately $\pi d^2 h_{cr}$. As discussed in Art. 10 the value of h_{cr} cannot be considerably smaller than h^0. Using Eq. 11-8 we may therefore write approximately

$$H_{cr} \cong \pi d^2 h^0 \cong \pi d^2 \frac{\lambda}{S_u^0} (T_b^0 - T_u) \tag{12-1}$$

This equation has been tested on mixtures of methane-oxygen-nitrogen and hydrogen-air at atmospheric pressure [*37*, pp. 390–418]. Results are shown in Table G,12a. Additional tests of Eq. 12-1 recently made by Calcote [*46*] for a large number of fuels show similar agreement between experiment and theory as in Table G,12a.

The occasional surprisingly close agreement between the experimental and calculated values of the minimum ignition energy must be considered fortuitous; however, the fact that experiment and theory agree with respect to order of magnitude and trend cannot be dismissed and may be accepted as proof that the theory, though strongly simplified, is basically correct.

Comparison of Eq. 12-1 with the equation $H_{cr} = Cd^2$ shows that the factor C is equal to $\pi(\lambda/S_u^0)(T_b^0 - T_u)$. It can be shown that this factor is indeed quasi-constant in the sense that it varies no more than is consistent with the scattering of points in a diagram of $\ln H_{cr}$ versus $\ln d$.

A review of other literature on spark ignition indicates that previously held views, according to which various electric parameters of the spark discharge rather than the total energy play a determining role in spark ignition, were based on erroneous interpretation of experimental observations. In particular the role of quenching distance was not previously recognized. A critical review of the earlier literature is found elsewhere [*37*, pp. 419–425].

Ignition by other sources. Fig. G,10b illustrates schematically relations among source energy, critical source temperature, and critical heating period for the ideal case of an ignition by a heated wire of zero heat capacity. In actual ignition experiments the energy input comprises not only the source energy as defined in Art. 10 but also the heat absorbed by the wire material. Let us imagine that the heat absorbed by the wire is much larger than the source energy. In this case the minimum energy input for ignition corresponds to a source temperature on the low quasi-constant branch of the T_{cr} curve; that is, the source temperature is substantially independent of the critical heating period. The apparent minimum ignition energy is thus larger than the true value by orders of magnitude. A case of this kind has been studied by Jones [*47*; also *37*, pp. 429–437]. This investigator experimented with fine wires embedded in

Table G,12a. Comparison of experimental and calculated minimum ignition energies. Atmospheric pressure and room temperature.

Mixture composition, per cent			S_u^0, cm/sec	T_b^0, °K	$\lambda \times 10^5$, cal/cm sec °C	d, cm	Minimum ignition energy H_{cr}, cal $\times 10^6$	
CH_4	O_2	N_2					Exptl.	Calc.
10	90	—	80	2200	6 3	0.072	19	24
15	85	—	175	2650	6.4	0.050	3.1	6.7
25	75	—	304	3000	6.5	0.028	1.4	1.4
40	60	—	305	3000	6.7	0.051	5.0	4.9
50	50	—	122	2650	6.9	0.165	110	109
52.5	47.5	—	87	2500	6.9	0.38	430	790
10	18.8	71.2	36	2200	6.3	0.28	170	830
16.3	29.3	54.4	110	2620	6.4	0.079	20	26
21.5	39.75	39.75	170	2810	6.5	0.053	9	8.4
26.4	49.1	24.5	240	2940	6.5	0.045	5.3	4.5
H_2	Air							
20	80		(100)*	1910	7.2	0.071	6.2	(18)*
30	70		195	2300	8.0	0.064	4.5	10
40	60		265	2240	9.0	0.076	6.6	12
57	43		190	1850	9.0	0.165	49	63

* This value uncertain.

matchhead compositions. From data of energy input versus critical heating period under various conditions of wire length and diameter and wire material it was possible to determine an equation of the energy input E of the form

$$E = (K + La)l + (H + Jl)t \tag{12-2}$$

In this equation the first term represents the energy input for zero critical heating period (corresponding to a very large current applied for a very short time). The length and cross-sectional area of the wire are denoted by l and a, respectively. The heat absorbed by the explosive medium, per unit length of wire, is K, and L is the heat content per unit volume of the wire at the quasi-constant source temperature. The second term of Eq. 12-2 comprises the contribution of the length of the critical heating period to the energy input. If the critical heating period t is larger than zero, some heat is lost to the wire leads, and since, under such conditions, the temperature wave penetrates farther into the explosive medium than under conditions of instantaneous heating, the explosive medium absorbs an amount of heat exceeding Kl. The term H represents the rate of heat loss to the wire leads and J represents rate of heat absorption by the medium, per unit length of wire. For further details of this illuminating investigation the reader is referred to the literature. Table G,12b shows values of source temperature T_{cr} and source energy K for three matchhead compositions. It is seen that the critical source temperatures are of rather low order.

A very different type of ignition source is illustrated by the experiments of Bowden and coworkers [*48*] on the impact sensitivity of nitroglycerin containing a minute gas bubble. The experiments demonstrate that ignition is due to the energy imparted to the gas bubble by adiabatic compression. Gas-free nitroglycerin is otherwise very insensitive to impact. From the data of critical bubble volume and compression ratio one obtains a minimum ignition energy of the order of 10^{-8} calories. This

Table G,12b. Values of source temperature T_{cr} and source energy K for various matchhead compositions.

Matchhead*	1	2	3
T_{cr}, °C.	380	440	740
K, cal/cm	0.5×10^{-3}	0.6×10^{-3}	1.0×10^{-3}

* Composition 1. 4 parts lead mononitroresorcinate to 1 part potassium chlorate by weight.
Composition 2. Copper acetylide.
Composition 3. 1 part charcoal to 5 parts potassium chlorate by weight.

represents an absolute minimum ignition in the sense discussed in Art. 10. From data of bubble diameter, predetonation burning velocity [*49*], adiabatic combustion temperature, and specific heat and heat conductivity of nitroglycerin, we compute from Eq. 12-1 a value of minimum ignition energy of the order of 10^{-8} calories.

Experiments of Stout and Jones [*3*, p. 329] on the ignition of a rich methane-air mixture with fine wires of the type employed in the matchhead experiments lead to a result very different from the latter experiments. The length of the wires did not exceed the dimensions of the minimum flame volume and the results indicate that for very short critical heating periods the energy required for ignition is simply the sum of the absolute minimum ignition energy and the heat absorbed by the wire material [*37*, pp. 443–444]. Under such conditions the critical source temperature is found to be somewhat in excess of the adiabatic flame temperature. The complicating conditions that can be incurred in such experiments are illustrated by data of the investigators on hydrogen-air mixtures. For such mixtures the minimum flame diameter was always much smaller than the wire length and the evidence points to ignition from tiny arcs that were forming at the instant that the current-carrying wire was melting and collapsing into droplets [*37*, pp. 445–446]. It is

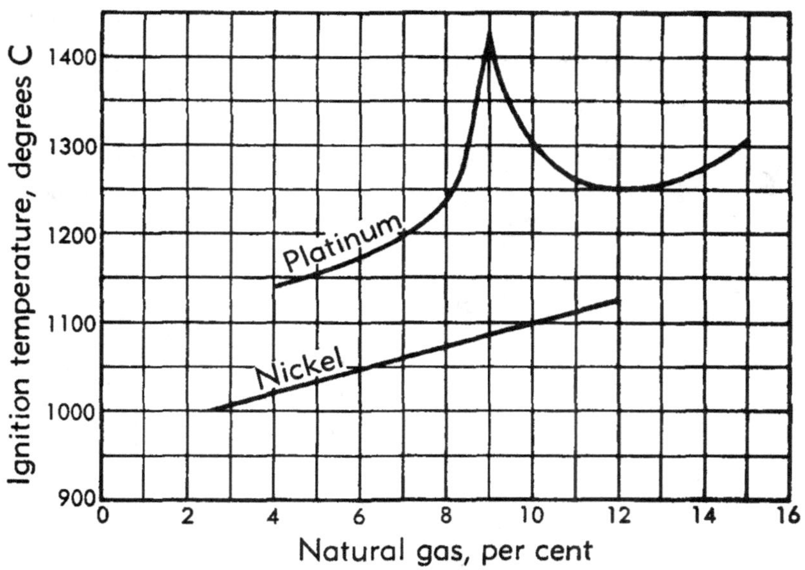

Fig. G,12d. Ignition of natural gas-air mixtures by heated bar of nickel and platinum (Coward and Guest).

evident that in experiments on ignition, even though experimental procedures are seemingly identical, much attention should be given to the possible change in the nature of the ignition process due to a change of some experimental parameter.

The experiments of Coward and Guest [*50,51*] on ignition of natural gas-air mixtures by heated bars of nickel and platinum are instructive. The former is a noncatalytic and the latter a catalytic metal. In agreement with the discussion in Art. 10 it was found that for the platinum bar the ignition temperature is always substantially higher than for the nickel bar. Representative data are shown in Fig. G,12d. Of interest is the sharp peak in the platinum curve which occurs at stoichiometric composition. This suggests that ignition is less readily obtained if the hot product mixture that surrounds the bar contains no excess oxygen or fuel.

G,13. Observations on the Propagation of Combustion Waves.

Propagation in tubes of large diameter. The gas flow through a combustion wave in a tube described in Art. 8 and depicted in Fig. G,8b can be studied experimentally in the following manner. A tube with one end closed and the other end open to the atmosphere is filled with explosive gas in which dust particles of suitable material, such as magnesium oxide, are suspended. The gas is ignited at the open end and as the combustion wave travels toward the closed end a camera moves parallel to the tube at the same rate, so that the image of the wave remains stationary. By means of a narrow sheet of light that bisects the tube, the flow lines are

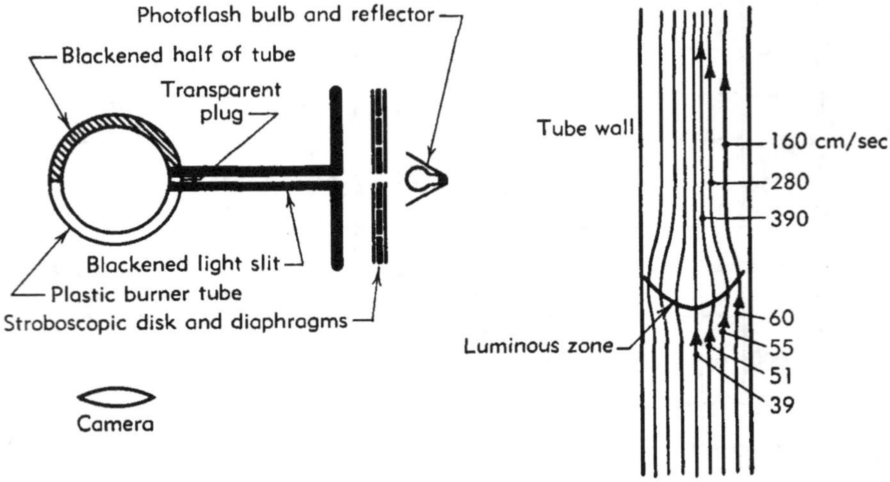

Fig. G,13. Investigation of combustion wave propagation in tubes by the particle-track method. Left, scheme of optical system. Burner tube closed at bottom and filled with explosive gas containing suspended MgO particles. Camera moving parallel to tube axis at the velocity of the combustion wave. Right, tracing of an experimental record. Tube diameter, 18 mm; gas mixture, 4.7 per cent ethane in air; burning velocity approximately 32 cm/sec. Arrows on flow lines indicate particle velocity with respect to wave.

made visible by the tracks of the dust particles on the film. A stroboscopic arrangement interrupts the particle tracks at constant time intervals, thus providing a measure of the particle velocity with respect to the combustion wave. Such experiments have been carried out by Grumer [*52*]. The apparatus used is shown schematically in Fig. G,13. The photographic records were not sufficiently contrasted to permit reproduction here. The original films clearly show the expansion and subsequent contraction of the central stream tubes and the corresponding contraction and subsequent expansion of the stream tubes near the wall as described in Art. 8. One of these records has been analyzed and drawn in Fig. G,13.

As the tube diameter is increased, the axial depth of the paraboloidlike surface of the combustion wave increases more than proportionally to the tube diameter, and thus the ratio of the areas of the combustion wave and tube cross section increases. Consequently, the rate of propagation of the wave along the tube increases with increasing tube diameter. Above a critical diameter the flow becomes turbulent but records of flame movement in nondetonable mixtures show [*40*] that the average rate of propagation of the wave may remain sensibly constant. On curves of the rate of uniform movement versus tube diameter a break is noted in the transition region from the laminar to the turbulent regime. In tubes of all sizes the acoustic vibrations of the gas column may be induced which are coupled with the combustion wave propagation causing the wave to oscillate. In the laminar regime these oscillations may be observed coincident with the arrival of the combustion at the antinodes, and between such zones of disturbance the movement is uniform. Experiments have established no limit to the increase of rate of flame propagation with tube diameter. In large tubes or ducts the initial uniform movement gives way to an accelerating movement resulting from increased turbulence. The mass flow may approach sound velocity and destructive effects, resembling those of detonation waves, are observed.

Isotropic and nonisotropic propagation. In Art. 4 it was pointed out that a concave curvature of the wave surface with respect to the unburned gas may in general be expected to increase the burning velocity because of the convergence of the heat flow into the unburned gas, but that in certain cases the opposite effect is possible. The latter effect is attributable to the divergence of the diffusion stream that carries molecules of reactants from the unburned to the burned side of the wave, and arises when the diffusion processes cause a shift of the composition of the unburned gas in the direction of the limits of inflammability. Experiments of Manton, von Elbe, and Lewis [*53*; also *37*, pp. 308–313] have shown that this always occurs in either fuel lean or rich mixtures when the deficient component is also the component of larger diffusivity. In these experiments the development of a combustion wave from an electric spark in a quiescent mixture was observed by high speed schlieren motion pictures. It was found that propagation was spherically symmetric for mixtures such as carbon monoxide and air and ethylene and air, whose components have about equal diffusivities. The propagation was also spherically symmetric in lean mixtures of higher hydrocarbons, propane, butane, etc., with air, and in rich mixtures of hydrogen or methane with air. On the other hand, in rich mixtures of higher hydrocarbons and lean mixtures of hydrogen or methane, spherical symmetry was never obtained, the wave surface developing more or less regular protrusions and recesses. In such mixtures the deficient component is also the component of larger diffusivity. The two types of flame propagation may be distinguished as isotropic and nonisotropic. For the details of the observations and photographs illustrating the effect the reader is referred to the literature [*37*, pp. 308–313].

There are a number of other manifestations of the phenomenon [*37*, pp. 276–279]. In burner flames of mixtures that give rise to nonisotropic propagation, the flame tip is apt to open up above some critical gas flow, disrupting the continuity of the combustion wave and permitting the passage of unburned gas. The inner cone of a Bunsen cone may develop ridges along which the burning velocity is reduced and, in the case of rich hydrocarbon mixtures, the change of mixture composition is visibly demonstrated by the change of color of the luminescent radiation and by carbon streamers originating from the ridges. In such instances the mixture is locally enriched so that solid carbon is formed. A flame of hydrogen-oxygen-carbon dioxide mixture in a stream issuing from a sintered glass plate disintegrates into filaments, and rich flames of hydrocarbons and air exhibit fluctuating discontinuities. In mixtures of hydrogen and inert gases near the lower limit of inflammability the breakup of the combustion wave produces many striking phenomena. This is particularly well demonstrated when the mixture is enclosed in a vertical tube and ignited at the bottom. Systematic investigations of the cellular breakup of rich hydrocarbon flames as a function of the molecular weight of the fuel and the pressure were made by Markstein [*54*; also *37*, pp. 330–333].

Inasmuch as the phenomenon may be expected to play a determining role at the limits of inflammability, it is probably chiefly responsible for the change of the limits with experimental conditions such as upward and downward propagation in tubes. Except for some observations on hydrogen mixtures this aspect of the phenomenon has not yet been studied.

G,14. Stability and Quenching Limits, and Structure of Burner Flames.

Stability limits. The principles of flame stabilization discussed in Art. 5 are readily applicable to the experimental determination of critical boundary velocity gradients of blowoff and flashback as a function of the parameters of the explosive gas, particularly mixture composition. In such experiments cylindrical burner tubes are used and the gas flow V of the mixture is increased or decreased to the point at which blowoff or flashback, respectively, occurs. The boundary velocity gradient, g, is found by determining the differential quotient $-dU_u/dr$ at the tube wall from the Poiseuille equation (6-1). The result is

$$g = 2nR \qquad (14\text{-}1)$$

where R is the radius of the tube and n is a constant. The flow V is equal to

$$\int_0^R 2\pi U_u r dr$$

so that

$$V = \tfrac{1}{2}\pi n R^4 \tag{14-2}$$

It follows that

$$g = \frac{4}{\pi}\frac{V}{R^3} \tag{14-3}$$

so that g_B and g_F are readily calculable from experimental values of tube diameter and critical flows of blowoff and flashback. Experiments have shown that g_B and g_F are indeed independent of tube diameter provided the measurements are made in the laminar regime and the tube diameter is large compared to the quenching diameter. As expected, g_B is found to be strongly dependent on the nature of the surrounding atmosphere, whereas, of course, g_F is unaffected by it. Rich flames burning in air are very stable and the stability is further increased by an oxygen surrounding. On the other hand, in an atmosphere of inert gas, values of g_B are small for both rich and lean mixtures. Data available in the literature [*37*, pp. 279–308] include mixtures of commercial gases and pure hydrocarbons with air, hydrogen with air and oxygen, and acetylene with oxygen.

In the turbulent regime the flow at the boundary may remain sufficiently laminar so that the stabilization mechanism is not affected by velocity fluctuations. It has been shown by Wohl, Kapp, and Gazley [*3*, p. 3; *55*] and Bollinger and Williams [*56*] that values of g_B obtained from the semiempirical equations for turbulent flow in the engineering literature permit excellent correlations with experiments in the laminar range. This subject is discussed further in Sec. I.

When the boundary velocity gradient exceeds the critical value for blowoff the flame may find a position of stability at some distance above the rim of the burner which may greatly exceed the burner diameter. Such so-called lifted flames are formed by fuel rich mixtures burning in air and, as fully discussed under Sec. I, they owe their stability to the diffusion of fuel gas into the surrounding slow-moving atmosphere.

For inverted flames anchored at the end of a wire mounted in the axis of a cylindrical tube, the boundary velocity gradient at the wire is given by

$$g = \frac{(2V/\pi R^3)[-2 + (1 - 1/b^2)/\ln b]b}{[1 - b^4 + (b^2 - 1)^2/\ln b]} \tag{14-4}$$

where $b = R_W/R$, R_W being the wire radius. The values of g_B are found to be independent of tube diameter but dependent on wire diameter in the manner discussed in Art. 5.

Calculation of quenching, flashback and blowoff—comparison of experiment and theory. The equations developed in Art. 11 may be applied to calculate values of d_P (cf. Eq. 11-41) and $d_\parallel$ and d_0 representing depth of penetration of quenching from a single surface, the quenching distance between plane-parallel plates, and the quenching diameter of tubes, respectively. According to Eq. 11-42, 11-43, and 11-44, the calculation requires a knowledge of a^0 defined by (see Eq. 11-24)

$$a^0 = \frac{\rho_u S_u^0 c_p}{\lambda} \tag{14-5}$$

and of the dimensionless quantities δ_P, $\delta_\parallel$, and δ_0. The latter are obtained from Fig. G,11e if the value of τ is known. τ is given by Eq. 11-32 and requires a knowledge of T_b^0 and T_1^0. T_b^0 is obtained from thermodynamic calculations and T_1^0 may be obtained from values of minimum ignition energy and burning velocity by the method outlined in Art. 11. A comparison of calculated and experimental values of d_P, $d_\parallel$, and d_0 is shown in Table G,14 for a number of methane-oxygen-nitrogen mixtures. The

Table G,14. Calculated and experimental flashback and quenching limits of mixtures of methane, oxygen, and nitrogen.

Mixture composition, per cent			Using Eq. 11-27						Using Eq. 11-27a	
			d_P, mm		$d_\parallel$, mm		d_0, mm		d_P, mm	$d_\parallel$, mm
CH_4	O_2	N_2	Calc.	Exptl.	Calc.	Exptl.	Calc.	Exptl.	Calc.	Calc.
10	90	—	0.43	0.40	0.86	0.86	1.9	1.18	0.22	0.47
15	85	—	0.19	0.14	0.39	0.46	0.94	0.68	0.10	0.21
25	75	—	0.096	0.066	0.19	0.28	0.39	0.48	0 035	0.09
40	60	—	0.16	0.11	0.33	0.43	0.65	0.69	0.05	0.13
50	50	—	0.48	0.41	0.90	1.5	1.9	2.15	0.12	0.30
52.5	47.5	—	0.74	1.0	1.4	2.2	3.40	3.25	0.25	0.50
10	18.8	71.2	0.85	0.90	1.7	2.8	3.54	3.35	0.40	0.90
16.3	29.3	54.4	0.28	0.44	0.56	0.79	1.17	1.42	0.12	0.29
21.5	39.8	39.7	0.17	0.19	0.38	0.53	0.75	0.92	0.07	0.17
26.4	49.1	24.5	0.15	0.11	0.39	0.45	0.61	0.70	0.05	0.12

δ, τ curves in Fig. G,11e have been computed using the hypothesis concerning the change of τ_1 embodied in Eq. 11-27. If the hypothesis embodied in Eq. 11-27a is adopted, the numerical values change. It is seen from the table that the values calculated using Eq. 11-27 agree considerably better with experimental data than when using Eq. 11-27a. However, even the latter results are of the correct order of magnitude, showing that the calculation is not very sensitive to the specific assumption concerning the change of temperature along the τ_1 surface.

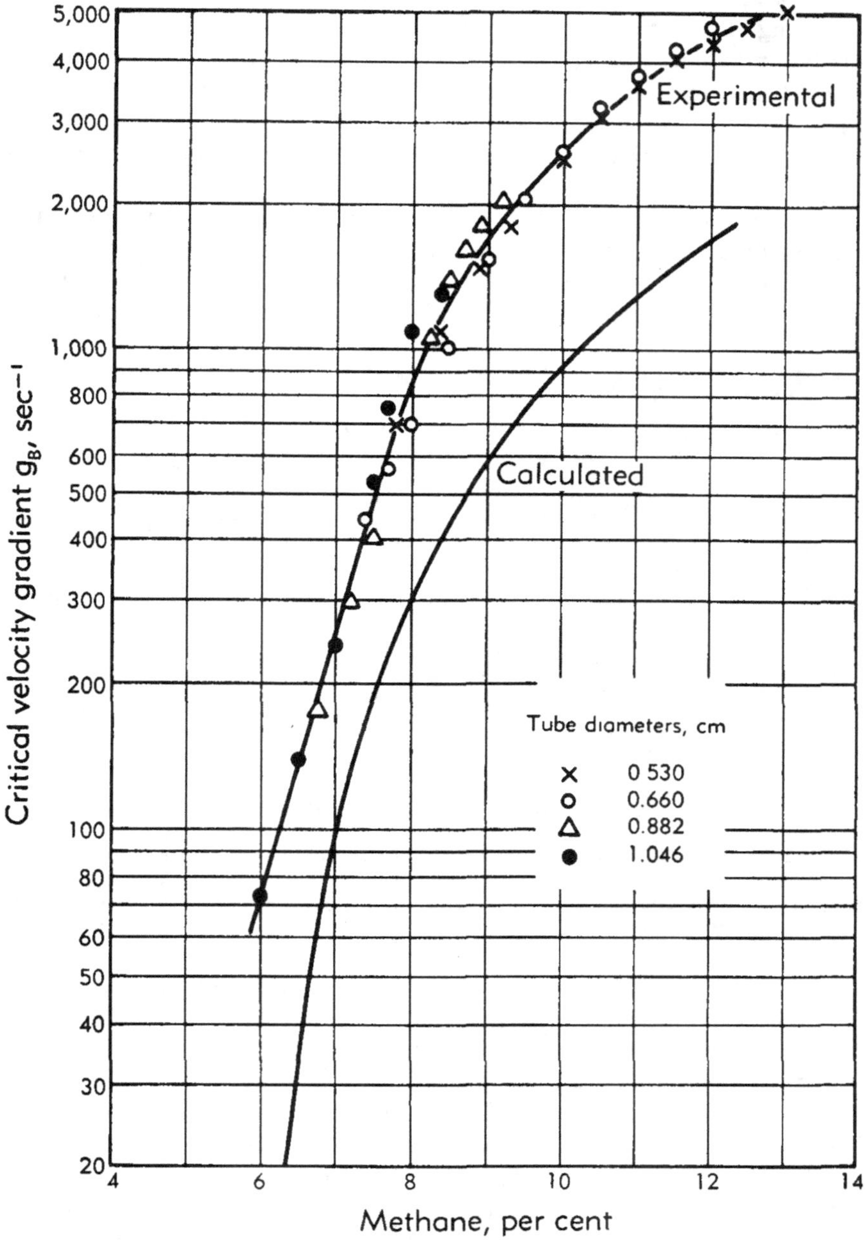

Fig. G,14a. Calculated and experimental critical boundary velocity gradients g_B for blowoff. Methane-air mixtures at room temperature, atmospheric pressure and laminar flow. Experimental determinations with cylindrical tubes of different diameters.

Blowoff calculations have been made for methane-air mixtures according to the method outlined in Art. 11. Fig. G,14 shows experimental and calculated values of g_B versus percentage of methane in the original mixture. The calculated curve follows the trend of the experimental curve and the numerical values agree within the order of magnitude which, in view of the approximations of the theory, may be considered satisfactory.

Structure of symmetrical flames. The flow pattern of an ordinary Bunsen flame can be investigated by the particle-track method as described in Art. 13 (see also IX,J). A comparison of an experimental and calculated flow pattern and combustion wave shape using the equations

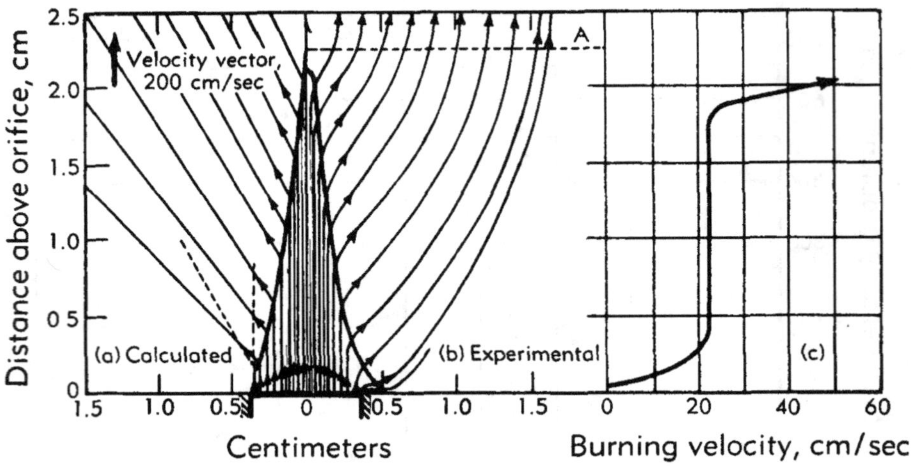

Fig. G,14b. Diagram of vertical center plane of natural gas-air flame on rectangular burner tube of 0.755 × 2.19-cm cross section. Mixture composition, 7.5 per cent natural gas; flow, 204 cm^3/sec. (a) and (b) Combustion zone and flow pattern. (c) Burning velocity.

given in Art. 6 is shown in Fig. G,14b. As seen from the right-hand diagram the experimental burning velocity is constant over a considerable range of the combustion wave between the base and the tip and, as shown in the left-hand diagram, in this range very good agreement of flow pattern and wave shape is obtained. Since, in the theoretical model, the burning velocity is taken to be constant over the entire wave surface the calculated area at the base is too small and at the tip it is too large. The gas velocities at the burner port and beyond the combustion wave are indicated by vector arrows. As a consequence of the refraction of the stream, the maximum gas velocity in the burned gas is shifted away from the axis; this is indicated by the vectors in the plane *A* [see *37*, pp. 250–258].

The temperature distribution in the flame of Fig. G,14b is shown in Fig. G,14c. The measurements were made by the sodium line reversal

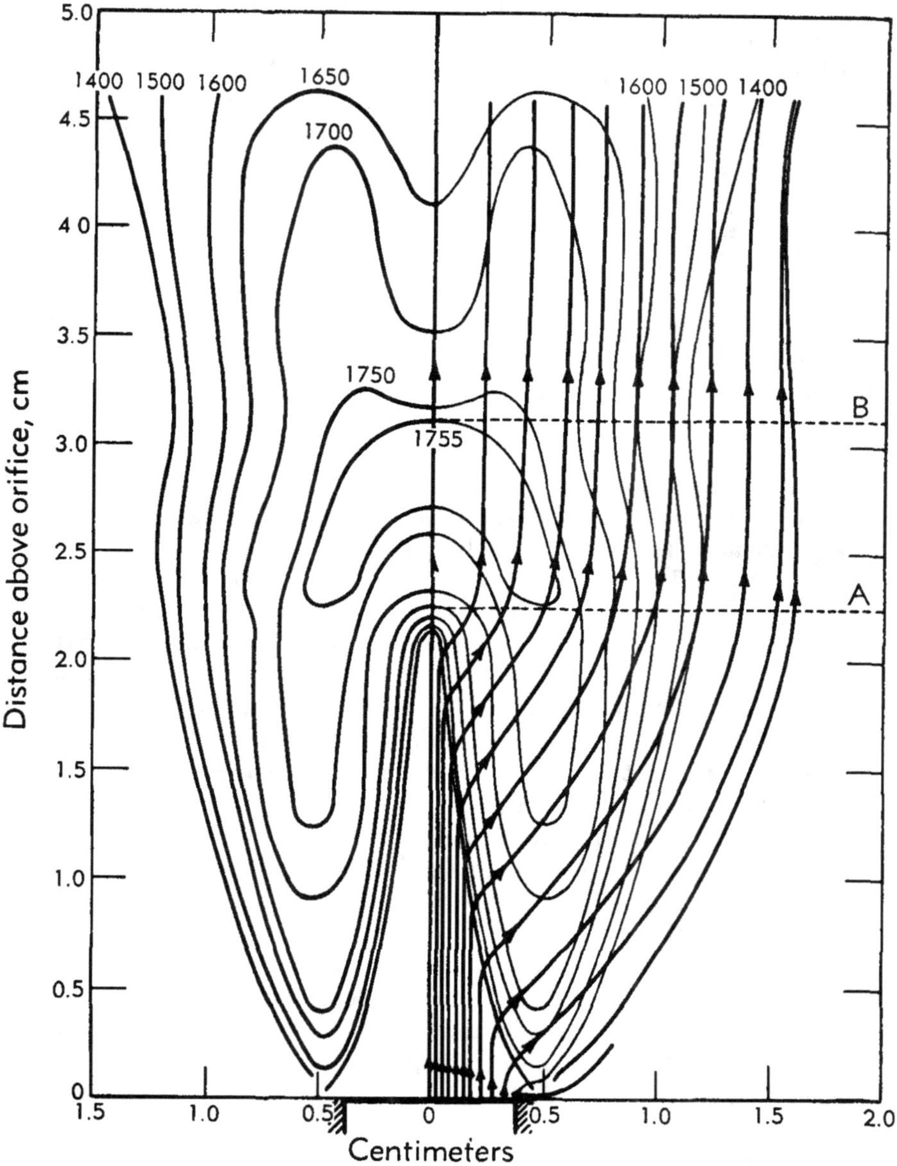

Fig. G,14c. Temperature distribution and flow pattern in the flame of Fig. G,14b. Temperatures in °C.

method [37, pp. 258–262; also IX,I]. In order to confine the observations to a central plane, sodium salt was added to only a portion of the gas mixture through an inner concentric tube.

It should be remarked that the data of Fig. G,14b refer to a flame on a slot burner viewed from the narrow side. The flame thus consists essentially of two inclined, mildly curved surfaces. In a flame of cylindrical symmetry the curvature of the wave increases from the base to the tip, which may make an appreciable contribution to the burning velocity. Whereas in the example illustrated in Fig. G,14b the burning velocity was found constant over a considerable range, it is noted that measurements on cylindrically symmetric flames by other investigators [*3*, p. 80; *57*; *37*, pp. 470–471] have yielded continuously increasing burning velocities along the cone surface. For the present status of determination of burning velocities see IX,K.

Burner flames of asymmetrical, polyhedral, and other structures. A flame form frequently observed when the gas flow is gradually reduced toward the flashback condition is typified by Plate G,14 [*58*; *37*, pp. 270–272]. The dynamic symmetry of the gas stream is destroyed; the combustion wave has partly entered the burner tube and exerts a back pressure normal to the tube axis which causes a deflection of the flow lines to the right in the unburned gas, leaving a wider spacing between flow lines near the wall. This is equivalent to a flattening of the velocity gradient at the wall; since the boundary velocity gradient of the stream, before a flame was superimposed upon it, was only slightly higher than the critical gradient g_F, the decrease of the gradient due to tilting results in local penetration of the flame into the tube. This penetration cannot continue indefinitely because the confining effect of the tube gradually increases the resistance of the flow lines to the deflecting flame pressure. Finally, a position of equilibrium is reached in which the boundary velocity gradient at the point of deepest penetration has become equal to the critical velocity gradient for flashback but exceeds the critical value at lower points in the burner tube. Obviously, the equilibrium position is maintained only when the burner tube below the point of penetration is kept cool. Otherwise, the temperature of the unburned gas increases, and both the burning velocity and critical flashback gradient g_F increase, resulting in continued downward travel of the combustion wave. If this precaution is taken, for example, by water-cooling the tube, such tilted flames are perfectly stable. The point on the circumference of the rim at which penetration occurs is fortuitous, and consequently slight changes in the circumferential temperature distribution can cause the flame to rotate. The diagram of Plate G,14 shows the velocity profile of the burned gas somewhat above the combustion wave. It is seen that the flow line of maximum velocity does not pass through the tip but is displaced in the direction of the point of entry, and the largest part of the combustion gases passes between the point of entry and the tip. This is also reflected in the limit of persistence of luminosity in the burned gas.

When the gas flow is gradually increased toward the blowoff condition, the combustion wave often remains attached to the burner rim at a single point and from there extends upward at a steep angle across the gas stream. The phenomenon is analogous to the foregoing description of the tilted flame: the back pressure exerted by the combustion wave deflects the flow lines in the unburned gas, changing the velocity distribution and causing a decrease of the boundary velocity gradient at the point of attachment to below the critical gradient for blowoff, g_B. Elsewhere along the burner rim, the critical velocity gradient for blowoff is exceeded. On further, but usually slight, increase of the gas flow the flame blows off completely.

The foregoing is just one example of modification of flame structure by back pressure from the combustion wave. As shown by Wohl, Kapp, and Gazley [*3*, p. 3; *55*], in certain cases the flow lines of a cylindrical jet are bent outward symmetrically about the axis, and characteristic flame shapes appear that are referred to as flat and button-shaped flames. The formation of ridges along the surface of the flame cone, mentioned in Art. 13, has the further consequence of altering the flow pattern, and complicated relations exist between the ridge angle and the flow pattern. Smith and Pickering [*59*] have shown that for certain critical adjustments of gas flow and mixture composition the flame cone assumes a stable polyhedral structure, the number of ridges obtainable being from three to as many as seven. For intermediate adjustments no steady state is obtained and the ridges shift rapidly, giving the appearance of a rotating flame.

CHAPTER 3. COMBUSTION WAVES IN CLOSED VESSELS

G,15. Dependence of Pressure Rise and Temperature Distribution on Fraction of Gas Burned. In a closed vessel propagation of a combustion wave is attended by pressure rise and by mass flow which is first directed away from, and later toward, the point of ignition. As the wave progresses, the temperature, as well as the pressure of the unburned gas, rises in accordance with the law of adiabatic compression, and the burning velocity increases. A peculiar feature of the process is the establishment of a temperature gradient in the burned gas, the temperature being highest at the point where ignition occurred, and lowest at the wave surface. The explanation of the temperature gradient is readily understood by considering the initial and final stages of the process. At the point of ignition the gas burns by expanding at practically constant pressure, which is the initial pressure p_i. It is subsequently compressed to nearly its original volume by the combustion of the rest of the gas in the vessel. The latter work of compression exceeds the former work of expansion because the compression of the gas at the point of ignition takes place at a steadily increasing pressure from p_i to the final pressure p_e, whereas the expansion took place at the lowest pressure p_i. On the other hand, the last portion of the gas to burn is first compressed under a pressure which rises from p_i to p_e and then expands on burning at the pressure p_e to approximately its original volume. The latter work of expansion is obviously larger than the former work of compression. Thus, the gas that burns last loses some of its energy while the gas that burns first gains energy in excess of the chemical energy released within it. The result is the establishment of a temperature gradient which rises from the portion burned last to the portion burned first, the temperature difference amounting to several hundred degrees centigrade [*37*, pp. 651–655]. Similar considerations apply to the volume of burned gas at any stage of the process when only a fraction of the contents has burned.

From thermodynamic principles it is possible to correlate the pressure at any instant with the amount and volume of the burned gas and to calculate the temperature distribution at any stage of the process. The treatment which was first given by Flamm and Mache [*60*] is described elsewhere [*37*, pp. 451–455]. Here it will suffice to state that the relation between the fraction n of gas burned and p, p_i, and p_e, the instantaneous, initial, and final pressures, respectively, is, in the first approximation, given by

$$n = \frac{p - p_i}{p_e - p_i} \tag{15-1}$$

This equation, which states that the fraction of gas burned is equal to the fraction of the total pressure rise, is sufficiently accurate for many practical purposes. It is strictly valid for small pressure rise in the early stages of the process, provided that the value of p_e is not taken as the true maximum pressure of the explosion but is calculated thermodynamically under the assumption that the heat capacities and dissociation equilibria of the burned gas in the early stages remain unchanged during the course of the combustion process.

The temperature in any given mass element rises by adiabatic compression from T_i to T_u, which is the temperature just prior to arrival of the combustion wave. As the wave passes over the mass element the temperature rises to T_b which is calculable from thermochemical data, and later to T_e, owing to further adiabatic compression.

G,16. Correlation of Rate of Pressure Rise with Burning Velocity. For the case of a spherical vessel with central ignition and isotropic propagation, a pressure-time record can be used to determine the burning velocity at any instant during the process. The fraction n of gas which, on burning, gives rise to a pressure p, occupies, before ignition, a sphere of radius r_i so that

$$n = \left(\frac{r_i}{a}\right)^3 \tag{16-1}$$

a being the radius of the vessel. r_i is thus known as a function of pressure from Eq. 15-1. The volume occupied by the fraction n when burned is a sphere of radius r_b. Its volume is obtained by subtracting the volume of the remaining unburned gas from the volume of the vessel. From this, r_b, as a function of pressure, is found to be

$$\left(\frac{r_b}{a}\right)^3 = 1 - \frac{p_i}{p}\frac{T_u}{T_i}\frac{p_e - p}{p_e - p_i} \tag{16-2}$$

where

$$T_u = T_i\left(\frac{p}{p_i}\right)^{\frac{\gamma_u - 1}{\gamma_u}} \tag{16-3}$$

To obtain burning velocities it is necessary to plot values of r_i corresponding to any given pressure on the pressure-time record as a function of time and to determine the slopes dr_i/dt. If, during the travel of the flame through the vessel, the burned gas had not expanded, then an element dr_i on this curve would represent the thickness of a shell at the temperature T_i and the pressure p_i about to be traversed by the wave in the time element dt. Its volume would have been

$$4\pi r_i^2 dr_i \tag{16-4}$$

However, by expansion, the radius of the shell increased from r_i to r_b, the pressure to p, and the temperature of the shell to T_u. Therefore, its volume is actually

$$4\pi r_i^2 dr \left(\frac{T_u p_i}{T_i p}\right) \tag{16-5}$$

Since the thickness of the shell is equal to $S_u dt$, its volume is also equal to

$$4\pi r_b^2 S_u dt \tag{16-6}$$

By equating the terms of Eq. 16-5 and 16-6 and substituting the value of T_u/T_i in Eq. 16-3, one obtains for the burning velocity

$$S_u = \left(\frac{dr_i}{dt}\right)\left(\frac{r_i}{r_b}\right)^2\left(\frac{p_i}{p}\right)^{\frac{1}{\gamma_u}} \tag{16-7}$$

In Fig. G,16 an analysis is given of a time-pressure record of a combustion wave in a mixture of ozone and oxygen containing 40.06 per cent ozone [*37*, p. 458]. The upper octant shows the conditions during wave propagation in increments of 0.1 of the total explosion time, and the lower octant the temperature and mass distribution at the end of the process. The radii of the circles 1*b*, 2*b*, etc., are the radii r_b of the volumes of the burned gas for the above time increments; 1i, 2i, etc., are the radii r_i of the volumes corresponding to the r_b's; and 1e, 2e, etc., correspond to volumes occupied by the gas after the combustion process is complete.

For the practical determination of burning velocity from experiments in spherical vessels with central ignition it is more advantageous to confine the observations to the early stages of the process corresponding to only a few per cent of the fraction n burned. The pressure p_e must then be

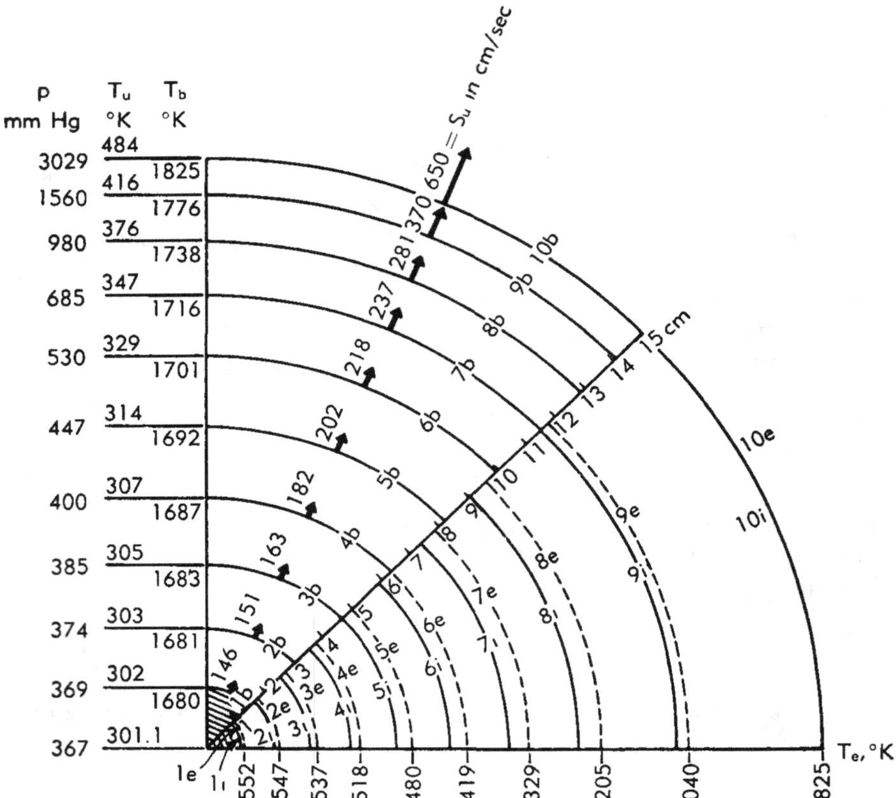

Fig. G,16. Combustion wave in a spherical vessel with central ignition. Data calculated from an experimental pressure-time record. Mixture, 40 per cent O_3 + 60 per cent O_2. Vessel radius, 15 cm. Upper octant shows wave positions 1*b*, 2*b*, etc., in intervals of 1.65 milliseconds. Lower octant shows spherical volumes 1i, 2i, etc., and 1e, 2e, etc., of fractions n corresponding to wave positions 1*b*, 2*b*, etc. Series *b* shows the process of burning; series i, the initial state; and series e, the end state.

calculated assuming constancy of heat capacities and dissociation equilibria as mentioned in Art. 15 in connection with Eq. 15-1. If simultaneously with recording of pressure the diameter of the flame is recorded photographically, then the radii r_b calculated from Eq. 16-2 can be compared with the experimental radii, thus providing a test of the experimental procedure. Discrepancies may arise from several causes. Since the calculation is based on the assumption that the combustion wave is a mathematical surface there arises the recurrent problem of selecting an experimentally determinable reference surface in the actual combustion. Evidently, direct and schlieren photography will give somewhat different results. Furthermore, the flame may not be an ideal sphere. A possible third cause may be a lag period in the establishment of thermodynamic equilibrium in the burned gas. Evaluation of experimental records of carbon monoxide-oxygen flames has yielded consistent data of burning velocity. The scattering due to errors of measuring the records was found to be small. The agreement between calculated and experimental values of r_b was remarkably good. Details are given elsewhere [*37*, pp. 473–479].

If, on the other hand, the experimental value of r_b determined for some reference surface, is accepted a priori as identical with the r_b defined by Eq. 16-2, the thermodynamic calculation of p_e can be dispensed with and the burning velocity can be determined from the records of r_b and p with time. For this purpose r_i is expressed as a function of p and r_b by the equation

$$\frac{r_i}{a} = \left[1 - \left(\frac{p}{p_i}\right)^{\frac{1}{\gamma_u}}\left(1 - \frac{r_b^3}{a^3}\right)\right]^{\frac{1}{3}} \tag{16-8}$$

which is obtained from Eq. 16-1, 16-2, and 16-3; thus dr_i/dt is determinable since p and r_b are known as functions of time, and S_u is obtained from Eq. 16-7. Another form of the equation for S_u is obtained by differentiating Eq. 16-8 with respect to r_b and combining with Eq. 16-7. This leads to

$$S_u = \left(1 - \frac{a^3 - r_b^3}{3p\gamma_u r_b^2}\frac{dp}{dr_b}\right)\frac{dr_b}{dt} \tag{16-9}$$

This equation can also be derived from the differential form of the equation of adiabatic compression as has been done by Fiock and coworkers [*61*; also IX,K].

Application of Eq. 16-9 requires the determination of two differential quotients from the experimental records. Close to the origin $dr_b/dt = S_b$, and since S_u/S_b is the density ratio ρ_b/ρ_u, the bracketed term is small compared to unity. Therefore, S_u is obtained as the difference of two quantities of comparable magnitude, and any error in the differential quotients and the other data is greatly magnified, and the results may be expected to scatter widely as has been found to be the case [*37*, pp. 473–479]. The same applies to the method of determining S_u using Eq. 16-8 in undifferentiated form as outlined above. Therefore, the method of utilizing simultaneous records of r_b and p and avoiding the thermodynamic calculation of p_e is not recommended.

A further discussion of the theory of steady state combustion waves in homogeneous gas mixtures (see Art. 2) is given by Hirschfelder, Curtiss, and Campbell [*62*, pp. 190–211]. Their treatment includes the conditions for transition to detonation. Another development of the system of combustion wave equations has been presented by von Kármán and Penner [*63*, p. 5], which with the aid of suitable simplifying assumptions has been used to obtain numerical solutions for flame propagation in hydrazine vapor and ozone.

G,17. Cited References.

1. Damköhler, G. *Der Chemieingenieur, 3,* (1931). (See [*2*, p. 17].)
2. Jost, W. *Explosions- und Verbrennungsvorgänge in Gasen.* Springer, Berlin, 1939.
3. *Third Symposium on Combustion, Flame and Explosion Phenomena.* Williams & Wilkins, 1949.
4. Hirschfelder, J. O., and Curtiss, C. F. *Univ. Wisconsin Rept. CM-529,* 1949.
5. Curtiss, C. F., and Hirschfelder, J. O. *J. Chem. Phys. 17,* 550 (1949). See also [*3*] and [*4*].
6. Mallard, E., and Le Chatelier, H. *Ann. mines 8 (ser. 4),* 274 (1883).
7. Crussard, L. *Compt. rend. 158,* 125, 340 (1914).
8. Jouguet, E. *Compt. rend. 156,* 872 (1913); *179,* 454 (1924).
9. Jouguet, E., and Crussard, L. *Compt. rend. 168,* 820 (1919).
10. Nusselt, W. *Z. Ver. deut. Ing. 59,* 872 (1915).
11. Eucken, A. *Müller-Pouillets Lehrbuch der Physik,* Vol. 3. Barth, Leipzig, 1926.
12. Daniell, P. J. *Proc. Roy. Soc. London A126,* 393 (1930).
13. Mache, H. *Die Physik der Verbrennungserscheinungen.* Veit & Co., Leipzig, 1918. Correction by Jost, W. [*2*, p. 106].
14. Zeldovich, Y. B., and Frank-Kamenetsky, D. A. *Compt. rend. Acad. Sci. U.R.S.S. 19,* 693 (1938), in English.
15. Zeldovich, Y. B., and Semenov, N. *J. Exptl. Theoret. Phys. U.S.S.R. 10,* 1116 (1940). (English transl. in *NACA Tech. Mem. 1084,* 1946).
16. Semenov, N. *Progr. Phys. Sci. U.S.S.R. 24,* 433 (1940). (English transl. in *NACA Tech. Mem. 1026,* 1942).
17. Zeldovich, Y. B. *J. Phys. Chem. U.S.S.R. 22,* 1 (1948).
18. Bartholomé, E. *Naturwissenschaften 36,* 171, 206 (1949).
19. Bartholomé, E., and Hermann, C. *Z. Elektrochem. 54,* 165 (1950).
20. Bartholomé, E., Dreyer, H. J., and Lesemann, K. J. *Z. Elektrochem. 54,* 246 (1950).
21. Sachsse, H., and Bartholomé, E. *Z. Elektrochem. 53,* 183 (1949).
22. Bechert, K. *Z. Naturforsch. A3,* 584 (1948).
23. Bechert, K. *Portugaliae Phys. 3,* 29 (1949), in French; *Ann. Physik* (6) *5,* 349 1950; *Naturwissenschaften 37,* 112 (1950); *Z. Elektrochem. 54,* 239 (1950).
24. Lewis, B., and von Elbe, G. *J. Chem. Phys. 2,* 537 (1934).
25. Boys, S. F., and Corner, J. *Proc. Roy. Soc. London A197,* 90 (1949).
26. Corner, J. *Proc. Roy. Soc. London A198,* 388 (1949).
27. Hirschfelder, J. O., and Curtiss, C. F. *J. Chem. Phys. 17,* 1076 (1949).
28. Hirschfelder, J. O. and Curtiss, C. F. *Univ. Wisconsin Nav. Research Lab. Bull. CM-398,* 1950.
29. Hirschfelder, J. O., Curtiss, C. F., Bird, R. B., and Spotz, E. L. *The Properties of Gases.* Wiley, 1954.
30. Stevens, F. W. *NACA Tech. Mem. 372,* 1931.
31. Tanford, C., and Pease, R. N. *J. Chem. Phys. 15,* 431, 861 (1947).
32. Tanford, C. *J. Chem. Phys. 15,* 433 (1947).
33. Hoare, M. F., and Linnett, J. W. *J. Chem. Phys. 16,* 747 (1948).
34. Dugger, G. L. *J. Am. Chem. Soc. 72,* 5271 (1950).
35. Evans, M. W. *Project Squid Tech. Rept. 27,* 1952; *Chem. Revs. 51,* 363 (1952).
36. Lewis, B., and von Elbe, G. *J. Chem. Phys. 11,* 75 (1943); *Trans. Am. Soc. Mech. Engrs. 70,* 307 (1948).
37. Lewis, B., and von Elbe, G. *Combustion, Flames and Explosions of Gases.* Academic Press, 1951.

38. Forsythe, J. S., Davies, D. G. R., and Townend, D. T. A. *Inst. Gas Engrs. Communs. 244*, 1941.
39. Silsbee, F. B. Cf. Stevens, F. W. *NACA Rept. 176*, 1923.
40. Coward, H. F., and Hartwell, F. J. *J. Chem. Soc.* 1996, 2676 (1932).
41. Frank, F., and von Mises, R. *Differentialgleichungen der Physik*, Vol. 2. Rosenberg, New York, 1943.
42. Peirce, B. O. *A Short Table of Integrals.* Ginn & Co., 1910.
43. Blanc, M. V., Guest, P. G., von Elbe, G., and Lewis, B. *J. Chem. Phys. 15*, 798 (1947).
44. Roth, W., Guest, P. G., von Elbe, G., and Lewis, B. *J. Chem. Phys. 19*, 1530 (1951).
45. Wheatcroft, E. L. E. *Gaseous Electrical Conductors.* Oxford Univ. Press, 1938.
46. Calcote, H. F. Private communication from Experiment, Inc.
47. Jones, E. *Proc. Roy. Soc. London A198*, 523 (1949).
48. Bowden, F. P., Mulcahy, M. F. R., Vines, R. G., and Yoffe, A. *Proc. Roy. Soc. London A188*, 291 (1947).
49. Mulcahy, M. F. R., and Vines, R. G. *Proc. Roy. Soc. London A191*, 210 (1947).
50. Coward, H. F., and Guest, P. G. *J. Am. Chem. Soc. 49*, 2479 (1927).
51. Guest, P. G. *U.S. Bur. Mines Tech. Paper 475*, 1930.
52. Grumer, J. Unpublished Bureau of Mines Reports.
53. Manton, J., von Elbe, G., and Lewis, B. *J. Chem. Phys. 20*, 153 (1952).
54. Markstein, G. H. *J. Chem. Phys. 17*, 428 (1949); *Project Squid Tech. Rept. 24*, 1950.
55. Wohl, K., Kapp, N. M., and Gazley, C. *Project Meteor Rept. UAC-26*, 1948.
56. Bollinger, L. M., and Williams, D. T. *NACA Tech. Note 1234*, 1947.
57. Garner, F. H., Long, R., and Ashforth, G. K. *Fuel 28*, 272 (1949); *J. Chem. Phys. 18*, 1112 (1950).
58. von Elbe, G., and Mentser, M. *J. Chem. Phys. 13*, 89 (1945).
59. Smith, F. A., and Pickering, S. F. *Bur. Standards J. Research 3*, 65 (1929).
60. Flamm, L., and Mache, H. *Wien Ber. 126*, 9 (1917).
61. Fiock, E. F., Marvin, C. F., Caldwell, F. R., and Roeder, C. H. *NACA Rept. 682*, 1940.
62. *Fourth Symposium on Combustion.* Williams & Wilkins, 1953.
63. von Kármán, Th., and Penner, S. S. *Selected Combustion Problems.* Combustion Colloquium, Cambridge Univ., Dec. 1953. Butterworths, London, 1954.

SECTION H

COMBUSTION WAVES IN TURBULENT GASES

BÉLA KARLOVITZ

The importance of the effect of turbulent motion on flame propagation was recognized early in the systematic study of flames. Mallard and Le Chatelier in their fundamental paper on flame propagation [*1*] attributed the high propagation velocity of flames, observed under certain circumstances, to the effect of turbulence. Similarly the fact that the entire charge in the cylinders of internal combustion engines could be burned during the short time interval available was explained by the effect of turbulence. It was believed that turbulence could increase the burning velocity by a factor of 10 to 100, but the mechanism of flame propagation in turbulent explosive mixtures was not understood, and even an empirical relationship between turbulence and burning velocity was unknown until 1941 when the work of Damköhler marked the beginning of a series of studies directed toward a fundamental understanding of turbulent flame phenomena.

To give a true picture of the growth of ideas, and of our present knowledge of turbulent flames, the important theoretical and experimental work done in this field will be reviewed first; a theory of turbulent flame phenomena then will be presented. This theory, although in need of further experimental confirmation, in the author's opinion can explain the observed facts and is in good agreement with all known measured data. To reduce the complexity of the problem, only flames in premixed gases will be treated in this section.

H,1. Phenomenological Description of Turbulent Flames. The effect of turbulence on flames may be best illustrated by two examples that are easily observed. Flames moving into a quiescent explosive mixture in tubes of a few centimeters diameter show a well-defined smooth flame surface and a propagation velocity close to the normal burning velocity. In channels above a certain critical diameter, fluctuations set in, the flame surface becomes rippled, and the velocity of flame propagation is greatly increased. Fig. H,1 shows the flame propagation velocity in methane-air mixtures as a function of the channel diameter, according to measurements by Coward and Hartwell [*2*]. The increased flame propagation velocity is assumed to be caused by the effect of turbulence, which also causes the fluctuation and distortion of the flame surface. This explanation is substantiated by the fact that the burned gas flow attains the critical Reynolds number at the channel diameter where the discontinuous increase of the flame velocity appears on the diagram.

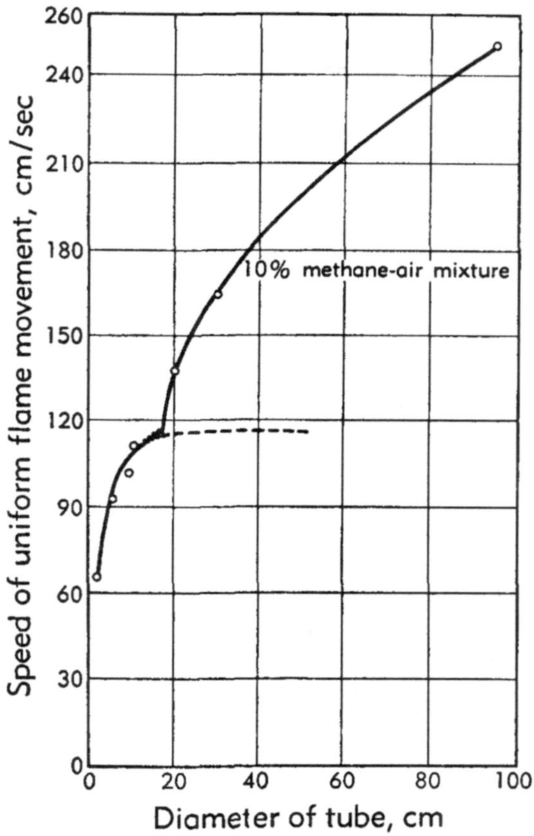

Fig. H,1. Flame velocity in tubes according to Coward and Hartwell [*2*].

As a second example the effect of turbulence on Bunsen-type flames may be considered. In laminar flow the flame front is smooth and thin (Plate H,1a). As the approach flow passes over the critical Reynolds number and becomes turbulent the flame suddenly becomes shorter, thicker, and diffuse in appearance (Plate H,1b). The shorter flame indicates an increase of the apparent burning velocity caused by the effect of turbulence.

H,2. The Work of Damköhler. Damköhler's work, which marks the beginning of serious studies of turbulent flames, was an experimental and theoretical study of the effect of turbulence [*3*].

For his experimental study Damköhler used Bunsen flames that burned on the top of long vertical burner tubes. The tubes were sufficiently long to produce fully developed turbulent flow. Three different tube diameters were used: 1.38 mm, 2.18 mm, and 2.71 mm. The flow velocity of the propane-oxygen mixture could be varied within wide limits. The Reynolds number of the flow could be increased up to $Re = 18{,}000$. Below $Re = 2300$ the flames were laminar, very steady, and quiet. Above 2300 the flames became turbulent, produced loud whistling noise, and were no longer sharply defined but blurred.

Damköhler determined the turbulent burning velocity by the formula of Gouy,

$$S_{\mathrm{T}} = \frac{\dot{V}}{A} \tag{2-1}$$

where S_{T} is the turbulent burning velocity, V is the volume of the gas burned per second, and A is the surface area of the flame.

The surface area, A, of the flame was taken as the inner area of the visible turbulent flame cone, and was measured from flame photographs. Damköhler makes the statement that this inner boundary of the turbulent flame is the locus of the fastest combustion, the outer boundary that of the slowest combustion. Having found from his experimental data that the burning velocity calculated from the outer boundary surface is approximately equal to the laminar burning velocity, he uses the ratio $A_{\mathrm{outer}}/A_{\mathrm{inner}}$ as a measure of the effect of turbulence on the flame propagation velocity. The measured value of $A_{\mathrm{outer}}/A_{\mathrm{inner}} = S_{\mathrm{T}}/S_u$ is shown in Fig. H,2a as a function of the Reynolds number of the flow.

To characterize the turbulence of the gas flow, Damköhler used the turbulent exchange coefficient introduced by Boussinesq, and the mixing length concept of Prandtl. The mixing length l represents the approximate average distance traversed by a small volume of fluid moving as a coherent unit during its random turbulent motion. The turbulent exchange coefficient ϵ is a quantity similar to the diffusion coefficient used in kinetic theory. The relationship between the mixing length l, the average velocity of the turbulent fluctuations $\overline{|w|}$, and the turbulent exchange coefficient is given by the equation

$$\epsilon = l\overline{|w|}$$

Damköhler assumed that the turbulence of the flow is not altered by the flame, and calculated the values of the turbulence parameters from the experimental data of J. Nikuradse for the pipe flow, and from the theoretical formula of W. Tollmien for the mixing of the jet with the surrounding still air.

In his theoretical considerations about the possible influence of turbulence on flame propagation, Damköhler recognizes that there are two fundamentally different effects possible, namely, whether the mixing length is large or small as compared with the thickness of the laminar flame front. The coarse, or large scale, turbulence distorts the otherwise smooth flame front, as shown schematically in Fig. H,2b. In this manner large scale turbulence increases the flame surface per unit cross section of the flow, and by it the apparent burning velocity, without altering the

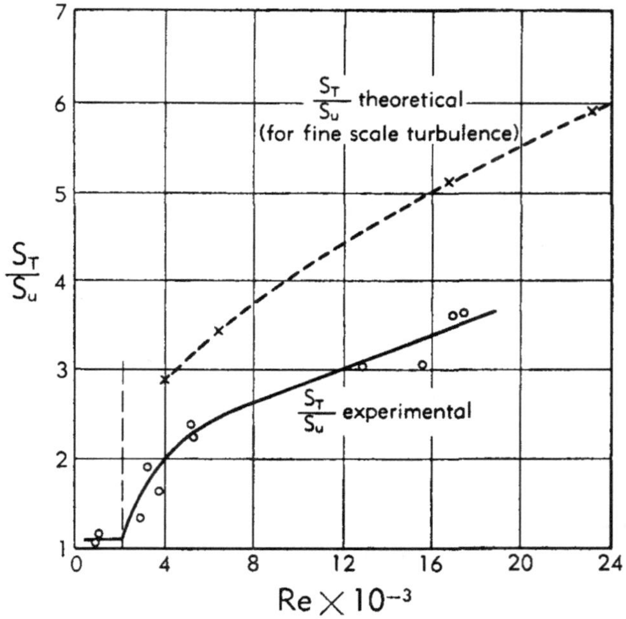

Fig. H,2a. Turbulent burning velocity as a function of the Reynolds number of the approach flow, according to Damköhler [3].

structure and the laminar burning velocity of the instantaneous combustion wave. Damköhler estimated the increased flame surface area, and from it the increased burning velocity, for not too large intensity of turbulence, by assuming that the flame surface is composed of small cones directed with or against the flow according to the direction of the instantaneous value of the local turbulent velocity component. From this picture Damköhler drew the conclusion that, for not too strong turbulence, the increased burning velocity should be proportional to the average of the absolute value of the turbulent fluctuating velocity component, which is normal to the average turbulent flame front; that is

$$S_T \sim \overline{|U|}$$

or for the same geometrical arrangement to the Reynolds number Re,

$$S_T \sim Re \tag{2-2}$$

However, experimentally he found the values of S_T/S_u to be proportional to $\sqrt{Re}$ at low values of Re and proportional to $aRe + b$ at higher values of Re, where a and b are constants, and $Re = \bar{U}d/\nu$ for the pipe flow. From this experimental result Damköhler concluded that fine scale turbulence must have had an important influence on the burning velocity under the experimental conditions.

Damköhler gave a more complete solution for the effect of fine scale turbulence. The effect of fine scale turbulence is to increase the intensity

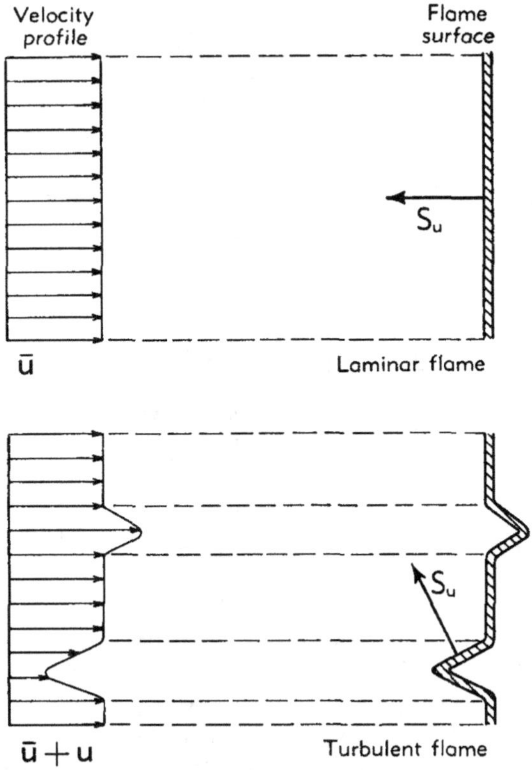

Fig. H,2b. Schematic representation of the effect of large scale turbulence on the flame surface, according to Damköhler [3].

of transport processes within the combustion wave. Instead of the molecular diffusion coefficient D, the transport of heat and active particles is proportional to the turbulent exchange coefficient ϵ. According to the theory of laminar flame propagation of Mallard and Le Chatelier [1], given in extended form in the paper, the laminar burning velocity is proportional to the square root of the quantity $\lambda/c_p\rho$, where λ is the heat conductivity, c_p the specific heat, and ρ the density of the unburned gas. The heat conductivity is $\lambda = c_p\frac{1}{3}\bar{v}\bar{L}$, making $\lambda/c_p\rho = \frac{1}{3}\bar{v}\bar{L} = D$, where $\bar{v}$ = molecular velocity, $\bar{L}$ = mean free path. Also, $D = \nu = \eta/\rho$, where η

is the viscosity and ν the kinematic viscosity; therefore fine scale turbulence should affect the burning velocity according to the relation

$$\frac{S_T}{S_u} = \sqrt{\frac{\epsilon}{\nu}} \tag{2-3}$$

To compare the experimental results with this theory of fine scale turbulence, Damköhler calculated the theoretical shape of the flames using the local values of the turbulent exchange coefficient of the approach flow. The average turbulent burning velocity could be calculated from the calculated flame shapes and then compared with the measured values. Burning velocities calculated in this manner are indicated in Fig. H,2a by the dashed line. The theoretical values are considerably higher than the experimental values. Damköhler explained this by the fact that the mixing length in the experiments was not small but comparable to the thickness of the flame, and therefore turbulent transport could not be effective with its full value.

Although Mallard and Le Chatelier already recognized that the effect of (large scale) turbulence is to increase the flame surface, Damköhler made the first attempt to derive a numerical relationship between the intensity of turbulence and turbulent burning velocity from this picture. While Damköhler's theory for the large scale turbulence has only limited validity, the clear distinction between the effect of large scale and fine scale turbulence is of fundamental importance. His experimental results are not sufficient to test the theory because the turbulence was not predominantly coarse or fine but intermediate. Also the experimental determination of turbulent burning velocity is open to criticism. The statement that the inner surface of the diffuse turbulent flame corresponds to the fastest and the outer to the slowest burning velocity cannot be accepted. There is no reason why the outer surface of the turbulent flame should reach the position which a laminar flame would occupy in a flow with the same mean velocity in the absence of turbulence. Furthermore the turbulent flame has no definite boundaries; its apparent thickness depends on how much light emission is considered still to be part of the flame front. Therefore the equality of the burning velocity calculated by Damköhler from the outer surface area of the flame with the laminar burning velocity should be considered as a coincidence.

H,3. The Work of Shelkin. Shelkin [4] extended the theoretical considerations of Damköhler. For the effect of fine scale turbulence ($l < \delta$) on the flame he took into account not only transport due to turbulence, but the molecular transport process, and gave the following formula for the turbulent burning velocity for fine scale turbulence:

$$S_T = S_u \sqrt{1 + \frac{\epsilon}{\alpha}} \tag{3-1}$$

where $\alpha = \lambda/c_p\rho$, thermal diffusivity; ϵ turbulent exchange coefficient; δ = thickness of the flame front.

Shelkin also considered the case where the scale of turbulence is not small but comparable to the thickness of the laminar flame. He reasons as follows. With turbulence of a scale comparable to the thickness of the flame, the temperature of a layer within the flame front is no longer constant but fluctuates about an average value. The reaction rate is an exponential function of the temperature, therefore the reaction rate does not correspond to the average temperature but is much larger. The reaction rate may become so high that not it but the rate of turbulent mixing would determine the velocity of flame propagation. Shelkin did not develop this idea further.

In the case of large scale turbulence, Shelkin used different approaches to calculate the increased burning velocity for strong and for weak turbulence.

According to the picture given by Shelkin, strong turbulence distorts the flame surface so much that elements of the flame start to break away

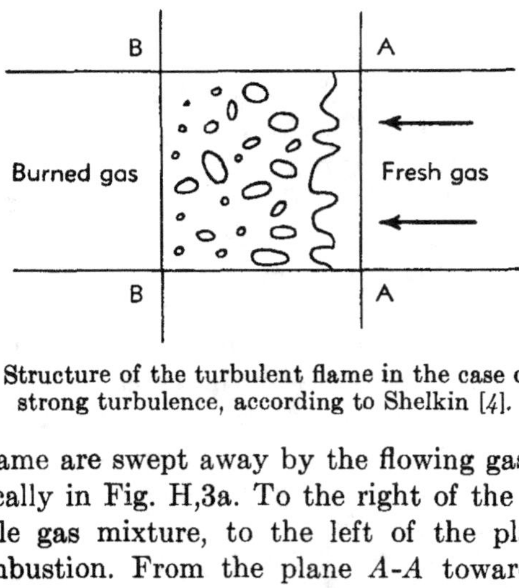

Fig. H,3a. Structure of the turbulent flame in the case of large scale, strong turbulence, according to Shelkin [*4*].

and islands of flame are swept away by the flowing gas. Such a flame is shown schematically in Fig. H,3a. To the right of the plane *A-A* is the fresh combustible gas mixture, to the left of the plane *B-B* are the products of combustion. From the plane *A-A* toward the plane *B-B* the average concentration of the fresh gas decreases and the concentration of the products of combustion increases gradually. The distance between *A-A* and *B-B* could be considered as the width of the turbulent flame front. Shelkin assumes by analogy to the laminar flame that the burning velocity is proportional to

$$S_T = \sqrt{\frac{\alpha}{\tau}} \tag{3-2}$$

where α is the thermal diffusivity and τ the reaction time.

For the turbulent flame the thermal diffusivity α has to be replaced by the turbulent exchange coefficient $\epsilon = lu'$. The reaction time τ is derived by the following argument. The time required to burn the small volume l^3 is

$$\tau = \frac{l^3}{AS_u}$$

where A is the actual flame surface within the small cube with the side length l. The flame starts with an elementary surface of l^2, but it becomes distorted and broken up by the fluctuations. The breaking up continues as long as the flame moves with its normal burning velocity over a distance equal to the scale of turbulence l. This time interval is equal to l/S_u. During this interval the total path traversed with the fluctuating velocity u' reaches the value of $L = (l/S_u)\, u'$. The ratio L/l shows how many times the volume l^3 is traversed by the fluctuations during combustion. *Each such traversal leads to the formation of a new flame area of* l^2. Therefore the new flame area for the combustion of the volume element l^3 is proportional to

$$A = l^2 \frac{u'}{S_u}$$

With this value the reaction time is

$$\tau = \frac{l}{u'}$$

and the turbulent burning velocity for *strong* turbulence

$$S_T = \sqrt{\frac{\epsilon}{\tau}} = \sqrt{\frac{lu'}{l/u'}} = u' \tag{3-3}$$

For weak turbulence Shelkin follows the argument of Damköhler, but he expresses the area of the small cones which represent the distorted flame surface as

$$A = l^2 \sqrt{1 + b\left(\frac{u'}{S_u}\right)^2}$$

instead of making it simply proportional to u' as Damköhler did. With this value of the surface area the increased burning velocity is given by the formula

$$\frac{S_T}{S_u} = \sqrt{1 + b\left(\frac{u'}{S_u}\right)^2} \tag{3-4}$$

where b is a coefficient of the magnitude of 1. For strong turbulence the formula reduces to

$$S_T = u'$$

in agreement with the result found above for strong turbulence.

According to Shelkin's formula for large scale turbulence, represented graphically in Fig. H,3b, the turbulent burning velocity should be independent of the scale of turbulence and it should increase at first very slowly with increasing turbulence intensity, but at larger turbulence intensities the turbulent burning velocity should be almost proportional to the turbulent intensity and independent of the normal burning velocity. This latter improbable result is the consequence of the assumption that each fluctuating traversal of a small volume l^3 by the flame should lead to the formation of a new flame area of the order of l^2.

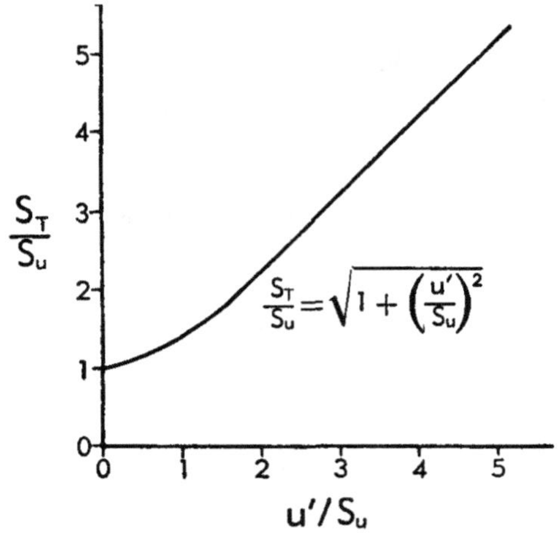

Fig. H,3b. Turbulent burning velocity as a function of turbulence intensity, according to Shelkin [*4*].

Shelkin carried out no experiments to test his theoretical predictions, and the experimental results to which he refers are not sufficient to support his theoretical predictions.

H,4. Turbulent Burning Velocity Measurements of the National Advisory Committee for Aeronautics. At the NACA Laboratory in Cleveland, L. M. Bollinger and D. T. Williams carried out turbulent burning velocity measurements [*5*] to test experimentally the theories of turbulent burning velocity described above. The experimental flames burned on the top of vertical burner tubes, long enough to insure fully developed turbulent pipe flow. They realized that the inner and outer boundaries of turbulent Bunsen flames are not well defined, and that the average position of the instantaneous flame front should be used in preference to an arbitrary boundary for the determination of turbulent burning velocity. Accordingly the average turbulent burning velocity was defined as

$$S_T = \frac{\dot{V}}{A} \tag{4-1}$$

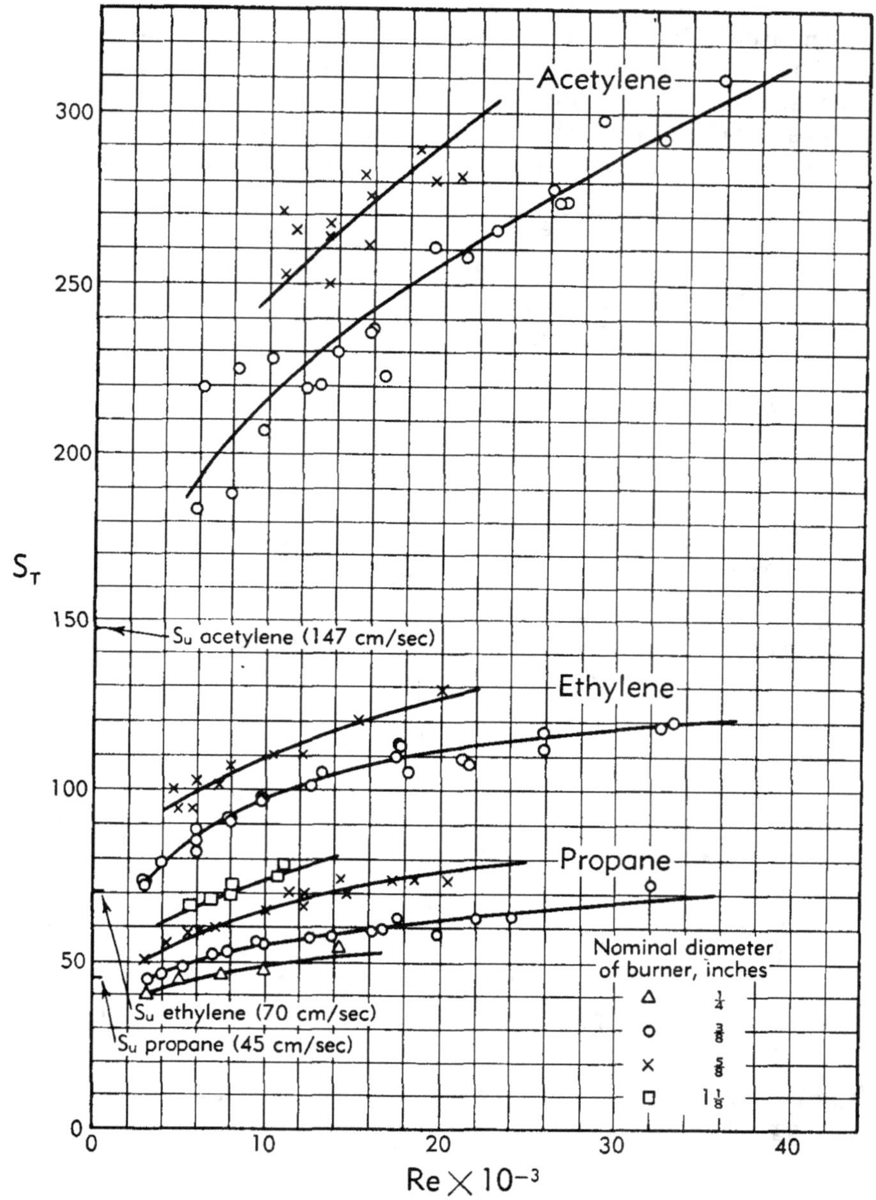

Fig. H,4a. Variation of the turbulent burning velocity with the Reynolds number of the flow for propane, ethylene, and acetylene flames, according to Bollinger and Williams [*5*].

where $\dot{V}$ is the volume rate of flow through the flame and A is the area of the average flame surface. Experimentally the average flame surface was measured from flame photographs on which the average surface was drawn in approximately halfway between the inner and outer boundaries of the flame (Plate H,4).

The turbulent burning velocities obtained by this method are shown in Fig. H,4a for propane-air, ethylene-air and acetylene-air mixtures.

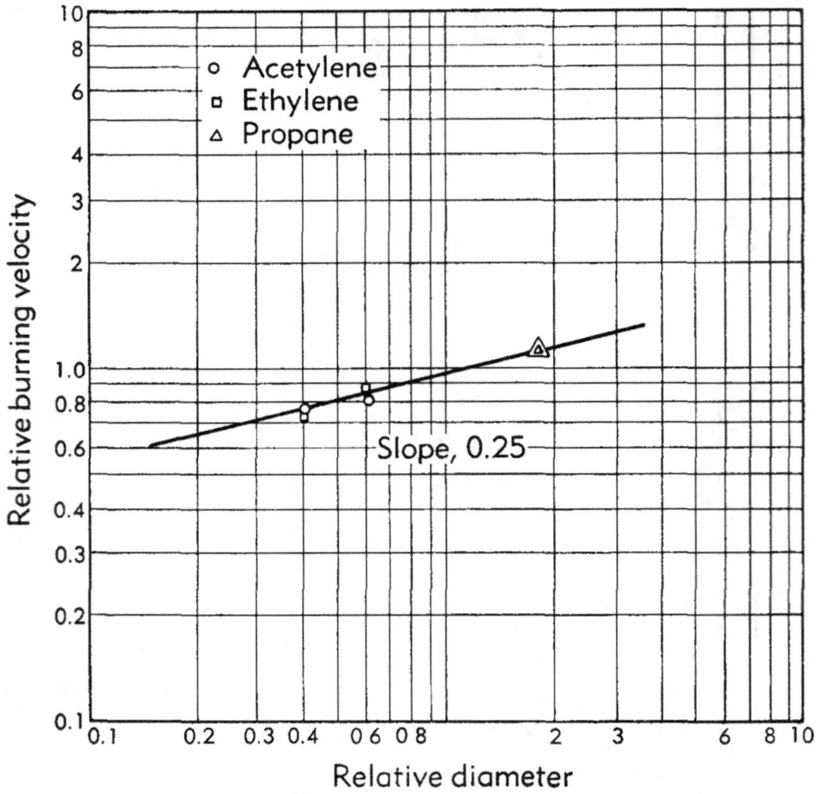

Fig. H,4b. Variation of relative burning velocity with relative burner diameter, based on ⅜-inch burner tube, according to Bollinger and Williams [5].

The mixture ratio used was always the one which gave the maximum burning velocity. These data show that the turbulent burning velocity increases with increasing Reynolds number, but not linearly. The increase is strong at first and much smaller at large Reynolds numbers. The measured turbulent burning velocity also increases with increasing normal burning velocity, and with increasing burner tube diameter. These experimental results are at variance with the theory of Damköhler, according to which the flame speed should be proportional to Re and independent of the scale of turbulence. The experimental results are also in disagreement with the formula of Shelkin

$$S_T = S_u\sqrt{1 + b\left(\frac{u'}{S_u}\right)^2} \tag{4-2}$$

which would require independence from the scale of turbulence, an almost linear increase with turbulence intensity, and independence from the normal burning velocity at high turbulence intensity.

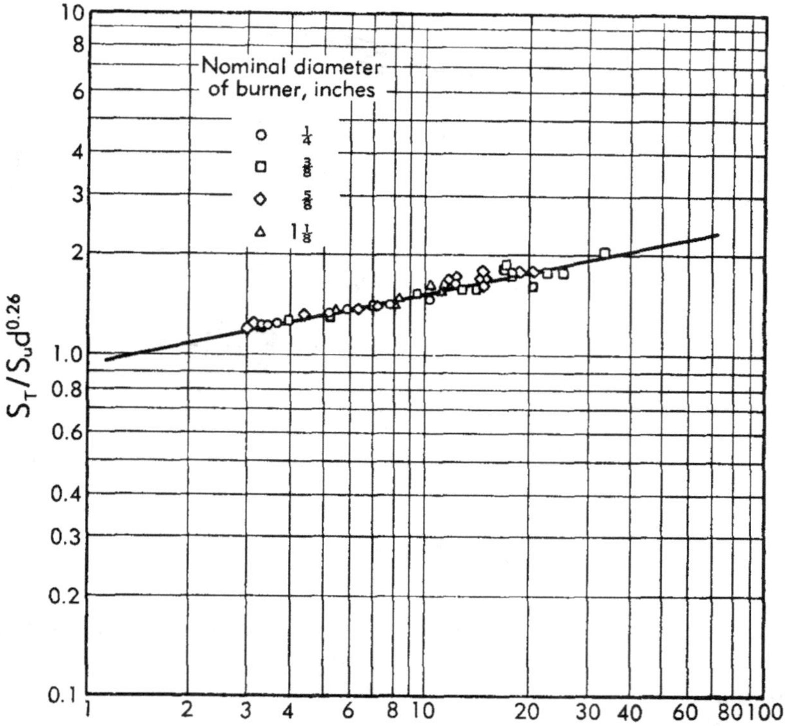

Fig. H,4c. $S_T/S_u d^{0.26}$ as a function of the Reynolds number for propane-air mixtures, according to Bollinger and Williams [5].

Bollinger and Williams found that the representation of their experimental data by an empirical formula required three independent parameters. They chose the normal burning velocity S_u, the burner tube diameter d, and the Reynolds number of the approach flow $Re = \bar{U}d/\nu$. The exponent of the normal burning velocity was chosen to be 1. The exponents of the tube diameter and of the Reynolds number were determined from a logarithmic plot of the experimental data. Fig. H,4b shows the variation of the relative flame speed with the relative tube diameter. From this diagram the exponent of the tube diameter was determined to be 0.256. Fig. H,4c shows $(S_T/S_u)d^{0.26}$ as a function of the Reynolds

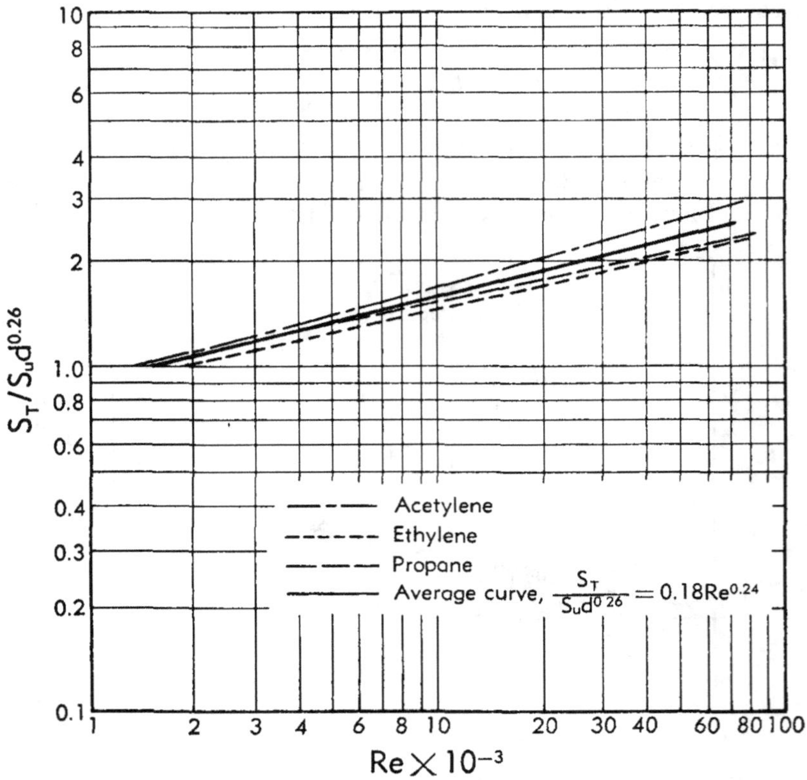

Fig. H,4d. Average value of $S_T/S_u d^{0.26}$ as a function of the Reynolds number, according to Bollinger and Williams [5].

number of propane-air mixtures. Similar plots were made for acetylene-air and ethylene-air mixtures. In Fig. H,4d the three curves obtained from these diagrams are shown together with a curve obtained by averaging. This average curve is represented by the empirical formula for the turbulent burning velocity:

$$S_T = 0.18 S_u d^{0.26} Re^{0.24} \text{ cm/sec} \tag{4-3}$$

where d is measured in centimeters.

H,5. Turbulent Flames Confined in Channels. Extensive experiments were carried out in the Fuels Research Laboratory of Massachusetts Institute of Technology on flames burning in high velocity streams, with the aim of finding the effect of turbulence on flame propagation and the conditions of flame stability. The flames in these experiments were confined in channels, which circumstance presents additional complications as compared to the open Bunsen type flames because the increased volume of the hot combustion products increases the flow velocity and substantially alters the velocity distribution in the combustion zone of the channel [*6*; *7*, pp. 21–40].

The experimental arrangement consisted of a channel with rectangular cross section of 3 × 1 in., in which the flame was anchored on a flame holder rod which was normal to the direction of flow. The combustion channel had Vycor side walls through which the flame could be observed and photographed. The explosive mixture entered into the combustion channel from a calming section through a nozzle of 25 to 1 cross-section

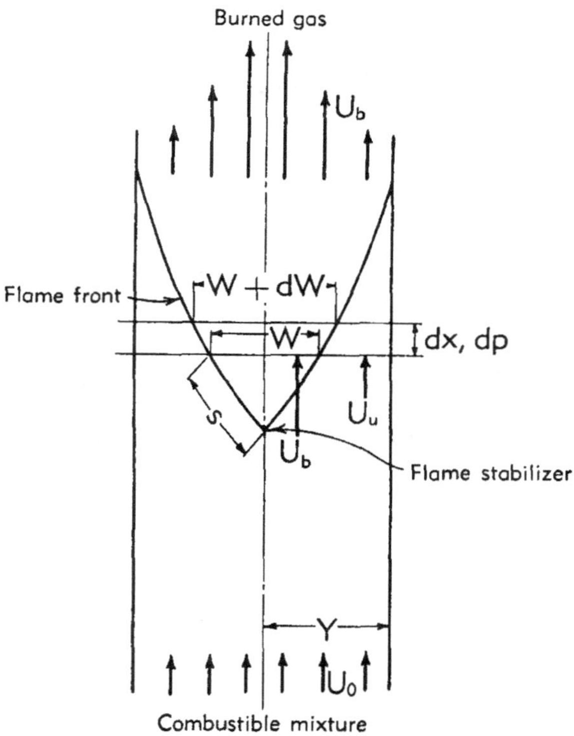

Fig. H,5a. Schematic representation of the flame configuration and of the flow conditions in the combustion zone of the experimental burner of the MIT Fuels Research Laboratory [*6*].

ratio. By this arrangement the turbulence intensity of the approach flow was reduced to 0.4 per cent of the mean velocity. The turbulence intensity could be increased by inserting screens of various mesh size before the entrance of the combustion channel. The intensity and the scale of the approach flow turbulence were measured in the absence of a flame with a hot wire anemometer [*8*; see also IX,F].

Two different fuels were used in these experiments—Cambridge city gas and propane. The fuel-air mixture ratio was varied on both the lean and rich side to the blowoff limits. The combustion chamber entrance velocity was varied between 20 and 350 feet per second.

The flow velocity in the combustion section of the channel was calculated with the simplifying assumptions that the entrance velocity is constant over the entire cross section, that the pressure is nearly constant over any cross section of the combustion chamber normal to its axis, and that the lateral velocity components and the friction between layers of flow with different velocities and on the walls can be neglected. The configuration of the flame and the flow conditions are shown schematically in Fig. H,5a. Y is the half width of the flow channel; W the width of

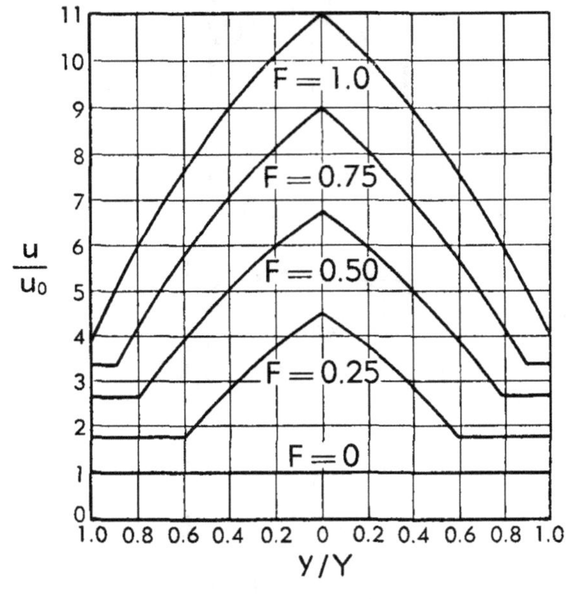

Fig. H,5b. Calculated velocity distribution in the MIT combustion channel for $\rho_0/\rho_l = 8$, according to Scurlock [6].

the flame (the distance between the two branches of the flame); s the length of the flame from the origin up to the cross section of the channel under consideration; p the gas pressure; ρ_b the density of the burned gas; U_u the flow velocity of the unburned gas; U_b the flow velocity of the burned gas; and F the fraction of burned gas flowing through any cross section of the duct.

Let the flame width at an arbitrary cross section of the channel be W. As the flow moves downstream a distance dx, an additional fraction of the gas is burned and the flame width increases to $W + dW$. At the same time the pressure drops by an amount dp to produce the increased flow velocity. The flow obeys the three relations:

Continuity

$$\rho_0 U_u(2Y - W) + \rho_b U_b W = \rho_0(U_u + dU_u)(2Y - W - dW) + \rho_b(U_b + dU_b)(W + dW) \quad (5\text{-}1)$$

Momentum equation for the volume element 2Ydx

$$-2Ydp = \rho_0(U_u + dU_u)^2(2Y - W - dW) + \rho_b(U_b + dU_b)^2(W + dW) - \rho_0 U_u^2(2Y - W) + \rho_b U_b^2 W \quad (5\text{-}2)$$

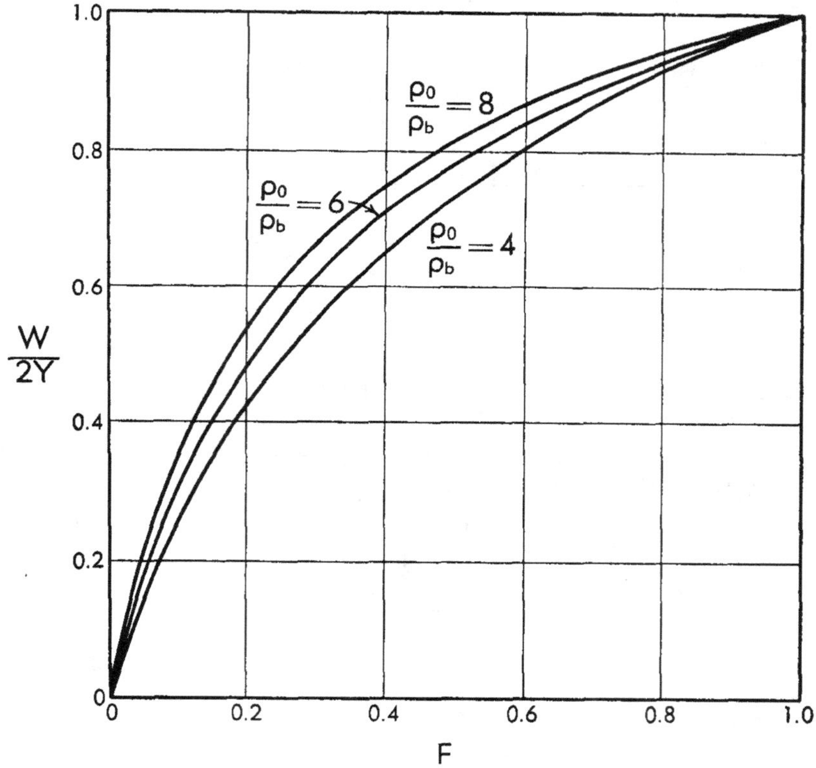

Fig. H,5c. Calculated flame width W versus fraction F of gas burned for the MIT combustion channel, according to Scurlock [6].

Bernoulli's equation for a streamline of the unburned gas

$$-dp = \rho_0 \tfrac{1}{2} dU_u^2 \quad (5\text{-}3)$$

No general solution of these equations was given, but assuming a finite step for the pressure drop dp the three unknown quantities dU_u, dW, and dU_b can be calculated from the above three equations. The fraction of burned gas at any cross section is given by the formula

$$F = \frac{\rho_b U_b W}{\rho_0 U_0 2Y} \quad (5\text{-}4)$$

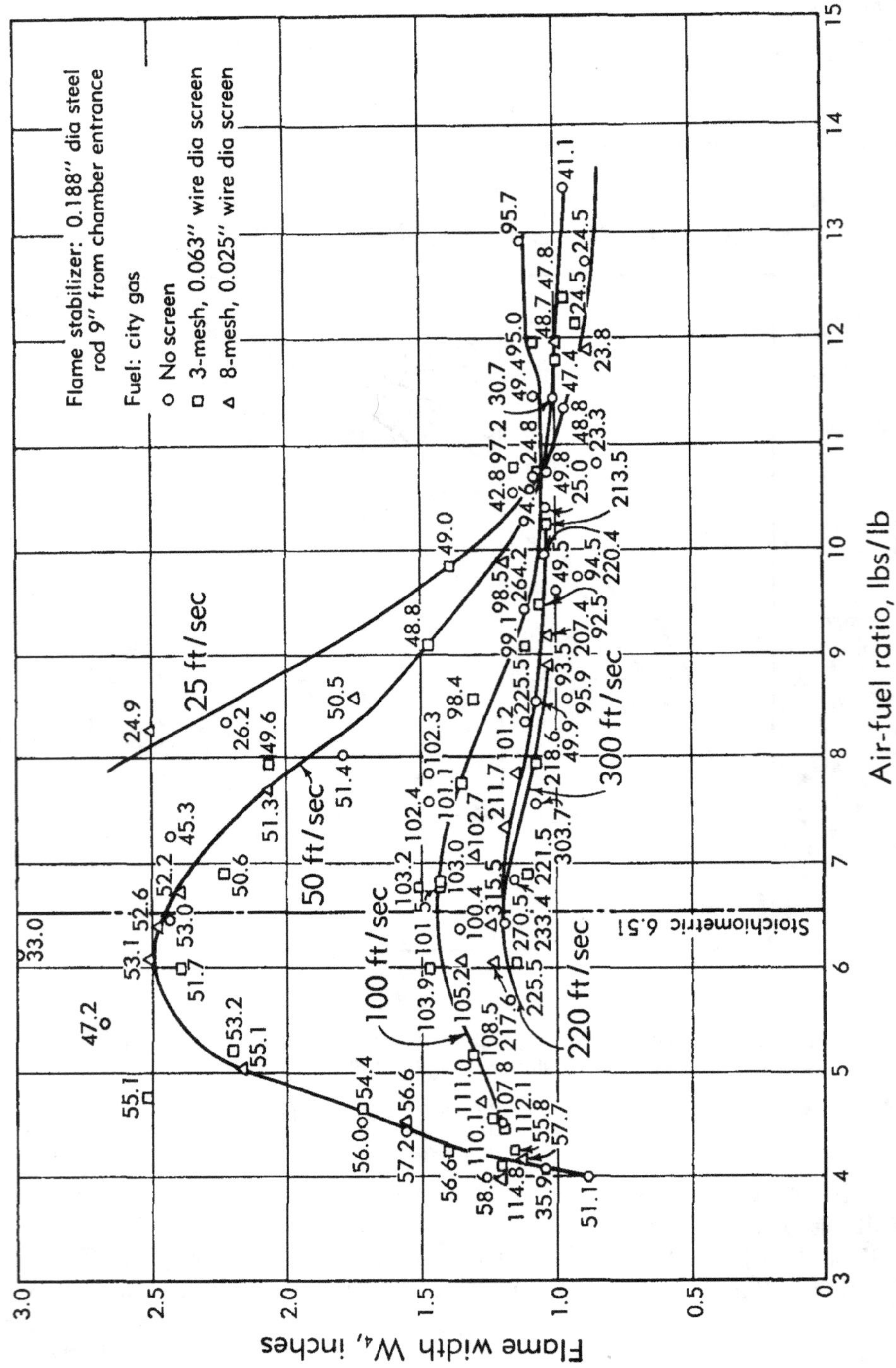

Fig. H,5d. Experimentally measured flame width W_4 4 inches from the stabilizer rod, as a function of the air-fuel ratio, for various chamber entrance velocities, according to Scurlock [6].

The calculations were carried out stepwise. The calculated velocity distribution in the combustion zone is shown in Fig. H,5b for a gas mixture with $\rho_u/\rho_b = 8$. The hot combustion products have a much lower density than the cold unburned gas and therefore the same pressure drop accelerates them to higher velocities than the unburned gas. Consequently the velocity distribution which is constant over the entrance cross section of the combustion chamber is disturbed by the effect of the flame, and strong velocity gradients appear in the flow. The flame width W versus the fraction of gas burned is shown in Fig. H,5c.

The burning velocity was measured by measuring the length of the flame s from flame photographs and the width of the flame W. The average turbulent burning velocity for the flame section of length s is defined by the relation

$$S_T = \frac{\dot{V}}{A} \tag{5-5}$$

where $\dot{V} = U_u YF$ is the volume of gas flow per second through the flame and $A = s \cdot 1$ is the flame surface area. Therefore the burning velocity is

$$S_T = \frac{\dot{V}}{A} = \frac{U_u YF}{s} \tag{5-6}$$

The experimentally measured flame width W at 4 inches from the stabilizer rod is shown in Fig. H,5d for the approximate chamber entrance velocities of 25, 50, 100, 220, and 300 ft/sec as a function of the air-fuel ratio. The exact values of the entrance velocity are indicated at any point in the figure. The turbulence intensity of the approach stream was 0.4 per cent without a turbulence generating screen, and 2.3 per cent with the 3-mesh screen. Surprisingly, the turbulence of the approach flow had very little influence on the flame width.

Fig. H,5e shows the turbulent burning velocity S_T calculated from these data. For comparison the normal burning velocity S_u of the mixture is shown in the lower part of the diagram. Also shown, in dotted lines, are the turbulent burning velocities calculated from the theoretical formula of Shelkin:

$$S_T = S_u \sqrt{1 + \left(\frac{u'}{2S_u}\right)^2} \tag{5-7}$$

for 2.3 per cent turbulence of the approach flow. The experimentally measured turbulent burning velocities are considerably higher than the values calculated from the approach flow turbulence. It was concluded from this result that the probable reason for the negligibly small effect of the approach flow turbulence on the burning velocity is caused by the production of very strong additional turbulence by another disturbance, which completely overshadows the effect of the approach flow turbulence.

The flame stabilizer was ruled out as a possible source of this additional turbulence because a variation of the stabilizer rod diameter from 0.016 in. to 0.498 in. had only a negligible effect on the flame width. It was also observed that no eddies are shed from the stabilizer during combustion.

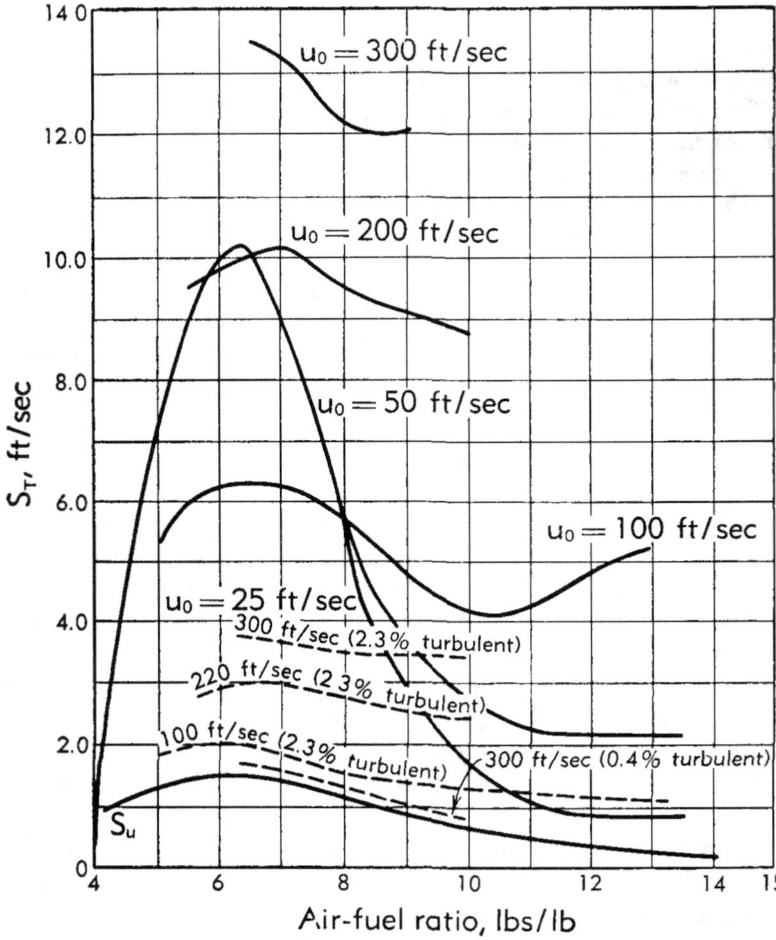

Fig. H,5e. Measured turbulent burning velocities (solid lines) compared with values calculated from Shelkin's formula $S_T = S_u\sqrt{1 + (u'/2S_u)^2}$ (dashed lines), according to Scurlock [6].

The source of additional turbulence was thought to be the strong velocity gradients produced by the flame itself. As shown in Fig. H,5f in the region immediately behind the flame holder the flow velocity of the gas mixture is higher than the velocity of the eddy flow behind the flame holder. Farther downstream the flow velocity of the burned gases is much higher than the velocity of the unburned gas. These strong velocity

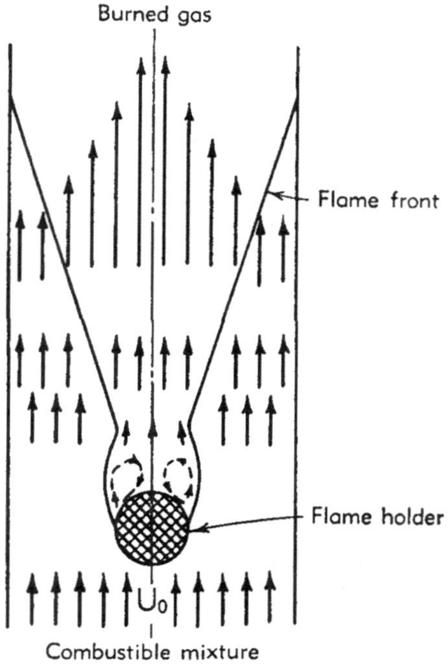

Fig. H,5f. Schematic diagram of flow through the MIT Fuels Research Laboratory combustion chamber, showing high velocity gradients at the flame front, according to Scurlock [*6*].

gradients produced by the flame generate eddies and finally turbulence. This was the first case in the study of turbulent flames where turbulence generation by the flame was recognized from careful measurements, and a mechanism was found by which the flame could produce turbulence. Unfortunately it is not possible to calculate the intensity of turbulence produced by the velocity gradients, and so the proposed theories of turbulent flame propagation cannot be tested experimentally by these measurements.

H,6. Turbulent Flame Measurements at California Institute of Technology. Extensive measurements of turbulent burning velocity and approach flow turbulence were carried out at the Jet Propulsion Laboratory of the California Institute of Technology [*9*]. The burner arrangement used in these experiments is shown in Fig. H,6a. The burner port was rectangular with a length of 12 cm and a width of 1, 2, or 3 cm. Thereby a two-dimensional flow was approximated. The flow velocity ranged from 100 cm/sec to 500 cm/sec in these experiments. The turbulence was generated by grids of various shapes and dimensions that could be inserted into the approach stream. The longitudinal and the transversal components of the turbulence intensity and their distribution

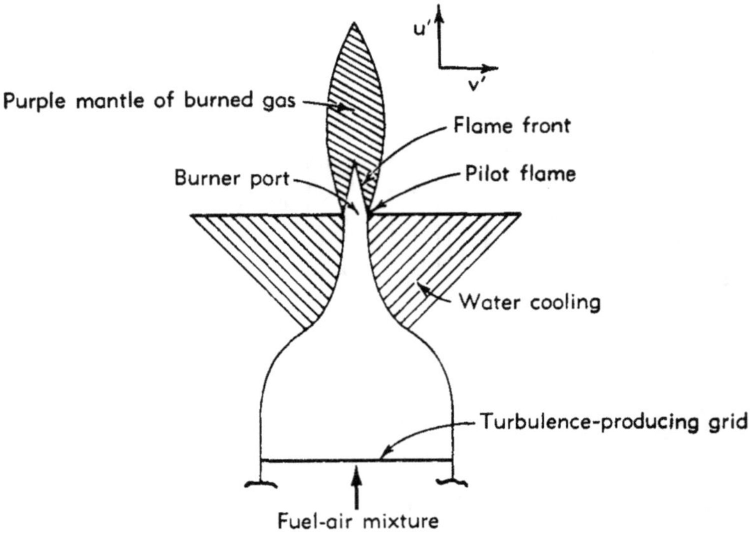

Fig. H,6a. Schematic drawing of the experimental burner used by Wright, California Institute of Technology [9].

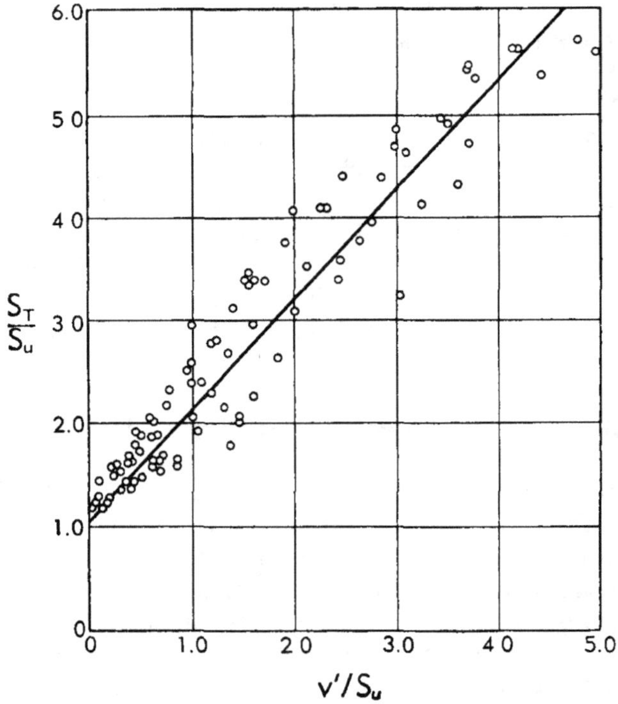

Fig. H,6b. Turbulent burning velocity versus turbulence intensity measured at the center of the burner port, according to Wright, California Institute of Technology [9].

in the flow, as well as the scale of turbulence, were measured with a hot wire anemometer in the absence of the flame.

The normal burning velocity was varied by varying the fuel-air ratio. Both the laminar and the turbulent burning velocity were determined by measuring the flame height, or the flame surface area and the mean flow velocity at the burner port. The burning velocity data are average values for the entire flame.

The ratio S_T/S_u of the turbulent to the laminar burning velocity was correlated with the ratio v'/S_u, where v' is the lateral turbulent velocity component measured at the center of the burner port in the absence of a flame. The experimental results are shown in Fig. H,6b. It was concluded that these data did not agree well with Shelkin's formula and that a linear relationship between S_T/S_u and v'/S_u represented the experimental data reasonably well.

It was also found in these experiments that although the spectrum of turbulence and the scale of turbulence varied greatly with the type of grid used for the production of turbulence and with the grid Reynolds number, no distinction could be made between the effect of different kinds of grids or different sizes of burner on the flame; that is, a change in the scale of turbulence did not alter the turbulent burning velocity if the intensity of turbulence was kept constant.

H,7. Other Experimental Studies of Turbulent Flames. Particularly detailed observations were carried out on turbulent flames traveling in tubes at New York University by M. W. Evans and collaborators [7, pp. 168–176]. The flames were initiated near the closed end of 8-foot long tubes of 4-inch diameter, that were filled with an explosive butane-air mixture. The apparent flame velocity was measured from streak photographs and the structure of the flames studied from high speed (5000 to 9000 frames per second) schlieren motion pictures.

It was found that the apparent flame velocity, i.e. the velocity with which the flame moves relative to the tube, was increased about tenfold when a grid was placed in the tube near the closed end, and the flame was initiated in the compartment between the closed end and the grid. As revealed by the high speed schlieren motion pictures, jets of flame shot through the holes in the grid, igniting the gas in the tube and producing a fine grained turbulent flame front in contrast to the rather smooth "normal" flame observed in a tube without grids.

Although the effect of turbulence was very pronounced in these experiments and the high speed schlieren motion pictures permit a rather close examination of the structure of the flame, a direct comparison of the experimental data with the proposed theories of turbulent burning velocity is not possible because of the great complexity of periodic flame phenomena.

Other interesting experiments were carried out by G. H. Markstein at the Cornell Aeronautical Laboratory [7, pp. 162–167]. Markstein questioned the correctness of the assumption made in the proposed theories of turbulent flames, according to which the turbulence in the combustion zone is equal to the turbulence of the approach flow. To study this problem under the simplest possible conditions, he made his observations not on turbulent flames but on laminar flames disturbed by acoustical vibrations or by vibrating wires. The motion of the disturbed flame front could be followed in great detail on high speed schlieren motion pictures.

It was found that the area of the distorted flame was equal to the area of the undisturbed flame. From this it follows that the normal burning velocity was not affected by the disturbances. The waves produced on the flame front by the disturbances moved up with the speed of the unburned gas, and their amplitude increased as they moved up. From this observed amplification of the disturbances by the flame front it was concluded that the intensity of turbulence in the unburned gas may be considerably amplified by the turbulence created in the burning zone of a turbulent flame.

Observations of flames burning in dust clouds led to results similar to those obtained for flames in premixed gases. Turbulence can increase the burning velocity many times above the value found in quiescent dust clouds [7, pp. 185–190].

H,8. Turbulence. As an introduction to the theory of turbulent flames, a brief description of the terms used to characterize turbulence is given here. For an extensive treatment of turbulence reference is made to Vol. V of this Series.

A flow in which adjacent layers of fluid slide past each other without mixing is called laminar flow. Such flow is realized for example in small diameter tubes at low flow velocities. As the flow velocity is increased a strong mixing motion suddenly appears, superimposed on the mean motion of the flow, and simultaneously the characteristic features of the flow, such as velocity distribution and pressure drop, are substantially altered. The superimposed random vortex motion is called turbulence. The randomness is essential. An organized system of eddies, or regular periodic fluctuations of velocity, do not possess the transport properties of random motion, and are not included in the term turbulence. As no general solutions of the Navier-Stokes equations, which describe turbulent motions, are known, accurate quantitative information about turbulent flow can be gained only from experiments. The extent of required experiments is greatly reduced by the similarity law of Reynolds. According to this law the flow in geometrically similar systems is similar if the dimensionless Reynolds numbers

$$Re = \frac{Ud}{\nu}$$

of the flow systems are equal. Here U = flow velocity; d = characteristic dimension of the system, for example, tube diameter; and $\nu = \eta/\rho$ = kinematic viscosity of the fluid.

The Cartesian components of the instantaneous flow velocity at a point may be written as the sum of a constant, or slowly changing, mean velocity component and the fluctuating velocity component

$$U = \bar{U} + u$$
$$V = \bar{V} + v$$
$$W = \bar{W} + w$$

where the mean value of any fluctuating velocity component is zero:

$$\bar{u} = 0$$
$$\bar{v} = 0$$
$$\bar{w} = 0$$

The root mean square value of the fluctuating velocity component is not zero; it is customary to denote this value by

$$u' = \sqrt{\overline{u^2}}$$
$$v' = \sqrt{\overline{v^2}}$$
$$w' = \sqrt{\overline{w^2}}$$

u', v', and w' are usually called the components of the intensity of turbulence.

The random fluctuating velocity components transport momentum across the fluid. For example the u and v components transport an amount $-\rho\overline{uv}$ of x momentum per unit area in the y direction in unit time. This momentum transport is equivalent to a shearing stress, the so-called Reynolds stress. The other shearing stress components are $-\rho\overline{vw}$ and $-\rho\overline{wu}$; the Reynolds normal stress components are $-\rho\overline{u^2}$, $-\rho\overline{v^2}$, and $-\rho\overline{w^2}$.

In the early so-called "mixing length" theories of turbulence, the turbulent motion was described by analogy with the molecular motion of a gas. Corresponding to the motion of a gas molecule between two collisions, a small volume of gas was pictured moving on the average a distance l, the mixing length, before losing its identity by mixing with the surrounding fluid. The product of the mixing length l and the mean fluctuating velocity u' is called the turbulent exchange coefficient

$$\epsilon = lu'$$

corresponding to the diffusion coefficient

$$D = \tfrac{1}{3}\bar{L}v_{\mathrm{m}}$$

of the kinetic theory.

By further analogy with the kinetic theory the apparent viscosity due to the turbulent motion is expressed as $\rho\epsilon$, and the increased heat conductivity as $c_p\rho\epsilon$. The numerical value of ϵ can be obtained only from experiments. Usually the value of ϵ is different for heat or for momentum transport.

The usefulness of the analogy of molecular motion is seriously limited because, in contrast to the mean free path of the kinetic theory, the mixing length is not well defined and not amenable to direct measurement.

In the newer "statistical" theories of turbulence, the statistical correlations between various velocity components are used to characterize the structure of the turbulent flow field. For example, the correlation coefficient between the u and v velocity components at one point is given by

$$R_{uv} = \frac{\overline{uv}}{\sqrt{\overline{u^2}}\,\sqrt{\overline{v^2}}}$$

Similar correlation coefficients exist between other velocity components, or between velocity components measured at different points.

A characteristic length, or scale, of turbulence can be defined by using the correlation coefficients in various ways. Using Lagrange's description of flow, that is, following the motion of a fluid particle, a characteristic length can be defined by the relation

$$l_1 = u' \int_0^\infty R_t dt$$

where R_t is the correlation coefficient between the u velocity component of the particle at any instant and of the same particle after the time interval t. This definition of scale is useful in the study of turbulent diffusion, and is rigorously defined only for homogeneous turbulence.

In the Eulerian description of flow the scale of turbulence may be characterized by the length

$$l_2 = \int_0^\infty R_y dy$$

where

$$R_y = \frac{\overline{u_1 u_2}}{u_1' u_2'}$$

is the correlation between the u component of the fluctuating velocity at a point 1 and the same velocity component at another point 2, separated by the distance y. This correlation factor is a function of the distance y between the two points; $R_y = 1$ for $y = 0$, and drops to $R_y = 0$ at a value of y beyond which there is no statistical correlation between the u components of the fluctuating velocity. No relationship is known between R_t and R_y, and correspondingly between the length l_1 and the length l_2. From experimental data, G. I. Taylor found that l_2 is approximately twice l_1.

In general the components of the turbulence u', v', and w', as well as the various correlation functions, may have different values. The simplest case of turbulence is where, at every point of the turbulent field, the direction in which the velocity component or a correlation function is measured has no effect on the result. Such turbulence was given the name of isotropic turbulence by G. I. Taylor. In isotropic turbulence $u' = v' = w'$ and $\overline{uv} = \overline{vw} = \overline{wu} = 0$; therefore the shear stresses are zero. Isotropic turbulence may be produced experimentally by grids in a flow. Such an arrangement is well suited for the measurement of the decay of turbulence. Isotropic turbulence can be sufficiently characterized for many purposes by two quantities, intensity and scale.

The variable eddies that constitute the turbulent motion are not all of the same size and velocity, their dimensions and velocities being distributed over a wide range. The function $F(\kappa)$ that describes the distribution of turbulent energy as a function of the wave number κ, the number of waves per unit length, is called the spectrum of turbulence. Energy is communicated to the turbulent motion from the mean flow through the largest eddies. From the large eddies it is transmitted to smaller and smaller eddies and finally absorbed by viscous friction.

H,9. Turbulent Diffusion. The diffusive properties of turbulent motion have been investigated in detail by G. I. Taylor, and others [*10,11,12,13,14,15,16,17*]. The problem treated by Taylor is the mean square displacement $\overline{X^2}$ of a particle from its original position by turbulent fluctuations in a given X direction.

The isotropic turbulent motion shall be characterized by the intensity of turbulence u' and by the correlation function $R_t = \overline{uu_t}/\overline{u^2}$ where u is the fluctuating velocity component of a particle in the X direction at a time t_0, and u_t is the velocity component of the same particle after the time interval t. Then the scale of the turbulence is defined by

$$l_1 = u' \int_0^\infty R_t dt \tag{9-1}$$

and a characteristic time by

$$T = \int_0^\infty R_t dt \tag{9-2}$$

A particle which at the time t_0 is at $X = 0$ is carried about by the random turbulent fluctuations, and after the time interval t it will be found at some distance X from the origin. If a great number of cases is considered, then after the same time interval the particles will be distributed at random in both positive and negative directions. The mean value of the displacement X is zero because positive and negative velocity components are equally probable. But the particles do not return necessarily to their starting point, and therefore the root mean square value of the displacement $\sqrt{\overline{X^2}}$ is not zero. To calculate the value of $\sqrt{\overline{X^2}}$, first we calculate its variation with time

$$\frac{d}{dt}\overline{X^2} = \overline{2X\frac{dX}{dt}} = 2\overline{uX} \tag{9-3}$$

From the definition of the correlation factor R_t, we have the relation

$$\int_0^t \overline{uu_t}dt = \overline{u^2}\int_0^t R_t dt \tag{9-4}$$

The left-hand side of this equation may be written as

$$\int_0^t \overline{uu_t dt} = \overline{u\int_0^t u_t dt} = \overline{uX} \tag{9-5}$$

Therefore

$$\frac{d}{dt}\overline{X^2} = 2\,\overline{uX} = 2\overline{u^2}\int_0^t R_t dt \tag{9-6}$$

For time intervals t which are small as compared with the characteristic time T, that is when $t/T \ll 1$, the value of the correlation function is nearly equal to 1 during the entire time interval. Therefore the root mean square displacement is

$$\sqrt{\overline{X^2}} = u't \tag{9-7}$$

proportional to the time interval.

For time intervals which are large as compared with the characteristic time T, that is when $t/T \gg 1$, the integral $\int_0^t R_t dt$ has a definite limiting value which is equal to the characteristic time T. Therefore

$$\frac{d}{dt}\overline{X^2} = 2\overline{u^2}T$$

and

$$\overline{X^2} = 2\overline{u^2}T\int_0^t dt = 2\overline{u^2}Tt$$

$$\sqrt{\overline{X^2}} = \sqrt{2u'l_1t} \tag{9-8}$$

For time intervals which are of the order of the characteristic time T, that is when $t \cong T$, the value of the integral $\int_0^t R_t dt$, and therefore the displacement, depends on the shape of the correlation curve $R(t)$. For simplicity, assuming a correlation function of the form $R(t) = e^{-t/T}$,

$$\frac{d}{dt}\overline{X^2} = 2\overline{u^2}\int_0^t e^{-t/T}dt \tag{9-9}$$

and

$$\overline{X^2} = 2\overline{u^2}\int_0^t\int_0^t e^{-t/T}dtdt$$

from which

$$\overline{X^2} = 2\overline{u^2}T[t - T(1 - e^{-t/T})]$$

and

$$\sqrt{\overline{X^2}} = \sqrt{2u'l_1t}\left[1 - \frac{T}{t}(1 - e^{-t/T})\right]^{\frac{1}{2}} \tag{9-10}$$

H,10. Theory of Flame Propagation by Large Scale Turbulence. Recent experimental studies [*18,19*] confirm the statement of Damköhler concerning the structure of flames in a flow with large scale turbulence. Plate H,10a shows a turbulent flame of $Re = 10{,}000$ as it appears to the eye. Its structure is revealed by the instantaneous schlieren photograph, Plate H,10b. Within the thick, diffuse, turbulent flame front there is a thin, sharp, wrinkled, instantaneous flame front. Small portions of this instantaneous flame front are carried by small masses of gas, which for short time intervals move as more or less coherent units. The average extension of these gas masses is characterized by the scale of turbulence $l_2 = \int_0^\infty R_x dx$, where $R_x = \overline{u_1u_2}/\overline{u^2}$ is the correlation coefficient of the u velocity components of two adjacent points 1 and 2 separated by the distance x. Superimposed over the random fluctuating motion of the small gas masses is the laminar burning velocity of the flame front by which it moves relative to the unburned gas. As the flame front propagates into the unburned gas it arrives, in the average, after the time interval

$$t_1 = \frac{l_2}{S_{\text{lam}}} \tag{10-1}$$

in a new mass of gas, the motion of which has no longer any correlation with the motion of the mass of gas which previously contained the flame. Therefore the further average displacement of the flame front by the random turbulent fluctuations has to be calculated as if it would start anew from this new position.

During the time interval t_1, the flame front is displaced by the turbulent fluctuations in the average by the amount $\sqrt{\overline{X^2}}$, forward and backward. The backward displacement is compensated by a process which will be described later. The average velocity of the displacement of the flame front by the large scale turbulent fluctuations is given therefore by

$$S_t = \frac{\sqrt{\overline{X^2}}}{t_1} \tag{10-2}$$

For weak turbulence, as compared with the normal burning velocity, where $u'/S_{lam} = t_1/T \ll 1$, the turbulent burning velocity is

$$S_t = \frac{\sqrt{\overline{X^2}}}{t_1} = \frac{u't_1}{t_1} = u' \tag{10-3}$$

that is, it is equal to the root mean square velocity of the turbulent fluctuations, and independent of the scale of the turbulence and of the normal burning velocity. This result already had been found by Damköhler by another line of reasoning. The same result could be obtained from simple considerations, for example, by comparing flame propagation to a relay race, in which the flame is always carried forward somewhere by the forward component of the fluctuating velocity; but as we have seen, this result is true only for weak turbulence. S_t is only a part of the total turbulent burning velocity, as will be shown later.

For strong turbulence, when $u'/S_{lam} \cong t_1/T \gg 1$, the average velocity of the displacement of the flame by the turbulent motion is

$$S_t = \frac{\sqrt{\overline{X^2}}}{t_1} = \frac{\sqrt{2l_1u't_1}}{t_1} = \frac{\sqrt{2l_1u'(l_2/S_{lam})}}{l_2/S_{lam}} = \sqrt{2u'S_{lam}} \tag{10-4}$$

if we assume that l_1 is approximately equal to l_2. According to the equation (10-4), at high turbulence intensity S_t increases only with the square root of the turbulence intensity u'; it is independent of the scale of turbulence, but is a function of the normal burning velocity.

For turbulence of intermediate intensity, when $u'/S_{lam} \cong 1$, the turbulent burning velocity is

$$S_t = \frac{\sqrt{\overline{X^2}}}{t_1} = \sqrt{2u'S_{lam}}\left[1 - \frac{S_{lam}}{u'}\left(1 - e^{-u'/S_{lam}}\right)\right]^{\frac{1}{2}} \tag{10-5}$$

One point of the above derivation has to be clarified still further. Displacement of particles by turbulent diffusion proceeds with equal probability in the positive and in the negative direction. In contrast to this, stationary turbulent flames move with constant velocity, relative to the gas, and maintain a constant average width of the flame brush. The reason for this behavior is shown in Fig. H,10a. In this figure, starting from an arbitrary wavy flame front, which is thought to have been produced by turbulent fluctuations, the position which the flame front would assume by normal propagation is constructed for subsequent short time intervals, during which turbulent motion is thought to be absent. In those regions of the flame which were moved back toward the burned gas the orientation of the flame front becomes oblique toward the mean direction of flame propagation, and portions of the flame front soon intersect each other causing sharp edges to appear on the back part of the flame. In those regions of the flame which were carried forward by the turbulent motion the flame becomes nearly perpendicular to the direction of mean propagation, and smooth, gently curved flame surfaces develop. This process explains the characteristic shape of turbulent flames, observed on all instantaneous flame pictures. The reason for the constant thickness of the flame and uniform propagation velocity also becomes apparent. The flame front elements with oblique orientation move forward with increased velocity and thereby not only compensate for the backward diffusion, but acquire on the average the same forward velocity as those elements of the flame front which are moved forward by turbulent diffusion. That the compensation is exact on the average follows from the constant width of the flame brush.

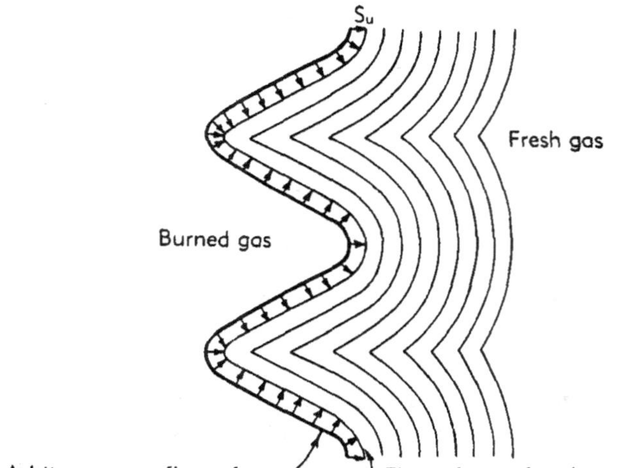

Fig. H,10a. The effect of normal flame propagation on the development of the turbulent flame front [*19*].

Even in those regions in which the flame front is normal to the direction of propagation of the entire flame system, the flame moves forward with its normal burning velocity in addition to the velocity with which the flame is transported by the turbulent motion. Therefore to obtain the total turbulent burning velocity S_T, the laminar burning velocity S_{lam} must be added to the velocity S_t produced by turbulent diffusion.

$$S_T = S_{lam} + S_t \tag{10-6}$$

The theoretical total turbulent burning velocity is shown in Fig. H,10b, as a function of the turbulence intensity. In this diagram the laminar burning velocity S_{lam} is used as the unit velocity, and accordingly the diagram is valid for any fuel and mixture ratio. The two limiting cases of very weak and very strong turbulence also are shown in the diagram.

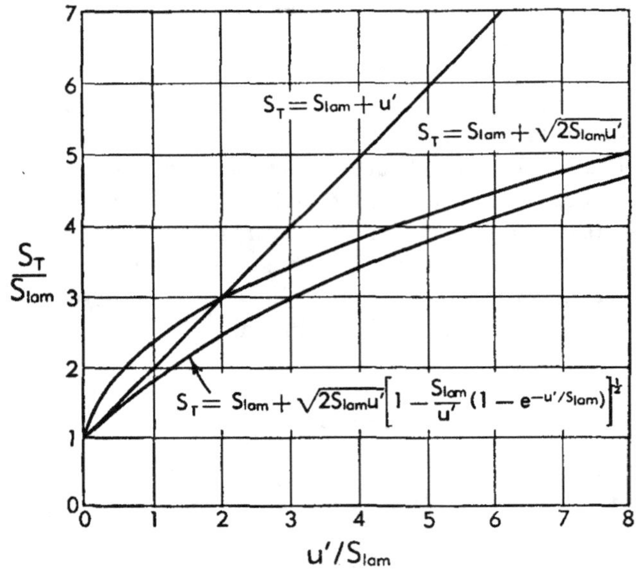

Fig. H,10b. Theoretical total turbulent burning velocity as function of the turbulence intensity, in units of the laminar burning velocity S_{lam} [*19*].

According to this theoretical result the effect of large scale turbulence on flame propagation should be independent of the scale of the turbulence; the total turbulent burning velocity should increase at first linearly, later only with the square root of the turbulence intensity; furthermore the laminar burning velocity S_{lam} should remain the most important controlling factor of flame propagation even at very high turbulence intensity.

H,11. Experimental Measurement of the Turbulent Burning Velocity. To test the theory of turbulent burning velocity described in Art. 10, experimental turbulent flames were produced and burning velocity measurements carried out [*19*]. The experimental Bunsen-type flames burned on the top of long vertical burner tubes, and were enveloped by a coaxial air flow to prevent eddy formation and mixing with the surrounding still air. The arrangement of the burner head is shown in Fig. H,11a. A stable flame was maintained even at high flow velocities by a narrow annular pilot flame burning at the edge of the burner tube. The velocity of the protective air flow was kept equal to the mixture flow velocity. This velocity is very nearly the same as the velocity of the combustion gases because of the almost free expansion of the flow around the flame. Considerable deviation from this matched air flow velocity had no observable effect on the shape and size of the flame.

The turbulence of the approach flow was the turbulence of the fully developed pipe flow. No grids were used to generate turbulence in these measurements. To characterize the turbulence, the radial component of the turbulence intensity, v', was used, which for a Bunsen-type flame is almost identical with the turbulence intensity component normal to the turbulent flame front. The turbulence intensity data were obtained from extensive turbulence measurements carried out by the Langley Field Laboratory of the National Advisory Committee for Aeronautics [*20*].

The burning velocity measurements were made on stoichiometric natural gas-air flames with a normal burning velocity of $S_u = 40.6$ cm/sec, and on stoichiometric acetylene-air flames with a normal burning velocity of $S_u = 175$ cm/sec.

The flow velocity and the intensity of turbulence vary along the radius of the pipe, and correspondingly along the height of the flame.

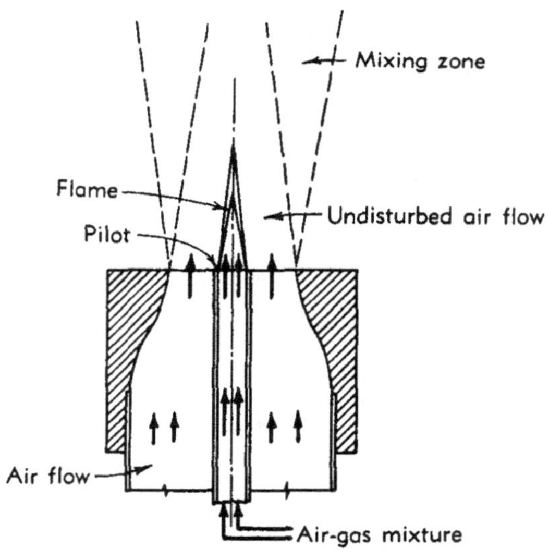

Fig. H,11a. Burner arrangement used for turbulent burning velocity measurements at the U.S. Bureau of Mines, Pittsburgh, Pa. [*19*].

Therefore an exact comparison between the experimental values and the predictions of the theory requires local measurement of the turbulent burning velocity. Values averaged over the entire surface of the flame do not give sufficient information. The local turbulent burning velocity was measured by the angle method, using the relation

$$S_T = U \sin \varphi$$

where U is the local mean flow velocity and φ the angle included by the average turbulent flame front and the direction of the approach velocity. The diffuse turbulent flame has no definite boundaries which could represent the average turbulent flame front. Furthermore it seems more appropriate to consider the most probable position of the instantaneous flame front as characteristic of the location and orientation of the average turbulent flame front than the more or less indefinite flame boundaries. The most probable position of the instantaneous flame front is marked with good approximation by the maximum brightness of the turbulent flame; the latter was therefore chosen to represent the average turbulent flame front. The location of the maximum brightness was measured on

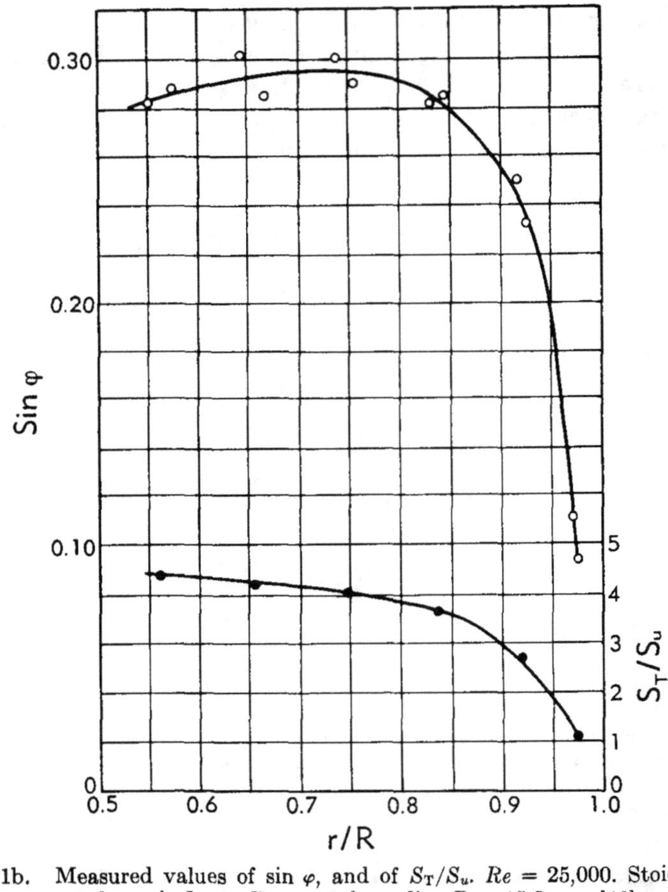

Fig. H,11b. Measured values of sin φ, and of S_T/S_u. Re = 25,000. Stoichiometric acetylene-air flame. Burner tube radius R = 15.8 mm [*19*].

long time (about 1 min) exposure flame negatives with a spectroscopic densitometer on equidistant horizontal lines, and the angle φ calculated from these measurements. Plate H,11 shows a negative of the type used for the burning velocity measurements.

Fig. H,11b, shows the measured values of sin φ and of S_T/S_u as a function of the radius for a stoichiometric acetylene-air flame. The Reynolds number of the approach flow was 25,000 and the burner tube radius was 1.58 cm. More such data are given in [*19*].

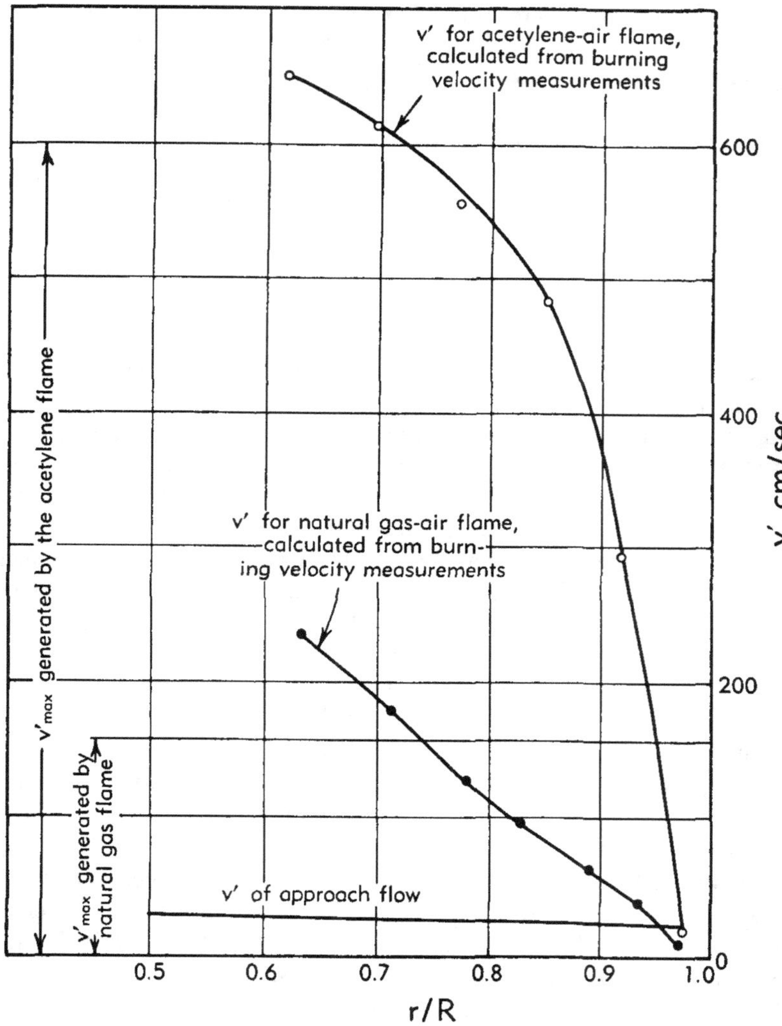

Fig. H,11c. Radial turbulence intensity v' in the flame calculated by the theory from measured turbulent burning velocity data for $Re = 50{,}000$, stoichiometric natural gas-air, and acetylene-air flames. Burner tube radius $R = 15.8$ mm. Mean approach flow velocity 2450 cm/sec [*19*].

The measured turbulent burning velocity data are in general higher than the values calculated from the theory described in Art. 10. For a closer study, the intensity of turbulence required by the theory in the approach flow to fit the measured turbulent burning velocity data was calculated. Such turbulence intensity values are shown in Fig. H,11c, together with the radial turbulence intensity values of the approach flow. The turbulence intensity which corresponds to the measured turbulent burning velocity is nearly zero in the immediate vicinity of the tube wall, but it increases very rapidly with decreasing values of r (corresponding to increasing height in the flame), and soon reaches values many times greater than that of the approach flow.

These experimental results can be explained only by the assumption that additional turbulence is generated by the turbulent flame itself, even without the intermediary step of producing strong velocity gradients in the mean flow first. Part of this flame-generated turbulence diffuses back into the approach flow and thus can affect the flame, especially the upper part of a Bunsen-type flame.

H,12. Turbulence Generation by the Turbulent Flame. The process of turbulence production by the turbulent flame may be pictured as follows [*19*].

As the explosive mixture passes through the instantaneous flame front with a velocity equal to the laminar burning velocity S_{lam}, its density decreases from ρ_1 to ρ_2 due to heating of the gas. Correspondingly the combustion gas leaves the flame front with the normal velocity component $(\rho_1/\rho_2)S_{lam}$, measured relative to the flame front. One may consider the flame front as a flow source that introduces an additional velocity component of the magnitude $[(\rho_1 - \rho_2)/\rho_2]S_{lam}$ into the gas flow. This velocity component is always normal to the instantaneous flame front. In a steady laminar flame the orientation of this additional velocity is constant in time. In contrast to this, the instantaneous flame front of a turbulent flame is in a state of random fluctuating motion which causes the orientation of small portions of the flame surface to fluctuate. Correspondingly the orientation of the flame-introduced velocity vector fluctuates also at random.

Over an extended time interval this velocity vector of constant magnitude but fluctuating orientation will have an average forward velocity component $\bar{U}_t$ normal to the average turbulent flame front. $\bar{U}_t$ is the mean value of $[(\rho_1 - \rho_2)/\rho_2]S_u \cos\varphi$ and not the velocity with which the hot combustion gas leaves the turbulent flame front. The magnitude of this average forward velocity component $\bar{U}_t$ will be large for slightly turbulent flames, and will decrease with increasing turbulence of the flame. After this average forward velocity is subtracted, the rest of the flame-introduced velocity components are distributed truly at random and constitute turbulence generated by the turbulent flame.

A schematic section of the turbulent flame front is given in Fig. H,12a. The undulated, fluctuating, instantaneous flame front has an average area A per unit area of the average turbulent flame front. The forward velocity component of the flame-introduced velocity for the element dA is

$$\frac{\rho_1 - \rho_2}{\rho_2} S_{lam} \cos\varphi \tag{12-1}$$

The average value of this forward component of the flame-introduced velocity vector is:

$$\bar{U}_f = \left[\frac{\rho_1 - \rho_2}{\rho_2} S_{lam} \cos \varphi\right]$$

$$= \frac{\rho_1 - \rho_2}{\rho_2} S_{lam} [\cos \varphi] \tag{12-2}$$

where the brackets [ ] denote average quantities.

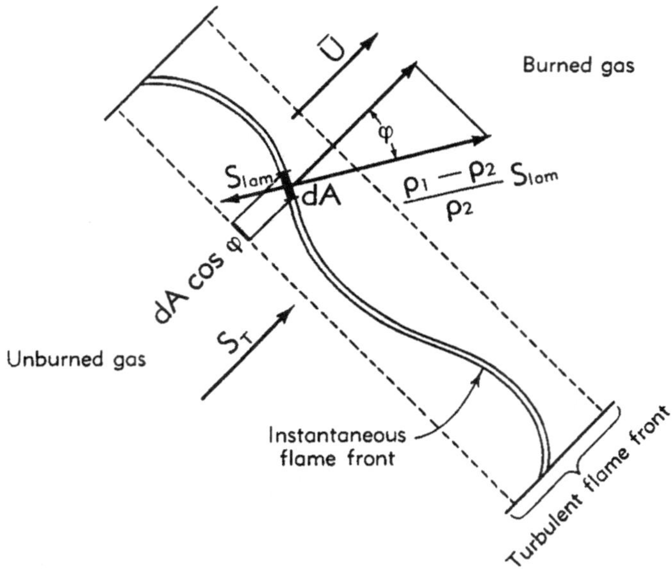

Fig. H,12a. Schematic cross section of the turbulent flame front showing the velocity vector $[(\rho_1 - \rho_2)/\rho_2]S_{lam}$ introduced by the instantaneous flame into the gas flow, and the average forward component $\bar{U}_f$ of this velocity [*19*].

From Fig. H,12a it can be seen that $[\cos \varphi] = 1/A$ since

$$\int_0^A \cos \varphi dA = 1$$

Therefore

$$\bar{U}_f = \frac{\rho_1 - \rho_2}{\rho_2} S_{lam} \frac{1}{A} \tag{12-3}$$

The area ratio A equals the ratio of the turbulent and laminar burning velocities because large scale turbulence does not affect the structure of the instantaneous flame front. Thus $A = S_T/S_{lam}$, and therefore

$$\bar{U}_f = \frac{\rho_1 - \rho_2}{\rho_2} S_{lam} \frac{S_{lam}}{S_T} \tag{12-4}$$

The intensity of the turbulence generated by the flame may be calculated from the amount of mechanical work developed by the gas as it flows across the pressure drop Δp of the instantaneous flame front, because this work is the source of the additional turbulent energy. In a coordinate system in which the unburned gas is at rest (Fig. H,12b) that part of the expansion work which is converted into kinetic energy of the flame-introduced velocity is given by

$$\int_{\rho_1}^{\rho_2} pd\left(\frac{1}{\rho}\right) - p_2\left(\frac{1}{\rho_2} - \frac{1}{\rho_1}\right) = \frac{1}{2g}\left(\frac{\rho_1 - \rho_2}{\rho_2} S_{\text{lam}}\right)^2 \tag{12-5}$$

Here

$$\int_{\rho_1}^{\rho_2} pd\left(\frac{1}{\rho}\right) = \text{total expansion work}$$

and

$$p_2\left(\frac{1}{\rho_2} - \frac{1}{\rho_1}\right) = \text{work done by the expanding gas on the surrounding.}$$

Part of this expansion work is added to the kinetic energy of the mean flow, corresponding to the mean forward velocity $\bar{U}$; the rest provides energy for the flame-generated turbulence. Therefore the total

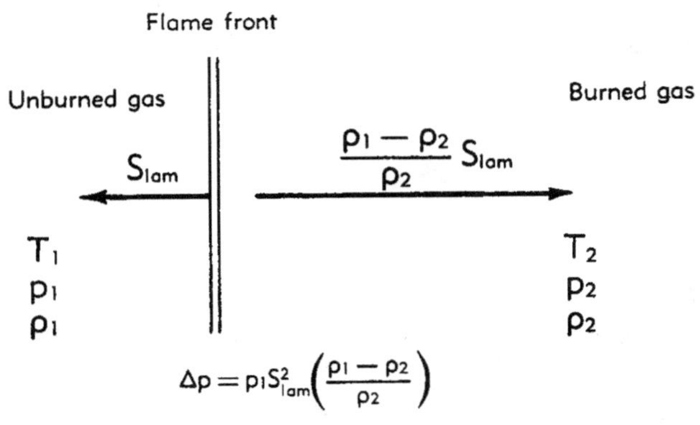

Fig. H,12b. Motion of the laminar flame. The laminar flame front moves into the unburned gas with the velocity S_{lam}, the burned gas leaves with the velocity $[(\rho_1 - \rho_2)/\rho_2]S_{\text{lam}}$, measured relative to the unburned gas [*19*].

intensity of the flame-generated turbulence is given by the energy equation

$$\overline{u^2} + \overline{v^2} + \overline{w^2} = \left(\frac{\rho_1 - \rho_2}{\rho_2} S_{\text{lam}}\right)^2 - \left(\frac{\rho_1 - \rho_2}{\rho_2} S_{\text{lam}} \frac{S_{\text{lam}}}{S_{\text{T}}}\right)^2 \tag{12-6}$$

where $\overline{u^2}$, $\overline{v^2}$, and $\overline{w^2}$ are the mean square values of the flame-generated turbulent velocity components. In the case of a laminar flame, $S_{\text{T}}/S_{\text{lam}} = 1$, the two terms on the right side of Eq. 12-6 are equal, and no turbulence is generated by the flame. With increasing values of $S_{\text{T}}/S_{\text{lam}}$, that is, as the flame becomes more turbulent, the flame-generated turbulence approaches an upper limit

$$\overline{u^2} + \overline{v^2} + \overline{w^2} = \left(\frac{\rho_1 - \rho_2}{\rho_2} S_{\text{lam}}\right)^2 \tag{12-7}$$

or if uniform energy distribution is assumed among the three components,

$$u' = v' = w' = \frac{1}{\sqrt{3}} \left(\frac{\rho_1 - \rho_2}{\rho_2}\right) S_{\text{lam}} \tag{12-8}$$

In Fig. H,11c this calculated maximal value of flame-generated turbulence is indicated on the left side, for natural gas-air and acetylene-air flames at $Re = 50{,}000$. The turbulence intensity which corresponds to the measured burning velocity of the flame approaches and even exceeds somewhat the maximal calculated value of flame-generated turbulence for both natural gas and acetylene flames. In this calculation the laminar flame velocity was taken to be equal to the normal burning velocity. The excess of the turbulence intensity, calculated from the measured turbulent burning velocity, above the theoretical maximum may be explained by a slight increase in the laminar burning velocity due to the effect of fine scale turbulence. The v' values of the acetylene flame seem to approach a maximal value not much larger than the last measured values; the v' values of the natural gas flame rise almost linearly even at the last measured points, but the shape of the flame indicates that the burning velocity and the corresponding v' values calculated from it do not reach much higher values than the last measured values.

These experimental results therefore seem to indicate the correctness of the above theories of turbulent burning and flame-generated turbulence. For a final experimental check the intensity of turbulence in the turbulent flame, or immediately after the turbulent flame, must be measured directly.

According to Eq. 12-8 the maximal intensity of flame-generated turbulence is proportional to the expansion ratio $(\rho_1 - \rho_2)/\rho_2$ of the combustion products and to the laminar burning velocity of the mixture. The laminar burning velocity, which, as was shown in Art. 10, has a controlling influence on the turbulent burning velocity, retains its fundamental importance also in the process of turbulence generation by the turbulent flames.

The generation of turbulence described here is an inherent property of the turbulent flame and is operative independent of other circumstances. Naturally the flame may affect the mean flow in a way which leads to turbulence production by shear flow as described in Art. 5. In such cases the turbulent energies produced by the various processes are additive.

Turbulence generation by the flame throws more light on the phenomena observed on flames propagating in tubes as described in Art. 1. When the flow of combustion products leaving the tube becomes turbulent, some of this turbulence diffuses back to the flame. The disturbed flame itself generates more turbulence, which is responsible for the sudden strong increase of the flame propagation velocity.

The strong turbulence generated by the turbulent flame is demonstrated by the shadow picture of a natural gas-air flame of $Re = 35{,}000$ (Plate H,12). This flame was surrounded by a protective coaxial air flow; therefore no turbulence could be produced by a velocity difference between the hot combustion products and the surrounding air.

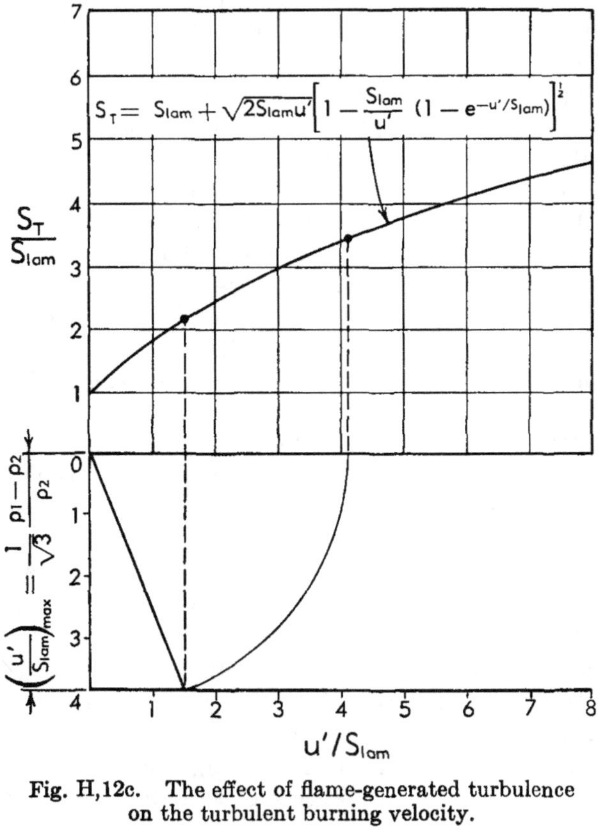

Fig. H,12c. The effect of flame-generated turbulence on the turbulent burning velocity.

The effect of flame-generated turbulence on the burning velocity is shown in Fig. H,12c where the theoretical turbulent burning velocity S_T is plotted as a function of the turbulence intensity u', in units of laminar burning velocity. In this representation the maximal intensity of the flame-generated turbulence is given by

$$\left(\frac{u'}{S_{\text{lam}}}\right)_{\text{max}} = \frac{1}{\sqrt{3}}\frac{\rho_1 - \rho_2}{\rho_2}$$

This expression is dependent on the mixture composition only. Its numerical value is nearly the same for most stoichiometric air-fuel mixtures. For example, for stoichiometric natural gas-air flames,

$$\left(\frac{u'}{S_{\text{lam}}}\right)_{\max} = \frac{1}{\sqrt{3}} \frac{\rho_1 - \rho_2}{\rho_2} = 3.88$$

for stoichiometric acetylene-air flames,

$$\left(\frac{u'}{S_{\text{lam}}}\right)_{\max} = \frac{1}{\sqrt{3}} \frac{\rho_1 - \rho_2}{\rho_2} = 4.11$$

The energies of the two random motions, that of the turbulence of the approach flow and that of the flame-generated turbulence, are additive. Therefore the resulting turbulence intensity which affects the burning velocity is given by the relationship

$$u'_r = (u'_a + u'_f)^{\frac{1}{2}}$$

where u'_r is the resulting turbulence intensity; u'_a the approach flow turbulence intensity; and u'_f the flame-generated turbulence intensity, assuming that the flame-generated turbulence is operative with its full intensity. An example of the effect of flame-generated turbulence is shown in Fig. H,12c for an approach flow where $u'/S_{\text{lam}} = 1.5$.

H,13. Pressure Drop Across the Flame, Diffusion of Turbulence, Thickness of the Turbulent Flame. In a strongly turbulent flame most of the kinetic energy produced by the pressure drop across the instantaneous laminar flame front,

$$\Delta p_{\text{lam}} = \rho_1 \left(\frac{\rho_1 - \rho_2}{\rho_2}\right) S_{\text{lam}}^2 \tag{13-1}$$

is converted into turbulent energy. The value of the mean forward velocity $\bar{U}_f$ is negligible. Therefore an additional pressure drop is required to produce the mean velocity increase across the turbulent flame. If we consider such a turbulent flame in a coordinate system where the turbulent flame front is at rest (Fig. H,13), we find that the velocity component normal to the turbulent flame front is increased by the flame from $U_1 = -(S_T + u_1)$ to $U_2 = -[(\rho_1/\rho_2)S_T + u_2]$, where u_1 and u_2 are the superimposed turbulent velocity fluctuations. The pressure drop required for this velocity increase is given by the momentum theorem:

$$\overline{\Delta p_t} = \rho_2 \overline{\left(\frac{\rho_1}{\rho_2} S_T + u_2\right)^2} - \rho_1 \overline{(S_T + u_1)^2} \tag{13-2}$$

or

$$\overline{\Delta p_t} = \rho_2 \left(\frac{\rho_1}{\rho_2}\right)^2 S_T^2 + \rho_2 \overline{u_2^2} - \rho_1 S_T^2 - \rho_1 \overline{u_1^2}$$

$$\overline{\Delta p_t} = \rho_1 \frac{\rho_1 - \rho_2}{\rho_2} S_T^2 + \rho_2 \overline{u_2^2} - \rho_1 \overline{u_1^2}$$

Neglecting $\rho_1\overline{u_1^2}$ as compared to $\rho_2\overline{u_2^2}$ and introducing the intensity of the flame-generated turbulence

$$\overline{u_2^2} = \frac{1}{3}\left(\frac{\rho_1 - \rho_2}{\rho_2}\right)^2 S_{\text{lam}}^2$$

$$\overline{\Delta p_t} = \rho_1\frac{\rho_1 - \rho_2}{\rho_2}\left[S_T^2 + \frac{1}{3}\left(1 - \frac{\rho_2}{\rho_1}\right)S_{\text{lam}}^2\right] \tag{13-3}$$

or in most cases with good approximation:

$$\overline{\Delta p_t} = \rho_1\left(\frac{\rho_1 - \rho_2}{\rho_2}\right)S_T^2 \tag{13-4}$$

The total pressure drop $p_1 - p_2$ is the sum of Δp_{lam} and Δp_t.

$$p_1 - p_2 = \Delta p_{\text{lam}} + \Delta p_t = \rho_1\left(\frac{\rho_1 - \rho_2}{\rho_2}\right)(S_{\text{lam}}^2 + S_T^2) \tag{13-5}$$

This expression is valid for flames which burn in the open, where only the normal component of the flow is affected by the flame. For flames

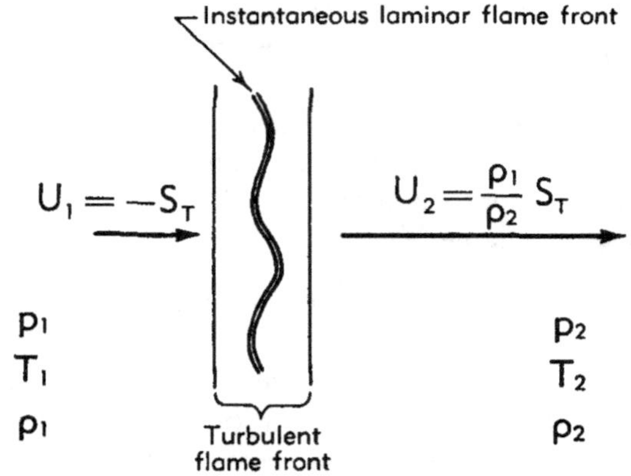

Fig. H,13. Gas flow across the turbulent flame front.

enclosed in channels, a further pressure drop is required to increase the mean flow velocity, corresponding to the decreased density of the combustion products.

The turbulence generated by the turbulent flame is carried away by the flow. However, because of the sudden increase of the turbulence intensity some turbulence diffuses back into the approach flow and into the neighboring flow tubes. It is by this diffusion of turbulent energy that flame-generated turbulence can affect the burning velocity of the flame which produces the turbulence. This diffusion process also accounts for the fact that the lower part of a turbulent flame is only slightly affected by the turbulent fluctuations, and the full effect of turbulence develops only at a considerable height on the flame (see Plate H,10b).

The turbulent flame front appears thick because the instantaneous, thin, laminar combustion wave moves about rapidly in the space occupied by the turbulent flame. Turbulence is generated in the immediate vicinity of the laminar combustion wave, therefore the average intensity of turbulence increases gradually across the turbulent flame front from the turbulence intensity of the approach flow to its maximal value. In the regions close to the unburned gas, the turbulence intensity is much smaller than the mean turbulence intensity in the flame. The turbulent burning velocity in these regions is therefore smaller than the normal component of the flow velocity. The instantaneous flame front is swept along by the flow, and only seldom for short intervals of time can momentarily strong fluctuations move the flame back into these regions. The opposite is true for the regions of the turbulent flame on the burned gas side. Here the turbulence intensity, and with it the turbulent burning velocity, is higher than the normal flow velocity, and the flow can carry the instantaneous flame front relatively seldom into these regions.

As a result, the probability of finding the instantaneous flame at a definite point will be large in the middle region of the turbulent flame front and will decrease on both sides. This qualitative reasoning shows that flame-generated turbulence tends to stabilize the thickness of the turbulent flame front, which otherwise would increase continually under the influence of the turbulent diffusion.

H,14. Propagation of Flames in Turbulent Explosive Mixtures. In the case of high speed turbulent Bunsen-type flames stabilized by a small annular pilot flame, it has been observed that as the flow velocity increases, the lower part of the flame becomes less luminous. At further increased flow velocity, holes appear on the sides of the flame through which explosive mixture can escape unignited. During all these changes the top portion of the flame remains unaltered to any observable degree (Plate H,14a).

If the flow velocity is increased further, the propagation of the flame is completely interrupted at a short distance above the pilot flame, and the top part of the flame disappears. There remains only a short, open, collarlike flame burning above the pilot flame. At sufficiently high flow velocities this phenomenon happens even in stoichiometric mixtures. Plate H,14b shows such a broken-off flame; a stoichiometric mixture of natural gas-air issues from the burner tube, in contact with the strong ring-shaped flame, without being ignited. This shows that under certain conditions flames do not propagate further in explosive mixtures even though a strong flame is burning in part of the mixture.

The interruption of flame propagation always occurs in a region of the flow where the velocity gradient is very high. Observation shows further that flames are more easily interrupted in gas mixtures which have a lower normal burning velocity, and on tubes of smaller diameter. Increasing the pilot flame can prevent the disruption of the flame.

A similar phenomenon was observed by A. C. Scurlock [6] on flames burning in channels and stabilized on flame holder rods. Under certain conditions the flame ceases to propagate across the entire flow of explosive mixture, although a shorter flame continues to burn behind the flame holder.

These phenomena of disruption of flame propagation may be explained by the reduction of the burning velocity as the flame moves across a steep velocity gradient. Fig. H,14a shows schematically a

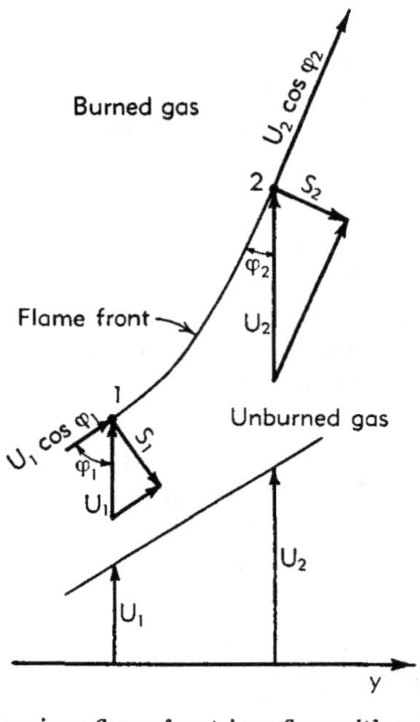

Fig. H,14a. Laminar flame front in a flow with velocity gradient.

laminar flame in a flow with moderate velocity gradient. At point 1, where the flow velocity is small, the velocity component parallel to the flame front, $U_1 \cos \varphi_1$, is small; at point 2, where the flow velocity is larger, the velocity component parallel to the flame front, $U_2 \cos \varphi_2$, is much larger than at point 1. Therefore the length of every small section of the flame front is increased; the flame front is "stretched" as it moves from the lower into the higher velocity regions of the flow. Consequently the heat flowing from the reaction zone into the unburned gas is distributed over an increased gas volume, and therefore the burning velocity must decrease.

The flow and heat flow conditions in the preparatory zone of a laminar flame existing in a flow with strong velocity gradient are shown in detail in Fig. H,14b. In this figure two coordinate systems are used. One is x, y with the x coordinate oriented in the direction of the mean flow velocity U, and another is ξ, η with the η coordinate oriented normal to the combustion wave.

The rate of change of the heat content of the elementary volume $d\xi \cdot d\eta \cdot 1$ of the preparatory zone of the combustion wave is given by the equation

$$\frac{dq}{dt} d\xi d\eta = \lambda \frac{\partial^2 \vartheta}{\partial \eta^2} d\xi d\eta + c_p \frac{\partial(\rho W_\eta \vartheta)}{\partial \eta} d\xi d\eta - c_p \frac{\partial(\rho W_\xi \vartheta)}{\partial \xi} d\xi d\eta \quad (14\text{-}1)$$

where q is the heat content per unit volume; t the time; λ the heat conductivity; ϑ the gas temperature above the approach flow temperature; c_p the specific heat of the unburned gas; W_η the gas flow velocity component normal to the combustion wave; and W_ξ the gas flow velocity component parallel to the combustion wave.

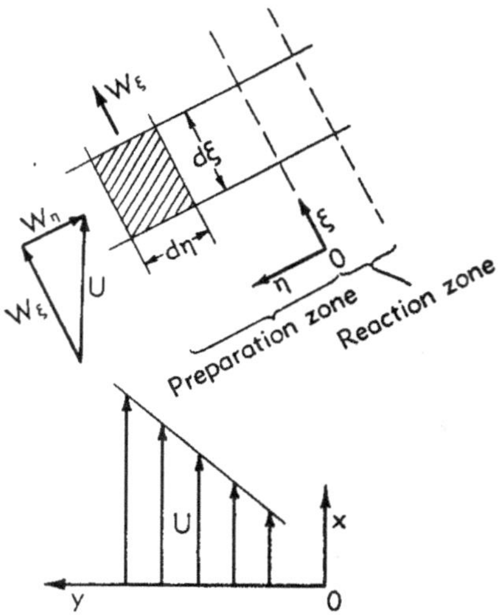

Fig. H,14b. Schematic representation of a volume element of the preparation zone of a combustion wave in a velocity gradient field.

In the steady state $dq/dt = 0$, and the equation reduces to

$$\lambda \frac{\partial^2 \vartheta}{\partial \eta^2} + c_p \frac{\partial(\rho W_\eta \vartheta)}{\partial \eta} - c_p \frac{\partial(\rho W_\xi \vartheta)}{\partial \xi} = 0 \qquad (14\text{-}2)$$

The equation of continuity is

$$\frac{\partial(\rho W_\eta)}{\partial \eta} - \frac{\partial(\rho W \xi)}{\partial \xi} = 0 \qquad (14\text{-}3)$$

Assuming $\partial\vartheta/\partial\xi = 0$, Eq. 14-2 and 14-3 can be combined into

$$\lambda \frac{\partial^2 \vartheta}{\partial \eta^2} + c_p (\rho W_\eta) \frac{\partial \vartheta}{\partial \eta} = 0 \qquad (14\text{-}4)$$

In the case of a normal, unstretched, combustion wave $\rho W_\eta = \text{const} = \rho_0 S_u$, where ρ_0 is the density of the unburned gas mixture a large distance before the flame front, and S_u is the normal burning velocity. In the case of a flame in a flow with velocity gradient, the mass flow rate ρW_η of the approach flow increases with increasing η. By assuming linear increase of the flow velocity we have

$$\rho W_\eta = \rho_0 S(1 + a\eta) \tag{14-5}$$

where the coefficient a characterizes the velocity increase. The value of the coefficient a can be calculated from the distribution of the mean flow velocity U as follows:

$$\rho W_\eta = \rho_0 U \sin\varphi = \rho_0 U_0 \sin\varphi + \rho_0 \left(\frac{dU}{dy}\cos\varphi \sin\varphi\right)\eta \tag{14-6}$$

where U_0 is the mean flow velocity at $\eta = 0$. With $\sin\varphi = S/U_0$, and $\cos\varphi \cong 1$, where S is the reduced burning velocity,

$$\rho W_\eta = \rho_0 U_0 \frac{S}{U_0} + \rho_0 \frac{dU}{dy}\frac{S}{U_0}\eta \tag{14-7}$$

or

$$\rho W_\eta = \rho_0 S(1 + a\eta) \tag{14-8}$$

where

$$a = \frac{1}{U_0}\frac{dU}{dy} \tag{14-9}$$

Therefore $1/a$ is the thickness within which the flow velocity increases by an amount equal to U_0.

With this relationship for mass flow, the equation for temperature distribution in the preparatory zone becomes

$$\lambda \frac{\partial^2\vartheta}{\partial\eta^2} + c_p\rho_0 S(1 + a\eta)\frac{\partial\vartheta}{\partial\eta} = 0 \tag{14-10}$$

By integrating once, Eq. 14-10 is transformed into

$$\frac{d\vartheta}{d\eta} = \left(\frac{d\vartheta}{d\eta}\right)_0 e^{-\frac{c_p\rho_0 S}{\lambda}\left(\eta + \frac{a}{2}\eta^2\right)} \tag{14-11}$$

where $(d\vartheta/d\eta)_0$ is the temperature gradient at the ignition point, whose value is determined by the reaction zone of the flame.

For the normal combustion wave in a constant velocity field, Eq. 14-11 reduces to

$$\frac{d\vartheta}{d\eta} = \left(\frac{d\vartheta}{d\eta}\right)_0 e^{-\frac{c_p\rho_0 S_u}{\lambda}\eta} \tag{14-12}$$

To maintain the flame permanently, the temperature of the approach flow must be increased in the preparation zone to the ignition temperature ϑ_0. Therefore

$$\int_0^{\infty} d\vartheta = \left(\frac{d\vartheta}{d\eta}\right)_0 \int_0^{\infty} e^{-\frac{c_p\rho_0 S_u}{\lambda}\eta}\, d\eta = \vartheta_0 \tag{14-13}$$

and

$$\left(\frac{d\vartheta}{d\eta}\right)_0 \frac{\lambda}{c_p\rho_0 S_u} = \vartheta_0$$

from which

$$S_u = \frac{\lambda}{c_p\rho_0}\frac{1}{\vartheta_0}\left(\frac{d\vartheta}{d\eta}\right)_0 \tag{14-14}$$

which is the familiar expression for the normal burning velocity.

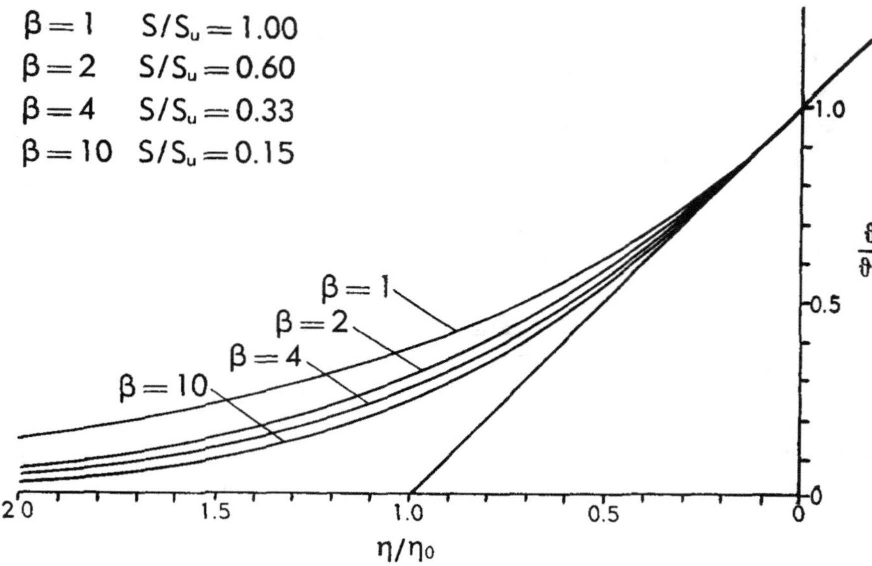

Fig. H,14c. Temperature distribution in the preparatory zone of a combustion wave in a flow with velocity gradient, for various values of β.

The temperature distribution in the preparatory zone of the normal combustion wave is shown in Fig. H,14c in nondimensional form. The length η_0 is defined by the relationship

$$\frac{1}{\eta_0} = \frac{1}{\vartheta_0}\left(\frac{d\vartheta}{d\eta}\right)_0 = \frac{c_p\rho_0 S_u}{\lambda} \tag{14-15}$$

For the flame in a velocity gradient field we have again the condition that

$$\int_0^{\infty} d\vartheta = \left(\frac{d\vartheta}{d\eta}\right)_0 \int_0^{\infty} e^{-\frac{c_p\rho_0 S}{\lambda}\left(\eta+\frac{a}{2}\eta^2\right)}\, d\eta = \vartheta_0 \tag{14-16}$$

The integration cannot be carried out in closed form, but inspection of the equation shows that with increasing value of a, the burning velocity S must decrease, otherwise the gas could not reach the ignition temperature.

The numerical integration of Eq. 14-16 was carried out in the nondimensional form

$$\int_0^\infty \frac{d\vartheta}{\vartheta_0} = \int_0^\infty e^{-\frac{S}{S_u}\left[\left(\frac{\eta}{\eta_0}\right)+\frac{\beta-1}{2}\left(\frac{\eta}{\eta_0}\right)^2\right]} \frac{d\eta}{\eta_0} = 1 \qquad (14\text{-}17)$$

where $\beta = 1 + a\eta_0$ characterizes the velocity field. The temperature distribution in the preparatory zone of the flame is given in Fig. H,14c for $\beta = 1$ (normal combustion wave), where $S/S_u = 1$; $\beta = 2$, where $S/S_u = 0.6$; $\beta = 4$, where $S/S_u = 0.33$; and $\beta = 10$, where $S/S_u = 0.15$.

These results show that the burning velocity may be seriously reduced if the flow velocity increases considerably within the preparatory zone of the flame. If the burning velocity is reduced to a small fraction of its normal value, the propagation of the flame may be entirely interrupted by small velocity fluctuations.

It is interesting to calculate the β value for a stoichiometric natural gas-air flame of $Re = 100{,}000$, on which the interruption of flame propagation is observed, depending on the size of the pilot flame. With $\lambda = 0.578 \times 10^{-4}$ cal/cm°C sec; $c_p\rho = 0.31 \times 10^{-3}$ cal/cm³°C; and $S_u = 40$ cm/sec,

$$\eta_0 = \frac{\lambda}{c_p\rho S_u} = 0.466 \times 10^{-2} \text{ cm}$$

The velocity distribution for the turbulent pipe flow near the wall is shown in Fig. H,14d [*21*, p. 79]. The velocity gradient reaches very high values in the laminar sublayer at the wall. Using the data of Fig. H,14d, the value of $1/a$ is calculated at the middle of this laminar sublayer as

$$\frac{1}{a} = U\frac{1}{dU/dy} = \frac{1}{2}R\frac{68.4}{Re^{\frac{7}{8}}}$$

where R is the radius of the tube. For $R = 1.58$ cm and $Re = 100{,}000$,

$$\frac{1}{a} = 0.228 \times 10^{-2} \text{ cm}$$

With these values

$$\beta = 1 + a\eta_0 = 1 + \frac{0.466}{0.228} = 3.04$$

at which value the burning velocity is reduced to about half its normal value. Closer to the wall, the value of β increases and the burning velocity is reduced further.

It is not possible from the above approximate theory to predict the exact conditions under which flame propagation will be interrupted, but the theory is sufficient to show that the steep velocity increase is responsible for the interruption of flame propagation. Furthermore, the theory shows definitely that flame propagation can be disturbed by a shear flow only if the flow velocity increases several-fold within the preparatory zone of the combustion wave. Turbulence may play a role in the final interruption of flame propagation, but its main importance is that sufficiently high velocity gradients can appear in turbulent flow to interrupt the flame. The theory also explains correctly the other experimental

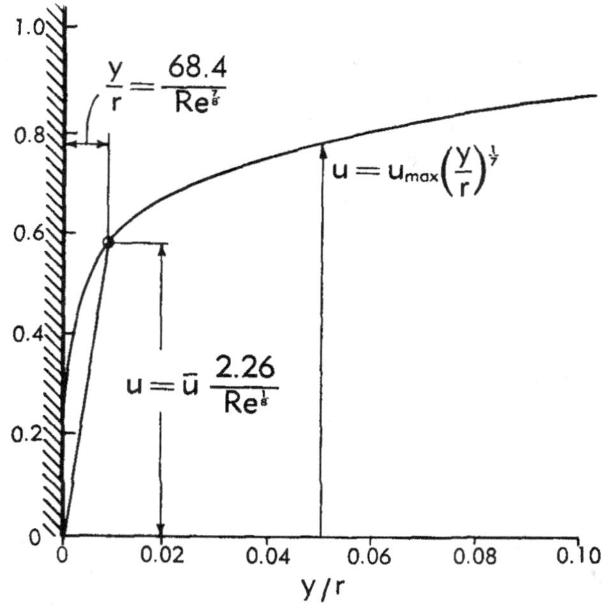

Fig. H,14d. Velocity distribution in turbulent pipe flow, showing the laminar sublayer near the wall [21].

observations. In leaner mixtures the thickness of the preparatory zone is larger, therefore flame propagation is more easily interrupted in such mixtures. Increased size of the pilot flame increases the smallest value of U_0, and thereby greatly reduces the effect of the velocity gradient on the burning velocity.

H,15. Stability of Turbulent Flames. The fundamental conditions of flame stabilization were found by B. Lewis and G. von Elbe [22], and are described in detail in Sec. G of this volume. Lewis and von Elbe found that the flashback and blowoff of laminar flames, stabilized on the rim of the burner tube, are controlled by the velocity gradient of the flow at the wall. L. M. Bollinger and D. T. Williams extended the stability determinations into the turbulent region [23] and found that the blowoff of turbulent flames occurs at the same velocity gradient, measured at the wall, as the blowoff of laminar flames. This result is to be expected since the lower edge of turbulent flames is hardly affected by turbulence, as shown, for example, in Plate H,10b. A detailed description of the experimental observations concerning the stability of turbulent flames stabilized on the rim of the burner tube, together with the stability data of lifted turbulent flames and turbulent diffusion flames, is given in Sec. I of this volume.

The velocity gradient of turbulent pipe flow at the wall as shown in Fig. H,14d is very high. Therefore, flames can be stabilized on the burner tube rim only at comparatively low flow velocities. In high speed flow, flames may be stabilized by pilot flames that burn outside of the high velocity flow, and provide a continuous source of ignition that is independent of the main flow as described, for example, in [*19*] and [*24*]. The effectiveness of pilot flames is limited only by the failure of the flame to propagate into the main flow across a boundary layer with steep velocity gradient, as described in Art. 14.

Another method for flame stabilization in high speed flow was described by Lewis and von Elbe [*22*]. In the wake of blunt objects placed in the flow, a recirculating eddy system is formed, that can act as a continuous source of ignition for an inverted flame, a flame which originates behind the blunt body and spreads from there obliquely across the explosive flow. The recirculating eddy system receives heat from the flame, carries it back upstream, and ignites the flow of explosive mixture with which it comes into contact. Considerable experimental work was carried out by various investigators [*6*; *7*, pp. 21–44; *25*] to determine the relationship between the dimensions and shape of the flame holder and the blowoff velocity. The results of these studies are described in Sec. J and in XI,E.

Blowoff occurs if, at high flow velocity, the heat received by the recirculating eddy from the hot combustion gases becomes insufficient to maintain a high enough temperature required for ignition. Sometimes blowoff is observed to occur in two steps. First the flame ceases to propagate across the entire explosive flow, although there is still a residual flame burning behind the flame holder; at somewhat increased flow velocity this residual flame also disappears.

Experiences with flame holders and pilot flames show that the problem of flame stabilization in high speed flow is really twofold. A continuous source of ignition must be provided by the flame holder or pilot flame, *and* propagation of the flame from the low velocity region of flow into the high velocity region must be assured. According to the results described in Art. 14, to fulfill this latter requirement the relative velocity increase within the preparation zone of the combustion wave must have a moderate value.

H,16. Numerical Data of Turbulent Flames. It is evident that our present knowledge of turbulent flame phenomena is not suffi-

Table H,16. Numerical data of turbulent flames.

Reynolds number	Tube radius, cm	Flow velocity U, cm/sec	Heat liberated Q, BTU/min	Gas mixture	Flame height h, cm	Flame diameter D, cm	Width of brush W, cm	Normal burning velocity S_u, cm/sec
10,000	1.58	490	740	Natural gas-air	12.62	3.28	1.15	40
25,000	1.58	1220	1,850		23.14	4.00	1.72	40
50,000	1.58	2450	3,680		34.28	4.11	1.61	40
75,000	1.58	3680	5,480		43.08	4.10	1.54	40
100,000	1.58	4900	7,470		58.82	4.71	2.06	40
75,000	2.54	2320	9,025		38.40	6.05	2.10	40
100,000	2.54	3085	12,200		53.08	7.10	2.63	40
160,000	2.54	4940	19,600		60.00	6.78	2.58	40
25,000	1.03	1810	1,450	Acetylene-air	6.73	2.21	0.708	175
35,000	1.03	2540	2,120		9.38	2.30	0.708	175
50,000	1.03	3630	2,970		12.30	2.36	0.908	175
75,000	1.03	5450	4,720		14.43	2.57	1.06	175
100,000	1.03	7250	5,980		16.22	2.52	0.910	175
25,000	1.58	1220	2,560		8.00	3.57	1.07	175
35,000	1.58	1720	3,480		9.71	3.43	1.07	175
50,000	1.58	2450	4,800		12.93	3.53	1.18	175

cient for the exact prediction of the size and shape of the turbulent flame, even if the data of the approach flow, and the value of the normal burning velocity, are given. We can calculate the maximal value of turbulent burning velocity to be expected, but not the distribution of the burning velocity along the lower part of the flame. Therefore, to help designers of combustion apparatus to estimate the size and shape of open turbulent flames, a series of photographs has been taken of natural gas-air and of acetylene-air flames. The height and maximal diameter of the visible flame and the thickness of the flame brush have been measured from these photographs, and the data are assembled in Table H,16. These flames were all surrounded by a coaxial air flow whose velocity was equal to that of the approach flow velocity of the explosive mixture. The normal burning velocity of the stoichiometric natural gas-air flames was 40 cm/sec; that of the stoichiometric acetylene-air flames 175 cm/sec.

H,17. The Status of Turbulent Flame Research. The theory of turbulent flames presented in Art. 12–13 was extended recently to include a more detailed description of the turbulent flame brush [*26*, pp. 248–262]. An electronic probe method [*27*, pp. 613–620] was developed by Karlovitz, Denniston, Knapschaeffer, and Wells, which permits the determination of the probability of finding the combustion wave at any point of the turbulent flame brush [*26*, pp. 613–620]. This makes it possible to define exactly the width of the turbulent flame brush, and to calculate the heat release rate in turbulent flames. The electronic probe also permits the direct measurement of the "fullness" of the flame, which is an important new concept in the study of flame stability.

The predictions of the theory are in good agreement with the available experimental results. For a complete experimental test of the theory and exploration of the limits of its validity, further measurements are required. These should include the measurement of turbulence intensity in the turbulent flame brush, or immediately downstream of the flame brush. This would require the development of new methods of measurement since available methods are not applicable to the high temperature flow. Another alternative is the measurement of the turbulent burning velocity or the volume of the turbulent flame brush, at such high turbulence intensity of the approach flow that the intensity of the flame-generated turbulence can be neglected.

A very important result concerning turbulent flames was obtained recently by J. P. Longwell, who showed that, at extremely high turbulence intensity, turbulent combustion approaches homogeneous volume combustion, where the combustion waves disappear and the kinetic reaction rate limits the heat release rate [*26*, pp. 508–510]. Longwell found experimentally 400×10^6 BTU/ft^3 atm^2 hr for the reaction-rate-limited heat release rate in stoichiometric fuel-air mixtures, and a dependence of the heat release rate on nearly the second power of the absolute pressure. This value is orders of magnitude higher than heat release rates found in practical combustion systems; it rises so rapidly with pressure that it will limit combustion rates only at low subatmospheric pressures.

Another approach was proposed recently for the theoretical study of turbulent flames [*28*] which dispenses with the idea of the wrinkled combustion wave, and considers the turbulent flame as a zone of distributed reaction. The experimental evidence cited in favor of this treatment can be explained on the basis of combustion waves, and hence there is no compelling reason to discard this very useful concept which eliminates the necessity to consider the chemical kinetics of combustion for the study of the effect of turbulence on flame propagation.

Other recent studies of turbulent flames include a theory of turbulent flame propagation by Scurlock and Grover [*26*, pp. 215–247] based on the turbulent displacement of the combustion wave; and an experimental work by Bolz and Burlage [*29*], who measured the growth of a flame globule in a turbulent stream by high speed photography.

There still exists considerable difference of opinion about the theory of turbulent flames between various workers as expressed, for example, in the Round Table Conference on Turbulent Flames of the Fourth Symposium on Combustion [*27*, pp. 923–926] and in the Discussions of the AGARD Combustion Colloquium [*26*, pp. 263–274]. It is hoped that further experiments and measurements will bring about the desired unification of ideas and will provide the designer of practical combustion systems with a powerful theory of turbulent flames.

H,18. Cited References.

1. Mallard, M. E., and Le Chatelier, H. L. *Ann. mines 4*, 343 (1883).
2. Coward, H. F., and Hartwell, F. J. *J. Chem. Soc. Part II*, 1996–2004 (1932).
3. Damköhler, G. *Z. Elektrochem. 46*, 601–626 (1940). English transl., *NACA Tech. Mem. 1112*, 1947.
4. Shelkin, K. I. *J. Tech. Phys. U.S.S.R. 13*, 520–530 (1943). English transl., *NACA Tech. Mem. 1110*, 1947.
5. Bollinger, L. M., and Williams, D. T. *NACA Tech. Note 1707*, 1948.
6. Scurlock, A. C. Flame stabilization and propagation in high velocity gas streams. *Project Meteor Rept. 19, Mass. Inst. Technol. Fuels Research Lab.*, 1948.
7. *Third Symposium on Combustion, Flame and Explosion Phenomena.* Williams & Wilkins, 1949.
8. Levine, R. S. Construction and use of a hot wire turbulence measuring device. *Project Meteor Rept. 5, Mass. Inst. Technol. Fuels Research Lab.*, 1947.
9. Wright, F. H. Measurements of flame speed and turbulence in a small burner. *Jet Propul. Lab. Calif. Inst. Technol. Progress Rept. 3-21*, 1950.
10. Taylor, G. I. *Proc. Roy. Soc. London A20*, 196–212 (1921).
11. Dryden, H. L. *Quart. Appl. Math. 1*, 7–42 (1943).
12. Batchelor, G. K. *Australian J. Sci. Research 2*, 437 (1949).
13. Batchelor, G. K. *Proc. Cambridge Phil. Soc. 48*, 345 (1952).
14. Brier, G. W. *J. Meteorol. 7*, 283 (1950).
15. Batchelor, G. K. *Proc. Roy. Soc. London A213*, 349 (1952).

16. Townsend, A. A. *Proc. Roy. Soc. London A209*, 418 (1951).
17. Uberoi, M. S., and Corrsin, S. *NACA Tech. Note 2710*, 1952.
18. Karlovitz, B. *Phys. Rev. 77*, 574 (1950).
19. Karlovitz, B., Denniston, D. W., Jr., and Wells, F. E. *J. Chem. Phys. 19*, 541 (1951).
20. Little, B. H., Jr., and Wilbur, S. W. *NACA Tech. Note 2361*, 1951.
21. Prandtl, L., and Tietjens, O. G. *Applied Hydro- and Aeromechanics*. McGraw-Hill, 1934.
22. Lewis, B., and von Elbe, G. *J. Chem. Phys. 11*, 75–97 (1934).
23. Bollinger, L. M., and Williams, D. T. *NACA Rept. 913*, 1948.
24. Buttner, H. J., and Teasel, R. C. Flame holding requirements for stable combustion of gaseous mixtures flowing at high velocities. *Project Squid Tech. Mem. PUR-12*, 1949.
25. Williams, G. C. *J. Aeronaut. Sci. 16*, 714–722 (1949).
26. *AGARD Combustion Colloquium*, Cambridge Univ. 1953. Butterworths, London, 1954.
27. *Fourth Symposium on Combustion*. Williams & Wilkins, 1953.
28. Summerfield, M., Reiter, S. H., Kebely, V., and Mascolo, R. W. *Jet Propulsion 24* (4), 254 (1954).
29. Bolz, R. E., and Burlage, H., Jr. The influence of turbulence on flame propagation rates. *Am. Rocket Soc. Paper 153* given Dec. 1954, New York.

SECTION I

DIFFUSION FLAMES

K. WOHL
C. W. SHIPMAN

I,1. Introduction. The phenomenon of flame is brought about by a complex interplay of chemical and physical processes. The mode and rate of supply of fuel and oxygen and the interaction between them determine, for a given configuration, flame structure and behavior. The net rate of interaction is the result of numerous elementary processes. According to conditions, one or the other of these processes or groups of processes is mainly responsible for the observable characteristics of a flame. In order to react chemically, fuel and oxygen must be mixed. In Diesel engines and in numerous types of gas turbines, jet engines, and furnaces, mixing occurs inside the combustion chamber and thus becomes an inseparable part of the combustion process. If liquid fuel is used, mixing of the gaseous components is preceded by atomization, spreading, and evaporation of the fuel; no consideration is given to the latter processes in the following survey.

In order to analyze the burning process it seems advisable to concentrate on a few relatively simple, well-defined situations. The case which is best understood is that of the burning of a gaseous, homogeneous mixture of fuel and oxygen or air in laminar flow. Here, too, the rate of burning is codetermined by mixing, namely, by the mixing of the burning gases with the unburned gas mixture. Mixing in this case occurs in such a manner that some reaction products diffuse from the flame front in an upstream direction against the approach stream until they are swept downstream by the gas flow, together with the bulk of the reaction products. This process is accompanied by heat conduction. Through both, a flame can be established as a steady state phenomenon in one-dimensional flow. Turbulence of the large scale type which is usually encountered seems to leave the process which happens in the immediate vicinity of a flame front element rather undisturbed; its effect on premixed flames is outside the scope of this survey.

The mixing process just described has to be distinguished from that which is operative in the so-called diffusion flames with which this survey is concerned. Here fuel and oxygen meet, form a combustible mixture, and, once ignition has taken place, start burning immediately upon creation of such a mixture. In this case the flame appears at the border between the fuel and oxygen zone. Combustion products created by the flame spread to either side, and fuel and oxygen have to diffuse against these streams in order to mix and to react. This over-all description fits the laminar diffusion flame in which mixing is brought about by molecular diffusion, as well as the turbulent diffusion flame in which the rapid creation of a coarse mixture through large scale transport processes greatly accelerates the formation of the final molecular mixture which is combustible. No steady state diffusion flame in a one-dimensional flow system is feasible; a steady diffusion flame of relatively simple structure can be established at the boundary of a fuel and oxygen zone by discharging a stream of fuel gas into an oxidizing atmosphere or by producing two parallel streams of fuel and oxidant.

The process of mixing fuel and oxygen is, of course, quite different according to whether a flame is lit or not, since the chemical reaction in the burning zone causes a change of temperature and composition. No rigorous analytical treatment of a diffusion flame has so far been attempted. It seems, however, that one can arrive at a description of certain features of diffusion flames with the help of radically simplified theories. This is especially true with the height of diffusion flames burning above a tube or nozzle in air, the air either streaming parallel to the fuel jet or being at rest [*1*; *2*, pp. 254, 266, 288; *3*; *4*]. The most important of these assumptions is that the rate of the chemical process is infinite so that the actual zones of combustion become infinitesimally thin, and the mixing process alone becomes responsible for the rate of burning and for flame location and size.

A flame is characterized morphologically not only by the shape of the zones in which chemical processes occur and which can be seen directly by the eye and recorded by ordinary photography, but also by the density pattern inside and outside the visible flame region which can be made visible by shadow and schlieren photography [*2*, p. 288]. Other important characteristics are the instantaneous and average distribution patterns of the reactants and main products of combustion [*2*, pp. 254, 266; *4*], of temperature, and of velocity. The two latter parameters have, e.g., been measured for the special case of a diffusion flame which is provided by the secondary burning zone of a Bunsen flame [*5*,*6*].

Recently a study of the chemical processes and the temperature in a diffusion flame has been initiated by Wolfhard and coworkers [*7*,*8*,*9*] who investigated the radiation from flat diffusion flames at various pressures. This research goes beyond the scope of the work mentioned so far, since it is concerned with the structure of a laminar combustion zone of finite thickness and yields insight into the sequence of chemical processes caused by interdiffusion of fuel and oxidant.

I,2. Laminar Diffusion Flames.

Flame Height. A reasonable measure of the progress of burning of a diffusion flame is its height. A jet of fuel is assumed to issue from a burner (tube or nozzle) of known diameter in a vertical direction upward into an ambient medium containing an oxidant (air). For a given fuel and burner, beginning at a low flow rate, the flame height increases with increasing flow rate, reaches a maximum, and then shortens. Shortly before the maximum height is reached the flame begins to flicker (becomes turbulent) at the top. This occurrence separates "laminar diffusion flames" from flames of the "transitional type." With further increase of flow rate, flickering spreads in a downward direction. This spreading stops rather suddenly when only a few diameters above the port are left free of fluctuations [*2*, p. 254]. At this point, the flame height becomes independent of flow rate, and the "transition flame" passes over into the "turbulent flame." The discussion in this section is confined to "laminar diffusion flames," i.e. to flames which show no or very little flickering [*1*; *2*, pp. 254, 288; *3*].

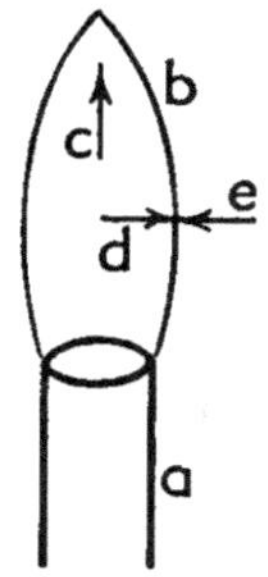

Fig. I,2a. (a) Burner tube, (b) flame front, (c) velocity of fuel jet in axis, (d) direction of diffusion of fuel towards flame front, and (e) direction of diffusion of oxygen towards flame front.

The situation in case of a laminar flame is illustrated by Fig. I,2a. The rate of chemical reaction can be assumed to be large compared with the rate of diffusion. Consequently, combustion is confined to a thin layer between the fuel and the oxidant, and the progress of combustion is determined primarily by the rates of molecular diffusion of the fuel and oxidant through the products of combustion to the flame zone. According to Burke and Schumann [*1*], the flame front is defined by the locus of those points where fuel and oxygen (diluted by combustion products) meet in stoichiometric proportions, or "where the rate of diffusion of combustible gas outward and the rate of diffusion of oxygen inward have the ratio required by the stoichiometric equation for complete combustion of the combustible gas. Thus, no unburned fuel exists on the oxidant side of the flame and there will be no unreacted oxygen on the fuel side unless the fuel jet contains primary oxygen."

Burke and Schumann studied a fuel jet surrounded by an annular stream of air confined in a tube, the two streams having the same initial velocity which was always low (2 ft/sec at most). In this case the shape of the flame depends on whether the annular stream carries more oxygen (overventilated flame) or less oxygen (underventilated flame) than is required to burn the fuel or fuel-air mixture in the inner jet to completion. The observed shapes of flame front for these two conditions are shown in Fig. I,2b.

For purposes of analytical treatment, Burke and Schumann stripped the flame model of nearly all features which are typical of a flame, and treated the laminar flame as if it were an isotherm mixing phenomenon within a homogeneous gas in which two different stream tubes had been labeled by tracer molecules. The relationship to a flame is given only by

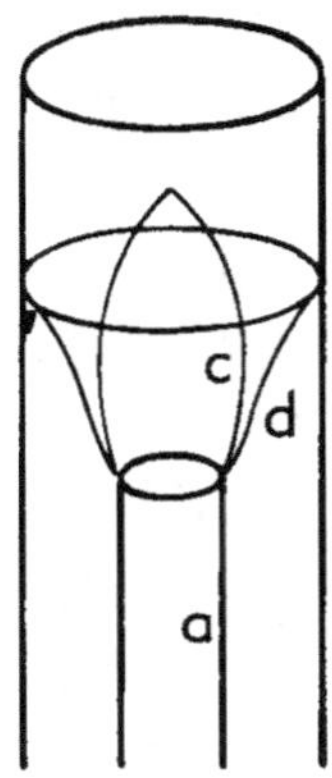

Fig. I,2b. (a) Inner tube, (b) outer tube, (c) overventilated diffusion flame, and (d) underventilated diffusion flame (Burke and Schumann [*1*]).

the provision mentioned above that a flame front is established where the original contents of the stream tubes meet in a fixed proportion (namely, the stoichiometric proportion which depends on the fuel and oxidant). It is remarkable how many essential features of diffusion flames can be described by this crude model, though it is not surprising that the application of the model is strongly limited.

The following specific assumptions have been made by Burke and Schumann: (1) The chemical reaction rate is infinite so that the burning zone degenerates to a geometrical plane, (2) the combustion reaction takes place with no change in number of moles, (3) diffusion takes place normal to the bulk flow only, (4) the velocities of the gas and air in the region of the flame are constant and equal to those of the approach streams, (5) the diffusivities of all the components are equal and constant. Assumptions (4) and (5) would, strictly speaking, imply that combustion causes no change of temperature as was indicated before. Actually, the

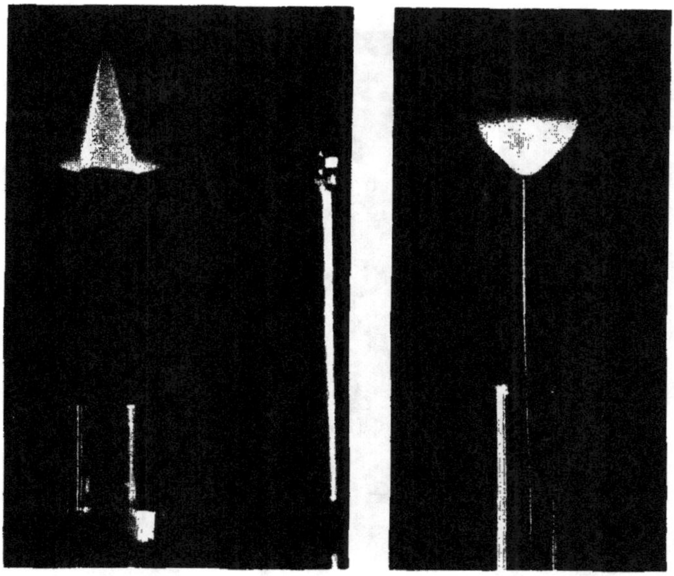

Plate G,5. Flame stabilization by wires.

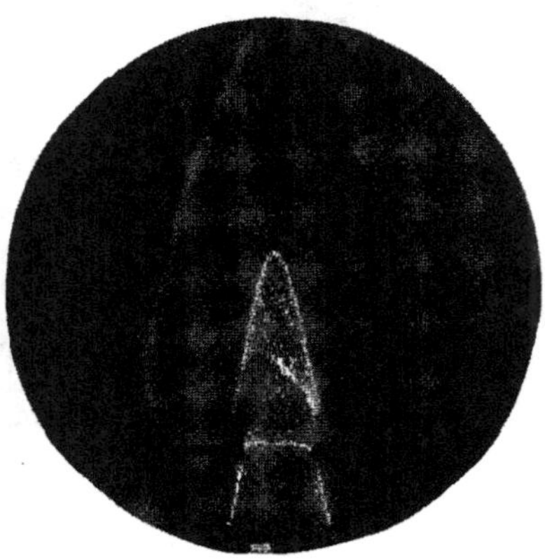

Plate G,6. Schlieren photograph of partially developed flame (Dery). Photograph made with open shutter. $\frac{1}{16}$-inch electrode spheres, 5.5 mm apart. Burner tube 8 mm i.d. Successive events recorded are:

1. Igniting discharge, which may be seen bridging electrodes, occurred.
2. Instantaneous schlieren photograph recorded by means of illuminating spark.
3. Flame subsequently grew to steady-state configuration, which appears in the form of double cone broken by electrodes and superimposed on schlieren photograph.

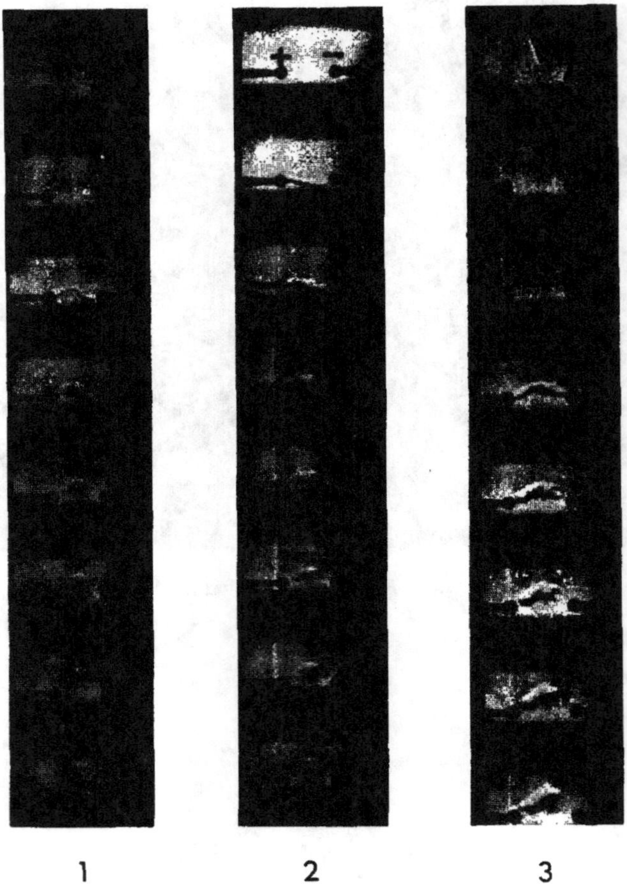

Plate G,12. Schlieren photographs of the early stages of flame propagation from sparks. Electrodes are dipping into a stream of methane and air at atmospheric pressure and average velocity of 160 cm/sec. The stream is laminar and originates from an 8-mm i.d. tube below the electrodes. Distance between electrodes is 6 mm. Electrodes are tipped by $\frac{1}{16}$-inch stainless steel spheres. 8-mm Fastax moving picture camera.

Series	1	2	3
Methane, per cent	8.5	8.5	10.8
Spark energy, millijoules	≅0.7	≅0.7	≅1.1
Time per frame, milliseconds	0.185	0.100	0.085
Time between frames, milliseconds	0.095	0.050	0.045

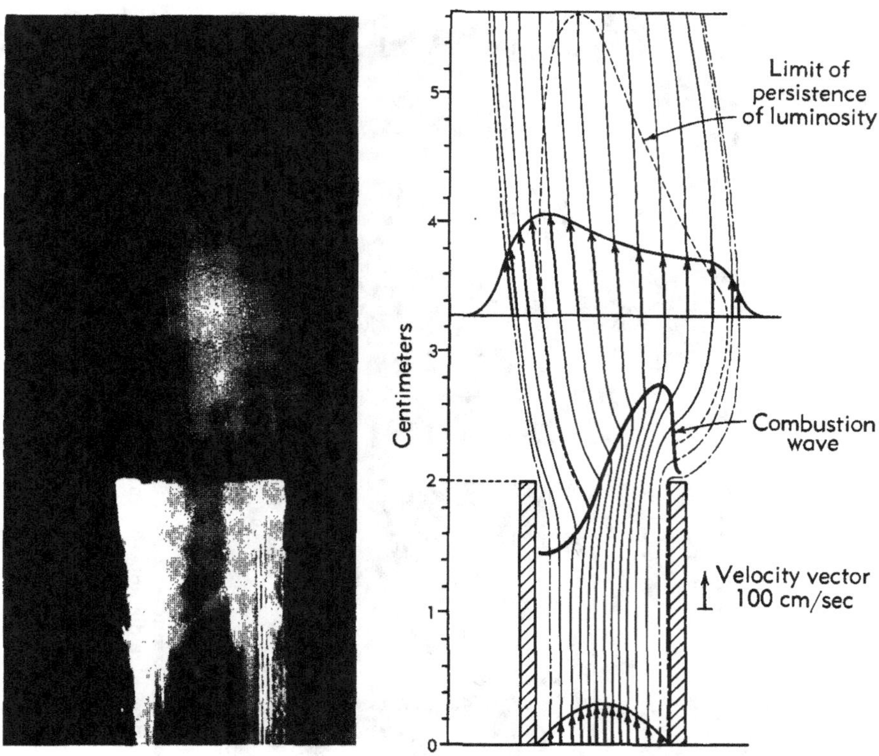

Plate G,14. Flame of natural gas and air, partially drawn into burner tube at gas flow just exceeding critical flow for flashback. Mixture composition, 8.1 per cent natural gas in air; gas flow, 52 cm^3/sec; inside diameter of tube, 1.068 cm. Left, photograph with particle tracks in vertical center plane. Right, diagram of flow pattern and velocity distribution. Solid flow lines are so located that the mass flow between any two adjoining lines is the same.

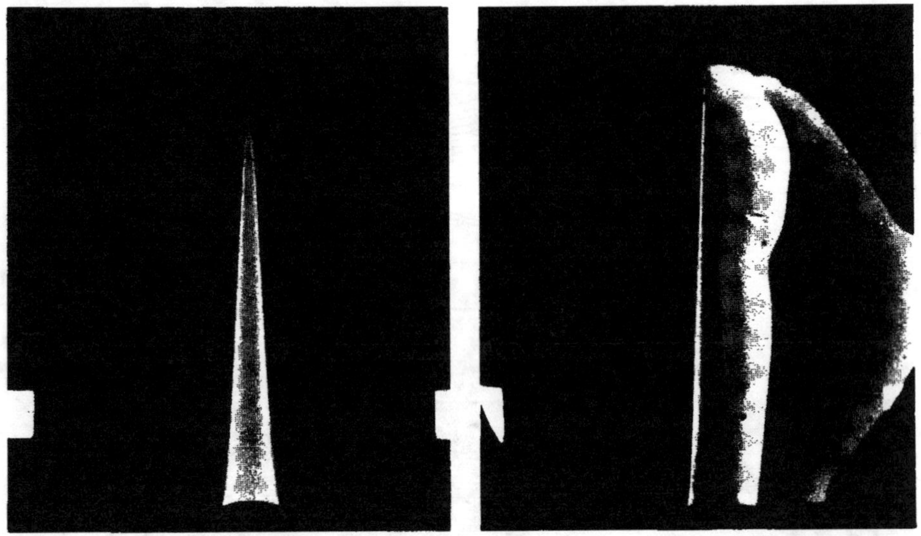

Plate H,1a. Time exposure (left) and instantaneous (right) schlieren photograph of a laminar Bunsen flame. Burner tube diameter 1.43 cm. Flow velocity 346 cm/sec, $Re = 3200$. Marker edge distance 10 cm.

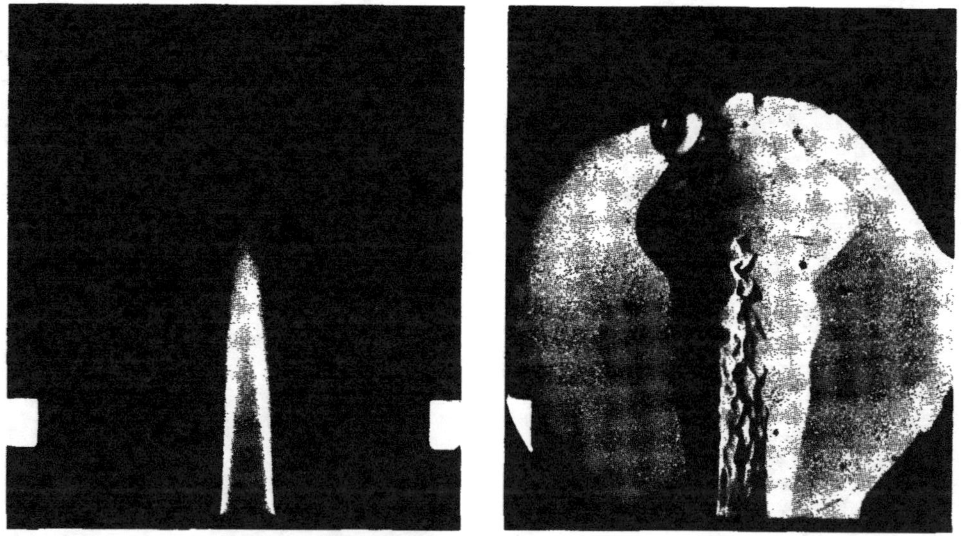

Plate H,1b. Time exposure (left) and instantaneous (right) schlieren photograph of a turbulent Bunsen flame. Burner tube diameter 1.43 cm. Flow velocity 496 cm/sec, $Re = 4600$. Marker edge distance 10 cm.

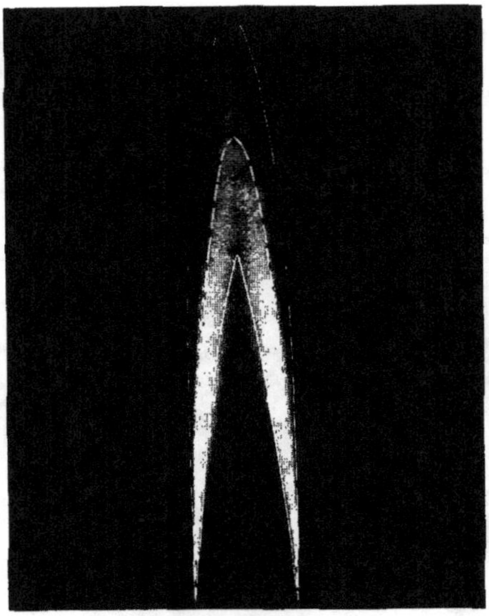

Plate H,4. Flame photograph with location of approximate mean flame surface outlined, as used for turbulent burning velocity determination by Bollinger and Williams.

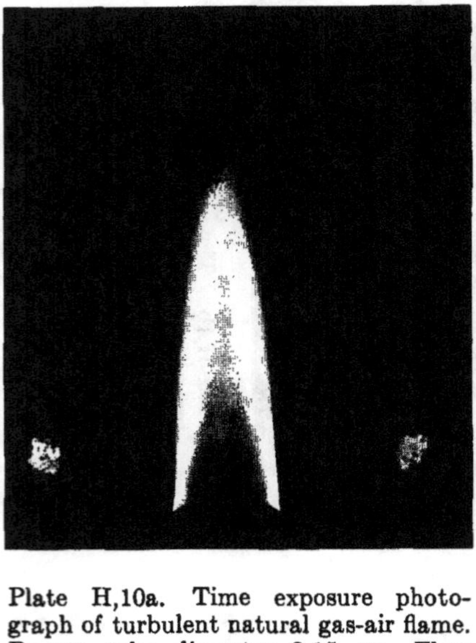

Plate H,10a. Time exposure photograph of turbulent natural gas-air flame. Burner tube diameter 3.15 cm. Flow velocity $U = 490$ cm/sec, $Re = 10{,}000$. Marker distance 10 cm.

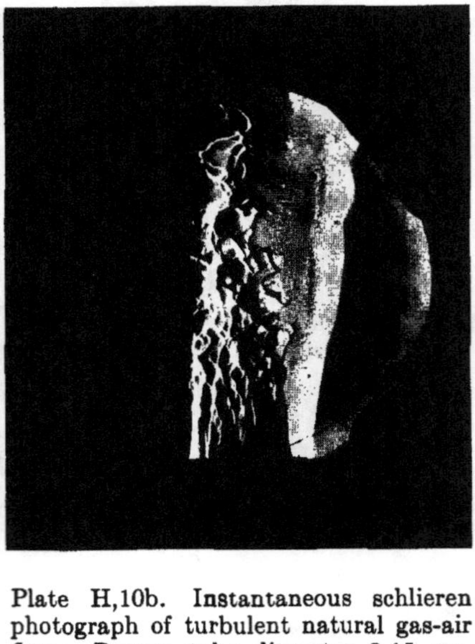

Plate H,10b. Instantaneous schlieren photograph of turbulent natural gas-air flame. Burner tube diameter 3.15 cm. Flow velocity $U = 490$ cm/sec, $Re = 10{,}000$.

Plate H,11 (*at left*). Flame negative used for turbulent burning velocity measurements.

Plate H,12 (*at right*). Shadow picture of natural gas-air flame. $Re = 35{,}000$. Burner tube radius $R = 1.58$ cm. Note the shape of the contact suface of hot combustion gas and cool air, which indicates strong turbulence of the combustion gas. Marker edge distance 10 cm.

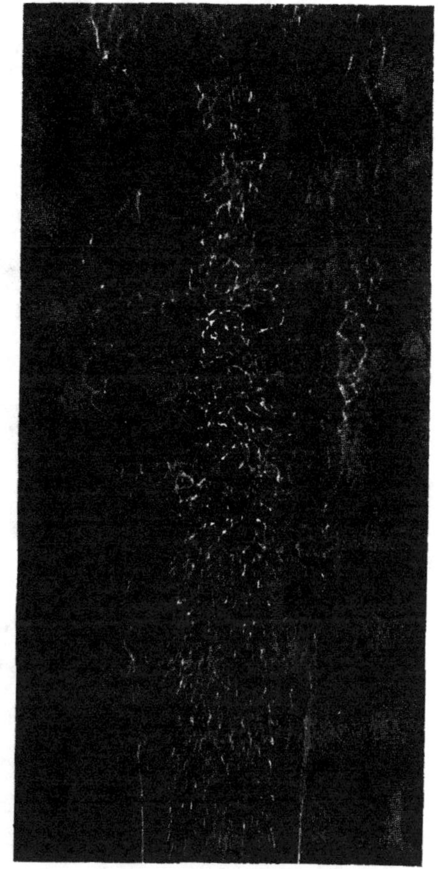

Plate H,14a. High speed Bunsen-type flame in stoichiometric mixture of natural gas-air. Burner tube radius $R = 1.58$ cm. Flow velocity 4900 cm/sec, $Re = 100{,}000$. Marker edge distance 10 cm. Note that the connection between the upper and the lower portion of the flame is almost interrupted.

Plate H,14b. Broken off flame in a high speed jet of stoichiometric natural gas-air mixture. Burner tube radius $R = 1.58$ cm. Flow velocity 6360 cm/sec, $Re = 130{,}000$. Marker edge distance 10 cm. Flame stabilized by annular pilot flame. Note that the flame propagation in the stoichiometric mixture is entirely interrupted at a short distance above the burner port.

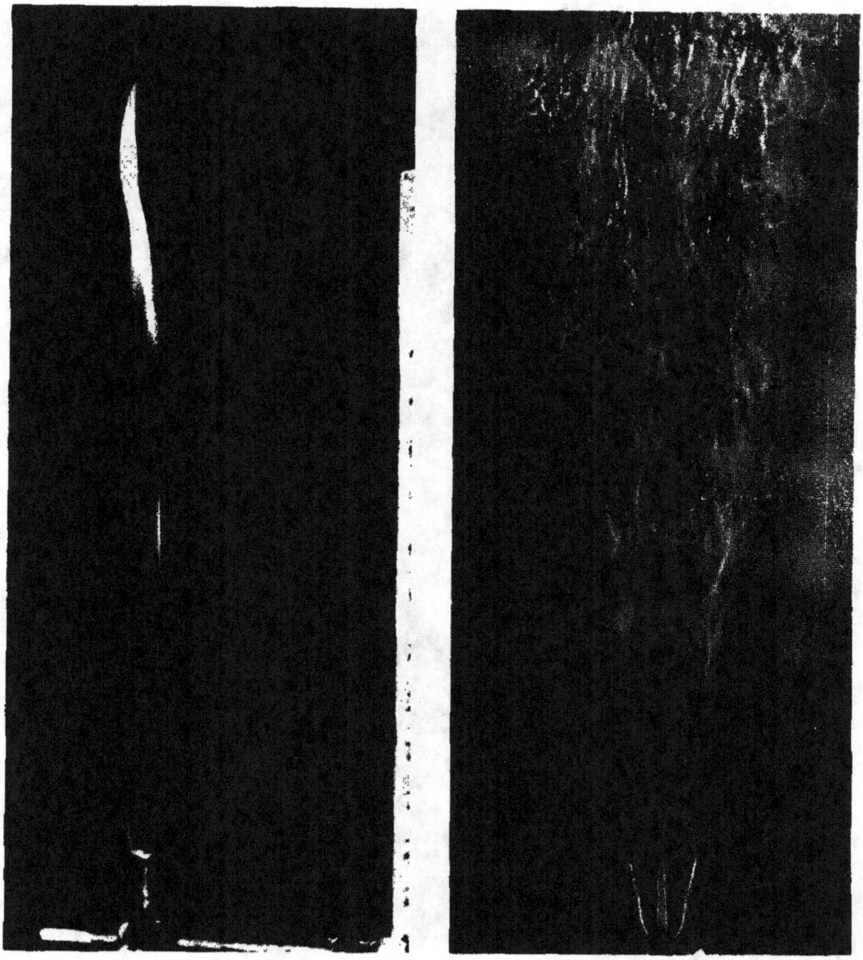

Plate I,2a. Direct (left) and shadow (right) photographs of laminar diffusion flame (50 per cent city gas). 0.40-inch tube. Velocity in tube 9 ft/sec.

Plate I,2b. Direct (left) and shadow (right) photographs of laminar diffusion flame (100 per cent Newark city gas). 0.40-inch tube. Velocity in tube 12.5 ft/sec.

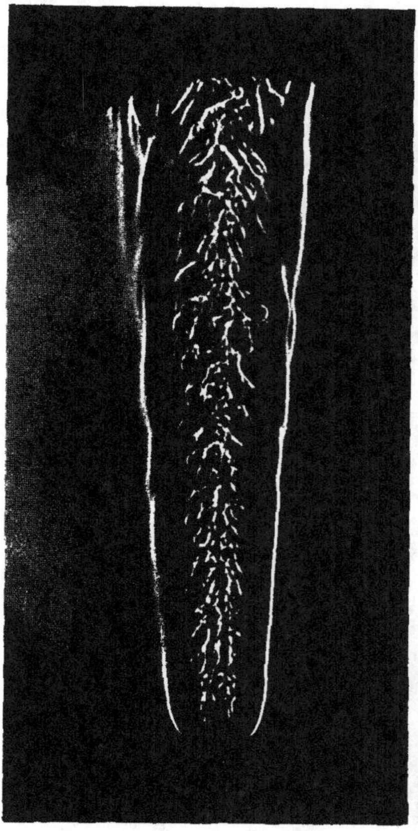

Plate I,3a. Shadow photograph of turbulent diffusion flame (100 per cent Newark city gas). 0.4-inch tube. Velocity in tube 51.5 ft/sec.

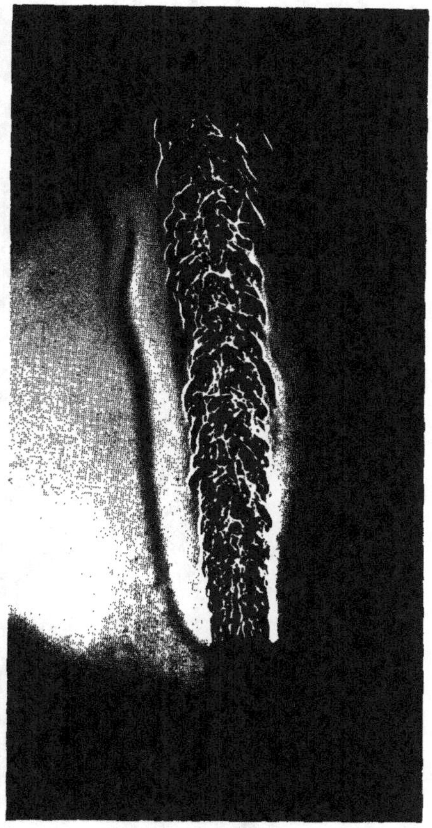

Plate I,3b. Shadow photograph of butane diffusion flame. 0.4-inch tube. Tube Reynolds number 8640.

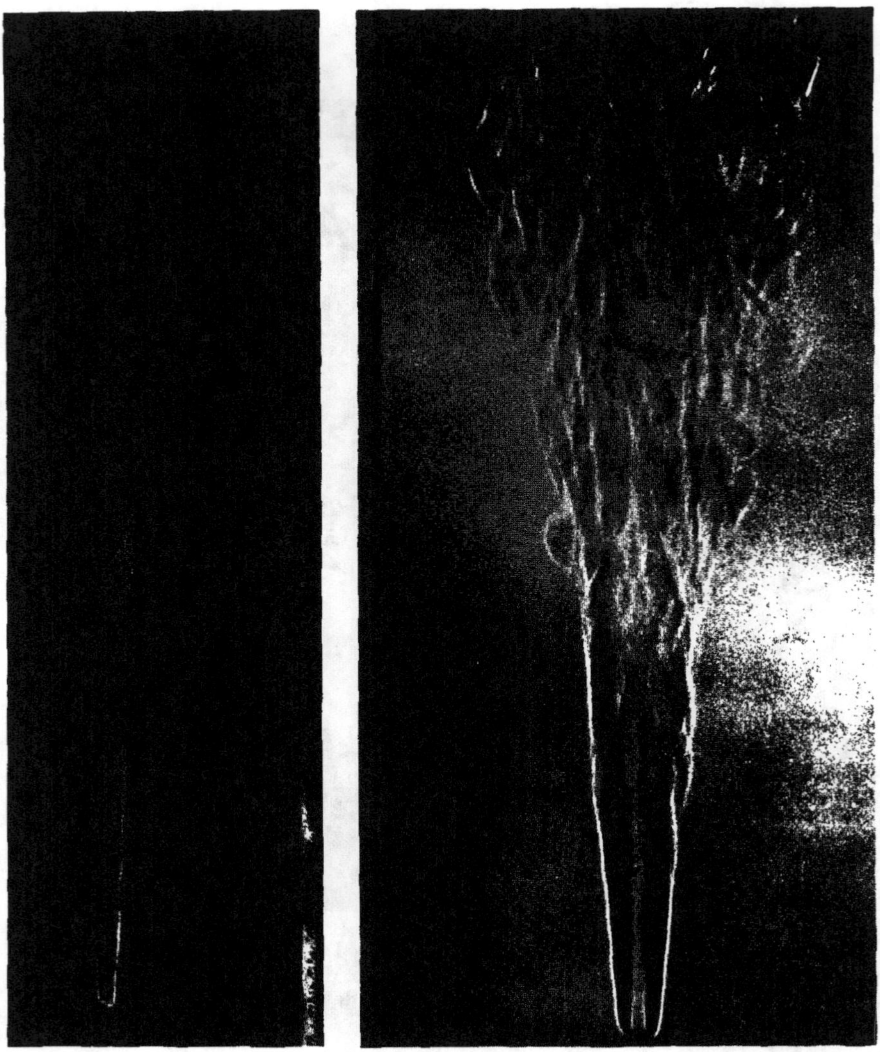

Plate I,4. Direct (left) and shadow (right) photographs of turbulent diffusion flame (50 per cent Newark city gas). 0.40-inch tube. Velocity in tube 66 ft/sec.

Plate J,1. Photograph of spray from swirl-type pressure atomizing nozzle. Pressure drop, approximately 2 lb/in.^{2}; flow rate, 0.018 lb/sec; fluid, kerosene.

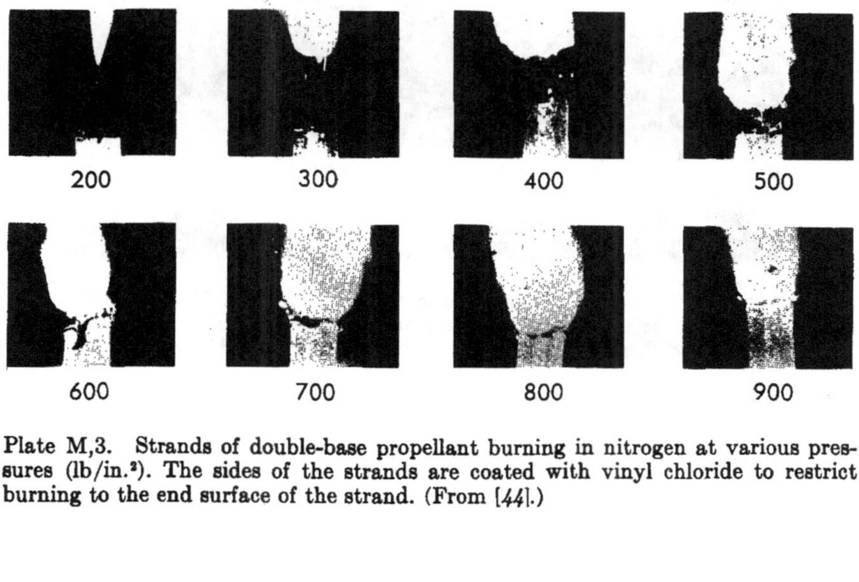

Plate M,3. Strands of double-base propellant burning in nitrogen at various pressures (lb/in.^{2}). The sides of the strands are coated with vinyl chloride to restrict burning to the end surface of the strand. (From [44].)

Plate N,2a. Fading to failure in a cartridge of TNT. Note the tendency for the velocity to remain constant first at high velocity (3800 m/sec) and then at low velocity (1800 m/sec). From [21] by permission.

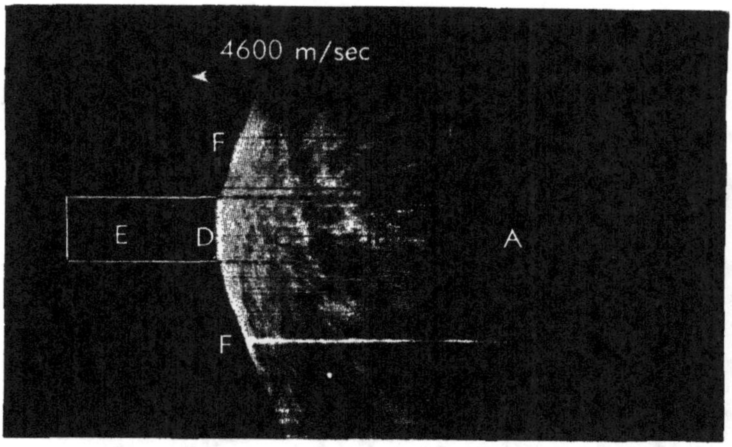

Plate N,2b. Detonation wave in column of TNT. From [*14*] by permission.

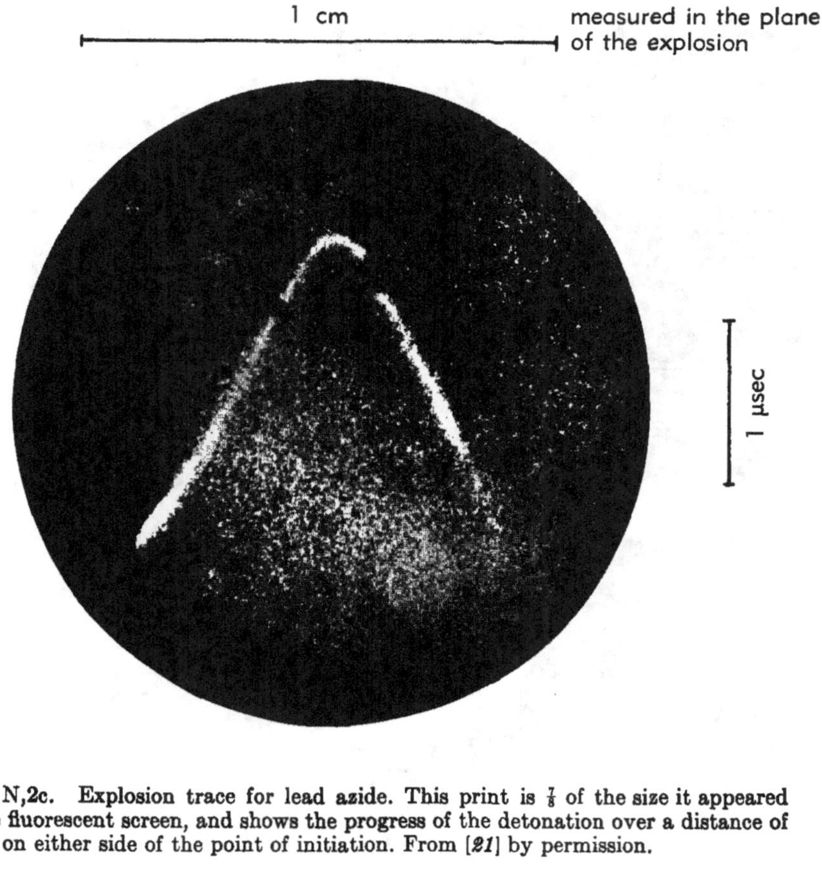

Plate N,2c. Explosion trace for lead azide. This print is $\frac{7}{8}$ of the size it appeared on the fluorescent screen, and shows the progress of the detonation over a distance of 4 mm on either side of the point of initiation. From [*21*] by permission.

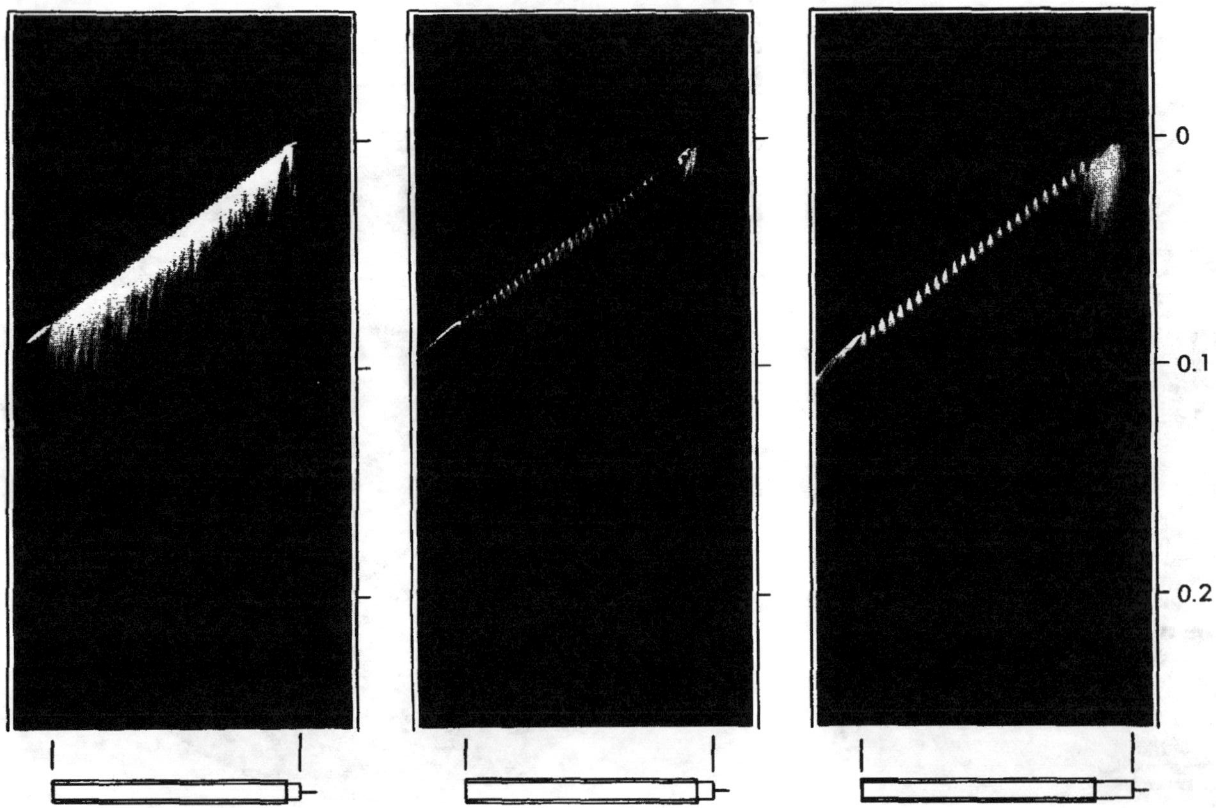

Plate N,2d. Photographic record of detonation rate using a series of small holes drilled in the opaque case. From [22] by permission.

authors presume that the effect of a change of temperature on velocity and diffusivity cancel each other (see below).

The mathematical formalism was simplified by Burke and Schumann by treating the fuel concentration as equivalent negative oxygen concentration, as is illustrated by Fig. I,2c which was taken from [*2*, p. 254]. The integrated diffusion equation is able to describe the flame shapes shown in Fig. I,2b quite well. Flame heights are defined as the vertical distances above the port of the inner tube at which combustion is complete. This is, according to Fig. I,2b, achieved for overventilated flames in the axis and for underventilated flames at the wall of the outer tube. A detailed account of the work of Burke and Schumann (and of diffusion flames in general) is given in Lewis and von Elbe's book on combustion [*10*,

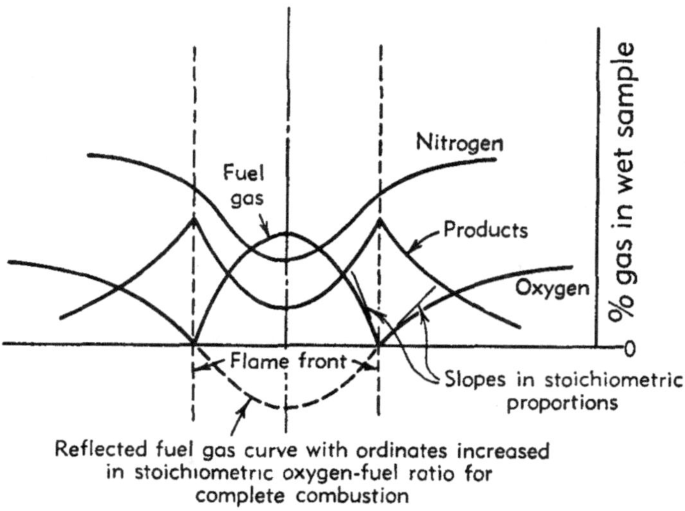

Fig. I,2c. Simplification of diffusion process in a typical flame [*2*, p. 258].

Chap. IX]. It is shown there that for low flow rates and flame heights the theory accounts satisfactorily for the dependence of flame height on flow rate, tube diameter, pressure, initial temperature, addition of inert gas to the fuel, and differences of air requirement for combustion of one mole of fuel.

Jost [*11*, p. 212ff.] showed that Einstein's diffusion equation $y^2 = 2Dt$ (y = mean distance of diffusion, D = diffusivity, t = time) leads, under Burke and Schumann's suppositions, to an approximate equation for the height of an overventilated flame. He introduced $t^* = h_{\text{low}}/u$ (t^* = time required for completion of combustion, h_{low} = flame height at low flow rates, u = average velocity in inner tube), and $y_{t^*} = r$ (r = radius of inner tube) and found

$$r^2 = \frac{2Dh_{\text{low}}}{u} \quad \text{or} \quad h_{\text{low}} = \frac{\Omega}{2\pi D} \tag{2-1}$$

(Ω = volume flow rate through burner tube). This equation accounts for the effect on flame height of all the variables mentioned above, except for the last two, about as well as the complicated equation of Burke and Schumann. In case of methane-air flames, Eq. 2-1 yields the right order of magnitude of the absolute flame height. This, however, cannot be generally true since the equation does not include consideration of the air requirement for combustion (see below).

Hottel and Hawthorne [*2*, p. 254] and Wohl, Gazley, and Kapp [*2*, p. 288; *3*] considered the case of a primary jet of fuel of higher velocity issuing into an infinite atmosphere of air with allowance for primary air in the fuel jet. If the same simplifying assumptions are used which were introduced by Burke and Schumann, it follows that the flame is always terminated when the flame front reaches the axis of the fuel jet. At this point, the concentration of fuel in the axis has dropped to zero. The boundary conditions are different from those for a confined air stream surrounding the jet. As a result, a very simple equation can be derived for the time t^* required for complete combustion [*2*, p. 254; *3*].

First, the treatment which has been given to this equation in [*2*, p. 288; *3*] is discussed. The equation is

$$t^* = \frac{d^2}{-16\,D \ln\left[1 - (X_{st}/X_0)\right]} \tag{2-2}$$

where d is the port diameter, D the diffusivity, X_{st} the mole fraction of fuel in a stoichiometric mixture, and X_0 is the original mole fraction of fuel in the fuel jet. This time is related to flame height h_0 as follows:[1]

$$h_0 = \int_0^{t^*} u\,dt = \bar{u}t^* = \frac{d^2\bar{u}}{-16D \ln\left[1 - (X_{st}/X_0)\right]} \tag{2-3}$$

where u is the velocity in the jet axis at any height and $\bar{u}$ is the average jet velocity. Similar to assumption (3) of Burke and Schumann, $\bar{u}$ is identified with the velocity in the burner mouth. This assumption is, however, more justified for jets which spread freely into an unconfined medium than for jets moving in a confined parallel gas stream (see below). The product $(\pi d^2\bar{u})/4$ is then equal to the volumetric flow rate through the burner. This yields

$$h_0 = \frac{\Omega}{-4\pi D \ln\left[1 - (X_{st}/X_0)\right]} \tag{2-4a}$$

or approximately

$$h_0 \cong \frac{\Omega X_0[1 - X_{st}/2X_0)]}{4\pi D X_{st}} \tag{2-4b}$$

[1] The symbol h_0 indicates that a simplified equation is used to calculate this value of flame height, (see next page).

For fuels for which X_{st} is reasonably small, especially for higher hydrocarbons, the equation can be further simplified, giving

$$h_0 \cong \frac{\Omega X_0}{4\pi D X_{st}} \tag{2-4c}$$

Eq. 2-4c bears a close resemblance to Eq. 2-1; it takes account, however, of the existing cylindrical symmetry and, more important, of the fuel concentration in the jet and of the air requirement for combustion of fuel. For a given fuel, the flame height is approximately proportional to ΩX_0, i.e. to the flow rate of fuel alone, irrespective of whether primary air is admixed to the fuel. The absolute flame height is approximately determined by $\Omega X_0/X_{st}$, i.e. by the volume of fuel and air which is consumed by combustion per second, and by the diffusion coefficient. Both approximations, Eq. 2-4b and 2-4c, fail if X_0 approaches X_{st}; before this happens, however, the equation given for h_0 breaks down because the flame has become self-propagating.

The proportionality of flame height and flow rate is empirically restricted to very small flame heights [*1*]. For larger flow rates, still remaining in the laminar range, the observed flame height increases less than proportionally to the flow rate, although conclusive experiments on this point exist only for fuel jets discharged into a quiet atmosphere of air. Fuel jet flames in a confined air stream behave differently according to Barr [*25*]. It appears hopeless to treat rigorously the case of a large fuel jet burning in air; thus only semiempirical formulas have been proposed to cover this range [*2*, p. 254; *3*]. For any diffusion flame, large or small, on traveling from the base to the tip, the mean temperature of the gases involved, the gas volume, and the diffusivity increase. If the flame is confined, the average velocity increases proportionally to the volume; if the flame burns freely, expansion in a sidewise direction occurs so that the velocity increase is less. Since the volume increase is proportional to T, and the diffusivity is about proportional to $T^{3/2}$, the flame height, according to Eq. 2-4b or 2-4c, will be proportional to $(\bar{T}/T_0)^{-1/2}$ if $\bar{T}$ is a proper average temperature of the gases involved and T_0 is the original temperature. As long as the flame height is small $\bar{T}$ will not depend on it. With increasing flame height, combustion products accumulate increasingly in the low velocity region around the flame. This hampers diffusion (tends to increase flame height) through increase of the distance to be covered by diffusion, but favors diffusion (tends to decrease flame height) through increase of T and consequently of D. Because of the strong dependence of D on $\bar{T}$, the increase of D overcompensates the effect of increase of diffusion distance so that flame height will become smaller. This then is the over-all effect, and it can be expected that the deviation of the observable flame height h_f from the value h_0 predicted by Eq. 2-4b will itself depend on flame height. This can be expressed in the simplest manner by dividing the right sides of Eq. 2-4a, 2-4b, and 2-4c, by the factor $1 + kh_t$ (k being a constant), and writing for the left side h_t instead of h_0. This then gives

$$h_t = \frac{h_0}{1 + kh_t} \tag{2-5}$$

For small flame heights, the equations thus obtained pass over into the well-established Burke and Schumann equations modified for open flames (Eq. 2-4a, 2-4b, and 2-4c). It is found that for medium and large heights of laminar flames kh_t is considerably larger than 1. The combination of Eq. 2-4b and 2-5 can thus be written with good approximation[2]

$$h_t = \frac{1}{\sqrt{\frac{4\pi X_{st} k D_0}{\Omega X_0 [1 - (X_{st}/2X_0)]}} + \frac{2\pi X_{st} D_0}{\Omega X_0 [1 - (X_{st}/2X_0)]}} \tag{2-6}$$

where D_0 = diffusivity at room temperature.

This equation has been verified for diffusion flames of city gas[3] (Newark, Delaware), issuing from tubes into an atmosphere of air, over a wide range of flow in the laminar region, and for tube diameters varied from 0.11 to 1.0 cm. Since the first term of the denominator is found to predominate, the flame height is nearly proportional to the square root of the volume flow rate of fuel over a wide range of flow rates (see Eq. 3-6 below). Fig. I,2d has been plotted especially to demonstrate the consequence of Eq. 2-6 that the flame height of a laminar flame at a given volume flow rate is independent of tube diameter. The lines in this figure represent average heights of city gas flames at a given volume flow rate. On moving from right to left, i.e. from large to small tube diameters or from low to high Reynolds numbers, the lines run horizontal, or the flame heights are constant, up to a point where the flames begin to flicker (see below). The points marked by the symbol ⊙ close to the right border of Fig. I,2d show the flame heights which follow from Eq. 2-6. The volume flow rate for any of these points is indicated at the horizontal line which runs closest to it. The agreement between calculated points and coordinated experimental lines appears satisfactory.

For a jet of 50 per cent city gas, the flame height was found to vary with jet diameter at constant volumetric flow rate, which is at variance with the theoretical result. This deviation was quantitatively accounted for by attributing it to a self-inflammability of 50 per cent city gas at elevated temperature. The values of the diffusivities computed from the data in all these cases are of the correct order of magnitude [*2*, p. 288].

[2] In [*2*, p. 288] and [*3*], the factor k in Eq. 2-5 and 2-6 had been given the designation $k/2D_0$.

[3] Composition: 35 per cent H_2, 15 per cent CH_4, 5 per cent C_2H_6, 6 per cent $C_{2.5}H_{4.2}$, 12 per cent CO, 0.8 per cent O_2, 5 per cent CO_2, 21 per cent N_2.

As was mentioned before, Hottel and Hawthorne [2, p. 254] also treat the laminar diffusion flame, starting with Eq. 2-2. They introduce the abbreviation

$$\Theta \equiv \frac{1}{-4 \ln [1 - (X_{st}/X_0)]} \tag{2-7}$$

and write Eq. 2-2 in the form

$$\frac{\Theta \Omega}{\pi D} = t^* u_0 = \int_0^{h_f} \frac{u_0 dh}{u} \tag{2-8}$$

where h is the vertical distance above the port, u_0 the velocity in the

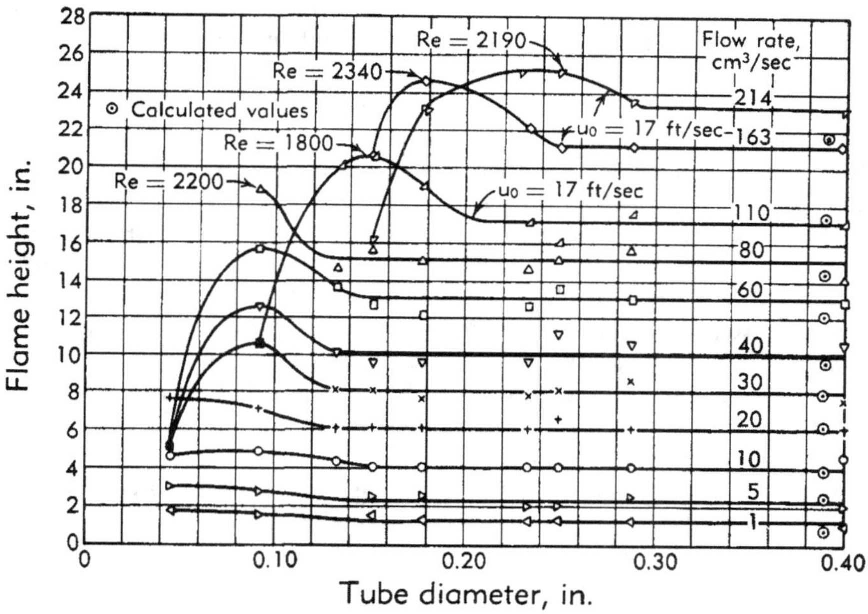

Fig. I,2d. City gas flames in free air. Effect of tube diameter on flame height at constant flow rate in laminar region, 100 per cent gas. Average density 0.052 lb/ft³.

nozzle, and u the velocity at h. The Grashof number is introduced as the decisive parameter for u/u_0, and it is stated that the empirical equation

$$\Theta \Omega = \phi(h_f, X_0) \tag{2-9}$$

is compatible with such dimensional considerations. The latter equation is given the specific form:

$$h_f = A \log (\Theta \Omega) + B \tag{2-10}$$

where A and B are constants depending on the fuel and the primary air-fuel ratio. Eq. 2-10 does not pass over into the experimentally confirmed Eq. 2-4a at low flow rates. The equation fits the data of Rembert and

Haslam [*12*] well. The data of Wohl, Gazley, and Kapp are also roughly described by an equation of the type of Eq. 2-10 though they show a somewhat different trend than the data of Rembert and Haslam, possibly (see [*2*, p. 254]) because the former authors have used tubes, the latter, nozzles.

Other Studies of Physical and Chemical Flame Characteristics.

Theory. Zeldovitch [*13*] analyzed the structure of the burning zone of a laminar diffusion flame. He assumed that the coefficients of diffusion are equal for all species and equal to the thermal diffusivity $\alpha = \lambda/\rho c_p$ (λ = heat conductivity, ρ = density, c_p = specific heat), and that heat loss by radiation is zero. The conclusion was reached that if the chemical reaction velocity is infinite, the mixture burning is of stoichiometric proportions and that the temperature of the flame front is that of a homogeneous stoichiometric mixture burning adiabatically. For further conclusions the NACA report cited may be consulted.

Experiments. At the stage of development indicated by the preceding discussion, it seems that further efforts should be directed toward an experimental examination of the physical and chemical characteristics of the flame. Such an examination has already begun. Chemical analysis of the gases from various parts of the diffusion flame, temperature measurements, the particle track method for velocity determinations, flame photography, both of the direct and shadow types, and spectroscopic measurements, have all been employed toward this end.

1. Chemical analysis. Burke and Schumann [*1*] took samples from the axis of the jet of a carbon monoxide flame at various points along the flame and found that nowhere was there any free oxygen which would indicate its having passed through the flame front. Hottel and Hawthorne [*2*, p. 254] analyzed samples taken at several radial locations for three distances from the port. The fuel was hydrogen. In no case, provided the flame did not flicker, were hydrogen and oxygen found to exist together. Concentrations of both dropped to zero at the flame front as is shown in Fig. I,2c.

2. Temperature measurements. Brody [*14*] measured the average temperature and absorptivity of the soot particles of a laminar butane-air diffusion flame at various heights by an absorption-emission method making use of the radiation in the region of the sodium lines. The jet had a very low velocity (down to 0.015 ft/sec) and was surrounded by an annular stream of air of a velocity chosen such as to keep the flames from flickering (about 0.27 ft/sec). These flames were overventilated and thus terminated in the jet axis. The visible flame emerging from a 1-inch tube was about 4.5 inches long. The maximum temperature occurred about 1.3 inches above the burner port. Maximum absorptivity (soot concentration) was observed for the same flame at a larger height, about 2.2 inches above the port. This is explained by the gradual spreading of the

temperature profile with increasing height. Near the port, soot can form only in the narrow, high temperature region close to the burning zone. Further up the tube, the widening of the temperature profile makes soot formation possible in the interior of the fuel jet.

The highest temperature observed (with a 0.43-inch tube) was 1830°K as compared with an adiabatic flame temperature of the stoichiometric butane-air mixture of 2240°K. A similar discrepancy was observed by Zeldovitch [*13*] in the case of fuel-air diffusion flames, whereas Wolfhard and Parker [*7*] found that the theoretical temperature was reached in the burning zone of fuel-oxygen flames. Zeldovitch ascribes his experimental results to heat losses by radiation. It seems more likely, however, that the difference between the observations of Brody and Zeldovitch on the one hand, and Wolfhard and Parker on the other hand, is essentially explained by the fact that the former are concerned with the average temperature of a cylindrical luminous flame zone and the latter with the maximum temperature occurring in a plane of a flat flame zone (see Fig. I,2f).

The temperature distribution in a rich, self-propagating laminar flame produced by a mixture of natural gas (Pittsburgh, Pennsylvania) and air was measured by Lewis and von Elbe [*5*]; see also [*10*, Fig. 92, p. 264]. The authors point out that the data are indicative of secondary burning, i.e. burning by a diffusion mechanism, in two respects: the maximum temperature is found to the sides of the primary cone in the region of the visible secondary burning zone, at a height above the port where a protective outer layer of combustion products has developed; and the temperature profile at heights about two to three times the height of the primary cone is nearly flat in the interior of the flame, since in this part of the flame the heat is produced only by secondary burning in the flame boundary.

3. Velocity measurements. Troy [*6*] measured the gas velocities within laminar, rich, Newark (Delaware) city gas-air flames and butane-air flames by use of stroboscopic illumination of small particles, a technique previously worked out by Lewis and von Elbe [*5*] (see also IX,J). The burner was a square duct of 0.37-inch inner width, the primary cone was about 1 in. high, and the measurements extended to 8 in. above the port, i.e. they were mainly concerned with the range of secondary burning. Since the flow lines were nearly vertical and no strong accelerations occurred in the diffusion flames, the deviations between powder velocity and gas velocity were small and could readily be estimated. Fig. I,2e, pertaining to city gas (Newark, Delaware), gives one example of the velocity fields observed. It is seen that the zone of relatively high velocities (in the case of Fig. I,2e of velocities above 7 ft/sec) remains nearly cylindrical for the full measured height, and that the velocity in the axis stays nearly constant over this range. The value of the axial

velocity (10.5 to 11.5 ft/sec) is nearly equal to the velocity of the unburned gas in the duct axis (2 × 5.35 = 10.7 ft/sec). These observations also hold for butane flames of the same velocity, and for city gas flames of higher velocity, though in the latter case the velocities measured begin to decrease through increased momentum transfer with the environment

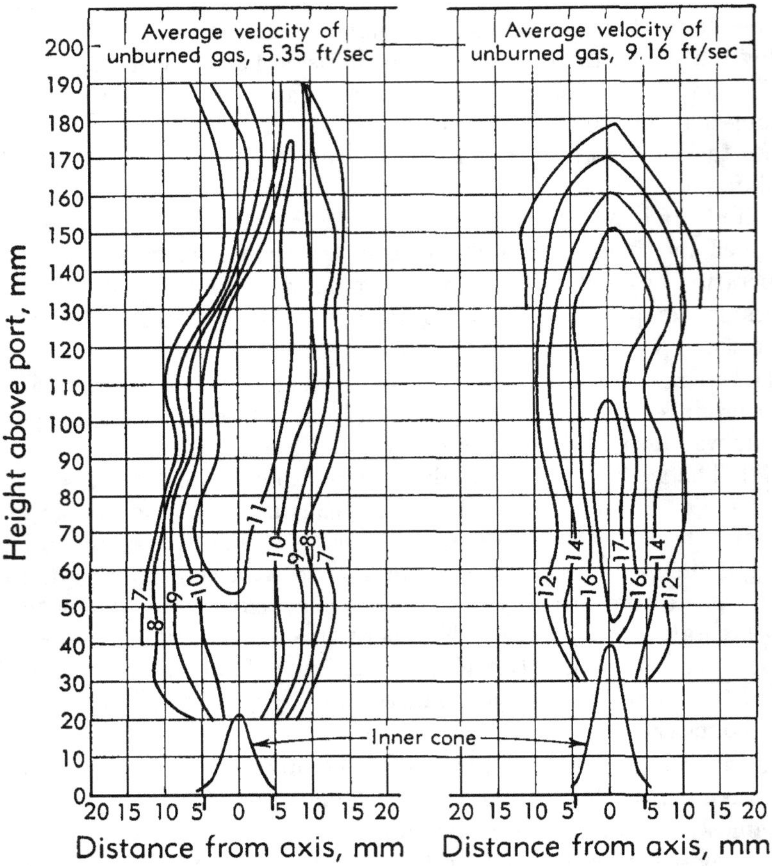

Fig. I,2e. Lines of constant velocity in secondary zone of self-propagating Newark (Delaware) city gas-air flames with 62 per cent of stoichiometric air (left) and 70 per cent of stoichiometric air (right).

somewhat earlier (5.5 in. above the port). It is noteworthy that the increase of velocity by expansion of the burning gases and the buoyancy force is balanced by retarding shear forces to produce no net acceleration of the burning stream. These results lend support to the assumption used in [*2*, p. 288; *3*] that the average velocity of unconfined diffusion flames in the axis is not very different from the port velocity (see the discussion of Eq. 2-3).

4. Photographic studies of flame structure. The structure of

diffusion flames in still air has been studied in [*2*, p. 288; *3*] by using both direct and shadow photography. Plates I,2a and I,2b are given as examples. It is seen from Plate I,2b that the boundary between the fuel jet and the combustion products which surround it, and the boundary between the ambient air and the envelope of hot gases are characterized by very steep density gradients. The burning zone lies between these two regions. Its position does not show on the shadow photograph (though it is, of course, visible on the direct photograph). This is in strong contrast to self-propagating flames where the burning zone is the place of the steepest density gradient. The reason for the difference is that, in self-propagating flames, the hot combustion products spread essentially only in a downstream direction while in diffusion flames they spread toward either side of the burning zone, creating an extended region of high temperature around it. The fuel jet in Plate I,2b remains straight and well defined until it is hidden by the schlieren in the outer flame envelope. The smooth continuation of the flame front in Plate I,2a is evidence that the inner fuel jet continues to still larger heights without being broken off. Neither the regular waves in the lower part of the flame envelope nor the irregular schlieren in the upper part of the flame envelope which occur far outside the flame zone (Plate I,2b), substantially affect the diffusion process responsible for combustion, as follows from the smooth appearance of the flame front (Plate I,2a). These flames are therefore designated as laminar.

This also applies to Plates I,2c and I,2d, which represent a more common type of laminar flame. Here the inner fuel jet is laminar in character as in Plate I,2b, and the flame front is slightly deformed (but does not show rapid fluctuations). On the basis of the molecular diffusion mechanism one would expect that the tops of the flames should be pointed as in Plate I,2a. Laminar flames of higher flow rate, like the one shown in Plate I,2c, however, broaden with increasing height and show a flattened top. This is probably due to the fact [*3*] that hot combustion products accumulate increasingly with increasing height in the space between the fuel jet and the burning surface (where the velocity is smaller than in the axis). This causes a lateral expansion of the flame front.

The waves in the outer flame envelope which are visible in the shadow photograph, Plates I,2b and I,2d, seem to be generally characteristic of laminar flames in still air. The occurrence of such waves is probably related to the fact that surfaces of discontinuity in an ideal fluid are unstable [*15*, p. 374]. In the case of steadily renewed surfaces between viscous fluids, a steady source of disturbances and some amplification mechanism is necessary to produce observable instability phenomena. In [*3*] the initial disturbance causing the waves is attributed to the sudden lateral expansion of the hot gases immediately above the burner port, which is visible on all flame shadow pictures. This expansion is the result of the intense burning of undiluted fuel and oxidant in that region. The expansion will be reduced if, at some phase of a wave motion, combustion products are transported into this region, and will be increased if at another phase the combustion products are replaced by fresh air. This may be the mechanism by which a resonance between wave formation and burning and an amplification of the wave motion is brought about.

The upward motion of the waves is most likely due to the motion of the medium in which they occur, so that relative to this medium the wave motion would be strictly radial. The vertical velocity of the gases which constitute the flame front of a diffusion flame has never been ascertained. At higher velocities (where the flames are turbulent) the lower portion of the flame envelope is free of waves. At larger heights, waves of small amplitude seem to develop, but since the height at which the flame envelope begins to show irregular fluctuations decreases with increasing velocity, no waves are observed at all in many cases. Plates I,2e and I,2f [*3*] which show a turbulent flame represent a borderline case. It seems that the resonance mechanism described above does not operate at high velocity where the percentage of fuel burning per unit of height is very small.

5. Spectroscopic measurements. The burning zone of a diffusion flame has been examined spectroscopically by Wolfhard and Parker [*7, 8, 9*]. The apparatus is so constructed that streams of fuel and oxidant are made to flow side by side, each having the same initial velocity, and the boundary between the streams is initially a plane surface. The systems examined were ammonia-oxygen, ethylene-oxygen, methane-nitric oxide, carbon monoxide-oxygen, and hydrogen-oxygen. The results for the ethylene-oxygen flame are shown in Fig. I,2f. The temperatures were measured both by the sodium D line reversal and OH line reversal techniques. The gases in the main reaction zone are in chemical equilibrium, and the peak temperature corresponds closely to the adiabatic flame temperature for stoichiometric mixtures—which confirms the theoretical conclusion by Zeldovitch [*13*]. For hydrocarbon-oxygen flames, there are two distinct zones, a yellow zone on the fuel side which is recognized as the result of emission from incandescent carbon particles, and a blue zone on the oxygen side. There is a dark space between them. The concentration of oxygen has dropped to zero at some distance from the yellow zone toward the oxygen side. It thus follows that soot formation in these flames is the result of hydrocarbon pyrolysis in the absence of oxygen. Water vapor, carbon monoxide, and carbon dioxide have diffused into this zone from the blue oxidation zone. Similarly separated zones of decomposition of fuel and of oxidation reactions appear in the ammonia-oxygen diffusion flame. The appearance of OH radiation on the fuel side, which is noted in Fig. I,2f, is attributed to scattering of light from the reaction zone by carbon particles. This radiation appeared only

when there was carbon formation. The CH radicals are believed to be the result of a reaction of OH and C_2 molecules. The explanation advanced by these authors for the presence of C_2 in the yellow flame zone is that those radicals are produced principally by vaporization of the carbon particles and that they are thermally excited afterward.

Wolfhard and Parker [7] draw the following comparisons between diffusion and premixed flames: " . . . all premixed gas flames start with

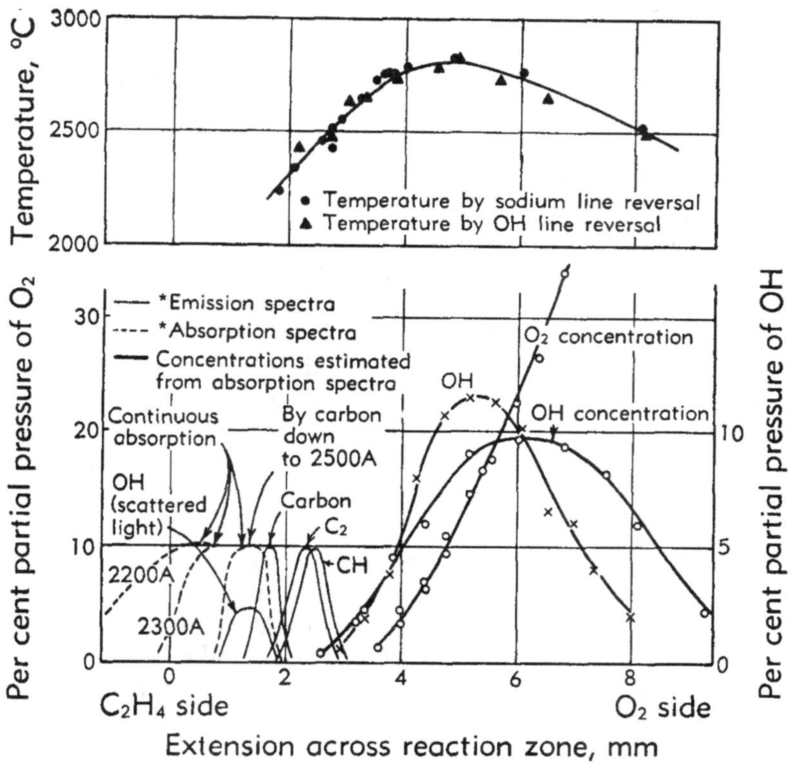

Fig. I,2f. Measurement of temperature and free radical concentrations in an ethylene-oxygen flat diffusion flame [7]. *Quantitative estimates of intensity only.

a high chemical potential which breaks down in the reaction zone . . . the diffusion flame, on the other hand never develops large chemical potentials in the main reaction zone. Diffusion endeavors to change the gas composition but chemical reaction occurs at once and restores the equilibrium. This is probably true only in the main reaction zone and some chemical potential may develop during decomposition in the preheating zones . . . "

The same authors found that as the pressure of the diffusion flame was decreased, the main reaction zone and the zone of hydrocarbon decomposition slowly merged while carbon formation decreased, and, at sufficiently low pressure, ceased altogether. As the pressure is lowered, the reaction rate decreases and the diffusivity increases so that a chemical potential can develop, and some of the characteristics of a premixed flame will appear. This latter picture coincides with the observation of Wohl and Peterson [*16*] that the primary and secondary burning zones of a fuel-rich premixed flame coalesce at low pressure ($\frac{1}{8}$ atmosphere and below).

Comparing these results with the assumption of an infinitely fast chemical reaction used in the preceding article, it may be said that this assumption is a reasonable approximation as far as the oxidation reactions are concerned, but that the preparatory reactions require a zone of finite width. It is, however, not likely that the slow preparatory reactions have to proceed to a certain degree of completion before the oxidation can take place; in other words, the rate of fuel consumption will not be determined by these slow reactions but by diffusion, in agreement with the flame model used before. Wolfhard and Parker assume that in air-fuel diffusion flames, as contrasted to oxygen-fuel diffusion flames, the reaction rates are reduced by the lower combustion temperature to such a degree that "some chemical potential may develop in the reaction zone," so that "diffusion will cease to be the sole rate-determining process." If this were true to a considerable degree, it would be expected that oxygen and fuel exist simultaneously in the combustion zone in measureable amounts. Judging from the few existing experiments (see above), this does not seem to be the case.

Synopsis of Laminar Diffusion Flames. A highly simplified theory of laminar diffusion flames has been developed by Burke and Schumann [*1*] which accounts very well for the effect of a change of variables on the dimensions of very small flames, and moderately well for the absolute dimensions of such flames. These flames were produced by fuel jets burning in a parallel annular stream of air of equal velocity. Larger flames have mainly been produced by fuel jets burning in still air [*2*, p. 254, 288; *3*]. For such flames Burke and Schumann's theory fails. A comparative study of laminar fuel jets burning in a parallel air stream and in quiet air is still missing.[4] Semiempirical corrections have been applied to Burke and Schumann's theory in order to describe the height of these larger flames. The corrections can be interpreted as taking into account the change of diffusivity with temperature and an accumulation of combustion products in the regions of low velocity surrounding the fuel jet. An integration of the exact equations describing the interaction of flow, chemical reaction (heat production), and diffusion of reactants and products in a diffusion flame is unlikely to succeed. However, attempts to develop a theory of the shape and over-all structure of the laminar

[4] See, however, [*25*].

diffusion flame, which is based on less crude simplifications than have been used so far, should definitely be encouraged.

At the same time additional experiments are needed. Flame height measurements have not sufficiently been corroborated by other types of investigation such as the measurement of composition, temperature, and velocity in diffusion flames (see, however, [*2*, p. 254] concerning composition). Further attention should also be given to the periodic motions which occur outside the burning zone of laminar diffusion flames in the boundary between the hot gases surrounding the flame and the outer atmosphere, and which can be ascertained by the shadow and schlieren methods and other means.

For optical investigations, two-dimensional flames are recommended. The relationship between chemical reaction rates and rates of diffusion can be varied by extending the studies to flames with various oxidants and diffusion flames at low pressure. Work in these directions has been initiated by Wolfhard and Parker [*7*, *8*, *9*]. The specific goal of the spectroscopic studies of these authors is an analysis of the kinetic characteristics of the processes occurring within and closely around the burning zone proper. They found that, in the main reaction zone of fuel-oxygen diffusion flames, equilibrium is very nearly established.

Recently, a candle flame has been subjected to a detailed chemical analysis with the help of a mass spectrometer by Smith and Gordon [*26*], and flat diffusion flames of organo-metallic vapors and air burning above the Wolfhard-Parker burner, just described, have been studied spectroscopically and by chemical analysis [*27*].

I,3. Transition Region Between Laminar and Turbulent Diffusion Flames. Diffusion flames in the transition and turbulent region are characterized by rapid fluctuations (flickering) of at least parts of the visible flame front and by inapplicability of the equations previously developed for flame height, the latter criterion indicating a change in the mechanism of mixing. In the "transition region," flickering spreads with increasing velocity of the fuel jet from the top of the flame to a distance of a few diameters from the port, as was mentioned in Art. 2 [*2*, p. 102, 254; *17*]. Flame turbulence seems to be mainly the result of the shear forces at the boundary between the—laminar or turbulent—fuel jet and the slower and usually less dense combustion products, and at the boundary between the flame envelope and the denser atmosphere. These boundaries coincide at the flame base, and separate more and more by the expanding region of high temperature combustion products.

Unignited jets. Garside and Whitehead [*17*], and Garside and Scholefield [*2*, p. 102] have examined both ignited and unignited ethylene jets emerging in still air, by means of shadow photographs. They used long and short tubes, and nozzles with a straight section of 3.8-mm length downstream of the constriction, the diameters being in the range from 0.42 to 1.54 mm. No data exist for tubes and nozzles of the larger size which have been used in [*2*, p. 254, 288], and in the flame stability studies of Bollinger and Williams [*18*] and of Wohl, Kapp, and Gazley [*2*, p. 3].

Within the rather large limit of irregular behavior of the jets it may be stated that, in the laminar range, the height above the port at which the jet becomes turbulent (height to turbulence h_t) increases for a given diameter with increasing length of the tube or straight downstream section of the nozzle to a limiting value. For relatively small ratios of tube length to diameter (or straight downstream section of nozzle to diameter), h_t varies more than inversely proportionally to the average velocity u at the port. For long tubes, h_t varies less than inversely proportionally to u. For short tubes, h_t at a given velocity u seems to be independent of diameter; with long tubes it seems to increase slightly with increasing diameter. At a given Reynolds number, h_t increases with increasing diameter in any case. The largest ratio (h_t:d) value observed was about 36 (d = port diameter).

This behavior of jets can be accounted for qualitatively by the assumption that the distance h_t at which the jet becomes turbulent decreases with increasing velocity gradient g at the rim of the tube or nozzle. A dimensionally plausible formula would be

$$h_t = k' \sqrt{\frac{\nu}{g}} \tag{3-1}$$

where ν = kinematic viscosity and k' = dimensionless constant.

For long tubes in laminar flow, g equals $8u/d$. There follows

$$h_t = k' \sqrt{\frac{\nu d}{8u}} \tag{3-2}$$

or

$$\frac{h_t}{d} = k' \sqrt{\frac{\nu}{8ud}} = k'' Re^{-\frac{1}{2}} \tag{3-3}$$

Eq. 3-2 and 3-3 describe roughly the trends observed with large tubes: with decreasing length of tube or straight nozzle section, the velocity gradient g increases; it tends to become independent of tube diameter and to vary more than proportionally to the velocity u (see, e.g. [*19*, Vol. I, p. 136]). These features, in conjunction with Eq. 3-1 explain qualitatively the change of behavior of the jet if one passes from long tubes to short tubes and nozzles. In the extreme the velocity gradient g becomes infinite so that events in the jet boundary become independent of the orifice configuration. In this case h_t might be determined by the critical value of a Reynolds number based on the length

$$Re_{\text{length}} = \frac{h_t u}{\nu} \tag{3-4}$$

With long tubes h_t drops abruptly to a few diameters at about $Re = 3000$, i.e. somewhat above the critical Reynolds number for tube flow as should be expected. With short tubes the drop of h_t in the transition zone between laminar and turbulent flow is more gradual. In case of tubes and nozzles with the smallest straight length-to-diameter ratio, the h_t curves show no anomaly in the transition zone at all and lie above the other h_t curves at tube Reynolds numbers above 3000, the reason obviously being that little or no turbulence develops in these short straight sections.

Flames. The height to turbulence for ignited jets, i.e. the height above the port at which the visible diffusion flame surface begins to flicker, has been measured by Garside and Whitehead [*17*], Garside and Scholefield [*2*, p. 102], and Gaunce [*20*]. The dependence of height to turbulence on flow velocity is of the same type as was observed for unignited jets. The heights to turbulence are larger for the ignited than for the unignited jet. This is in agreement with the fact that the fuel jets within flames are much more stable than the unignited jet, due to the higher viscosity within the flame. The starting point of flickering is frequently not well defined. Observations of flickering of the flame zone have therefore occasionally been supplemented by observations of flickering phenomena in the boundaries which appear in shadow pictures (see [*2*, p. 262]), though these two phenomena are rather independent of each other (see Art. 2).

The transition from laminar to turbulent flames is also indicated in Fig. I,2d. This figure shows the conditions at which the heights of city gas flames (Newark, Delaware), at constant volume flow rate through tubes, become a function of tube diameter. The points at which this occurs mark the end of the range for which the laminar diffusion mechanism applies, and thus indicate the beginning of turbulence. On moving at constant volume flow rate toward smaller diameters (higher Reynolds numbers for tube flow) the flame height first increases and then decreases. It thus appears that the first occurrence of flickering in the flame tip region prolongs the flame, and that the well-known reduction of flame height through turbulence sets in after a substantial part of the flame has become turbulent. The reason for the primary increase of flame length seems to be that the first fluctuations are of a very large scale so that they result partly in a transport of large lumps of burning material to larger heights without substantially affecting the combustion process as such. At the higher flow rates of city gas the maximum which is characteristic of the transition to a turbulent flame occurs at about the critical Reynolds number for tube flow.

The beginning of flickering at a certain flow rate can be qualitatively understood if it is realized that the height of laminar flames h_f increases, and the height to turbulence h_t decreases, with increasing flow rate. At low flow rates, h_t is larger than h_f. At a certain higher flow rate both heights become equal so that flickering starts. If Eq. 3-2 is used for h_t and Eq. 3-6 (which is an approximation of Eq. 2-6) for h_f, it follows that the critical volume flow rate for the starting point of flickering is proportional to the $\frac{3}{2}$ power of the tube diameter d (or that $u \sim d^{-1/2}$). This relation points in the right direction. A more quantitative treatment of the height to turbulence of *burning* jets is a matter of future analysis.

Hottel and Hawthorne [2, p. 254] observed the transition from laminar to turbulent flames for a number of fuels burning from nozzles of $\frac{1}{8}$- to $\frac{1}{4}$-inch diameter in air and give the following approximate values for the cold-flow nozzle Reynolds numbers of the transition range:

Table I,3.

Fuel	H_2	Cambridge (Mass.) city gas	CO	H_2 Cambridge city gas propane, all with primary air	Propane acetylene
Re	2000	3600	4900	5500–8500	9500

The dependence of the transition point on the type of burner (tube or nozzle) has so far not been clearly established.

Wohl, Gazley, and Kapp [2, p. 288; 3] observed that the butane diffusion flame burning from tubes showed no typically turbulent flame front up to a Reynolds number of 11,000 for the pure fuel, 6700 for a fuel-air mixture with 17 per cent fuel, and 4200 for 10 per cent fuel (the experiments were not carried to higher Reynolds numbers). The shadow pictures of all these flames showed a straight, central fuel jet extending to large heights, and wave formation in the flame boundary, which is characteristic of laminar flames. The main evidence for the essentially laminar character of these flames is provided by the fact that the heights of these flames are nearly perfectly described by the equation

$$h_f = k' \sqrt{\Omega X_0 \left(1 - \frac{X_{st}}{2X_f}\right)} \tag{3-5}$$

The factor k' is the same for pure butane and the two butane-air mixtures mentioned, and its values for 0.18-inch and 0.40-inch tubes are 10.4 and 9.7 respectively. An expression of the form of Eq. 3-5 is produced if the two equations characteristic of laminar flow (2-4b and 2-5) are combined, with the provision that in Eq. 2-5 the number 1 is neglected in comparison with kh_f. The resulting equation, which is also obtained from Eq. 2-6 by neglecting the second term in the denominator is

$$h_t = \sqrt{\frac{\Omega X_0[1 - (X_{st}/2X_0)]}{4\pi X_{st} k D_0}} \tag{3-6}$$

The reason that the linear term in Eq. 2-5 can be neglected for butane is that, according to the discussion preceding Eq. 2-5, the quadratic term is largely due to the increase of diffusivity with increasing temperature and to the amount of combustion gases produced by combustion of a given fuel volume. Since the diffusivity of butane at room temperature is smaller than that of city gas and since the air requirement for combustion of butane is larger than that of city gas, it can be expected that the kh_t term will be larger in the case of butane.

A fundamental difference between the structure of a city gas flame and a butane flame at a tube Reynolds number above 8000 is revealed by a comparison of Plates I,3a and I,3b. Plate I,3a shows a structure of the jet boundary which gives evidence of intense turbulent mixing. Plate I,3b shows a surface of round smooth eddies which are quite stable (as was ascertained with the help of high speed motion photography [*2*, p. 288]) and contribute little to mixing which thus has to occur mainly through molecular diffusion across the boundary of these eddies. The apparent laminarity of the butane jet up to a tube Reynolds number of 11,000 can be explained by the fact that the kinematic viscosity of butane at room temperature is excessively small (0.035 cm^2/sec against 0.155 for air). When the butane stream emerges into the zone of combustion products, the kinematic viscosity in the stream boundary which is responsible for the structure of the boundary and for mixing will increase by a large factor, so that the effective Reynolds number drops below the critical value. It is an open question whether the data of Hottel and Hawthorne [*2*, p. 254] which were collected in Table I,3 can be interpreted in a similar manner.

Synopsis of diffusion flames in the transition region. A number of experiments by Garside, et al. [*17*; *2*, p. 102] on the height above the port at which unignited laminar jets break up and become turbulent seems to indicate that the velocity gradient at the rim of the port is greatly responsible for the value of this height. The distances from the port at which laminar flames begin to flicker are much higher than the "heights to turbulence" observed with unignited jets, and the tube Reynolds numbers at which flame flickering begins are frequently in the turbulent range [*2*, p. 254]. The critical tube Reynolds number for flickering seems to be larger when the kinematic viscosity of the fluid leaving the port at the initial temperature is smaller [*2*, p. 288]. The reason probably is that the kinematic viscosity in the jet boundary increases strongly through mixing with the hot combustion products, so that a Reynolds number lower than the tube Reynolds number applies to the fuel jet inside the flame.

I,4. Turbulent Flames.

FLAME HEIGHT.

Introduction. As noted in the discussion of laminar flames, when the velocity of the fuel jet from a given burner reaches a certain value, the flame length remains essentially constant as the velocity is increased further. This behavior is characteristic of turbulent diffusion flames. The space requirement for combustion (flame length) in such circumstances has been studied by Hawthorne, Weddell and Hottel [*2*, p. 266] and by Wohl, Gazley, and Kapp [*2*, p. 288; *3*]. Hawthorne, et al. used a number of fuels (see Table I,4) burning from nozzles and orifices. They measured the flame height by inspection and by determining the point on the axis at which 99 per cent of the fuel was burned. The latter height was usually somewhat lower than the former. Their correlation was based on the visible flame height. In the case of orifices, the vena contracta was used for defining diameter and average velocity at the jet origin. Little difference was noted between data obtained with nozzles and orifices on this basis of correlation. Wohl, et al. measured the visible height of city gas flames burning from tubes (in which the flow was turbulent).

The fact that the flame length is independent of volumetric flow rate means that the higher the jet velocity, the faster oxygen from the surroundings is mixed with the fuel, so that the distance along the jet axis required for mixing of sufficient air to burn the fuel is more or less constant. This can be explained by assuming that the rate-determining process is the mixing of the reactants by eddy diffusion, since normally the eddy diffusivity is proportional to the jet velocity. It has also been found that the flame length measured in port diameters remains about constant for turbulent flames burning from ports of varying diameter; this can be considered a consequence of the fact that normally the eddy diffusivity is proportional to the characteristic dimensions of the flow system [*2*, p. 266, 288]. The constancy of flame length measured in port diameters can also be understood from general dimensional considerations [*2*, p. 266]. The actual value for the flame length in terms of port diameter depends mainly on the fuel used. Hawthorne, et al. found flame heights up to nearly 300 port diameters (see Table I,4); Wohl, et al. found heights of 70 to 100 diameters.

The approach of Hawthorne, Weddell, and Hottel. Hawthorne, Weddell, and Hottel [*2*, p. 266] give a summary of observations of various authors on the structure of turbulent jets emerging into a medium of similar density and viscosity. Three characteristics are emphasized:

1. The concentration of fluid from the jet at any point (height h above the port, horizontal distance r from the axis) is independent of

port diameter and velocity if the coordinates are measured in port diameters (r/d, h/d).

2. The ratio of the concentration of jet fluid at any point to the concentration of jet fluid in the axis at the same height varies with radial position in nearly the same way for all values of height, port diameter, and initial jet velocity, if comparisons are made at radial distances which are a constant fraction of the "width of the jet."

3. The boundary of the jet, defined as the point where the concentration is a small constant fraction of the axial concentration at the same height, is an inverted circular cone.

For the treatment of the turbulent diffusion flame, i.e. of a fuel jet which undergoes a chemical reaction and produces heat as a result of mixing with the surrounding air, these three characteristics have been retained, and the situation has been simplified further by assuming that the velocity and composition are constant over any cross section of the flame jet and change abruptly into the constant velocity and composition of the surrounding air. The latter assumption has been made for the purpose of calculating flame heights notwithstanding the fact that the authors themselves have ascertained the gradual change of composition over the flame cross section. It has been demonstrated for the case of unignited jets that the assumption leads to reasonable results. The flame height is then obtained by applying the laws of conservation of mass and momentum and an equation of state as follows:

1. Continuity. The moles of "port fluid" leaving the burner equal the moles of port fluid which pass through any cross section of the jet if the moles of constituents, irrespective of chemical transformation, are always counted as unreacted. There follows

$$\begin{aligned}\pi r^2 u\rho\chi_N &= \frac{\pi r^2 u \mathfrak{M}_0 p X_N \gamma}{RT} = \frac{\pi r^2 u \mathfrak{M}_0 p X \gamma}{RTX_0} \\ &= \pi r_0^2 u_0 \rho_0 = \frac{\pi r_0^2 u_0 \mathfrak{M}_0 p}{RT_0}\end{aligned} \tag{4-1}$$

where r = radius of the jet at some height h, u = velocity, ρ = density, T = temperature, χ_N = mass fraction of jet port fluid, X_N = mole fraction of port fluid counted as unreacted (χ_N and X_N are equal to 1 at the burner port), $\mathfrak{M}_0$ = molecular weight of port fluid in the unreacted state, X = mole fraction of fuel, counted as unreacted, and γ = moles of reactants per moles of product; all values without subscript are averages at the height h; the subscript $_0$ refers to burner port conditions.

2. Momentum law. No forces acting, i.e. the momentum of the jet is constant. (A derivation was also given, taking the buoyancy force

into consideration. The resulting equation, however, was not applied quantitatively to experimental results, although it was shown that the buoyancy reduces the flame height at relatively low Froude numbers.)

$$\pi r^2 u^2 \rho = \pi r_0^2 u_0^2 \rho_0 \tag{4-2}$$

3. Equation of state, i.e. the ideal gas law written for the varying composition.

$$\frac{\rho}{\rho_0} = \left[\frac{X}{X_0} + \frac{\mathfrak{M}_s}{\mathfrak{M}_0}\left(1 - \frac{X}{X_0}\right)\right]\frac{\gamma T_0}{T} \tag{4-3}$$

$\mathfrak{M}_s$ = the molecular weight of the surrounding fluid (air)

At the height where the combustion is just completed it is assumed that the temperature T is the adiabatic flame temperature (T_f), the mole fraction that of the stoichiometric mixture (X_{st}), and the density that of the stoichiometric mixture at $T_f(\rho_f)$. Eq. 4-1 and 4-2 can each be solved for $(u/u_0)^2$ and equated to each other to yield an expression for the jet radius. For the maximum radius r_f occurring at the height of completion of combustion there is obtained

$$\frac{r_f}{r_0} = \frac{1}{\chi_f}\sqrt{\frac{\rho_0}{\rho_f}} = \frac{X_0}{X_{st}}\sqrt{\frac{\rho_f}{\rho_0}}\left(\frac{T_f}{\gamma T_0}\right) \tag{4-4}$$

(χ_f = mass fraction of port fluid at this height). Substitution of Eq. 4-3 into Eq. 4-4 gives

$$\frac{r_f}{r_0} = \frac{X_0}{X_{st}}\sqrt{\left[\frac{X_{st}}{X_0} + \frac{\mathfrak{M}_s}{\mathfrak{M}_0}\left(1 - \frac{X_{st}}{X_0}\right)\right]\frac{T_f}{\gamma T_0}} \tag{4-5}$$

The height h is obtained as a function of the radius r with the help of the assumption that the angle of spread of a flame jet is the same for all flames. This is expressed in the form

$$h = k_1(2r) \tag{4-6}$$

where k_1 is a constant. Eq. 4-5 and 4-6 yield the following formula for flame height h_f:

$$\frac{h_f}{d} = k_1 \cdot \frac{X_0}{X_{st}} \cdot \sqrt{\left[\frac{X_{st}}{X_0} + \frac{\mathfrak{M}_s}{\mathfrak{M}_0}\left(1 - \frac{X_{st}}{X_0}\right)\right]\frac{T_f}{\gamma T_0}} \tag{4-7}$$

As noted in the discussion of flames in the transition range, there is always some distance s above the port for which the flame appears free of turbulence. Since, in that region, mixing probably occurs by molecular diffusion, and since molecular diffusion is a slower mixing process than eddy diffusion, it is assumed that little combustion occurs in this region. Eq. 4-7 thus obtains the final form:

$$\frac{h_f - s}{d} = \frac{k_1 X_0}{X_{st}} \sqrt{\left[\frac{X_{st}}{X_0} + \frac{\mathfrak{M}_s}{\mathfrak{M}_0}\left(1 - \frac{X_{st}}{X_0}\right)\right] \frac{T_f}{\gamma T_0}} \tag{4-8}$$

Eq. 4-8 complies with the general predictions mentioned in the introduction of this article. The rate of mixing enters the equation only indirectly in the form of the assumption that the angle of spread has a given constant value.

The value of k_1 has been determined empirically from a graph in which the measured height of several flames was plotted versus the maximum flame radius calculated for these flames according to Eq. 4-5. This graph is reproduced here as Fig. I,4a (Fr = Froude number). The best straight line corresponds to a value of $k_1 = 5.3$. The assumed half angle of spread of the flames can be computed from the value of k_1 and is found to be 5.4° (see below). A more detailed comparison of Eq. 4-8 with the data is given in the following table:

Table I,4.

	H_2	Cambridge (Mass.) city gas	C_3H_8	C_2H_2	CO
$\left(\frac{h_f - s}{d}\right)_{calc}$ (Eq. 4-8)	152	101	300	205	46.5
$\frac{h_f}{d}$ (max exptl values)	147	136	296	188	76
$\frac{s}{d}$ (max exptl values)	2	6	—	11	35
$\left(\frac{h_f - s}{d}\right)_{exptl}$ (max values)	145	130	—	177	41
$\frac{(h_f - s)_{calc}}{(h_f - s)_{exptl}}$	1.05	0.78	1.0	1.09	1.13

Recently, Thring and Newby [*28*, p. 789] have predicted the lengths of turbulent jet flames burning in enclosures by means of cold jet models and similarity theory.

The approach of Wohl, Gazley, and Kapp. Wohl, Gazley, and Kapp [*2*, p. 288; *3*] developed an equation for the height of turbulent diffusion flames in which they retain the concept of a time for mixing t^* derived for the laminar flame, but use the eddy diffusivity ϵ, instead of the molecular diffusivity D. The time for mixing in the laminar case was given by Eq. 2-2, which may be approximated by

$$t^* = \frac{d^2 X_0}{16 D X_{st}} \tag{4-9}$$

For turbulent mixing there follows instead

$$t^* = \frac{d^2 X_0}{16 \epsilon X_{st}} \tag{4-10}$$

It is assumed, as in the case of the laminar flame discussed previously, that the average velocity does not greatly differ from the initial velocity of the jet, so that there follows for the flame height:

$$h_f = u_0 t^* \tag{4-11}$$

In [2, p. 288; 3] it was assumed that the eddy diffusivity is proportional

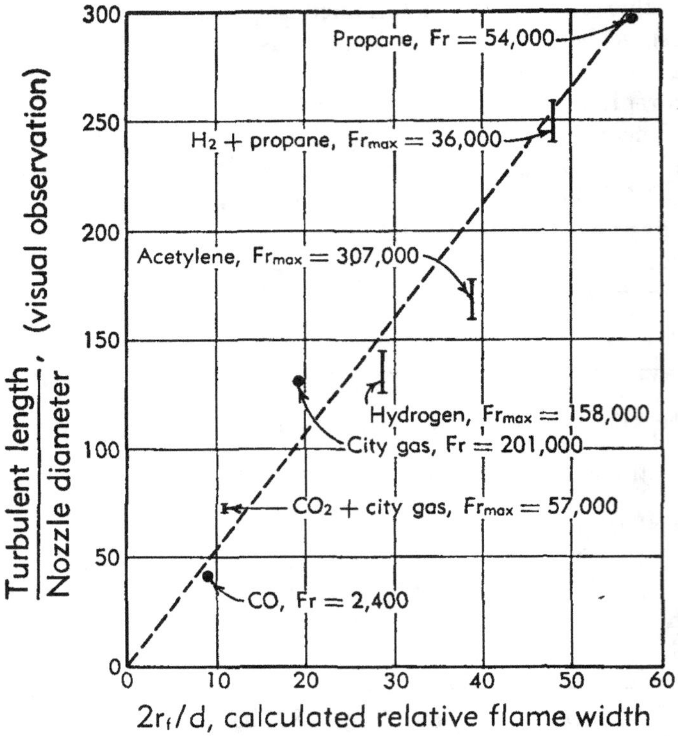

Fig. I,4a. Correlation of turbulent free flame lengths with calculated flame spread. Abscissa is calculated width at point of completion of combustion, ignoring buoyancy [2, p. 280].

to the product of the burner diameter and the jet velocity, in analogy to pipe flow. There follows

$$\epsilon = 0.00255 f u_0 d \tag{4-12}$$

where the numerical factor refers to pipe flow and f is a constant. Combination of Eq. 4-10, 4-11, and 4-12 leads directly to

$$\frac{h_f}{d} = \frac{25.6 X_0}{X_{st} f} \tag{4-13}$$

This equation describes the main characteristics of turbulent diffusion flames, namely, the proportionality of flame height with port diameter, and its independence of flow rate. For city gas issuing from a tube (turbulent flow in the tube) f was not greatly different from unity. However, this cannot be generally true since additional factors which were taken into consideration by Hawthorne, Weddell, and Hottel (Eq. 4-8) affect the flame height. The f factor must thus depend on the fuel used.

Another treatment of turbulent flames which is also based on a diffusion mechanism and makes use of an expression for eddy diffusivity similar to Eq. 4-12, has been given by Yagi and Saji [*28*, p. 771]. Concerning earlier work by Yagi, see [*28*, p. 97].

Derivation of the equation (4-8) of Hawthorne, Weddell, and Hottel with the help of a diffusion mechanism. By approximate considerations of the jet structure, and by introducing the momentum equation and equation of state derived by Hawthorne, Weddell, and Hottel, the diffusion equations given above (Eq. 4-10 to 4-13) can be extended so that they lead to the same result which Hawthorne, Weddell, and Hottel had derived (Eq. 4-8).

Eq. 4-10 is rewritten

$$t^* = \frac{\overline{r^2} X_0}{4\epsilon X_{st}} \tag{4-14}$$

in which $\overline{r^2}$ is the mean square of a reference radius which replaces the port radius r_0, and accounts for the expansion of the original jet through temperature increase. This results from the fact that it is not realistic to treat the system as if diffusion occurred between a cylindrical fuel column of constant radius r_0 and the surrounding air. Rather, the cylinder itself expands in time through temperature increase, quite apart from the diffusion effect. This expansion is roughly described by

$$\frac{r_f^2}{r_0^2} = \frac{T_f}{\gamma T_0} \tag{4-15}$$

where r_f is the reference radius at the flame tip. The average value of r^2 lies between r_f^2 and r_0^2 and may be written

$$\frac{\overline{r^2}}{r_0^2} = \frac{T_f}{k_2 \gamma T_0} \tag{4-16}$$

where $k_2 > 1$. Eq. 4-11 for flame height is retained, but u_0 is now given the meaning of velocity in the flame *axis*. Instead of using Eq. 4-12, the eddy diffusivity is expressed by

$$\epsilon = k_3 \overline{ur} = k_3 u_0 r_0 \frac{\overline{ur}}{u_0 r_0} \tag{4-17}$$

where $\overline{ur}$ is the average product of the velocity and radius of the jet. This implies that the velocity over a cross section of a jet is not uniform, u_0 being constant and $\bar{u}$ decreasing with increasing height.

The momentum equation used by Hawthorne, Weddell, and Hottel (Eq. 4-2) gives

$$\frac{\overline{ur}}{u_0 r_0} \cong \sqrt{\frac{\rho_0}{\bar{\rho}}} \tag{4-18}$$

The r term in the product $\overline{ur}$ is not necessarily the same as in Eq. 4-16.

The equation of state (Eq. 4-3), written for the average temperature, is

$$\frac{\bar{\rho}}{\rho_0} = k_4 \left[\frac{X_{st}}{X_0} + \frac{\mathfrak{M}_s}{\mathfrak{M}_0} \left(1 - \frac{X_{st}}{X_0} \right) \right] \frac{\gamma T_0}{T_f} \tag{4-19}$$

where $k_4 > 1$. Substitution of Eq. 4-14, 4-16, 4-17, 4-18, and 4-19 into Eq. 4-11 gives the following equation for the flame height:

$$h_f = \frac{r_0 T_f X_0 \sqrt{k_4 \left[\frac{X_{st}}{X_0} + \frac{\mathfrak{M}_s}{\mathfrak{M}_0} \left(1 - \frac{X_{st}}{X_0} \right) \right] \frac{\gamma T_0}{T_f}}}{4 k_2 k_3 \gamma T_0 X_{st}} \tag{4-20}$$

Dividing both sides of Eq. 4-20 by $2r_0$, and ascribing to the constants k_4 and k_2 the value 2, the final equation reads

$$\frac{h_f}{d} = \frac{X_0 \sqrt{\left[\frac{X_{st}}{X_0} + \frac{\mathfrak{M}_s}{\mathfrak{M}_0} \left(1 - \frac{X_{st}}{X_0} \right) \right] \frac{T_f}{\gamma T_0}}}{8\sqrt{2}\, k_3 X_{st}} \tag{4-21}$$

This equation is essentially the same as Eq. 4-7. Modification of this equation for the height over which the flame is not turbulent gives Eq. 4-8.

In comparing the two derivations of this equation, it is noted that Hawthorne, Weddell, and Hottel have made use of the assumption that the angle of spread of the flame is constant. The present derivation assumes instead a diffusion mechanism and takes into consideration the expansion of the whole system by increase of temperature (an expansion which is superimposed over the diffusion process), and the increase of eddy diffusivity with spreading of the jet.

It will be seen that the factor k_1 in Eq. 4-7 corresponds to the term $1/(8\sqrt{2}k_3)$ in Eq. 4-21. The average empirical value of 5.3 for k_1 which was given by Hawthorne, Weddell, and Hottel thus corresponds to $k_3 = 1.66 \times 10^{-2}$. The significance of k_3 is that, according to Eq. 4-17, the eddy diffusivity is proportional to k_3 at otherwise equal circumstances. (Consequently, the flame height is inversely proportional to k_3.) The values of k_3, determined from the data of Wohl, Gazley, and Kapp, are 1.12×10^{-2} for 100 per cent city gas and 0.825×10^{-2} for 50 per cent city gas. This means that the flame heights obtained by Wohl, et al. lie above the average line of Fig. I,4a. It would be difficult to trace the cause for this discrepancy.

Recently, Baron [21] has given a semiempirical equation for the height and shape of a vertical turbulent diffusion flame. The approach bears resemblance to that of Hawthorne, Weddell, and Hottel though it is considerably refined; the result as to flame height is again identical with Eq. 4-7 derived by the latter authors. Baron considers first the mass and momentum transfer for an unignited vertical jet of a combustible fluid emerging into quiet air. He disregards molecular diffusion and, in the development leading to the final explicit equation for flame height, also disregards buoyancy effects. Turbulent mixing is introduced only implicitly, by making use of the empirical differential equations for the horizontal component of mass flow and momentum which has been proposed by Reichhardt [22] and Baron and Alexander [23]. For the case of a jet emerging from a point source, these relations lead to a simple error function for the radial distribution of mass flow and momentum as shown by the following equations:

$$\overline{\rho\chi_N u} = \frac{(\overline{\rho u})_0 r_0^2}{C_\chi^2 h^2} e^{-\frac{r^2}{C_\chi^2 h^2}} \tag{4-22}$$

$$\overline{\rho u^2} = \frac{(\overline{\rho u^2})_0 r_0^2}{C_m^2 h^2} e^{-\frac{r^2}{C_m^2 h^2}} \tag{4-23}$$

where h = distance from burner port in the direction of the jet axis, r = radial distance, u = velocity component in the direction of the jet axis at h and r, χ_N = mass fraction of port fluid (the initial value of χ_N is 1), and r_0 = "radius of the point source"; the subscript $_0$ refers generally to burner port conditions. C_χ and C_m are empirical constants. The terms $C_\chi h$ and $C_m h$ in these equations define characteristic radii (r_χ and r_m, respectively) of the jet at the height h. If these are introduced into the above equations, it is realized that Hawthorne, Weddell, and Hottel's equations (4-1 and 4-2) represent a kind of integration of Eq. 4-22 and 4-23 respectively, since they deal with the total flow through a jet cross section of characteristic radius r, average velocity u, and average mass fraction of nozzle fluid χ_N.

Dividing Eq. 4-22 by the square root of Eq. 4-23 yields according to Baron,

$$\frac{\overline{\rho\chi_N u}}{(\overline{\rho u^2})^{\frac{1}{2}}} \cong \overline{\chi_N \bar{\rho}^{\frac{1}{2}}} \cong \frac{\rho_0^{\frac{1}{2}} C_m r_0}{C_\chi^2 h} e^{-\frac{r^2}{h^2}\left(\frac{1}{C_\chi^2} - \frac{1}{2C_m^2}\right)} \tag{4-24}$$

This is the same procedure by which Eq. 4-4 was obtained from Eq. 4-1 and 4-2. If the characteristic jet radius $r_{\chi m} = (C_\chi^2/C_m)\, h$ is introduced, it is realized that Eq. 4-4 is a kind of integrated form of Eq. 4-24.

The mass fraction χ_N can be replaced by the fuel mole fraction X

with the help of the equation

$$\frac{1}{\chi_N} = 1 + \frac{\mathfrak{M}_s}{\mathfrak{M}_0}\left(\frac{X_0}{X} - 1\right) \tag{4-25}$$

The author realizes that in turbulent flow the average of the products in Eq. 4-22 and 4-23 does not equal the product of the average factors, so that the factoring by which Eq. 4-24 has been obtained involves errors. The estimation of these errors does not appear conclusive.

If C_χ^2/C_m and $\bar{\rho}/\rho_0$ are known, Eq. 4-24 yields the coordinates of the surface (and specifically, for $r = 0$, the axial distance from the port) at which χ_N has a given value. For nonisothermal jets without a chemical reaction, one might attempt to determine $\bar{\rho}/\rho_0$ from χ_N, the equation of state, and a heat transfer equation which is analogous to Eq. 4-22 and 4-23. According to Baron, such a procedure is not feasible because of the uncertainties introduced by the turbulent fluctuations. The method would not be applicable to a diffusion flame in any case.

Baron characterizes a diffusion flame in the usual manner, assuming that the surface at which the mole fraction of fuel (counted as unreacted) has dropped to the value in the stoichiometric mixture represents the flame front, and that this surface is a source of heat such that its temperature is equal to the adiabatic combustion temperature of the stoichiometric mixture. This, and the change of mole numbers, is the only way in which the chemical reaction in the jet is taken into consideration. The density in the flame front is thus obtained from Eq. 4-3 and it is assumed that the temperature and density distribution throughout the jet—which is radically different from that in a jet without chemical reaction—does not cause a substantial error in Eq. 4-24. By introducing Eq. 4-25 and 4-3 into Eq. 4-24, and by making $X = X_{st}$ and $r = 0$, the following equation is obtained for the flame height $h = h_f$:

$$\frac{h_f}{d} = \frac{C_m X_0}{2C_\chi^2 X_{st}}\sqrt{\frac{X_{st}}{X_0} + \frac{\mathfrak{M}_s}{\mathfrak{M}_0}\left(1 - \frac{X_{st}}{X_0}\right)\frac{T_f}{\gamma T_0}} \tag{4-26}$$

(d = "diameter of the point source"). This equation is formally identical with Hawthorne, Weddell, and Hottel's equation (4-7) which is not surprising. Baron finds empirically, from experiments on jet spreading [24], $C_m = 0.075$, $C_\chi = 0.0855$, i.e. $C_m/2C_\chi^2 = 5.13$, while Hawthorne, Weddell, and Hottel found 5.3 from flame heights. The agreement is astonishingly good.

Application of Eq. 4-24 to an arbitrary point of the flame front and to the point at $r = 0$, $h = h_f$, yields the following equation for the flame contour produced by a fuel jet emerging from a point source:

$$r_{st} = 0.145h\sqrt{\ln\frac{h_f}{h}} \tag{4-27}$$

where r_{st} = distance of flame from jet axis at the height h.

The treatment of a turbulent diffusion flame with appreciable buoyant forces and of intermingling parallel flames has been outlined by Baron in general terms. It is stated that "buoyancy increases the rate of mixing thereby decreasing the height and diameter of the flame." Consideration of special states and boundary conditions of fuel jets of finite width (turbulent or streamline flow at the burner mouth, long tube, nozzle, or orifice) seems to go beyond the scope of the proposed procedure.

STRUCTURE OF TURBULENT DIFFUSION FLAMES. The structure of turbulent diffusion flames has been investigated by Wohl, Gazley, and Kapp [*2*, p. 288; *3*] who used direct and shadow photography, and by Hawthorne, Weddell, and Hottel [*2*, p. 266] who used chemical analysis. Sage, et al. also examined velocity, temperature, and composition distribution [*4a,4b*].

Direct and shadow photographs. Plates I,4a, I,4b, and I,3a [*2*, p. 288; *3*] are examples of direct and shadow photographs of turbulent diffusion flames. The direct picture, Plate I,4a, shows that the flame has the shape of an inverted cone with straight sides, as was claimed by Hawthorne, Weddell, and Hottel [*2*, p. 266]. The boundary is somewhat fuzzy compared with that of laminar flames (Plates I,2a and I,2c), due to the fluctuations of the turbulent flame front. The flame envelope visible in the shadow picture, Plate I,4b, has already been discussed in Art. 2. In addition, it may be stated that the average boundary of the upper turbulent part of the envelope is also strictly conical. The jet of unburned fuel which rises inside the flame and which is gradually consumed by combustion has a decidedly turbulent appearance in the shadow picture, Plate I,4b, as is evident from a comparison with the laminar fuel jets visible in the flame shadow pictures, Plates I,2b and I,2c. A more detailed picture of the structure of a fuel jet in a turbulent diffusion flame is given in Plate I,3a. The downward bending of the striations is due to the fact that for every cross section the velocity decreases with increasing distance from the axis.

Plate I,4b may also be used to measure the angle of spread of the turbulent portion of the flame. The half angle between axis and one jet boundary is found to be about 12°. Plate I,4a gives a half angle of about 3.2°, which, however, does not measure the spread of the jet, but that of the luminous reaction zone. As was noted earlier, the angle determined by the value of the constant in Eq. 4-8 is 5.4° and thus lies in between the two measured angles, as is plausible.

Concentration patterns in turbulent diffusion flames. Hough, Sears, Wilcox, and Sage [*4a*] studied the mixing of two turbulent coaxial cylindrical streams, natural gas inside, air outside, without and with burning, by making measurements of composition, total pressure, and static pressure at numerous axial distances downstream of the port of the inner tube, and at numerous radial distances from the axis. The diameter of the inner tube was 0.941 in., that of the outer tube 3.82 in. The common average inlet velocity was 25 ft/sec. Ignition was effected by a continuous spark at an axial distance from the gas nozzle of 7.25 in. and a radial distance below the center line of 0.625 in. The writers presume that the flame burned not from the spark but from the port. Hough, et al. conclude tentatively from their data that the eddy diffusivity under burning conditions is more than ten times that prevailing in the unreacted state.

Berry, Mason, and Sage [*4b*] continued these studies, varying the velocity, but keeping the initial velocities of both streams equal as before. It is stated that the ratio of the two streams was such as to provide stoichiometric air for the complete combustion of the natural gas to CO_2 and H_2O. (From the dimensions of the two coaxial ducts one might have presumed that there was some excess of oxygen.) This time the spark holder was located close to the fuel port. The positioning of the spark holder in the boundary between the streams caused asymmetric patterns of temperature, velocity, and composition.

The relation used in [*4b*] for prediction of the mole fraction X_m of fuel in the axis (or rather the maximum mole fraction at any cross section) as a function of distance h from the starting point of mixing (at which point $X_m = X_0 = 1$), is

$$X_m = \frac{X_0 r_0^2 u_0}{4(\epsilon + D)(h + h')} \tag{4-28}$$

where r_0 is the radius of the inner tube and h' is a length added to h to give a linear relationship between X_m and a reciprocal distance. At the flame tip, $h = h_f$ and $X_m = X_{st}$ so that one obtains

$$h_f + h' = \frac{X_0 r_0^2 u_0}{X_{st} 4(\epsilon + D)} \tag{4-29}$$

If the molecular diffusivity D is small in comparison with the eddy diffusivity ϵ, and h' is small in comparison with h, this reduces to Eq. 4-10 and 4-11 which apply to the experiments of Wohl, Gazley, and Kapp.

In their work on quasi-confined and free flames, Hawthorne, Weddell, and Hottel [*2*, p. 266] took numerous samples at different heights and distances from the jet axis for hydrogen-air flames surrounded by an enclosure. The latter materially affected the flame even though large quantities of excess air were admitted to the enclosure. The results were evaluated to give the concentration of the nozzle fluid, and correlated [*2*, p. 266] with the help of dimensionless coordinates as shown in Fig. I,4b. It is seen that the concentration profiles which were obtained at various conditions are similar to each other and to profiles measured for air jets by Ruden. The insensitivity of the profile to jet velocity is especially noteworthy. The absolute height of the concentration profiles decreases, of course, with increasing distance from the port.

Plots of the reciprocal of port fluid concentration (calculated from the samples as unburned reactants) against distance from the port, measured in port diameters, gave straight lines, which is similar to the findings of Berry, Mason, and Sage [4b] (see Eq. 4-28).

Hawthorne, et al. furthermore studied the concentration in the axis of free turbulent diffusion flames for hydrogen, hydrogen-propane, propane, Cambridge city gas and a city gas-carbon dioxide mixture. For all these flames, the reciprocal of the concentration of the nozzle fluid, plotted against the distance from the point of beginning flame turbulence yielded, in contra-distinction to confined flames, curves the slopes of

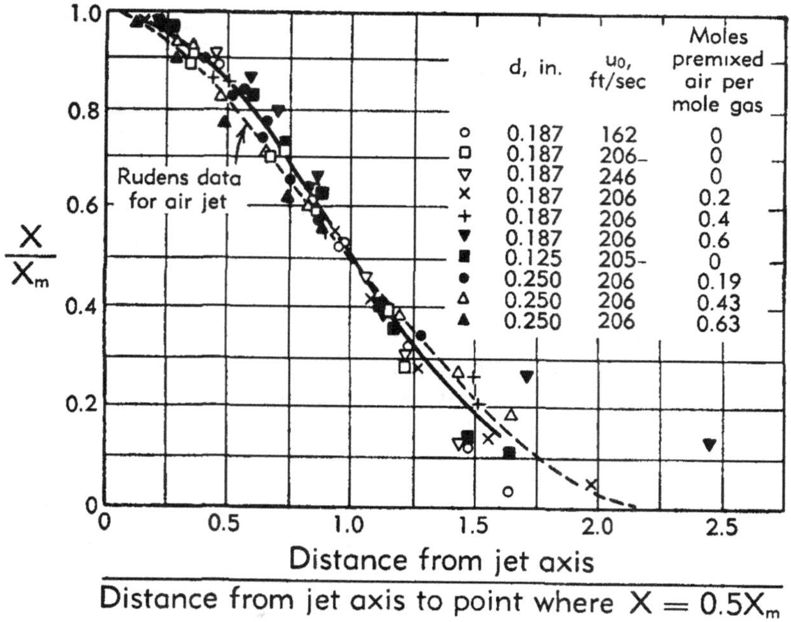

Fig. I,4b. Correlation of radial concentration profiles in confined hydrogen flames [2, p. 276].

which increased with increasing distance. For confined and free hydrogen flames the curves run closely together up to 40 to 50 nozzle diameters from the point of beginning flame turbulence and diverge from each other strongly beyond this distance. It follows that the free flame is shorter than the confined flame. The reason probably is that the velocity is higher in the upper parts of the confined flame.

By analyzing the gas composition in the axis of a free turbulent flame, it was found that the average sample contains substantial amounts of unburned fuel gas and free oxygen simultaneously. This is due to the coarseness of turbulent mixing. Primarily, eddies of air are drawn in from the surrounding atmosphere. Reaction with fuel requires molecular mixing which has to spread from the surface into the interior of the eddy. This process is aided by resolution of the eddies and formation of new surfaces. These events take time, and as a result the combustion process does not keep pace with the air induction. Obviously, the lag between mixing and chemical reaction increases with increasing scale or eddy size.

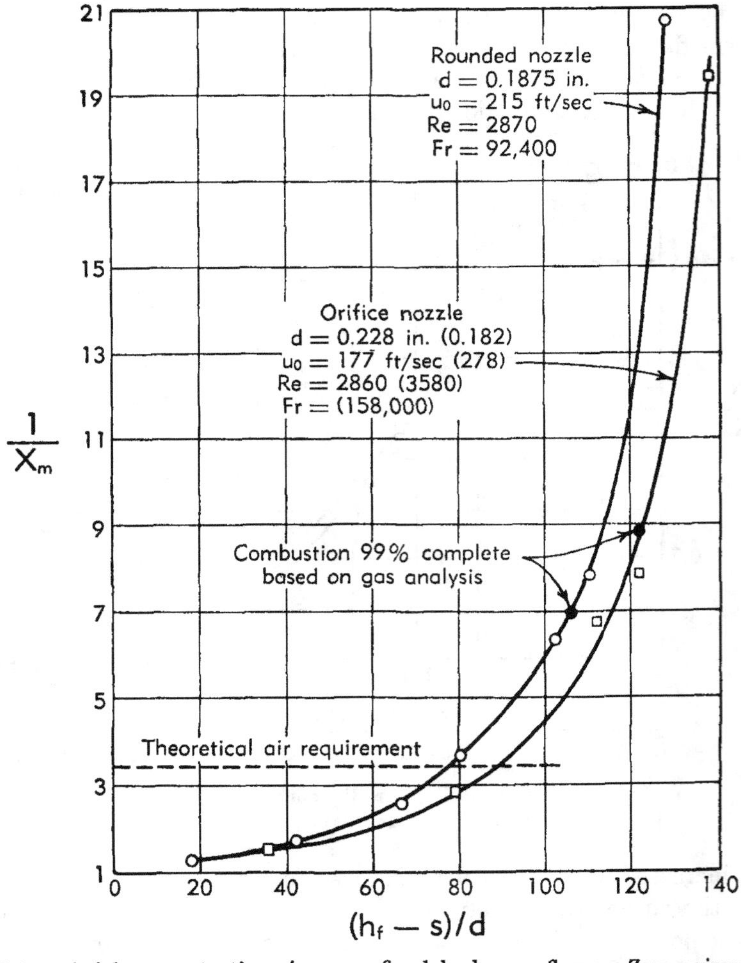

Fig. I,4c. Axial concentrations in unconfined hydrogen flames. Zero primary air. Numbers in parentheses are based on diameter of vena contracta of sharp-edged orifice nozzle [*2*, p. 277].

This lag is experimentally evident from Fig. I,4c [*2*, p. 266] according to which the oxygen and fuel atoms in the jet axis are in stoichiometric proportion at a flame height which is much lower than the one at which the fuel is nearly completely burned.

Hawthorne, Weddell, and Hottel [*2*, p. 266] treat this situation analytically by introducing the unmixedness factor which is defined as the mean deviation of the nozzle fluid mole fraction from its average value $\sqrt{\overline{X'^2}}$. By applying the normal error law to the fluctuations of mole fraction, a relationship of the form

$$\frac{\sqrt{\overline{X'^2}}}{-\beta \Delta X} = \phi \left(\frac{\text{actual oxygen in sample}}{O_2 \text{ equivalent of unburned gas}}\right)^{\beta} \tag{4-30}$$

is derived. This relationship is shown in Fig. I,4d. ΔX is explained in the caption of Fig. I,4d. β has the value $+1$ when the average mixture is

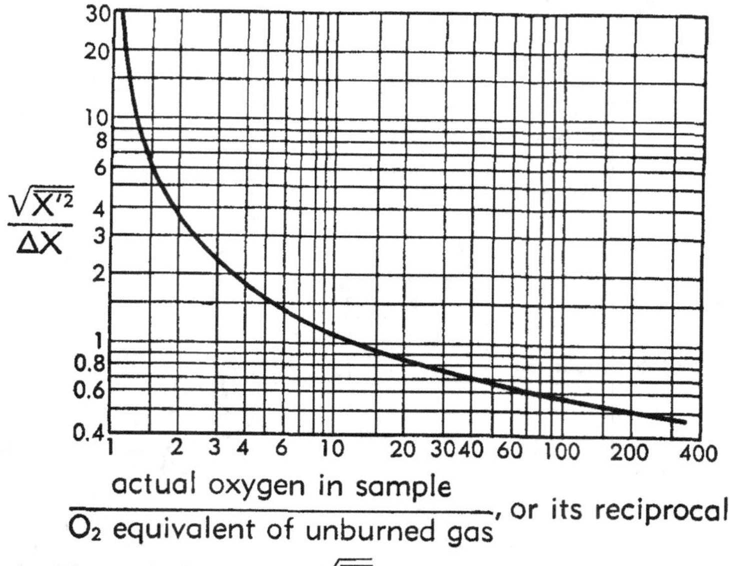

Fig. I,4d. The unmixedness factor $\sqrt{\overline{X'^2}}$ in turbulent flames. Derivation assumes that fluctuating component X' follows the normal error law. Use (actual oxygen in sample)/(oxygen equivalent of unburned gas) or its reciprocal, whichever is greater than unity. $\Delta X = |X - X_T|$, where X = mole fraction of nozzle fluid in sample reduced to unreacted condition, X_T = mole fraction of nozzle fluid in unreacted stoichiometric mixture. When $\Delta X = 0$, $\sqrt{\overline{X'^2}} = \sqrt{2\pi} \times$ (moles unburned nozzle fluid)/(total moles of nozzle fluid + moles of air in unreacted sample) [2, p. 282].

lean, and -1 when it is rich. Since ΔX is given by the nozzle fluid used, and since the actual oxygen and the oxygen equivalent to the unburned gas are found by chemical analysis, the unmixedness factor can be obtained from Fig. I,4d. Results are shown in Fig. I,4e. The abscissa, $1/X_m$, is the reciprocal of the mole fraction of the port fluid in the axis. With the help of curves like those shown in Fig. I,4c the abscissa can be converted into flame height.

The largest possible value of the unmixedness factor is 1. The largest values recorded come close to this limit. They pertain to a distance of 25 or more nozzle diameters from the port. It should be expected that the unmixedness factor will decrease when the port is approached more closely. It is seen that the relative unmixedness $\sqrt{\overline{X'^2}}/X_m$ in the jet axis does not vary strongly since the experimental lines in Fig. I,4e have a slope close to -1. The average value of the relative unmixedness is about 0.5. The relative unmixedness is usually smallest at medium heights. The increase toward lower heights may be due to a relatively low temperature, i.e. slow molecular diffusivity, and the increase toward larger heights may be due to increased turbulent mixing.

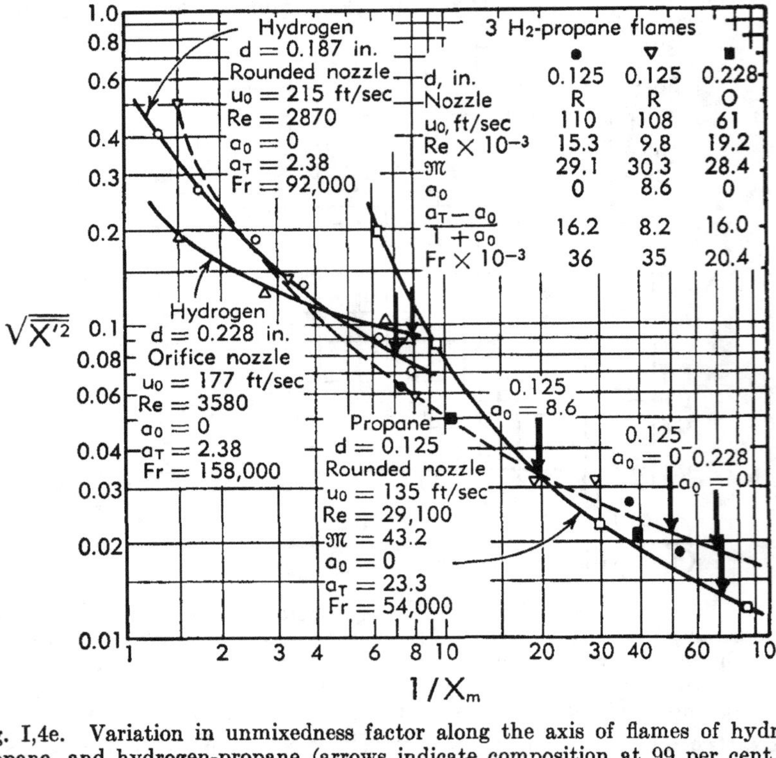

Fig. I,4e. Variation in unmixedness factor along the axis of flames of hydrogen, propane, and hydrogen-propane (arrows indicate composition at 99 per cent completion of combustion) [*2*, p. 282]. a_0 = moles of primary air per mole of fuel gas; a_T = moles of air required for complete combustion of one mole of fuel gas.

Synopsis of Turbulent Diffusion Flames. Hawthorne, Weddell, and Hottel [*2*, p. 266] developed a very useful formula for the height of turbulent diffusion flames, which is based on the principle of conservation of momentum and mass flow and on the assumption of a universal constant angle of spread and of uniform properties of the jet over every cross section. The formula can be shown to be compatible with the assumption of a turbulent diffusion mechanism in which the eddy diffusivity is proportional to the velocity and local diameter of the burning jet. Baron [*21*] arrived at the same formula, and at the same time at an equation for the shape of the burning surface, by assuming an error function for the distribution of fuel flow and momentum across the jet. In all these approaches the chemical reaction in the diffusion flame has been treated essentially by assuming that the burning surface is the locus where fuel and oxygen meet in stoichiometric proportions, and that the temperature of this surface is the adiabatic flame temperature of the stoichiometric mixture. Chemical analysis of the turbulent diffusion flame [*2*, p. 266] has yielded important information on the relationship between the coarse turbulent mixing and the molecular mixing which has to precede chemical reaction. A study of composition and velocity in a turbulent diffusion flame which is produced by a fuel jet burning in the axis of a confined annular air stream has been initiated by Sage, et al. [*4a,4b*].

It would be desirable if the chemical processes and the various mixing processes occurring in a turbulent diffusion flame between the center line of the fuel stream and the surrounding air were to be studied further by measuring the patterns of composition, temperature and velocity, and possibly the fluctuations of these quantities, throughout the flame for a number of characteristic cases. These might include the cylindrical fuel jet emerging into an atmosphere of air, coaxial jets of fuel and air, and, for optical studies, corresponding two-dimensional systems. The state of turbulence of the streams prior to mixing should be varied independently of their velocity. Comparison with the mixing of nonreacting, nonisothermal streams or nonreacting isothermal streams of different density, and with premixed turbulent flames would be helpful. Such data should present a challenge to the theoretician to go beyond a semiempirical treatment of height and shape and to explain physically the events occurring in the different regions of a turbulent diffusion flame.*

Several important papers have appeared recently dealing with the diffusion flame, its geometry as affected by flow pattern, and its behavior in situations where the kinetics of the combustion reactions may be important.

Barr [*28*, p. 765] and Barr and Mullins [*29*] have studied the industrially important case of confined diffusion flames for the case of laminar flow. The geometry of the systems studied was similar to that of Burke and Schumann (see Fig. I,2b). Air and fuel were admitted in separate streams (the air in an annulus around the fuel port) at rates varying over a wide range, both as measured in total gas flow and relative velocities of the two jets. This wide range of flow conditions and confinement of the flame results in several modifications of the flame behavior as discussed earlier. For example, in the extreme case of the overventilated flame (Fig. I,2b line c) (very low fuel flow rate and high air flow rate) the flame behaves as if the fuel were injected into the wake of a bluff object (the central tube). The extreme case of the underventilated flame (line d of Fig. I,2b) results in several interesting phenomena of pulsation and extinction by "burnback" into the air stream. In addition, as is painfully common with all types of confined flames, vibratory behavior was observed but no satisfactory explanation found.

Barr and Mullins present a solution of the diffusion equation for the case of variable air and fuel velocities with the usual assumptions of Burke and Schumann and the modification that, instead of assuming equality and constancy of air and fuel velocities, only constancy of air and fuel velocities is assumed but the shear stress due to inequality of the velocities is neglected. They include the possibility that the air stream is "vitiated," i.e. diluted by combustion products. The solution contains the dimensionless groups

$$\frac{D\lambda^2 h}{u_2}, \quad \frac{iu_1}{au_2}, \quad \text{and} \quad \frac{D\lambda^2 h}{u_1}$$

h = flame height
D = diffusivity
λ = m^{th} root of $J_1(\lambda R)$; R = radius of the outer tube
u_1 = velocity of the fuel stream
u_2 = velocity of the air stream
i = stoichiometric ratio of oxygen to fuel
a = mole fraction of oxygen in the vitiated air stream

(For a given geometry, fuel, and oxidant composition, i, λ, D and a will be constant, and the first and third groups are related by the second.) The first and second of these groups were used to correlate the heights of flames with the fuel flow rates for given air flow rates. The first group is effectively flame height per unit of air velocity; the second, the fuel-air ratio. These coordinates gave a satisfactory correlation of the flame heights. Barr [*25*] later examined the influence of several variables neglected in the derivation of the theoretical equation in [*29*].

Another interesting study has been undertaken by Powell [*30*, p. 154]. The unmixedness in turbulent diffusion flames, referred to earlier under *Structure of Turbulent Diffusion Flames*, leads to a dispersion of parcels of fuel and oxidant mixed with each other, but not sufficiently mixed for reaction. The final stage is dependent on molecular mixing. The scale of unmixedness seems frequently to have the same order of magnitude as the thickness of the reaction zone (for diffusion flames with oxygen about 2 mm., see Fig. I,2f and [*33*]). These circumstances require cognizance of both diffusion and chemical reaction rates. The problem is also of practical interest in the burning of liquid fuels, since 100-micron droplets result in mixing zones of comparable dimensions. The physical situation analyzed by Powell is that of a laminar diffusion flame burning above a "sandwich burner," i.e. a burner consisting of long, thin fuel and air ports placed alternately, each being of such dimensions that the fuel-air ratio at equal velocity of the fuel and air streams is stoichiometric. The scale of unmixedness is represented by the thickness of one pair of fuel and air ports. This thickness is chosen to be of the same order of magnitude as the thickness of the reaction zone. The problem considered is two-dimensional, the pertinent dimensions being the height above the port and the distance normal to the plane of the sheets. In the work cited, a mathematical analysis of the problem was presented. The principal result showed that the height required for 90 per cent completion of the combustion is equal to the product of the initial velocities of the jets and the sum of two terms which are proportional to the characteristic reaction time t_r and the characteristic mixing time t_d, respectively.

$$h_{90\text{ per cent}} = u_0(at_d + bt_r) \tag{4-31}$$

The variation in velocity and diffusivity of the gases with height is accounted for by assuming the velocity to be proportional to the temperature and the diffusion coefficient to be proportional to $T^{1.75}$. Experimental work on the problem is in progress.

Soot formation in diffusion flames has been treated, e.g. by Behrens [*31*], Arthur and Napier [*30*, p. 156] and Schalla and McDonald [*30*, p. 157]; see also [*33*].

Mention should also be made of a number of experiments in which more complex configurations of technical interest have been investigated. Sawai, Kunugi, and Jinno [*28*, p. 805], following the pioneer work of Rummel [*32*] considered the case where gas and air emerge as separate jets into the confined space of a furnace.

Hottel and Person [*28*, p. 781] used a vortex system which was intended to promote mixing of the fuel and oxidant in a controlled manner rather than the random manner as in conventional high output combustion chambers. However, they made the observation—which is important as such—that the vortex established was of the Rankine type so that the shear generated was not sufficient to provide intense mixing.

Schmidt and Schoppe [*30*, p. 98] described two systems in which rapid mixing and high heat release were obtained as a result of the turbulence caused by the imposed flow pattern. In the first of these, air was introduced at the outer periphery of a circular chamber with a tangential velocity component. The fuel was admitted through slits on the downstream side of tubes placed in a circle and parallel to the axis of the chamber. In a second experiment, two air streams of equal speed (one of which contained the fuel) were directed adjacent to one another in opposite directions, producing a turbulent zone, the position of which was independent of the total gas flow. A flame was established and maintained in this region without benefit of a flameholder.

Two excellent surveys of diffusion flames have appeared since the writing of the main body of this article. Gaydon and Wolfhard [*33*, pp. 123, 162] put special emphasis on the problems of the structure of the reaction zone and of soot formation. Hottel [*28*, p. 97] gives a thorough discussion of mixing in laminar and turbulent jets and jet flames.

I,5. Cited References.

1. Burke, S. P., and Schumann, T. E. W. *Ind. Eng. Chem. 20*, 998 (1928).
2. *Third Symposium on Combustion, Flame and Explosion Phenomena.* Williams & Wilkins, 1949.
3. Wohl, K., Gazley, C., and Kapp, N. *Univ. Delaware Eng. Exptl. Sta. Bull. 1*, 1948.

4a. Hough, E. W., Sears, R. E., Wilcox, O. W., and Sage, B. H. *Calif. Inst. Technol. Jet Propul. Lab. Progress. Rept. 4-82*, Aug. 1949.

4b. Berry, V. J., Mason, D. M., and Sage, B. H. *Calif. Inst. Technol. Jet Propul. Lab. Progress. Rept. 20-148*, Aug. 1951.

5. Lewis, B., and von Elbe, G. *J. Chem. Phys. 11*, 75 (1943).
6. Troy, D. J., Jr. *Chem. Eng. M.S. Thesis*, Univ. Delaware, 1948; carried out under a subcontract sponsored by United Aircraft Corp. and Project Meteor.
7. Wolfhard, H. G., and Parker, W. G. *Proc. Phys. Soc. London A65*, 2 (1952).
8. Wolfhard, H. G., and Parker, W. G. *Proc. Phys. Soc. London A62*, 722 (1949).
9. Wolfhard, H. G., and Parker, W. G. *J. Chem. Soc.*, 2038 (1950).
10. Lewis, B., and von Elbe, G. *Combustion, Flames and Explosions of Gases.* Academic Press, 1951.
11. Jost, W. *Explosion and Combustion Processes in Gases.* McGraw-Hill, 1946.
12. Rembert, E. W., and Haslam, R. T. *Ind. Eng. Chem. 17*, 1236 (1925).
13. Zeldovitch, Y. B. *NACA Tech. Mem. 1296*, 1951. (Transl.).
14. Brody, C. *Project Squid Semi-Annual Progress. Rept.*, Apr. 1951. Also *Chem. Eng. M.S. Thesis*, Univ. Delaware, 1952; carried out under sponsorship of Project Squid.
15. Lamb, H. *Hydrodynamics.* Dover, 1945.
16. Wohl, K., and Peterson, C. *Project Meteor Rept. UAC-54*, Feb. 1952.
17. Garside, J. E., and Whitehead, A. E. *Inst. Gas Engrs. Communs. Repts. 274/110*, London, 1946.
18. Bollinger, L. M., and Williams, D. T. *NACA Tech. Note 1234*, 1949.
19. Goldstein, S. *Modern Developments in Fluid Dynamics.* Clarendon Press, 1938.
20. Gaunce, H. Unpublished research at Mass. Inst. Technol. mentioned in [*2*, Fig. 1].
21. Baron, T. *Chem. Eng. Progress. 50*, 73 (1954).
22. Reichhardt, H. *Z. angew. Math. u Mech. 21*, 1941. Translated in *J. Roy. Aeronaut. Soc. 47*, 167 (1943).
23. Baron, T., and Alexander, L. G. *Chem. Eng. Progress. 47*, 181 (1951).
24. Hinze, J. O., and Van Der Hegge Zignen, B. G. *Appl. Sci. Research A1*, 433 (1949).
25. Barr, J. *Fuel 33*, 51 (1951).
26. Smith, S. R., and Gordon, A. S. *J. Chem. Phys. 22*, 1150 (1954).
27. Egerton, A., and Rudrakanchana, S. *Proc. Roy Soc. London A225*, 427 (1954).
28. *Fourth Symposium on Combustion.* Williams & Wilkins, 1953.
29. Barr, J., and Mullins, B. P. *Fuel 28*, 181, 200, 205, 225 (1949).
30. *Fifth Symposium on Combustion. Pittsburgh, Pa.* Abstract of Papers, 1954 (in print).
31. Behrens, H. *Brennstoff—Chem. 34*, 135 (1953).
32. Rummel, K. *Arch. Eisenhuttenw. 10*, 505, 541; *11*, 17, 67, 113, 163, 215 (1937).
33. Gaydon, A. G., and Wolfhard, H. G. *Flames, Their Structure, Radiation, and Temperature.* Chapman and Hall, 1953.

PART FOUR

COMBUSTION OF LIQUIDS AND SOLIDS

SECTION J

COMBUSTION OF LIQUID FUELS

J. P. LONGWELL

Fuels for air-breathing jet propulsion engines are commonly in liquid form when introduced into the combustion equipment. If the fuel is of sufficient volatility, it can be evaporated and mixed with the air before burning by breaking the liquid into small particles and allowing sufficient space for the above processes to take place. Since it is very important to keep the volume of combustion equipment to a minimum in jet propulsion engines, the volume necessary for complete evaporation of the fuel is not generally available and drops of liquid fuel enter the flame zone. The process of burning a liquid fuel can be subdivided into the processes of atomization, mixing of the resulting spray with air, evaporation of the spray, and combustion of the resulting mixture. None of these processes is understood with the degree of precision necessary to design combustors for practical application. Information currently available on the various processes, however, is of value in understanding the development problems encountered in building equipment and in arriving at a first approximation in combustor design.

J,1. Atomization. Atomization refers to the process of breaking a large mass of liquid fuel into small drops. Successful analysis of this process should result in expressions describing the size distribution of drops as a function of physical properties of the fluids, geometry of the system, and operating conditions. Such a complete analysis is not available; however, discussions of possible mechanisms of atomization are found in the literature.

The process of atomization of a liquid stream entering a combustion chamber consists of some combination of the following mechanisms:

1. Spreading of the fuel jet into a cone, disk, or sheets by virtue of its own velocity.
2. Disruption of the liquid by virtue of its velocity relative to the air.
3. Breakage of thin sheets and ligaments into several drops by surface tension forces.
4. Agglomeration of drops by collision.

The importance of these several mechanisms depends strongly upon the details of the atomizing equipment, the fluid properties, and the velocity of the liquid and the gas.

Plate J,1 illustrates the process of atomization under conditions where the relative velocity between gas and liquid is small. In this case a swirl-type atomizing nozzle, as commonly used in turbojet practice, discharges oil into a chamber of quiescent air. The fuel jet before leaving the nozzle orifice rotates rapidly so that it spreads out into a conical sheet on leaving the nozzle. This sheet becomes thinner as the distance from the orifice increases. The effect of surface tension in breaking up the sheet is apparent from the photograph and it is seen that separated ligaments and sheets are formed which are drawn into drops. On collecting drops formed under these conditions one finds many with entrapped air bubbles indicating that the sheets of liquid are folded over, thus trapping air bubbles. At high liquid velocities the conical sheet breaks up nearer the nozzle, and finally, the liquid appears to be torn apart as soon as it contacts air. Under these conditions few entrapped air bubbles are found and it appears from photographs that drops are primarily formed from ligaments, or from pieces short enough to form one drop rather than breaking into several, as is common with the long thin ligaments.

If fuel is introduced into the air stream through a tube at low liquid velocity, one finds [*1*,*2*] that when the air stream has very low velocity the column of liquid is broken up by surface tension forces into drops having a diameter of the same order of magnitude as the diameter of the liquid column. As air velocity is increased, pieces are torn from the surface of the column and these pieces are stretched out into thin ligaments which eventually break into drops. This mechanism is nicely illustrated by Scheubel [*1*,*2*] in spark photographs with an air velocity of approximately 80 ft/sec. As air velocity increases, the ligaments become much less predominant (150 ft/sec), and the drops appear to form very near the surface of the liquid column which is completely broken up a few injection-tube diameters below the point of introduction into the air stream.

Castleman [*2*] has discussed the formation of drops from these ligaments and has shown, on the basis of Rayleigh's [*3*] calculation of the time of collapse of a cylindrical jet of liquid, taking surface tension and inertia forces into account, that these forces would cause collapse of the ligaments in a time of the same order of magnitude as that estimated from Scheubel's photographs.

A similar analysis of breakage of a liquid column, which takes both viscosity and surface tension into account, has been made by Weber [*4*,*5*]. The time required for a column of diameter d to break up is

$$t = C\left[\left(\frac{8\rho}{\sigma}\right)^{\frac{1}{2}} d^{1.5} + \frac{6\mu}{\sigma d}\right] \tag{1-1}$$

while the optimum length-to-diameter ratio β for breakage is

$$\beta = \sqrt{2}\,\pi \left[1 + \left(\frac{2\mu^2}{\sigma \rho d}\right)^{\frac{1}{2}}\right]^{\frac{1}{2}} \tag{1-2}$$

(See Eq. 1-4 for meaning of symbols.) Both the time of breakup and the length of particle are reduced if the column is being stretched, as would probably be the case in a high velocity air stream. Eq. 1-1 shows that the higher the surface tension the more quickly a ligament will break into drops. Since the effect of the air stream is probably to stretch ligaments, thereby reducing the size of drops formed from them, high surface tension would seem to promote formation of larger drops by this mechanism. High viscosity has the opposite effect; however, high viscosity and also high surface tension resist the deforming action of the air stream, and probably result in larger diameter ligaments being formed initially. Eq. 1-2 shows that increasing viscosity increases the size of particle formed from a ligament of given diameter, while increased surface tension tends to reduce it somewhat. Experiment shows that increasing either surface tension or viscosity increases drop size.

A study of the breakup of single, large, water drops is given by Lane [*12*]. In this study, drops were dropped in a steady flow wind tunnel and were also subjected to sudden air blasts. At low air velocities the drops were forced into the shape of a hollow bag which ultimately broke into smaller drops of various sizes. Below a certain critical velocity the drops did not break. This critical velocity was given by the following formula which describes results obtained with 0.5- to 5.0-mm diameter (d) drops placed in a wind tunnel with constant air velocity

$$\Delta u^2 d = 612$$

where Δu is the relative velocity (m/sec) between the drop and the air at the instant of breaking. Investigation of fluids other than water also indicated that

$$\Delta u \sim \sqrt{\frac{\sigma}{d}}$$

If this relationship holds over a wide range of drop diameters one would expect a 45-micron hydrocarbon drop to withstand a relative velocity of approximately 200 ft/sec, so it is apparent that small drops are quite stable. Experiments in which sudden blasts of very high velocity air struck a drop showed that small pieces of fluid were torn from the surface of the drop before the drop was badly deformed; the proportion of disruption taking place in this manner increased with increasing velocity.

In jet engine applications, flow in the vicinity of an atomizing system is generally very turbulent and it is to be expected that drops once formed would tend to collide if their number per unit volume is large. On collision these drops agglomerate into a single larger drop, and since their relative velocity with respect to air has probably diminished they may not break again. An expression for the distance between drops can be derived as follows:

$$\text{the volume of air per drop} = \frac{\pi \dot{V}_a d^3}{6\dot{V}_f}$$

where $\dot{V}_a$ is the volume rate of flow of air and $\dot{V}_f$ is the volume rate of flow of fuel. Assuming that the drop volume is small relative to air volume and assuming cubic spacing of the drops, the distance between drops in terms of drop diameter is

$$l = \left(\frac{\pi \dot{V}_a}{6\dot{V}_f}\right)^{\frac{1}{3}}$$

It is seen that the distance between drops in terms of drop diameters depends only on the volume ratio of air to fuel in the region of interest. Nukiyama and Tanasawa [8] in their study of air atomization found that if this volume ratio was less than 5000 an increase in average drop size was observed. This would correspond to a distance between drops of approximately 14 drop diameters. Under practical operating conditions the volume ratio is often such as to give an average distance between drops of less than 10 drop diameters; so it is believed that recombination of drops is an important phenomenon.

Expressions for drop size distribution. A compact method of reporting the distribution of drop sizes is desirable, and since no analytic method of describing the spray distribution has been developed, various investigators have resorted to empirical formulas for this purpose. A number of such formulas have been used, and detailed comparison and discussion of several of these equations have been presented [6,13]. In Table J,1 the

Table J,1. Summary of distribution equations.

Equation	*Rosin-Rammler*	*Nukiyama-Tanasawa*
$\dfrac{d\chi}{d(d/d_m)}$	$-0.693\kappa \left(\dfrac{d}{d_m}\right)^{\kappa-1} e^{-0.693(d/d_m)^\kappa}$	$ad^5e^{-bd^\kappa}$
χ	$e^{-0.693(d/d_m)^\kappa}$	$1 - \dfrac{\Gamma_{bd^\kappa}(6/\kappa)}{\Gamma(6/\kappa)}$

two formulas that have been given most extensive use in describing atomization data of interest in combustion are given for χ, the fractional mass above a certain drop diameter d, and for $d\chi/d(d)$. (The last term in the Nukiyama-Tanasawa relations is the ratio of the incomplete and complete gamma functions.) Formulas for the various means can be calculated from these equations or can be found in the literature [6,13]. These mean values are sometimes useful in characterizing a spray, but one must generally use the distribution function in calculation of evaporation or combustion rates.

The equation identified as Nukiyama and Tanasawa was used by these investigators [*7,8*] to describe the drop distribution obtained in a carburetor-type air-atomizing nozzle, while the Rosin-Rammler equation has been used in describing the spray from vortex-type atomizing nozzles [*9,10,11*]. In both of these equations the uniformity of the distribution is characterized by the quantity κ, the higher the value of κ the more uniform the distribution.

Distribution equations in which $y = \ln \phi(d)$ is substituted in the normal probability distribution curve have been suggested [*6,13*] but have not been extensively applied to experimental data. One form suggested [*13*] gives a maximum size drop which is more realistic than the prediction of a few very large drops encountered in the other formulas. This equation would be expected to fit the data better for cases where coarse atomization was encountered; however, with the fine atomization generally encountered in jet propulsion engines it is not believed that this refinement is necessary since the other forms, which are simpler to

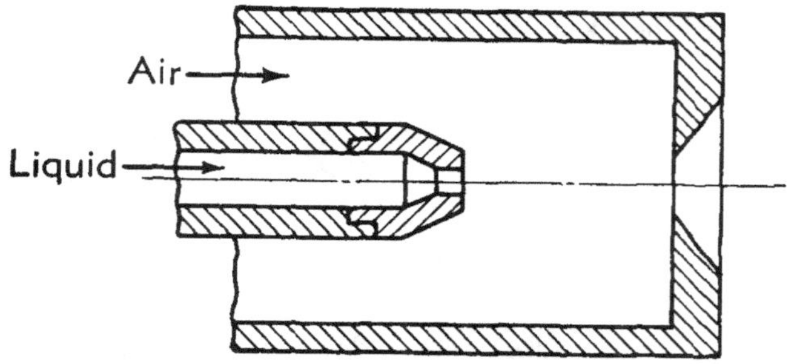

Fig. J,1a. Air blast atomizing nozzle used by Nukiyama and Tanasawa.

handle mathematically, appear to give an adequate description of experimental data for the conditions of interest. Bevans [*6*] found that the Rosin-Rammler equation fitted the experimental data somewhat better for large drops than the Nukiyama and Tanasawa equation, and is probably more convenient to use when mass fractions are of interest. This equation, however, is not satisfactory when the number of drops is being considered since for $\kappa < 3$ it predicts an infinite number of drops of zero radius.

Air atomization. Nukiyama and Tanasawa [*7,8*] have carried out an extensive series of measurements of the drop size distribution obtained with the type nozzle shown in Fig. J,1a. The liquid was introduced into the air stream at a relatively low velocity. The liquid (usually water) and the air then passed through a nozzle. Air velocity was varied from approximately 150 to 1000 ft/sec and the sizes of liquid tube and air nozzle varied from 0.2 to 2 mm and 2 to 5 mm respectively. Samples were generally taken on the axis of the nozzle at a distance of 150 to 250 mm.

The samples were collected on glass slides coated with a viscous oil when water drops were being collected. The slides were exposed for approximately $\frac{1}{300}$ sec, and at least 500 drops were measured for each operating condition. The effects of exposure time and location of samples were studied and the values were felt to give representative results.

Assuming that the diameter of the device holding the collecting plate was 1 cm, it is possible to estimate the accuracy of sampling in these experiments by means of Fig. J,2f. The collection efficiency for a mean drop size of 20 microns was found to be 0.67 when using an air nozzle diameter of 3.71 mm and an air velocity of 226 m/sec at the nozzle. This collection efficiency varies with drop size, greater for the larger drops and smaller for the smaller ones. When this effect is considered, one finds that the mean drop size reported is approximately 10 per cent too large and that the correct distribution was not as narrow as reported. At lower air velocities the drops are larger and one finds that the sampling error is smaller. Since most of the data refer to drops greater than 20 microns it is not felt that this source of error is serious.

The results of their studies are summarized by the following empirical formula:

$$d_0 = \frac{586\sqrt{\sigma}}{u\sqrt{\rho}} + 597\left(\frac{\mu}{\sqrt{\sigma\rho}}\right)^{0.45}\left(\frac{1000\dot{V}_{\text{liq}}}{\dot{V}_{\text{a}}}\right)^{1.5} \tag{1-3}$$

where

$$\frac{d\chi}{d(d/d_{\text{m}})} = 286\left(\frac{d}{d_{\text{m}}}\right)^{5} e^{-5.7(d/d_{\text{m}})} \tag{1-4}$$

where d_0 = the surface/volume mean drop diameter, microns
χ = fractional mass of drops having diameters larger than d
u = air velocity at nozzle exit, m/sec
$\dot{V}_{\text{liq}}$, $\dot{V}_{\text{a}}$ = volume rates of flow of liquid and gas respectively at discharge conditions
σ = surface tension, dynes/cm (19–73)
ρ = liquid density, g/cm^3 (0.7–1.2)
μ = liquid viscosity, poises (0.003–0.5)

The mass median drop diameter d_{m}, the size of drop at which the mass of a classified sample is divided into two equal parts, is 13 per cent greater than d_0.

It was found that when the volume ratio of air to liquid exceeds 5000, drop size was a function of the surface tension, liquid density, and the relative velocity between air and liquid. If less air than this was used, the second term becomes of importance and a larger drop size is obtained. A sharp-edged orifice and a cylindrical air tube were also used with no apparent effect on drop size if the same air velocity at the vena contracta was used. The position of the water outlet with respect to the sharp-edged orifice was varied, and no effect was found as long as the water tube did not protrude far enough through the air orifice to place the water in a zone of low air velocity.

Variations of the air orifice diameter and the water inlet tube diameter did not affect the drop size unless $\dot{V}_a/\dot{V}_{liq}$ became less than 5000, in which case the difference was accounted for by the correction given in Eq. 1-3.

In jet propulsion equipment the fuel stream is commonly injected into a large stream of high velocity air. This situation differs from that found in the above measurements in that the air velocity is sustained after injection, as contrasted to the rapid reduction in velocity of the relatively small jet of air exhausting into a still chamber. It is possible that smaller drops would be obtained by maintaining the air velocity as in a jet engine,

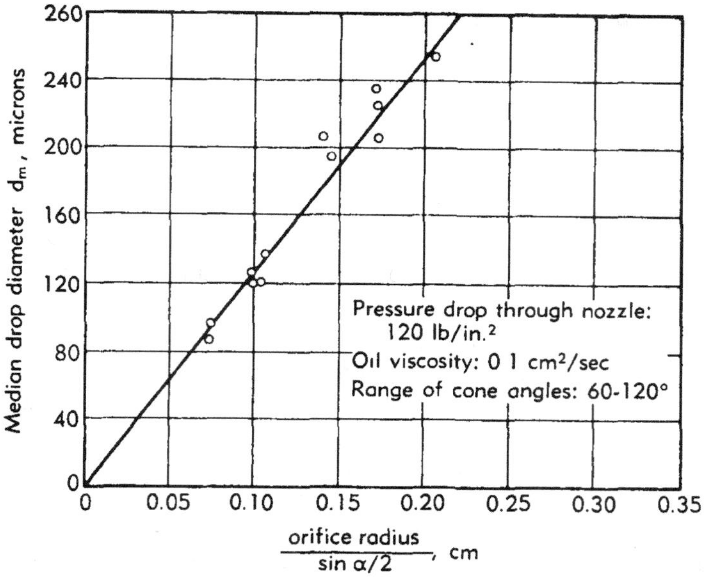

Fig. J,1b. Median drop diameter vs. orifice radius/sin $\alpha/2$.

and measurements taken under these conditions would be of considerable interest.

The effect of air density and air temperature were not studied; however, a few runs made at the University of Illinois [7] indicate that drop size increases with decreasing air density and decreasing air temperature. More data are needed on these variables as well as on atomizers mounted in a large duct.

As an example of the size of drop predicted by Eq. 1-3 it is estimated that a mean drop size of 45 microns would be obtained in a 300-ft/sec air stream when kerosene is the liquid. This is considerably smaller than the drop size to be expected from pressure-atomizing nozzles operating in still air where mean drop sizes of 100 to 200 microns are commonly found.

Pressure-atomizing nozzles. Atomization by swirl-type pressure-atomizing nozzles has been studied in some detail [*9,10,11*]. These results were obtained discharging the nozzles into still room temperature air and, therefore, represent conditions different from those obtained in jet engines where a high velocity hot air stream generally impinges on the nozzle.

Longwell [*10*] studied atomization with this type of nozzle by spraying hydrocarbon fuels and freezing part of the spray. The size distribution was then determined by sieving. As shown in Fig. J,1b the mass median drop diameter increased linearly with nozzle orifice diameter divided by the sine of half the cone angle, when pressure and viscosity are held constant. The results of the investigation are summarized by the following empirical formulas and Fig. J,1c.

$$\frac{d_m \sin \alpha/2}{r_0} = \frac{0.72e^{0.70\mu/\rho}}{\Delta p^{0.37}} \tag{1-5}$$

$$\chi = e^{-0.693(d/d_m)^\kappa} \tag{1-6}$$

where

d_m = mass median drop diameter
d = drop diameter
r_0 = radius of the nozzle orifice—units same as chosen for drop diameter—(0.04–0.14 cm)
Δp = pressure drop through nozzle, psi (50–300)
μ = viscosity, cm²/sec (0.08–0.8)
χ = fractional mass of drops having a diameter greater than d
κ = diversity factor (see Fig. J,1c)
α = cone angle (60–120°)

Viscosity has very little effect at low values of d_m but increases drop size 10 per cent when it reaches 0.13 poises. While jet fuels such as kerosene and the wide cut type have low viscosities at room temperature, under extremely cold conditions their viscosity can sometimes become sufficiently large to adversely affect atomization. When fuel is atomized by spraying into hot gases it is heated as the drops are formed, which reduces the viscosity. Spraying into colder air has the reverse effect. This is illustrated by measurements made with oils of identical viscosity, 0.22 poises, except that one of the oils had its low viscosity by virtue of being heated to 200°F. When sprayed into room temperature air the heated oil gave a median drop size of 186 microns while the unheated oil gave a drop size of 144 microns.

A similar study has been made by Joyce [*11*] by spraying molten paraffin and sieving the solidified drops. These data showed essentially the same dependence of drop size on pressure as found in Longwell's investigation; however, drop size was found to depend on orifice diameter to the 0.6 power as contrasted to the first power dependence found in

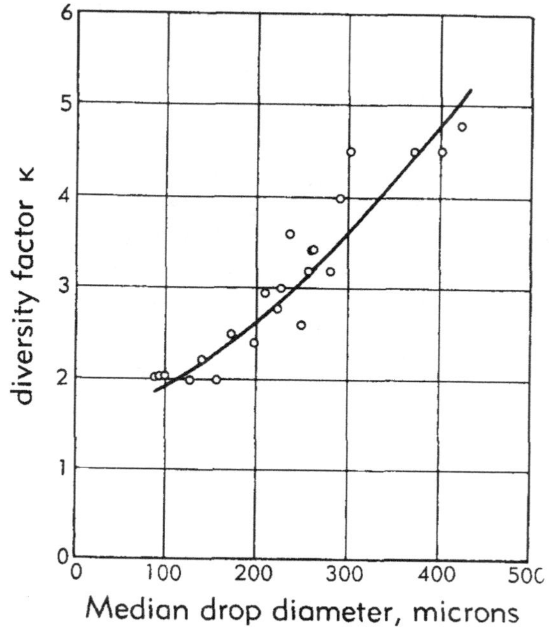

Fig. J,1c. Diversity factor for pressure-atomizing nozzles.

Eq. 1-5. In the range of 75 microns, drop size predictions by Joyce were found to be about 60 per cent too large; for nozzles that gave 200 to 300 microns, agreement was fairly good. It is suspected that the increase in viscosity of the paraffin as it cooled down on contacting the air made atomization more difficult and increased the drop size.

J,2. Mixing and Precipitation of Fuel Sprays. In order for combustion to take place, air and fuel molecules must be intimately mixed on a molecular scale. Molecular diffusion, which ultimately accomplishes this mixing, is sufficiently rapid only if the diffusion must take place across a very short distance. It is not practical to have fuel injection points spaced this closely; hence one must rely on spraying the fuel into the air stream, and on mixing caused by the turbulent flow encountered in jet engines, to spread the fuel drops throughout the air stream in such a manner that molecular diffusion can take place with sufficient rapidity.

Quantitative knowledge of the mixing process is of particular importance in cases where a large excess of air, over that required for stoichiometric proportions of fuel and air, is available. Combustion will probably not take place at all if more than twice the stoichiometric amount of air is mixed with the fuel before burning; so it is important in such a case to restrict the mixing of fuel and air until combustion is complete.

Fuel drops carried by the air stream tend to collect on objects placed in the air stream and on walls, so that this mechanism of "unmixing" is also of practical interest.

Mixing by turbulent diffusion. Diffusion by the random motion of turbulent velocity fluctuations is closely analogous to diffusion as encountered on a molecular scale or in Brownian movement of small particles suspended in a liquid or gas. A basic difference is that molecular diffusion is considered to be made up of small excursions suddenly interrupted by a collision with another molecule, while the path of a small volume of gas in turbulent flow is a continuous though random motion. The theory of diffusion by continuous movements has been discussed by G. I. Taylor [*14,15*]. It is concluded that diffusion in isotropic turbulent flow can be described by the following equations:

$$-\frac{\partial N}{\partial t} = E\frac{\partial^2 N}{\partial x^2} \tag{2-1}$$

where

$$E = L\sqrt{\overline{u^2}}$$

$$L = \int_0^T R_T dT$$

$$R_T = \frac{\overline{u_t u_{t+T}}}{\sqrt{\overline{u_t^2}}\sqrt{\overline{u_{t+T}^2}}}$$

It is seen that diffusion is described by the same differential equation used for molecular diffusion and that the eddy diffusivity E is proportional to a length depending on the scale of turbulence, L, times a velocity (turbulent intensity), which is quite analogous to expressions derived from kinetic theory for molecular diffusion where the diffusion coefficient is proportional to the mean free path times the root mean square molecular velocity. The magnitude of eddy diffusivity E is generally more than 100 times the magnitude of molecular diffusivity.

The above relations apply over a time interval sufficient to allow a fairly large number of fluctuations [*15*]. If a very short time interval is considered, no changes in direction of motion of the diffusing material occur, and under these conditions the average distance traveled by the diffusing material will depend only on the intensity of the turbulence.

These considerations apply to the diffusion of a small volume of material of density comparable to that of the main stream fluid. A drop of fuel carried by a turbulent air stream has a density approximately 500 times that of the air, and it is to be expected that the inertia of the drop will interfere with its ability to follow the fluctuations in air flow. The importance of the effect can be very roughly evaluated by assuming that the velocity fluctuations in turbulent flow are sinusoidal in nature and that the drag on the particle follows Stokes' law. The equation of motion of a drop under such circumstances would be:

$$m\frac{d^2s}{dt^2} = 3\pi\mu d_0\left(u_0 \cos \omega t - \frac{ds}{dt}\right)$$

where s is the distance of the drop from its mean position and u_0 is the maximum air velocity relative to its time average velocity. If one lets $k = 3\pi\mu d_0/2m$, a solution of this equation is

$$s = \frac{2ku_0(2k \sin \omega t - \omega \cos \omega t)}{\omega^3 + 4k^2\omega} \tag{2-2}$$

If s_0 is the amplitude of the gas motion relative to its average position then

$$s_0 = \int_0^{\pi/2} u_0 \sin \omega t \, dt$$

where

$$u = u_0 \sin \omega t$$

so

$$s_0 = \frac{u_0}{\omega} \tag{2-3}$$

Dividing Eq. 2-2 by Eq. 2-3 we have

$$\frac{s}{s_0} = \frac{4k^2 \sin \omega t - 2\,k\omega \cos \omega t}{\omega^2 + 4k^2} \tag{2-4}$$

From Eq. 2-4 one obtains by setting $d(s/s_0)/dt = 0$

$$\frac{s_{max}}{s_0} = \left(\frac{4k^2}{\omega^2 + 4k^2}\right)^{\frac{1}{2}} \tag{2-5}$$

The above equation gives the ratio of amplitude of drop motion to the amplitude of the motion of the gas. The frequency of the motion of the drop is the same as the frequency of the air velocity fluctuation so that the velocity of the drop is reduced in proportion to the reduction in amplitude and, since the diffusion coefficient is proportional to the product of velocity and amplitude of motion,

$$\frac{E}{E_0} = \left(\frac{s_{max}}{s_0}\right)^2 = \frac{4k^2}{\omega^2 + 4k^2} \tag{2-6}$$

where E is the diffusivity of the drop and E_0 is the eddy diffusivity of the gas in the turbulent stream.

As an example of the magnitude of this effect, a 300-ft/sec air stream at 1 atmosphere pressure and 45-micron kerosene drops were assumed. The frequency ω was estimated to be 300 radians/sec for a 6-inch diameter duct assuming fully developed pipe turbulence. For these conditions $E/E_0 = 0.35$, showing a large reduction in diffusivity for conditions of interest in jet engines. This conclusion is supported by experimental data discussed later in this section.

The distribution of fuel given by solution of Eq. 2-1 can give a time average fuel concentration at a given point which is useful for many purposes; however, it does not give information on the fluctuation of fuel concentration with time at a given point which would be necessary to judge accurately the quality of mixing for purposes of combustion. One would expect that the larger the scale of the turbulence the less intimate the mixing; however, one would also expect that increasing the turbulent intensity would tend to improve the intimacy of mixing.

Turbulence under the complex flow conditions encountered in jet engines is generally not isotropic which adds to the complexity of rigorous treatment. The lack of information on turbulence in ducts, as found in jet engines, prohibits more than qualitative use of turbulence theory at the present time; however, a number of measurements of eddy diffusivity have been made under conditions of interest to designers of jet engines. These measurements [*16,17,18*] were made by studying the spreading of vapors or liquid drops, downstream from a single injection point located in the center of a circular duct through which air was flowing at a high velocity. By expressing the number of drops or amount of fuel vapor in terms of fuel-air ratio f, a material balance on a small volume of air gives in cylindrical coordinates using Eq. 2-1

$$\frac{\partial f}{\partial x} = \frac{E}{u}\left(\frac{1}{r}\frac{\partial f}{\partial r} + \frac{\partial^2 f}{\partial r^2}\right) \tag{2-7}$$

It is assumed that diffusion in the axial, x, direction can be neglected, that f depends on r and x only, and that E, the eddy diffusivity, and u, the main stream velocity, are constant.

Integration of this equation for a duct of infinite diameter gives [*19*]

$$f = \frac{\dot{w}_f u}{4\pi \dot{w}_a' E x} e^{-r^2 u/4Ex} \tag{2-8}$$

where $\dot{w}_f$ is the weight rate of fuel flow and $\dot{w}_a'$ is the weight rate of air flow per unit duct area.

This equation describes quite accurately the measured distributions [*16,17,18*]. An example of this agreement is shown in Fig. J,2a where log of fuel-air ratio is plotted as a function of R^2. This should give a straight line with a slope of $-u/4Ex$ and it is seen that the ata points fit a straight line over a considerable portion of the duct, indicating that in spite of variations of E and u with radius the assumption of constant u/E gives an accurate description of the distribution. Eddy diffusivities are therefore frequently determined by the use of this equation; however, in doing this the source of diffusing material must be a reasonable approximation to a point source.

Unfortunately, measurements of correlation coefficients and intensities of turbulence were not made in these experiments [*16,17,18,19,20*]. However, some of the tests [*16,17*] were made in long ducts which should have given turbulence approaching equilibrium pipe turbulence. Sherwood and Towle [*16*] and McCarter, et al. [*20*] found that if a fine mesh screen were placed upstream of the point of injection of diffusing material, the diffusivity was reduced by as much as a factor of 2, and 30 to 40 duct diameters were required for it to return to its original value. Although one

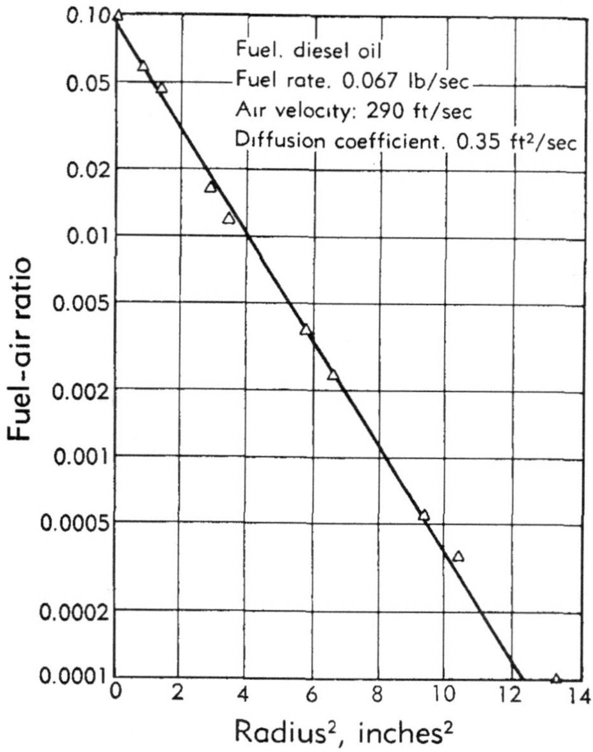

Fig. J,2a. Fuel distribution 38 inches downstream from a fuel injection point in a 10-inch duct.

might expect some increase in intensity immediately downstream from the screen, the pronounced reduction in scale appears to be predominant in its effect. In this case, although the fuel might not be spread as far radially, the intimacy of mixing might be better despite the lower diffusivity.

According to measurements by Simmons [*21*], the scale of turbulence in the center of a pipe is approximately 0.17 times the pipe radius, and the intensity is approximately 3 per cent of the main stream velocity. One would then expect that the diffusivity would be proportional to the duct diameter. Sherwood and Towle [*16*] find that this is the case for their tests in long 6- and 12-inch ducts; the diffusivity is twice as large for the 12-inch duct. The scale of turbulence, however, is very sensitive to the length of the duct (for short ducts) and to the nature of the flow preceding the sections of interest. One should not assume that changing the diameter of an engine combustion chamber will necessarily make a proportionate change in diffusivity without giving careful consideration to the other changes in flow conditions.

If intensity is a constant fraction of the average flow velocity one would expect the diffusion coefficient to be proportional to velocity, and the ratio E/u which appears in the distribution equation (Eq. 2-8) to be independent of velocity. This seems to be approximately the case for the situations on which data are available [*16,17,18,20*]. At the high Reynolds numbers encountered in jet engines (10^5 to 10^6) one would not expect changes in air density or temperature to have a large effect on diffusivity. This proved to be the case [*17*] where a change in air density by a factor of 10 produced only minor changes in diffusivity. It then appears that for a given geometry, E/u is nearly constant which, according to Eq. 2-8, means that the fuel distribution patterns will be essentially independent of operating conditions, assuming that the manner in which the fuel is injected does not change.

The above conclusions apply to a vaporized fuel, but, as shown by Eq. 2-6, one does not expect drops to follow the same rules since the rate of spreading will also depend on drop size, frequency of turbulence, etc. The larger the drops and the higher the frequency of the turbulent velocity fluctuations the lower the diffusivity will be, compared to that of vaporized fuels. For the same drop size one might therefore expect E/E_0 to increase with increasing duct diameter and to become smaller with increasing main stream velocity. Since the air stream velocity, density, and air-fuel ratio all affect atomization, detailed knowledge of both atomization and turbulence are necessary for prediction of diffusivity of fuel drops. A few data are available to compare the diffusivity of sprays with that of a vaporized fuel. Typical data [*17*] are given in Fig. J,2b where E/u is plotted as a function of air velocity for a light hydrocarbon fuel which vaporized a short distance downstream from the point of injection, and for diesel fuel which remained essentially unevaporated. It is seen that the two diffusivities approach each other at the lower velocities, but that at high velocities the diffusivity of diesel fuel drops becomes much lower than for the vaporized fuel. Atomization in these runs was accomplished by flowing the liquid fuel at low velocity through a tube pointed upstream. The impact of the air stream on the fuel jet was responsible for atomization.

For purposes of prediction of fuel spray characteristics, a value of E/u of 0.001 ft^{-1} has been found reasonable for smooth flow through a smooth duct 6 inches in diameter using a fuel that vaporizes. When flow is disturbed by obstructions or for other reasons, a value more than twice that for smooth flow is sometimes observed. From Fig. J,2b it is apparent that lower values of E/u should be used when the fuel does not vaporize.

If the diffusion coefficient E is known, the spreading of a spray in an air stream can be described provided the initial distribution of the spray in the air is known. The simplest case is the point source for which Eq. 2-8 applies. The point source is a good approximation if the fuel enters the air stream at a very low velocity through a small tube. However, if the velocity at which the fuel enters the air stream is large, the

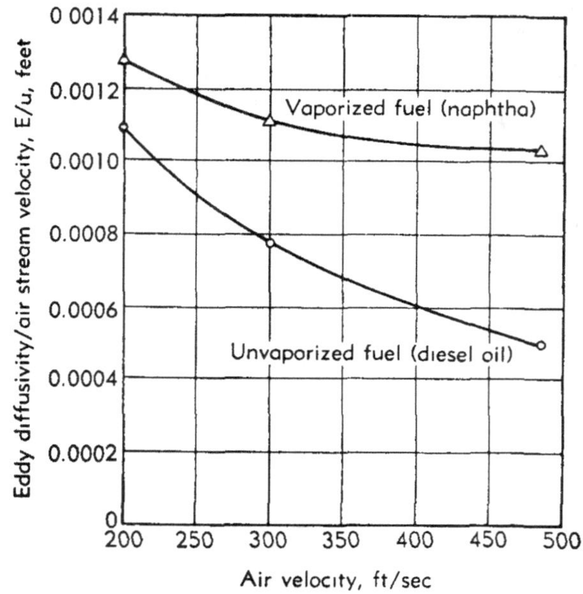

Fig. J,2b. Effect of air velocity on diffusion parameter, E/u. Distance from point of injection, 17 in.; duct diameter, 6 in.; air temperature, 300°F.

fuel jet penetrates the air stream, and the initial fuel distribution will depend on fuel velocity and the configuration of the injection device. The simplest device for injecting fuel into an air stream is a straight cylindrical tube. Where such a device is pointed downstream a point source is a good approximation, but when it is pointed upstream and the fuel has a high enough velocity to penetrate the air stream the spray is turned back and spread out radially. This is also the case for swirl-type atomizing nozzles or pintle-type nozzles which send out a cone-shaped spray. A fuel spray can be described by assuming that it issues from a number of point sources (Eq. 2-8) properly distributed to approximate the initial distribution [17]. The source of fuel for these cases can be approximated by assuming that these point sources are located on the circumference of a circle. The fuel distribution downstream from such a "ring" source is given by the following equation which is derived by adding the contribution of an infinite number of point sources evenly distributed around the circumference of a ring of radius R_0.

$$f = \frac{\dot{w}_f}{\dot{w}'_a R_0^2} \frac{K}{\pi} \left[I_0 \left(2K \frac{r}{R_0} \right) \right] e^{-K(1+r^2/R_0^2)} = \frac{\dot{w}_f}{\dot{w}'_a R_0^2} \phi \tag{2-9}$$

$$K = \frac{uR_0^2}{4Ex} \tag{2-10}$$

The function ϕ is plotted in Fig. J,2c vs. r/R_0 and K, and the fuel distribution downstream from a ring source can be found from this figure

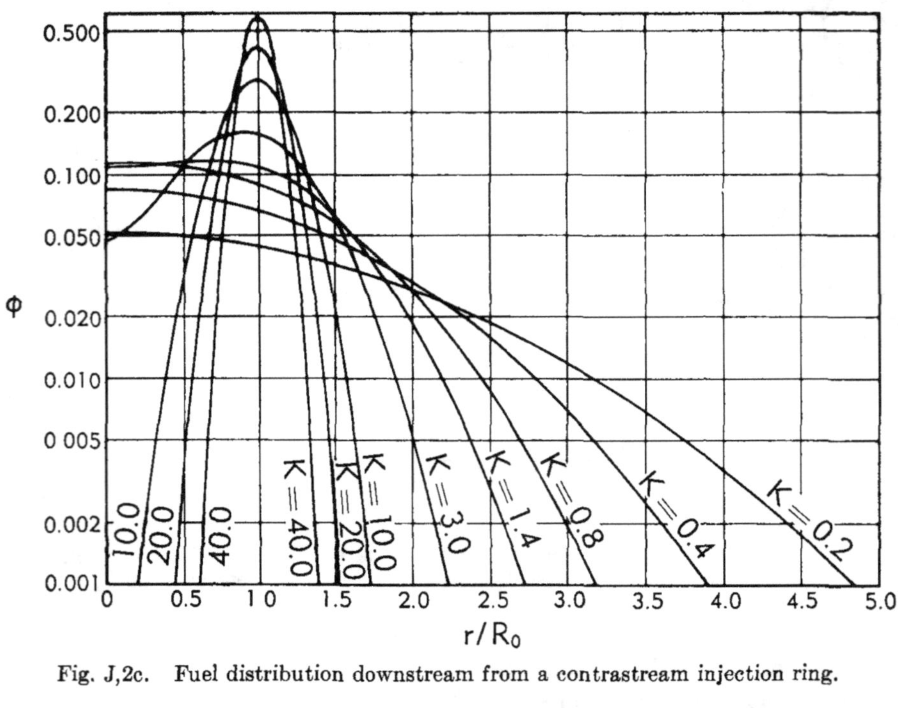

Fig. J,2c. Fuel distribution downstream from a contrastream injection ring.

and Eq. 2-10 if the various constants are known. It has been found that the radius of the spray observed visually or scaled from a photograph can be used for R_0. A number of measurements of R_0 have been made for contrastream injection tubes, and the results of these measurements are given in Fig. J,2d. It is seen that the value of R_0 for a wide variety of conditions can be correlated by plotting the ratio of R_0 to the fuel jet radius r_1 as a function of the momentum of the air stream flowing through an area equal to the initial fuel jet area divided by the momentum of the fuel jet $\rho_a u^2/\rho_f u_f^2$. The spray from other types of injection equipment can

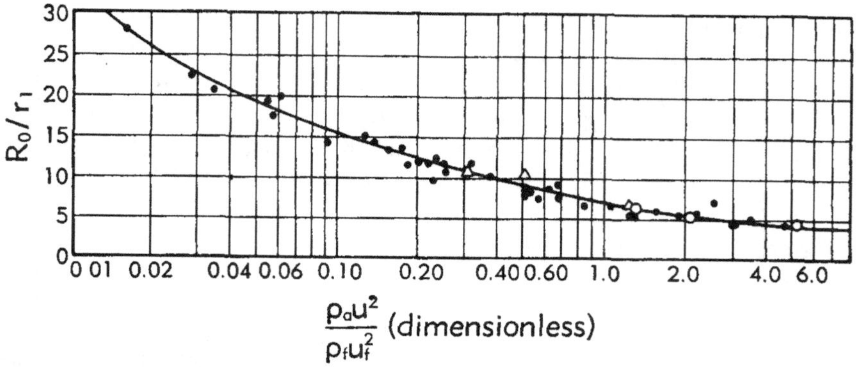

Fig. J,2d. Initial radial penetration of a contrastream "open tube" fuel spray. Open points were determined from measured distributions. Solid points were scaled from photographs (maximum diameter not more than 2 in. downstream from tip of tube).

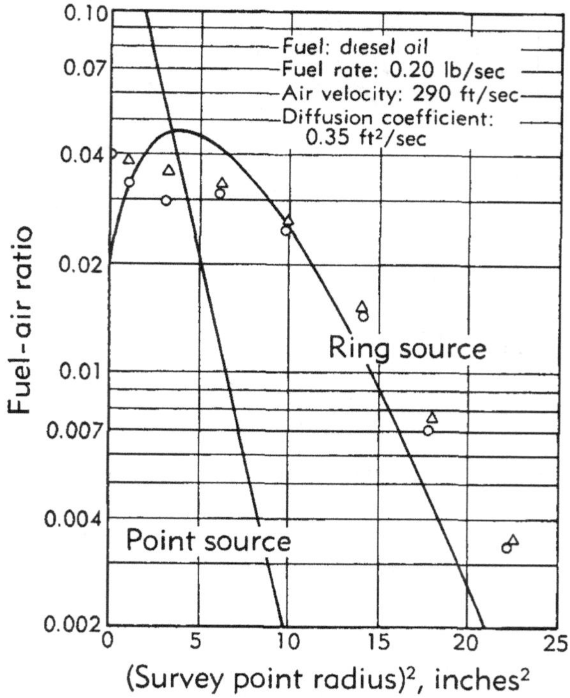

Fig. J,2e. Fuel distribution 35 inches downstream from a contrastream spray nozzle, 10-inch duct.

probably be similarly described; however, if data are lacking, a few observations of the size of the spray in a high speed air stream will give the desired information.

The need for this means of describing a fuel spray is shown in Fig. J,2e where the fuel distribution downstream from a contrastream injection tube is shown. The curve described by Eq. 2-10 is seen to describe the distribution fairly well, but a point source is seen to be unsatisfactory, although it would probably do if a fictitiously high diffusivity were used.

If a jet engine using a straight tube injector were to climb to a high altitude so that air density was reduced but air-fuel ratio held constant, it would follow from Fig. J,2d that a smaller value of R_0 would be obtained at high altitudes. This would result in a narrower distribution of fuel since E/u would probably remain constant.

Precipitation of fuel sprays. It is frequently desirable to know the amount of liquid that will collect on an object mounted in a flowing mixture of fuel drops and air. This is of interest in research where one is sampling such a mixture to determine drop size or fuel distribution, and

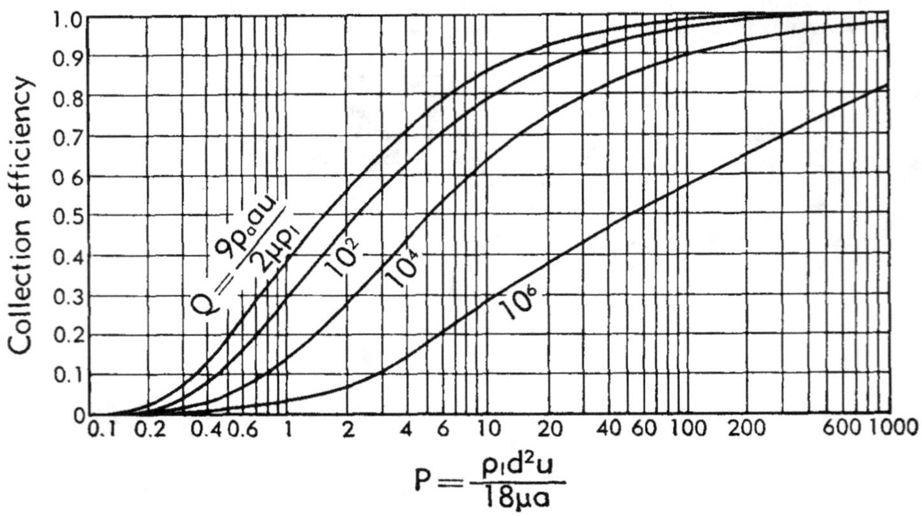

Fig. J,2f. Collection efficiency of spherical particles on cylinders.

in the development of combustors, since the amount of liquid collecting on spark plugs, walls, and flame-stabilizing elements sometimes has an important effect on the combustion. The calculations of drop trajectories made by Langmuir and Blodgett [*22*] are useful in such problems. The fraction of drops, within a volume swept out by a cylinder moving through air, that will strike the leading edge of the cylinder has been presented in generalized terms so that one can easily determine this collection efficiency for many situations of interest.

In calculating the collection efficiency, potential flow around the cylinder was assumed. For the moderately high velocity conditions encountered within jet engines, this is felt to be a good approximation of the flow around the leading edge of the cylinder, which is the region of interest. The velocity field was assumed uniform an infinite distance upstream from the cylinder and in that region the drops were assumed to have no motion with respect to the air. When the streamlines curve because of the cylinder, a relative velocity between the drop and the air exists, and a drag force is exerted on the drop. This force is expressed as follows:

$$F = \frac{\pi C_D Re \mu \Delta u d}{8} \tag{2-11}$$

$$\frac{C_D Re}{24} = 1 + 0.197 Re^{0.63} + 2.6 \times 10^{-4} Re^{1.38} \tag{2-12}$$

$$Re = \frac{d \Delta u \rho_a}{\mu}$$

At very low values of Reynolds number, Eq. 2-11 reduces to Stokes' law. The equations of motion were solved on a differential analyzer and solution for collection efficiency is given in Fig. J,2f where the collection efficiency η is plotted as a function of the two dimensionless groups:

$$P = \frac{\rho_{liq} d^2 u}{18 \mu a} \tag{2-13}$$

$$Q = \frac{9 \rho_a a u}{2 \mu \rho_{liq}} \tag{2-14}$$

where u is the velocity of the undisturbed air stream relative to a cylinder of radius a.

Collection efficiency increases with increasing P and decreases with increasing Q. The larger drops are therefore more easily collected, and the larger the cylinder the lower the collection efficiency. Similar calculations have also been made for drop deposition on spheres and ribbons [*22*].

J,3. Evaporation of Fuel Drops.

Equilibrium vaporization. The mechanism of combustion and its rate depend on whether the fuel is vaporized or unvaporized. It is therefore of interest to determine the extent to which fuel injected into the air stream of an engine will vaporize. The amount of vaporization is ultimately limited by the equilibrium conditions as determined by fuel volatility, the enthalpy of the air, and the fuel. While sufficient time for attainment of equilibrium is usually not available, calculation of equilibrium vaporization is of interest in approximating the maximum amount of vaporization that can be anticipated.

If the constituents of a fuel are known, the equilibrium vaporization can be calculated by assuming an adiabatic mixing process of air and fuel of known enthalpy. For this case one can write the following equations:

Conservation of energy

$$w_iH_i + (w_a)_i(H_a)_i = (w_{liq})_1(H_{liq})_1 + (w_{liq})_2(H_{liq})_2 + \cdots + (w_{liq})_n(H_{liq})_n + (w_f)_1(H_f)_1 + (w_f)_2(H_f)_2 + \cdots + (w_f)_n(H_f)_n + w_aH_a$$

$$(H_f)_1 = \psi_1(T), \qquad (H_f)_2 = \psi_2(T), \qquad \text{etc.} \tag{3-1}$$

$$(H_{liq})_1 = \phi_1(T), \qquad (H_{liq})_2 = \phi_2(T), \qquad \text{etc.}$$

where all symbols with subscript $_a$ refer to air, w is the number of moles, and H is the molar enthalpy of the respective constituents at equilibrium, and subscripts $_{liq}$ and $_f$ designate liquid and vapor phase respectively.

Conservation of matter

$$w_1 = (w_{liq})_1 + (w_f)_1 \qquad w_2 = (w_{liq})_2 + (w_f)_2 \qquad \text{etc.} \tag{3-2}$$

Equilibrium relationship

$$p_1 = K_1 \frac{(w_{liq})_1}{(w_{liq})_1 + (w_{liq})_2 + \cdots + (w_{liq})_n} \qquad \text{etc.} \tag{3-3}$$

$$p_a + p_1 + p_2 + \cdots + p_n = p \tag{3-4}$$

$$\frac{p_1}{p} = \frac{(w_f)_1}{w_f + w_a}, \qquad \frac{p_2}{p} = \frac{(w_f)_2}{w_f + w_a} \qquad \text{etc.} \tag{3-5}$$

One has $5n + 4$ equations and $5n + 4$ unknowns, where n is the number of components in the fuel; so, if the necessary physical properties and fuel analysis are available, it is possible to determine the composition of the vapor and the liquid phase of such a mixture. For hydrocarbon fuels a very large number of compounds are present in the fuel, and an analysis giving the constituents of these fuels is seldom available. Even if such analysis were available, solution of this large number of equations would be a laborious task. One is therefore interested in simplified methods of determining equilibrium vaporization. Distillation data for fuels are usually available or are easily obtained and can be used as a basis for estimating equilibrium vaporization if it is assumed that the fuel is arbitrarily divided into several portions of reasonably narrow boiling range. These portions can then be treated as pure compounds, using average properties and determining the equilibrium constants by means of Raoult's law which sets K of Eq. 3-3 equal to the vapor pressure of the fraction of interest at the mixture temperature. Average physical properties for such a treatment can be obtained from [*23*]. This technique is also quite laborious because of the large number of equations; so, empirical methods relating the fraction of fuel vaporized as a function of mixture temperature (flash vaporization curve) to the ASTM distillation have been developed [*23*]. The method recommended in [*23*] makes use of several empirical charts which are available in the above reference. Typical results of a calculation making use of this method are shown in Fig. J,3a where fraction vaporized is plotted as a function of pressure and air temperature for a stoichiometric mixture. It is seen that, for this kerosene, vaporization is complete at stoichiometric air-fuel ratios above a temperature of 260°F at a pressure of one atmosphere. At lower pressures, lower temperatures are required for complete vaporization. An interesting phenomenon is observed when one considers flight at constant Mach number or compression ratio and estimates fraction vaporized as a function of altitude. As one climbs from sea level the effect of reduced temperature is greater than the effect of reduced pressure and the ability of fuel to vaporize decreases, but when one reaches the stratosphere

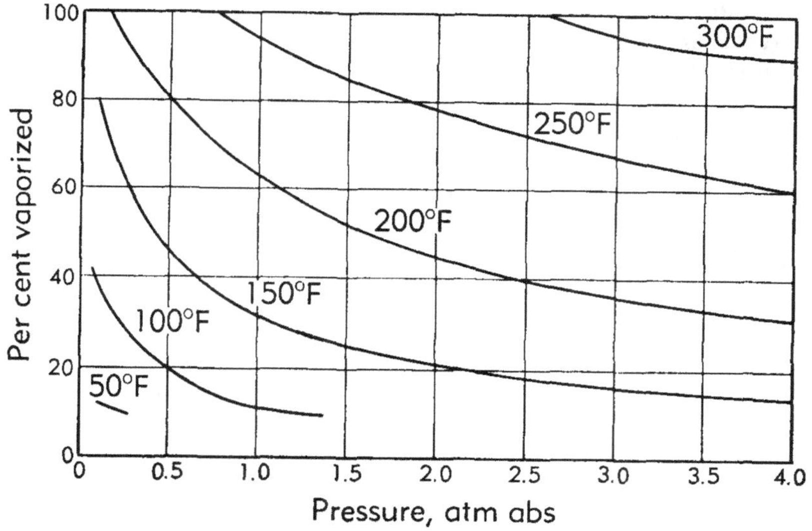

Fig. J,3a. Equilibrium vaporization of kerosene as a function of pressure and temperature.

where the temperature is constant, the pressure continues to decrease which allows an increase in the amount of fuel vaporized. One finds therefore that poorest vaporization conditions generally occur in the region of 30,000 to 40,000 ft altitude.

Evaporation rate. Calculation of evaporation rate is considerably more difficult than calculation of equilibrium vaporization. One needs to know the size distribution of drops and details of the flow of surrounding gas past the drop, in addition to accurate information on physical characteristics of the hydrocarbon liquid and vapors. It is, in principle, possible to calculate the evaporation rate; however, rigorous solution of this problem would be very complex. The approximate treatment which follows, however, illustrates many of the major effects of interest in combustor design.

It is first assumed that steady state conditions exist. This assumes that the concentration and temperature gradients around the drop can adjust themselves to the steady state condition rapidly, compared to their rate of change with time as the drop evaporates. Under these circumstances the rate of evaporation is determined by the rate at which heat can be supplied to the drop, this heat being used to supply the latent heat of evaporation. Since it is necessary for the evaporated material to diffuse away from the surface of the drops, the rate of diffusion depends on the partial pressure of fuel vapors at the drop surface and therefore on the drop temperature. Here it is further assumed that the evaporation rate is sufficiently slow so that the vapors in the immediate vicinity of the drop surface are in equilibrium with the drop liquid. The drop reaches an equilibrium temperature below the boiling point of the drop, this temperature being determined by requirements of heat transfer and diffusion.

An important simplification is made by the assumption that concentration and temperature are functions only of distance from the center of the drop. This is not an accurate description of conditions in the vicinity of the drop; however, this assumption has been successfully used by chemical engineers [*24*] to describe heat and mass transfer phenomena for many conditions, and will be employed here.

Net heat transfer rate, dq/dt, at a given radius is described as follows:

$$\frac{dq}{dt} = -kA\frac{dT}{dr} + m_0 c_p T \tag{3-6}$$

The first term is the usual expression for heat transfer by conduction while the second term gives the amount of heat carried by the diffusing fuel vapors, a consideration of importance at high rates of evaporation. A heat balance across a differential change in r gives for steady state conditions:

$$\frac{m_0}{4\pi}\frac{d(c_pT)}{dr} = r^2\frac{d}{dr}\left(k\frac{dT}{dr}\right) + 2rk\frac{dT}{dr} \tag{3-7}$$

While c_p and k change appreciably with temperature, it is assumed that they are constant. Integration then gives the temperature

$$\frac{T_1 - T}{T_1 - T_0} = \frac{e^{\alpha r_0/r_1} - e^{\alpha r_0/r}}{e^{\alpha r_0/r_1} - e^{\alpha}} \tag{3-8}$$

$$\alpha = \frac{-m_0 c_p}{4\pi k r_0}$$

The heat entering the liquid drop serves to supply latent heat of evaporation and is therefore related to the evaporation rate as follows:

$$\lambda m_0 = 4\pi k_0 r_0^2 \left(\frac{dT}{dr}\right)_{r=r_0}$$

By differentiating Eq. 3-8, substituting for $(dT/dr)_{r=r_0}$, and solving for the evaporation rate one obtains

$$m_0 = \frac{4\pi r_0 r_1 k}{c_p(r_1 - r_0)} \ln\left[1 + \frac{k_0}{k}\frac{c_p}{\lambda}(T_1 - T_0)\right] \tag{3-9}$$

c_p and k are average values over the range of temperatures from T_0 to T_1, while k_0 is evaluated at temperature T_0.

Eq. 3-9 gives the rate of evaporation as a function of temperature, taking into account the effect of the mass transfer which is counter to the direction of heat flow. When $(k_0/k)(c_p/\lambda)(T_1 - T_0) \ll 1$, then the effect of mass transfer can be neglected and Eq. 3-9 reduces to

$$m_0 = \frac{4\pi k_0 r_0 r_1}{\lambda(r_1 - r_0)}(T_1 - T_0) \tag{3-10}$$

or for $r_1 \gg r_0$,

$$m_0 = \frac{4\pi r_0}{\lambda} k_0(T_1 - T_0) \tag{3-11}$$

According to kinetic theory, diffusion of a binary mixture of gases is described as follows [*25*, Chap. 4]:

$$u_f - u_a = \frac{n^2}{n_f n_a} D \frac{d(n_a/n)}{dr} - \frac{1}{\rho}\Gamma_f \frac{d \ln T}{dr} \tag{3-12}$$

This equation assumes that total pressure gradients are small compared to the magnitude of the pressure and that external forces, such as centrifugal fields, are small. The first term describes the normal diffusion rate and the second the rate of thermal diffusion where Γ_f is the thermal diffusivity. Thermal diffusion is generally quite small compared with normal diffusion so that this term can be neglected. u_f and u_a are velocities of fuel vapor and air, respectively, relative to a common reference—the average velocity of the mixture. Therefore $u_f - u_a$ is the velocity of fuel with respect to air. In this system the air remains stationary with respect to the drop as the fuel diffuses so that $u_f - u_a$[1] also represents the quantity of direct interest—the velocity of fuel with respect to the drop. The result is

$$u_f - u_a = \frac{n^2}{n_f n_a} D \frac{d(n_a/n)}{dr} \tag{3-13}$$

Fuel velocity can also be expressed in terms of the evaporation rate as

$$u_f - u_a = \frac{m_0}{\rho_f A} = \frac{m_0}{\rho_f 4\pi r^2} \tag{3-14}$$

[1] The more commonly used expression $\Delta u_f = -AD[d(n_f/n)/dr]$ applies to the case of two counterdiffusing components with no net flow and follows from Eq. 3-12 when $n_f u_f = -n_a u_a$ and, using the relations $dn_a = -dn_f$, $\Delta u_f/A = u_f(n_f/n)$. Eq. 3-13, which applies to the case of one component diffusing through a stagnant film, reduces to the other case when the concentration of fuel vapors is very low and $n_a \to n$.

Assuming the perfect gas law, both mole fractions and densities can be expressed in terms of partial pressures or

$$\frac{p_f}{\rho_f} = \frac{RT}{\mathfrak{M}_f} \quad \text{and} \quad \frac{n_f}{n} = \frac{p_f}{p}$$

Substituting these relations and Eq. 3-14 into 3-13, we have

$$m_0 = \frac{4\pi r^2 p D \mathfrak{M}_f}{RT} \frac{d(\ln p_a)}{dr} \tag{3-15}$$

If one equates the mass flow, m_0, at r and $r + dr$, then

$$0 = \frac{r d^2(\ln p_a)}{dr^2} + \frac{2d(\ln p_a)}{dr}$$

On integration between r_0 and r_1

$$\ln p_a = \frac{r_0 \ln (p_a)_0[(r_1/r) - 1] + r_1 \ln (p_a)_1[1 - (r_0/r)]}{(r_0 - r_1)} \tag{3-16}$$

Differentiation of this equation and substitution of the resulting expression for $d(\ln p_a)/dr$ into Eq. 3-15 gives

$$m_0 = \frac{4\pi p D \mathfrak{M}_f}{RT} \frac{r_0 r_1}{(r_1 - r_0)} \ln \frac{(p_a)_1}{(p_a)_0} \tag{3-17}$$

If the concentration of fuel vapors is small, Eq. 3-17 reduces to

$$m_0 = \frac{4\pi D \mathfrak{M}_f}{RT} \frac{r_0 r_1}{(r_1 - r_0)} [(p_a)_1 - (p_a)_0] \tag{3-18}$$

and if it is further assumed that $r_1 \gg r_0$, we have:

$$m_0 = \frac{4\pi \mathfrak{M}_f D r_0}{RT} [(p_a)_1 - (p_a)_0] = \frac{4\pi \mathfrak{M}_f D r_0}{RT} [(p_f)_0 - (p_f)_1] \tag{3-19}$$

Eq. 3-19 is Langmuir's equation [27] which is commonly used for describing the evaporation rate of spherical particles in still air under conditions giving low evaporation rates. This equation is analogous in form to Eq. 3-11 which describes the rate of heat transfer under similar conditions illustrating the well-known analogy between mass and heat transfer. It is seen by comparison of Eq. 3-17 and 3-9 that this analogy is not valid for high rates of evaporation, but the effect of drop radius is the same for both cases, the difference being the effect of temperature or concentration.

The temperature T_0 of the drop is necessary to find the rate of evaporation from the heat transfer equations (Eq. 3-9 to 3-11) or from the mass transfer equations (Eq. 3-17 to 3-19). Since $(p_f)_0$ is the partial pressure of fuel in equilibrium with the fuel at temperature T_0, these two sets of equations can be related by knowledge of the effect of temperature on fuel vapor pressure. Detailed data on temperature-vapor pressure relationships can be found in handbooks such as [23].

To aid in evaluating the importance of the effect of mass transfer on heat and mass transfer rates, a calculation of the ratio of evaporation rate to that obtained if this effect is neglected has been made. This ratio τ is defined below by dividing Eq. 3-9 by Eq. 3-10 letting $k_0 = k$.

$$\tau = \frac{\lambda}{c_p(T_1 - T_0')} \ln \left[1 + \frac{c_p}{\lambda} (T_1 - T_0) \right] \qquad (3\text{-}20)$$

T_0' is the temperature of the drop which one would calculate assuming a low rate of evaporation (Eq. 3-9 and 3-19) as contrasted to T_0, the drop temperature obtained by taking this effect into account.

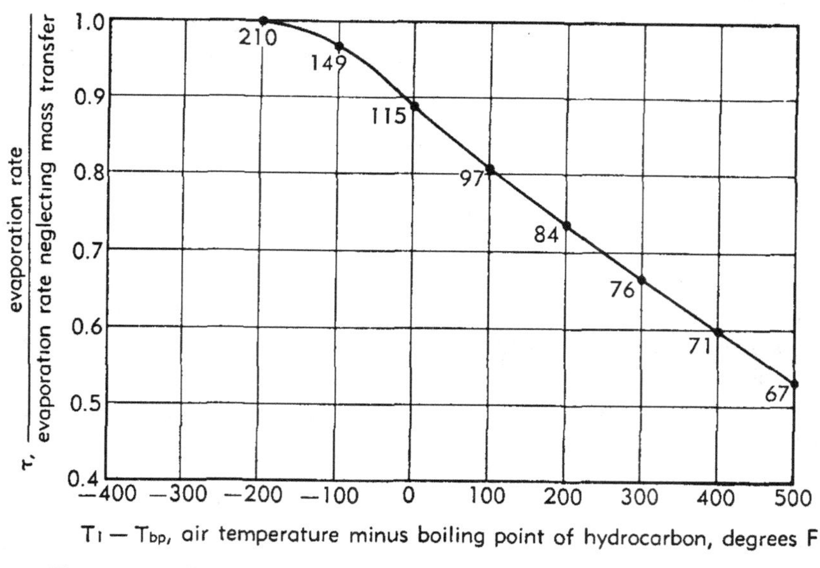

Fig. J,3b. Effect of mass transfer on the evaporation rate of kerosene.

Fig. J,3b gives τ as a function of the temperature difference between air and the average boiling point of a kerosene-type fuel. It is seen that the effect on evaporation rate becomes important when the air temperature is greater than the fuel boiling point; so, it is apparent that the more complex equation should be used in applications such as high speed ramjets and turbojet afterburners where high gas temperatures are encountered. The difference between the boiling point of the fuel and the drop temperature T_0 is marked on the curve. These temperatures are for the case where evaporation rate is taken into account and the total pressure is 1 sea level atmosphere. Total pressure has very little effect on this correction factor; similar curves for $\frac{1}{3}$ and 3 atmospheres are almost indistinguishable from the 1-atmosphere curve.

At very high gas temperature the error involved in using the boiling point of the fuel for T_0 becomes small compared with the uncertainty in the proper value of physical properties, r_1, etc., so that one can then find the evaporation rate directly from Eq. 3-9.

The various expressions for m_0 can be used to calculate the time required for evaporation of a drop. If the weight w of a drop is expressed as follows:

$$w = \tfrac{4}{3}\pi r_0^3 \rho_{\text{liq}}$$

then

$$dw = m_0 dt = 4\pi r_0^2 \rho_{\text{liq}} dr_0 \tag{3-21}$$

The drop temperature T_0 is independent of drop radius for given ambient conditions, so that Eq. 3-21 can be written in terms of radius and time. The radius r_1 at which conditions correspond to ambient conditions will, in general, be a function of r_0 and the ambient conditions; however, if $r_1 \gg r_0$, using Eq. 3-9 to eliminate m_0 one can immediately integrate Eq. 3-2 to obtain

$$t = \frac{c_p \rho_{\text{liq}}[(r_1^2)_0 - r_0^2]}{2k \ln\left[1 + \dfrac{k_0}{k}\dfrac{c_p}{\lambda}(T_1 - T_0)\right]} \tag{3-22}$$

where $(r_1)_0$ is the initial drop radius and r_0 is the radius at time t. The time t for complete evaporation is seen to be a function of the square of the initial drop radius, showing the importance of drop size in determining the evaporation time.

Since a wide variety of fuel drop sizes is normally injected into a practical combustion chamber, it is of interest to consider the evaporation of such a collection of drops. The Rosin-Rammler equation (Table J,1) describes the distribution of drop sizes before the beginning of evaporation. Differentiation of the expression gives the weight fraction $\Delta\chi_{R_0}$ of material originally found in size range Δr_1 at average size $\bar{r}_1$.

$$\Delta\chi_{R_0} = -\frac{0.693\kappa}{\bar{r}_1}\left(\frac{r_1}{\bar{r}_1}\right)^{\kappa-1} e^{-0.693(r_1/\bar{r}_1)^{\kappa}}\Delta r_1 \tag{3-23}$$

The fraction of this material remaining after the average size has been reduced to r by evaporation is

$$\frac{\Delta r}{\Delta\chi_{R_0}} = \left(\frac{r}{r_1}\right)^3 \tag{3-24}$$

Eq. 3-22 gives the r as a function of time for evaporation in still gas:

$$C_1 t = r_1^2 - r^2 \quad \text{where} \quad C_1 = \frac{2k \ln\left[1 + \dfrac{k_0}{k}\dfrac{c_p}{\lambda}(T_1 - T_0)\right]}{C_p \rho_l}$$

Elimination of r from Eq. 3-24 followed by summation of all unevaporated drops gives

$$\chi = -\frac{0.693\kappa}{\bar{r}_1^{\kappa}} \int_{\sqrt{C_1 t}}^{\infty} r_1^{\kappa-4}(r_1^2 - C_1 t)^{\frac{3}{2}} e^{-0.693(r_1/\bar{r}_1)^{\kappa}} dr_1 \tag{3-25}$$

where χ is the fraction unevaporated. Probert [*29*] has graphically integrated this expression for several values of κ. The results of this integration

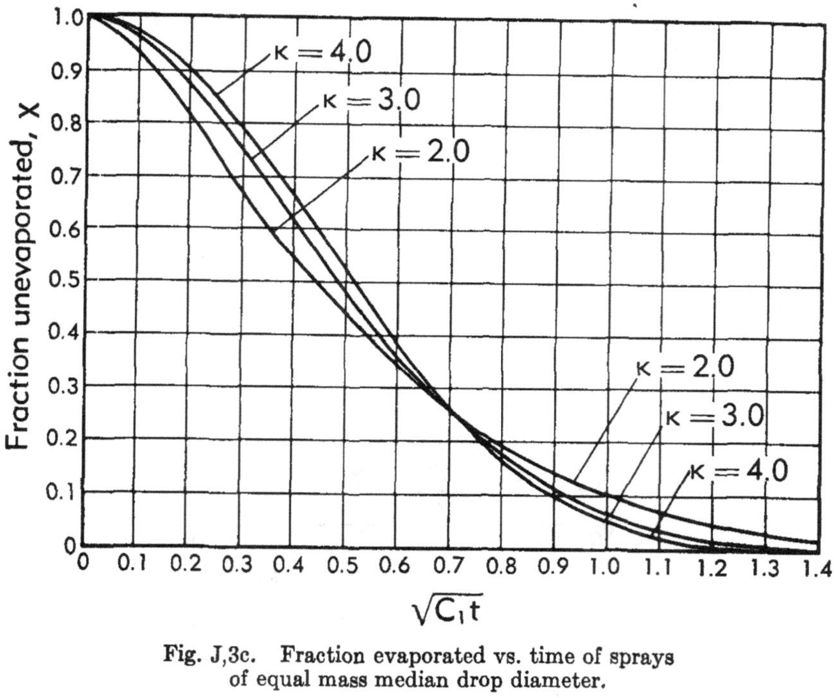

Fig. J,3c. Fraction evaporated vs. time of sprays of equal mass median drop diameter.

are given in Fig. J,3c. Since κ characterizes the width of the drop size distribution and small values of κ denote a broad distribution of sizes, it is seen that the highest initial rate of evaporation is given by the spray of broadest distribution. This is generally advantageous in obtaining ignition of the spray; however, after about 75 per cent of the material has evaporated, the greater number of large drops in the wide distribution spray results in a decreased rate of evaporation. Since burning the last bit of fuel is generally more of a problem than burning the part that evaporates quickly, this residue of large drops is quite important from a practical point of view. These curves illustrate the importance of knowing the distribution of drop sizes in a study of spray evaporation, since some important characteristics are overlooked by use of an average drop size.

When the relative velocity of the main air stream and the drop is sufficiently high to increase the rate of evaporation, one can no longer assume that r_1, the radius where ambient conditions can be assumed to exist, is large. Measurements by Frössling [*28*] plotted in Fig. J,3d show the rate of evaporation of a small naphthalene sphere at various points on its surface expressed in terms of Nusselt's number

$$Nu' = \frac{2k_g r_0 \mathfrak{M}_f p_f}{\rho_f D} = \frac{2r_1}{r_1 - r_0} \tag{3-26}$$

where k_g is defined by the relation $m_0 = \mathfrak{M}_f k_g \pi r_0^2 (p_f - p_{f_l})$. It is seen that for very small Reynolds number, $Nu' = 2$ and $r_1 \gg r_0$. As relative air velocity and Re increase, a quite high transfer coefficient occurs on the upstream portion of the drop. This rate decreases to a low value 120 to

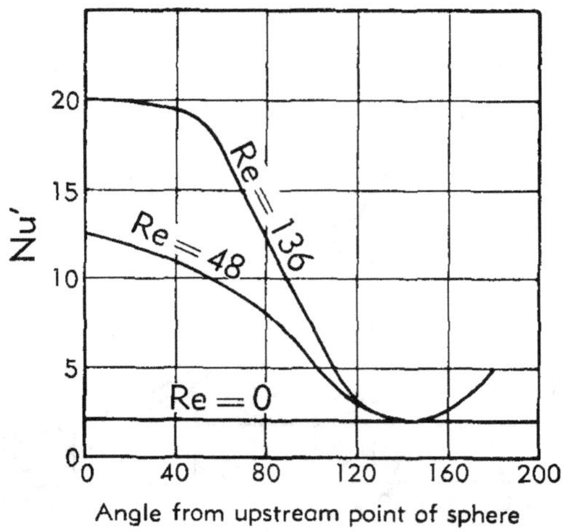

Fig. J,3d. Effect of Reynolds number on local mass transfer rates at the surface of a sphere.

140° away from the upstream end, and then increases again at the rear of the drop where reverse flow and better mass transfer rates are again encountered. This rather complex situation is usually simplified by postulating that mass and heat transfer rates to spheres can be correlated by assuming a constant average value of r_1 around the drop. One such correlation [*31*] is given by the following empirical formulas

$$Nu = 2.0 + 0.60 Pr^{\frac{1}{3}} Re^{\frac{1}{2}} = \frac{2r_1}{r_1 - r_0} \tag{3-27}$$

$$Nu' = 2.0 + 0.60 Sc^{\frac{1}{3}} Re^{\frac{1}{2}} = \frac{2r_1}{r_1 - r_0} \tag{3-28}$$

where

$$Nu = 2\frac{hr_0}{k}$$

$$Nu' = \frac{2k_g r_0}{D}\frac{\mathfrak{M}_f p_f}{\rho_f}$$

$$Pr = \frac{c_p\mu}{k}$$

$$Sc = \frac{\mu}{\rho D}$$

$$Re = \frac{2\Delta u r_0 \rho}{\mu}$$

where h is defined by

$$m_0 = \frac{4\pi r_0^2 h}{\lambda}(T_1 - T_0)$$

The data for Eq. 3-27 and 3-28 were taken under conditions such that u_r, the velocity of evaporated material, could be neglected. Data for rapid evaporation are not available; however, as a first approximation, it might be assumed that the effective film thickness $r_1 - r_0$ remains unchanged by the rate of evaporation. It is known [*30*] that in the case of a flat surface, mass transfer away from the surface will increase the boundary layer thickness, which would cause an increase in the value of r_1 giving a lower rate of evaporation. Measurements of droplet evaporation for large temperature differences would be of value in determining the magnitude of this effect. In developing the preceding equation it was assumed that an average value of physical constants could be used. These constants change with temperature, and where a large temperature gradient exists it is suggested that a logarithmic mean of the property for the two temperatures T_1 and T_0 be used. The logarithmic mean is defined for property B as follows:

$$B_{\text{mean}} = \frac{B_0 - B_1}{\ln (B_0/B_1)} \tag{3-29}$$

In order to determine the rate of heat or mass transfer it is necessary to know the relative velocities of the drop and the air. Consideration of this problem can be divided into two parts: the period when the drop is approaching the main stream velocity after injection, and the period during which the turbulent velocity components are primarily responsible for the relative velocity between the air and the drop.

The relative velocities of drop and air can be estimated by using Eq. 2-11 to make a force balance on the drop:

$$F = \frac{4}{3}\pi r_0^3 \rho_1 \frac{d\Delta u}{dt} = 6\pi\mu\Delta u r_0(1 + 0.197Re^{0.63} + 2.6 \times 10^{-4}Re^{1.38}) \tag{3-30}$$

Examination of probable conditions to be encountered in jet engines shows that the error due to neglecting the third term in the expression for C_D is small; so, integration of Eq. 3-30 gives

$$\left(\frac{Re}{Re_0}\right)^{0.63}\left(\frac{1 + .197Re_0^{0.63}}{1 + .197Re^{0.63}}\right) = e^{-2.84\mu t/r_0^2\rho_{liq}} \tag{3-31}$$

The above equations apply to single spherical drops. At high relative velocities it is probable that the drops are sufficiently deformed to give them a somewhat higher drag and acceleration. At the point of injection the drops are quite close together and cannot be expected to follow accurately the formula developed for a single drop, since the air in an accelerating cloud of drops is slowed down relative to the air traveling outside the spray. For purposes of illustration, however, it is felt that Eq. 3-31 is of sufficient accuracy.

A sample calculation assuming a 45-micron drop of kerosene in a 300-ft/sec air stream at 1-atmosphere pressure showed that starting from rest this drop would travel 1.2 ft before reaching 97 per cent of the main-stream speed. At this point the relative velocity between the air and the drop is equal to the turbulent intensity in the center of the duct, assuming fully developed pipe turbulence, and the Reynolds number would be 6. The above calculation assumes a drop of constant size; if the drop were evaporating it would approach the main stream velocity more rapidly as its size decreased.

Eq. 2-4 can be used to indicate the effect of turbulence on the relative velocity between the drop and the air. For the case of the 45-micron drop in the 300-ft/sec air stream, assuming an intensity of 3 per cent, the maximum Reynolds number for the flow of air past the drop due to turbulence would be 4.7. From Eq. 3-26, which gives the effect of Reynolds number on the heat transfer rate, one finds that the ratio $r_1/(r_1 - r_0)$ is increased 60 per cent at this Reynolds number.

It is apparent from the preceding discussion that the effect of relative air velocity, during the acceleration period and while the drop is being carried by a turbulent stream, must be considered in a calculation of evaporation rate. In addition to the effects previously discussed, the heat absorbed by the drop while it is being heated to its equilibrium temperature, the effect of changing composition of the fuel drop as it evaporates, and the effect of fuel vapors in suppressing evaporation when air temperature is low must be considered. These effects can be taken into account by proper application of equations previously presented.

The importance of the various factors determining evaporation rate vary widely depending on the specific application; however, for purposes of illustration some data on the evaporation of hydrocarbons injected contrastream into a high velocity air stream are presented in Fig. J,3e and J,3f [*17*]. In these experiments the amount of unvaporized fuel was determined at a given distance downstream from the point of injection, by sampling at various points in the cross section with a small tube pointed upstream. Air flowed through the tube at a sufficiently low velocity so that the drops of fuel that entered the tube were not appreciably evaporated, once they had entered the tube. The collection efficiency as calculated from Fig. J,2f or measured by injecting a fuel of low

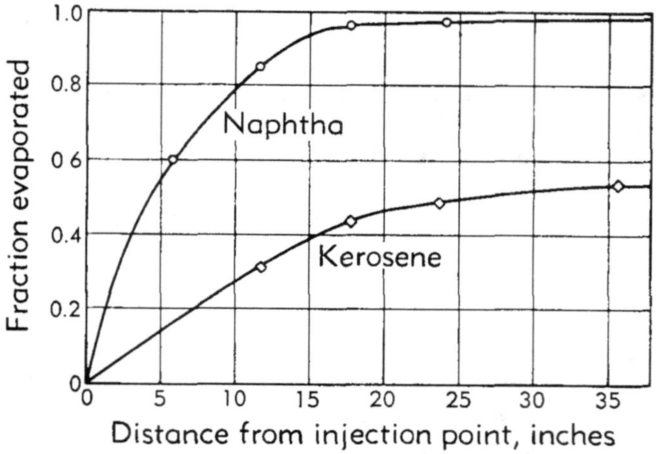

Fig. J,3e. Rate of evaporation of a fuel spray. Air velocity 500 ft/sec; air temperature, 200°F; fuel rate, 0.20 lb/sec.

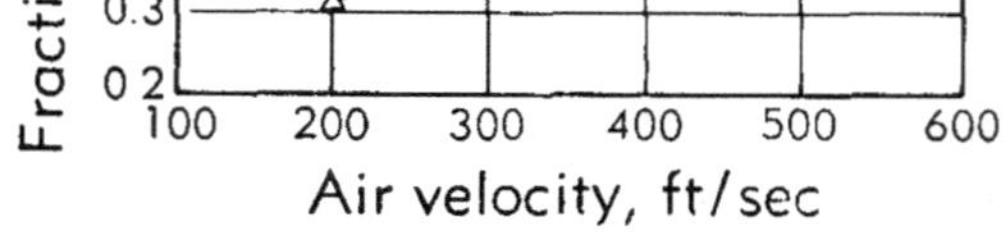

Fig. J,3f. Rate of evaporation of a fuel spray. Distance from point of injection 17 in.; fuel, 330°F boiling point, narrow cut; air temperature, 235°F; constant fuel-air ratio (0.0243 in a 6-in. duct).

volatility was essentially 100 per cent. It is seen from Fig. J,3e that, with an air velocity of 500 ft/sec, evaporation takes place quite rapidly in the first 20 inches downstream from the point of injection. In the case of the more volatile fuel, naphtha, evaporation was essentially complete in 20 inches, while in the case of the kerosene, which was barely capable of evaporation, assuming perfect mixing in the duct, only 50 per cent was evaporated at this point and evaporation proceeded rather slowly at greater distances. At a distance of 20 inches, with the naphtha, combustion is essentially that of a fuel vapor, while for the kerosene, combustion initiated at a distance of 20 inches would take place in the presence of considerable quantities of liquid fuel.

Fig. J,3f shows the effect of air velocity on evaporation when the fuel rate was increased in proportion to the air velocity in order to maintain a constant average ratio of fuel to air. It is seen that the fraction of material evaporated at a distance of 17 inches downstream from the point of injection increases as air velocity is increased. The fact that a greater proportion of the fuel is evaporated in a shorter time with increasing velocity is probably a result of both increased heat and mass transfer coefficients, and reduced drop diameter. The effect of air pressure has also been studied, and results indicate that holding constant air-fuel ratio the fraction evaporated may increase somewhat as pressure is reduced. It is believed that this may indicate a decrease in drop size with decreasing fuel flow in spite of the probable tendency toward larger drops with decreasing air density.

J,4. Combustion of Fuel Drops. When fuel drops enter a flame zone or are contacted by oxygen containing gas of sufficiently high temperature, the vapors may be ignited in the immediate vicinity of the drop, forming a very high temperature zone surrounding the drop. Observation of the burning of single drops under such circumstances [*32,33*] has shown that a drop in motion, relative to the surrounding atmosphere, will have a rather thin zone of flame around its upstream face and sides, and a long "tail" of flame in its wake. The surrounding flame may or may not be luminous depending on the oxygen pressure, the composition of the fuel, and the size of the drop. Aromatic fuels tend to give a more luminous flame than paraffinic fuels, and low oxygen pressure and very small drops tend to give less luminous flames.

In jet propulsion equipment, times of less than 0.01 sec are generally allowed for evaporation and combustion of the fuel. Since most fuels of interest in jet propulsion will evaporate before a fuel temperature of 700°F is reached, cracking or oxidation of the liquid fuel will not occur to any appreciable extent in the very short time available.

Fig. J,4a gives a highly simplified model of combustion near the surface of a drop [*33*], as proposed by Spalding. Starting at the surface of the drop (point r_0) the gas film is at the drop temperature T_0 and the pressure of fuel vapors is in equilibrium with the liquid drop. These fuel vapors diffuse through a stagnant film of combustion products to r_1 where combustion takes place. It is assumed that all fuel is consumed at this point so that the vapor pressure of fuel is zero. Oxygen is assumed to diffuse from the surrounding air to this point, and is also assumed to be completely consumed so that the ratio of amount of fuel and oxygen diffusing to r_1 is the stoichiometric ratio. Both heat and combustion products travel out to r_2 where conditions correspond to the temperature and composition of the surrounding atmosphere. This simple model is capable of mathematical treatment using the equations developed in the preceding article. The following discussion follows closely that given by Spalding [33] except that the equations are written in spherical coordinates and assume that the drop has already reached its equilibrium temperature. The equations used are as follows:

$$m_0 = \frac{4\pi r_0 r_1 k_f}{(r_1 - r_0)c_p} \ln\left[1 + \frac{k_0}{k_f}\frac{c_p}{\lambda}(T_1 - T_0)\right] \tag{4-1}$$

$$m_a = C_1 m_0 \tag{4-2}$$

$$m_a = \frac{4\pi r_1 r_2 \mathfrak{M}_a D_a}{(r_2 - r_1)RT_a}[(p_a)_2 - (p_a)_1] \tag{4-3}$$

$$m_0 \Delta H = \frac{4\pi r_1 r_2}{(r_2 - r_1)} k_a(T_1 - T_2) \tag{4-4}$$

Eq. 4-1 gives the rate of fuel evaporation due to the heat transferred from r_1 to r_0. This equation is Eq. 3-4 of the previous article and assumes

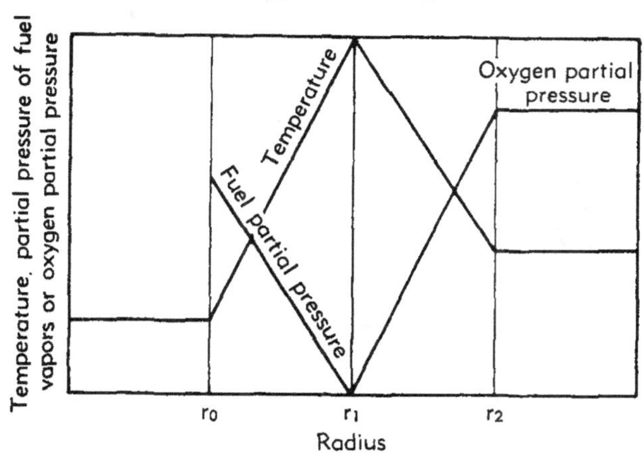

Fig. J,4a. Model for study of combustion near the surface of a drop.

that all the heat transferred to the drop is used to supply the latent heat of evaporation λ. If the drop is below its equilibrium temperature, some of the heat will go into increasing the temperature of the drop, and the evaporation rate will be lower than that given by Eq. 4-1. Proper treatment of this unsteady state condition, while feasible, would complicate the problem considerably, and it is believed that many of the important features can be illustrated by assuming that the fuel enters the combustion zone at its equilibrium temperature which is somewhat below its normal boiling point. Eq. 4-2 is the stoichiometric relation between fuel and oxygen where C is the stoichiometric ratio. For the high temperature difference encountered in this case, very little accuracy is lost by assuming that T_0 is the boiling point of the fuel. This makes it possible to avoid using the diffusion equation between r_0 and r_1. Eq. 4-3 gives the amount of oxygen diffusing into the combustion zone. In this case we have fuel diffusing into the combustion zone, and combustion products diffusing out of the combustion zone through a stagnant film of nitrogen. The equations for this case have been derived by Gilliland [26] and are quite inconvenient to use. A rough calculation has indicated that the error due to using the simpler Eq. 4-3, which assumes that the concentration of the diffusing components is low, is probably less than 15 per cent. Eq. 4-4 states that all the heat released by combustion of the fuel ($m_0\Delta H$) is transferred to the surrounding air. A calculation of the effect of mass transfer on heat transfer from r_1 to r_2 showed that Eq. 3-11 can be used with an error less than 10 per cent. The correct equation is similar in form to Eq. 4-1 and use of this form would have complicated the algebra; the simple form was used in view of the relatively small error involved.

By eliminating m_0 from Eq. 4-2, 4-3, and 4-4 one obtains the following expression for the temperature of the combustion zone:

$$T_1 = \frac{\mathfrak{M}_a D_a \Delta H[(p_a)_2 - (p_a)_1]}{CRT_a k_a} + T_2 \tag{4-5}$$

$$= B + T_2$$

By using Eq. 4-4 and 4-5 to eliminate r_1 and T_1 from Eq. 4-1, the following expression for m_0 is obtained:

$$m_0 = \frac{4\pi}{\Delta H}\left(\frac{r_0 r_2}{r_2 - r_0}\right)\left\{k_a B + \frac{k_f \Delta H}{c_p} \ln\left[1 + \frac{k_0}{k_f}\frac{c_p}{\lambda}(B + T_2 - T_0)\right]\right\} \tag{4-6}$$

Knowing the various quantities in this expression, m_0, the burning rate for a given drop radius r_0, can be found.

The first term in the parenthesis is generally small compared with the second, so Eq. 4-6 can be simplified to

$$m_0 = \frac{4\pi k_f r_0 r_2}{c_p(r_2 - r_0)} \ln\left[1 + \frac{k_0}{k_f}\frac{c_p}{\lambda}(B + T_2 - T_0)\right] \tag{4-7}$$

This equation (4-7) is quite similar to the heat transfer equation (4-1) except that the temperature difference is made up of two terms. The first term, B, depends on the product of partial pressure of oxygen in the surrounding atmosphere and the diffusivity of oxygen in the outer film. This product is independent of total pressure for a given atmosphere composition since diffusivity is inversely proportional to pressure. The other terms are quite insensitive to pressure; so, it is seen that the combustion rate, as predicted by Eq. 4-7, will not depend on pressure. The relative importance of the diffusion term B and $T_2 - T_0$ in Eq. 4-7 depends on the composition of the atmosphere surrounding the drop. If very little burning has taken place in this atmosphere, the partial pressure of oxygen, and thereby B, will be large and $T_2 - T_0$ will generally be small, while if most of the oxygen has been consumed by combustion, B is small but $T_2 - T_0$ is large. These two effects tend to balance each other as shown in Fig. J,4b, where the time for complete evaporation of a 45-micron kerosene drop is plotted as a function of temperature of the surrounding atmosphere, assuming that the initial temperature of the air was 200°F and that its temperature has been raised by combustion so that at 3640°F no oxygen remains. The time required for consumption of the drop is approximately 10^{-3} sec and it decreases slightly as the atmospheric temperature increases.

For the conditions assumed for Fig. J,4b it is estimated that the velocity of hydrocarbon vapors at the surface of the drop is, at 1 atmosphere

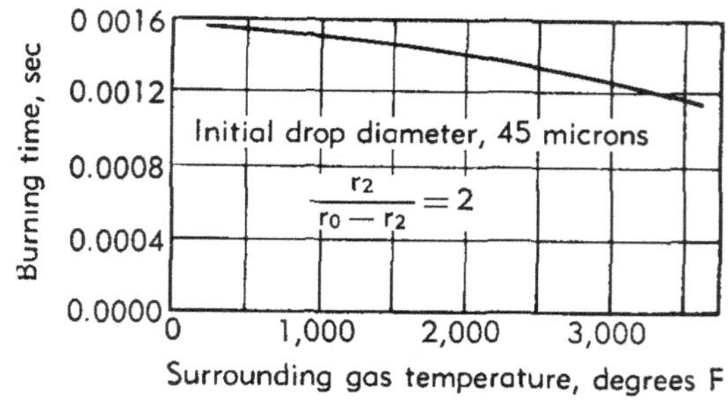

Fig. J,4b. Time for combustion of a fuel drop.

total pressure, of the same order of magnitude as the laminar burning velocity of premixed hydrocarbon vapors and air. Since the conditions in the vicinity of the drop are generally less favorable for rapid combus tion than for the case of a premixed flame, it would appear that the assumption of instantaneous reaction employed in the preceding treatment may not be sufficiently accurate for drops this size or smaller, or for lower pressure where the velocity is higher since the mass flow does not change with pressure. However, it would appear that for larger drops and higher pressures that this assumption may give useful results.

A few measurements made at the Standard Oil Development Company illustrate the effect of fuel evaporation on burning rate [*17*]. In these tests the rate at which a flame spreads through a flowing mixture of either vaporized fuel and air, or fuel drops and air, was measured. The equipment used is shown diagrammatically in Fig. J,4c. Mounted in the center of the duct is a pilot where fuel and air are burned completely, and the hot combustion products exhausted parallel to the flow of the main stream of fuel and air. The velocity of the gases leaving the pilot was generally somewhat greater than the main-stream velocity of 200 ft/sec. Fig. J,4d shows the point combustion efficiency along a duct diameter 18 inches downstream from the end of the pilot. These combustion efficiencies were determined by gas analysis and are therefore a time average of the gas composition flowing past the point of interest. The flow is highly turbulent, and the gas analyses indicate that the low combustion efficiencies at the larger distances from the center of the duct are the consequence of sampling an occasional eddy of completely burned material, rather

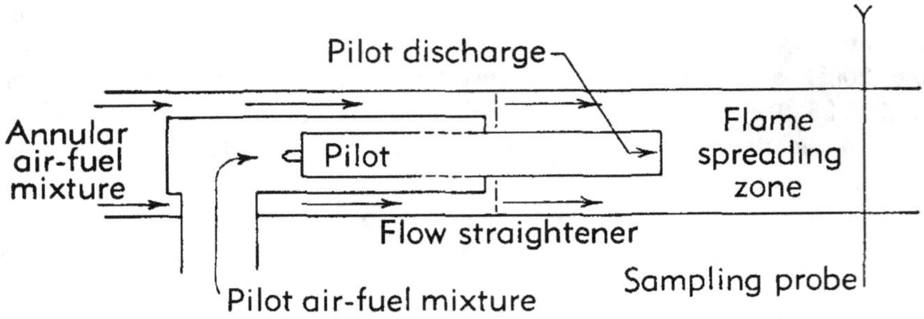

Fig. J,4c. Apparatus for determining amount of flame spreading from a controlled heat source.

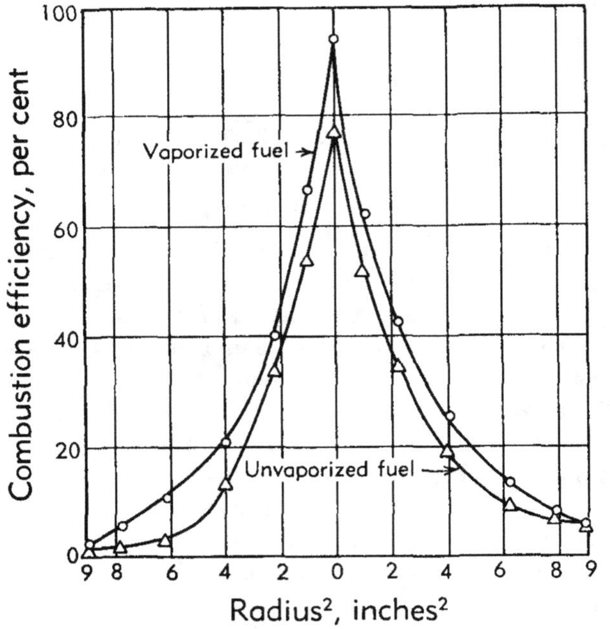

Fig. J,4d. Effect of degree of fuel vaporization on point combustion efficiency.

than the mixture being uniformly but partially oxidized. Almost half of the burned material shown in Fig. J,4d comes from the pilot, and calculation shows that between the pilot and the sampling point approximately 50 per cent more fresh mixture was burned using a vaporized and premixed fuel, than when a mixture of atomized diesel oil drops and air was burned. This lower rate of flame propagation corresponds to a time delay in burning the drops of the order of 10^{-3} sec, which corresponds roughly to the time of evaporation of a drop calculated in the preceding example. The importance of having good atomization is apparent from these considerations. Atomization by the main air stream as used in these tests is apparently adequate to give combustion without a very serious increase in required combustion chamber volume; however, if much larger drops were used one might encounter an undesirable delay in combustion due to the time necessary for the evaporation of the fuel drops.

J,5. Cited References.

1. Scheubel, F. N. *Wissenschaftliche Gesellschaft fur Luftfahrt (WGL), Jahrbuch*, 140 (1927).
2. Castleman, R. A., Jr. *J. Research Natl. Bur. Standards 6*, 369 (1931).
3. Rayleigh, Lord (J. W. Strutt). *Proc. Roy. Soc. London A10*, 4 (1879). (Sci. Papers, Art. 58, Chap. XX.)
4. Weber, C. *Z. angew. Math. u. Mech. 11*, 136–154 (1931).
5. Schweitzer, P. H. *J. Appl. Phys. 8*, 513–521 (1937).
6. Bevans, R. S. Mathematical expressions for drop size distribution in sprays. *Conference on Fuel Sprays*, Univ. Mich., Mar., 1949.
7. Lewis, H. G., Edwards, D. G., et al. *Ind. Eng. Chem. 40*, 67–74 (1948).
8. Nukiyama, S., and Tanasawa, Y. *Trans. Soc. Mech. Engrs. Japan 4*, 86, 138, (1938); *5*, 63, 68 (1939); *6*, II-7, II-8 (1940).
9. Kalupaev, P. G. *Chem. Eng. Sc.D. Thesis*. Mass. Inst. Technol., 1941.
10. Longwell, J. P. *Chem. Eng. Sc.D. Thesis*. Mass. Inst. Technol., 1943.
11. Lloyd, P. *Power Jets Special Rept. 510*, or *War Emergency Issue 12, Proc. Inst. Mech. Engrs. 153*, 1945.
12. Lane, W. R. *Ind. Eng. Chem. 43*, 1312–1317 (1951).
13. Mugele, R. H., and Evans, H. D. *Ind. Eng. Chem. 43*, 1317–1324 (1951).
14. Dryden, H. L. *Ind. Eng. Chem. 31*, 416 (1939).
15. Taylor, G. I. *Proc. Roy. Soc. London A20*, 196 (1921).
16. Sherwood, T. K., and Towle, W. L. *Ind. Eng. Chem. 31*, 416 (1939).
17. Combustion studies done at the Standard Oil Development Co. under *Navy Contract Nav. Ord.-9233*. See also Longwell, J. P., Weiss, M. A., et al. *Ind. Eng. Chem. 45*, 667–676 (1953).
18. Brown, H. E., and Amstead, B. H. Droplet dispersion in high velocity air streams. *Conference on fuel sprays*, Univ. Mich., Mar. 1949.
19. Longwell, J. P. *J. Aeronaut. Sci. 16*, 707 (1949).
20. McCarter, R. J., et al. *Ind. Eng. Chem. 41*, 1290–1295 (1949).
21. Taylor, G. I. *Proc. Roy. Soc. London A157*, 537 (1936).
22. Langmuir, I., and Blodgett, K. B. A mathematical investigation of droplet trajectories. *Army Air Forces Tech. Rept. 5418*, 1946.
23. Maxwell, J. B. *Data Book on Hydrocarbon*. Van Nostrand, 1950.
24. Walker, W. H., Lewis, W. K., McAdams, W. H., and Gilliland, E. R. *Principles of Chemical Engineering*. McGraw-Hill, 1937.
25. Hirschfelder, J. O., and Curtiss, C. F. *Univ. Wisc. CF-1501*, 1952.
26. Sherwood, T. K. *Absorption and Extraction*. McGraw-Hill, 1937.
27. Langmuir, I. *Phys. Rev. 12*, 368–370 (1918).
28. Frössling, N. *Gerlands Beitr. Geophys. 52*, 17 (1928).
29. Probert, R. P. *Phil. Mag. 37*, 95–167 (1946).
30. Stewart, W. E. *Sc.D. Thesis*. Mass. Inst. Technol., 1950.
31. Ranz, W. E., and Marshall, W. R., Jr. *Chem. Eng. Progr. 48*, 141–146 (1952); *49*, 173–180 (1953).
32. Chang, T. Y. *Chem. Eng. Sc.D. Thesis*. Mass. Inst. Technol., 1941.
33. Spalding, D. B. *Fuel 29*, (*1*), 2–7; (*2*), 25–32 (1950).

SECTION K

COMBUSTION OF SOLID FUELS

MELVIN GERSTEIN
KENNETH P. COFFIN

K,1. Introduction. The combustion processes in aircraft propulsion systems considered in the past have involved almost exclusively gaseous and liquid fuels. This has been due primarily to the nature of the propulsion systems, which were not easily adaptable to the injection techniques required for solid fuels. The advent of the jet engine with combustion chambers free from moving parts permits the consideration of all combustible materials which are capable of the necessary energy release levels and rates. Thus fuels in solid form may be considered from the same point of view as the more conventional liquid and gaseous fuels. While many of the combustion characteristics of solid fuels are similar to those of liquid and gaseous fuels, there are sufficient differences to necessitate a separate discussion. The term *solid fuel* will be applied to any solid material capable of releasing energy under the conditions encountered in jet engines. Not included are fuel-oxidant propellant mixtures, which are discussed in Sec. M.

K,2. Properties of Solid Fuels and Combustion Products. In evaluating those solids holding the most promise as fuels, important considerations are the quantity of energy released by the chemical reaction and the temperature of the resulting flame, as well as the nature of the exhaust products, since the propulsive thrust depends on the momentum of the exhaust stream.

The heat of combustion is a measure of the maximum energy which can be imparted to the working fluid. The production of high temperature, which is related to high exhaust momentum, is dependent upon a high exothermic heat of reaction and a high thermal stability of the combustion products [*1*, pp. 559–582]. Both Leonard [*10*] and Grosse [*11*] have compiled lists of substances which have relatively high heats of combustion. These are compared in Table K,2a with some of the conventional carbonaceous fuels and with a typical liquid fuel. Fuel density has also been included in the table since, for propulsion applications, not only is the quantity of heat per unit weight of fuel important but also the space necessary to contain the fuel. It is evident that some solid fuels have high heats of combustion both on a weight basis and on a volume basis. For rocket propulsion, Leonard [*10*] indicates that the final velocity of a rocket can be increased by using high density propellents at the start of firing and a high energy combination at the end.

It has already been indicated that a high heat of combustion associated with stable combustion products is important in producing high flame temperatures. For substances producing only products gaseous at ordinary temperature the principal energy absorption process will be thermal dissociation. The presence of condensed products at the temperature at which the heat of combustion is reported will absorb considerable

Table K,2a. [*12,13*]

Fuel	Combustion product	Fuel density, g/cm³	Heat of combustion	
			kcal/g fuel	kcal/cm³ fuel
Coal	CO	1.3–1.7	1.8–2.3	2.3–3.9
	CO_2		6.4–8.1	8.3–13.8
Gasoline	H_2O, CO_2	0.74	11.5	8.5
Lithium	Li_2O	0.53	10.3	5.5
Beryllium	BeO	1.84	16.2	29.8
Boron	B_2O_3	2.34	13.8	32.3
Magnesium	MgO	1.74	5.88	10.2
Aluminum	Al_2O_3	2.70	7.35	19.7
Silicon	SiO_2	2.35	7.22	16.9
Calcium	CaO	1.55	3.79	5.9
Zirconium	ZrO_2	6.44	2.83	18.2
Lanthanum	La_2O_3	6.16	1.65	10.2
Hafnium	HfO_2	13.3	1.52	20.2
Thorium	ThO_2	11.3	1.26	14.2

heat of fusion and vaporization, thus minimizing to some extent the effect of a high heat of combustion. Table K,2b contains some of the transition temperatures for the products of combustion of the fuels listed in Table K,2a. The heat of vaporization of 77 kcal/mole for B_2O_3 [*12*] and the heat of sublimation of 145.4 kcal/mole for BeO [*12*] indicate that the volatilization process involves appreciably greater energy than does the fusion process. Wolfhard and Parker [*14*] quote boiling points of 2980°C for aluminum oxide and 2800°C for magnesium oxide.

Due to the difficulties of measuring high flame temperatures and the lack of thermodynamic data for the calculation of flame temperatures, there has been little work done for the fuels considered here. Several measurements have been made of the flame temperature of aluminum and magnesium. Scartazzini [*16*] obtained a temperature of 3200°C for aluminum and 2500°C [*17*] for magnesium. Wolfhard and Parker [*14*] obtain "true" temperatures by a line reversal technique and by plotting the absolute intensity of the radiation for comparison with theoretical black body radiation curves. They report "true" temperatures of 3000°C for aluminum and 2800°C for magnesium. Zirconium is reported in [*11*] to have a flame temperature greater than 3000°C.

Table K,2b. [*12,13*]

Compound	Molecular weight	Melting point, °C	Boiling point, °C	Heat of fusion, kcal/mole
Li_2O	29.88	>1700		
BeO	25.02	2550	≅3900	
B_2O_3	69.64	450	1250	5.27
MgO	40.32	2800		18.5
Al_2O_3	101.94	2040	2250	26.
SiO_2	60.06	≅1700	2230	≅2.
CaO	56.08	2600	2850	12.
ZrO_2	123.22	2677	4300	20.8
La_2O_3	325.84	2305	4200	
HfO_2	210.60	2777		
ThO_2	264.12	3050	4400	

K,3. Fuel and Air Contact. The first stage of the combustion process involves bringing the fuel and air into contact so that reaction can take place. In the case of dispersible fuels, this process involves atomization and mixing. For volatile fuels, evaporation may occur so that part of the mixing may involve fuel vapor and air.

Obviously, those solid fuels which have melting points below their flame temperature could be liquefied and treated as liquid fuels (Sec. J). One technique for handling solid fuels in this way has been demonstrated for aluminum [*15*]. The aluminum, in the form of a wire, was passed through a hydrocarbon flame and the liquid metal atomized by an air jet. Since this is essentially an air-atomizing nozzle, the Nukiyama and Tanasawa equations (Sec. J) can be used to obtain drop size and drop size distribution. The fact that the resulting spray may contain some solid particles should not affect the applicability of the equations.

Solid fuels may also be used in conjunction with a liquid fuel carrier. The combustion of carbon emulsions has been considered [*8*], and Bowman [*9*] and Leonard [*10*] have suggested the use of certain solid rocket fuels in this form. Presumably, with due regard to the surface tension, viscosity, and density of the fuel, the atomization of these emulsions would be treated as for conventional liquid fuels.

The atomization step can be eliminated completely if the solid is injected into the combustion chamber as a dust. Coal has been used in this way in a high heat-release combustor for a gas turbine [*4*]. The transport of particles in a gas stream has been studied extensively [5,6,7]. When a jet of air is blown upward through particles of varying size, those particles having a terminal velocity (measured in free fall in still air) less than the air velocity will be carried by the gas stream. The terminal velocity of a particle w can be estimated [52]. For laminar flow

$$w = K_{\mathrm{lam}} d_{\mathrm{p}}^2 \left(\frac{\rho_{\mathrm{p}} - \rho_g}{\rho_g \nu_g} \right) \tag{3-1}$$

where d_{p} is the diameter of the particle, ρ_{p} the density of the particle, ρ_g the density of air, ν_g the kinematic viscosity of air, and K_{lam} is a constant which varies with particle shape. For turbulent flow

$$w = K_t d_{\mathrm{p}}^{\frac{1}{2}} \left(\frac{\rho_p - \rho_g}{\rho_g} \right)^{\frac{1}{2}} \tag{3-2}$$

For spheres, $K_{\mathrm{lam}} = 54.5$ and $K_t = 25.4$ for d in cm and w in cm/sec.

In actual transport through a duct, the particles will collide with each other and with the walls of the duct. The air velocity required must therefore exceed the terminal velocity. Dalla Valle [52] has developed the following equation for particle transport:

$$u = K \left(\frac{\rho_{\mathrm{p}}}{\rho_{\mathrm{p}} + 1} \right) d_{\mathrm{p}}^n \tag{3-3}$$

For air velocity u in cm/sec and d in cm, the constants are for horizontal transport, $K = 2.10 \times 10^3$ and $n = 0.4$; for vertical transport, $K = 3.86 \times 10^3$ and $n = 0.6$.

Once the particles are suspended in the air stream, either as a result of atomization or elutriation, the particles and air undergo turbulent mixing. The motion of solid particles in turbulent air will follow the laws discussed in Sec. J for liquid particles. The equations derived there apply even more closely to the solid case since it is usually assumed in trajectory calculations that the particles are solid and nonevaporating.

The problems of atomization and mixing are eliminated if a solid fuel bed is used. Shaped fuel charges similar to those used in solid fuel rockets (Sec. M) have been suggested for ramjet application in [53]. The charge may be made by compacting fine particles with the aid of heat, pressure, and adhesive additives such as oil or pitch. A comprehensive bibliography on briquetting has been prepared by the U. S. Bureau of Mines [2]. Inasmuch as there can be no simple control of the fuel feed in such a system, the shape and size of the charge is varied to give a predetermined burning rate. In general, the rate of heat release will depend on the exposed surface area and the burning rate per unit of surface. The latter is discussed in Art. 6.

Evaporation of solid fuel is usually considered as part of the combustion process. In most cases, evaporation is negligible in the atomization and mixing stages, because the temperature is not sufficiently high for the fuel to exhibit an appreciable vapor pressure. Solids, such as naphthalene, which have a relatively high vapor pressure, may be assumed to evaporate according to the equations established for liquid fuels (Sec. J).

K,4. Oxidation Mechanisms. It has been stated previously that evaporation of a solid fuel prior to burning will be negligible in most cases. Then the early stages of reaction, particularly the steps leading to ignition, will depend on the mechanism and rate of the surface reactions. These reactions will raise the temperature of the surface until burning and/or evaporation can occur. Here, again, is a major difference between solid and liquid fuels, since apparently all of the oxidation of hydrocarbon fuels takes place in the vapor phase.

A great deal of work has been done on the mechanism of carbon combustion and on the low temperature oxidation of metals, which are being considered here as solid fuels. Very little has been done on the high temperature oxidation mechanisms of any solid fuels except carbon. Some studies have been made on the reaction of metals with nitrogen to form nitrides which may be of importance in the combustion of metal fuels in air.

Jones and Townend [*18*, pp. 359–365; *19*, pp. 348–352] have investigated the oxidation of carbonaceous material at temperatures below 350°C. Gulbransen and Andrew [*3*, pp. 1034–1051] have studied the oxidation of graphite in the 425–575°C range and have found that the rate constant has an exponential temperature dependence

$$k = Ze^{-E/RT} \tag{4-1}$$

For an oxygen pressure of 76 mm Hg, an activation energy E of 36.7 kcal/mole was found while Z had a value of 6×10^{24} in terms of carbon atoms reacting per cm^2 per sec. The effect of pressure on the rate of reaction has been studied at temperatures in the range 450–500°C, and the oxidation was found to be first order with respect to oxygen pressure above 100 mm and zero order below 10 mm. The rate of reaction is very dependent on the nature of the graphite and its pretreatment.

A review of work on the mechanism of carbon oxidation prior to 1933 has been given by Mayers [*20*] and much of the more recent work is included in a symposium on carbon combustion [*19*, pp. 313–378, 523–585]. Both the low temperature and high temperature oxidations of carbon are included as well as some efforts to determine the over-all mechanism of the active burning of carbonaceous materials.

In the high temperature regime (above 900°C), Meyer [*19*, pp. 328–330] has summarized his experimental work on filaments at low pressure (below 0.1 mm), which indicated a first order reaction with respect to oxygen in the temperature range 730–1230°C with the primary products of carbon monoxide and carbon dioxide in a 1:1 ratio, independent of pressure and of temperature over the specified ranges. He reports an activation energy of 20–30 kcal/mole. In the range 1530–2230°C, he reports a zero order reaction with 2 parts of carbon monoxide being formed per part of carbon dioxide, and an activation energy of 90 kcal/mole; again in this region, the CO-CO_2 ratio was independent of pressure and temperature. Strickland-Constable [*21*; *19*, pp. 322–327], attempting to clarify the work of earlier investigators on carbon filaments at low oxygen pressures and interpreting his own experimental results (0.01 to 0.5 mm Hg), agrees only in part with Meyer. Strickland-Constable reports a first order reaction in the 900–2000°C range but reports only carbon monoxide as the primary product of the surface reaction. The rate increases with temperature from 900 to 1100°C and is independent of temperature from 1600 to 2000°C. Between 1100°C and 1600°C the rate appears to depend on the temperature reached by the carbon filament during the preceding run.

Studies of carbon combustion at atmospheric pressure, which is more closely related to the conditions of importance in jet propulsion, have shown that both carbon monoxide and carbon dioxide are formed. The ratio of oxides depends on the precise nature of the fuel and the conditions of temperature and gas flow under which combustion occurs. There are a number of possible mechanisms:

1. Formation of both CO and CO_2 in a surface reaction between C and O_2
2. Formation of CO_2 from C and O_2 at the surface, with some CO_2 later reduced to CO
3. Formation of CO at the surface from C and O_2 or C and CO_2, the CO being oxidized to CO_2 later in a gas-phase reaction

In a theoretical discussion of the combustion of fuel particles, Spalding [*22*] discussed the effects of diffusion on the combustion of solid fuel particles and cited experimental work to support a mechanism employing a carbon and carbon dioxide surface reaction at high temperature with subsequent gas-phase oxidation of the carbon monoxide. Arthur, Bangham, and Bowring [*18*, pp. 466–474], Mertens and Hellinckx [*18*, pp. 474–475], and Mertens [*19*, pp. 353–356] have definitely shown that the third mechanism, the reaction of carbon to form primarily carbon monoxide, prevails. There appears to be a general agreement then between the research at high pressure and the low pressure studies of Strickland-Constable that carbon monoxide is the primary product of the surface oxidation of carbon. Whether the oxygen reaches the surface as oxygen or carbon dioxide has not yet been resolved, although it seems likely that both may occur, depending on the conditions involved. Although considerable progress has been made, the exact nature of carbon combustion is still uncertain with regard to the quantitative rate of the oxidation process and to the nature of the details of the mechanism at the carbon surface.

The slow noncombustive oxidation of metals has been studied extensively but much of this research has been conducted at low temperatures in corrosion studies. Laidler [23] has discussed the mechanism of some single-stage surface processes. Gulbransen [24] has reviewed some of the theoretical and experimental work on oxide film growth and has applied the transition state theory of diffusion to the problem [25]. Cabrera and Mott [26] have given a more extended theoretical review. In a nonmathematical survey of his earlier papers, Evans [23, pp. 71–76] has reconciled various physical mechanisms of oxide growth with four mathematical expressions for the thickening of oxide films with time at a fixed temperature, that is, the linear, parabolic, logarithmic, and asymptotic relations which have been found experimentally. The relative physical and crystallographic properties of the metal and its oxide determine which growth rate governs the oxidation. Because of the temperature dependence of these properties, a shift of the controlling growth law may occur with temperature.

Gulbransen and coworkers have studied the growth of oxide films on iron [27], aluminum [28], beryllium [29], tungsten and magnesium [23, pp. 222–236], columbium and tantalum [30], titanium [31, pp. 741–748], and zirconium [31, pp. 515–525] at reduced pressures and at temperatures generally well below the ignition point; reactions with nitrogen are included in some of the references. Terem has studied the oxidation of aluminum [32], magnesium [33], and beryllium and nickel [34] in air and oxygen near the ignition temperature and at atmospheric pressure. Some general statements can be made about the temperature dependence of the oxidation of metals. The temperature dependence of the rate constant for oxide formation is an exponential following to, within experimental error, either the Arrhenius equation [35]

$$k = Ae^{-E/RT} \tag{4-2}$$

or the transition state theory [25]

$$\frac{k}{T} = Ce^{-E/RT} \tag{4-3}$$

where

$$C = ce^{\Delta S^*/R} \tag{4-4}$$

Some typical values of activation energy and the temperature range to which they apply have been taken from Gulbransen's work and are listed in Table K,4. Although some of the oxidation experiments have been conducted close to the ignition temperature of the metal, the relationship between the slow oxidation and the rate of active combustion has not yet been clarified. The slow oxidation process occurs at temperatures rarely exceeding 1000°C, and for many metals ignition occurs appreciably lower; the active combustion occurs at temperatures ranging upward of 2500°C. The most optimistic point of view is that the rate of the chemical process in active combustion can hardly be slower than the rate obtained by extrapolation from the slow oxidation studies.

Table K,4

Metal	Temperature, °C	Activation energy E, kcal/mole
Aluminum [*28*]	350–475	22.8
Beryllium [*29*]	350–950	50.3
Columbium [*30*]	200–375	22.8
Tantalum [*30*]	250–450	27.4
Titanium [*31*, pp. 741–748]	350–600	26.0
Zirconium [*31*, pp. 515–525]	200–425	18.2

K,5. Ignition. The ignition of solid fuels depends upon many of the same basic factors which are important in the ignition of liquid and gaseous fuels. Essentially the fuel and oxidant receive energy from the ignition source until a temperature is reached at which the oxidation reaction becomes self-sustaining. The temperature at which ignition occurs will be governed to a large extent by the rates of the reaction mechanisms discussed in Art. 4. The ignition process can be accomplished either by the use of a local source of energy of high intensity such as a spark, heated rod, or chemical igniter, or by the use of a large amount of heat at a minimum energy level such as the heating of the fuel-air mixture by the walls of a furnace. The energy required for ignition will be a function of the fuel type, configuration, and preignition conditioning as well as of the ignition source itself. Because the observed ignition energy is so characteristic of the experiment and is markedly influenced by trace materials present in the fuel, no attempt has been made to compile a large list of absolute ignition energies and temperatures.

Minimum electric energies for the ignition of dusts suspended in air and in layers have been determined in connection with dust explosion hazards. Boyle and Llewellyn [*36*, pp. 45–49, 173–181] measured the ignition energies of aluminum and magnesium by the discharge of capacitance sparks through dust clouds of the metal in air. They found that the ignition energy varies with the series resistance which determines the duration of the discharge. In general, there is an optimum discharge time for ignition. The discharge resistance shown in Fig. K,5a [36, pp. 173–181] is inversely related to the discharge time and indicates such an optimum. The ignition energy of a dust cloud is a function of particle size, the smaller particles requiring less energy. Typical values for magnesium and aluminum are given in Table K,5a [36, pp. 173–181] and for cellulose acetate dust in Fig. K,5b [37]. The ignition energy of carbonaceous materials will depend on the composition of the material and, in particular, the quantity and nature of the volatile components.

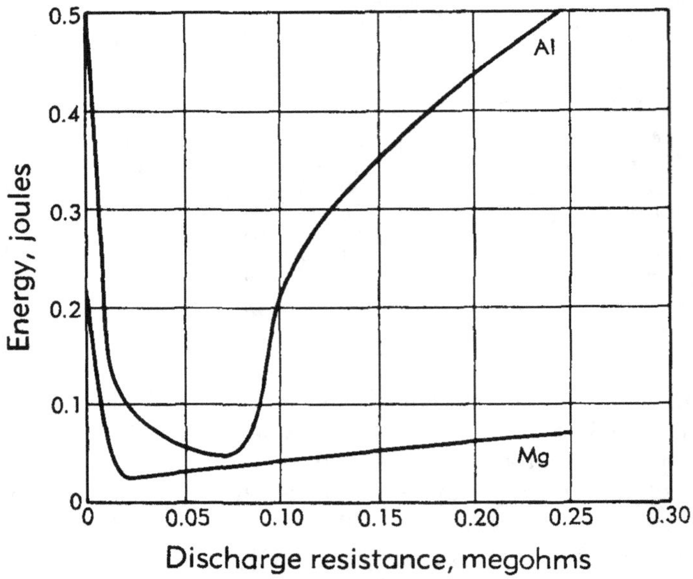

Fig. K,5a. Effect of discharge resistance (which determines discharge time) on the minimum ignition energy of magnesium and aluminum dust clouds. From [36, pp. 173–181] by permission.

The ignitability of a dust cloud containing particles of various sizes will in general be determined by the smallest particles present since these are the most easily ignited. Llewellyn used the Mg-Al column in Table K,5a to illustrate this effect. The energy required for the electric ignition of aluminum and magnesium dusts, of the order of 0.02 to 0.05 joules, is high when compared to 10^{-3} to 10^{-4} joules for conventional fuels in gaseous form. The large resistances and long spark durations required for ignition suggest that the energy is expended in heating up the suspended particles. Llewellyn states that the addition of a resistance of a few hundred ohms, or over, to a spark source reduces it to a thermal source. Hartmann, Nagy, and Jacobson [38] working with titanium, zirconium, uranium, and their hydrides found that the ignition energy values for dust

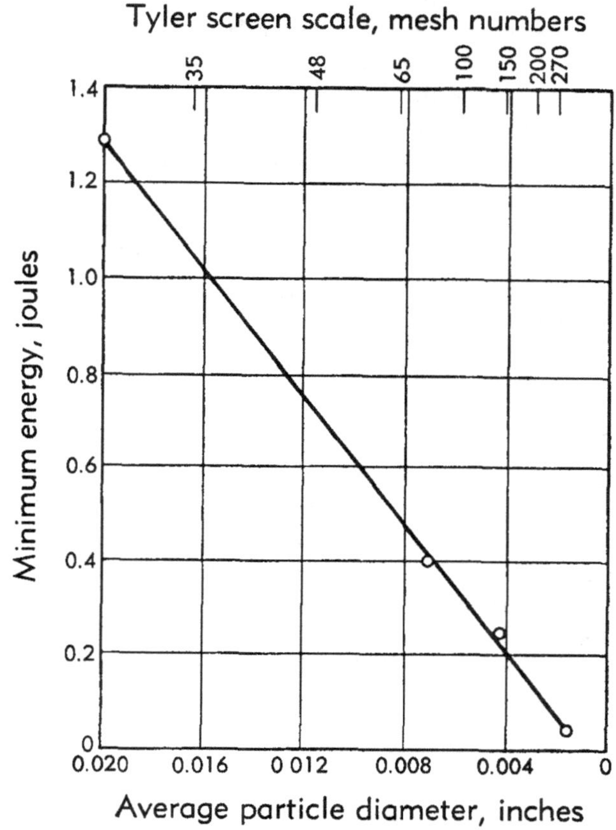

Fig. K,5b. Effect of particle size on the minimum ignition energy of dust clouds of cellulose acetate [*37*].

Table K,5a. After [*36*, pp. 173–181] by permission.

Resistance, ohms	Ignition energy, joules				
	Magnesium			Aluminum	Mg-Al 50:50
	90μ	70μ	40μ	10μ	Mg-90μ Al-10μ
50×10^3	0.20	0.033	0.025	0.055	0.060
10^6	1.20	0.14	0.09	0.55	0.80

clouds ranged from 3 to 200 millijoules, while the energy required for dust layers of the same materials ranged from 0.001 to 3 millijoules.

The ignition energy remains constant over a wide range of concentrations. At very low concentrations ignition is erratic because a single particle may ignite if caught by the discharge. Nevertheless there is a concentration below which the flame will not spread to adjacent particles, essentially a lower propagation limit. For magnesium and aluminum this occurs at about 50 mg/liter, while no upper limit of propagation has been observed [*36*, pp. 173–181]. Hartmann, Nagy, and Brown [*39*] report the

Table K,5b. After [*39*].

Minus 200 mesh, per cent	1	15	40	92
Milligrams per liter	125	60	40	25

minimum explosive concentration limits for a wide variety of metals and for various forms of the metal. For each dust, the minimum explosive concentration varies with particle size, uniformity of dispersion, impurities, strength and nature of the ignition source, and time of contact of the cloud with the ignition source. The effect of particle size on the minimum explosive limit is illustrated in Table K,5b for magnesium [*39*]. The effect of the nature of the ignition source is illustrated by Table K,5c from

Table K,5c. After [*40*].

Dust	Milligrams per liter of air		
	Glower	Arc	Induction spark
Aluminum	7.0	7.0	13.7
Coal	17.2	24.1	No ignition
Starch	7.0	10.3	13.7

Trostel and Frevert [*40*]. On the basis of the percentage of solid present to that required for complete consumption of the oxygen, Hartmann [*39*] reports that the lower explosive limit of magnesium is 5 to 6 per cent and of aluminum 11 to 13 per cent. In general, the lower the reactivity of the metal toward oxygen, the higher the percentage composition necessary to permit reaction.

The ignition of solid fuels by heat radiated and conducted from the walls of the combustion chamber is important not only as a means of ignition but also in the burning process itself. One of the first steps in burning is the heating of the cold fuel to the temperature at which the reaction becomes self-sustaining. For this reason considerable research has been done, both theoretical and experimental, on what might be called thermal ignition. A complete theoretical solution of the thermal ignition problem has not yet been achieved but much can be learned from the work done on simplified models of the process. The discussion will consist principally of the ignition process of a single particle as distinct from the more complicated problem of the ignition of dust clouds. In the case of carbonaceous fuels which may contain appreciable quantities of volatile constituents, only the residual solid is considered since the ignition characteristics of the volatile constituents resemble those of liquid fuels. While most of the theoretical work has been done with carbon, the results are probably applicable to the other solid fuels.

The first step in the thermal ignition process is the absorption of heat by the particle. Nusselt [*41*], considering spherical carbon particles suspended in air, developed equations which govern the rate of heat absorption. The absorption of heat by the particle (which acts to raise its temperature) is equated to the difference between the heat absorbed by radiation from the walls of the combustion chamber and the heat lost by conduction to the gas in which the particle is suspended. Nusselt's equation is

$$\left(\frac{4}{3}\pi r^3 c\rho \frac{dT}{dt}\right)_{\mathrm{p}} = \varphi\sigma\left[\left(\frac{T_{\mathrm{w}}}{100}\right)^4 - \left(\frac{T_{\mathrm{p}}}{100}\right)^4\right] 4\pi r_{\mathrm{p}}^2 - 4\pi r_{\mathrm{p}} k(T_{\mathrm{p}} - T_0) \qquad (5\text{-}1)$$

The subscript $_{\mathrm{p}}$ refers to the particle while $_{\mathrm{w}}$ refers to the furnace wall and $_0$ to the initial conditions of the system. The radiation angle factor is φ, the radiation constant is σ, and the thermal conductivity of air is k.

Neglecting the fourth power of the absolute particle temperature T_{p} in relation to the fourth power of the combustor wall temperature T_{w} and integrating Eq. 5-1 with the boundary condition $T_{\mathrm{p}} = T_0$ at $t = 0$, Nusselt obtains

$$T_{\mathrm{p}} = T_0 + \frac{\varphi\sigma r_{\mathrm{p}}}{k}\left(\frac{T_{\mathrm{w}}}{100}\right)^4 \left(1 - e^{-\frac{3k}{c\rho r_{\mathrm{p}}^2}t}\right) \qquad (5\text{-}2)$$

If the residence time in the combustion chamber is sufficiently large, the particle temperature reaches an equilibrium value given by

$$(T_{\mathrm{p}})_{t\to\infty} = T_0 + \frac{\varphi\sigma r_{\mathrm{p}}}{k}\left(\frac{T_{\mathrm{w}}}{100}\right)^4 \qquad (5\text{-}3)$$

Eq. 5-3 indicates that the difference between final particle temperature and its initial temperature is proportional to the particle radius and intensity of radiation, but inversely proportional to the thermal conductivity of the air. It is obvious that ignition cannot occur if the required ignition temperature is greater than $(T_{\mathrm{p}})_{t\to\infty}$. It is evident from Eq. 5-2, since time occurs in the exponent, that the particle temperature increases rapidly as residence time increases. To reach a given particle temperature, therefore, the combustor wall temperature must be increased as the time lag is decreased. For very short time lags, taking only the first two terms in the binomial expansion of the exponential, Eq. 5-2 reduces to

$$(T_p)_{t\to\infty} = T_0 + \varphi\sigma\left(\frac{T_w}{100}\right)^4 \frac{3t}{(c\rho r)_p} \tag{5-4}$$

Eq. 5-4 indicates that the initial rate of heating of the particle is inversely proportional to the particle size and is independent of the thermal properties of the gas. In other words, for very short time lags, Nusselt's equation predicts that smaller particles will reach a given temperature at lower combustor wall temperatures than larger particles; thus the wall temperature necessary for ignition may be a function of particle size.

In a brief review of some of the earlier theories of the heating rate of particles, Orning [42] points out shortcomings in Nusselt's simplified equations. Nusselt assumed that the particle is surrounded by an infinite air supply so that the air temperature does not change. If the quantity of air is limited, the equations must be corrected for the change in air temperature. According to Orning, Traustel has done this by expressing the average air temperature as a function of time and a dimensionless quantity N, the ratio of the heat capacities of the air and the particle. Traustel's equations for very short times reduce approximately to Nusselt's, but for long heating times the factor N becomes important. Orning also points out that Nusselt has assumed an infinite thermal conductivity for the particle so that the surface and interior are at the same temperature, and that Nusselt has used steady state heat transfer coefficients although the heating of a particle is a transient process. Using a more rigorous treatment of the heat transfer process, Orning showed that Nusselt had underestimated the rate of heat conduction from the particle to the air by as much as 11 per cent, the magnitude of the correction depending on particle size, residence time, and pressure. While Nusselt indicated no effect of pressure, Orning's more rigorous treatment shows that increasing pressure will have an adverse effect on the heating rate of a particle, presumably because the heat loss term rises as pressure rises while the radiation term which accounts for the heating of the particle is unaffected.

All the above treatments of the heating rate of particles have assumed that the particle is at rest with respect to the air stream. This is a reasonable assumption for small particles in a laminar air stream [7] but does not correctly describe the process under turbulent conditions. The introduction of a relative velocity between the particle and air complicates the equations so that a rigorous treatment cannot be made. Johnstone, Pigford, and Chapin (according to Orning [42]) gave an approximate estimate of the heat flow from the particle to the surrounding air to be

$$Q_0 \sim (2r_{\mathrm{p}})^{\frac{3}{2}}(\vartheta_{\mathrm{p}} - \vartheta_0)\sqrt{ku(\rho c_p)_g} \tag{5-5}$$

which shows the rate of heat loss to be proportional to the square root of the gas density and to the square root of the relative velocity u.

Suspension of particles in air represents only one of the methods for the utilization of solid fuels discussed in Art. 3. Many of the same concepts will apply to other systems although the form and quantitative aspects of the equations will be different. In the heating of bulk forms, such as in furnace beds, the thermal conductivity of the material will become increasingly important. Mayers [*43*] has given an excellent discussion of the heat transfer associated with fuel beds in a flowing gas stream.

The heating rate of the fuel is, of course, only the first step in the thermal ignition process; it determines the time necessary for the particle temperature to reach its ignition temperature under a given set of conditions. The actual particle temperature necessary for ignition will depend on the mechanism and rate of oxidation of the substance as discussed in Art. 4. The rate of the reaction which causes the particle to liberate more heat than it loses to the surroundings will determine the ignition temperature. Unfortunately it is not possible, because of the complexity of the heat transfer process and the oxidation process, to compute absolute ignition temperatures even under ideal conditions. Considerable experimental work has been done on the determination of the furnace wall temperature necessary to ignite solid fuels in air. Minimum ignition temperature (furnace wall temperature) will depend on such things as particle diameter, residence time of the fuel in the furnace, the density of the dust cloud or fuel bed, and the nature of the material. Some of these factors are associated with the heating rate while others are associated with the chemical oxidation process discussed in Art. 4. Minimum ignition temperatures vary from 315 to 900°C under normal circumstances while some substances such as zirconium powder have been known to ignite at room temperature, although Hartmann [*39*] indicates that this may be due to static sparks.

The effect of particle size on ignition temperature is illustrated by the data of Hartmann [*39*] for powders suspended in air. Although the effect of particle size is not clear-cut, the general tendency is for the smaller particles to ignite at a lower furnace temperature.

The thermal ignition of settled powders, determined experimentally, generally occurs at a lower furnace wall temperature than the ignition of the same powder suspended in air.

Since dust layers and suspended dusts will absorb heat at different rates depending on experimental conditions, such as furnace design, air temperature, and flow, it is possible that the particle temperature may be the same for ignition in each case. This suggestion is similar to the one made earlier that the temperature of the furnace wall necessary for ignition of particles of different size may be due to differences in heating rate as a function of particle size and not the actual particle temperature.

Some typical comparisons from Hartmann's work [*39*] are shown in Table K,5d.

Table K,5d

Material	Ignition temperature, °C	
	Dust cloud	Quiescent powder
Aluminum	645	585
Magnesium (coarse)	790	560
Magnesium (fine)	540	480

K,6. Combustion Processes. Once a particle has been ignited, the burning rate determines the rate at which energy will be released by the fuel. Considerable theoretical and experimental work has been done on the burning of single particles of solid fuel. The burning of dust clouds or fuel beds represents a more complex problem. The work on single particles, however, has supplied much background for the determination of burning rates in the more complex systems.

Nusselt [*41*], considering the diffusion of oxygen to the surface of a spherical particle and assuming immediate conversion of carbon to carbon dioxide at the surface as well as an unlimited supply of air, derived an expression for the burning time,

$$t_0 \sim \frac{(\rho r_0^2)_p V_{min}}{D_0 T_m} \tag{6-1}$$

where r_0 is the initial radius of the particle, V_{min} is the volume of air necessary to burn a unit weight of the particle, and T_m is the mean reaction temperature defined by the equation

$$T_m = \frac{(T_s)_p - T_0}{\ln [(T_s)_p/T_0]} \tag{6-2}$$

where $(T_s)_p$ is the surface temperature of the particle and T_0 is the initial air temperature. According to Eq. 6-1, the burning time is proportional to the square of the particle radius. If it is assumed that the air supply is limited, then the oxygen concentration decreases as the burning progresses. Nusselt included this variation of oxygen concentration during the combustion process and obtained an equation which multiplies Eq. 6-1 by a complicated function of the oxygen concentration. For a system containing 20 per cent more air than stoichiometric, the burning time is

2.67 times as long as with unlimited air, according to Nusselt. Other studies of the burning rate of carbon particles in a stagnant gas in which diffusion was considered the rate-controlling process are found in [*44,45,46*].

Hurley [*7*] has also computed the burning times of coal particles on a very simplified basis in which diffusion is assumed to be the controlling process. In all cases his predictions were well above the measured burning times of coal particles. Hurley points out, however, that in burning coal containing volatile matter, as differentiated from carbon or graphite, the escaping volatile matter can either split the coal particle into finer particles having a greater burning rate, or produce hollow latticed spheres known as "cenospheres." Considering these hollow spheres, Hurley computes burning times about $\frac{1}{50}$ of the burning time of the original coal particle and thus brings his results into agreement with experiment. Orning [*47*] has verified the existence of cenospheres in burning coal. The formation of such hollow spheres will not be discussed further because it has no significance in the burning of most solid fuels, since they do not contain volatile material.

Hottel and coworkers have made a comprehensive study of the burning rate of carbon spheres in moving air in which they have considered both the reaction rate and the diffusion rate [*48,49,50*]. They have treated the burning rate of a sphere as inversely proportional to the sum of two resistances, (1) a chemical resistance at the carbon surface, and (2) a resistance to the counterdiffusion of oxygen and the products of combustion through the gas moving in a laminar film near the surface of the sphere. Consider a burning sphere with the gaseous medium flowing around it; if the combustion is slow compared to the gas flow, the gas composition over the outer film boundary can be taken as the same as in the main stream. The diffusion of oxygen through the film to the sphere is given by [*48*]

$$N_O = \frac{p}{p_i} \frac{(p_O)_g - (p_O)_s}{RT} D \left(\frac{1}{x} + \frac{2}{d}\right) \tag{6-3}$$

where x is the film thickness, d the particle diameter, p the total pressure, p_i the partial pressure of inert gas, $(p_O)_g$ the partial pressure of oxygen in the ambient gas, $(p_O)_s$ its partial pressure at the surface of the sphere, T the mean film temperature, and D is the diffusion coefficient. The rate of oxygen consumption at the surface of the sphere, if the reaction is first order, is given by

$$N_O = \beta(p_O)_s = \frac{c}{\sqrt{T_s}} e^{-E/RT_s}(p_O)_s \tag{6-4}$$

T_s is the surface temperature; $(p_O)_s$ can be eliminated from Eq. 6-3 and

6-4 to give

$$N_O = \frac{(p_O)_g}{\dfrac{p_1 RT}{pD\left(\dfrac{1}{x} + \dfrac{2}{d}\right)} + \dfrac{\sqrt{T_s}}{ce^{-E/RT_s}}} \tag{6-5}$$

Further substitutions can be made to put Eq. 6-5 in terms of measurable variables at the beginning of the reaction [*50*]. If carbon dioxide is the primary product, the molal rate of combustion of carbon K_m is equal to the rate of oxygen consumption. If only carbon monoxide is formed and it is not consumed in a secondary gas-phase reaction, then $K_m = 2N_O$. The number of grams of carbon burned will be given by

$$K = 12K_m = 12N_O\xi \tag{6-6}$$

where ξ represents the coefficient in the equation $K_m = \xi N_O$ and is determined by the relative quantities of carbon monoxide and carbon dioxide produced at the surface with a value $1 \leqq \xi \leqq 2$. The probable value of ξ in terms of the CO-CO_2 ratio has been discussed in Art. 4. From Eq. 6-4 and 6-6 the burning rate of carbon in g/cm sec is given by

$$K = \frac{12\xi(p_O)_g}{\dfrac{p_1 RT}{pD\left(\dfrac{1}{x} + \dfrac{2}{d}\right)} + \dfrac{\sqrt{T_s}}{ce^{-E/RT_s}}} \tag{6-7}$$

The denominator of Eq. 6-7 contains two terms, a diffusion resistance term to the left and a chemical resistance term to the right. Agreement with an equation of this type for the burning of carbon spheres was excellent as shown in Fig. K,6a [*50*].

Hottel, after replacing x by a function obtained by dimensional analysis, found the following limiting conditions to exist:

1. If E/T_s is large the chemical resistance controls and

$$K \sim [\xi(p_O)_g/\sqrt{T_s}]e^{-E/RT_s} \tag{6-8}$$

2. If E/T_s is small then diffusion controls and two cases appear:

(i) for small diameter spheres

$$K \sim \frac{\xi(p_O)_g p T^{n-1}}{p_1 d} \tag{6-9}$$

(ii) for large diameter spheres

$$K \sim \frac{\xi(p_O)_g p T^{n-1-a/2}(du\rho)^a}{p_1 d} \tag{6-10}$$

where a and n are constants. The equations indicate that diffusion will control at high temperatures where the reaction rate is rapid, the exact temperature depending on the reactivity indicated by the magnitude of E. Under conditions where diffusion is the controlling factor, the burning rate will be inversely proportional to the particle diameter for very small particles. In all cases the temperature dependence of the over-all reaction rate is small when diffusion controls and large when the chemical activity controls. Hottel's experimental data verify these conclusions as illustrated in Fig. K,6b [*48*]. At low temperatures, the rate increases rapidly with temperature, the effect of temperature decreasing as temperature increases. The effect of air velocity is small at low temperatures where

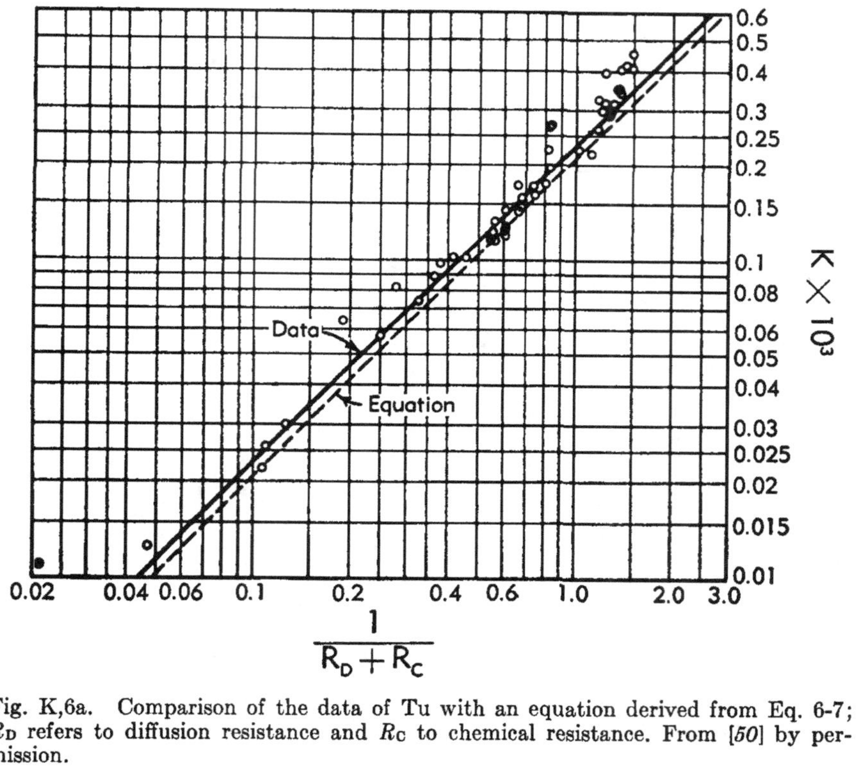

Fig. K,6a. Comparison of the data of Tu with an equation derived from Eq. 6-7; R_D refers to diffusion resistance and R_C to chemical resistance. From [*50*] by permission.

chemical reactivity controls and becomes significant at higher temperatures where diffusion is important.

While the equations were developed for a single particle, Hottel [*50*] has shown how they can be modified for the combustion of powdered coal where the flame temperature is high and the particle size small. He obtains

$$K \sim \frac{(p_O)_g T_s^{0.5}}{d} \tag{6-11}$$

for a single particle burning in air with carbon dioxide as the surface product. The burning time of the particle

$$t \sim \frac{\rho_p d_0^2}{T_s^{0.5}} \tag{6-12}$$

for infinite excess air. This shows the same dependence on particle size as Nusselt showed (Eq. 6-1), but uses the square root of the surface temperature instead of the first power of a mean reaction temperature, although Hottel suggests the use of a mean temperature. The more complicated equation for the case of finite excess air also agrees with Nusselt's equation.

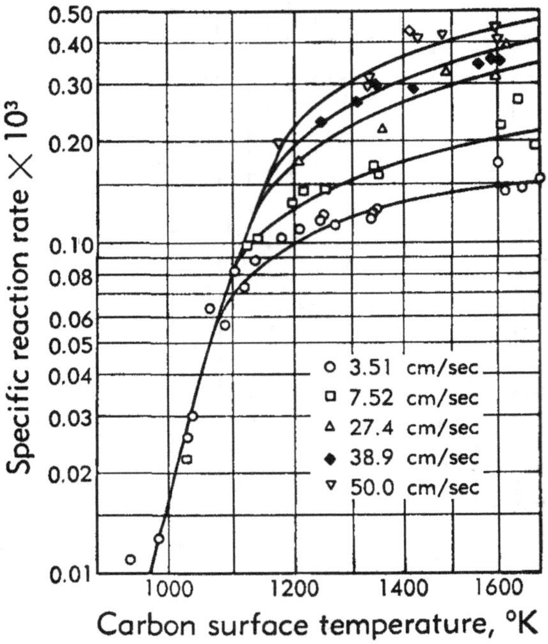

Fig. K,6b. Effect of temperature and velocity on the rate of combustion of 1-inch electrode carbon spheres in dry air. From [*48*] by permission.

Mayers [*20*], summarizing a detailed mathematical treatment of the burning rate of solid carbon, suggests an empirical equation

$$K \sim \frac{(p_O)_g}{0.21}\left(\frac{T}{1273}\right)^n u^{0.4} r_p \tag{6-13}$$

where the value of n is probably between $\frac{1}{2}$ and $\frac{3}{2}$. This equation is of interest because the effect of air velocity on burning rate agrees well with some recent work of Damon [*3*] at high air velocities. Damon studied the kinetics of the combustion of carbon cylinders placed in an air stream flowing at velocities of from 28 to 540 ft/sec and air temperatures of from

900 to 1200°C. The results indicated a reaction rate proportional to the 0.47 power of the velocity and relatively independent of the air temperature. Both the velocity and temperature effects are in good agreement with Mayers' empirical results (Eq. 6-13).

Recently, Spalding [*22*] has calculated the burning time of suspended carbon powders based on equations developed for the burning rate of liquid fuel droplets, and taking into account the reaction of the carbon surface with carbon dioxide as well as the dissociation of carbon dioxide. He compares the burning time of a carbon particle with that of a liquid

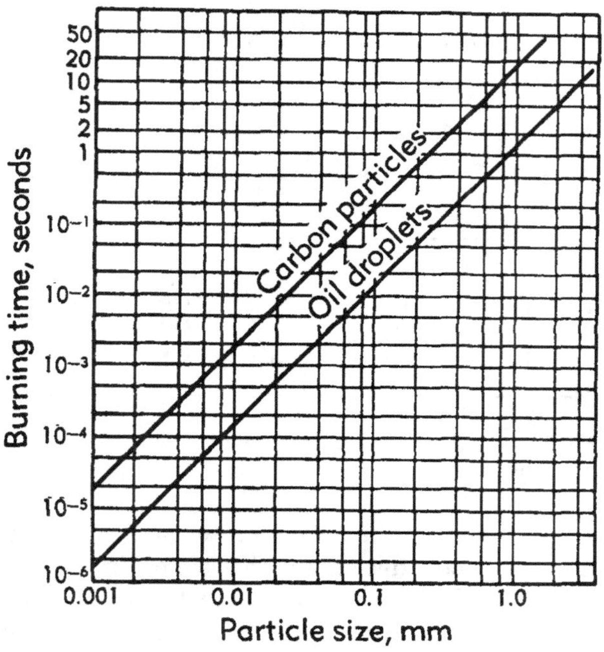

Fig. K,6c. Burning times of fuel particles under average engine conditions; upper limit. From [*22*] by permission.

fuel droplet, and obtains the results shown in Fig. K,6c. Spalding considers the carbon dioxide dissociation important because it reduces the total diffusion time, since carbon in the form of carbon monoxide diffuses out to meet the oxygen. Consideration of the dissociation doubles the burning rate relative to the case of no dissociation, but the oil burning rate is still 13 times as fast as that of the carbon. Spalding predicts that coal containing volatile material will have an intermediate burning rate because the volatile material diffusing outward will hasten the reaction. Spalding also demonstrates that the burning time of a fuel particle is proportional to the square of the initial diameter of the particle, which is in agreement with the equations of Nusselt and Hottel discussed earlier in this article.

Cassel, Das Gupta, and Guruswamy [*18*, pp. 185–190] studied the rate of flame propagation (flame speed) through dust clouds. They modified an equation derived by Mallard and le Chatelier for flame propagation in gas by adding a term for radiant heat transfer. They assumed that the diffusion of oxygen was rate-controlling and calculated a time of 0.01 seconds for the burning of a 25-micron aluminum particle and a burning zone thickness of 2.5 cm, a value of 100 to 1000 times greater than values estimated for premixed hydrocarbon-air flames. Experimentally they found the results shown in Table K,6 [*18*, pp. 185–190].

Table K,6. After [*18*] by permission.

Atomized aluminum				
Particle size, μ	Concentration, mg/liter	Gas	Reynolds number	S_u, cm/sec
6–40	120	Air	1420	19.2
	160		1280	22.0
	190		1430	23.8
0–40	210	Air	1280	25.0

The above table indicates a slight dependence on particle size and concentration. The flame velocities are slightly below those measured for hydrocarbon-air flames.

While the equations for the burning rates of single particles of solid fuel appear to be in good agreement with some of the experimental data, the equations cannot be applied exactly to the more complex process which occurs in the combustion chamber of a jet propulsion system. Nevertheless some work has been done to show the type of results which might be expected in such systems.

Hazard and Buckley [*4*] studied the effect of pressure on the burning of pulverized coal in various types of combustion chambers adaptable to gas turbines. They measured the fraction of combustible burned as a function of residence time in the combustion chamber. The residence time was calculated from the measured air flow by assuming a combustion-gas temperature of 1370°C (2500°F). The coal particles were assumed to travel through the combustion chamber at the velocity of the gas. Typical results are shown in Fig. K,6d [*4*]. Little effect of pressure on the fraction of fuel consumed was noted in the tests. The relatively small pressure effect found in the gas turbine type combustors is not in agreement with an appreciable effect of pressure in decreasing the fraction of fuel consumed in laboratory combustion chambers noted by Omori and Orning

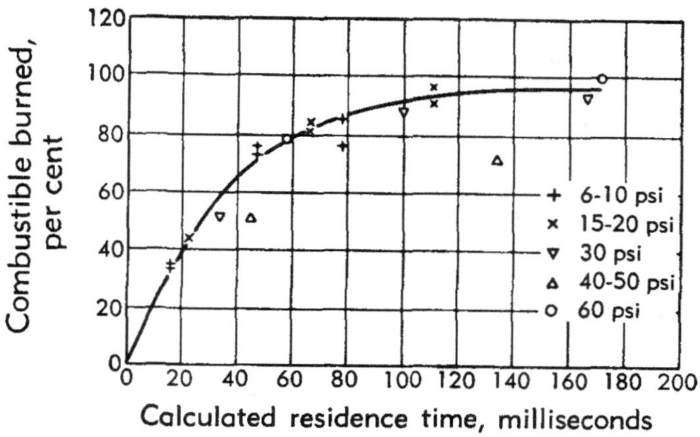

Fig. K,6d. Per cent of coal dust burned as a function of residence time and pressure in a pressure furnace [4].

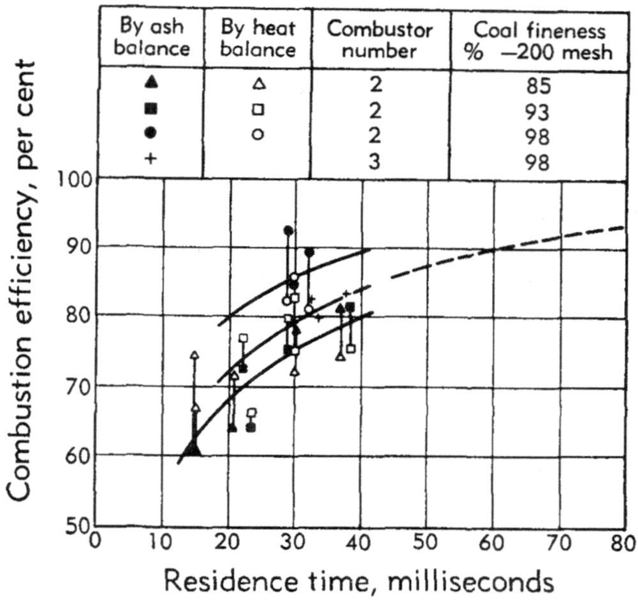

Fig. K,6e. Combustion efficiency as a function of residence time in a TG-180 jet engine combustor (no. 2) and an unperforated flame tube (no. 3) [4].

[42]. Hazard and Buckley also studied the effect of preheat up to 315°C (600°F) on the burning rate and found no measurable effect, although preheat did improve ignition and flame stability. The importance of residence time in increasing the fraction of fuel consumed emphasizes the advantages of the balanced-vortex combustor which circulates the combustion air so that the large particles are retained in the combustor and only very fine particles of a predetermined size escape, thus reducing

the loss in efficiency due to unburned fuel. Heat release rates as high as 7.4 cal/cm^3 sec (3 million BTU/ft^3 hr) were obtained with a balanced-vortex combustor [*4*].

Hazard and Buckley investigated the burning of pulverized coal in a single combustor from a tubular turbojet engine, replacing the spark plug by a hydrogen igniter. Coal rates of 8.8 to 20.2 gm/sec (70 to 160 lb/hr) corresponding to heat release rates from 3.1 to 6.2 cal/cm^3 sec (1.25×10^6 to 2.5×10^6 BTU/ft^3 hr) were studied. Their results are shown in Fig. K,6e [*4*]. Their studies indicate that coal fineness and residence time are important variables, combustion efficiency increasing with both.

K,7. Exhaust. In gas turbine and jet propulsion applications, the momentum of the exhaust determines the thrust developed. While the burning of carbonaceous materials, particularly those with low ash contents, produces an exhaust gas similar to that of conventional fuels, the combustion of solid metallic fuels introduces new exhaust problems. The use of fuels containing metallic elements introduces into the exhaust stream under some operating conditions a large quantity of solid and/or liquid products. In addition to the problems of slagging and abrasion mentioned briefly by Leonard [*10*], Bowman [*51*] points out that an exhaust stream containing agglomerated particles has a lower momentum than a totally gaseous exhaust of the same mass and at the same temperature. Consider an exhaust containing two types of molecules, neither of which can agglomerate. Assume that one mole of fuel gives particles Am_1 and Bm_2 where A and B are the number of moles of particles of mass m_1 and m_2 respectively. These particles have average velocities u_1 and u_2 and an average kinetic energy given by

$$\frac{m_1u_1^2}{2} = \frac{m_2u_2^2}{2} = \text{const} \tag{7-1}$$

The total kinetic energy is given by

$$E_{\text{total}} = \frac{Am_1u_1^2}{2} + \frac{Bm_2u_2^2}{2} \tag{7-2}$$

and the total momentum by

$$I_0 = Am_1u_1 + Bm_2u_2 \tag{7-3}$$

Now consider the case where m_1 cannot agglomerate but m_2 does agglomerate to give solid particles. There will be a new mass $m_{12} = m_2Z$ where Z is the agglomeration factor. Since the total kinetic energy does not change and since there are fewer particles, the average kinetic energy per particle will be higher and the particle velocities will change. Let the new velocities be called u_{11}, and u_{12}, corresponding to masses m_1 and m_{12}. The average kinetic energy per particle is now given by

$$\frac{m_1 u_{11}^2}{2} = \frac{m_{12} u_{12}^2}{2} = \text{const} \tag{7-4}$$

and the total kinetic energy by

$$E_{\text{total}} = \frac{A m_1 u_{11}^2}{2} + \frac{B m_{12} u_{12}^2}{2Z} \tag{7-5}$$

and the momentum by

$$I_{\text{ag}} = A m_1 u_{11} + \frac{B m_{12} u_{12}}{Z} \tag{7-6}$$

The ratio of momenta in the two cases becomes

$$\frac{I_{\text{ag}}}{I_0} = \frac{u_{11}(A m_1 \sqrt{Z} + B \sqrt{m_1 m_2})}{u_1(A m_1 + B \sqrt{m_1 m_2}) \sqrt{Z}} \tag{7-7}$$

From Eq. 7-2 and 7-5

$$\frac{u_{11}}{u_1} = \sqrt{\frac{Z(A + B)}{B + AZ}} \tag{7-8}$$

From Eq. 7-7 and 7-8

$$\frac{I_{\text{ag}}}{I_0} = \sqrt{\frac{A + B}{B + AZ}} \frac{A m_1 \sqrt{Z} + B \sqrt{m_1 m_2}}{A m_1 + B \sqrt{m_1 m_2}} \tag{7-9}$$

As Z approaches infinity

$$\left(\frac{I_{\text{ag}}}{I_0}\right)_{Z \to \infty} = \frac{m_1 \sqrt{A(A + B)}}{A m_1 + B \sqrt{m_1 m_2}} = \left[\frac{A^2 + AB}{A^2 + 2AB \sqrt{m_2/m_1} + B^2(m_2/m_1)}\right]^{\frac{1}{2}} \tag{7-10}$$

If m_2 is of the order of m_1 or greater, which is usually the case, the ratio expressed in Eq. 7-10 will be less than unity, indicating a loss in momentum due to agglomeration. [*51*]

K,8. Concluding Remarks. For many years solid fuels have been successfully employed in stationary engines and in recent years have been used in gas turbines. The advantages of solid fuels, including high densities, high heats of combustion, and high flame temperatures, and in many cases, ready availability, low cost, and ease of handling and storage, suggest that solid fuels have a place in aeronautical power plants. The previous articles have reviewed the fundamental information on the combustion of solid fuels as a basis for its extension to the conditions found in aircraft combustion chambers. The extension of knowledge of solid fuel combustion to jet engine power plants will require solutions of the practical engineering problems involved in the design of solid fuel systems, slag removal, and high temperature materials. While much of the current knowledge concerns carbon and carbonaceous materials, it has been shown that the solid metals may also be of importance as fuels. These, however, present even greater problems in combustion systems and a critical evaluation must be made to determine if their advantages outweigh the problems involved in their use, since it is the practicality of the ultimate system which will determine the extent to which solid fuels are utilized in jet propulsion.

K,9. Cited References.

1. Symposium on production of synthesis gas. *Ind. Eng. Chem. 40*, 558–641 (1948).
2. Fisher, P. L. *U. S. Bur. Mines Inform. Circ. 7469*, 1948.
3. Symposium on theoretical aspects of combustion and gasification. *Ind. Eng. Chem. 44*, 1028–1082, 1547–1575 (1952).
4. Hazard, H. R., and Buckley, F. D. *Trans. Am. Soc. Mech. Engrs. 70*, 729–737 (1948).
5. Perry, H., Corey, R. C., and Elliott, M. A. *Trans. Am. Soc. Mech. Engrs. 72*, 599–610 (1950).
6. Foley, R. B. *Trans. Am. Soc. Mech. Engrs. 69*, 101–108 (1947).
7. Hurley, T. F. *J. Inst. Fuel London 4*, 243–254 (1930–1931).
8. Audibert, E., and Raineau, A. *Chem. Abstracts 43*, 3591 (1949).
9. Bowman, N. J. *J. Space Flight 1*, (7), 1–8 (1949).
10. Leonard, A. S. *J. Am. Rocket Soc. 68*, 12–16; *70*, 20–31; *72*, 10–23 (1947).
11. Grosse, A. V., and Conway, J. B. The combustion of metals. *First Tech. Rept., High Temperature Project Contract N9-Office of Nav. Research 87301, Temple Univ.*, Oct. 1951.
12. Rossini, F. D., Wagman, D. D., Evans, W. H., Levine, S., and Jaffe, I. Selected values of chemical thermodynamic properties. *Natl. Bur. Standards Circ. 500*, 1952.
13. *Handbook of Chemistry and Physics*, 33rd ed. Chemical Rubber Publishing Co., Cleveland, 1951–52.
14. Wolfhard, H. G., and Parker, W. G. *Proc. Phys. Soc. London B62*, 523–529 (1949).
15. Branstetter, J. R., Lord, A. M., and Gerstein, M. *NACA Research Mem. E51B02*, Mar. 1951.
16. Cueilleron, J., and Scartazzini, H. *Compt. rend. 228*, 489–490 (1949).
17. Scartazzini, H. *Compt. rend. 230*, 97–98 (1950).
18. *Third Symposium on Combustion, Flame and Explosion Phenomena*. Williams & Wilkins, 1949.
19. International symposium on combustion of carbon. *J. chim. phys. 47*, 313–378, 523–585 (1950).
20. Mayers, M. A. *Chem. Revs. 14*, 31–52 (1934).
21. Strickland-Constable, R. F. *Chemistry & Industry 67*, 771–774 (1948).
22. Spalding, D. B. *Fuel 30*, 121–130 (1951).
23. *Pittsburgh International Conference on Surface Reactions*. Corrosion Publishing Co., Pittsburgh, 1948.
24. Gulbransen, E. A. *Ind. Eng. Chem. 41*, 1385–1391 (1949); and in more detail, *Rev. Mét. 45*, 181–204, 287–300 (1948).
25. Gulbransen, E. A. *Trans. Electrochem. Soc. 83*, 301–313 (1943).
26. Cabrera, N., and Mott, N. F. *Repts. Progr. in Phys. 12*, 163–184 (1948–1949).
27. Gulbransen, E. A., and Ruka, R. *J. Metals 188, Trans. Am. Inst. Mining Met. Engrs.*, 1500–1508 (1950).
28. Gulbransen, E. A., and Wysong, W. S. *J. Phys. Chem. 51*, 1087–1103 (1947).
29. Gulbransen, E. A., and Andrew, K. F. *J. Electrochem. Soc. 97*, 383–395 (1950).
30. Gulbransen, E. A., and Andrew, K. F. *J. Metals 188, Trans. Am. Inst. Mining Met. Engrs.*, 586–599 (1950).

31. Gulbransen, E. A., and Andrew, K. F. *J. Metals 1, Trans. Am. Inst. Mining Met. Engrs.*, 1949.
32. Terem, H. N. *Rev. fac. sci. univ. Istanbul A12*, 299–307 (1947).
33. Terem, H. N. *Rev. fac. sci. univ. Istanbul A13*, 147–159 (1948).
34. Terem, H. N. *Rev. fac. sci. univ. Istanbul A8*, 9–22 (1943).
35. Dunn, J. S. *Proc. Roy. Soc. London A111*, 203–209 (1926).
36. Boyle, A. R., and Llewellyn, F. J. *J. Soc. Chem. Ind. London 69*, 1950.
37. Hartmann, I., and Nagy, J. *U. S. Bur. Mines Rept. Invest. 3751*, 1944.
38. Hartmann, I., Nagy, J., and Jacobson, M. *U. S. Bur. Mines Rept. Invest. 4835*, 1951.
39. Hartmann, I., Nagy, J., and Brown, H. R. *U. S. Bur. Mines Rept. Invest. 3722*, 1943.
40. Trostel, L. J., and Frevert, H. W. *Chem. Met. Eng. 30*, 141–146 (1924).
41. Nusselt, W. *Verfahrenstechnik Beih. A. Ver. deut. Ing. 68*, 124–128 (1924).
42. Omori, T. T., and Orning, A. A. *Trans. Am. Soc. Mech. Engrs. 72*, 591–597 (1950).
43. Mayers, M. A. *Trans. Am. Soc. Mech. Engrs. 79*, 279–288 (1937).
44. Burke, S. P., and Schuman, T. E. W. *Ind. Eng. Chem. 23*, 406–413 (1931).
45. Lewis, W. K. *Ind. Eng. Chem. 15*, 502–503 (1923).
46. Wentzel, W. *Fuel 11*, 177–196, 222–228 (1942).
47. Orning, A. A. *Trans. Am. Soc. Mech. Engrs. 64*, 497–508 (1932).
48. Tu, C. M., Davis, H., and Hottel, H. C. *Ind. Eng. Chem. 26*, 749–757 (1934).
49. Davis, H., and Hottel, H. C. *Ind. Eng. Chem. 26*, 889–892 (1934).
50. Parker, A. S., and Hottel, H. C. *Ind. Eng. Chem. 28*, 1334–1341 (1936).
51. Bowman, N. J. *J. Space Flight 3*, (*6*), 1–15 (1951).
52. Dalla Valle, J. M. *Micromeritics*. Pitman, 1943.
53. Chandler, E. F. *Aero Digest 61*, (*3*), 19 (1950).

SECTION L

COMBUSTION OF LIQUID PROPELLANTS

DAVID ALTMAN
S. S. PENNER

CHAPTER 1. IGNITION PHENOMENA IN BIPROPELLANT AND MONOPROPELLANT SYSTEMS

L,1. Introduction. Stable combustion of liquid propellants in jet engines, rockets, etc., occurs at a measurable finite rate under steady state or quasi-steady state conditions. Prior to the establishment of steady state combustion, any propellant system must pass through one or more transitory stages commonly designated as the ignition process. A measure of the time required to initiate combustion is the ignition delay, the precise definition of which is a function of the empirical studies which are used for its measurement. The initial stages of steady state combustion are profoundly affected by heat and/or mass transport to the reactants, whereas these factors are of less importance during the transient pre-steady state reactions.

It is now commonly agreed that what the experimenter calls stable combustion can be described adequately only by considering both hydrodynamic and chemical phenomena, the relative influence of the two processes varying widely from the diffusion flame to the turbulent combustion flame. A similar situation exists for the ignition phenomena. The interplay of physical rate processes such as evaporation rates, gas and liquid phase diffusion, and mixing processes on the one hand; and gas and liquid-phase chemical reaction rates with or without (surface) catalysis, on the other hand, is in general so complex that the notion of a single rate-determining step is generally not applicable, although it may still be useful in some cases. The various possible pre-steady state combustion paths may be summarized conveniently by the diagram shown in Fig. L,1 for a bipropellant system. The modifications required for a description of monopropellants are obvious. The diagram shown in Fig. L,1 is not meant to include every conceivable physical and chemical process which is of importance in defining the measured ignition delay. Thus heat transfer to and from the motor wall is not considered; nor is explicit allowance made for the importance of chamber size in determining partial pressures and hence chemical reaction and diffusion rates, or of heats of reaction in determining rates of heat transfer, diffusion, chemical reaction, and hence the rate of establishment of a steady state. It is to be recognized, of course, that the reaction scheme shown in Fig. L,1 remains equally valid for steady state combustion as for the ignition processes, the

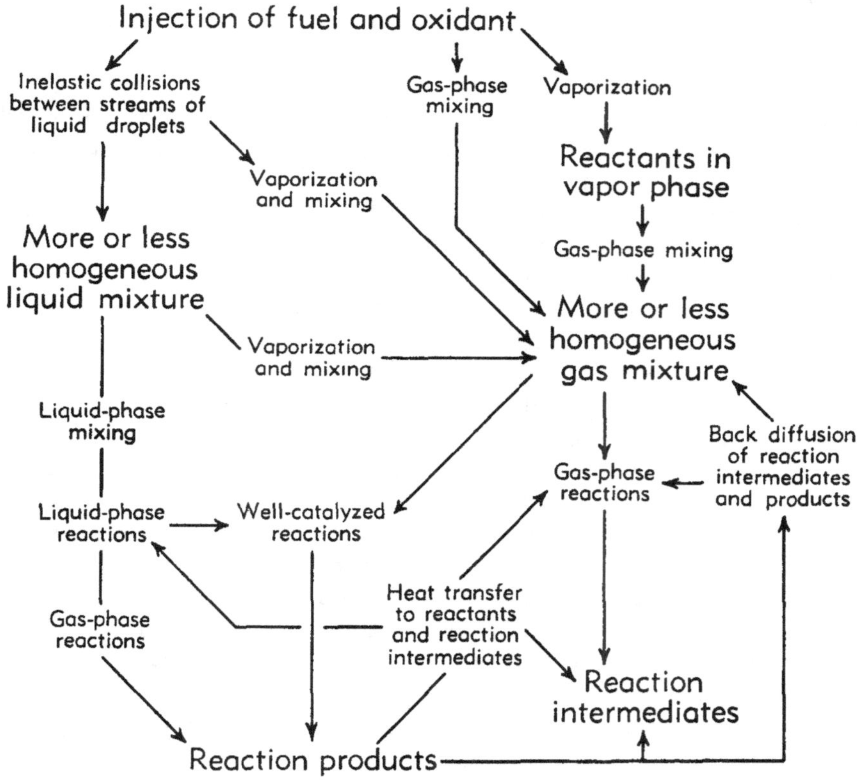

Fig. L,1. Schematic representation of preignition processes.

essential difference being that for steady state combustion the various rates of reaction are no longer changing with the time. Reference to Fig. L,1 clearly indicates the importance of both physical and chemical processes in determining ignition delay. Hence it follows that the ignition delay is in general a function not only of the propellant system used, but also of the injection procedure and the combustion chamber configuration.

At this stage of the discussion it may appear that the measurement of the ignition delay is meaningless since the ignition delay is defined through the measuring technique as well as through the physical and chemical properties of the propellant system. Although the truth of this remark can hardly be disputed, it is nevertheless known that rocket engineers have found it convenient and useful to classify propellants as (1) spontaneously-igniting or self-igniting propellants[1] and (2) non-spontaneously-igniting or non-self-igniting propellants. The mere fact that a classification of this sort is possible indicates that the injection patterns and combustion chamber configurations are sufficiently similar in many cases to permit a useful classification of propellant systems for practical purposes, or else that for many reactants relatively uncomplicated processes such as liquid-phase mixing and reactions are rate-controlling.

From the point of view of satisfactory motor performance, the use of spontaneous propellants[2] is highly desirable. Thus the temperature and pressure in the combustion chamber should increase rapidly and continuously to their equilibrium values. For nonspontaneous propellants the pre-steady state reactions may not occur at all or may occur so slowly that the propellant mixture flows through the chamber without the liberation of enough heat to initiate and sustain steady state burning. For practical purposes the ignition delay associated with this type of behavior is infinite. More frequently, undesirable motor performance is the result of excessively long ignition delays resulting in the accumulation of abnormally large concentrations of combustible gases and/or liquids in the combustion chamber before the rate of liberation of heat by the ignition processes has become sufficiently rapid to sustain steady state combustion. As the result of accumulation of abnormally large amounts of propellants prior to steady state combustion, pressure transients of disastrous magnitudes may result. This phenomenon is normally referred to as a "hard start." As a remedy for this sort of situation, the previous considerations suggest either a modification of the propellant system by the addition of a suitable catalyst or else installation of appropriate physical changes such as a better injection system to give smaller liquid droplets, more intimate liquid, and/or gas-phase mixing, etc.

The use of bipropellant systems with very short ignition delays is particularly important for service applications requiring close timing. Good ignition characteristics are also important for bipropellants used in motors which require repetitive operation.

A conservative upper limit for the transient chamber pressure during ignition, as a function of equilibrium chamber pressure, ignition delay, and motor design characteristics, can be obtained as follows. The initial mass flow rate $\dot{m}$ of propellant into the combustion chamber is given by

[1] The term "hypergolic," adopted from the German literature, is frequently used for "self-igniting."

[2] The use of "spontaneous propellant" for "spontaneously-igniting propellant" is a common abbreviation of the expression.

the relation

$$\dot{m} = A\sqrt{2p_f'\rho} \tag{1-1}$$

where A is the cross-sectional area of the injector, p_f' represents the injector gauge pressure, and ρ is the density of the injected fluid.[3] Actually, Eq. 1-1 overestimates the effect of $\dot{m}$ slightly since it corresponds to the equilibrium injection rate, thereby neglecting the approach to steady state flow. If it is assumed that all of the propellant remains in the chamber until ignition occurs at the time t_{ign},

$$\dot{m}t_{ign} = At_{ign}\sqrt{2p_f'\rho} \tag{1-2}$$

It is apparent that all of the assumptions made thus far tend to maximize the product $\dot{m}t_{ign}$.

During steady state combustion the required equilibrium flow rate of propellant $\dot{m}_0$ is

$$\dot{m}_0 = A\sqrt{2(p_f - p_c)\rho} = \frac{p_c f_t}{c^*} \tag{1-3}$$

where p_f and p_c represent, respectively, the absolute equilibrium values of the pressures on the injector and in the chamber; f_t is the cross-sectional area of the nozzle throat; and c^* represents the characteristic velocity. On the other hand, the maximum transient pressure p_{tr} must be less than or equal to the pressure corresponding to instantaneous combustion of all of the propellant in the chamber at the time t_{ign}.[4] Thus

$$p_t \leqq \frac{\dot{m}t_{ign}RT_c}{\mathfrak{M}V_c} = At_{ign}\sqrt{2p_f'\rho}\,\frac{RT_c}{\mathfrak{M}V_c} \tag{1-4}$$

where T_c is the chamber equilibrium temperature, $\mathfrak{M}$ represents an average molecular weight of the combustion products, and V_c is the chamber volume. Combining Eq. 1-3 and 1-4 leads to the result

$$\frac{p_t}{p_c} \leqq \frac{t_{ign}}{c^*}\frac{RT_c}{\mathfrak{M}L^*}\sqrt{\frac{p_f'}{p_f - p_c}} \tag{1-5}$$

Here the value of L^* is defined in the usual manner as the ratio of chamber volume to nozzle throat area. Eq. 1-5 can be transformed to the equivalent form

$$\frac{p_t}{p_c} \leqq \frac{c^*\Gamma(\gamma)}{L^*}\,\frac{t_{ign}}{\sqrt{(p_f - p_c)/p_f'}} \tag{1-5a}$$

[3] The discussion used to estimate the ratio of transient preignition pressure to equilibrium chamber pressure will be restricted, for convenience, to monopropellants. Extension to bipropellant systems may be made without difficulty. For bipropellants, $\dot{m}$ is the sum of terms corresponding to the injection of oxidizer and fuel.

[4] Instantaneous combustion will evidently be approximated closely if the pressure rise at time t_{ign} occurs in a time interval Δt which is small compared to the equilibrium residence time in the chamber during steady state combustion.

where $\Gamma(\gamma)$ is defined as

$$\Gamma(\gamma) = (\gamma)^{\frac{1}{2}} \left(\frac{2}{\gamma + 1}\right)^{\frac{\gamma+1}{2(\gamma-1)}}$$

and γ represents the specific heat ratio.

The problem of ignition delay should not be confused with that of incomplete combustion. Thus it is well known that in some cases of steady state combustion in rocket chambers the experimentally observed performance is appreciably lower than the performance predicted theoretically from calculations assuming that thermodynamic equilibrium is reached in the combustion chamber. Although the chemical reactions occurring during the transitory ignition delay period may be in some way related to rocket performance, incomplete combustion is not considered an ignition delay problem. Nevertheless, interesting correlations have been observed by various investigators between short ignition delay and good combustion efficiency.

In conclusion it appears desirable to call attention to the fact that it is not difficult in principle to conceive an experiment in which the "true chemical ignition delay," uncomplicated by mixing phenomena, is measured. Thus it might be possible to produce essentially homogeneous mixtures of unreacted propellants at very low temperatures and then to measure reaction rates as a function of temperature for the premixed system. Studies of this type would undoubtedly be of interest to the reaction kineticist but would probably not be of immediate value to the rocket engineer interested in ignition delay measurements. At least these data would not be of practical value until the complicated interaction between physical and chemical phenomena during the ignition period is completely understood.

L,2. Experimental Methods Used for Measuring Ignition Delay in Bipropellant Systems. Experimental techniques which have been developed for differentiating between spontaneous and nonspontaneous propellants, and for estimating quantitatively the ignition delay in spontaneous propellants, fall into three general classes. These three types of tests will be designated as laboratory tests on unconfined systems, laboratory tests on semi-enclosed chambers, and motor tests, respectively.

In the laboratory tests on unconfined systems the ignition delay is usually defined by a measuring technique which emphasizes the contributions to the total ignition delay of liquid-phase mixing and reaction. Only if these liquid-phase processes occur sufficiently slowly in a rocket combustion chamber is it to be expected that laboratory estimates will agree reasonably well with observed motor performance. This type of correlation, in fact, has been observed for a number of bipropellant mixtures.

Laboratory tests on semienclosed chambers are characterized by injection systems and chamber configurations very similar to those used in actual service units. They differ from service units in that they have larger openings, and therefore maintain lower peak pressures during the ignition delay period than correspond to the equilibrium pressures which are usually encountered in combustion chamber operation.

Motor tests for determining ignition delays are carried out under conditions approximating those existing in service rocket motors. Nevertheless, scale effects on ignition reactions may be of sufficient importance to invalidate extrapolation of measurements made in small combustion chambers to performance in service units.

LABORATORY TESTS FOR MEASURING IGNITION DELAYS IN UNCONFINED SYSTEMS.

Cup tests. Among the earliest attempts made in this country at classification of propellants into spontaneous and nonspontaneous pro-

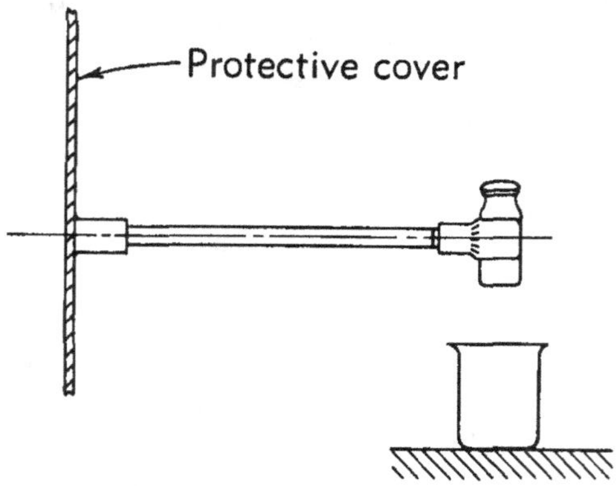

Fig. L,2a. Experimental arrangement used for cup tests. The oxidizer may be poured into the fuel or conversely.

pellant combinations, and assigning ignition delays to the spontaneously reacting mixtures, may be mentioned the simple cup-test experiments (Fig. L,2a). These tests [*1*,*2*] are performed by pouring one component of a bipropellant system into another and noting the time required for the appearance of a flame. The relative amounts of the two propellants mixed are not controlled precisely nor is there a uniform procedure for deciding which of the two components is to be poured into the other. It is usually expedient to perform several modifications of the cup-test ignition delay experiment on a given propellant mixture. In some cases the occurrence of a rapid reaction with a flame may depend critically on the manner in which the cup-test experiment is performed. More frequently, however, substantially the same results are obtained for several different procedures, thus permitting a more or less unique characterization of the propellants. In the experiments of Powell and Kaplan [*1*], the ignition lag in mixtures of red fuming nitric acid and aniline was found to be between 0.08 and 0.12 seconds for more than tenfold variations in the relative proportions of the added propellants. Results of this sort may be understood qualitatively in terms of an energy balance between heat liberated during reaction at the interface between the two components and the heat absorbed primarily by the propellant which is present in large excess. Both of these terms are of similar order of magnitude regardless of relative composition since the physicochemical characteristics are determined mostly by the propellants used and only to a lesser extent by the mixture ratio.

In cup test experiments the flame may be observed visually or else may be recorded by means of a high speed motion picture camera. In spite of the extreme simplicity of the cup tests, the results obtained correlate well in many cases with performance in a rocket motor under actual operating conditions. In the absence of facilities for more refined estimates of ignition delay, the cup test remains to this day one of the simplest useful methods for obtaining qualitative information on ignition delay measurements for propellant mixtures in which liquid-phase mixing and reaction contribute materially to the observed ignition delay.

Falling-drop tests. Closely related to cup tests for making preliminary estimates of ignition delay are the drop tests in which the disparity in concentration of the two components of a bipropellant system may be greatly magnified. Thus a drop of one component is allowed to fall into a cup containing an arbitrary amount of the second component.

In one type of apparatus [*3,4,5,6*] the falling drop interrupts the light reaching one of two photocells, thereby furnishing the reference point from which the length of the ignition delay is counted. The transitory ignition reactions are considered to be completed when the chemical reaction has become sufficiently vigorous to trigger a second photocell as the result of light emitted from the reaction mixture. The standard German drop tester was designated as the LFM Tester J. It consisted of a fuel receptacle and a calibrated orifice to release the oxidizer. The distance through which the oxidizer was allowed to fall was kept constant in order to assure similar impact conditions.

The performance of a two-photocell type of ignition delay apparatus may not be satisfactory since the second photocell is occasionally triggered by spurious radiation effects which are not related, in any obvious way, to the progress of the chemical reactions being studied. For this reason a modification of the two-photocell technique has been introduced [*4,6*], in which the significant physical variable determining the length of ignition delay is the time required to generate a fixed total pressure in a given combustion chamber (Fig. L,2b). Thus, Clark and Sappington define the ignition delay as the time interval between passage of the drop at the position A-A' and the time required to build up a pressure of 2 inches of water in the closed ignition chamber used for study [*6*]. This definition for the ignition delay represents an unfortunate choice, in that the chosen pressure is so small that the measured ignition delays appear to be abnormally short. Nevertheless the effect of various factors on ignition delay, such as composition and temperature of propellants, could still be investigated.

Another variation of the drop-test method for measuring ignition delay involves introducing one drop of oxidizer from a pipette placed a given distance above an aluminum plate containing one drop of fuel. The delay from the time of contact to ignition is measured by use of a high speed camera. Ignition is considered to occur when vigorous chemical

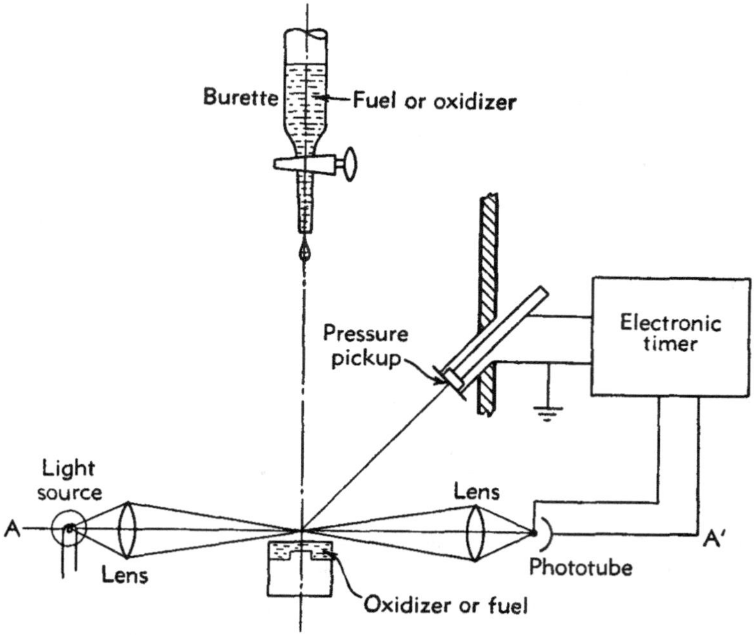

Fig. L,2b. Drop test apparatus used by Clark and Sappington [6].

reaction takes place. Whereas this method has been found to work well for the system H_2O_2-N_2H_4, CH_3OH, H_2O, satisfactory ignition does not occur for the system RFNA[5]-aniline, even though the latter system is known to be spontaneous in rocket combustion chambers. For RFNA-aniline systems, surface catalysis and liquid-phase mixing appear to be of importance in shortening ignition delay.

It is apparent that the physical variables determining the duration of the measured ignition delay in drop tests are essentially the same as the variables involved in cup tests. Considering the tenuous relation between the quantity measured in cup tests or drop tests and the empirical ignition delay which is of significance in the combustion chamber, one wonders if

[5] RFNA = red fuming nitric acid.

the refinements in measuring technique involved in the drop-test apparatus as compared with the cup tests are warranted. It appears reasonable to assume that the physical reasons leading to any significant correlation between laboratory tests and motor performance are, in large measure, the result of the fact that the rate-controlling steps in bipropellant ignition of many propellant systems involve primarily liquid-phase mixing and liquid-phase reactions. If this suggestion were not valid one would certainly expect greater disparity between measured ignition delays for changes in relative concentrations or in available surface than are actually observed.

Ignition delay measurements in the twin-jet apparatus. An apparatus for measuring ignition delay, in which the ignition delay is defined as the time required for liquid-phase mixing and sufficiently vigorous liquid-phase reaction to lead to the production of a luminous flame, has been developed by German[6] and British [*4*] workers. This work may be considered to carry cup tests and drop tests to their logical conclusion.

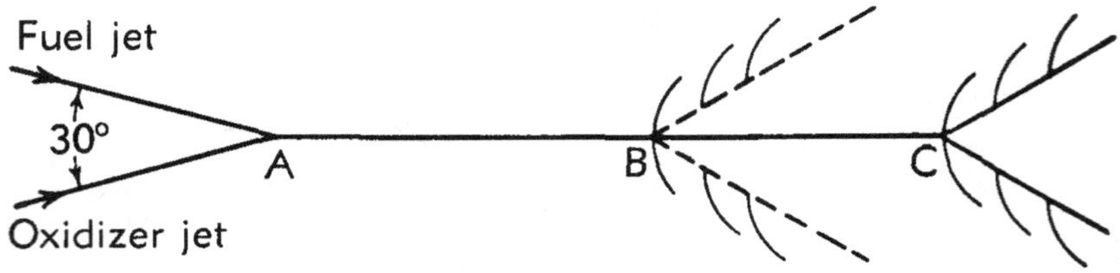

Fig. L,2c. Twin jet apparatus (schematic, cf. [*4*]). *C* represents the point where ignition is first observed; *B* represents the point beyond which rapid exothermic reaction occurs during the steady state. The points *B* and *C* may not coincide. The combined fuel and oxidizer jets beyond *A* are allowed to fall freely under gravitational forces.

In the twin-jet apparatus the oxidizer and fuel streams projected from two separate glass capillaries are allowed to impinge in air or nitrogen at atmospheric pressure (Fig. L,2c). The angle between the coalescing liquid propellants is chosen sufficiently small so that a single continuous stream results after collision, with relatively little carry-through. The resulting liquid mixture falls freely under gravitational acceleration until liquid phase mixing and chemical rate processes lead to vigorous exothermic reaction and vaporization of the propellant mixture. The ignition delay, as measured by the twin-jet apparatus, should then be defined as the time elapsed between impingement of the separate oxidizer and fuel streams and the *first* occurrence of rapid chemical reaction. In practice it has proved to be desirable to emphasize the point at which the ignition delay period ends by placing a strong light behind the stream of propellant mixture and measuring the length of the liquid path, which now appears as a dark silhouette against a lighter background [*4*].

The use of the twin-jet apparatus for the measurement of ignition

[6] The German development was under way at Trauen and is quoted by Broatch [*4*].

delay is somewhat similar to its application for steady state combustion studies. In this case it is not the *first* occurrence of vigorous exothermic reaction which is taken as the end point; rather, a steady state measurement is made in which the position of the resulting point of vigorous exothermic reaction is dependent, to some extent, on heat and mass transfer to the liquid from the combustion products and intermediates. It is obvious that use of the twin-jet apparatus for steady state measurements will be related to ignition delay studies only if the point of vigorous exothermic reaction does not travel upstream to an appreciable extent, as the steady state is approached.

As mentioned earlier in this discussion, the twin-jet apparatus is essentially a device for measuring quantitatively the time required for the occurrence of vigorous chemical reaction as dependent on liquid-phase mixing and chemical reactions. In view of this fact it is not surprising to note that the measured ignition delay appears to increase as the angle between the two impinging propellant-component streams is decreased, thereby decreasing the rate of liquid-phase mixing. On the other hand, if the angle between the two converging liquids is made too large, considerable penetration occurs which may again lead to improper mixing and an increase in the length of the ignition delay period. For measurements on the system using 80 per cent H_2O_2-20 per cent H_2O (containing sodium pyrophosphate as stabilizer) as oxidizer and 30 per cent $N_2H_4 \cdot H_2O$-13 per cent H_2O-57 per cent CH_3OH (containing 0.6 g Cu per liter in the form of $KCu(CN)_2$ for ignition catalysis) as fuel, an angle of 30° between the impinging liquid streams was selected.

The twin-jet apparatus has also been incorporated in a large chamber in order to permit quantitative study of the effect of low pressure and temperature on ignition delay defined through the occurrence of liquid-phase mixing and reactions. These measurements are useful in assessing the importance of liquid-phase rate processes at high altitudes.

Ignition Delay Measurements in Semiconfined Chambers. Several devices for measuring ignition delays in semienclosed chambers have been developed. In general, one would expect reasonably good correlation between this type of ignition delay test and motor performance since the various physical and chemical factors influencing the length of the ignition delay are similar.

The MIT semienclosed chamber. In an ignition delay apparatus developed at the Massachusetts Institute of Technology, a high speed motion picture camera is used to record the reaction between streams of oxidizer and fuel impinging at high velocity (approximately 150 ft/sec) in a semienclosed chamber [7]. The chamber has an opening to facilitate the photographing of the ignition reactions. The ignition delay is defined as the time elapsed between first impingement of the liquid streams and the first observation of incandescence. The end of the ignition delay period is also estimated by recording the noise accompanying the vigorous chemical reaction, with the aid of a microphone. The use of high injection velocities leads to rapid liquid-phase mixing and reactions, resulting in relatively short ignition delays.

IGNITION DELAY MEASUREMENTS IN SMALL SCALE AND SERVICE ROCKET CHAMBERS. Ignition delay measurements, published and unpublished, have been made on a great many propellant systems in a wide variety of rocket motors. A critical evaluation and compilation of the results which have been obtained is outside the scope of the present discussion. Instead, a brief review of representative apparatus for ignition delay measurements in simulated motor tests is presented.

Microrocket measurements. Ignition delay measurements in the microrocket are carried out under conditions closely approximating those occurring in service combustion chambers, the principal difference being a much larger surface-to-volume ratio and fewer injection nozzles. As with the cup-test and twin-jet measurements, results will obviously be useful if physical and chemical changes occurring in the liquid phase are rate-controlling. Otherwise, a scale effect of unknown magnitude is involved whenever an extrapolation to a larger combustion chamber is performed. This scale effect may be the result of decreased surface-to-volume ratio. Furthermore, in large combustion chambers the time required to reach the motor wall is increased, particularly for the propellants injected near the middle of the combustion chamber. As a result, increased ignition delays will be observed in large motor chambers if surface catalysis is important.

Microrockets [*8*] are convenient devices not only for measuring ignition delay but also for motor performance studies requiring minimal consumption of the propellants being tested. Representative microrockets have chamber lengths from $1\frac{1}{2}$ to 4 inches and nozzle throat diameters from 0.144 to 0.179 inches. A variety of injectors (Fig. L,2d) has been used. The ignition delay is defined as the time elapsed between injection into the combustion chamber and the first occurrence of a rapid increase in chamber pressure. Ignition lag measurements in the microrocket on the system H_2O_2-N_2H_4 have shown an interesting correlation between short ignition delay and high combustion efficiency.

Facilities are generally available for studying the effect of temperature, injection pressure, and initial chamber pressure on ignition lag in the microrocket.

Ignition lag measurements in transparent combustion chambers. A glass combustion chamber has been constructed in order to permit photographing of the transient chemical reactions preceding steady state combustion. The entrance of propellants, impingement, and onset of luminosity are recorded with a high speed camera. The transparent motor tests have been correlated with results obtained in metal combustion chambers in which the electrical conductivity of the liquids was used to record the time of the first impingement. The pressure rise in the combustion chamber was followed with a suitable gauge and was used to estimate the end of the ignition delay period. The injectors used for these ignition lag measurements yielded two impinging liquid streams and required between 5 and 20 cm^3 of propellant per run. An apparatus of this type has been used to study the effect of temperature and of chemical additives on the measured ignition delay.

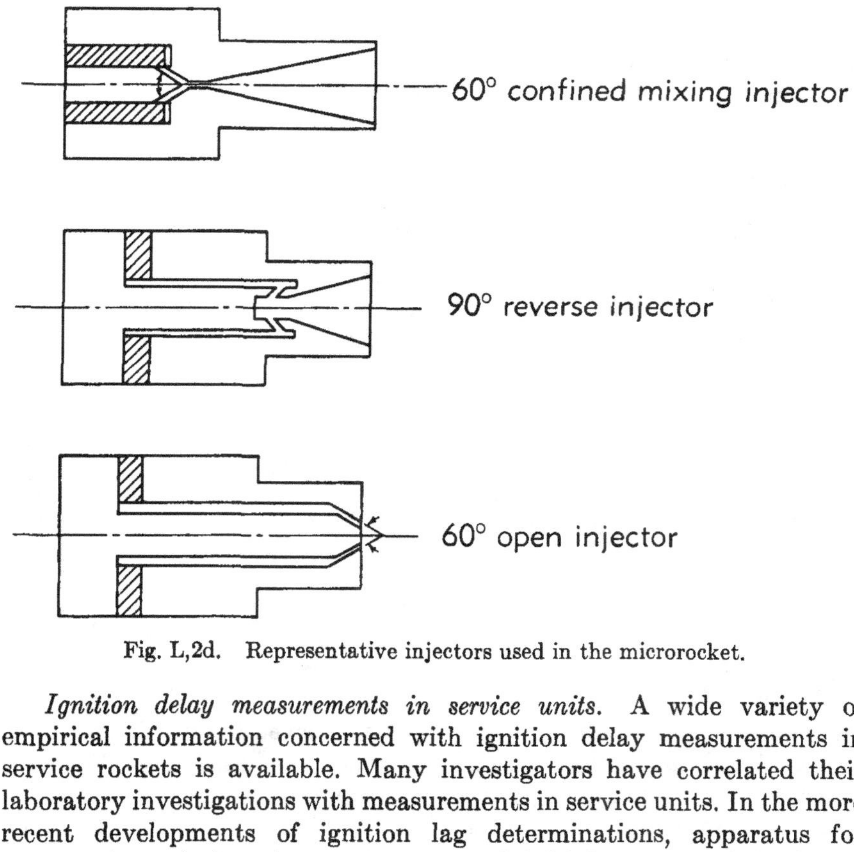

Fig. L,2d. Representative injectors used in the microrocket.

Ignition delay measurements in service units. A wide variety of empirical information concerned with ignition delay measurements in service rockets is available. Many investigators have correlated their laboratory investigations with measurements in service units. In the more recent developments of ignition lag determinations, apparatus for systematic variations in temperature and initial pressure (in order to simulate high altitude operation) has been included. The rocket motors used for ignition lag measurements are usually small, developing thrusts in the neighborhood of 50 pounds. For this reason the problem of scale effects in designing large motors remains as an important research program to which relatively little attention has been devoted thus far.

L,3. Representative Ignition Delay Measurements on Spontaneous Bipropellant Systems. A tabulation of experimentally observed ignition delay measurements is of interest since it may aid our understanding of the transient processes preceding steady state combustion. Seven topics will be considered, viz. (1) the effect of experimental technique on observed ignition delay, (2) the effect of soluble catalysts and composition on ignition delays observed in a given apparatus, (3) the effect of initial temperature, (4) the effect of initial chamber pressure, (5) the effect of surface catalysts, (6) the relation between ignition delay and chemical structure, and (7) the effect of injection pressure on ignition delay.

Of coordinate practical importance with any of the factors enumerated above, is the relation between injection technique and ignition delay. Although the study of this problem has received considerable attention

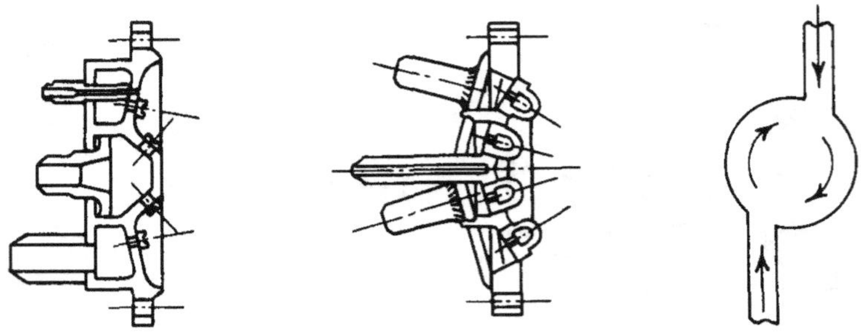

Fig. L,3a. Injectors with intersecting liquid streams which emphasize liquid-phase reactions. These injectors are useful for such systems as RFNA-AN, RFNA-$N_2H_4 \cdot H_2O$, H_2O_2-N_2H_4, etc. Left, intersecting stream injector; center, multiple-cone injector; right, premixing injector (the mixed fuel-oxidizer stream passes out in a direction normal to the page as a spray).

from engineers associated with motor development projects, the quantitative classification and tabulation of this type of information does not appear to be feasible. Hence a few general statements will have to suffice.

In many bipropellant systems a decrease in ignition delay, and also an increase in motor performance, may be achieved by utilizing methods for rapid and complete liquid-phase mixing. In recognition of this fact, injection techniques utilizing the type of injectors shown in Fig. L,3a and L,3b have been used in the past. Furthermore, liquid-phase premixing injectors[7] have been developed (Fig. L,3a(right)). These injectors are useful for propellant systems containing components of low volatility. On the other hand, for systems involving very volatile components such as liquid hydrogen, oxygen, fluorine, or even ammonia, short ignition delays and adequate motor performance are not critically dependent on liquid-phase mixing. In this case the injector designs drawn in Fig. L,3b(left) and L,3a(center) are useful since they give fine sprays of drop-

[7] The premixer can evidently be used for any bipropellant system provided the residence time in the injector is appreciably less than the observed ignition delay.

lets which will evaporate rapidly, thereby accelerating gas-phase reactions. Although some investigators [*12*] have claimed that the efficiency of the system ethyl alcohol-liquid oxygen is improved with liquid-phase mixing, this result is probably caused by the increased atomization.

Comparison of Observed Ignition Delays for Given Propellant Systems. In order to indicate the concordance between experimental results obtained by various methods for a given propellant system, a tabulation of observed data is given in Table L,3a for the reactants

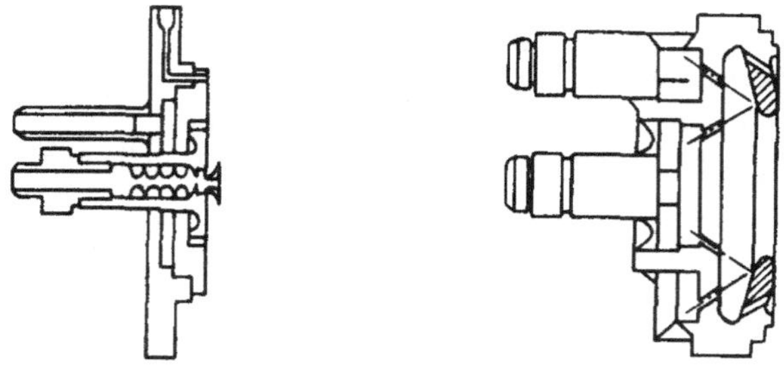

Fig. L,3b. Injectors which yield fine sprays of fuel and oxidizer, emphasizing gas-phase reactions. These injectors are useful for such systems as O_2-NH_3, H_2-O_2, etc. Left, concentric hollow cone injector; right, target type injector.

RFNA and AN. Reference to Table L,3a shows large differences in results reported by different observers although the more reliable cup and drop tests [*1*,*5*] are seen to agree well with results obtained in a semienclosed chamber and a small motor. The absolute values of the observed I.D. (ignition delay) reported in [*6*] are too small because a poor definition of I.D. was used (cf. Art. 2). The cup tests reported in [*4*] probably yielded small values for the I.D. because the reactions were catalyzed by meta

Table L,3a. Ignition delay determined by various methods for the RFNA (red fuming nitric acid)-AN (aniline) system at room temperature and an initial pressure of 1 atm.

Observed I.D., sec	*Reference*	*Type of apparatus*	*Remarks*
0.08 to 0.12	[*1*]	Cup test	The observed values of the I.D. cover a one-hundredfold variation in concentration
0.09	[*5*]	Drop test	
0.003 to 0.008	[*6*]	Drop test	The observed I.D. was roughly 2 msec larger when AN was dropped into excess RFNA, than when RFNA was dropped into excess AN
0.06	—	Semienclosed chamber and small scale motor tests	
0.004 to 0.008	[*2*]	Cup test	Ignition may have been catalyzed by metal salts

salts dissolved in the nitric acid. Comparing all the values obtained with the different sets of apparatus, it appears that I.D. values in the range 0.06 to 0.12 probably are most typical for this system. Some scale factor is expected when extrapolating to large service units since surface catalysis affects the initiation of the ignition reaction and surface-to-volume ratios decrease with size in geometrically similar chambers.

Effect of Composition on Ignition Delay. One of the important bipropellant systems for which studies of suitable ignition catalysts have been carried out is the system H_2O_2 (containing NH_4NO_3)-$N_2H_4{\cdot}H_2O$. The ignition delay for this system was found to increase as the amount of ammonium nitrate added to H_2O_2 was increased [7]. Ammonium nitrate is not generally considered as a combustion catalyst, and since it does not give up oxygen as readily as does H_2O_2 its action is mainly that of a diluent.

The system H_2O_2-$N_2H_4{\cdot}H_2O$ was used by the Germans in rocket aircraft and as a spontaneous combination for ignition in rocket torpedoes, with decalin employed as the fuel in the latter application for steady state combustion. In carrying out these developments the Germans found that a small amount of copper or iron catalyst, added to hydrazine hydrate in the form of complex cyanides, materially decreased the observed ignition delay. The German work has been reviewed by Darling and Kavanagh [7], who have also supplemented the available ignition delay data by carrying out measurements in semienclosed ignition chambers using photographic records to follow the progress of ignition. Representative results taken from this work are reproduced in Table L,3b [7].

Table L,3b. The effect of copper and iron catalysts on the ignition delay of 100 per cent $N_2H_4{\cdot}H_2O$ reacting with 90 per cent H_2O_2 at room temperature [7].

Catalyst added	*Concentration of Cu or Fe in g/l*	*Ignition delay, sec*
$K_3Cu(CN)_4$	0.01	no ignition
	0.1	0.090
	1.0	0.021
$Na_2Fe(CN)_5NO$	0.19	0.087
	0.47	0.067
	0.93	0.036
	1.87	0.022

Although the absolute values of the I.D. listed in Table L,3b are certainly dependent on the type of ignition delay apparatus in which measurements are made, experience has shown that laboratory tests are generally useful in evaluating ignition catalysts for given bipropellant systems. Thus the use of furfuryl alcohol-aniline mixtures with RFNA and WFNA (white fuming nitric acid) is the result of laboratory measurements on the I.D. [*2*]. Similar remarks apply to the use of ferric chloride, vanadium pentoxide, and ammonium metavanadate used as ignition catalysts in WFNA reacting with aniline which may contain copper formate as ignition catalyst [*2*]. These experimental results suggest that oxidation catalysts appear to have an accelerating effect on the ignition process of these systems.

The chemical rate processes occurring either during ignition or during steady state combustion are not understood well enough to provide a complete explanation for the role of catalysts during ignition. Accordingly the development of ignition catalysts must be viewed as an art rather than as a science.

Effect of Temperature on Ignition Delay. Reference to the general scheme for describing ignition reactions given in Fig. L,1 suggests immediately that an increase in temperature should decrease the ignition delay since the physical and chemical factors which might conceivably be rate-controlling are all known to change as the temperature is raised. This supposition has been amply verified by studies on various bipropellant systems.

In drop-test experiments, Gunn [*5*] found a marked increase in I.D. for the system RFNA-AN as the temperature was decreased below about 15°C. Whereas the observed I.D. was more or less constant between 15 and 50°C at 0.09 sec, it increased to nearly 0.20 sec as the temperature was lowered to 0°C. Qualitatively similar results were obtained for the same system in drop-test experiments [*6*].

For a given composition of ignition catalysts used in hydrazine hydrate, Darling has observed a nearly tenfold increase in I.D. with H_2O_2 containing various amounts of NH_4NO_3, as the temperature was decreased from 25 to -35°C. These measurements were performed in semienclosed chambers. A much less pronounced dependence of I.D. on temperature has been reported for other propellant systems.

In tests performed in a 50-lb motor, an increase in ignition delay from about 0.015 to 0.040 sec has been observed as the temperature was decreased from -10 to -30°C for the system RFNA reacting with 2 parts by volume of FA plus 1 part by volume of AN. The ignition delay was observed to be constant between -10 and $+10$°C. Broatch [*4*] observed a linear decrease of I.D. in the twin-jet apparatus as the temperature was increased from 15 to 25°C for reaction between 80 per cent H_2O_2 (containing sodium pyrophosphate as stabilizer) and a fuel mixture consisting of 30 per cent $N_2H_4 \cdot H_2O$, 13 per cent H_2O, 57 per cent CH_3OH (containing 0.6 g of Cu per l in the form of $KCu(CN)_2$).

It is not possible to predict the quantitative effect of temperature on the I.D. at the present time. The data appear to indicate that not only do the rate constants influence ignition, but also physical properties such as viscosity and surface tension which affect the rate of mixing.

Effect of Initial Pressure on Ignition Delay. In order to simulate ignition at high altitudes, extensive experimental work has been done to determine the effect of low initial pressures, with or without simultaneous cooling, on I.D. It has been found, using the twin-jet apparatus, that spontaneous systems may exhibit sufficiently large increases in I.D. to become nonspontaneous at low pressures.

In a semienclosed chamber it has generally been found that the I.D. for systems such as RFNA-80 per cent AN plus 20 per cent FA increased as the pressure was reduced. It appears reasonable to suppose that the observed increase in I.D. for bipropellant combinations for which liquid-phase reactions are of importance is, in some measure, the result of increased droplet dispersion and vaporization at reduced pressures. Furthermore, if ignition requires the reaction of partially reacted vapors, it is clear that low pressures will inhibit this gas-phase reaction and hence increase the ignition delay.

Effect of Surface Catalysts on Ignition Delay. No extensive studies of surface catalysis on the I.D. of spontaneous bipropellant systems appear to have been carried out. However, this type of catalysis has been found to be effective for monopropellants such as H_2O_2 and N_2H_4 (cf. Art. 5). Accordingly one might expect that surface catalysis would also be of some value in bipropellant systems, particularly in those combinations using potential monopropellants. Failure to use surface catalysis for spontaneous bipropellants may be ascribed to the fact that other methods of accelerating reactions are more convenient and adequately efficient. As mentioned earlier in this discussion, ignition of the system RFNA-AN is facilitated by the presence of a surface. Combustion efficiency in this case is also improved by using turbulence rings.[8]

Surface catalysis has important applications in the use of nonspontaneous bipropellant and monopropellant systems (cf. Art. 4 and 5).

Relation Between Ignition Delay and Chemical Structure. The relation between ignition delay and chemical structure or composition is obviously complicated by the interplay between chemical and hydrodynamic factors. Nevertheless it is to be expected that for bipropellant systems in which liquid-phase reactions are of critical importance, useful correlations can be obtained [*9*].

Ignition delay measurements using RFNA for the oxidizer show that unsaturated groups, such as the allyl group, enhance ignition. Saturated hydrocarbons are normally nonspontaneous. On the other hand, the aromatic nucleus has an activating effect when connected to a nitrogen or sulfur atom. Ignition is also facilitated by replaceable hydrogen atoms connected to the nitrogen or sulfur atom. These conclusions are based on cup tests, drop tests, and motor performance tests on a great variety

[8] The Enzian injector for the RFNA-N_2H_4 system makes efficient use of surface catalysis.

of organic compounds, of which mercaptans and unsaturated amines are considered to be the most promising spontaneous fuels with RFNA. The unsaturated amines were found to tolerate extensive dilution with petroleum hydrocarbons and still remain spontaneous. Preliminary screening of fuels which ignite spontaneously with RFNA was carried out by determining how far a self-igniting material may be diluted with a combustible but nonspontaneous diluent before satisfactory ignition fails to occur in drop tests. Investigations of this type are closely related to the material discussed under the heading EFFECT OF COMPOSITION ON IGNITION DELAY.

EFFECT OF INJECTION PRESSURE ON IGNITION DELAY. Relatively little quantitative work has been done on the effect of hydrodynamic factors on the ignition delay. From the earlier discussion on the mechanism of ignition delay, one might expect that high injection velocities would decrease the ignition delay because of better atomization, mixing, and frictional heating of the streams. This effect has been observed over a limited pressure range for several systems employing RFNA as oxidizer and also for the H_2O_2-N_2H_4 system in closed chambers at very high injection pressures [*9*]. A quantitative explanation of these effects has not been given but it appears that more experiments of this type should be useful in clarifying the influence of hydrodynamic factors on the ignition process.

L,4. Ignition of Representative Nonspontaneous Bipropellants. The discussion presented in Art. 3 suggests that it may be possible to obtain satisfactory ignition with nonspontaneous propellants by using soluble or surface ignition catalysts, by changing the injector design appropriately, or by using an extraneous heat source to start reaction. If these methods fail, it is still generally possible to use a given nonspontaneous combination operationally by utilizing a small amount of a self-igniting combination in order to initiate combustion. An example of this procedure was quoted in Art. 3 in the use of $N_2H_4 \cdot H_2O$ to start the reaction between decalin and hydrogen peroxide in one type of German rocket torpedo. Ignition experience obtained to date indicates that this method can be successfully employed for nearly all nonspontaneous combinations containing nitric acid or H_2O_2 because these oxidizers form spontaneous combinations with many fuels.

An outstanding example of a nonspontaneous bipropellant system, which is attractive operationally because of wide availability and low cost of components, is the system RFNA-gasoline. Many soluble ignition catalysts have been developed for this system. Thus the addition of suitable amines or mercaptans to the gasoline and, alternately, the addition of a powerful oxidizing agent such as $KMnO_4$ to RFNA, results in spontaneous combinations. Unfortunately the concentrations of amines and mercaptans in gasoline required to obtain satisfactory ignition are fairly large (usually in excess of 10 per cent by volume). Furthermore the activity of $KMnO_4$, added to RFNA as an ignition catalyst, decreases rapidly with time as the result of precipitation of MnO_2. It is expected that further developments of the art will ultimately lead to the invention of more satisfactory soluble catalysts than are available at the present time.

Many bipropellant systems using liquid oxygen as oxidizer are nonspontaneous. Ignition of the system, liquid oxygen-liquid ammonia, is brought about conveniently by using a spark plug to supply the energy necessary to initiate the exothermic reactions which are characteristic of the oxidation of ammonia. High performance of this nonspontaneous system may then be obtained by using a suitable injection technique. The bipropellant combinations, oxygen-hydrazine, RFNA-alcohol, and oxygen-alcohol, can be ignited readily by using an independent energy source such as a powder squib or pyrotechnic igniter. Ignition of the oxygen-alcohol propellant in the German V-2 was initiated in this manner.

The system, RFNA-ammonia, can be made spontaneous by dissolving a small amount of an alkali metal (e.g. Li) in the ammonia.

Ignition of the nonspontaneous combination, 80 per cent H_2O_2-kerosene, may be accomplished by spraying the H_2O_2 on an appropriately heated noncatalytic surface. Catalytic surfaces for this system fall into two classes, viz. surfaces which catalyze the decomposition of H_2O_2 alone and those which catalyze both the decomposition of H_2O_2 and the oxidation of kerosene. Platinum, palladium, and hopcalite are examples of the latter while MnO_2 pellets and silver-plated copper gauze are typical of the former.

L,5. Ignition of Monopropellants. Since all useful monopropellants are intrinsically unstable, they can always be decomposed by suitable application of heat or energy. The practical requirement of short ignition delay places definite limitations on the manner in which the requisite energy is applied. A method of ignition which is suitable for one combustion chamber geometry may not be suitable for another. Consider, for instance, the ignition of a monopropellant by catalysis on the preheated walls of the combustion chamber. Heat transfer from the wall, and reaction kinetics will determine a characteristic time of ignition. This "ignition delay time," when multiplied by the injection rate of monopropellant, will give a definite quantity of propellant in the chamber leading to pressure transients which are characteristic of the chamber volume. Consider a second motor similar to the first but of shorter length, and hence smaller volume. Since the ignition geometry is substantially the same, the quantity of propellant accumulated at ignition is similar, and therefore a transient pressure greater than that occurring in the larger motor will result. Such transient pressures have been known to rupture rocket motors at starting. Thus it must be concluded that for monopropellants as well as for bipropellants not only the chemistry of the monopropellant but also the geometry of the injector and combustion chamber will influence the ignition.

The methods used for initiating the thermal decomposition of monopropellants are similar to those employed for nonspontaneous bipropellants. For convenience they are summarized explicitly below.

Heated surface. Ignition is effected by heating the combustion chamber walls or, better, by introducing into the chamber a resistance element heated electrically to a suitable temperature. This technique of ignition is applicable to nearly all monopropellants.

Pyrotechnic ignition. Pyrotechnic ignition involves the use of a small quantity of combustible placed near the injector. The heat generated by the igniter is sufficient to initiate steady state decomposition of the monopropellant.

Spark. This method, although quite commonly used for the ignition of gas mixtures, can be employed effectively with only a limited number of monopropellants. The spark is an intense temperature source confined to a small region and is usually of limited energy output. Because of the local confinement of the spark, there is danger of drowning by large droplets. Good atomization, therefore, is usually necessary for successful application of this technique.

Bipropellant ignition. This mode of ignition is restricted to those monopropellants which form spontaneously ignitable combinations with another propellant, a small quantity of which is used to initiate steady state combustion.

Surface catalysis. Chemical catalysis on surfaces may be of two types, viz. one in which the surface functions as a true catalyst and is not consumed during ignition and the other in which it reacts chemically to produce ignition. The former type of ignition is obviously more desirable since it can be used repeatedly.

Detailed information concerning specific methods for initiating the decomposition of monopropellants will be presented in connection with the discussion of the performance of these compounds.

CHAPTER 2. MOTOR PERFORMANCE OF SELECTED MONOPROPELLANTS

L,6. General Characteristics of Monopropellants. A monopropellant is a single fluid capable of undergoing exothermic reaction to yield gaseous products. The rate of decomposition of a useful monopropellant must obviously be negligible at storage temperatures. During steady state decomposition, heat and mass transport from the reaction zone occur at a rate sufficient to maintain the decomposition processes. The operation of a monopropellant rocket motor has several distinct advantages over a bipropellant motor. These are: (1) For operation of a monopropellant, one propellant tank with a single pumping arrangement is required. The simplicity so achieved may be desirable for applications where high performance is not required, as in gas generators and assisted take-off units. (2) The injection is simplified since exact impingement of propellant streams for purposes of uniform mixing is unnecessary. (3) The operation of a monopropellant motor is less apt to vary with ambient temperatures. For bipropellants, changing ambient temperatures may cause unequal density changes in the fuel and oxidizer. For a given fluid volume injected, this density change affects the mixture ratio at which the propellant operates because rates of injection of fuel and oxidizer may change differently. As a result, one propellant tank will empty prior to the other, causing an unnecessary waste of the remaining component. (4) Finally, the use of a single propellant will simplify field operations, since only one tank needs to be filled. For prefilled rockets this consideration is of little importance but it is a significant factor for units requiring tank filling in the field.

In order that a fluid be suitable for monopropellant operation, it must satisfy certain thermodynamic and kinetic requirements. The thermodynamic requirement that the substance be capable of undergoing exothermic decomposition to produce gases has already been mentioned. The kinetic requirements are generally dictated by operational procedures. The kinetic requirements are: (1) the decomposition rate at storage temperatures must be negligibly small or else capable of inhibition by chemical or other means; (2) the decomposition should be readily initiated (some of the techniques used have been discussed in Art. 5); (3) the release of energy should occur smoothly and reproducibly so that explosion or detonation can be avoided; and (4) the rate of decomposition at the steady state should be reasonably fast so that excessively large combustion chamber volumes need not be employed. The last two kinetic factors should, in principle, be capable of analysis by a theoretical treatment similar to that used to describe the burning of solid propellants. However, our knowledge regarding high temperature kinetics is so inadequate that at best only semiquantitative correlations are feasible. Since the thermodynamics of combustion is well understood, it is evident that future advances of fundamental significance in combustion will lie in expanding our knowledge of rate processes, particularly of the interplay of hydrodynamics and chemical reactions. Important steps in this direction have already been made by several investigators [*10,11,12*].

L,7. Classification of Monopropellants. Depending on their chemical composition, monopropellants may be conveniently divided into three separate classes.

Class A. Members of class A contain both oxidizer and fuel constituents in the same molecule. Class A embraces the large field of conventional liquid explosives, only a small portion of which has the necessary stability and kinetic requirements for use as monopropellants. Examples of class A compounds which have been considered for use as monopropellants are nitromethane (CH_3NO_2), methyl nitrate (CH_3NO_3), nitroethane ($C_2H_5NO_2$), ethyl nitrate ($C_2H_5NO_3$), nitroethanol ($HOC_2H_4NO_2$), diethylene glycol dinitrate or DEGN $(C_2H_4ONO_2)_2O$, ethylene oxide (C_2H_4O), nitroglycerin $C_3H_5(ONO_2)_3$, and hydrogen peroxide (H_2O_2).

Class B. This class contains, in general, only the separate oxidizer or fuel constituents. The decomposition energy is derived from a chemi-

Table L,7. Physical properties and heats of formation of selected monopropellants.*

Monopropellant	Density, g/cm^3	Melting point, °C	Boiling point, °C	Standard heat of formation $\Delta H^\circ_{f\,(298)}$, kcal/mole
Class A compounds				
CH_3NO_2	1.130^{20}_{4}	−29	101	−26.7, −27.6 [*14*], −21.3 [*16*]
CH_3NO_3	1.203^{25}_{4}	—	65, explodes	−41 (estimated)
$C_2H_5NO_2$	1.052^{20}_{20}	−90	115	−30
$C_2H_5NO_3$	1.105^{20}_{4}	<−102	88	−44.3
$HOC_2H_4NO_2$	1.270^{15}_{4}	−80	194	—
$(C_2H_4ONO_2)_2O$	1.377^{25}_{4}	−11.3	161	−103 (estimated)
C_2H_4O	0.887^{7}_{4}	−111	10.7	−18.4
$C_3H_5(ONO_2)_3$	1.601	13.3	260, explodes	—86 (calc)
H_2O_2	1.465^{0}_{4}	−0.9	152	−44.84
Class B compounds				
N_2H_4	1.011^{15}_{4}	1.4	113.5	+12.05
C_2H_2	0.621^{-84}_{4}	−82	−84, sublimes	+54.19
C_2H_4	0.566^{-102}_{4}	−169	−104	+12.50
NO	1.269^{-150}_{4}	−163.6	−151.8	+21.6
Class C compounds				
$NH_{3(l)}$	0.817^{-79}	−77.7	−33.4	−16.07 [*14*]
CH_3OH	0.792^{20}_{4}	−97.8	64.7	−57.04
$C(NO_2)_4$	1.650 [*41*]	13	126	+8.9 [*15*]

* Unless indicated to the contrary, all data in Table L,7 have been taken from the compilation of physical properties of chemical compounds prepared by the National Bureau of Standards [*13*].

cally unstable arrangement of the atoms in the molecule. For this reason all of the compounds in this group have positive values for the standard heats of formation (cf. Table L,7). Examples of compounds in this group, some of which may be usable monopropellants, are hydrazine (N_2H_4), acetylene (C_2H_2), ethylene (C_2H_4), and nitric oxide (NO).

Class C. A propellant of this class is a synthetic mixture of two or more compounds of fuel and oxidizer. It is evidently necessary that the components used to prepare a class C compound are not spontaneous with one another but are capable of smooth and continuous reaction once ignition has occurred. These restrictions narrow down the apparently limitless number of combinations which might otherwise be possible. Class C compounds contain the high performers among the monopropellants. Members of this class are: NH_3(l)-NH_4NO_3 (Diver's solution), NH_4OH-NH_4NO_3, CH_3NO_3-CH_3OH (Myrol), H_2O_2-CH_3OH, $C(NO_2)_4$-fuel.

Densities, melting points, boiling points, and heats of formation of representative monopropellants are summarized in Table L,7.

L,8. Performance Characteristics of Monopropellants. This article deals with theoretical performance values and operational characteristics of several monopropellants. The evaluation of monopropellants is concerned with performance, thermal and shock sensitivity, rate of thermal decomposition at temperatures experienced in the regenerative cooling coils, ease of combustion, and general physical properties. Next to performance, the most important of these items are the sensitivity and the thermal decomposition rate. The apparently contradictory requirements of low thermal decomposition rates at moderate temperatures, on the one hand, and ease of combustion, on the other hand, can only be realized in monopropellants with relatively high activation energies for decomposition. A suitable compromise between sensitivity and ease of combustion is not always feasible. Thus the use of a sufficiently insensitive monopropellant usually requires rather large combustion volumes leading to values for L^* which are generally considerably larger than those used for bipropellants.

PERFORMANCE OF CLASS A COMPOUNDS. A considerable amount of work has been done on the heats of combustion and the explosive properties of class A compounds, which are widely used as explosives. A comprehensive review of these compounds, as well as of class B compounds, containing an interesting correlation between oxygen balance and physicochemical properties such as power or brisance, heat of combustion, etc., has been given by Lothrop and Handrick [*18*]. We shall utilize the results and methods of these authors in order to correlate composition and properties for monopropellants.

The oxygen balance is defined as $-1600[2x + (y/2) - z]/($molecular

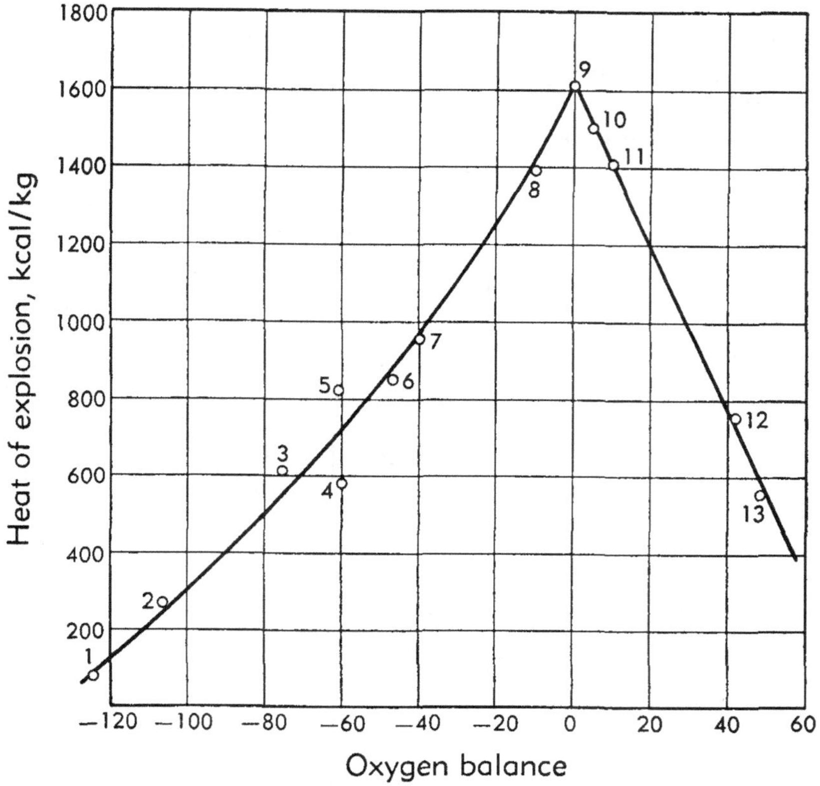

Fig. L,8a. Heat of explosion of organic nitro and nitrate compounds as a function of oxygen balance.

Legend

1. 2,3 dinitro-2,3 dimethyl butane
2. 1,3 dinitro-2,3 dimethyl propane
3. TNT
4. 2,2 dinitro propane
5. ethyl nitrate
6. tetryl
7. nitromethane
8. pentaerythritol tetranitrate
9. ethylene glycol dinitrate
10. nitroglycerin
11. manitol hexanitrate
12. hexanitro ethane
13. tetranitro methane

weight), where x represents the number of atoms of carbon, y equals the number of atoms of hydrogen, and z represents the number of oxygen atoms in the molecule. A zero balance therefore indicates that there is just sufficient oxygen to convert all the carbon and hydrogen to CO_2 and H_2O, respectively; it represents the condition for maximum heat release per gram.[9]

[9] The use of the oxygen balance ignores differences in the heats of formation of the compounds, which causes some scatter in graphs showing the desired correlation (Fig. L,8a to c). The scatter is not serious for class A compounds since the heat of combustion is much larger than the heat of formation.

The relation between heat of explosion and oxygen balance for several representative compounds of class A is shown in Fig. L,8a. The expected behavior of maximum heat release at zero oxygen balance is evident in this graph, which also shows a relatively small slope for a negative oxygen balance. This effect is far more pronounced for rocket performance parameters like I_{sp} and c^* which have a broad maximum extending toward negative values of the oxygen balance. Fig. L,8b shows the relation between explosive power as measured in the lead-block expansion test

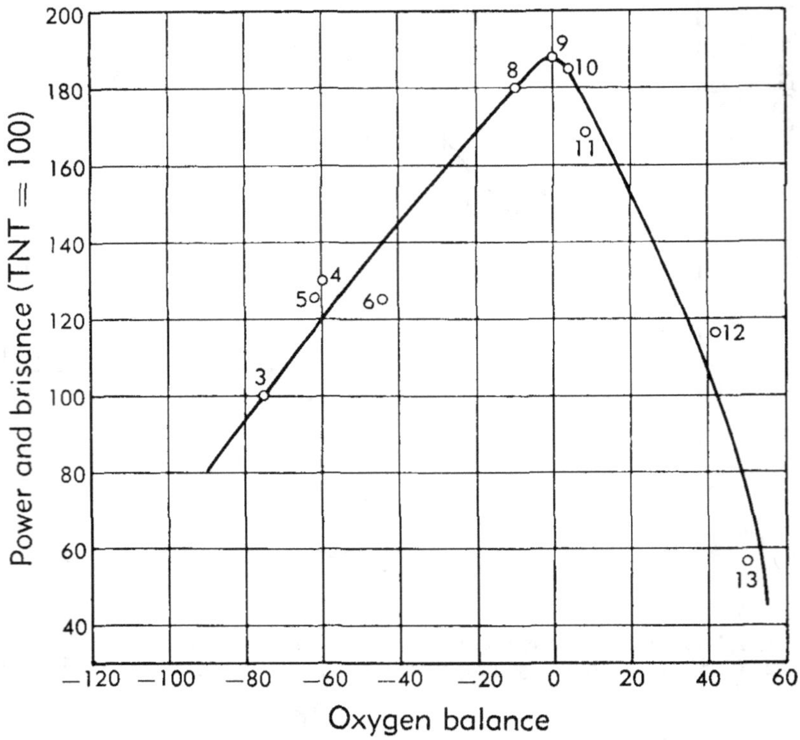

Fig. L,8b. Power and brisance as a function of oxygen balance. Legend: cf. Fig. L,8a.

and oxygen balance; the curve is seen to be very similar to that given in Fig. L,8a. The similarity in shape of these two curves shows that heat of explosion and explosive power are related for class A compounds. This relation is brought out in Fig. L,8c. It is of fundamental importance in evaluating class A monopropellants to recognize that high performance, which is directly related to the heat of explosion, will generally be obtained at the expense of increased explosive power.

Nitromethane. This compound is the first member of the nitroparaffins. Thermodynamic calculations show that at the adiabatic flame temperature, the equilibrium composition is very closely represented by the following equation:

$$CH_3NO_2 = 0.2CO_2 + 0.8CO + 0.8H_2O + 0.7H_2 + 0.5N_2 \quad (8\text{-}1)$$

Using a heat of formation, $\Delta H^0_{f(298)}$ of -27.6 kcal/mole for CH_3NO_2 gives a flame temperature of 2450°K and an I_{sp} of 218 seconds at 300 psia chamber pressure exhausting to one atmosphere. The recently reported $\Delta H^0_{f(298)}$ value of -21.3 kcal/mole [*16*] raises the flame temperature and I_{sp} by about 4 and 2 per cent, respectively. Because of the very favorable

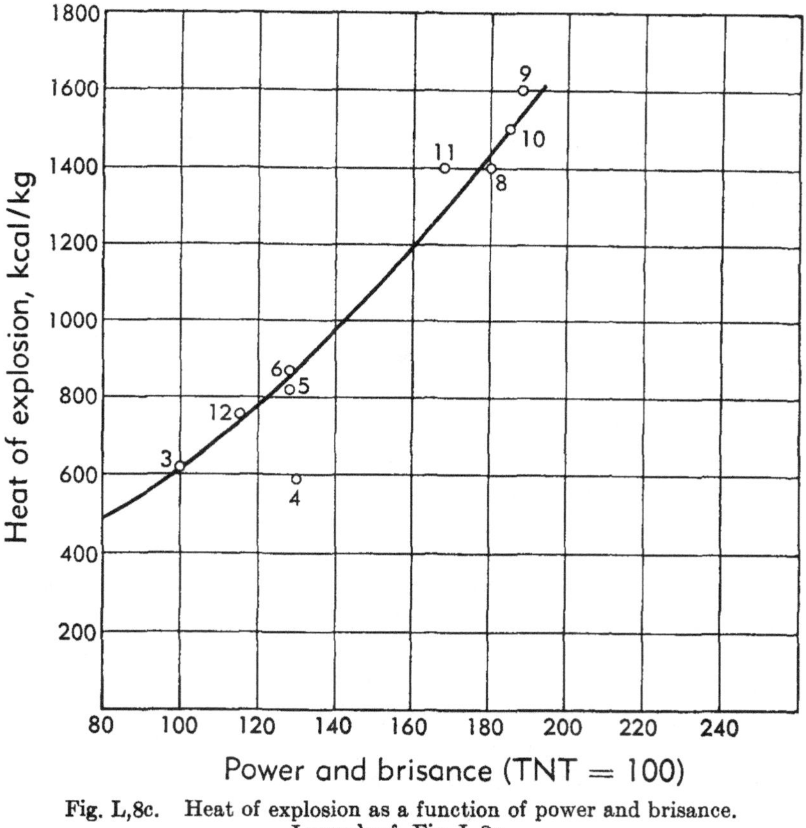

Fig. L,8c. Heat of explosion as a function of power and brisance. Legend: cf. Fig. L,8a.

oxygen balance in this molecule, CH_3NO_2 represents one of the higher-performing monopropellants of class A.

A method which has been developed for ignition of nitromethane involves the simultaneous injection of oxygen gas and the use of a spark [*17*]. It is probable that the large L^* values ($\cong$500 inches) required for continuous combustion are the result of slow gas- or liquid-phase chemical reactions rather than of slow vaporization. This supposition is supported by the fact that the use of certain combustion catalysts decreases the minimum L^* for combustion. It is generally considered, however, that such catalysts are still not sufficiently active to reduce the combustion volume to reasonably low values, say of the order of 100 inches or less. The occurrence of sporadic explosions in the regenerative cooling coils [19] is related to the use of large-L^* motors which cause a large temperature rise of the coolant. Thus the problems of increasing the combustion rate and eliminating the explosion hazard in the cooling coils are closely related for nitromethane.

Methyl nitrate. Methyl nitrate, CH_3NO_3, has a still more favorable oxygen balance than nitromethane and therefore yields higher rocket performance. Very little work has been done with this compound, however, because of its great thermal and shock sensitivity. The fact that the liquid explodes at 65°C (Table L,7) makes its use as a regenerative coolant almost impossible. Desensitizing CH_3NO_3 with CH_3OH (see Myrol in class C compounds) has been considered, but the amount of alcohol necessary for desensitization results in considerable loss in performance.

Diethylene glycol dinitrate. The oxygen balance of DEGN is about the same as that of nitromethane and hence it has very similar performance. Thermodynamic calculations [19] indicate a chamber temperature of 2500°K and an I_{sp} of 213 sec at a 20-to-1 pressure ratio. The somewhat lower performance of DEGN as compared to nitromethane results from the lower percentage of hydrogen in the molecule.

Small scale tests show that DEGN explodes more readily and at lower temperatures than does nitromethane. For this reason it is not considered possible to use this compound as a regenerative coolant.

Nitroglycerin. Nitroglycerin is only slightly overoxidized and therefore is capable of yielding high performance as a monopropellant. Its extreme thermal and shock sensitivity, however, prohibits its use as a monopropellant. Motor tests in which as much as 30 per cent nitrobenzene has been added indicate that the mixture is still too sensitive for general use. Because of its sensitivity, little development work has been carried out on this compound.

Ethylene oxide. Ethylene oxide has been considered as a monopropellant mainly because of its potential availability and low adiabatic flame temperature. The low oxygen balance (−182) indicates that the performance should be quite low. The low boiling point (10.7°C) would necessitate the use of refrigeration in moderate climates if the propellant were to be used as a liquid. Thermodynamic calculations show that the decomposition at equilibrium can be described mainly by the following reactions:

$$C_2H_4O\ (l) = CO + CH_4 + 26.6\ \text{kcal} \tag{8-2}$$

$$C_2H_4O\ (l) = CO + 2H_2 + C + 8.4\ \text{kcal} \tag{8-3}$$

$$C_2H_4O\ (l) = CO + \tfrac{1}{2}C_2H_4 + H_2 + 3.1\ \text{kcal} \tag{8-4}$$

$$C_2H_4O\ (l) = CO + \tfrac{1}{2}C_2H_6 + \tfrac{1}{2}H_2 + 18.9\ \text{kcal} \tag{8-5}$$

The relative extents to which the reactions proceed are given by the equilibria between the product species. However, because of the relatively low temperature of the reaction, it is likely that one or more of the reactions listed in Eq. 8-2 to 8-5 are inhibited as a result of slow reaction rates.

Because of the large heat release, maximum performance should result if reaction (Eq. 8-2) occurs exclusively, which would lead to a chamber temperature of 1450°K and an I_{sp} of 165 sec at a 20-to-1 pressure ratio. Laboratory experiments on the mechanism of the thermal decomposition of C_2H_4O show that free radicals are formed as intermediates and that the end products listed above are reasonable expectations [*20,21*]. It is apparent that the lower heat releases corresponding to reactions (Eq. 8-3, 8-4, and 8-5) will reduce performance. Because of the low performance and low adiabatic flame temperature it appears that the main field of application for this compound might be as a gas generant rather than as a rocket propellant.

Hydrogen peroxide. With suitable catalysts it is possible to decompose H_2O_2 according to the equation:

$$H_2O_2\ (l) = H_2O\ (g) + \tfrac{1}{2}O_2 + 12.96 \text{ kcal} \qquad (8\text{-}6)$$

Thermochemical calculations [*22*] on 100 per cent H_2O_2 show the chamber temperature to be 1250°K and the specific impulse 146 sec at a 20-to-1 pressure ratio. Because of this low performance, H_2O_2 cannot compete with the other members of the class A monopropellants, but it is interesting to note that as a result of the simplicity of operation of H_2O_2 motors and because of the low chamber temperature, H_2O_2 was actually used in military weapons by the Germans in World War II. Examples of systems utilizing H_2O_2 catalyzed by calcium permanganate are (1) the Focke-Wolf ATO Fw56 which developed 650 lb thrust for 30 seconds, (2) pilot-controlled units for the Henkel He 112 and He 126 rocket motors, and (3) the Messerschmitt Me 163-A rocket airplane with a thrust of 1650 lb [*22*]. More recently a group of American engineers has developed, as a hobby [*23*], a small monopropellant rocket of 5 lb thrust using H_2O_2. In spite of its low theoretical I_{sp}, useful rocket application can be made of this monopropellant because of its high efficiency (as measured by c^*) and its high liquid density (1.39 g/cm^3 for commercial 90 per cent H_2O_2 at 20°C). Considerable application has also been made of the gas-generating properties of H_2O_2 (as for instance in the German V-2) because of ease of ignition and low temperature of the product gases.

Hydrogen peroxide is normally manufactured as an aqueous solution. Prior to World War II, concentrations no higher than 30 to 35 per cent were generally available. As a result of interest in H_2O_2 as a rocket propellant, however, commercially available solutions of H_2O_2 containing up to 99 per cent are now produced by means of vacuum distillation of the more dilute solutions. A complete review of the thermodynamic properties of aqueous solutions of H_2O_2 has been given by Kavanagh [*24*] and Williams, et al. [*25*]. For high performance it is desirable to employ as high a concentration of H_2O_2 as possible, whereas in certain gas-generating devices it may be advantageous to add water in order to lower the decomposition temperature. Calculations show that an 87 per cent solution will yield a temperature of 930°K (320°K decrease for 13 per cent H_2O) and an I_{sp} of only 126 sec (20 sec decrease). The use of aqueous solutions of H_2O_2 may be desirable because of their low freezing points.

A considerable amount of work has been done on the kinetics of the catalytic decomposition of dilute H_2O_2 solutions. It is known that permanganate solutions are effective catalysts for the decomposition of H_2O_2 and, in fact, calcium permanganate has been used very successfully for the ignition of peroxide motors [*19,22*]. Once the decomposition has been started, the reaction proceeds smoothly with a nonluminous exhaust. Calcium permanganate can be used in two ways, i.e. as a concentrated aqueous solution injected together with the H_2O_2, and also as a catalytic surface on alundum pellets which are soaked with the catalytic solution, dried, and then packed into the motor chamber. Both methods have been found satisfactory.

The disadvantages in the use of H_2O_2 either as a monopropellant or in a bipropellant combination are its thermal sensitivity, which requires vented containers for shipping and storage, and its high freezing point (cf. Table L,7). Concentrated H_2O_2 reacts with many metals, and the oxides formed catalyze the decomposition reaction. If poured on oxidized iron, concrete, dust, or clothing, the exothermic decomposition becomes so rapid that an explosion may result if the liquid is partially confined. This property constitutes a hazard and great care should be exercised in handling H_2O_2 to avoid spillage. Concentrated H_2O_2 is practically inert, however, in contact with 2S aluminum alloy and polyethylene, teflon and Kel-F plastics [*26*]. A detailed description of the equipment suitable for handling concentrated H_2O_2 has been given by Davis and Keefe [*26*]. With reasonable handling care and with the use of proper materials for containers these authors conclude that concentrated peroxide solutions can be handled on a large scale with safety.

Performance of Class B Compounds. The compounds in this class depend upon an unstable arrangement of the atoms for liberation of energy during decomposition. Rocket performance values for members of this group are generally lower than those of class A monopropellants.

Hydrazine. In closed-bomb tests [*27*] it has been shown that the thermal decomposition of N_2H_4 proceeds according to the two consecutive over-all reactions

$$3N_2H_4 = 4NH_3 + N_2 + 80.15 \text{ kcal} \tag{8-7}$$

$$4NH_3 = 2N_2 + 6H_2 - 44.00 \text{ kcal} \tag{8-8}$$

The NH_3 is only negligibly dissociated for temperatures below 400°K, and therefore the gases contain only NH_3 and N_2 according to the reaction given in Eq. 8-7. However, if the temperature is raised appreciably above 400°K, the NH_3 decomposes thermally according to Eq. 8-8. The equilibrium decomposition becomes essentially complete above 800°K as indicated by the data in Table L,8a which shows the equilibrium fraction x of NH_3 dissociated, originating from the decomposition of N_2H_4 at a pressure of 20 atmospheres.

Table L,8a. Equilibrium fraction of NH_3 dissociated as a function of temperature (total pressure 20 atm).

Temperature, °K	400	500	600	700	800
Fraction of NH_3 dissociated	0 05	0.25	0.60	0.88	0.96

If N_2H_4 decomposes to NH_3 and N_2 adiabatically according to Eq. 8-7, sufficient heat is liberated to heat the gases to 1649°K. If, however, equilibrium dissociation of NH_3 occurs according to Eq. 8-8, the temperature of the gases is only 867°K. Since the equilibrium concentration of NH_3 at 867°K is negligibly small, it follows that if N_2H_4 is decomposed adiabatically and permitted to reach equilibrium, the gases essentially contain N_2, H_2, and only a trace of NH_3. Suppose now that the decomposition occurs within a short period of time, of the order of milliseconds, which is characteristic of transient times in conventional rocket motors. It would then be reasonable to expect that only part of the NH_3 could decompose, depending upon the rate of the over-all reaction given in Eq. 8-8. Experimental results indicate [*27*] that the reaction of Eq. 8-8 proceeds more slowly than the reaction of Eq. 8-7. If the reaction of Eq. 8-7 proceeds completely and the reaction of Eq. 8-8 incompletely, then the over-all decomposition reaction can be represented as a function of the fraction x of NH_3 decomposed:

$$3N_2H_4 = 4(1 - x)NH_3 + (1 + 2x)N_2 + 6\,xH_2 + (80.15 - 44.00x)\text{kcal} \tag{8-9}$$

The performance of N_2H_4 as a monopropellant or a gas generant, therefore, is dependent on the amount of NH_3 decomposed. Some control over this factor can be had in practice through the design of motor chambers giving various L^* values.

Performance calculations [*28*], showing the variation of I_{sp}, T_c, and c^* with the arbitrary percentage of NH_3 decomposed, are given in Fig. L,8d. Because of the low temperatures involved and because of the high rate of travel through the nozzle, composition changes according to Eq. 8-8 would not be expected to occur. Hence constant-composition flow may be assumed. It is instructive to note that I_{sp} and c^* are quite constant over the range $0 < x < 0.5$ even though T_c is steadily diminishing. This result follows from the fact that the decomposition of NH_3 produces H_2, causing the average molecular weight to decrease about as rapidly as T_c decreases. For a monopropellant system, therefore, it is best to design the motor so that x falls in the range 0.3 to 0.5 in order to obtain maximum I_{sp} at the lowest T_c.

It is very difficult to manufacture anhydrous hydrazine because N_2H_4 forms a relatively stable hydrate $N_2H_4 \cdot H_2O$. A complete discussion of the thermodynamic and chemical properties of N_2H_4 can be found in the works of Audrieth, et al. [*29*].

The thermal decomposition rate of liquid N_2H_4 has been studied in closed-bomb tests [*27*]. It has been found that the reaction is surface-catalyzed in the gas and liquid phase, and that the hydrazine salts of HNO_3, HCl, and H_2SO_4 catalyze the decomposition in the liquid phase.

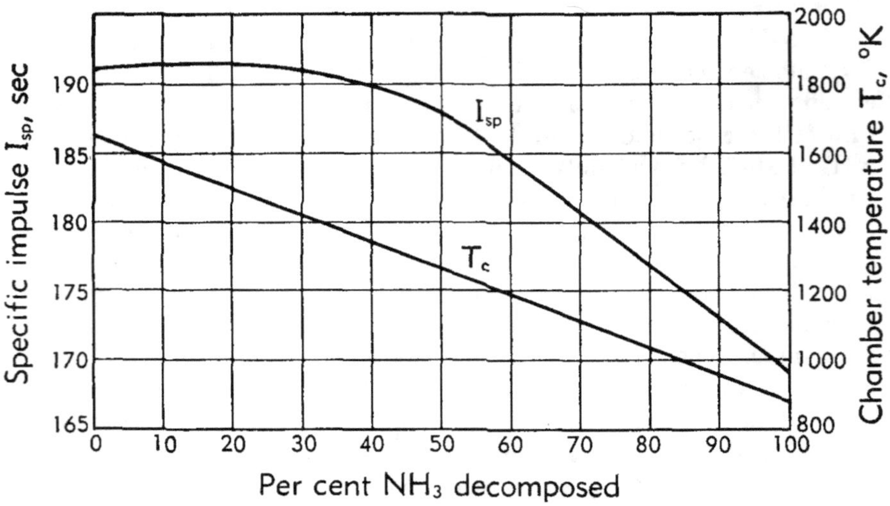

Fig. L,8d. Specific impulse and chamber temperature of N_2H_4 monopropellant as a function of per cent NH_3 decomposed.

For uncatalyzed N_2H_4 in the constant volume system, the rate of decomposition did not vary with moderate changes in the ratio of liquid to vapor volumes, indicating that the major portion of the decomposition probably occurred in the vapor phase [*27*]. It has not been possible to detonate liquid N_2H_4 by ordinary means, and it is therefore believed that explosions experienced with N_2H_4 are probably thermal in character. This important fact makes possible the design of regeneratively cooled motors using N_2H_4, in spite of its thermal instability, since cooling passages can be designed with the maximum temperature of N_2H_4 remaining at a temperature below the value at which rapid decomposition occurs. This remark is also true for bipropellant motors using N_2H_4 as coolant, since these systems require low-L^* motors and hence lead to relatively small temperature rises of the coolant.

Acetylene, ethylene, and nitric oxide. No serious consideration has been given to the use of any of these compounds as monopropellants mainly because they are gases at normal temperatures and have several undesirable properties. Because of its large positive standard heat of formation, C_2H_2 would be expected to yield high performance. Unfortunately this compound has the serious drawback of being rather easily detonated. Both C_2H_2 and C_2H_4 would be expected to deposit carbon during decomposition. Nitric oxide has been successfully decomposed when preheated at atmospheric pressure in a flame holder, but thermodynamic calculations do not show this compound to be promising. Assuming $\Delta H_f^0(\mathrm{NO}) = +21.6$ kcal/mole, it is found that $T_c = 2800°\mathrm{K}$ and $I_{sp} = 190$ sec for a 20-to-1 pressure ratio employing NO gas at 300°K

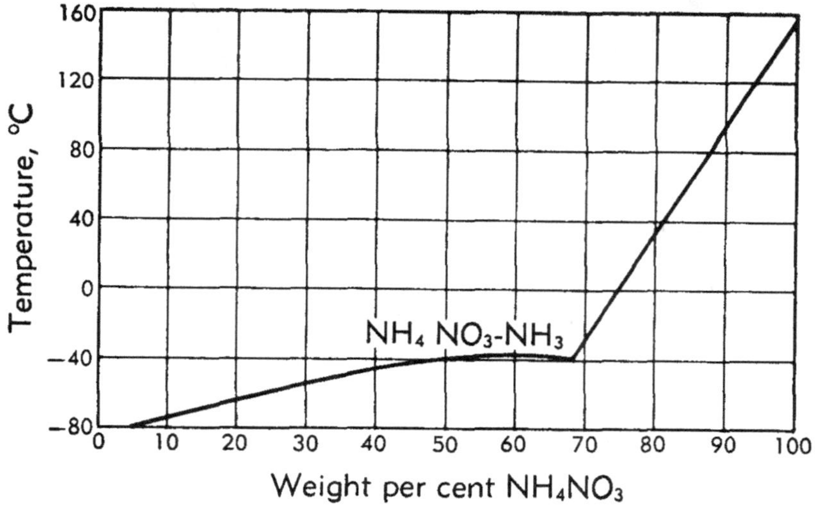

Fig. L,8e. Freezing point and solubility data for the system NH_4NO_3-NH_3.

initially. The low performance for such a high flame temperature results from the high molecular weight of the product gases.

Performance of Class C Monopropellants.

Diver's solution. Diver's solution is a solution of NH_4NO_3 in liquid NH_3. Interest in this mixture was stimulated by the fact that the components are cheap and readily available commercially, and also because of the high solubility of NH_4NO_3 in NH_3. The freezing point or solubility curve for the system NH_3 (l)-NH_4NO_3 is shown in Fig. L,8e. Since the stoichiometric concentration occurs at 87.5 per cent NH_4NO_3, it is obvious from the solubility curve that the practical application of these solutions is restricted to the underoxidized side. The solubility at 0°C is about 75 per cent NH_4NO_3 by weight; the I_{sp} calculated at this point is approximately 175 sec at a 20-to-1 pressure ratio; the density at 20°C is about 1.15 g/cm³.

Practical application of Diver's propellant, however, has been retarded by problems of ignition and stable combustion. It is known that propellants containing NH_3 or NH_4NO_3 are difficult to ignite. Since the combustion occurs only when both components are heated to a temperature considerably above the boiling point of NH_3, it is obvious that rapid liquid-phase reaction cannot occur without catalysis. Evaporation and heating of the separate components appear to be prerequisites for both satisfactory ignition and combustion processes. Successful applications of Diver's solution depend on the development of a powerful catalyst for the combustion and a suitable arrangement of surfaces to effect a large heat transfer rate to the propellant.

Methyl nitrate and methanol (Myrol). As mentioned earlier in discussing methyl nitrate as monopropellant, the addition of CH_3OH to CH_3NO_3 has been considered in order to desensitize the methyl nitrate. The Germans did a considerable amount of work with these propellant mixtures during World War II (the term Myrol was invented by them) but the general conclusion appears to be that organic nitrates such as CH_3NO_3 cannot be desensitized sufficiently to lead to hazard-free operation, particularly in regeneratively cooled motors.

Peroxide and methanol. The purpose of adding a fuel to H_2O_2 is to bring the system closer to stoichiometric mixture ratio, thereby improving performance. Thus the addition of 21.4 per cent by weight of CH_3OH to 87 per cent H_2O_2 will raise the performance from 126 sec to 225 sec [*19*]. A serious drawback, however, is the great sensitivity of mixtures such as these. The presence of a fuel greatly increases the exothermic heat release during thermal decomposition, thereby making thermal explosions highly probable. The stoichiometric mixture referred to above can be detonated in a beaker by the addition of a few drops of permanganate solution which initiates the thermal decomposition of the peroxide. Because of this great sensitivity, monopropellants containing peroxide and a fuel cannot be recommended.

Tetranitromethane and fuel. Tetranitromethane, $C(NO_2)_4$, because of its high oxygen balance is a poor monopropellant. When mixed with a fuel, however, the heat of combustion is increased as is also the explosive power (cf. Fig. L,8b and c). Table L,8b shows some results obtained by

Table L,8b. Explosive power of tetranitromethane mixtures.

Detonator cap no. (German standard)	TNM	TNM:N_2O_4 70:30	TNT	Nitroglycerin	TNM:benzene 87:13
1	12	0	0	171	413
2	44	0	208	172	404
3	65	13	218	379	404
5	86	48	268	407	404
8	71	37	332	397	445

the lead-block expansion test on the explosive power of several TNM mixtures [*30*], compared with TNT and nitroglycerin.

Reference to Table L,8b shows that the presence of only 13 per cent by weight of benzene causes the explosive power of TNM to surpass even that of nitroglycerin, a result which is in accord with the correlation presented in Fig. L,8b. Hence mixtures of TNM and a fuel cannot be considered to be suitable monopropellants. Pure TNM, however, is a relatively low order explosive which can be further desensitized by addition of N_2O_4.

CHAPTER 3. COMBUSTION OF SELECTED BIPROPELLANT SYSTEMS

L,9. Classification of Oxidizers and Fuels. Bipropellant combinations for use in rocket engines are derived from a relatively small group of oxidizers and a large group of fuels.

An oxidizer is a substance containing an excess of oxygen or halogen; fluorine and chlorine are the most useful halogens from an energetic point of view. The fuels generally contain carbon, hydrogen, and light metals. An element such as sulfur occupies a more or less neutral position and is considered an oxidizer in the presence of excess fuel (where it forms hydrogen or metal sulfides), but acts as a fuel in excess oxygen (where it forms SO or SO_2 during combustion).

Oxidizers which have been considered for rocket applications include O_2, O_3, H_2O_2, HNO_3, N_2O_4, NO, N_2O, $C(NO_2)_4$, F_2, ClF_3, and OF_2. Of these oxidizers, O_2, O_3, NO, N_2O, F_2, and OF_2 are gases at ambient tem-

Table L,9. Properties of representative oxidizers.

Oxidizer	Formula	Density,* g/cm³	Melting point, °C	Boiling point, °C	Standard heat of formation $\Delta H^0_{f(298)}$, kcal/mole [*13*]
Oxygen	O_2	1.14^{-182}	−218.7	−183.0	0
Ozone	O_3	1.71^{-111}	−251.4	−111.5	34.0
Hydrogen peroxide	H_2O_2	1.44	−0.1	150.5	−44.84
Nitric acid	HNO_3	1.50	−41.6	86.0	−41.40
Nitrogen tetroxide	N_2O_4	1.44	−11.2	21.2	2.3
Nitric oxide	NO	1.27^{-150}	−163.6	−151.8	21.6
Nitrous oxide	N_2O	1.23^{-89}	−102.4	−88.5	19.5
Tetranitromethane	$C(NO_2)_4$	1.65	13	126	8.9 [*15*]
Fluorine	F_2	1.55^{-187}	−217.9	−188.0	0
Chlorine trifluoride	ClF_3	1.77^{12}	−82.6	12.1	37.0
Oxygen bifluoride	OF_2	1.90^{-224}	−223.8	−144.8	5.5

* At 20°C, unless otherwise specified by the superscript.

peratures and require refrigeration or high pressures for liquefaction. The physical properties of these oxidizers are summarized in Table L,9.

The number of compounds containing carbon and hydrogen which can act as fuels is almost limitless and includes practically all of the oxygen-deficient organic chemicals. In addition, a large group of fuels containing inorganic hydrogen compounds can be employed. Thus fuels could be subdivided conveniently into organic and inorganic compounds. The organic fuels may be classified further according to the conventional groupings of organic chemistry, i.e. saturated and unsaturated hydrocarbons, alcohols, amines, ethers, nitroparaffins, aromatic compounds, etc. Important inorganic fuels include H_2, NH_3, N_2H_4, PH_3, H_2S, B_2H_6, Si_2H_6, $LiNH_2$, etc., and some of the light metals, their alloys, and solutions. The highest performing members in this group are H_2, N_2H_4, NH_3, and the hydrides and amides of the low atomic weight metals.

L,10. Combustion of Bipropellants in Rocket Engines. The physicochemical phenomena involved in the analysis of combustion processes in bipropellant motors are so involved that a quantitative discussion does not appear to be feasible at the present time. However, it may be instructive to classify the more important processes which must take place and to present simplified physical pictures for some of the complicated phenomena associated with heterogeneous combustion.

In general, bipropellant propulsion systems involve injection of liquid oxidizers and fuels. The physicochemical processes leading to the formation of the combustion products may be represented by a reaction scheme which is similar to that used in categorizing the rate processes involved in ignition (cf. Fig. L,1). The major difference between the ignition and steady state burning processes arises from the fact that heat transfer from regions of active combustion and diffusion of combustion intermediates play a more important role during steady state burning than during ignition. This difference has been emphasized previously in the discussion of measurement of ignition delay by the twin-jet apparatus (cf. Art. 2). Only a small change with time in flame position is observed during ignition for the system N_2H_4-H_2O_2. On the other hand, for the bipropellant system acid-aniline, the flame position changes markedly after ignition occurs on a surface. Hence it follows that the course of combustion during ignition and steady state burning need not be the same. This result is contrary to the idea that ignition delays are necessarily related to steady state combustion times or motor sizes required to obtain optimum performance.

During ignition of spontaneously-igniting propellants, an initial exothermic reaction generates the heat necessary for the propellant to proceed through the combustion reactions. During steady burning the flame front must be considered as an additional source of heat. In the case of a nonspontaneous system such as HNO_3-hydrocarbon, for instance, an injector which can rapidly vaporize the propellant is found to be efficient since the liquid-phase mixing can contribute only negligible heating. On the other hand, a spontaneous system such as HNO_3-aniline is found to combust more efficiently with liquid stream impinging-jet injectors. Still another situation arises with a system such as HNO_3-N_2H_4 which is spontaneously-igniting. Combustion efficiency in this system is highest when an Enzian-type injector (Fig. L,3b) is used where the propellant impinges on a surface. In this case, the initial exothermic reactions are hastened by the presence of a surface, an observation which is in agreement with the fact that the thermal decomposition of N_2H_4, which is exothermic, is surface-catalyzed [*27*].

In a general way, these principles concerning the path of combustion lead to some conclusions concerning the influence of chemistry on the burning process. If combustion is initiated in the liquid phase as a result of spontaneous chemical reaction, then chemical additives which alter the ignition delay may be expected to influence the over-all combustion time. Thus for HNO_3 and aniline, the presence of water in the acid increases the ignition delay and also the motor length (L^*) for steady state combustion, whereas the presence of metallic additives has an opposite effect (Table L,3a). Thus it has been observed that the presence of the corrosion products of steel (Fe, Ni, and Cr nitrates) in the acid increases the efficiency of combustion. In the case of the nonspontaneous systems such as O_2 and alcohol, however, as much as 25 per cent H_2O can be added to the alcohol without seriously altering the necessary motor dimensions. A small effect can be attributed mainly to the lower flame temperature and hence lower rate of heat transfer back to the initial combustion phase. One may further conclude from these remarks that the use of detergents in the propellants would have a more pronounced effect with systems requiring liquid-phase mixing since the ignition delay may be lowered by this means.

The theoretical calculation of combustion rates in bipropellant motors must be regarded as one of the most important unsolved problems in rocket engineering. If it were obvious that any particular process is slow enough to control the over-all reaction rate, then it would not be difficult to obtain approximate analytic expressions for the combustion rate. For example if the slow step involved diffusion of reactants, reasonable approximations could be derived from the kinetic theory of gases. If the chemical reactions between liquid propellants are important, an analysis of liquid-phase mixing and liquid-phase chemical reaction rates is indicated. If the liquid-phase reaction processes are slow compared with gas-phase reactions, evaporation rates may become rate-controlling and useful estimates of over-all reaction rates can be derived from application of formulas for droplet burning in reacting gases.

The a priori classification of rate-controlling steps for bipropellant systems is, at best, hazardous. An attempt to analyze heterogeneous combustion between gaseous oxygen and liquid hydrocarbons under various simplifying assumptions has been made by Spalding [*33,34*]. The conditions in bipropellant rocket motors do not, in general, correspond to the idealized treatment given by this author. Nevertheless an extension of the ideas proposed by Spalding may be useful in some cases. For example, in an ammonia-oxygen motor one would expect relatively rapid evaporation of the oxygen, leading to combustion conditions which may be approximated by the heterogeneous reaction between liquid ammonia and gaseous oxygen. For a process of this type, the notion of a "combustion surface" or a "small combustion volume" may be useful. The over-all reaction rate is then determined by the following physico-chemical processes:

1. Diffusion of gaseous oxygen and ammonia to the combustion surface which has become established close to the droplet of liquid ammonia.
2. Heat transfer to the liquid ammonia droplets and to gaseous oxygen.
3. Evaporation of liquid ammonia, which depends on the heat transfer to the ammonia and on the evaporation coefficient of the liquid.
4. Chemical reactions between gaseous ammonia and oxygen at the "combustion surface." In the treatment of Spalding, the chemical reactions are assumed to occur so rapidly that they do not contribute appreciably to the reaction time.

The diffusion of reaction products from the "combustion surface" will influence the over-all reaction rate indirectly through its effect on the distances from the "combustion surface" to the surface of the liquid and to the unreacted oxygen. The preceding picture is based on the assumption that reactions in an ammonia-oxygen motor can be represented by the heterogeneous combustion of liquid ammonia in an atmosphere of gaseous oxygen. Injection conditions also enter into the formulation of the problem indirectly insofar as they affect the distances from the "combustion surface."

Without analytic study of the results derivable from these assumptions it is not possible to assess the validity of the proposed picture. However, in a field as complicated as the combustion processes in rocket motors, it is perhaps justifiable to commence analytic treatment by using somewhat drastic simplifying assumptions. Combustion models for other types of heterogeneous combustion can be obtained by an obvious extension of the model proposed for the ammonia-oxygen system. In concluding this discussion, mention should be made also of important contributions to the problem of heterogeneous combustion which have been made recently by workers at the National Gas Turbine Establishment of Pyestock, Hants, England. For example, Barr and Mullins [*35*] have studied the effect of addition of combustion products to air in air-fuel combustion, Godsave has investigated the burning of single drops of fuel both theoretically [*36*] and experimentally [*37*], and Mullins has studied the ignition of fuels injected into a hot air stream [*38,39*]. Of particular interest is an experimental study by Topps [*40*] on the evaporation and combustion of falling fuel droplets which shows that the mass evaporated at any given time, divided by the product of the frictional drag on the droplet and the diameter raised to the -0.3 power, is a unique function of the temperature for a wide variety of hydrocarbon fuels.

L,11. Performance Characteristics of Several Bipropellant Combinations. In order to illustrate the behavior of representative bipropellant combinations, the performance of the following systems will be considered: RFNA-aniline; RFNA-hydrazine; RFNA-gasoline; hydrogen-fluorine. Before proceeding with the discussion of individual propellants, it may be of interest to summarize the desirable properties of propellants used in liquid-fuel rocket engines. Such a summary is presented in Table L,11.

RFNA-aniline. Theoretical performance data, assuming a 20-to-1 pressure ratio with unit atmospheric pressure, indicate a maximum I_{sp} for equilibrium flow of 226 sec for RFNA-aniline. The stoichiometric mixture ratio r (weight of oxidizer/weight of fuel) occurs at a value of 4.17 whereas the maximum value of I_{sp} (221 sec) for frozen nozzle flow occurs at a lower mixture ratio. The maximum value of I_{sp} (226 sec) for equilibrium expansion through the nozzle lies closer to the stoichiometric mixture ratio, a characteristic which is general for all systems investigated thus far.

Ignition delay measurements for this system give a value in the vicinity of 60 milliseconds in a small scale injection apparatus (cf. Table L,3a). The fact that a multiorifice-type injector gives good performance indicates that liquid-phase reactions are important initially. These reactions are probably of three types [*41*, pp. 610–616]: (1) neutralization of the amine group, (2) oxidation of the amine group to form nitroso- and diazo compounds which are highly colored, and (3) nitration of the benzene ring. These reactions are sufficiently exothermic to initiate steady state combustion and probably account for the observed spontaneous reaction between acid and aniline. It is known that the nitration reactions of aromatic nuclei are facilitated by the presence of NO_2^+ present in concentrated nitric acid [*42,43*] and that the presence of water in the acid suppresses the formation of NO_2^+. This fact is in accord with the observation that the addition of water to the acid materially increases the ignition delay and decreases the combustion efficiency.

RFNA-hydrazine. This spontaneous combination is representative of the class of noncarbonaceous propellants. The relatively high performance in this system (I_{sp} = 246 sec for equilibrium flow) results because of the high energy release in the reaction (N_2H_4 itself is a monopropellant) and because of the low average molecular weight of the combustion gases. The small value of r (1.56) for the stoichiometric mixture is also characteristic of most noncarbonaceous systems.

Table L,11. Desirable properties of propellants used in liquid-fuel rocket engines.

1. Small negative or preferably positive standard heats of formation, $\Delta H^0_{f(298)}$, of the propellants are desired for high performance. However, stability requirements may limit the use of compounds with large positive values for ΔH^0_f.
2. The reaction products should have low molecular weights and large negative heats of formation. If conditions (1) and (2) are met, then the reaction products will consist of low molecular weight compounds at high temperatures.
3. The propellants should have large densities in order to minimize the dead weight of storage tanks.
4. The oxidizers and reducing agents are best handled as liquids. Hence it is desirable to obtain propellants which are normally liquid in the operating temperature range of service units (i.e. from about −40 to +60°C). For substances such as liquid oxygen and hydrogen, special equipment must be provided which represents added dead weight which the propulsion unit must carry and is warranted only in the case of very high energy propellant mixtures.
5. For purposes of regenerative and sweat cooling of the combustion chamber, the propellant used for cooling should have a high specific heat and a high heat of vaporization; it should be thermally stable at elevated temperatures and also noncorrosive.
6. Since it may be necessary to store the propellants for long periods of time before use, good propellants should have high storage stability, i.e. they must not decompose or change chemically during storage so that their use as a propellant is impaired.
7. Since propellants are chemicals which have to be handled by service personnel it may be desirable for some applications to use propellants of relatively low toxicity.
8. For repetitive motor operation, it is desirable to consider bipropellant mixtures which are spontaneously combustible with minimum time lag. These may be used as the main propellants or in small quantities as starting fluids.

In addition to the items listed here, other specific considerations such as erosion of the combustion gases on the nozzle throat, deposition in the nozzle throat of solid deposits, effect of viscosity on line pressure losses, radar attenuation of the jet gases, and the economics of the propellant must be considered.

Although the ignition delay for this propellant system is quite short, the use of the multiorifice injectors employing liquid stream impingement does not result in optimum performance. However, if liquid stream impingement occurs on a surface, then the combustion efficiency is greatly improved (cf. Art. 10). The initial rapid exothermic reaction is probably the neutralization reaction which liberates sufficient heat to facilitate oxidation. When heated, hydrazine will decompose to NH_3 and the elements; and if insufficient mixing between oxidizer and fuel droplets occurs, the droplets with excess fuel will generate NH_3 which must subsequently be oxidized in the vapor stream. If the initial neutralization reaction occurs on a surface, however, then the probability of local inhomogeneities is diminished and the rate of oxidation of the N_2H_4 or NH_3 will be increased because of surface catalysis. The oxidation of NH_3, in particular, is known to be surface-catalyzed.

RFNA-gasoline. The propellant system RFNA-gasoline is of great practical utility because of the relatively large availability of the components. Gasoline is a mixture of various hydrocarbons in the range of five to twelve carbon atoms. The average composition is generally taken as C_8H_{18}. Performance calculations on this propellant combination indicate a slightly higher performance (by several seconds in I_{sp}) than for the acid-aniline system. The stoichiometric mixture ratio, $r = 5.5$, is rather high, but maximum performance occurs near $r = 4$. The propellant system is nonspontaneous and therefore requires an external ignition source [*44*].

If nitric acid and gasoline are mixed in a beaker, exothermic chemical reaction occurs. Depending upon the quality of the fuel, the temperature may either rise very rapidly with considerable fuming of the mixture or else the temperature rise may be practically nonexistent. This reaction is associated with nitration of the hydrocarbon molecules, which is exothermic and which proceeds more rapidly with aromatic and unsaturated hydrocarbons than with the straight-chain saturated molecules. The order of reactivity generally follows the scheme: straight-chain < branched-chain < aromatic < unsaturated compounds. The unsaturated hydrocarbons (generally olefins) undergo several types of reactions of the type [*45*, pp. 110–549]:

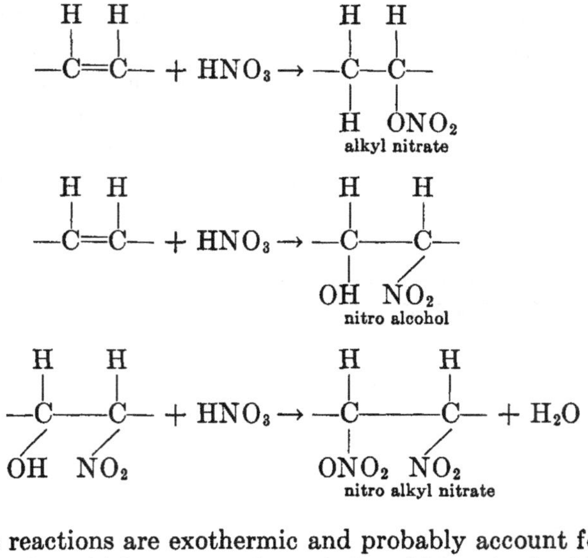

All of these reactions are exothermic and probably account for the large temperature rise which occurs when nitric acid is mixed with unrefined gasoline. In a study of the chemical behavior of solids formed in the initial reactions between WFNA and dicyclopentadiene, Trent and Zucrow [46] have concluded that the nitro and nitrate groups are formed. Although the nitration reactions probably constitute an initial step in combustion, their occurrence at low temperatures appears to be undesirable from an operational point of view. Thus, if a long ignition delay occurs, these reactions cause the accumulation of nitrated substances which can subsequently explode. With straight-chain saturated hydrocarbons, however, the amount of nitrated products formed during a long ignition delay is small, thereby minimizing this hazard.

Hydrogen-fluorine. The hydrogen-fluorine propellant combination is one of the highest performing systems available. Performance calculations for equilibrium expansion through the nozzle for a chamber pressure of 300 psia and an external pressure of 1 atm yield $I_{sp} = 365$ sec for $r = 3.3$, i.e. an appreciable excess of H_2 compared to the stoichiometric mixture ($r = 19$).

Although the adiabatic flame temperature at 20 atm for the H_2-F_2 system is very high (ca. 4800°K), the total equilibrium radiant heat transfer from the combustion products is not very large. The only molecule present in the combustion chamber which makes an important contribution to the equilibrium radiant heat transfer is HF. Rough estimates [47] indicate that the radiant heat transfer to the motor chamber from a H_2-F_2 motor will not exceed about 20 per cent of the total observed heat transfer.

The H_2-F_2 system is nonspontaneous, a result which has interesting chemical implications. Thus it has been shown, both from an experimental study [48] and on theoretical grounds [49], that the activation energy for the direct reaction between the two molecular species must be quite large. This work suggests that one of the functions of the igniter, such as a squib, is the production of minute concentrations of atomic hydrogen and/or fluorine. Thus the igniter may be considered to be a chain initiator, rapid reaction occurring once a small amount of the chain carriers has been produced.

The reaction kinetics of the H_2-F_2 system is relatively so simple that an approximate analysis of composition changes during nozzle flow can be carried out. The reaction scheme is

$$\begin{aligned} H + H + M &\rightarrow H_2 + M \\ F + F + M &\rightarrow F_2 + M \\ H + F + M &\rightarrow HF + M \\ H + F_2 &\rightarrow HF + F \\ H_2 + F &\rightarrow HF + H \end{aligned}$$

or for the over-all process:

$$H_2 + F_2 = 2HF$$

where M represents any third body in a collision process. This seems to be compatible with the maintenance of chemical equilibrium during expansion through a Laval nozzle [*50*].

L,12. Modern Trends in Combustion Research on Liquid-Fuel Rocket Engines. In recent years there has been increased activity on various phases of combustion research relating to rocket motor development. A review of some of this work is given in a recent survey paper [*51*] on fundamental studies of droplet burning, spray combustion, observations in motor chambers, and instability. In a series of four papers presented at the Fourth Symposium on Combustion [*52*, pp. 818–864], both theoretical and experimental data were presented for single fuel droplets burning in atmospheres containing oxidizer. These papers give equations for the mass burning rate as a function of the physical and thermodynamic properties of the fuel as a function of pressure up to about 20 atmospheres, and as a function of Reynolds number of the gas based on the droplet. Consideration is also given to the problem of the time required to ignite a droplet when subjected to a heated gaseous atmosphere. In several later papers [*53,54,55,56*], the droplet burning equations are examined in greater detail. Recognition is taken of the fact that the combustion surface is established at that point where the mass flow rates of fuel and oxidizer meet in stoichiometric proportions. This concept has also been expressed by Spalding [*33*; *34*; *52*, pp. 818–864]. In a recent treatment of this subject [*57*], an equation is derived for the burning rate of a monopropellant droplet, where kinetics have been explicitly introduced, thereby eliminating the necessity of arbitrarily describing a combustion surface. In an attempt to apply these equations of droplet burning to conditions actually occurring in a rocket motor, Miesse [*58*] has considered the burning of a droplet when superimposed on a flow field, which is taken to be representative of that expected in a motor chamber. In addition to these works on individual droplets there have been several papers in which these concepts have been applied to the problem of spray combustion [*59,60*].

The effect of nonuniformity in the combustion pattern in a rocket motor has also been analyzed from a thermodynamic point of view [*61*]. It was found that for some systems the existence of variations in mixture ratio over the cross section can lead to a higher performance than if the mixture ratio were uniform. There have also been other studies on the observation of combustion in transparent motor chambers [*62,63,64*]. Although these observations are of qualitative interest, the process is too complex to draw quantitative relations from them. A problem which has recently received much attention in large scale rocket motors is that of instability during combustion [*52*, pp. 880–885; *65*; *66*; *67*; *68*; *69*; *70*, pp. 352–376]. In this connection consideration has been given to the "chugging" frequency in the motor chamber, acoustical vibrations, and mechanical feedback from the propellant pumping system.

L,13. Cited References.

1. Powell, W. B., and Kaplan, N. *Calif. Inst. Technol. Jet Propul. Lab. Rept. 24*, July 1944.
2. Kaplan, N., and Boden, R. H. *Calif. Inst. Technol. Jet Propul. Lab. Progr. Rept. 1-40*, Jan. 1946.
3. Bilfinger. Zündverzugsmessgerät für R-Stoffe. *B. M. W. Rept. 857*, Munich, Dec. 1944.
4. Broatch, J. D. *Fuel 29*, 106 (1950).
5. Gunn, S. V. *M.S. Thesis*. Purdue Univ., 1949.
6. Clark, F. B., and Sappington, M. H. *Calif. Inst. Technol. Jet Propul. Lab. Mem. 9-3*, June 1947.
7. Darling, B., and Kavanagh, G. M. *Mass. Inst. Technol. Rept. 19, Office of Nav. Research Contract N5-ori-78*, July 1947.
8. Collins, A. S. *Sc.D. Thesis*. Mass. Inst. Technol., 1948.
9. Gunn, S. V. *J. Am. Rocket Soc. 22*, 33 (1952).
10. Hirschfelder, J. O., and Curtiss, C. F. *J. Chem. Phys. 17*, 1076 (1950).
11. Bechert, C. *Portugaliae Phys. 3*, 29 (1949).
12. Zeldovich, Y. B. *Tech. Rept. F-TS-1226-IA (GDAM A9-T-45)*, May 1949. (Transl. by Brown Univ.).
13. Rossini, F. D., Pitzer, K. S., Arnett, R. L., Braun, R. M., and Pimentel, G. C. *Selected Values of Physical and Thermodynamic Properties of Hydrocarbons and Related Compounds*. Carnegie Inst. Technol. Press, Pittsburgh, 1953.
14. Bichowsky, F. R., and Rossini, F. D. *The Thermochemistry of the Chemical Substances*. Reinhold, 1936.
15. Roth, W. A., and Isecke, K. *Ber. deut. chem. Ges. 77 B*, 537 (1944).
16. Holcomb, D. E., and Dorsey, C. L., Jr. *Ind. Eng. Chem. 41*, 2788 (1949).
17. Bellinger, F., Friedman, H. B., Bauer, W. H., Eastes, J. W., and Bull, W. C. *Ind. Eng. Chem. 40*, 1320, 1324 (1948).
18. Lothrop, W. C., and Handrick, G. R. *Chem. Revs. 44*, 419 (1949).
19. *Jet Propulsion*. Prepared by Staffs of Jet Propul. Lab. and Guggenheim Lab., Calif. Inst. Technol. for Air Tech. Service Command, 1946.
20. Steacie, E. W. R., and Folkins, H. O. *Can. J. Res. B17*, 105 (1939).
21. Mueller, K. H., and Walters, W. O. *J. Am. Chem. Soc. 73*, 1458 (1951).
22. Bloom, R., Davis, N. S., Jr., and Levine, S. D. *J. Am. Rocket Soc. 80*, 3 (1950).
23. Sanz, M. C. *J. Am. Rocket Soc. 75*, 122 (1948).
24. Kavanagh, G. M. *Mass. Inst. Technol. Rept. 28, D. I. C.*, 6552 (1949).
25. Williams, G. C., Satterfield, C. N., and Isbin, H. S. *J. Am. Rocket Soc. 22*, 70 (1952).
26. Davis, N. S., Jr., and Keefe, J. H., Jr. *J. Am. Rocket Soc. 22*, 63 (1952).
27. Thomas, D. D. *Calif. Inst. Technol. Jet Propul. Lab. Progr. Rept. 9-14*, 1947.
28. Altman, D., and Thomas, D. D. *Calif. Inst. Technol. Jet Propul. Lab. Progr. Rept. 9-36*, 1949.
29. Audrieth, L. F., et al. *N6-ori-71 Chemistry Task XX*, Chap. 1–10. Univ. Illinois, Oct. 1948–Feb. 1949.
30. Hager, K. F. *Ind. Eng. Chem. 41*, 2168 (1949).
31. Penner, S. S. *Calif. Inst. Technol. Jet Propul. Lab. Prog. Rept. 9-13*, Sept. 1947.
32. Hartwig, F. W. *A.E. Thesis*. Calif. Inst. Technol. Guggenheim Jet Propul. Center, 1952.
33. Spalding, D. B. *Fuel 29*, 2–7, 25–32 (1950).
34. Spalding, D. B. *Fuel 30*, 121–130 (1951).
35. Barr, J., and Mullins, B. P. Concerning combustion in vitiated atmospheres. *Natl. Gas Turbine Establishment (England) Rept. R44*, June 1949.
36. Godsave, G. A. E. The burning of single drops of fuel, Part I. *Natl. Gas Turbine Establishment (England) Rept. R66*, Mar. 1950.

37. Godsave, G. A. E. The burning of single drops of fuel, Part II. *Natl. Gas Turbine Establishment (England) Rept. R87*, Apr. 1951.
38. Mullins, B. P. Studies of the spontaneous ignition of fuels injected into a hot air stream, General Introduction and Part I. *Natl. Gas Turbine Establishment (England) Rept. R89*, Aug. 1951.
39. Mullins, B. P. Studies on the spontaneous ignition of fuels injected into a hot air stream, Part II. *Natl. Gas Turbine Establishment (England) Rept. R90*, Aug. 1951.
40. Topps, J. E. C. *J. Inst. Petroleum, 37*, 535–553 (1951).
41. Fieser, L. F., and Fieser, M. *Organic Chemistry*. D. C. Heath, 1944.
42. Hughes, E. D., Ingold, C. K., et al. *J. Chem. Soc.*, 2400–2473 (1950).
43. Ingold, C. K., Millen, D. J., et al. Parts I–VI, *J. Chem. Soc.*, 2576–2627 (1950).
44. Sage, B. H., Hough, E. W., and Green, J. *Calif. Inst. Technol. Jet Propul. Lab. Prog. Rept. 1-3*, Apr. 1942.
45. Gilman, H. *Organic Chemistry*, Vol. I. Wiley, 1938.
46. Trent, C. H., and Zucrow, M. J. *J. Am. Rocket Soc. 21*, 129 (1951).
47. Penner, S. S. *J. Appl. Phys. 21*, 685–695 (1950).
48. Eyring, H., and Kassel, L. S. *J. Am. Chem. Soc. 55*, 2796–2797 (1933).
49. Eyring, H. *J. Am. Chem. Soc. 53*, 2537–2549 (1931).
50. Penner, S. S. *J. Franklin Inst. 249*, 441–448 (1950).
51. Penner, S. S., and Datner, P. P. *Tech. Rept. 7, Contract DA-495-Ord-446*, Sept. 1954. Presented at the *Fifth Symposium on Combustion*, Sept. 1954.
52. *Fourth Symposium on Combustion*. Williams & Wilkins, 1953.
53. Goldsmith, M., and Penner, S. S. *Jet Propulsion 24*, 245–251 (1954).
54. Kumazai, S., and Isada, H. *Science of Machine 4*, 337–342 (1952).
55. Graves, C. C. Paper presented before the *Third Midwestern Conference on Fluid Mechanics*, Univ. Minn., 1953.
56. Wise, H., Lorell, J., and Wood, B. J. Paper presented at the *Fifth Symposium on Combustion*, Sept. 1954.
57. Lorell, J., and Wise, H. *Calif. Inst. Technol., Jet Propulsion Lab. Progress Rept. 20-237*, May 1954.
58. Miesse, C. C. *J. Am. Rocket Soc. 24*, 237 (1954).
59. Probert, R. P. *Phil. Mag. 37*, 94 (1946).
60. Graves, C. C., and Gerstein, M. Paper presented before the *AGARD Combustion Panel*, Scheveningen, The Netherlands, May 1954. Butterworths, London (in press).
61. Altman, D., and Lorell, J. *J. Am. Rocket Soc. 22*, 252–255 (1952).
62. Berman, K., and Cheney, S. H., Jr. *J. Am. Rocket Soc. 23*, 89–96 (1953).
63. Berman, K., and Logan, S. E. *J. Am. Rocket Soc. 22*, 78–85 (1952).
64. Altseimer, J. H. *J. Am. Rocket Soc. 22*, 86–91 (1952).
65. Summerfield, M. *J. Am. Rocket Soc. 21*, 108 (1951).
66. Crocco, L. *J. Am. Rocket Soc. 21*, 163 (1951); *22*, 7–17 (1952).
67. Crocco, L., and Cheng, S. I. *AGARD Combustion Instability*. Butterworths, London (in press).
68. Tsien, H. S. *J. Am. Rocket Soc. 22*, 256 (1952).
69. Marble, F. E., and Cox, D. W., Jr. *J. Am. Rocket Soc. 23*, 63 (1953).
70. Ross, C. C., and Datner, P. O. *AGARD Selected Combustion Problems*. Butterworths, London, 1954.

SECTION M

COMBUSTION OF SOLID PROPELLANTS

CLAYTON HUGGETT

M,1. General Characteristics of Solid Propellants. Modern solid propellants used in jet propulsion devices may be divided into two general classes, double-base propellants and composite propellants, according to their composition and physical structure. Both types of propellants contain sufficient oxygen within their structure to effect their transformation into gaseous products. By the combustion or burning of a solid propellant we shall mean here the series of chemical processes through which the propellant, when subjected to a suitable ignition stimulus, is converted into gaseous products without the addition of oxygen from an external source. This combustion results in the evolution of a large amount of heat and, when it occurs in a confined space, in the development of a considerable pressure.

Double-base propellants have as their principal components nitrocellulose and an explosive plasticizer, usually nitroglycerin. Other materials may be added in smaller proportion to serve as stabilizers, nonexplosive plasticizers, coolants, lubricants, opacifiers, and burning rate modifiers, or otherwise to confer desirable properties on the product. Typical compositions together with some of their ballistic properties which will be discussed in later articles are given in Table M,1. The explosive plasticizer and minor constituents, with the exception of inorganic materials which may be present in small quantities, are physically compatible with nitrocellulose. The propellant is prepared in the form of a rigid plastic having a nearly homogeneous structure. Double-base rocket propellant compositions derive largely from those of the ballistites and cordites which have been used as gun propellants for many years. The principal innovations in adapting them for rocket applications have been in the techniques of fabricating the large grains needed for this purpose.[1]

Composite propellants are made by embedding a finely divided solid oxidizing agent in a plastic, resinous, or elastomeric matrix. The matrix

[1] A single piece of rocket propellant is called a grain although it may have a length of several feet, a web thickness of several inches, and weigh several hundred pounds. The web thickness is the minimum burning distance through the propellant between two opposite surfaces of the grain.

material usually provides the fuel for the combustion reaction although solid reducing agents are sometimes included in the compositions. Explosive binders requiring no additional source of oxygen have also been used. Composite-type propellants have been made in a great variety of compositions. Oxidizing agents which have been used extensively include ammonium nitrate, sodium nitrate, potassium nitrate, ammonium perchlorate, and potassium perchlorate. Asphalt, natural and synthetic rubbers, aldehyde-urea and phenolic resins, vinyl polymers, polyesters, and nitrocellulose are among the matrix materials which have been

Table M,1. Compositions and ballistic properties of typical rocket propellants.

Constituent	JPN Ballistite	HES 4016	Cordite SC	Russian cordite	Composite A	Composite B	Composite C
Nitrocellulose	51.5	54.0	50.0	56.5	21.0		
(Per cent nitrogen)	(13.25)	(13.25)	(12.20)	(12.20)	(12.60)		
Nitroglycerin	43.0	43.0	41.0	28.0	13.0		
Ethyl centralite	1.0	3.0	9.0	4.5	1.0		
Diethyl phthalate	3.25						
Dinitrotoluene				11.0			
Carbon black	(0.2)				9.0		
Potassium sulfate	1.25						
Potassium perchlorate					56.0		
Potassium nitrate							50.0
Ammonium perchlorate						75.0	
Ammonium picrate							41.0
Resin binder						25.0	9.0
Ballistic properties							
Heat of explosion, cal/g	1230	1260	965	880	1200		
Adiabatic flame temperature, °K	3028	3087	2492	2340		2200	
Burning rate; in./sec at 2000 lb/in.^{2}, 25°C	1.02	0.90	0.49	0.47	1.95	0.3	1.4
Pressure index at 2000 lb/in.^{2}	0.73	0.75	0.69	0.73	0.45	0.0	0.42

employed as binders. Ammonium picrate, carbon black, and aluminum powder have been used as fuel fillers. Minor constituents may be added to modify the properties of the binder or to catalyze the burning process. The common characteristic of these varied compositions is a markedly heterogeneous structure with adjacent regions of oxygen-rich and oxygen-deficient materials. In this respect they may be considered to be related to black powder although their physical form derives from modern plastics technology.

The laws of burning. Composite and double-base propellants show certain gross similarities in their burning behaviors. It is these properties which they possess in common that largely determine their usefulness in rocket propulsion devices.

Solid propellants burn by parallel layers; that is, all burning surfaces regress in a direction normal to the original surfaces of the grain and the grain tends to retain its original configuration until the web has burned through. This generalization, known as Piobert's law, was first applied to granular black powder long before the discovery of smokeless propellants. It receives its most striking confirmation and has its greatest utility when applied to the large and geometrically intricate grains of modern rocket charges.

The linear rate of burning of a propellant depends on the pressure under which the burning takes place. The simplest form of this dependence states that the burning rate increases with some power of the pressure,

$$r = bp^n \tag{1-1}$$

where b and n are constants characteristic of the particular propellant composition under study. This generalization, known as Vieille's law, represents quite accurately the experimental results obtained with many simple double-base and some composite propellants. However, as indicated in Art. 4, the pressure dependence of the burning rate of many of the new propellant compositions is much more complex, particularly in the pressure region below 3000 lb/in.^{2}

A third generalization that will be found useful in considering the combustion of solid propellants may be stated as follows: The gaseous products from a propellant, burned under pressure in an atmosphere of its own combustion products, are in thermodynamic equilibrium at the temperature of the flame. Thus it is possible to calculate the composition and thermodynamic properties of the propellant gas from a knowledge of the composition of the propellant and the thermodynamic properties of its constituents, by methods which have been described in Sec. A. In any practical experiment the decrease in flame temperature due to heat loss to the walls of the combustion chamber, and to the expansion of the gas in the case of burning in a vented vessel such as a rocket motor, must be taken into account. More significant deviations from equilibrium will be observed when the burning takes place at low pressures (below perhaps 1000 lb/in.^{2}), when the burning takes place in a foreign atmosphere which may enter into or quench the reactions of the powder gas, and when the residence time of the reacting gas within the combustion chamber is too short for equilibrium conditions to be reached (Sec. A). Since these departures from equilibrium are due to the interruption of the reaction before it has had time to approach completion, they are indicative of conditions at some intermediate stage in the combustion process. The study of such nonequilibrium conditions, brought about by the deliberate manipulation of experimental conditions, has provided much of the direct evidence concerning the nature of these intermediate steps.

Interior ballistic considerations. Before considering in detail the chemical processes which take place during the combustion of a solid propellant, it will be desirable to discuss very briefly the interior ballistics of solid propellant rockets. In this way we will be better able to evaluate from a practical standpoint the importance of the various factors which influence the burning process. A more detailed discussion of the interior ballistics of solid fuel rockets will be found elsewhere in this series (XII,H).

Most modern solid fuel rockets are designed around a neutral burning charge; that is, one whose burning surface area remains nearly constant until the charge is completely burned.[2] A simple example of such a charge consists of a centrally perforated, cylindrical grain with the end surfaces covered with an inhibiting material in order to confine the burning to the lateral surfaces. In this case the area of the burning surface within the central perforation increases at exactly the same rate as the area of the external surface decreases. The mass rate of burning of the propellant, or the mass rate at which the propellant gas is being generated within the chamber, will be given by

$$\dot{m}_b = S\rho r \tag{1-2}$$

where r is the linear burning rate, ρ is the density of the solid propellant and S is the area of the burning surface. We wish to consider the effect of variations in the chamber pressure p_c and the initial temperature of the powder T_0 on $\dot{m}_b$. Assume that the pressure dependence of the linear burning rate is given by Vieille's law (1-1).

$$\dot{m}_b = S\rho b p_c^n \tag{1-3}$$

The pressure index n is nearly independent of the propellant temperature but b increases as the initial temperature of the grain is raised.

At the same time that the grain is burning within the chamber, gas will be discharged from the nozzle of the rocket at a rate

$$\dot{m}_d = p_c C_d A_t \tag{1-4}$$

where A_t is the cross-sectional area at the throat of the rocket and C_d is called the discharge coefficient, a quantity whose value depends only on the composition and temperature of the propellant gas.

At the start of burning, the rate of gas generation will exceed the rate of gas discharge and the pressure in the chamber will rise. However, since n has a value of less than 1 for all practical propellants, the rate of gas

[2] Progressive charges in which the burning surface increases during burning, and regressive charges in which the burning surface decreases, may be used for special purposes. A great variety of grain designs have been used to obtain desired variations in the area of the burning surface with time during burning (XII,H).

discharge will increase more rapidly with pressure than the rate of gas generation, and eventually a chamber pressure will be reached where the two rates become equal and the chamber pressure assumes a constant value. At higher pressures, the mass rate of discharge would be greater than the rate of gas generation and the chamber pressure would fall to the equilibrium value. Thus a steady state pressure is established whose magnitude is, combining Eq. 1-3 and 1-4,

$$p_{eq} = \left(\frac{S\rho b}{A_t C_d}\right)^{\frac{1}{1-n}} \tag{1-5}$$

The pressure index may have a value in the neighborhood of 0.67 to 0.75 (Table M,1) so that the exponent $1/(1 - n)$ has a value of about 3 or 4. The equilibrium pressure is then seen to be an extremely sensitive function of S, the area of the burning surface and b, the coefficient of the burning law (it is assumed that ρ, A_t, and C_d are fixed for a given case). Since it is usually desirable to have the pressure in the rocket chamber show a minimum of variation under all circumstances, it is apparent that a suitable propellant must burn regularly in accordance with Piobert's law to give a constant burning area S and that b must show the smallest possible variation with propellant temperature. It is also apparent that the effects of variations in S and b will be minimized if n has a sufficiently small value. Thus the magnitude of the pressure index n is one of the important factors in determining the suitability of a propellant for rocket propulsion applications. The low value of n associated with many composite propellant compositions and the small variation of b with temperature are among the major advantages of this type of propellant.

M,2. Thermal Decomposition of Propellant Components. The combustion of a solid propellant takes place through a complex series of chemical processes which start within the interior of the grain and continue at an accelerated rate through the burning surface, finally reaching completion with the attainment of thermodynamic equilibrium in the gas phase at some distance from the surface of the propellant. The reaction zone possesses a physical discontinuity as well as sharp gradients in temperature and composition. The reactions are rapid, particularly at the high pressures which are of greatest interest. Consequently the precise experimental investigation of the burning process presents a difficult problem.

In order to gain some insight into the nature of the reactions taking place, it has been desirable to isolate experimentally certain probable reactions or groups of reactions. In particular the low temperature thermal decompositions of the oxidizing or explosive propellant ingredients have been studied extensively. While there is some question as to the applicability of the results of these *in vitro* experiments to the more complex conditions of combustion and to the multicomponent mixtures encountered in propellants, it appears that the reactions observed are similar to the initial steps of the combustion reaction. Moreover, the products of these low temperature decompositions must resemble the products from the initial stages of combustion, which in turn form the reactants for succeeding steps in the burning process.

Nitrocellulose. Nitrocellulose undergoes a slow spontaneous decomposition at temperatures ranging from those normally encountered in storage up to those where the sample ignites and rapid combustion follows. The reaction is catalyzed by traces of acids which may remain from the manufacturing process if the nitrocellulose is not sufficiently purified. Since some of the products of decomposition are themselves acidic in nature, the reaction may assume an autocatalytic character.

Will [*1*], in 1901, made careful studies of the effect of impurities on the rate of the decomposition reaction. He succeeded in showing that by careful purification the rate of the reaction could be reduced to a minimum value which could not be lowered further by additional purification. This characteristic rate increased with temperature and, for a given temperature, was greater the higher the degree of nitration of the nitrocellulose.

Robertson and Napper [*2*] studied the low temperature reaction in an apparatus wherein the volatile products of decomposition were removed continuously from the sample by a stream of carbon dioxide. They found that nearly half of the nitrogen liberated during this slow decomposition was in the form of nitrogen dioxide. If the decomposition products were allowed to remain in contact with the nitrocellulose, the NO_2 disappeared and the rate of total nitrogen evolution was accelerated.

Roginskii [*3*] has examined the data of Will and of Robertson and Napper and concludes that between 90 and 175°C the reaction follows the simple first order law (Sec. D).

$$-\frac{dc}{dt} = Ace^{-E_a/RT} \tag{2-1}$$

with an activation energy $E_a \cong 50$ kcal/mole and a frequency factor $A \cong 10^{20}$ sec^{-1}. Daniels [*4*] found an activation energy of 46.7 kcal/mole. Although a "normal" value of 10^{13} sec^{-1} for A is assumed, an examination of the data [*81*] indicates that the actual value must be in the neighborhood of 10^{19} sec^{-1}.

This slow decomposition is accompanied by the evolution of heat. When the rate of heat production at any point in the sample is greater than the rate of conduction of heat away from that point the temperature will rise, further accelerating the rate of decomposition and leading ultimately to a thermal explosion [*5*, p. 536]. Because of the low thermal conductivity of nitrocellulose, large samples stored at temperatures as low as 100°C may explode by this mechanism.

When small samples of nitrocellulose are heated above about 180°C, the sample ignites after a short induction period. Ignition is preceded by liquefaction and an acceleration of the decomposition process. A reactive gas containing nitrogen oxides, formaldehyde, carbon monoxide, and hydrogen is given off by the accelerated decomposition reaction. In the presence of an inert gas these products accumulate in the neighborhood of the nitrocellulose surface and a gas-phase ignition can occur, leading to a surface combustion of the sample. At very low pressures this reactive gas diffuses away so rapidly that ignition does not occur. Under these conditions the accelerated decomposition in the liquid phase can lead to a thermal explosion.

At very low pressures, under about 5 mm, small samples of nitrocellulose will burn with a feeble flame, depositing a powdery residue that has been termed the "white substance" (WS). The composition of WS has been studied by Wolfrom [*6,7*] and his associates. According to Wolfrom, a major part of this material is made up of partially denitrated and oxidized nitrocellulose residues having an average molecular weight of about 1500. This is accompanied by low molecular weight products such as formic and oxalic acids, furfural, glyoxal and formaldehyde, oxides of nitrogen and carbon, hydrogen, and water.

The decomposition becomes more complete at higher pressures. At 100 mm, about 40 per cent of the sample can be recovered as a complex mixture of liquid products called the "red substance" (RS) [*6*]. Glyoxal, formic acid, formaldehyde, and water are principal constituents of the RS, together with small amounts of glyceraldehyde, mesoxaldehyde, and other more complex products. The balance of the nitrocellulose is converted to gaseous products.

The following scheme has been proposed by Wolfrom, Dickey, and Prosser [*7*] for the initial reactions leading to the formation of RS.

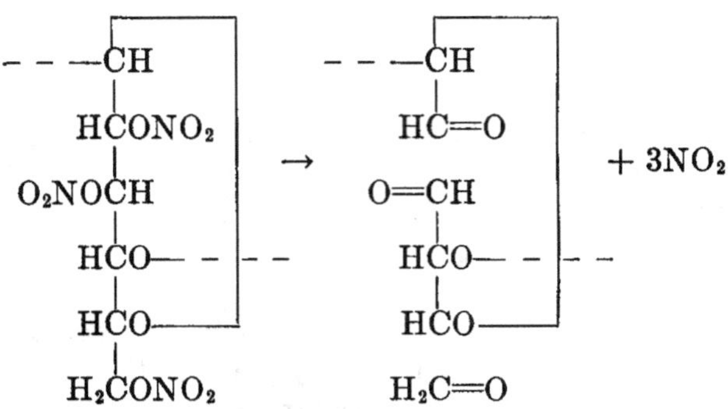

(2-2)

Cellulose nitrate

The first step is the breaking of the RO—NO$_2$ bonds, followed by rearrangement of the resulting free radical and scission of the 2—3 and 5—6 carbon—carbon bonds. By this scheme, glyoxal would originate from carbons 1 and 2 by hydrolysis of the acetal bonds. Formaldehyde would be formed from carbon 6 while the C—3,4,5 fragment would lead to the three-carbon aldehydes found among the products. Further reactions between NO_2 and these initial products would produce the more highly oxidized compounds found in the RS mixture.

Further increases in pressure cause a decrease in the amount of RS formed, but it is not until a pressure of several atmospheres is reached that all of the aldehydic products disappear. Fenimore [*8*] has formulated the final steps in the burning process as

$$HCO_2H,\ (HCO)_2,\ NO \rightarrow N_2,\ CO,\ CO_2,\ H_2O \tag{2-3}$$

Nitroglycerin. The behavior of relatively large samples of nitroglycerin when heated at atmospheric pressure has been described by Snelling and Storm [*9*]. They observed that their samples began to discolor at 135°C. The color deepened and ebullition started at 145°C. This ebullition was due to the formation of gaseous decomposition products rather than to boiling, since on continued heating the temperature of the liquid rose to 218°C where explosion occurred. The nonexplosive decomposition is exothermic, as in the case of nitrocellulose, since the temperature of the sample rose above that of the surrounding air bath during the decomposition. By continued heating below the explosion temperature, a nonexplosive residue having a low nitrogen content was obtained.

Robertson [*10*] studied the low temperature reaction by a method similar to that which he had used to study nitrocellulose (Art. 2). He concluded that practically all of the nitrogen liberated during the decomposition appeared first in the form of nitrogen dioxide.

Roginskii and Sapozhnikov [*11*] carried out the decomposition in a closed system where the products of reaction remained in contact with the sample. At temperatures between 120 and 150°C an induction period during which a slow first order decomposition takes place was terminated by an abrupt transition to an explosive reaction. This was attributed to the accumulation of a critical concentration of some product of decomposition, probably NO_2, leading to an autocatalytic reaction. At temperatures between 155 and 190°C the reaction was first order, presumably because of the decreased solubility of the autocatalytic agent and the occurrence of a unimolecular decomposition in the gas phase. The activation energy of this first order decomposition was estimated to be 47.8 kcal/mole.

A very careful study of the autocatalytic reaction was made by Lukin [*12*]. He obtained explosions in sealed ampoules at temperatures as low as 70°C after long isothermal induction periods. Water and nitrogen dioxide, and oxalic, nitric, and sulfuric acids were among the active catalysts for the reaction. The activation energy for the autocatalytic process was found to be about 26 kcal/mole.

The explosive decomposition of nitroglycerin, as in the case of nitrocellulose, thus appears to be due to a combination of autocatalytic and thermal effects. A branching-chain type of explosion mechanism in the liquid phase appears unlikely in view of the failure of known chain-breakers such as benzophenone to inhibit the reaction [*3*] and the failure of ultraviolet light or electron bombardment to initiate long reaction chains [*13*]. However, Gray and Yoffe [*14*] present strong evidence for a chain mechanism for the vapor-phase inflammation of nitroglycerin and other alkyl nitrates from the accelerating effect of inert gas diluents on the reaction.

The decomposition of organic nitrates. While nitrocellulose and nitroglycerin are almost the only nitrate esters to be used extensively in solid propellants,[3] the low temperature decompositions of other members of the series have been studied in order to elucidate the mechanism of the process. In particular, the simple structures of the low molecular weight compounds make them more amenable to kinetic investigation.

A number of workers have suggested that the primary step in the reaction is the breaking of the $RO—NO_2$ bond with the resulting formation of a nitrogen dioxide molecule and an alkoxyl radical. The latter gives rise to aldehydic and hydroxylic products which then react further with NO_2 to give the observed final products, CO, CO_2, NO, N_2O, N_2, H_2, H_2O, etc. (Table M,2a).

The decomposition of methyl nitrate was studied by Appin, Todes, and Khariton [*17*]. They found the reaction to follow first order kinetics between 210 and 240°C and proposed the reaction scheme:

$$CH_3ONO_2 \rightarrow CH_3O\cdot + NO_2 \tag{2-4}$$

$$2CH_3O\cdot \rightarrow CH_3OH + CH_2O \tag{2-5}$$

At higher temperatures, explosion occurred following an induction period. The explosion was supposed to have resulted from a thermal process, but O. K. Rice [*18*] has questioned this interpretation of the experimental results. Gray and Yoffe [*14*] found that the addition of argon or nitrogen facilitated the explosion of methyl nitrate vapor, indicating a branching-chain mechanism for the process. Adams and Bawn [*15*] found the slow decomposition of ethyl nitrate to be similar to that of methyl nitrate and suggested an alternate scheme for the stabilization of the alkoxyl radical,

$$CH_3CH_2O\cdot + C_2H_5ONO_2 \rightarrow CH_3CH_2OH + C_2H_4ONO_2\cdot \tag{2-6}$$

$$C_2H_4ONO_2\cdot \rightarrow CH_3CHO + NO_2 \tag{2-7}$$

More recently, Phillips [*19*], and Pollard, Wyatt, and Marshall [*20*] have shown, through the retarding effect of NO_2 on the decomposition, that

[3] Diethylene glycol dinitrate was used extensively in Germany during World War II. This was due in part to a shortage of glycerin for the manufacture of nitroglycerin, but certain of the DEGN powders showed improved physical properties.

Table M,2a. Composition of the volatile products from the slow decomposition of nitrate esters.

<table>
<tr><th rowspan="2">Compound</th><th rowspan="2">Temp., °C</th><th colspan="7">Gases, per cent</th><th rowspan="2">Liquid products</th><th rowspan="2">Reference</th></tr>
<tr><th>NO_2</th><th>NO</th><th>N_2O</th><th>N_2</th><th>H_2</th><th>CO</th><th>CO_2</th></tr>
<tr><td>Ethyl nitrate</td><td>185</td><td></td><td>60.0</td><td>3.2</td><td>1.2</td><td></td><td>8.9</td><td>23.4</td><td>CH_3CHO, H_2O</td><td>[*15*]</td></tr>
<tr><td>Nitroglycerin</td><td>165</td><td></td><td>37</td><td colspan="2">11</td><td></td><td>22</td><td>30</td><td>H_2O</td><td>[*12*]</td></tr>
<tr><td>Nitrocellulose</td><td>160</td><td>10.1</td><td>37.2</td><td>5.4</td><td>9.9</td><td>3.9</td><td>13.8</td><td>19.7</td><td>CH_2O, H_2O</td><td>[*5*, p. 536]</td></tr>
<tr><td>Pentaerythritol tetranitrate</td><td>210</td><td>11.8</td><td>47.7</td><td>9.5</td><td>1.6</td><td>2.0</td><td>21.1</td><td>6.3</td><td>CH_2O, H_2O</td><td>[*16*]</td></tr>
</table>

the initial dissociation step (2-4) is reversible,

$$RCH_2O\cdot + NO_2 \rightarrow RCH_2ONO_2 \tag{2-8}$$

Phillips has pointed out that the two-step mechanism (Eq. 2-6 and 2-7) is more probable than the simple disproportionation of alkoxyl radicals (Eq. 2-5) but Levy [22] has questioned the occurrence of this step because of the failure of diethyl peroxide to accelerate the decomposition of ethyl nitrate. The occurrence of free radicals during the thermal decomposition of nitrate esters has been discussed by F. O. Rice [82], but no experimental tests have been reported. The presence of free radicals has been demonstrated in the somewhat similar decomposition of ethyl nitrite [21]. In this case, the process

$$RCH_2O\cdot \rightarrow R\cdot + CH_2O \tag{2-9}$$

apparently occurs when the reaction takes place at low pressures or in the presence of an inert gas diluent.

Most kinetic studies of nitrate ester decomposition have made use of manometric methods to follow the course of the reaction, assuming that the rate of pressure increase was proportional to the rate of disappearance of the nitrate. Analytical studies, if made, have been confined to the examination of the final products, and the nature of the intermediate steps has been inferred from the kinetic data and the products of decomposition. Recently, Levy [22] has used the infrared spectrophotometer to study the vapor-phase decomposition of ethyl nitrate. By this method he was able to follow directly the disappearance of the ethyl nitrate and the formation of some of the intermediate and final products. Ethyl nitrite was found to be formed as an intermediate in nearly quantitative yield. This decomposed subsequently to give acetaldehyde, ethyl alcohol, and nitric oxide, the products required by the older mechanisms (Eq. 2-4 to 2-8); so the failure of the prior experiments to reveal the presence of the nitrite as an intermediate is not surprising. Rate data obtained from the direct observation of the disappearance of the nitrate were in good agreement with those obtained previously by the manometric method.

The following series of equations provides a satisfactory explanation for the experimental observations:

$$C_2H_5ONO_2 \rightleftarrows C_2H_5O\cdot + NO_2 \tag{2-10}$$

$$C_2H_5O\cdot \rightarrow CH_3\cdot + CH_2O \tag{2-11}$$

$$2CH_2O + 3NO_2 \rightarrow 3NO + 2H_2O + CO + CO_2 \tag{2-12}$$

$$2CH_3\cdot + 7NO_2 \rightarrow 7NO + 3H_2O + 2CO_2 \tag{2-13}$$

$$C_2H_5O\cdot + NO \rightarrow C_2H_5ONO \tag{2-14}$$

By carrying out the decomposition in the presence of added acetaldehyde or nitric oxide, the reverse step (Eq. 2-10) is suppressed and the rate of disappearance of ethyl nitrate follows simple first order kinetics.

If further work confirms the general occurrence of nitrites as intermediates in the thermal decomposition of organic nitrates, mechanisms concerned with the detailed chemistry of the decomposition process will have to be revised, although the over-all kinetic result as it affects the burning of propellants will probably be little changed.

Table M,2b. Thermal decomposition of organic nitrates.

Compound	Temp., °C	E_a, kcal/mole	$\log_{10} A$†	Ref.	Remarks
Methyl nitrate	210–240	39.5	14.3	[*17*]	gas
Ethyl nitrate	180–215	39.9	15.8	[*15*]	gas
	161–181	41.2	16.85	[*22*]	gas
n-Propyl nitrate			14.7	[*23*]	
Ethylene glycol dinitrate	85–105	39	15.9	[*23*]	
Trimethylene glycol dinitrate	85–110	38.1	15.2	[*23*]	
Propylene glycol dinitrate	80–100	37.4	15.2	[*23*]	
Nitroglycerin	75–105	40.3	17.1	[*23*]	liquid
	90–125	42.6	18.0	[*3*]	liquid
	125–150	45	19.2	[*3*]	liquid
	150–190	50	23.5	[*3*]	liquid
Pentaerythritol tetranitrate	161–233	47	19.8	[*16*]	liquid
	171–238	39.5	16.1	[*16*]	5 per cent solution in dicyclohexyl phthalate
	100–120	50.9	20.6	[*24*]	solid
	120–150	54.0	23.3	[*24*]	liquid, 50 per cent in TNT
Mannitol hexanitrate			20.6	[*23*]	
Nitrocellulose	90–135	49	21	[*3*]	solid
	145–155	48	20	[*3*]	
	155–175	56	24	[*3*]	
	130–155	46.7	19[1]	[*4*]	
Nitroxy-nitro-compounds					
Trimethylolnitromethane trinitrate	75–95	36.4	15.3	[*23*]	
Dimethylolnitromethane dinitrate	134–208	38.9	16.5	[*25*]	

† $\log_{10} A$ estimated from published data [*81*].

The slow thermal decompositions of a large number of organic nitrates have been found to follow the first order kinetic law when precautions were taken to avoid the complicating effects of autocatalysis and self-heating. Values of the activation energy E_a and the frequency factor A in the unimolecular rate equation which have been reported by various investigators are collected in Table M,2b.

While there is some lack of agreement among different investigators studying the same compound, due probably to the experimental difficulties noted above, virtually all agree in assigning a high energy of activation and an unusually large frequency factor to these reactions. While the evidence is less conclusive, it also appears that A and E_a increase with the temperature and with increasing structural complexity of the molecule.

Large values of A and E_a are characteristic of most explosive substances (Table M,2b and M,2c). In particular, a large activation energy is necessary to confer adequate stability at low temperatures in combination with a high order of reactivity at elevated temperatures. The energy of activation for the decomposition of nitrate esters has been found to compare reasonably well with estimates of the strength of the $RO—NO_2$ bond.

The interpretation of the large value of A is more difficult. The possible occurrence of long, stable reaction chains (chain lengths of the order of 10^5 to 10^6 would be necessary to explain some of the reported results) is not supported by the experimental evidence, except possibly in the case of simple nitrate esters in the vapor phase. The more complex molecules show similar rates of decomposition in the liquid phase, in solution, and in the vapor phase (Table M,2b).

In the more complicated molecules, energy from a number of degrees of freedom of molecular motion may contribute to the activation of the $RO—NO_2$ bond, thus increasing the probability of bond rupture. Phillips [*23*] has suggested that the formation of the activated complex which precedes the breaking of the bond is accompanied by a large positive entropy of activation, due to a lessening of the repulsion between neighboring nitrate groups. While these considerations may account for a value of A well above the classical one, it seems probable that some of the very large values reported result from the complexity of the decomposition process and the difficulty in eliminating thermal and autocatalytic effects in the experimental measurements.

The decomposition of nitramines and nitro-compounds. While organic explosives other than nitrate esters have not been used extensively in propellants in the past, they appear to offer interesting possibilities for the development of new compositions. It has been shown that a variety of high explosives including TNT, PETN, and cyclonite will burn without detonation at rates comparable to those of propellant explosives.

The slow thermal decompositions of nitramines and nitro-compounds are similar in many respects to the decomposition of nitrate esters. Many of these reactions have been found to follow first order kinetics. The published kinetic data are collected in Table M,2c. The large activation energies and high frequency factors appear to be characteristic of all explosive compounds. Autocatalytic and thermal effects complicate the simple unimolecular picture as in the case of nitrate ester decompositions.

This may account for some of the discordantly large values of A and E_a reported for TNT and tetryl. Robertson [*28*] attributes the decreased rate of decomposition of cyclonite in solution to the quenching of a chain reaction which occurs in the pure liquid.

An examination of the products of decomposition reveals a more significant difference between the mechanism of decomposition of nitramines and of nitrate esters. While the primary step in the latter reaction

Table M,2c. Thermal decomposition of nitramines and nitro-compounds.

Compound	Temp., °C	E_a, kcal/mole	$\log_{10} A$	Ref.	Remarks
Nitramines					
Ethylenedinitramine	100–120	37.3	12.2	[*26*]	solid
	120–135	41.5		[*26*]	solid
	135–145	51.5		[*26*]	solid
	184–254	30.5	12.8	[*27*, p. 677]	liquid
Cyclotrimethylenetrinitramine	213–299	47.5	18.5	[*28*]	liquid
	201–280	41	15.46	[*28*]	5 per cent solution in dicyclohexyl phthalate
	195–280	41.5	15.55	[*28*]	1 and 5 per cent solutions in TNT
Cyclotetramethylenetetranitramine	271–314	52.7	19.7	[*28*]	liquid
Nitro-nitramine					
2,4,6-Trinitrophenylmethylnitramine	211–260	38.4	15.4	[*27*, p. 677]	liquid
	140–150	55.5	24.5	[*3*]	liquid
Nitro-compounds					
Trinitrotoluene	270–310	34.4	11.4	[*27*, p. 977]	liquid
		53	19	[*3*]	liquid
Picric acid		58.6	22.5	[*3*]	liquid
Nitromethane	380–430	53.6	14.6	[*29*]	gas

appears to involve the liberation of NO_2, followed by reduction to NO, most of the nitrogen from the nitramines appears first in the form of nitrous oxide, N_2O. The breaking of the RN—NO_2 bond to give nitrogen dioxide is not an important process in the decomposition of nitramines. This difference is illustrated by the difference in the composition of the gaseous products from decomposed nitramines and nitrate esters as shown in Tables M,2a and M,2d. Since the final step in the combustion

Table M,2d. Composition of the products from the slow decomposition of nitramines.

Compound	Temp., °C	Gases, moles/mole						Liquid products	Reference
		NO	N_2O	N_2	H_2	CO	CO_2		
Ethylenedinitramine	195	0.10	1.4	0.39	—	0.028	0.065	CH_3CHO, H_2O	[*27*, p. 677]
Cyclotrimethylenetrinitramine	225	0.54	0.98	1.16	0.09	0.40	0.48	CH_2O, H_2O	[*28*]
Cyclotrimethylenetrinitramine 5 per cent in TNT	220	0.70	0.63	1.65	0.20	0.57	0.83	CH_2O, H_2O	[*28*]
Cyclotetramethylenetetranitramine	280	0.95	1.51	1.16	—	0.57	0.64	CH_2O, H_2O	[*28*]

of an explosive appears to be the reduction, by carbon monoxide and hydrogen, of the oxides of nitrogen formed during the early stages of the reaction, we may expect certain differences between the combustion of nitramine and nitrate ester propellants because of the preponderance of either N_2O or NO in the gas-phase reaction.

A comparison of the data in Tables M,2b and M,2c indicates that, as a class, nitramines have greater stability at low temperatures than nitrate esters. From this it may be inferred that a higher surface temperature will be necessary for the combustion of a nitramine propellant. As shown in Art. 4, this suggests that the temperature dependence of the burning rate of a nitramine propellant should be low. No experimental tests of this sort have been reported.

The thermal decomposition of solid oxidizing agents. Of the various solid oxidizing agents which have been used extensively in rocket propellants, ammonium nitrate is the only one whose thermal decomposition has been studied in detail [*30*]. Robertson [*16*] found that the decomposition followed the unimolecular equation

$$k = 10^{13.8}e^{-40,500/RT}\ \text{sec}^{-1}$$

over the temperature range 243 to 361°C. Nitrous oxide and water are almost the only products of decomposition of liquid ammonium nitrate at low temperatures. At higher temperatures NO_2, NO, and N_2 are found, probably as the result of a vapor-phase decomposition. By the use of N^{15} as a tracer, it has recently been shown that the nitrous oxide is formed from the reaction of an ammonium group and a nitrate group [*31*]. From the isotopic molecule $N^{15}H_4N^{14}O_3$, the isomer $N^{15}N^{14}O$ is formed exclusively. Friedman and Bigeleisen [*31*] suggest the dehydration mechanism,

$$NH_4^+ + NO_3^- \rightarrow H_2N\text{—}NO_2 + H_2O$$

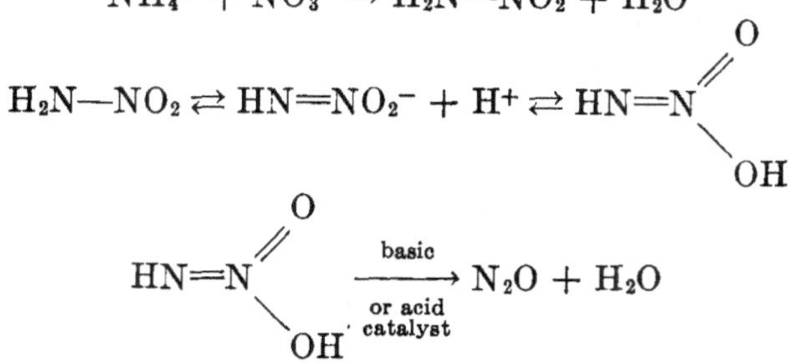

A variety of catalysts have been found to accelerate the decomposition of ammonium nitrate. Chromium compounds are particularly effective. This effect has been used to increase the burning rate of ammonium nitrate propellants, which otherwise are very slow burning [*5*, pp. 572–579].

The thermal decomposition of ammonium perchlorate has been studied by Dodé [*32*] and by Bircumshaw and Newman [*92*]. The reaction is complex; decomposition starts in the neighborhood of 200°C and is still slow enough to measure at 450°C. At temperatures below 300°C the major products of decomposition are indicated by the equation

$$4NH_4ClO_4 \rightarrow 2Cl_2 + 3O_2 + 8H_2O + 2N_2O$$

At temperatures above 350°C the following equation is more representative of the products:

$$2NH_4ClO_4 \rightarrow 4H_2O + Cl_2 + O_2 + 2NO$$

The interpretation of the experimental results is further complicated by the occurrence of an induction period whose duration is sensitive to the presence of impurities and by a change in crystal structure which takes place at 240°C. Manganese dioxide and ferric oxide catalyze the decomposition. Schultz and Dekker [*93*] have interpreted the experimental data of Bircumshaw and Newman in terms of a topochemical reaction starting at nuclei on the crystal surfaces and spreading hemispherically into the interior of the crystals.

Potassium perchlorate decomposes when heated to give potassium chloride and oxygen. The decomposition starts at about 500°C. Otto and Fry [*33*] report that the reaction is unimolecular over the temperature range from 536 to 617°C. From their results the rate equation

$$k = 10^{14.0} e^{-60,800/RT} \text{ sec}^{-1}$$

can be deduced. The large activation energy, combined with a normal frequency factor, is indicative of the stability of the material at low temperatures. A somewhat more complicated decomposition process is indicated by the more recent studies of Bircumshaw and Phillips [*94*]. The decomposition can be catalyzed by a variety of agents, ferric oxide and manganese dioxide being among the most effective.

Recently it has been shown that solid potassium perchlorate and carbon black react together at comparatively low temperatures, 320 to 385°C, at a rate much too fast for the normal dissociation of the perchlorate into potassium chloride and oxygen to be the rate-controlling step [*34*]. It is probable that similar processes take place in the burning of a composite-type propellant.

Sodium and potassium nitrate are also comparatively stable oxidizing agents. When heated they first lose a portion of their oxygen and are converted to the nitrites. This reaction is reported to start at 255°C in the case of $NaNO_3$ and at 286°C for KNO_3 [*35*]. At higher temperatures the nitrites decompose to form alkali oxides and peroxides. More oxygen and oxides of nitrogen are given off. The decomposition of KNO_3 is rapid at 1000°C but less rapid than the combustion of a KNO_3 propellant.

Gas-phase reactions. The varied observations on the slow decomposition of explosives suggest that the combustion of a propellant takes place through a surface zone decomposition leading to the formation of a combustible gas mixture. The combustion of this gas close to the propellant surface then provides much of the energy necessary to support the primary decomposition. The combustible mixture may contain reducing gases such as formaldehyde and other more complex organic fragments, carbon monoxide, and hydrogen; oxidizing agents such as nitrogen dioxide, nitric oxide, nitrous oxide, and oxygen; and the comparatively inert end products, nitrogen, water, and carbon dioxide. A knowledge of the kinetics of the various reactions that can occur in such mixtures would be of obvious value in attempting to arrive at an understanding of the complete combustion process. Unfortunately only a few of these reactions have been studied.

The reaction between nitrogen dioxide and formaldehyde has been studied by Pollard and Woodward [*36*, pp. 760, 767] and Pollard and Wyatt [*37*]. The reaction proceeds smoothly at temperatures between 118 and 160°C. The rate is proportional to $p_{CH_2O} \times p_{NO_2}$. The value of k is given by

$$k = 10^{7.1}e^{-15,100/RT} \text{ l/mole-sec}$$

Above 160°C the rate accelerates more rapidly, and at about 180°C the reaction becomes explosive. The slow reaction can be represented by the over-all equation

$$5CH_2O + 7NO_2 \rightarrow 3CO + 2CO_2 + 7NO + 5H_2O$$

The explosive reaction leads to the formation of large amounts of N_2 and CO_2 together with nitric oxide and a small amount of CO. The transition from slow reaction to explosion apparently occurs through a thermal mechanism.

The oxidation of CO by NO_2 is much slower than that of formaldehyde and thus is probably not important during the first stages of combustion where NO_2 is present [*38*].

In the later stages of the flame reaction NO, CO, and H_2 are the principal reactants. Nitric oxide is a relatively stable gas, decomposing only slowly into N_2 and O_2 at 1200°C [*39*]. It reacts slowly with carbon monoxide in a homogeneous third order process at 1380 to 1580°C [*40*]. At lower temperatures the reaction is heterogeneous and is catalyzed by water vapor [*41*]. The suggested mechanism is

$$CO + H_2O \rightarrow CO_2 + H_2$$
$$2NO + H_2 \rightarrow N_2O + H_2O$$
$$2N_2O \rightarrow 2N_2 + O_2$$

The reaction between NO and H_2 [*42*] takes place slowly at 900°C. It also follows a third order law, as do most reactions involving oxidation by NO.

M,3. The Burning of Double-Base Propellants. The many studies of the decomposition of pure explosives described in the preceding article suggest the types of chemical reactions that can be expected to occur in the combustion of a double-base propellant. This approach to the elucidation of the mechanism of the process has been stressed by Bawn [*25*] and by Adams [*95*, p. 277]. It must be kept in mind that these experiments deal with simple starting materials and, in general, with reaction in a single phase. Under these conditions we observe only a single reaction or closely related series of reactions. In the burning of a propellant, on the other hand, we encounter a series of reactions differing widely in chemical character and in the physical environment in which they occur. Any detailed

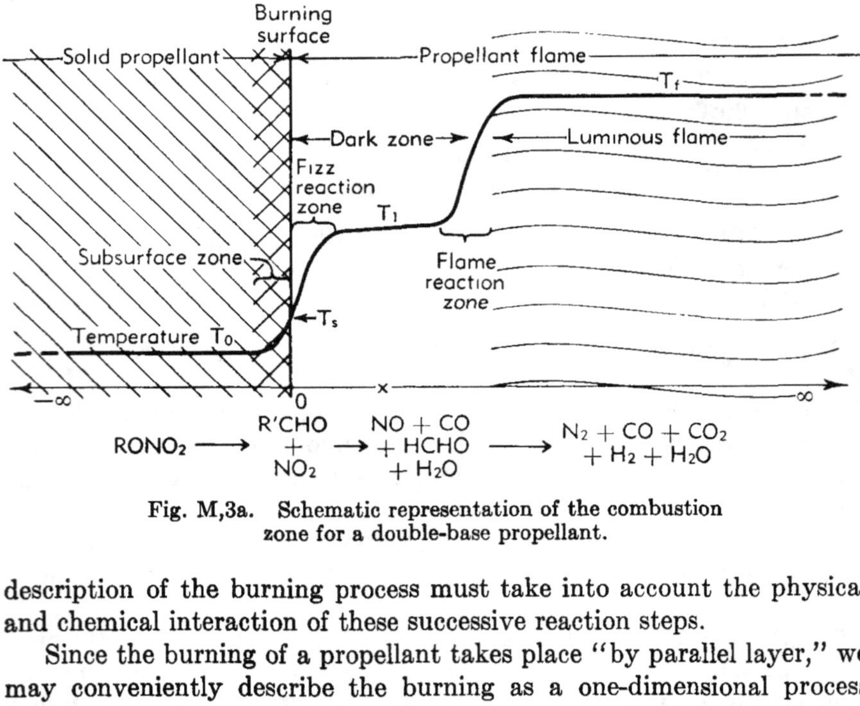

Fig. M,3a. Schematic representation of the combustion zone for a double-base propellant.

description of the burning process must take into account the physical and chemical interaction of these successive reaction steps.

Since the burning of a propellant takes place "by parallel layer," we may conveniently describe the burning as a one-dimensional process taking place along a spatial parameter normal to the burning surface. Because of the poor thermal conductivity of the material, the propellant at a short distance beneath the burning surface is largely unaffected by the combustion process. As the burning surface approaches, the propellant is heated by conduction and chemical reaction; the temperature rises, liquefaction may occur, and at the surface the propellant is completely decomposed to volatile fragments. These volatile products of decomposition react further in the gas phase and ultimately are converted to the end products of combustion at the high temperature of the propellant flame. This reaction sequence is illustrated schematically in Fig. M,3a.

It is convenient to choose a frame of reference so that the position of the burning surface remains fixed. As the propellant moves from left to right through the combustion zone, the products from each reaction step form the reactants for the following step until the end products, in thermodynamic equilibrium at the high temperature of the flame, are formed. Heat conduction in the opposite direction, from a region of higher to lower temperature, maintains the temperature of the reactants at a level necessary for reaction to occur. Thus a steady state is established, characterized by constant rates of mass flow and energy flow through the reaction zone and by stable concentration and temperature gradients. All derivatives with respect to time conveniently vanish. Such a one-dimensional model can be approximated experimentally in convenient fashion by a long cylindrical strand burning from one end. By coating the sides of the strand with some less combustible material, burning can be restricted to the end surface of the grain (Plate M,3). Such grains burn in a very regular and reproducible fashion and afford a convenient means of making a variety of observations on the burning process.

The subsurface zone. The propellant layers immediately beneath the burning surface are warmed by heat conducted from the relatively hot surface of the grain.[4] This rise in temperature starts the decomposition of the nitrocellulose and nitroglycerin a short distance below the surface. As has been shown (Art. 2), these decompositions are exothermic processes which cause further heating of the subsurface zone.

A second type of reaction which contributes to this subsurface heating is that which takes place between the nitrate ester and the stabilizer. These stabilizers are usually weakly basic aromatic compounds such as diphenylamine or ethyl centralite (*sym*-diethyldiphenylurea). It is their function to react with the oxides of nitrogen formed by the spontaneous decomposition of the nitrate esters during storage, thus preventing an accelerating autocatalytic reaction. At higher temperatures they can react directly with the nitrate esters in a highly exothermic reaction. The importance of this process is shown by experiments in which the temperature in the interior of a heated powder grain was measured by a thermocouple [*43*]. The temperature of the interior rose above that of the heating bath, and if the temperature of the bath was sufficiently high the sample ignited. The magnitude of this self-heating effect is greater for propellants containing relatively large amounts of stabilizer than for propellants with little or no stabilizer. Diphenylamine reacts more vigorously than centralite under these conditions and thus is a less satisfactory stabilizer

[4] We will neglect, for the present, the effect of radiation penetrating into the interior of the grain. This will be a good approximation in the case of a propellant containing an opacifying agent such as carbon black or in the case of a small end-burning strand burning at a low pressure where the volume of radiating flame is small and located at some distance from the burning surface. The effect of radiation in cases where it is of importance is discussed in Art. 4.

for high temperature storage. The importance of these reactions to the burning process is shown by the observation that small strands of propellants containing large amounts of stabilizers will burn smoothly in an inert atmosphere at atmospheric pressure and 35°C, whereas propellants containing only nitrocellulose and nitroglycerin will not burn at pressures below 200 lb/in.^{2} at 35°C [*44,84*]. It appears that at low pressures only a small amount of energy returns to the burning surface from the flame zone. These exothermic reactions, taking place within the solid phase beneath the burning surface, must contribute a significant part of the energy necessary to decompose the surface layers of the propellant and transport the products to the flame reaction zone.

The heat balance in any plane of the combustion wave (Fig. M,3a) is given by

$$\frac{\partial}{\partial x}\left(k\frac{\partial T}{\partial x}\right) - \dot{m}\frac{\partial}{\partial x}(c_pT) + \dot{q}_x = 0 \tag{3-1}$$

where x is the distance from the burning surface, k is the coefficient of thermal conductivity, $\dot{m}$ is the mass burning rate, c_p is the specific heat at constant pressure, and $\dot{q}_x$ is the rate of heat release due to chemical reaction in the x plane. The first term in the equation represents the rate of change in heat content per unit volume due to conduction, the second term is the rate of change of heat content due to mass flow, and the third term is the rate of heat release by chemical reaction. For purposes of simplification we assume that k and c_p are constant within the solid phase. With this approximation Eq. 3-1 becomes

$$k\frac{\partial^2 T}{\partial x^2} - \dot{m}c_p\frac{\partial T}{\partial x} + \dot{q}_x = 0 \tag{3-2}$$

Unfortunately $\dot{q}_x$ contains the rates of the exothermic decomposition reactions. These depend on the temperature through an exponential factor of the Arrhenius type. The equation then could not be solved in exact form, even if the appropriate kinetic and thermodynamic terms for the evaluation of $\dot{q}_x$ were known. However, because of the large activation energies of these exothermic processes (Art. 2), $\dot{q}_x$ will be small outside of the high temperature region close to the burning surface. Qualitatively the effect of $\dot{q}_x$ will be to increase the steepness of the temperature gradient in the neighborhood of the burning surface.

If we make a further approximation, setting $\dot{q}_x = 0$, Eq. 3-2 can be integrated to give

$$T - T_0 = (T_s - T_0)e^{\dot{m}c_px/k} \tag{3-3}$$

where T_0 is the initial temperature of the propellant and T_s is the temperature at the burning surface. A typical double-base propellant may have the following properties:

$$\text{density} = 1.6 \text{ g/cm}^3$$
$$c_p = 0.35 \text{ cal/g deg}$$
$$k = 5 \times 10^{-4} \text{ cal/cm sec deg}$$

If we take $T_0 = 25°\text{C}$ and assume a value of 300°C for T_s we can calculate T as a function of x for various values of the burning rate. The results of such calculations are shown in Fig. M,3b.

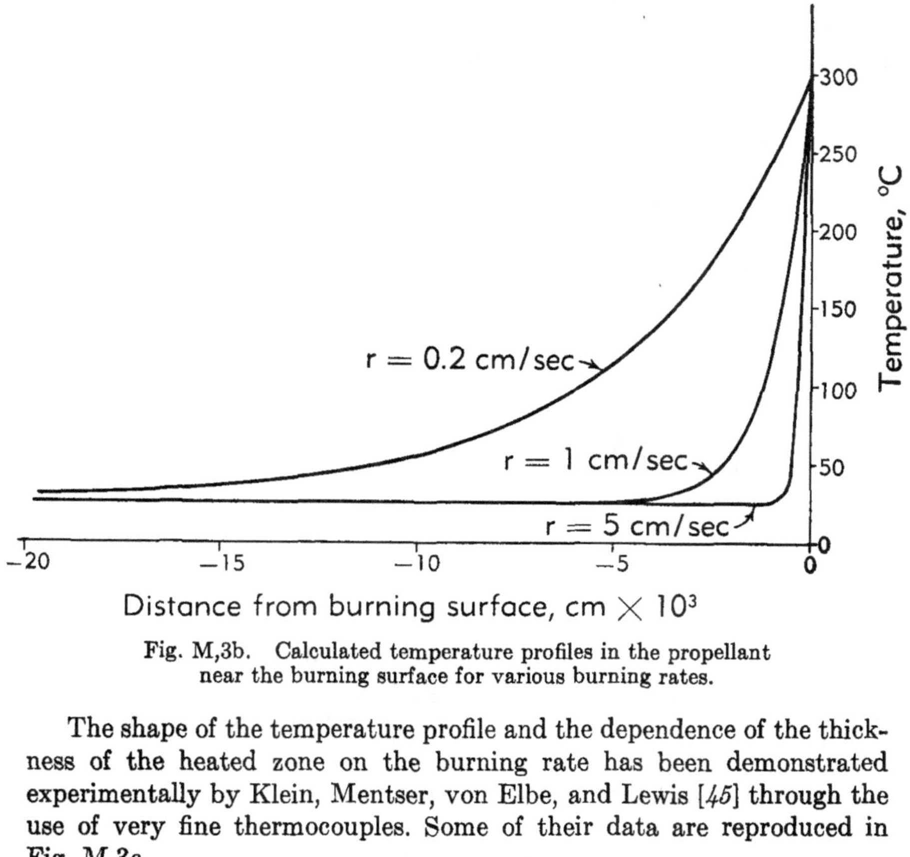

Fig. M,3b. Calculated temperature profiles in the propellant near the burning surface for various burning rates.

The shape of the temperature profile and the dependence of the thickness of the heated zone on the burning rate has been demonstrated experimentally by Klein, Mentser, von Elbe, and Lewis [*45*] through the use of very fine thermocouples. Some of their data are reproduced in Fig. M,3c.

It is apparent that only a very thin layer at the surface of the propellant grain takes part in the burning process. As the burning rate decreases, the thickness of this heated layer increases rapidly and the heat supplied by the exothermic reactions taking place therein makes up an increasingly large proportion of the total energy necessary to support combustion. In contrast to the transfer of energy from the flame region to the surface, the total amount of heat from this source may actually increase as the pressure decreases and the burning rate falls off. It is found experimentally that for most double-base propellants the pressure index of the burning rate becomes smaller as the burning rate decreases at low pressures. It is also generally observed that slow-burning propellants have smaller pressure indexes in the low pressure region than fast-burning ones. Since most jet propulsion devices operate at comparatively low pressures, below 2000 lb/in.^{2}, reactions in the subsurface zone may be of considerable practical importance and are deserving of more careful study.

The surface temperature. In the preceding paragraphs, reference was made to the temperature of the burning surface. It is apparent that this is a quantity which will be difficult to define and even more difficult to

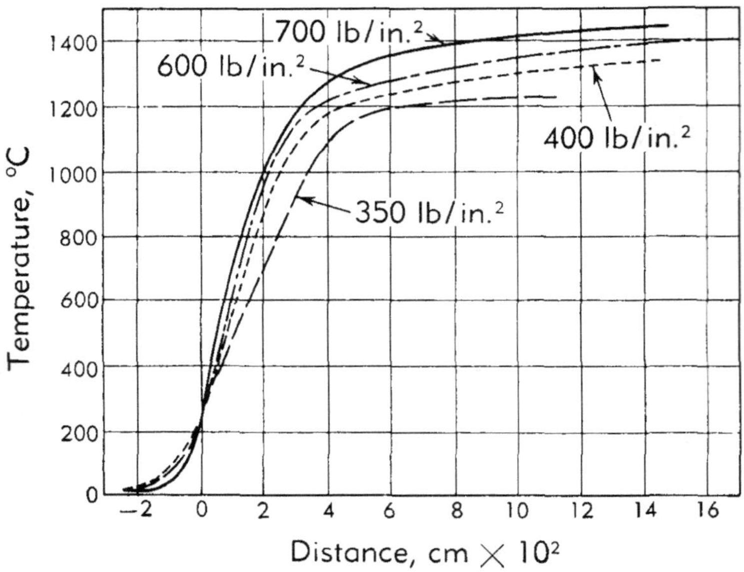

Fig. M,3c. Experimental temperature profiles in the neighborhood of the burning surface. Nitrocellulose (13.15 per cent N) plus 1 per cent ethyl centralite. The burning surface is assumed to be at $T = 250°C$ (from [*45*]).

measure experimentally. The temperature gradient is at its steepest in this region; thermal equilibrium between the various degrees of freedom of the reacting species may be upset by chemical excitation; and the location of the surface itself is rendered uncertain by liquefaction and foaming. Nevertheless, because of the utility of the concept of a surface temperature in discussing various characteristics of the combustion process, it is worthwhile to try to establish the approximate level of the temperature in the neighborhood of the burning surface. A number of experimental measurements of this quantity have been attempted.

From studies of the melting of small metal particles and the decomposition of alkaline earth carbonates when incorporated in propellants, Daniels, Wilfong, and Penner [*4*] estimated the surface temperature to be above 1000°C. This value appears to be too high since the surface of the burning grain is not luminous. The observed experimental results may be due to catalytic reactions at the surface of the solid particles, as has been observed with bare thermocouples in the propellant flame [*45*], or to variations in heat transfer to the solid particles which may project out from the propellant surface into the very steep temperature gradient of the flame.

The amount of heat in the surface layers of a suddenly extinguished propellant grain was measured calorimetrically by Aristova and Leipunskii [*46*]. If it is assumed that the temperature distribution within the burning surface is given by Eq. 3-3, the quantity of stored heat q is given by

$$q = \int_{-\infty}^{0} (T - T_0)\rho c_p dx = (T_s - T_0)\rho c_p \int_{-\infty}^{0} e^{\dot{m}c_p x/k} dx \tag{3-4}$$

Then

$$T_s = T_0 + \frac{q\dot{m}}{\rho k} \tag{3-5}$$

In this way a value of 330 ± 45°C was obtained for the surface temperature of a double-base propellant. A single-base (nitrocellulose) propellant gave 252 ± 48°C. The use of Eq. 3-3 assumes that no heat is released by chemical reaction within the solid phase. As this is not the case, the value of T_s given by Eq. 3-5 will be too small. Since these measurements were made at atmospheric pressure where the subsurface heating effect may be large, this may cause a significant error.

The direct measurement of the temperature profile by means of thermocouples, discussed in the previous article, also leads to a determination of T_s. Referring to Eq. 3-3 a plot of $\ln(T - T_0)$ vs. x should yield a straight line. This is found to be the case in the low temperature region of the temperature profile, but at higher temperatures a considerable curvature is observed [*45*]. This is due to the heat of reaction term which was neglected in obtaining Eq. 3-3. The point of departure from linearity represents the plane in the combustion zone where heat production becomes important. This is taken to be the burning surface. In this way a surface temperature in the neighborhood of 250°C was found for a single-base propellant. In this case also, a consideration of the heat of reaction term will lead to a somewhat higher value.

Finally it may be of interest to calculate an upper limit to the surface temperature from the measured rate of decomposition of nitrocellulose and nitroglycerin. Following Daniels, et al. [*4*], we estimate the number of nitrate groups in a typical double-base propellant to be $1.07 \times 10^{22}/\text{cm}^3$. For a burning rate of 1 cm/sec the surface reaction zone will have an effective thickness of perhaps 10^{-3} cm (Fig. M,3c). The rate of the unimolecular decomposition reaction necessary to give a burning rate of

1 cm/sec will be

$$-\frac{dn}{dt} = kn = 1.07 \times 10^{22}/\text{cm}^2\ \text{sec} \tag{3-6}$$

$$n = 1.07 \times 10^{19}/\text{cm}^2 \tag{3-7}$$

$$k = 10^3/\text{sec} \tag{3-8}$$

where n is the number of nitrate groups in the reaction zone and k is the specific rate constant. From Table M,2b

$$k = 10^{20}e^{-48,000/RT_s} \tag{3-9}$$

and from Eq. 3-8 and 3-9 the value of the surface temperature[5] is:

$$T_s = 617°\text{K} = 344°\text{C}$$

Increasing or decreasing the thickness of the reaction zone by an order of magnitude leads to values of 310°C and 383°C respectively. Since the surface decomposition probably takes place, in part at least, by the faster autocatalytic mechanism, these values calculated for the simple unimolecular process must give an upper limit for the surface temperature.

While none of these results is completely satisfactory, it is probable that the temperature of the surface of a typical double-base propellant burning under moderate pressure is in the neighborhood of 300°C.

The flame reaction zone. When an end-burning strand of double-base propellant burns in an inert atmosphere at pressures below about 200 lb/in.², the flame is nonluminous [*44,84*]. As the pressure of inert gas is increased, a luminous flame region appears at some distance from the burning surface (Plate M,3). Further increases in pressure cause the luminous flame to approach the surface more closely, until at 1000 lb/in.² it is difficult to detect the dark zone between the propellant surface and the luminous flame. For a typical double-base composition the length of this dark zone between burning surface and luminous flame is given approximately by

$$l = \frac{10^7}{p^3}$$

where l is the dark zone length in inches and p is the pressure in lb/in.²

The heat produced by the nonluminous reaction, some 500 cal/g, is less than half of that produced by complete reaction through the luminous flame zone (Fig. M,3d). The products of reaction show a similar change in going from the nonluminous to the luminous stage (Fig. M,3e). Large

[5] Daniels calculates a much higher value for T_s by a similar process [*4*]. However, he uses a "normal" value of 3×10^{13} sec^{-1} for A, the temperature independent term in the rate equation rather than the experimental value used here. While Daniels does not calculate a value for A, an examination of his data [*81*], making reasonable assumptions concerning the nature of the products of decomposition, indicates that the value would not be far from 10^{19} sec^{-1}.

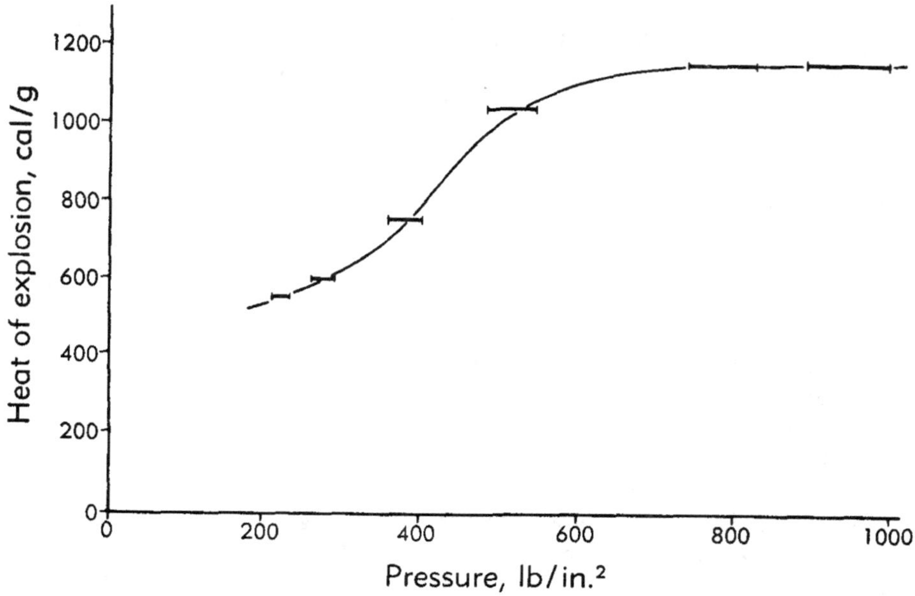

Fig. M,3d. Change in the heat of explosion with change in the initial pressure of inert gas, propellant HES 4016 (from [*44*]).

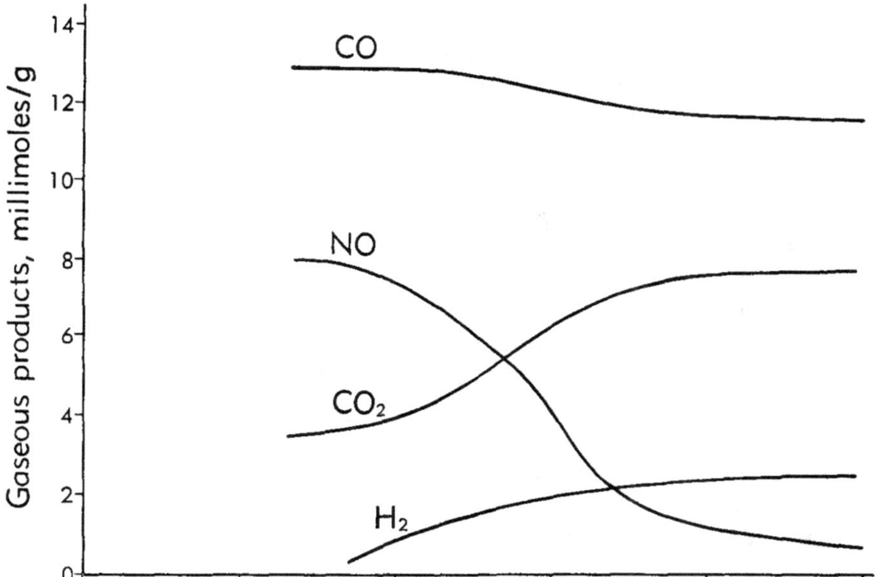

Fig. M,3e. Change in the products of combustion with change in the initial pressure of inert gas, propellant HES 4016 (from [*44*]).

amounts of carbon monoxide and nitric oxide are formed in the nonluminous reaction. In the luminous flame zone the nitric oxide is consumed. The amount of carbon monoxide decreases, carbon dioxide increases, and hydrogen is formed by the water gas reaction.

The reacting flame can thus be divided into two regions with distinctly different properties and in which different chemical processes take place. The nonluminous region has been called the fizz zone since the product gases are projected outward in a direction normal to the plane of the burning surface with a considerable velocity, making a fizzing sound. The manner in which the products are projected outward indicates that the principal site of reaction must be very close to the burning surface. The luminous region is commonly called the flame zone to differentiate it from the fizz zone, but this usage is an unfortunate one since it may be misinterpreted to mean the gross volume of the propellant flame.

The temperature profile through the fizz zone of a nitrocellulose propellant has been measured by the thermocouple method described previously. Fig. M,3c shows experimental curves of temperature vs. distance for a number of pressures. As might be expected, the temperature gradient near the burning surface increases with pressure since the gas-phase reactions are accelerated. The fizz reaction zone is very narrow, extending only a few hundredths of a centimeter beyond the burning surface. This distance is much shorter than the length of the dark zone observed at comparable pressures. Beyond this narrow reaction zone the temperature levels off to a value of about 1400°C through the rest of the fizz zone. This latter figure is in satisfactory agreement with values found for double-base propellant by spectroscopic measurement of the intensity of nitric oxide absorption bands in the flame [*85*]. This latter method while less precise than the thermocouple method has the advantage that no foreign object is introduced into the flame. Calculations based on the measured heat of the fizz reaction and an analysis of the products lead to similar values for the temperature of the fizz zone.

From the results discussed in the previous paragraph it appears that the luminous flame reaction follows a period of comparative inactivity in the outer regions of the fizz zone. The reason for this delay in the onset of the final series of reactions is not known. We may compare this delay to the induction period frequently observed in gas-phase reactions. The depletion of some chain-breaking agent, the formation of a critical concentration of some autocatalytic intermediate, or the attainment of a critical temperature leading to a thermal explosion may bring about a sudden increase in reaction. No temperature profile measurements have been made in this region, but from the sharpness of the edge of the luminous flame we may infer that the gradient is rather steep. The temperature of the completely reacted flame has been measured by optical methods [*87*]. Good agreement with calculated values is found when heat loss in the experimental setup is taken into account. The final products of the reaction are in equilibrium at the temperature of the propellant flame.

Fig. M,3a summarizes the main features of this descriptive and qualitative picture of the burning process.

M,4. The Burning Rate of Propellants. The linear burning rate of a propellant and the manner in which the burning rate depends on the intrinsic properties of the propellant and on the conditions under which it is burned are factors of the greatest importance in determining the suitability of the propellant for jet propulsion applications. The elucidation of the factors which affect the burning rate is the primary objective and the principal justification for the study of the mechanism of the burning process. Conversely, a study of the effect of a given variable on the rate of burning can give important information on the combustion mechanism. In this article we shall follow this dual course: first, to describe the dependence of the burning rate on the intrinsic and extrinsic properties of the propellant and where possible to relate this dependence causally to the mechanism of burning; and second, to make use of the results of empirical burning rate studies to develop further the qualitative picture of the burning process presented in the previous article.

Experimental measurement of burning rate. Double-base propellants were used in guns for many years before they were widely used in rockets. It is not surprising therefore that the early measurements of burning rates were made under conditions simulating gun performance. The measurement of the variation of pressure with time in a gun chamber or, more conveniently, in a closed pressure chamber where the variation of chamber volume with shot travel does not occur, permits a determination of the linear burning rate. However, the calculation of the burning rate from such a pressure-time record is a laborious process which requires a number of assumptions and approximations. Indeed this calculation is seldom made; instead, a quantity called the "quickness," which involves both the linear burning rate and the geometry of the charge, is used to characterize and compare gun propellants. Moreover, the closed-chamber method of burning rate measurement is least satisfactory in the low pressure region of greatest interest in rocketry.

The extensive development of solid fuel rockets at the start of World War II led to more suitable methods for the study of rocket propellants. In the vented vessel technique the propellant charge is fired in a chamber fitted with a venturi nozzle, simulating a rocket motor. The variation of pressure with time is recorded. With a neutral-burning charge the chamber pressure remains fairly constant during burning. The linear burning rate at the average chamber pressure during burning is calculated directly from the web of the grain and the measured time of burning. Such a measurement is most useful in ballistic design work since the conditions under which the charge is burned can be made to approach closely those which prevail in a rocket in flight.[6] The method is less useful for the basic measurement of burning rates since the relatively large propellant grains required remove it from the realm of laboratory scale operations and the comparatively poor precision tends to obscure the small trends that are of importance in fundamental investigations.

The necessity for a true laboratory method of determining burning rates was recognized very early in the course of the study of the burning of rocket propellants. Such a method is provided by the strand burning technique [47].[7] In this method a long cylindrical propellant strand is ignited at one end in a pressure vessel which has been prepressurized to the desired level with an inert gas. The sides of the strand are coated with a suitable nonexplosive material to prevent the flame's spreading along the surface of the grain. The burning surface then remains planar and normal to the axis of the strand. Burning takes place at a nearly constant pressure and the burning rate is determined from the direct measurement of the time necessary to burn a measured length of strand. The method is rapid and precise and the small samples required are easily prepared in small scale equipment. The burning rate measured by this method is determined almost exclusively by the pressure and temperature at which the measurement is made and by the composition and physical structure of the propellant sample. It is thus well suited to the investigation of these variables. On the other hand, the application of data obtained by this method to the prediction of the performance of a charge burning in a rocket motor must be made with some care, since under the latter conditions the burning is influenced by the geometry of charge and chamber (see the latter part of Art. 4) as well as by the fundamental variables listed above.

Dependence of the burning rate on pressure. We have already noted that the linear burning rate increases strongly with an increase in the pressure under which the propellant burns, and that this pressure dependence at constant temperature is represented approximately by Vieille's law (Eq. 1-1),

$$r = bp^n \tag{4-1}$$

Here b, n, and a (Eq. 4-3 below) are constants characteristic of the propellant composition. They may vary with the temperature.

[6] However, these static tests in which the rocket test motor is fastened securely to a static test stand do not duplicate those forces on the propellant charge which are due to the motion of the rocket in flight. The setback force, due to acceleration of the rocket during burning, and centrifugal force, due to rotation in the case of a spin-stabilized rocket, will not affect the intrinsic burning rate of the propellant but they may have a serious effect on the over-all motor-charge performance due to displacement or breakup of the propellant charge.

[7] A number of ingenious methods which may be suitable for laboratory-scale investigations were also developed by British workers. These have not yet been described in the open literature.

Alternate forms of the burning law have been proposed. In the past the simple form

$$r = bp \tag{4-2}$$

has been used in gun ballistics because it lends itself readily to ballistic calculations and because it represents the burning rate of simple single-base (nitrocellulose) gun propellants with some accuracy. However, it fails to describe the behavior of most rocket propellants adequately.

The linear form

$$r = a + bp \tag{4-3}$$

has been advocated, particularly by Muraour and his associates [*48*]. This equation describes the behavior of many double-base propellants as well or better than the index law (Eq. 4-1). The two forms (Eq. 4-1 and Eq. 4-3) give similar results over the range of pressures usually of interest in rocketry, and both are used extensively in rocket ballistics calculations.

Careful measurements extending down to a pressure of a few atmospheres have shown that for many double-base propellants the burning rate can be represented very accurately over a wide range of pressures by the combined form

$$r = a + bp^n \tag{4-4}$$

For most propellants a is small and n is not too different from unity so that Eq. 4-4 is approximated by either Eq. 4-1 or Eq. 4-3. For most heterogeneous propellants, for many homogeneous propellants containing various burning rate catalysts, and for most propellants at very low pressures, even Eq. 4-4 is inadequate to describe their complex behavior.

It is apparent that no single simple law can be put forth as *the* law of pressure dependence of the burning rate. However, the following generalizations can be made:

1. At high pressures (above a few hundred to one or two thousand $lb/in.^2$, depending somewhat on composition) the burning rate is a smooth function of the pressure and can be represented adequately by Eq. 4-1, 4-3, or more precisely by Eq. 4-4.
2. At intermediate pressures the pressure dependence may be complex and strongly dependent on composition. Regions of zero or even negative pressure dependence are sometimes found.
3. At very low pressures the burning rates of most propellants extrapolate to a finite rate at zero pressure although the propellant ceases to burn at pressures somewhere between a few pounds and a few hundred pounds per square inch.

In terms of the mechanism of the burning process, it appears that at high pressures most of the energy necessary to bring about the surface decomposition (which leads to a regression of the propellant surface and thus establishes the burning rate) comes from the high temperature flame region. The rate of energy transfer to the surface will be determined by the thermal conductivity of the flame and the proximity of the high temperature region to the burning surface. Since the rates of the flame reactions are strongly pressure dependent, this will lead to increased energy transfer and thus to a smoothly increasing burning rate as the pressure increases.

At intermediate pressures the flame zone is some distance removed from the burning surface or entirely absent. Much of the energy necessary to support combustion must come from the fizz zone. The reactions in this region are sluggish but susceptible to catalysis. They are strongly dependent on propellant composition, in contrast to the flame reactions which are virtually the same for all double-base propellants. In consequence, the pressure dependence of the burning rate is much less regular in this pressure region.

At very low pressures very little energy will return to the burning surface from the gas phase and the burning rate becomes nearly independent of pressure. Muraour attributes the pressure independent term in his burning rate law (Eq. 4-3) to thermal conductivity, independent of pressure. This can apply only to thermal conductivity in the condensed phase. We are then faced with the problem of accounting for the source of the energy necessary to maintain the high temperature of the burning surface, a source which must be *essentially independent of the pressure.* It is apparent that this energy necessary to support combustion at very low pressures (at atmospheric pressure or below) must come from the exothermic reactions taking place in the heated region below the burning surface.

It is necessary to point out that values of the limiting low pressure burning rate a, obtained by the application of Eq. 4-3 to burning rate data obtained in the high pressure regions, are not reliable because of the uncertainty of the pressure dependence at lower pressures. It is only by extending the burning rate measurements to very low pressures, preferably to atmospheric pressure or lower, that a value of a which may reasonably be interpreted in this manner is obtained.

Dependence of the burning rate on initial temperature. The change in burning rate with change in the initial temperature of the propellant is comparatively small for a chemical process, usually under 5 per cent per 10°C and frequently much less. Attempts to interpret this small temperature coefficient in terms of an activation energy for the rate-controlling process lead to implausibly low values for the energy of activation or improbably high values for the surface temperature. However, it is apparent from the nature of the burning process described in the preceding articles that such a simple interpretation cannot be expected to apply. The reaction does not take place in an isothermal region, but in a region of very steep temperature gradients where the precise application of the concept of an activation energy becomes difficult. The rate at which energy is supplied to the solid propellant may be considered to control the burning rate; the temperature in the reaction zone adjusts itself to maintain the steady state.

An increase in the initial temperature of the propellant will cause a similar rise, the exact value of which can be obtained from the specific heats of propellant and reaction products, in the final temperature of the flame. It might be assumed that the temperature profile in Fig. M,3a would be raised by an approximately constant amount at all points. However, the temperature of the burning surface cannot increase by this amount since, because of the increased burning rate, the surface layers have a shorter time to receive energy from the flame and subsurface reactions. The burning surface will then show a small increase in temperature but much less than the increase in T_0 or T_f.

For convenience in predicting the effect of an increase in T_0 we may relate the change in burning rate to the increase in energy transfer from the flame due to the increased differential between T_s and T_f or to the decreased amount of energy to be supplied to raise the powder from T_0 to T_s. Actually T_s assumes a value which represents a balance between these two factors, thus either method should lead to a similar result.

Corner [*49*] has developed the first method of approach. From the increase in T_f he calculates an increase in burning rate of 2.2 per cent per 10°C for cordite SC. The experimental value is between 3 and 4 per cent.

The second approach is implied in the use of the equation

$$r = \frac{c'p^n}{T' - T_0} \tag{4-5}$$

to represent the temperature dependence of the burning rate. Here T' is a constant, characteristic of the propellant composition and having the dimensions of temperature. This form for the temperature dependence, originally suggested by Crow and Grimshaw [*50*], has been applied by Avery and his associates [*88,89*] to a wide range of double-base propellants. Assuming for the moment that Eq. 4-5 represents the behavior of a propellant reasonably well, it can be seen that T' is the temperature (extrapolated) at which the burning rate would become infinite, i.e. the temperature at which burning would take place simultaneously throughout the propellant grain. This should be the surface temperature T_s. While it has not been possible to extend burning rate measurements to the neighborhood of T' because of the adiabatic heating of the sample, it is interesting to note that values of T' obtained by extrapolation from lower temperatures fall between 200 and 350°C, in reasonable agreement with estimated values of the surface temperature and the ignition temperature.

According to this argument a propellant with a high value of T_s should have a low temperature coefficient. It has been found experimentally that propellants which contain relatively stable coolants such as dinitrotoluene, and which might therefore be expected to have higher decomposition temperatures than a hot double-base composition, generally show a comparatively small dependence of burning rate on temperature. Many composite propellants, which because of the exceptional stability of their ingredients might be expected to have high surface temperatures, are also noteworthy for their small temperature coefficients.

The discussion in the preceding paragraphs has assumed that the burning rate is controlled by the final temperature of the flame T_f and that the effect of reactions in the solid phase can be neglected. As has been indicated, this is a good approximation at high pressures where the subsurface zone is narrow and the flame approaches close to the burning surface. It is not a satisfactory approximation at low pressures. Consequently it is not surprising that the temperature dependence of the burning rate at low and intermediate pressures is more complex than this simple picture would indicate. Propellants with zero or small negative temperature coefficients in this pressure region have been reported. The effects of compositional changes are often highly specific, but insufficient experimental data are available to permit useful generalizations. A careful study of the temperature dependence of the burning rate at low and intermediate pressures is very much needed.

Effect of propellant composition on the burning rate. The burning rate of a propellant at constant temperature and pressure is primarily a function of propellant composition. Changes in burning rate which accompany changes in composition may be divided into two classes: gross effects, which can be related to the change in the temperature of the propellant flame; and specific effects, which depend on a particular physicochemical action at some intermediate point in the burning process.

Muraour [*51*] was one of the first to point out the relationship between the burning rate and the flame temperature. He gives the equation

$$\log_{10} V = 1.36 + 0.27 \frac{T}{1000} \tag{4-6}$$

where V is the rate of decrease of the web (twice the burning rate as we have used the term) in mm/sec at a constant pressure of 1000 kg/cm^2 (14,223 lb/in.^{2}) and T is the calculated uncooled temperature of the propellant flame. Using the linear form of the burning law

$$V = a + bp \tag{4-7}$$

where V is again the decrease in web in mm/sec and p is given in kg/cm^2, Muraour [*52*] finds that

$$\log_{10}(1000b) = 1.214 + 0.308\frac{T}{1000} \tag{4-8}$$

while a is nearly independent of composition and has a value of about 10 mm/sec. Eq. 4-6 and 4-8 are claimed to predict correctly the burning rate over a range of T_f from 1500 to 4000°C and p from 25 to 4500 kg/cm^2.

For most double-base compositions the flame temperature is approximately a linear function of the heat of combustion. It is therefore not surprising that a relationship similar to Eq. 4-6 should exist between the heat of explosion and the burning rate. Gibson [*53*] was the first to point out such a relationship. Muraour [*54*] gives the equation

$$\log_{10} V = 1.47 + 0.846\frac{Q}{1000} \tag{4-9}$$

where Q is the heat of explosion at constant volume in kilocalories per kilogram. Eq. 4-6 might be expected to show slightly better agreement with experiment than Eq. 4-9 since the rate of energy transfer to the burning surface depends more directly on T_f than on Q.

These empirical relationships were obtained from burning rate data taken at high pressures. From the mechanism of the burning process we may expect such a result to hold in the high pressure region where the rate of burning is largely controlled by energy transfer from the flame zone. We should hardly expect them to hold in the low and intermediate pressure regions where the hot flame zone is either absent or at some distance from the burning surface. Experimentally it is found that formulas similar to Eq. 4-6 or 4-9 hold reasonably well at pressures above some 2000 lb/in.^{2} At lower pressures, propellants with similar heats of explosion may differ in burning rate by more than a factor of two. It is in these low pressure ranges that specific compositional effects become important.

Since at very low pressures the rate of burning is largely controlled by exothermic reactions taking place within the burning surface, we would expect compositions which favor such reactions to have relatively greater burning rates at these pressures. Propellants containing large amounts of stabilizers such as ethyl centralite or diphenylamine fall in this class and burn smoothly down to very low pressures. A related compound, *p*-phenylenediamine, is particularly active in this respect. A propellant to which 5 per cent of this material has been added burns faster at low pressures than the control sample without *p*-phenylenediamine despite the decreased heat of explosion. At higher pressures the burning rate falls below that of the control in accordance with Eq. 4-9.

More stable coolants such as dibutyl phthalate or triacetin do not enter into these low temperature reactions as readily; thus propellants containing these materials do not burn as readily at low pressures. A material which would react endothermally in the neighborhood of the burning surface would be expected to have an even greater retarding effect on the burning rate at low pressures. An experimental propellant containing paraformaldehyde is believed to illustrate this point [*55,56,57*]. It is probable that the paraformaldehyde dissociates to formaldehyde and then to carbon monoxide and hydrogen in the surface and fizz regions with the absorption of a large amount of energy. A propellant containing 5 per cent of paraformaldehyde failed to burn at low pressures but at higher pressures, as the luminous flame approached the burning surface, the burning rate increased rapidly with pressure and approached the value predicted from the heat of explosion.

Effect of radiation on propellant burning. Up to the present we have not considered the effect of radiation on the burning process. During the burning of a propellant, radiative energy transfer must occur between the hot flame and the propellant grain. According to the picture of the burning process which has been presented, this energy transfer must influence the burning rate. However, in contrast to the primary variables (pressure, temperature, and composition) which have been discussed previously, the radiation effect and other secondary effects to be discussed in this and the following articles depend on the geometry of the propellant charge and of the chamber in which it burns.

The effect of radiation on the burning process appears to be purely thermal in nature. No convincing evidence pointing to a specific photochemical reaction has been found.

Radiant energy absorbed by the propellant grain will raise the temperature of the propellant and thus increase its burning rate in a manner analogous to an increase in initial temperature. The effect of radiation on the burning rate can be represented by an equation similar to Eq. 4-5,

$$r = \frac{c'p^n}{T' - (T_0 + \Delta T_\lambda)} \tag{4-10}$$

where ΔT_λ is the increase in powder temperature due to absorption of radiation.

In case of a propellant having a large absorption coefficient at all wavelengths, the radiation will be absorbed in the surface layers and the effect will remain relatively constant (neglecting the effect of an increase in radiation path length) during the entire time of burning. If the propellant absorbs less strongly, radiation will penetrate into the interior of the grain. The propellant at the center of the web will be exposed to radiation for a longer time than that at the outside, so the burning rate will increase progressively during the entire time of burning. This effect has been discussed by Avery [*58,59*] and by Penner [*59,60*] who have made practical applications to the design of rocket charges. The increase in burning rate during burning, due to radiation, amounts to less than five per cent in typical rockets.

In extreme cases (transparent propellants with high flame temperatures) the intensity of radiation in the interior of the grain may become high enough to cause ignition at the surface of particles of absorbing impurities which are usually present [*59*]. This subsurface ignition causes fissuring and breakup of the propellant grain which may lead to dangerously high pressures in the rocket chamber. The addition of an opacifying agent such as carbon black to the propellant eliminates this source of erratic behavior.

The intensity of radiation at the burning surface is approximately

$$I = \sigma T_f^4(1 - e^{-k\rho_g l}) \tag{4-11}$$

where σ is the Stefan-Boltzmann constant, T_f is the temperature of the propellant gas, k is the mass emissivity of the gas, ρ_g is its density, and l is the average radiation path length. This simple equation neglects the effect of emission and reflection from the comparatively cool propellant surface and from the chamber wall. It also assumes that the propellant gas is uniform in temperature and composition throughout the combustion chamber, while in fact the burning surface is separated from the hot flame by the comparatively cool fizz gas layer (Art. 3).

From Eq. 4-11 it is apparent that, because of the radiation path length term, the magnitude of the radiation effect will depend on the initial geometry of charge and chamber for a given propellant formulation. The radiation effect is strongly dependent on the flame temperature of the propellant, and to a lesser extent on the density of the flame gas which is proportional to the chamber pressure.

The simple molecules in the propellant flame are poor emitters so that the mass emissivity of the gas from a double-base propellant is low. The presence of a small amount of inorganic matter increases the mass emissivity greatly, probably due to the presence of incandescent solid particles in the flame. Estimates of the mass emissivity are not too satisfactory. Penner [*60*, p. 278] has given a value of $k = 25\ \text{cm}^2/\text{g}$ for an unsalted propellant and $k = 40\ \text{cm}^2/\text{g}$ for a propellant containing 1.5 per cent potassium sulfate. Experimental measurements by Craig [*87*] indicate even lower values.

An estimate of the importance of radiation in supplying the energy necessary to support combustion can be made. Consider a propellant with the following properties:

$$T_0 = 300°\text{K} \qquad c_p = 0.35\ \text{cal/g deg}$$
$$T' = 600°\text{K} \qquad \rho = 1.6\ \text{g/cm}^3$$
$$T_f = 3000°\text{K} \qquad k = 25\ \text{cm}^2/\text{g}$$

From Eq. 4-7 and 4-8 the burning rate of a propellant with a flame temperature of 3000°K at a pressure of 2000 lb/in.^{2} will be approximately

1.5 cm/sec. The density of the flame gas at this pressure will be approximately 0.0125 g/cm³. We assume that all incident radiation is absorbed at the burning surface. Taking the mean radiation path length to be one centimeter, we get from Eq. 4-11

$$I = 29.6 \text{ cal/cm}^2 \text{ sec} \tag{4-12}$$[8]

and

$$\Delta T_\lambda = \frac{I}{r\rho c_p} = 35° \tag{4-13}$$

From Eq. 4-10

$$r = \frac{c'p^n}{600 - (300 + 35)} \tag{4-14}$$

As a crude interpretation of these results, we can say that in this case radiation has increased the burning rate of the propellant by a little more than ten per cent or that radiation has supplied a little more than ten per cent of the energy which must be furnished to the burning surface to support combustion at the observed rate. While this figure will vary with the properties of the propellant (or with the nature of the assumptions just made), we note that in the most favorable case (large values of k, ρ_g, or l) the radiation intensity cannot exceed a limiting value set by the intensity of black body radiation at the temperature of the propellant flame. In the present case this is

$$I = 1.364 \times 10^{-12} \times 81 \times 10^{12} = 110.5 \text{ cal/cm}^2$$

and

$$\Delta T_\lambda = 131°$$

While radiation may play a significant role in the burning process, particularly in the case of hot propellants, it appears that it cannot play a dominant role in determining the burning rate.

Chuffing. It is observed that, when the pressure in a rocket motor decreases below a certain critical value, the chamber pressure may fall suddenly to atmospheric and the charge apparently ceases to burn. At times, however, the charge reignites after a delay lasting from a fraction of a second to many seconds and a new period of normal burning follows. This cycle may be repeated a number of times, leading to a series of explosions or "chuffs." The mechanism of the burning process which has been described in previous articles provides a qualitative explanation for this phenomenon.

The final stages of the gas-phase reaction have been shown to be sluggish at low pressures and to take place at some distance from the burning surface (Art. 3). If this reaction ceases within the rocket motor, as a result of a decrease in pressure due to regressive burning or to some

[8] T_f is the calculated flame temperature. In a typical rocket the actual temperature will be lower by perhaps 10 per cent. This will cause the radiation intensity to be lower than the values calculated here by a factor of $0.9^4 = 0.656$.

random disturbance within the motor, both the flame temperature and the number of moles of gaseous products will decrease suddenly. This will cause a sudden drop in the chamber pressure as well as a decrease in energy transfer to the burning surface. The temperature of the burning surface will fall and burning will cease.

However, the temperature of the surface layers of the propellant will still be relatively high; moreover, the surface will receive additional energy from the hot motor walls. Exothermic reactions, taking place within the surface layers, will lead to the formation of a heated zone penetrating a significant distance into the interior of the propellant grain. The exothermic decomposition of the surface layers leads to the formation of a combustible gas mixture which, when a critical concentration is reached, ignites spontaneously. The heated surface layers of propellant burn rapidly, maintaining the chamber pressure at a level high enough for the flame reaction to take place. As soon as this heated layer has been burned away, however, the burning rate decreases, the chamber pressure falls, and the cycle repeats itself. While most double-base propellants will cease to burn in a rocket when the chamber pressure falls below a critical value, the ability to reignite, thus producing a chuff, is seen to be closely related to the ease with which the propellant undergoes an exothermic decomposition, i.e. to the chemical composition of the propellant.

Andreev [*61*] has observed a similar type of pulsating reaction in the combustion of nitroglycol and has advanced a similar explanation.

Erosive burning. When a propellant grain burns under conditions such that there is a rapid flow of propellant gas parallel to the burning surface, the burning rate is enhanced. This condition may occur within a perforation in the grain where the products of combustion flow outward into the larger volume of the combustion chamber, or in a rocket chamber where the propellant gases flow over the propellant charge toward the nozzle located at one end. This phenomenon has been called erosion because of its superficial resemblance to the mechanical wearing away of a solid body by a rapidly flowing fluid. It is possible that a portion of the material at the burning surface, which may have a fluid or foamlike structure, may be carried away in this fashion. However, the principal effect appears to be due to increased heat transfer to the propellant surface, brought about by turbulence in the gas stream near the surface caused by the gas flow. The effect of erosive burning on rocket motor performance is discussed in detail elsewhere in this series (XII,H).

The rate of burning of a propellant under erosive conditions can be represented by the equation [*90*]

$$r = r_0(1 + kv) \tag{4-15}$$

or perhaps better by

$$r = r_0(1 + k\rho_g v) \tag{4-16}$$

where r_0 is the burning rate at zero gas velocity, v is the gas velocity parallel to the burning surface, ρ_g is the gas density, and k is the erosion constant characteristic of the particular propellant composition. The erosion constant is greater for propellants having a low flame temperature than for those with a high flame temperature; experimental values range from 10×10^{-4} sec/ft for a relatively "cool" cordite to 3×10^{-4} sec/ft for a "hot" ballistite. The erosion constant appears to be relatively independent of pressure; the temperature dependence of k is not known.

The gas velocity parallel to the burning surface will vary along the length of the propellant grain, being greatest at the end closest to the nozzle, and in a typical rocket may reach values of 500 to 1500 ft/sec. This will lead to a variation of the burning rate along the charge, causing the grains to assume a tapered shape, and to a significant increase in the average burning rate, factors of great practical importance in rocket design.

Corner [*49,62*] has given a simple theory of propellant erosion in guns, based on a simple one-step model of the burning process. From hydrodynamic considerations it may be concluded that the propellant gases in the chamber will be very largely in a state of turbulence. Close to the burning surface, however, there will be a laminar layer having essentially viscous properties. Outside this region a transition layer occurs wherein turbulence becomes increasingly important as the distance from the surface increases. Estimates of the thickness of the reaction zone indicate that the greater part of the flame reaction takes place in this transition region. Here heat transfer, which largely determines the burning rate, takes place both by convection and by conduction. A high gas velocity parallel to the burning surface increases the turbulence in the reaction zone, thus increasing the rate of heat transfer and the rate of burning. A cool propellant, having a thicker reaction zone, will have a larger portion of reaction occurring in the transition zone and thus will show a greater erosion effect than a hot propellant having a thinner reaction zone. This is in accord with experimental observations.

No attempt has been made to apply this theory of erosion to the more detailed models of the reaction zone which have been used in studying the burning process at the relatively low pressures encountered in rocket operation. The erosive burning of composite propellants has been studied by Green [*96*].

Resonance burning. The term "resonance burning" has been applied to a type of unstable burning sometimes encountered in rocket motors. This condition is found most frequently with tubular charges although other grain configurations have been reported to show similar effects. Resonance burning causes sudden rises in the rocket chamber pressure; these pressure peaks may be accompanied by breakup of the propellant grains, which in extreme cases may cause motor blowups. The central perforations of partially burned grains recovered from shots exhibiting resonance burning are found to have a rippled surface resembling a standing wave pattern. This first led to the suggestion that the unstable burning was caused by some form of resonance effect.

The occurrence of resonance burning appears to be closely related to the location of the stagnation point, the point of zero gas velocity within the grain perforation where the combustion gases flow outward in opposite directions. Resonance appears to be observed only when the charge geometry is such that the stagnation point occurs within the perforation of the grain. Empirically it has been found that resonance burning can be prevented by drilling radial holes in the grain, by placing a small noncombustible rod in the perforation, or by the use of perforations of noncircular cross section.

A limited amount of study has been given to the mechanism of resonance burning. Since the rate of burning of most propellants is a strong function of pressure and propellant temperature, a small fluctuation in pressure in the gas stream near the burning surface will cause a corresponding fluctuation in the burning rate at the adjacent surface. This will have the effect of reflecting the disturbance in an amplified form. The question of the origin of the initial disturbance has not been considered; however, all observations agree that small fluctuations due to small variations in composition or physical characteristics occur in all types of flames. It is only necessary to provide a suitable mechanism for their amplification to explain the present phenomenon.

The problem of a resonance effect between oscillations in the gas stream and the burning rate of the propellant has been studied by Grad [*63*]. It was found that under certain conditions the amplitude of such a disturbance would increase with time, leading to the possibility of increased burning rates, large pressure oscillations within the perforation, and grain breakup. The modes most favorable to such a resonance effect were found to consist of waves spiraling around the inside of the perforation. Experimental evidence supporting the acoustical theory of resonance burning has been presented by Smith and Sprenger [*64*, pp. 893ff.] who found that large amplitude pressure oscillations occurred within the rocket chamber during resonance burning with frequencies corresponding to the fundamental tangential mode of acoustical oscillation or its lower overtones. Similar results have been reported recently by Cheng [*97*], and his conclusions have been further discussed by Green [*98*].

The use of a noncylindrical perforation or of holes drilled through the sides of the grain will tend to break up the wave patterns and thus inhibit resonance burning. The inhibiting effect of an axial rod is less easily explained on the basis of this model. If the gas velocity is high the disturbance may be swept out of the perforation before it can reach dangerous amplitudes; in the neighborhood of the stagnation point, however, the gas velocity will be low so that the buildup time may be relatively long. Finally it may be noted that resonance effects are most severe with fast-burning "hot" compositions and that the burning rates of such powders are most sensitive to variations in temperature and pressure. The simple mathematical theory thus affords an explanation for most of the empirical observations concerning resonance burning.

M,5. Theories of the Burning of Propellants. The preceding articles, largely descriptive in nature, have given a qualitative picture of the burning process from which useful generalizations may be drawn. It would be desirable, however, to have a theory which would relate the burning rate and the basic parameters of the process, composition, pressure, and initial temperature, in a more fundamental manner. A successful theory would provide a detailed model from which properties of the reaction zone inaccessible to direct experimental measurement might be predicted and would provide a rational explanation for such phenomena as chuffing, erosion, and resonance burning, which up to now have been treated in a very speculative fashion.

Double-base propellants have been used in guns for a much longer time than in rockets, so that the early attempts to formulate a satisfactory theory of propellant burning were based on observations of the performance of propellants under gun conditions. The high pressures encountered in guns and closed chambers have the effect of shortening the scale of time and distance in the reaction zone. When the difficulties encountered in making accurate experimental observations under these conditions are considered, it is not surprising that these early theories were based on very simple models of the burning process. More recent investigations carried out in the low pressure regions have revealed the complexity of the process and demonstrated the necessity for treating a much more detailed model.

Corner [*49*] has reviewed the older theories of propellant burning, and more recently Geckler [*95*, pp. 289ff.] has summarized modern developments with particular reference to their applicability to the burning of composite propellants. Corner divides the earlier proposals into surface- and gas-phase theories depending on whether attention is directed at the initial gasification of the propellant surface or at the subsequent reaction of the combustible gas mixture. A complete theory will have to consider both processes and the interaction between the two. Theories of this type have been developed only within the last few years.

Surface theories. Of the surface theories, only that of Muraour [*65*] and a more recent hypothesis suggested by Daniels [*4*] need be considered here.

Muraour has not given a complete quantitative theory, but rather has correlated the results of a great variety of experimental observations in terms of a qualitative picture of the burning process. The surface decomposition of the propellant to give a combustible gas mixture is considered to be the rate-determining step, and the effects of such variables as pressure, initial temperature, flame temperature, heat of explosion, and radiation are interpreted as they affect this initial decomposition. Energy transfer from the flame to the propellant surface occurs by means of a conductive process whose rate is proportional to the pressure and by radiation, independent of pressure. This leads to the burning rate law

$$r = a + bp$$

While the theory, lacking a quantitative form, does not provide an insight into the detailed mechanism of the process it has been valuable as a guide to experimentation and as a framework for the correlation of experimental results. Some of the empirical correlations obtained by Muraour have already been cited (Eq. 4-6, 4-8, and 4-9).

Daniels has called attention to the first order decomposition of nitrocellulose and suggests that this is the rate-determining step in the surface decomposition, thus determining the rate of the whole combustion process. The nitrocellulose decomposition is assumed to have the experimentally measured activation energy, 46,700 cal/mole, and the "normal" frequency factor, 3×10^{13} sec^{-1}. This leads to a burning rate equation at constant pressure of the form

$$r = \frac{nAe^{-E_a/RT_s}}{m} \tag{5-1}$$

where n is the number of nitrate ester groups per square centimeter of surface reaction zone and m is the number of nitrate ester groups per cubic centimeter.

The effect of varying the initial temperature of the propellant is assumed to be reflected directly in the temperature of the burning surface (after making a suitable allowance for the increase in heat capacity of the propellant with temperature). This permits a calculation of the surface temperature, making use of the energy of activation and the experimentally determined temperature coefficient of the burning rate. In this way a value of about 1000°C is found for the surface temperature. The experimental evidence presented to support this high value of the surface temperature has already been discussed (Art. 3). In order to bring the burning rate calculated by means of Eq. 5-1 into agreement with experimental values, it is necessary to assume that the effective surface reaction zone is approximately one hundred molecules (5×10^{-6} cm) thick.

The effect of pressure on the burning rate cannot be derived from this hypothesis. Qualitatively it is assumed that an increase in pressure will speed up the (bimolecular) gas-phase reaction, bringing the reaction zone closer to the burning surface and increasing the heat flow to the surface. The theory is thus incomplete, as any theory which confines itself solely to the condensed phase must be, in that it tells nothing about the most significant step in the process, the flow of energy to the burning surface.

As a picture of the surface decomposition process, the greatest weakness of this hypothesis lies in the very large value assumed for the surface temperature. This in turn is made necessary by the use of a normal value for the temperature independent term in the decomposition rate expression despite the fact that all experimental evidence indicates that the true value (without attempting to interpret its possible significance at this time) is some 10^6 times greater.

Gas-phase theories. The surface theories concentrate their attention on the decomposition of the solid propellant at the burning surface and consider the flame zone only as an energy reservoir, with the rate of energy transfer from flame zone to burning surface depending on pressure and flame temperature in a simple fashion. In contrast, Boys and Corner [*49,66,67*] have considered the propagation of a reaction zone through a homogeneous combustible gas mixture. Their theory is applicable to any flame reaction which can be represented by a single-reaction step of thermal origin. In applying the theory to solid propellants, the manner of interaction between the flame zone and the burning surface is not considered. It is assumed that the propellant surface will decompose to produce a combustible gas mixture at a rate just equal to that at which it is consumed by the flame reaction. The manner of arriving at this steady state is not covered by the theory nor is the stability of such a state demonstrated; instead, the experimental facts are assumed to prove that such steady states do exist.

As a model for the combustion process, Boys and Corner consider the propagation of a plane reaction zone through a homogeneous combustible gas. The gas is assumed to be capable of undergoing a single reaction process whose rate at any point is determined only by the temperature and concentration of reactants at that point. The effect of diffusion of reactants in the reaction zone was neglected at the start but this effect has been included in a more recent paper [*67*]. The pressure is assumed to be effectively constant throughout the reaction zone,[9] thus the case of a gaseous detonation is not covered by the theory.

Consider a plane reaction zone moving along an x axis perpendicular to the plane of reaction with the origin moving with the reaction zone. The reactants approach the reaction zone from the region of $-x$, pass through the reaction zone, and the products move on in the direction $x \to \infty$. The conditions at any point in the reaction zone are a function of the x coordinate alone.

The total energy flux per unit area across any plane parallel to the reaction surface must be a constant since in the steady state there can be

[9] This can be shown to be true for slow flames to a very high degree of accuracy. See for example appendix B of [*70*].

no accumulation of energy between successive planes. This energy flux will consist of the intrinsic energy of the gas transported by mass flow, the work done in increasing the volume of the gas at constant pressure, and the energy flux due to thermal conductivity across the plane. At large distances from the reaction zone the temperature will be constant and the transfer of heat by conduction will be zero. This leads to the energy flux equation

$$\dot{m}(e + pV) - k\frac{dT}{dx} = \dot{m}(e_f + pV_f) = \dot{m}(e_0 + pV_0) \tag{5-2}$$

or

$$\frac{k}{\dot{m}}\frac{dT}{dx} = h - h_f = h - h_0 \tag{5-3}$$

Here $\dot{m}$ is the mass rate of flow per unit area (the mass burning rate), k is the thermal conductivity coefficient, h is the total heat content per unit mass, and the subscripts $_0$ and $_f$ refer to the initial and final states respectively. The heat content at any time when a fraction ϵ of the reaction is completed is

$$h - h_f = (1 - \epsilon)q + (T - T_f)c_p \tag{5-4}$$

where q is the total heat of reaction per unit mass and c_p is the specific heat capacity, both at constant pressure. Since most of the reaction takes place in a comparatively small region at the high temperature end of the reaction zone, it is assumed that c_p, and thus q, is independent of temperature. Combining Eq. 5-3 and 5-4 we have

$$\frac{k}{\dot{m}}\frac{dT}{dx} = (1 - \epsilon)q + (T - T_f)c_p \tag{5-5}$$

A second relationship between ϵ and T can be obtained from the kinetic expression for the homogeneous reaction rate. This will have the form

$$\frac{d\epsilon}{dt} = f(\epsilon, V, T) \tag{5-6}$$

and since the rate at any point can be expressed as a function of x,

$$\frac{d\epsilon}{dt} = \dot{m}V\frac{d\epsilon}{dx} = f(\epsilon, V, T) \tag{5-7}$$

Since Eq. 5-5 and 5-7 contain x only as dx, it is possible to eliminate x between them and obtain

$$\frac{d\epsilon}{dT} = \frac{kf(\epsilon, V, T)}{\dot{m}^2V[(1 - \epsilon)q + (T - T_f)c_p]} \tag{5-8}$$

Finally, V can be eliminated by means of the equation of state

$$\frac{pV}{RT} = \frac{1 + n\epsilon}{\mathfrak{M}} \tag{5-9}$$

where a molecule of reactant of molecular weight $\mathfrak{M}$ gives $n + 1$ product molecules. Then

$$\frac{d\epsilon}{dT} = \frac{kp\mathfrak{M}f(\epsilon, T)}{\dot{m}^2RT(1 + n\epsilon)[(1 - \epsilon)q + (T - T_f)c_p]} \tag{5-10}$$

This is an equation in ϵ and T where the one variable parameter $\dot{m}$ must be so chosen that it satisfies the boundary conditions $\epsilon = 0$ at $T = T_0$ and $\epsilon = 1$ at $T = T_f$.

Boys and Corner consider three particular cases of the general equation wherein the reaction is represented by (i) a first order rate resulting from a unimolecular reaction, (ii) a second order rate from a bimolecular reaction, and (iii) a second order rate as the low pressure result of a unimolecular reaction. The corresponding rate expressions are:

(i) $$\frac{d\epsilon}{dt} = f(\epsilon, T) = A_1(1 - \epsilon)e^{-E_a/RT} \tag{5-11}$$

(ii) $$\frac{d\epsilon}{dt} = f(\epsilon, T) = \frac{A_2}{V}(1 - \epsilon)^2e^{-E_a/RT} \tag{5-12}$$

(iii) $$\frac{d\epsilon}{dt} = f(\epsilon, T) = \frac{A_3}{V}(1 - \epsilon)(1 + n\epsilon)e^{-E_a/RT} \tag{5-13}$$

By substituting these expressions in Eq. 5-10 and making use of Eq. 5-9, the corresponding equations for $d\epsilon/dT$ are obtained.

These latter equations are solved by a process of successive approximation. In cases (i) and (iii) explicit expressions for the mass burning rate are obtained.

Case (i) $$\dot{m}^2 = \frac{kA_1p\mathfrak{M}e^{-E_a/RT_f}}{c_pRT_f(1 + n)\left[\frac{(1 + n/2)}{(1 + n)}\left(\frac{E_a}{RT_f}\right)\left(\frac{q}{c_pT_f}\right) - 1\right]} \tag{5-14}$$

Case (iii) $$\dot{m}^2 = \frac{kA_3p^2\mathfrak{M}^2e^{-E_a/RT_f}}{c_pR^2T_f^2(1 + n)\left[\frac{(1 + n/2)}{(1 + n)}\left(\frac{E_a}{RT_f}\right)\left(\frac{q}{c_pT_f}\right) - 1\right]} \tag{5-15}$$

For case (ii) the solution has the form

Case (ii) $$\dot{m}^2 = \frac{kA_2p^2\mathfrak{M}^2\psi e^{-E_aR/T_f}}{c_pRT_f^2(1 + n)^2} \tag{5-16}$$

where ψ depends on the composition of the fuel. The value of ψ is determined by the auxiliary equation

$$g_1(\psi) - \frac{2n}{1 + n}g_2(\psi) + \frac{n}{1 + n}g_3(\psi) = \left(\frac{c_pT_f}{q}\right)\left(\frac{RT_f}{E_a}\right) \tag{5-17}$$

This equation is solved for ψ by trial and error for a particular case, making use of the $g(\psi)$ functions tabulated by Boys and Corner.

The value of ψ is independent of pressure so that for all three cases, if the reaction can be represented by a rate equation of order n, the burning rate is proportional to the pressure raised to the $n/2$ power. This is a general result obtained with all flame theories based on a purely thermal reaction mechanism. The burning rate should be a smoothly increasing function of pressure unless the mechanism of reaction changes as the pressure rises. The theory fails to predict the irregular dependence of the burning rate on pressure frequently observed at low pressures.

With the value of the mass burning rate determined, the integration of Eq. 5-10 leads to a relation between ϵ and T. This in turn can be used in Eq. 5-5 to find a relationship between T and x. Thus the theory provides an insight into the structure of the reaction zone. The reaction zone is shown to be very narrow, with a steep temperature gradient and with most of the reaction taking place near the upper temperature limit. For a typical propellant, the reaction is estimated to go from 10 to 80 per cent toward completion in a distance of $1.6/p$ cm where p is the pressure in lb/in.^2

In a later paper, Corner [*67*] has taken into account the diffusion of reactants and products in the reaction zone. The general features of the theory with diffusion are very similar to the simple theory in most respects, but it is found that inclusion of the diffusion effect can reduce the calculated flame speed by more than half in a typical case. Using the theory with diffusion and plausible values for the various reaction constants, including an activation energy of 26.4 kcal/mole, Corner obtained excellent agreement between observed and calculated burning rates for a series of propellants with flame temperatures ranging from 1900 to 3050°K.

A number of authors have developed theories of flame propagation in gases similar to that of Boys and Corner, but only the latter have attempted to apply their methods to the combustion of a solid propellant. While none of these gas-phase theories can be expected to give a complete account of the burning of a solid, they are applicable to that part of the propellant burning process which takes place in the flame zone. Theories of this type have been reviewed by Markstein and Polanyi [*68*] and by Evans [*69*]. Recently, Hirschfelder and Curtiss [*70*; *5*, pp. 121, 127, 135] and their associates have developed their theory along the lines followed by Boys and Corner, but in a more general form. They obtained solutions to the flame equations for a number of reactions by numerical integration.

The flame theory of Zeldovich [*71*] and Frank-Kamenetsky has been applied by Belayev [*72*] and Semenov [*73*] to the combustion of nitroglycol. The liquid is assumed to volatilize at the surface, thus providing a constant supply of fuel for the vapor-phase flame reaction. The surface temperature is the boiling point of the liquid. It is stated that the theory is not applicable to nonvolatile explosives but it would appear that there is little difference, insofar as the mathematical formulation is concerned, between a simple vaporization and a surface decomposition of the type assumed by Boys and Corner.

These vapor-phase theories must be regarded as incomplete when applied to the combustion of a solid propellant in much the same way that the surface theories discussed in the last article are incomplete. They consider only the later stages of the reaction process, passing over the initial stages of the reaction with a simplifying assumption which is not compatible with the complexity of the process. These deficiencies of the older theories, made increasingly evident by recent experimental work, have led to the development of what may be called combination theories wherein an attempt is made to consider the interaction of all of the successive stages of the reaction process.

Combination theories. Theories of the combination type have been put forward by Rice and Ginell [*55*] and by Crawford and Parr [*74*]. The two theories are essentially similar, differing only slightly in the model which is assumed for the reaction zone and in mathematical detail. Both are based on a three-stage reaction zone similar to that described in Art. 3 and shown in Fig. M,3a. It is assumed that the homogeneous reaction rates of the surface, fizz, and flame stages can be represented by single expressions of the Arrhenius type similar to Eq. 5-11, 5-12, and 5-13, although each of these stages may consist of a complex series of individual reaction steps.

The nearly complete separation in space of the three reaction stages, which appears to occur at least at low pressures, permits each stage to be treated separately before considering its interaction with the adjacent stage. The rate of the surface reaction, expressed as a function of a surface temperature T_s, is considered first. The effect of the fizz reaction, with a maximum temperature T_1, on T_s is considered next. Finally the effect of the flame reaction, leading to the final flame temperature T_f, on T_1 and thus on T_s and on the burning rate, is considered.

Both theories assume a first order rate law for the surface reaction. Rice considers a surface decomposition of the Daniels type (Art. 5). From experimental observations the limiting low temperature value of the surface temperature T_s is estimated to be about 700°K. Values for the other parameters which determine the burning rate are then found by fitting the theoretical equations to experimental burning rate data. The low value of T_s together with the observed temperature and pressure coefficients of the reaction leads to a value for the energy of activation of the surface decomposition of some 16,000 cal/mole in a typical case. The frequency factor A_s is found to have a "normal" value in the range 10^{11} to 10^{13} sec^{-1}. These low values of E_s and A_s may be compared with the high values indicated by thermal decomposition experiments (Art. 2).

Crawford has extended Rice's model to consider a surface reaction zone of finite thickness, extending from the burning surface back into the powder to where the temperature has fallen to such a low value that further reaction can be neglected. Since the propellant decomposes to give gaseous products, the reaction zone is considered to have a foamlike structure with the density decreasing as the burning surface is approached. This foam zone is treated as a thermal reaction zone of the Boys-Corner case (i) type (Eq. 5-10 and 5-11), neglecting diffusion in the condensed medium.

The fizz zone is represented by a second order thermal process. Crawford follows the method used by Boys and Corner to treat such a process, with appropriately altered boundary conditions to allow for the proximity of the foam and flame zones on either side. Rice uses a somewhat different mathematical method to arrive at a similar result.

Crawford has not extended his theory to include the effect of the luminous flame but suggests that this effect will become important at high pressures. The reaction could be treated as a thermal reaction of the Boys-Corner type but, according to Crawford, it would have to be of a higher order than the fizz reaction to prevent coalescing of the two reaction zones. Rice considers the flame reaction to be a branching-chain explosion. It acts as a heat source whose distance from the burning surface depends on the time necessary to reach a critical concentration of some active species in the flame. For the purpose of fitting the theory to the experimental data this distance is obtained from measurements of the dark zone length (Art. 3). The flame reaction begins to make a significant contribution to the burning rate at a relatively high pressure, in the neighborhood of 1000 $lb/in.^2$ for a typical composition.

Since the burning rate is determined by a series of reaction stages having different pressure dependences and making their greatest contribution to the burning rate in different pressure regions, it is possible to reproduce the complex pressure dependence observed in the low pressure region by means of these theories. Fig. M,5 shows a comparison of the experimental burning rate data and the theoretical curves obtained by the two methods for a typical double-base composition. Except in the very low pressure region (surface reaction controlling) the theories of Crawford and of Rice give similar results. In the very high pressure region, where the flame reaction plays a dominating role, these theories can be made to merge smoothly with the gas-phase theory of Boys and Corner.

All of the theories discussed here require the evaluation of a number of reaction parameters which are not readily susceptible to direct measurement. Values are selected to give the best agreement between calculated and experimental burning rates but, because of the number of parameters involved and the approximate nature of the model considered, the selec-

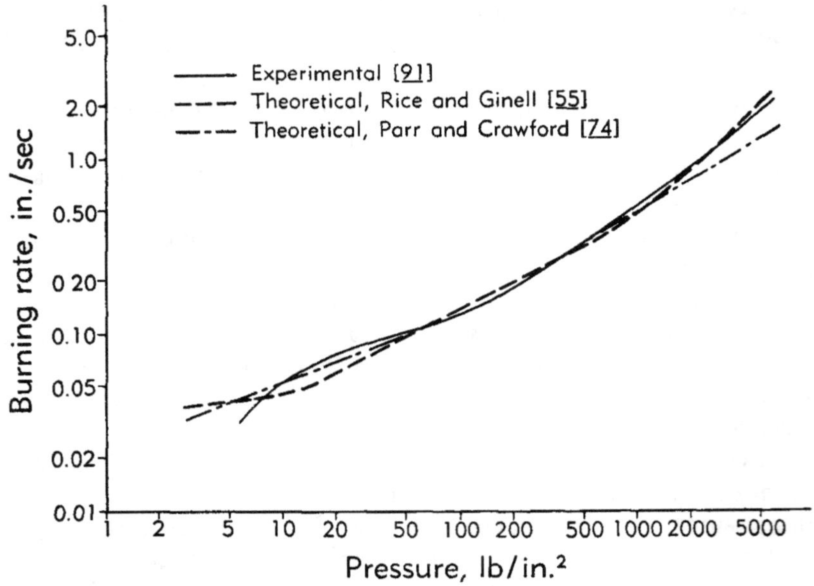

Fig. M,5. Comparison of experimental and theoretical burning rate curves, propellant HES 4016, 25°C.

tion of a unique set of values is not possible. Table M,5 gives values of some of the important reaction parameters used by various investigators to obtain satisfactory agreement with experimental data.

***Table M,5.* Values of reaction parameters.**

	Boys and Corner [*66*]	Daniels, Wilfong, and Penner [*4*]	Rice and Ginell [*55*]	Crawford and Parr [*74*]
T_s	750°K	1273°K	700°K	700°K
A_s		3×10^{13} sec^{-1}	1.9×10^{11} sec^{-1}	
E_s		46,700 cal/mole	16,000 cal/mole	20,000 cal/mole
T_1			1400°K	1100°K
A_1			$10^{-4}Z$	
E_1			5500 cal/mole	30,000 cal/mole
T_f	3050°K		3370°K	
A_f	$10^{-1}Z$			
E_f	26,400 cal/mole			

As an example of the difficulties that are encountered in attempting to obtain a consistent set of values in this fashion, we can consider the surface decomposition reaction. If its rate can be represented by a first order law of the Arrhenius type, three basic parameters, A_s, T_s, and E_s, must be evaluated. If two of these can be established through independent measurements, the third can be obtained by comparison with the experimental burning rate. We can obtain a second relationship between T_s and E_s by means of the temperature coefficient of the burning rate (provided we can find a satisfactory relationship between T_s and T_0). Then the evaluation of either T_s or E_s, together with the observed burning rate and temperature coefficient of the burning rate, enables us to fix all three parameters. It should be pointed out, however, that measurements of the temperature coefficient are among the least satisfactory of the available data, from the standpoint of precision and accuracy, on which to base an evaluation of the burning parameters.

Daniels has taken his value of E_s from independent kinetic measurements of nitrocellulose decomposition. The value of T_s is then found by comparison with the observed temperature coefficient of burning although experimental evidence is also cited to support this value. A_s is then determined by the observed linear rate of burning. Rice has fixed his value of T_s on the basis of experimental evidence. The evaluation of E_s from the temperature coefficient and A_s from the linear burning rate then follows as before. Crawford has used experimental attempts to determine the parameters directly only as a guide to the appropriate range of values, and has varied both T_s and E_s to obtain the best possible fit to the experimental burning rate curves. In this way the three distinctly different sets of parameters in Table M,5 are obtained. Each of the three sets is self-consistent and capable of reproducing the observed burning rates with reasonable accuracy when applied to the appropriate form of the theory.

Much of the difficulty in evolving a completely satisfactory theory of the burning process must be attributed to our lack of knowledge of the individual chemical processes which take place. The theoretical treatment, while sometimes laborious in application, appears to be satisfactory for the simple models which have been proposed. Future advances will come through refinement of the model as new experimental work furnishes more information on the structure of the reaction zone and the nature of the chemical reactions comprising the successive reaction stages.

All of the theories discussed in detail in this article postulate a thermal model for the reaction zone, wherein the reaction is propagated by the conduction of thermal energy from the hot region of the flame to the cooler reactants entering the reaction zone. The importance of autocatalytic processes in the decomposition of propellants and explosives has been pointed out repeatedly, while chain reactions, propagated by free radicals or other energy-rich particles, are recognized to occur in most flames. Rice has suggested that the flame reaction is of the chain type, but this assumption is not essential to his theory.

Flame propagation theories in which the flame velocity is controlled by the diffusion of active chemical species rather than by the transfer of thermal energy have been developed by Lewis and von Elbe [*75*] and by Tanford and Pease [*76*].

The extension of the present theories to include reaction processes of these types appears to present no fundamental difficulties. An autocatalytic rate law could be substituted for the simple first order or second order laws assumed for the surface reaction zone of Crawford, the fizz zone of Rice or Crawford, or the flame zone of Boys and Corner, with only superficial changes in mathematical detail. The diffusion treatment of Boys and Corner, or of Rice and Ginell, can be adapted to the case of propagation by diffusion of active particles. Future developments will probably consider such reaction schemes, but here again substantial progress will be aided by further experimental elucidation of the role which autocatalytic and chain reactions play in the combustion process.

M,6. The Mechanism of Burning of Composite Propellants. Composite propellants have come into prominence only since the latter part of World War II. Consequently they have been studied less than the older double-base types. A recent review by Geckler [*95*, pp. 289ff.] provides an excellent summary of the present status of the problem.

The variety of compositional types falling within the broad definition of the composite class is so great that any general treatment of the burning process is probably out of the question. Nevertheless most composite propellants show similarities in physical structure and ballistic behavior which make it desirable to group them together as a single class. A comparatively low rate of dependence of burning rate on temperature and pressure is characteristic of many composite types.

The reaction zone in the combustion of a double-base propellant, as a consequence of the homogeneous structure of the solid, will be homogeneous in composition throughout any plane parallel to the burning surface. In contrast, the heterogeneous structure of the composite types leads to a similar heterogeneity at the burning surface and in the flame reaction zone. The thermal decomposition of oxidizer particles and matriz at the burning surface will give rise to alternate oxidizer-rich and fuel-rich gas streams. The manner of mixing of these oxidizer and fuel streams to form the flame reaction zone will play an important role in the burning process and may be largely responsible for the distinctive characteristics of the burning of composite propellants.

Crawford [*91*] and his associates have carried out experiments on the burning of a propellant composed of a stoichiometric mixture of potassium perchlorate and carbon black in a matrix of double-base propellant. Rice [*57*] has discussed the mechanism of burning of such materials in qualitative fashion. In this type of composite, the matrix in the absence of the filler is capable of burning in the manner described in Art. 3. The effect of the carbon-perchlorate filler is superimposed on the normal behavior of the matrix. The burning rates of propellants of this type increase rapidly with pressure in the region below 300 lb/in.^{2}, greatly exceeding the burning rate of the double-base matrix alone (Fig. M,6a). At higher pressures the burning rates of the composites appear to approach that of the matrix asymptotically, giving rise to the low dependence of burning rate on pressure, which is desirable in rocket applications.

At very low pressures the perchlorate appears to act as an inert diluent, cooling the burning surface by absorbing heat from the surrounding propellant, so that the burning rate falls below that of the matrix.

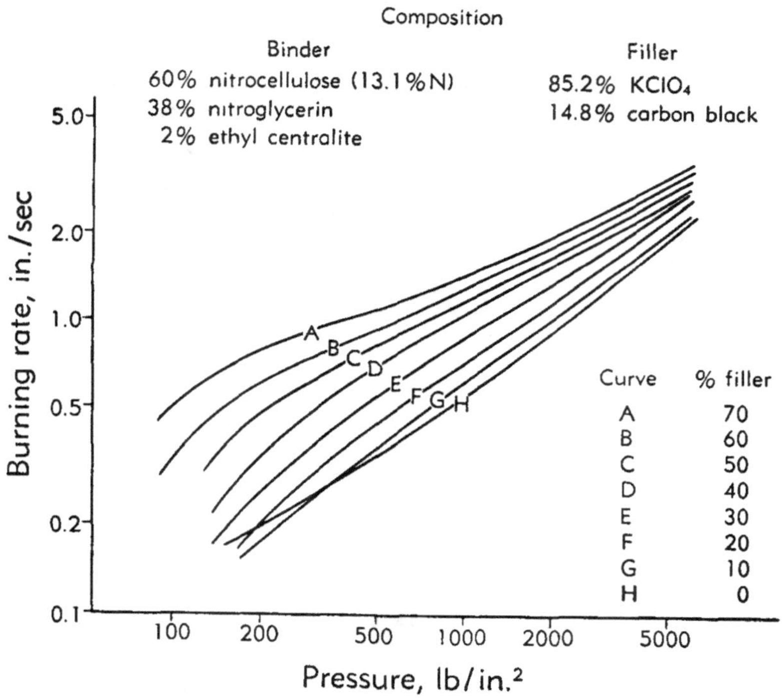

Fig. M,6a. Burning rates of composite propellants (from [*91*]).

Finely ground perchlorate is more effective than an equal weight of coarse material in lowering the low pressure burning rate, indicating that although the crystals do not reach thermal equilibrium in the burning surface the small crystals approach equilibrium more closely than the large ones.

As the pressure is increased, the perchlorate takes a more active part in the reaction. Photographs of burning grains show flashes of flame which appear to originate at perchlorate crystals. The perchlorate decomposes to give off streams of oxygen which can react with the carbon black and reducing gases in the flame. From the rate of decomposition of these crystals and published data on the rate of decomposition of potassium perchlorate (Art. 2), Rice has estimated the surface temperature of the crystals to be in the neighborhood of 1100°K. This is considerably higher than the probable surface temperature of the double-base binder. This high temperature may be caused by a reaction between perchlorate and the binder or binder decomposition products at the surface of the perchlorate particle. A more likely explanation is that the perchlorate particles extend out from the surface into the higher temperature region of the fizz flame. The diameters of the perchlorate crystals used are of the

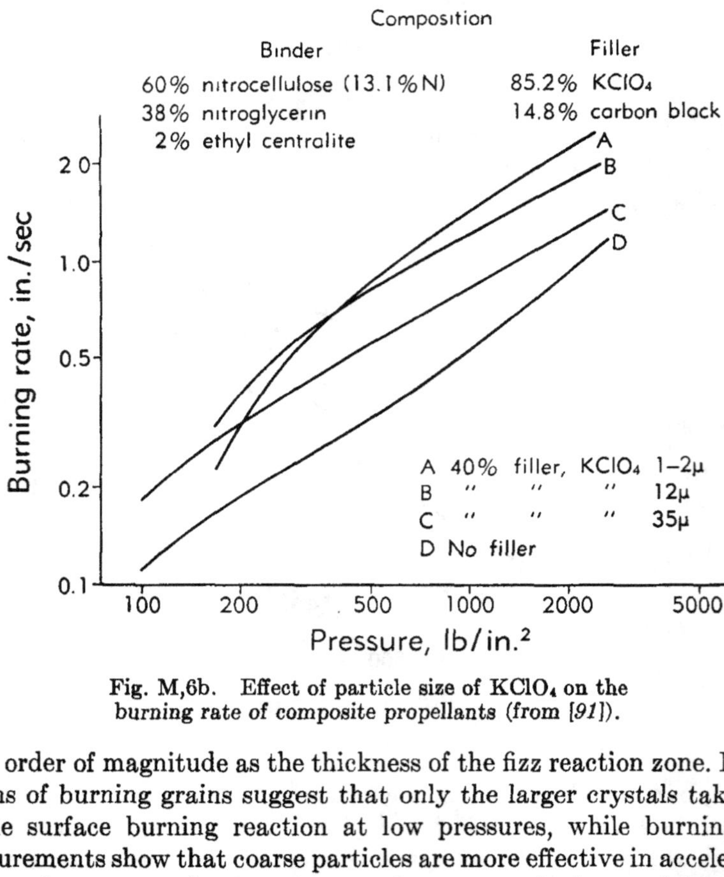

Fig. M,6b. Effect of particle size of $KClO_4$ on the burning rate of composite propellants (from [*91*]).

same order of magnitude as the thickness of the fizz reaction zone. Photographs of burning grains suggest that only the larger crystals take part in the surface burning reaction at low pressures, while burning rate measurements show that coarse particles are more effective in accelerating the burning rate at lower pressures than an equivalent weight of fine particles of perchlorate (Fig. M,6b). If reactions at the surface of the crystal play a dominant role, the particle size effect should be reversed with the more finely divided material showing the greater effect at low pressures.

The decomposing perchlorate will give off streams of oxygen and particles of perchlorate may be ejected from the surface by gases generated at the particle-matrix interface, while the surrounding matrix will give off a combustible gas mixed with particles of carbon. The rate of reaction between these, and hence the distance from the hot flame zone to the propellant surface, will depend on the rate of mixing of the streams. Finely ground perchlorate will provide a great many tiny oxygen sources, thus promoting mixing by diffusion, decreasing the mixing distance, and increasing the burning rate by promoting energy transfer to the burning surface (Fig. M,6b). Rice has shown that this mixing distance will increase as the pressure increases so that the effect of this diffusion process is to decrease the fraction of heat returning to the burning surface and to resist an increase in burning rate as the pressure increases. This compensates to some extent for the strong pressure dependence of the matrix burning rate and leads to a more favorable pressure index for this type of propellant.

Propellants containing potassium perchlorate, ammonium perchlorate, or ammonium nitrate as the oxidizer have been prepared. When the compositions are adjusted to give propellants with similar flame temperatures, the order of the burning rates is $KClO_4 > NH_4ClO_4 > NH_4NO_3$. Surprisingly, the oxidizer with the greatest thermal stability gives the fastest burning rate while the least stable oxidizer gives the slowest burning rate.

The thermal decomposition of potassium perchlorate at the burning surface presumably leads to a gas mixture containing two parts of oxygen to one of potassium chloride (at flame temperature above the boiling point of KCl). Ammonium perchlorate decomposes to give a gas mixture containing large amounts of water vapor, chlorine, hydrogen chloride, nitrogen, and nitric oxide in addition to a small amount of oxygen (Art. 2). The principal products from the thermal decomposition of ammonium nitrate are nitrogen oxides, water vapor, and nitrogen. The relative burning rates of these composite types thus are found to parallel the oxygen content of the gas mixtures resulting from the thermal decomposition of the oxidizer rather than the ease of decomposition of the oxidizer.

The examination of the surface of partially burned grains of composites of the crystalline oxidizer-resinous binder type indicates that the solid- or liquid-phase reaction is confined to a very thin layer at the burning surface. Solid-phase reactions of the type which play such an important part in the burning of double-base propellants are not important here. The possible significance of reactions at the surface between matrix and oxidizer particles has not been explored.

This relative freedom from complicating solid-phase reactions, the possibility of studying the thermal decomposition of oxidizer and matrix independently, and the control which can be exercised over type, proportion, and particle size of oxidizer, makes this an attractive field for experimental and theoretical study. Further significant development of the theory of the burning of these relatively new propellant types can be expected.

M,7. The Ignition of Solid Propellants. The ignition of a solid propellant is a transient process leading to steady state burning. A discussion of the ignition process has been deferred until the nature of steady state burning could be examined.

In simplest terms, ignition consists in supplying energy from an external source to the propellant surface to produce a chemical and thermal state approximating that characteristic of steady burning. If the ignition system is well matched to the requirements of the propellant, energy will be supplied at a rate and in a quantity sufficient to lead to a close approximation of the final steady state. With the removal of the ignition stimulus, the normal combustion processes will take over the control of the reaction without discontinuity. If the energy supply from the igniter is not matched to the requirements of the propellant, various abnormalities such as overignition, igniter peaks, ignition delays, hangfires, and chuffing may occur in going from the transient to the steady state.

The ignition of gun or rocket propellant charges is usually accomplished by means of a small charge of black powder set off with an electric match. Less frequently, a metal oxidizer or other pyrotechnic mixture, finely divided propellant, or nitrocellulose is used as the main igniter charge.

In such a system, energy transfer to the surface of the propellant grain can occur principally through conductive and convective transfer from the hot igniter gas, through radiative transfer from the hot gas and incandescent solid particles, and through the impingement of hot solid particles from the igniter onto the propellant surface.

Ignition has been accomplished by each of these mechanisms under conditions such that the alternate methods of energy transfer could not play a significant part. The relative importance which the various mechanisms will assume in a particular case will depend on the composition of igniter and propellant and on the geometric parameters of the charge and igniter system. The use of black powder affords conditions favorable to all three mechanisms of energy transfer. The use of "gasless" igniters of the metal oxidizer type largely precludes the possibility of conductive and convective transfer from the gas. The effectiveness of this type of igniter must be due to the presence of hot solid particles coming in contact with the propellant surface and perhaps to radiation from the hot particles. When nitrocellulose or, more strikingly, an explosive gas mixture is used as the source of ignition energy, solid particles will be absent and the intensity of radiation from the gas will be low. Conductive transfer from the hot gas will be the principal mechanism of energy supply to the propellant surface.

Propellants can be ignited by radiation alone but the long ignition times observed are such as to suggest that radiation does not play a major role in the usual case. A consideration of the secondary role which radiation plays in steady state burning (Art. 4) supports this conclusion. The effect of radiation appears to be purely thermal; no photochemical effects have been observed. In the case of transparent propellants, radiation will penetrate into the interior of the grain. The distribution of absorbed energy will depend on the absorption coefficient of the propellant. Opaque propellants are found to be most easily ignited by radiation since most of the energy is absorbed in a thin layer at the surface, while propellants with some degree of transparency would be expected to burn more vigorously once ignition has been accomplished since the propellant will be heated to a considerable depth below the ignition surface. In the case of transparent propellants, radiation may even bring about a subsurface ignition due to the heating of radiation-absorbing particles of foreign matter unavoidably present in the propellant. Such subsurface ignitions occur after normal combustion has started and are caused by radiation from the propellant flame as well as from the igniter.

As is to be expected, propellants ignite more readily the higher their initial temperature. A rough correlation may be obtained between the minimum energy necessary to produce ignition in a given case and the energy necessary to heat the propellant from its initial temperature to some arbitrary ignition temperature.

The effect of pressure on the ignition process is more complex. High pressures will increase the emissivity of the igniter gas, thus increasing radiant energy transfer, but should have little effect on conductive transfer or on transfer by solid particles. The development of pressure will be accompanied by gas turbulence which may have an important effect on convective energy transfer, depending somewhat on the geometry of igniter and charge. Perhaps the most important effect of pressure in the ignition process lies in the transition to steady burning. The final gas-phase combustion reactions have been found to be unstable below a well-defined threshold pressure. If this pressure is not established in the combustion chamber during ignition, the transition to steady burning cannot take place smoothly and hangfire, chuffing, or an ignition failure will result.

Propellants with large heats of explosion are usually easier to ignite than cooler compositions. Propellants based on nitrate esters ignite more readily than those containing nitro- or nitramino-compounds. Double-base propellants ignite more readily than most composite types. A difference in ignition behavior is to be expected in the latter case because of the difference in burning mechanism.

The ignition of nitrocellulose has been studied by Rideal and Robertson [*5*, p. 536]. Two mechanisms can occur. If the gaseous products of thermal decomposition are allowed to accumulate in the neighborhood of the hot surface, a gas-phase explosion will occur leading to ignition.

This gas-phase reaction may be similar to the thermal explosion of formaldehyde and nitrogen dioxide studied by Pollard and Woodward [*36*, p. 767]. If the gaseous decomposition products are removed as they are formed, a gas-phase ignition cannot occur but on continued heating a thermal explosion will occur in the condensed phase.

The ignition of double-base propellants probably follows along similar lines. The importance of the retention of the ignition products in the neighborhood of the propellant surface has been illustrated by the work of Brian and McDowell [*77*] and others on the ignition of single grains of propellant by a stream of hot gases. The observed ignition times are longer than those observed for nitrocellulose under static conditions, and at times the grain vaporizes and the products are carried away by the gas stream without the appearance of a luminous flame. The presence of oxygen in the gas stream reduces the ignition time markedly. High speed photographs of igniting grains indicate that ignition starts in the gas stream close to the propellant surface. The occurrence of exothermic reactions in the condensed phase, which could lead to ignition by the thermal explosion mechanism, has been discussed previously (Art. 3).

Frazer and Hicks [*78*] have studied the properties of a thermal model for the ignition process. The propellant in the form of a semi-infinite slab with a plane surface at $x = 0$ receives heat at the surface from a high temperature gas representing the igniter flame. The propellant is also assumed to undergo a zero order exothermic reaction. Within the propellant grain the heat flow equation then has the form

$$c_p\rho \frac{\partial T}{\partial t} = k \frac{\partial^2 T}{\partial x^2} + qe^{-E_a/RT} \qquad x > 0; t > 0 \tag{7-1}$$

in which c_p is the heat capacity, k is the coefficient of thermal conductivity, and q is the quantity of heat evolved per unit of reaction. At the surface being heated the boundary condition is

$$-k \frac{\partial T}{\partial x} = h(T_g - T) \qquad x = 0; t > 0 \tag{7-2}$$

while at a sufficient distance from the surface the temperature gradient disappears and the other boundary condition becomes

$$-k \frac{\partial T}{\partial x} = 0 \qquad x \to \infty ; t \geqq 0 \tag{7-3}$$

The behavior of the igniter is idealized to the form of a thermal square wave; the temperature of the gas in contact with the propellant surface rises abruptly to a high value, $(T_g)_1$ at zero time, remains constant for a time t_0, and then falls to a low value $(T_g)_2$ for the duration of the experiment. This set of equations has been solved by numerical integration for a range of values of the parameters.

Ignition is arbitrarily defined to occur when the temperature of the propellant surface reaches a high value such that $T_i > 0.05E/R$, since at this point the exponential term in Eq. 7-1 is large and the temperature rise due to self-heating has become very rapid.

In this semi-infinite model, self-heating of the interior of the grain due to exothermic decomposition will always lead to ignition after a time t_{ia}, the adiabatic ignition time, without the addition of heat from an external source. The value of t_{ia} will depend strongly on the initial temperature T_0 of the propellant. In a real case, heat loss from the surface of a propellant grain of finite size will prevent an adiabatic ignition at initial temperatures below some critical value which depends on grain geometry as well as on the thermal parameters of the system. Frank-Kamenetsky [*79*] has derived the limiting conditions for adiabatic ignition for a number of simple geometries.

When heat is supplied to the propellant surface from an external source, the igniter gas at a temperature $(T_g)_1$, the surface temperature will rise and a thermal gradient will be propagated within the grain. If the heating is continued, ignition will occur after a time $(t_i)_{min}$, the minimum ignition time for the given rate of heat input. If heating is discontinued before ignition occurs ($t_0 < (t_i)_{min}$), heat will be lost from the surface to the cool gas at temperature $(T_g)_2$. However, if the propellant has reached a sufficiently high temperature before the hot igniter gas is removed, the temperature may continue to rise due to the exothermic decomposition reaction and the grain may ignite after a time $t_i > (t_i)_{min}$. As t_0 decreases below $(t_i)_{min}$ the ignition delay $(t_i - (t_i)_{min})$ increases rapidly and t_i approaches the adiabatic ignition time, corresponding to complete decay of the ignition stimulus.

The theory of Frazer and Hicks affords a satisfying picture of the thermal behavior of the propellant surface layers up to the point of ignition. The work of Rideal and Robertson, and others, seems to indicate clearly, however, that the site of the principal ignition process is in the layer of gaseous decomposition products adjacent to the propellant surface. If the definition of ignition used by Frazer and Hicks is extended by the assumption that the attainment of the ignition temperature is accompanied by the rapid evolution of an explosive gas mixture from the surface, this objection to the thermal theory is largely removed. The lack of suitable quantitative experimental data has prevented a rigorous test of the theory up to the present time.

The thermal theory of Frazer and Hicks with its emphasis on the solid-phase decomposition reactions is not applicable to most composite propellants where such reactions are largely absent. Jones [*80*] has studied the ignition of pyrotechnic mixtures similar in composition to certain composite propellants and has noted the absence of any heat of reaction term in the ignition energy equation. The ignition of a composite propellant by means of hot wires was studied by Altman and Grant [*64*, pp. 158ff.]. Thermal conduction was found to be the governing factor in this ignition process, heat generation by chemical reaction being relatively unimportant. The less frequent occurrence of chuffing or hangfires in the case of composite propellants also points to this difference between the mechanisms of ignition of composite and of double-base propellants.

With the heat of reaction term eliminated, Eq. 7-1 is reduced to the well-known unsteady state heat transfer equation

$$c_p\rho \frac{\partial T}{\partial t} = k \frac{\partial^2 T}{\partial x^2} \tag{7-4}$$

solutions of which have been tabulated for a variety of cases. The development of a theory of ignition of composite propellants along these lines with an accompanying experimental study appears to be highly desirable.

M,8. Cited References.

1. Will, W. *Angew. Chem. 14*, 743, 774 (1901).
2. Robertson, R., and Napper, S. S. *J. Chem. Soc. 91*, 764 (1907).
3. Roginskii, S. *Physik. Z. Sowjetunion 1*, 640 (1932).
4. Wilfong, R. E., Penner, S. S., and Daniels, F. *J. Phys. Chem. 54*, 863 (1950).
5. *Third Symposium on Combustion, Flame, and Explosion Phenomena.* Williams & Wilkins, 1949.
6. Wolfrom, M. L., Dickey, E. E., and Maher, G. G. The thermal decomposition of cellulose nitrate under reduced pressures. *Ohio State Univ. Research Foundation*, 1950.
7. Wolfrom, M. L., Dickey, E. E., and Prosser, H. C. A possible free radical mechanism for the initial reactions in the thermal decomposition, at reduced pressure, of cellulose nitrate (and other organic nitrates). *Ohio State Univ. Research Foundation*, 1948.
8. Fenimore, C. The final reactions of the burning of nitrocellulose. *Ballist. Research Lab. R-464*, 1944.
9. Snelling, W. O., and Storm, C. G. *U.S. Bur. Mines Tech. Paper 12*, 1912.
10. Robertson, R. *Proc. Chem. Soc. 25*, 179 (1909).
11. Roginskii, S. Z., and Sapozhnikov, L. M. *J. Phys. Chem. U.S.S.R. 2*, 80 (1931).
12. Lukin, A. Y. *J. Phys. Chem. U.S.S.R. 3*, 406 (1932).
13. Muraour, H. *Bull. soc. chim. France 53*, 612 (1933).
14. Gray, P., and Yoffe, A. D. *Proc. Roy. Soc. London A200*, 114 (1949).
15. Adams, G. K., and Bawn, C. E. H. *Trans. Faraday Soc. 45*, 494 (1949).
16. Robertson, A. J. B. *J. Soc. Chem. Ind. London 67*, 221 (1948).
17. Appin, A., Todes, O., and Khariton, Y. *J. Phys. Chem. U.S.S.R. 8*, 866 (1936).
18. Rice, O. K. *J. Chem. Phys. 8*, 727 (1940).
19. Phillips, L. *Nature 165*, 564 (1950).
20. Pollard, R. H., Wyatt, R. M. H., and Marshall, H. S. B. *Nature 165*, 564 (1950).
21. Rice, F. O., and Rodowskas, E. L. *J. Am. Chem. Soc. 57*, 350 (1935).
22. Levy, J. B. *J. Am. Chem. Soc. 76*, 3254, 3790 (1954).
23. Phillips, L. *Nature 160*, 753 (1947).
24. Brewer, S. D., and Henkin, H. The stability of PETN and pentolite. *Office Sci. Research and Develop. Rept. 1414*, 1943.
25. Bawn, C. E. H. The decomposition and burning of nitric esters. *British Ministry of Supply Rept. A.C. 10068/LFC. 58*, 1948.
26. Henkin, H. Stability of ethylenedinitramine (Haleite) and related nitramines. *Office Sci. Research and Develop. Rept. 1734*, 1943.
27. Robertson, A. J. B. *Trans. Faraday Soc. 44*, 677, 977 (1948).

28. Robertson, A. J. B. *Trans. Faraday Soc. 45*, 85 (1949).
29. Cottrell, T. L., Graham, T. E., and Reid, T. J. *Trans. Faraday Soc. 47*, 585 (1950).
30. Doescher, R. N. *Calif. Inst. Technol. Jet Propul. Lab. Rept. PR 9-43*, 1950.
31. Friedman, L., and Bigeleisen, J. *J. Chem. Phys. 18*, 1325 (1950).
32. Dodé, M. *Bull. soc. chim. France 5 (5)*, 170 (1938).
33. Otto, C. E., and Fry, H. S. *J. Am. Chem. Soc. 45*, 1134 (1923).
34. Patai, S., and Hoffmann, E. *J. Am. Chem. Soc. 72*, 5098 (1950).
35. Lewchewski, K., and Degenhard, W. *Ber. deut. chem. Ges. B72*, 1763 (1950).
36. Pollard, F. H., and Woodward, P. *Trans. Faraday Soc. 45*, 760, 767 (1949).
37. Pollard, F. H., and Wyatt, R. M. H. *Trans. Faraday Soc. 46*, 281 (1950).
38. Brown, F. B., and Crist, R. H. *J. Chem. Phys. 9*, 840 (1941).
39. Wise, H., and Frech, M. F. Rate of decomposition of nitric oxide at elevated temperatures. *Calif. Inst. Technol. Jet Propul. Lab. Rept. PR 9-46*, 1950.
40. Fenimore, C. P. *J. Am. Chem. Soc. 69*, 3143 (1947).
41. Musgrave, F. F., and Hinshelwood, C. N. *J. Chem. Soc. London*, 56 (1933).
42. Hinshelwood, C. N., and Green, T. E. *J. Chem. Soc. London*, 730 (1926).
43. Crawford, B. L., Jr., and Isbin, H. S. Determination of ignition temperatures of double-base powders. *Office Sci. Research and Develop. Rept. 1713*, 1943.
44. Crawford, B. L., Jr., Huggett, C., and McBrady, J. J. *J. Phys. Chem. 54*, 854 (1950).
45. Klein, R., Mentser, M., von Elbe, G., and Lewis, B. *J. Phys. Chem. 54*, 877 (1950).
46. Aristova, Z. I., and Leipunskii, O. I. *Compt. rend. Acad. Sci. U.R.S.S. 54*, 503 (1946).
47. Crawford, B. L., Jr., Huggett, C., Daniels, F., and Wilfong, R. E. *Anal. Chem. 19*, 630 (1947).
48. Muraour, H. *Chimie & industrie 47*, 602 (1942).
49. Corner, J. *Theory of the Interior Ballistics of Guns*. Wiley, 1950.
50. Crow, A. D., and Grimshaw, W. E. *Trans. Roy. Soc. London A230*, 387 (1932).
51. Muraour, H. *Chimie & industrie 50*, 105 (1943).
52. Muraour, H. *Bull. soc. chim. France 9*, 511 (1942).
53. Gibson, R. E. The rate of burning of double-base powders and the possible effects of change in nitroglycerin and total volatiles content on the burning of jet propulsion tube powder. *Office Sci. Research and Develop. Rept. 943*, 1942.
54. Muraour, H., and Aunis, G. *Compt. rend. 229*, 173 (1949).
55. Rice, O. K., and Ginell, R. *J. Phys. Chem. 54*, 885 (1950).
56. Rice, O. K. The theory of the burning of double-base rocket powders. *Office Sci. Research and Develop. Rept. 5224*, 1945.
57. Rice, O. K. The theory of the burning of rocket powders. *Office Sci. Research and Develop. Rept. 5574*, 1945.
58. Avery, W. H. *J. Phys. Chem. 54*, 917 (1950).
59. Beek, J., Jr., Avery, W. H., Dreshner, M. J., McClure, F. T., and Penner, S. S. Studies of radiation phenomena in rockets. *Office Sci. Research and Develop. Rept. 5817*, 1946.
60. Penner, S. S. *J. Appl. Phys. 19*, 278, 392, 511 (1948).
61. Andreev, K. K. *J. Phys. Chem. U.S.S.R. 20*, 365 (1946).
62. Corner, J. *Trans. Faraday Soc. 43*, 635 (1947).
63. Grad, H. *Commun. on Pure and Appl. Math. 2*, 79 (1949).
64. *Fourth Symposium (International) on Combustion*. Williams & Wilkins, 1952.
65. Muraour, H. *Mém. Artillerie Franç. 24*, 586 (1950).
66. Boys, S. F., and Corner, J. *Proc. Roy. Soc. London A197*, 90 (1949).
67. Corner, J. *Proc. Roy. Soc. London A198*, 388 (1949).
68. Markstein, G. H., and Polanyi, M. Flame propagation—a critical review of existing theories. *Cornell Univ. Aeronaut. Lab. Bumblebee Series Rept. 61*, 1947.
69. Evans, M. W. *Chem. Revs. 51*, 363 (1952).
70. Hirschfelder, J. O., and Curtiss, C. F. *J. Chem. Phys. 17*, 1076 (1949).
71. Zeldovich, Y. B. *J. Phys. Chem. U.S.S.R. 22*, 27 (1948). English transl. in *NACA Tech. Mem. 1282*, 1951.

72. Belayev, A. T. *Acta Physicochim. U.R.S.S. 8*, 763 (1938).
73. Semenov, N. M. *Prog. Phys. Sci. U.S.S.R. 24*, 433 (1940). English transl. in *NACA Tech. Mem. 1026*, 1942.
74. Parr, R. G., and Crawford, B. L., Jr. *J. Phys. Chem. 54*, 929 (1950).
75. Lewis, B., and von Elbe, G. *Combustion, Flames, and Explosions of Gases.* Academic Press, 1951.
76. Tanford, C., and Pease, R. N. *J. Chem. Phys. 15*, 431, 433, 861 (1947).
77. Brian, R. C., and McDowell, C. A. *Trans. Faraday Soc. 45*, 212 (1949).
78. Frazer, J. H., and Hicks, B. L. *J. Phys. Chem. 54*, 872 (1950).
79. Frank-Kamenetsky, D. A. *J. Phys. Chem. U.S.S.R. 13*, 738 (1939); *Acta Physicochim. U.R.S.S. 10*, 365 (1939).
80. Jones, E. *Proc. Roy. Soc. London A198*, 523 (1949).
81. Daniels, F. Studies of the mechanism of burning of double-base rocket propellants. *Office Sci. Research and Develop. Rept. 6559*, 1949.
82. Rice, F. O. Presented at classified conference, 1950.
83. Roth, J. Presented at classified conference, 1950.
84. Crawford, B. L., Jr., Huggett, C., and McBrady, J. J. Observations on the burning of double-base powders. *Office Sci. Research and Develop. Rept. 3544*, 1944.
85. Crawford, B. L., Jr. Spectrographic studies of the powder flame. *Univ. Minn. MR UMN/S2,3,4,5,7*, 1946.
86. Crawford, B. L., Jr., Measurement of flame temperature. *Univ. Minn. MR UMN/S9,10,12,13,14,15*, 1947.
87. Craig, R. S. Flame temperature and radiation studies in rockets. *Office Sci. Research and Develop. Rept. 5832*, 1945.
88. Avery, W. H., and Hunt, R. E. Effect of pressure and temperature on the rate of burning of double-base powders of different compositions. *Office Sci. Research and Develop. Rept. 1993*, 1943.
89. Avery, W. H., Hunt, R. E., and Sachs, L. D. Revisions and corrections to Natl. Defense Research Comm. Formal Rept.a-225 (*O.S.R.D. Rept. 1993*, 1943). *Office Sci. Research and Develop. Rept. 4568*, 1944.
90. Thompson, R. J., and McClure, F. T. Erosive burning of double-base powders. *Office Sci. Research and Develop. Rept. 5831*, 1945.
91. Crawford, B. L., Jr., et al. Studies on propellants, 30, 35, 42. *Office Sci. Research and Develop. Rept. 6374*, 1945.
92. Bircumshaw, L. L., and Newman, B. H. *Proc. Roy. Soc. London A227*, 115 (1954).
93. Schultz, R. D., and Dekker, A. O. The kinetics of decomposition of ammonium perchlorate. *Paper presented at the American Chemical Society meeting, Atlantic City, Sept. 1952.* See also [95].
94. Bircumshaw, L. L., and Phillips, T. R. *J. Chem. Soc.*, 703 (*1953*).
95. *Selected Combustion Problems: Combustion Colloquium.* AGARD, Palais de Chaillot, Paris, 1953.
96. Green, L., Jr. *Jet Propulsion 24*, 9 (1954).
97. Cheng, S. I. *Jet Propulsion 24*, 27, 102 (1954).
98. Green, L., Jr. *Jet Propulsion 24*, 252 (1954).

PART FIVE

DETONATION PROCESSES IN GASES, LIQUIDS, AND SOLIDS

SECTION N

DETONATION PROCESSES IN GASES, LIQUIDS, AND SOLIDS

A. R. UBBELOHDE
JOHN COPP

CHAPTER 1. EXPERIMENTAL METHODS FOR OBSERVING DETONATION

N,1. Introduction. The discovery of that mode of chemical decomposition known as detonation, and the development of methods [*4,50*] for determining detonation rates, were followed by the formulation of a hydrodynamic theory of the phenomenon. General predictions of the theory as presented by Chapman [*15*] and Jouguet [*37*] and amplified by Becker [*2*] have been confirmed by the work, among others, of Dixon [*24*], Lewis and Friauf [*48*], and Payman and Walls [*59*]. Much of the early basic work was done on gaseous systems.

Throughout this period of theoretical and experimental development, continued progress in synthetic organic chemistry greatly increased the range of potential solid and liquid explosives. A multiplicity of ad hoc methods were developed for testing the technological usefulness of explosives by evaluating the power, brisance, and sensitiveness [*52*]. The results of such ad hoc tests have contributed largely to the elimination of accidental explosions arising from the mechanical and thermal hazards associated with manufacture and storage of explosive substances. However, few basic experiments for investigating detonation in solid and liquid explosives, particularly the physicochemical aspects, were developed until comparatively recent years.

For interpreting detonation, important characteristics are as follows:

1. Detonation velocity.
2. Mass flow velocity of the reaction products behind the shock front.
3. The temperature, pressure, and density distribution in the region immediately behind the shock front, which are determined by—
4. The thermodynamic properties of the explosive and of its reaction products.

Of these, (1) and (4) are most easily accessible to experiment.

N,2. Measurement of Detonation Velocities. For the measurement of detonation velocities the photographic method initiated by Mallard and Le Chatelier [*50*] has been much more widely used than the chronographic method of Berthelot and Vieille [*4*]. In addition to the advantage of recording detonation rates in more or less permanent form, photographs can provide quantitative information on the movement of flame, product gases, and shock disturbances set up in and around a detonating charge (Plate N,2a), and can be used to study the sequence of events that lead up to and follow on stable detonation. Developments in the photographic study of detonation in gases were due, among others, to Dixon [*25*], and further refinements to Bone and Frazer [*6*]. Laffitte [*41*], Jones [*32*], and especially Payman and his collaborators [*60,61,62,63*] extended the method to the study of shock waves in fluids and of detonation waves in solid and liquid explosives.

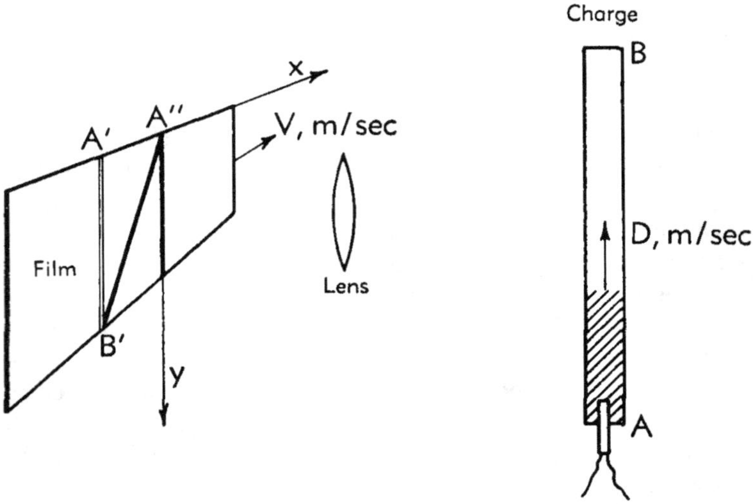

Fig. N,2a. Experimental layout of a rotating-drum camera for measuring detonation velocities.

Wave speed cameras. The principle of the wave speed camera is illustrated in Fig. N,2a. When an explosive charge, mounted with its axis vertical and in front of a lens and a vertical slit, is fired from one end, a luminous detonation front moves up the column from A to B with an instantaneous velocity D. The instantaneous velocity of the image of the flame front down a stationary film is

$$\frac{dy}{dt} = -\frac{D}{K}$$

where K is the ratio of the charge length to the length of its image on the film. If the film be moved past the slit with a constant horizontal velocity

V, this is compounded with the vertical motion of the flame-front image to produce an inclined trace $A''B'$. At any point the slope of this trace with respect to the x axis is dy/dx, where

$$\frac{dy}{dx} = \frac{dy/dt}{dx/dt} = -\frac{D}{KV}$$

or

$$D = -KV\frac{dy}{dx} \tag{2-1}$$

To measure D accurately it is necessary that KV be at least of the same order of magnitude, so as to give the flame trace a slope of about 45°.

In a recent camera [*14*] for studying detonation of high explosive charges of up to 1 kg in weight, a 35-mm film, 1 meter long, was fitted inside a balanced drum mounted horizontally on the shaft of a synchronous motor. A linear film velocity of the order of 2×10^4 cm/sec was obtained by rotating the drum at about 10^4 rpm. Alternative geometrical arrangements of the charge with respect to the slit were also used. Plate N,2b gives typical records of the luminous effects emanating from a detonating cylindrical charge.

A mechanically simpler method of spreading the image of a detonation front over the surface of a film is to concentrate the light onto a double-sided mirror rotating at the center of a fixed drum that carries a film on its inner surface [*7,14,22,31,61*]. Two independent slits at right angles to one another have been used by W. M. Evans [*64*] for simultaneously recording events as they develop in these two directions.

In the earliest cameras with the film mounted on the outside of the drum, the writing speed was limited by the strength of the film. With the film inside the rotating drum the writing speed is limited by the mechanical strength of the drum to about 200 m/sec. The higher writing speed attained with rotating-mirror cameras is limited (above about 350 m/sec) by the reduced intensity of illumination at high speeds.

Electronic velocity determinations. The main disadvantage of photographic determinations of detonation velocities is that the instantaneous D has to be evaluated from the tangent (2-1) to the luminous trace on the photograph. This limits accuracy. The chronograph method of Berthelot and Vieille [*4*], which measures the time of transit of a detonation wave over a finite segment of an explosive charge, provides an average value for D. For special purposes this is no disadvantage. Modern electronic devices, whose greater precision permits the use of a shorter charge, are being increasingly applied for this purpose [*28,77*]. A recent development in which all mechanical moving parts are eliminated is due to Courtney-Pratt ([*20,21*] and Fig. N,2b). A luminous point image from the detonating charge travels over a photosensitive surface. The electron beam thus generated is focused and deflected by a linearly varied field in an image converter, and the image trace recorded on a fluorescent screen is photographed (cf. Plate N,2c). No obvious limit to the writing speeds attainable has yet been found, and the resolution of events separated by as little as 10^{-8} seconds has already been achieved. For example, spark initiation of nitroglycerin shows that an initial slow burning transforms to fast explosive reaction in a time of the order of 30 microseconds. Again, in the spark initiation of lead azide, a typical salt, true detonation is immediately set up with no preliminary period of slow burning. Differences of this kind are discussed in Art. 15.

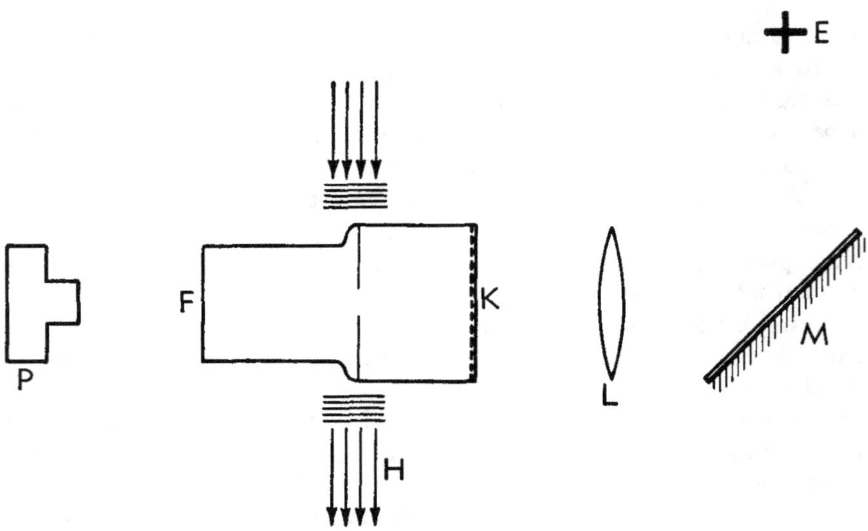

Fig. N,2b. The Courtney-Pratt high speed camera. Light from the explosion center E is concentrated by a mirror M and a lens L onto the photosensitive cathode K of an image converter tube. The electron beam is focused and deflected by an applied field H, is recorded on a fluorescent screen F, and the image is photographed with a plate camera P.

Some examples of the experimental evaluation of detonation velocities. The wave speed camera has been extensively employed to determine stable detonation velocities of gases and of condensed explosives in charges which are bare or are wrapped in transparent cases of negligible mass and strength. Much of this work has not been published in detail [*26,40*]. Although optical methods for determining detonation rates of solid explosives are less easily adapted to charges confined in heavy metal cases, Cybulski, Payman, and Woodhead [*22*] have reported precise detonation rates obtained photographically for such charges by drilling a series of small holes in the opaque cases (Plate N,2d). The less precise comparative method of Dautriche [*23*] has been extensively employed to study the effect of charge density [*27*] on the detonation velocity of solid explosives. More recently, Copp and Ubbelohde [*16,17*] have used this method with comparatively large charges of explosive up to 3.5 kg, and a sensitive detonating "cordeau" containing PETN to determine the effect of inert and slow-reacting fillers on the detonation velocity of solid explosives in cased charges.

Measurements [*22,26*] may be quoted to illustrate some effects of charge density, crystal size, and charge diameter.

In general the detonation velocity D of a solid explosive increases with its density ρ. For granular PETN, observed detonation velocities are 6110 m/sec at $\rho = 1.17$ g/cm^3 and 7520 m/sec at $\rho = 1.51$ g/cm^3 in unconfined charges of diameter 3.17 cm. When the effect of lateral venting is eliminated (see Art. 9), this dependence can be represented with fair accuracy by the general equation

$$D_0\rho = (D_0)_{1.0} + \mathbf{M}(\rho - 1.0) \tag{2-2}$$

where $(D_0)_{1.0}$ and $\mathbf{M}$ are constants of the explosive [*18*]. For TNT, $(D_0)_{1.0} = 4800$ m/sec and $\mathbf{M} = 3500$ (m/sec)/(g/cm^3).

The average crystal size and size distribution of the particles in a granular explosive can have an important effect on the velocity and stability of detonation. (Values of D quoted are not necessarily the same as limiting values D_0 in Table N,2a.)

Table N,2a. Effects on D of grain size and size distribution.

Explosive	Mean density, g/cm^3	Velocity D, m/sec
Coarse-grained TNT (unconfined charge 3.17-cm diameter)	1.605	6430
Fine-grained TNT	1.625	6850
Ammonium picrate	1.0	
Percentage of coarse-grained material mixed with fine-grained		
0	1.0	4800
16	1.0	4450
40	1.0	3850
60	1.0	2950
(Detonation failed between 60 and 90)		
90	1.0	3700
100	1.0	4150

With increasing charge diameter, the detonation velocity of an explosive such as TNT, at constant density and grain size, tends to a limiting value.

Table N,2b. Unconfined charges, fine-grained TNT. Effects of charge diameter on D.

Charge density, g/cm^3	1.605	1.605	1.610	1.615	1.615	1.620
Charge diameter, cm	1.26	1.66	1.90	2.20	2.54	3.17
Detonation velocity, m/sec	fails	6470	6740	6820	6890	6900

A similar conclusion can be drawn from results (Fig. N,10b) for charges confined in heavy metal cases. The interpretation of experimental measurements of this kind is discussed in Art. 9 and 10.

For *liquid explosives* such as nitroglycerin and its analogues, it appears [*38*] that two steady detonation rates are possible. A "high" value of about 8000–10,000 m/sec corresponds to the theoretical hydrodynamic detonation velocity, whereas a "low" velocity detonation of about 2000 m/sec is not much larger than sonic velocities. Weak priming charges (or microinitiation; cf. *Impact tests* in Art. 15) always initiate low order detonation, which can subsequently transform to high velocity detonation. The reverse transition from high to low order detonation has not been observed. A noteworthy feature of low order detonation in liquids is that with increasing charge radius the velocity decreases (see Art. 12). Photographic techniques have been applied by Bowden and his coworkers [*9,10,11*] to record the sequence of events in microinitiation of detonation in liquid explosives.

N,3. Detonation Temperatures. Every application of hydrodynamic theory involving equations of state for the detonation products requires calculation of the detonation temperature. A practical difficulty in such calculations is the need to know the composition of the gases when detonation energy is completed. Empirical rules have been proposed for estimating the composition of the reaction products in the wave front, and more refined calculations in which equilibrium is assumed to follow temperature and pressure changes have been made [*18,48*]. Experiments have been carried out to determine the composition of detonation products, by analyzing the gases formed when detonation occurs in a lead block enclosed in a sealed bomb [*70*]. The results obtained cannot be referred (cf. Haid and Schmidt [*30*]), to the zone immediately behind the detonation wave, since the various equilibria involving CO, CO_2, H_2, H_2O, etc. are only "frozen" after extensive expansion has occurred [*33*].

A small number of experiments have been described in which spectroscopic analysis of the flash from a detonating charge has led to direct estimates of detonation temperatures [*45,1*] (cf. data of Fox quoted by Cook [*18*]). A difficulty exists in that, according to Muraour and Michel-Levy [*55*], the origin of the most intense luminosity at the instant of detonation lies not in the product gases but in the associated shock wave. For this reason these authors claim that direct determination of detonation temperatures by spectroscopic methods is impossible.

N,4. Flow Velocities. The flow velocity U (cf. Art. 6) behind a pure shock front has been measured by means of schlieren photography with a wave speed camera and compared with calculated values [*63*]. Wave speed camera records of gaseous detonation rates also reveal the magnitude of the flow velocities. Estimates of the flow velocity U in the detonation of solids have been obtained by X-ray photography of special so-called "zebra" charges containing alternating layers of explosive with and without incorporated lead (cited in [*26*]).

N,5. Other Experiments. Measurements on the decay and failure of detonation, and on the build-up of detonation are best discussed after considering the effects of lateral venting on the propagation of detonation (Art. 9 and 12). Relative measurements on detonation pressures have been proposed from observations on relative brisance or hollow charge effects, but the figures are at present mainly of technological interest [*52*] and have not as yet been elucidated theoretically.

X-ray pulse radiography of detonating charges. By using X-ray generators giving high intensity pulses with a duration $\cong 10^{-7}$ sec [*12*] it is possible in principle to radiograph some of the rapid density changes due to detonation. These include the initial movement of the inner wall of the case around a charge, from which the detonation pressure can be calculated, and the mass flow in "layered" charges with alternate regions of high and low opacity to X rays. It is too soon to predict how much precise information about the physicochemical parameters of detonation will eventually be obtainable by such means.

CHAPTER 2. THE PHYSICAL CHEMISTRY OF DETONATION PROCESSES IN GASES, LIQUIDS, AND SOLIDS

N,6. Some Characteristic Differences Between Detonation and Explosion. Detonation has been frequently discussed from a hydrodynamic standpoint [*2,5,15,37,49,71,72,73,74*] which makes no explicit reference to the physical chemistry of the processes involved. To understand various recent lines of investigation concerned especially with this physical chemistry, it is important to grasp how detonation differs from explosion or flame (Sec. G and I). When a cylindrical system capable of undergoing chemical reaction is ignited at one end, a *chemical reaction* may in suitable systems be propagated along the cylinder either as a flame or as a (considerably faster) detonation. The propagation of flame does not involve any physicochemical processes substantially different from those in reacting systems at lower temperatures, except for the greater preponderance of free radical and free atom mechanisms of chemical reaction. Ignition of the layers of unburnt gas immediately ahead of an advancing flame front is effected by forward diffusion of "hot" molecules, radicals, and atoms. Detonation differs from flame in that an impact wave (III, E) passing through unreacted material *causes* the chemical reaction to take place in such a way that the hydrodynamic effects of the energy liberated can be fed into the propagation of the wave.

The momentary pressures set up are much higher than in the case of flames, and in solid high explosives may reach 10^5 atmospheres or more. This explains the shattering effect of detonations on containers of insufficient mechanical strength.

Since the front of a detonation wave, traveling with velocity of $y \times 10^3$ m/sec (where y usually ranges from 1 to 10), traverses distances of the order of a molecular diameter (2×10^{-8} cm) in $(2/y) \times 10^{-13}$ sec, it seems at first sight that the chemical reaction which maintains detonation should be completed in times of this order. Recent experiments show that this is not so at any rate in condensed explosives, but that certain chemical reactions may be complete in times as short as 10^{-6} sec. The mechanism of physicochemical activation which leads to such rapid chemical reactions can be discussed only after a consideration of the nature of shock or impact waves.

The propagation of shock waves through gaseous, liquid, or solid media has received comparatively rigorous mathematical analysis in two limiting cases. In case 1, the thermal properties of the medium can be expressed by a simple equation of state, and the passage of the shock wave does not induce chemical reaction in the material. In case 2, the passage of the shock wave is accompanied by and leads to chemical reaction of the material. If the energy of chemical reaction ΔQ per unit mass be released at a sufficiently high rate so that it can contribute to the propagation of the shock wave, then under favorable circumstances a stable "detonation," i.e. a shock wave involving chemical reaction, is set up.

The simplest hypothesis about the structure of an impact or detonation wave is that it involves a region of sudden increase in pressure and temperature [*65*]. This region travels forward into the undisturbed material with a velocity which is greater than the acoustic velocity of disturbances of small amplitude. Behind this steep rise there is a fall in pressure which explains the use of the term "wave" for this kind of disturbance. Equations characteristic of the propagation of such a wave front with a steady linear velocity D follow from expressions for the conservation of mass, momentum, and energy across a finite zone behind the shock front. These equations are not derived in this article but are quoted for reference (see III,D and G). The flow velocity U of material behind the shock front in the direction of the front is related to the velocity of propagation D_0 of the shock front by the two equations

$$\frac{U}{D_0} = \frac{v_1 - v_2}{v_1} \tag{6-1}$$

$$UD_0 = v_1(p_2 - p_1) \tag{6-2}$$

where v_1 and p_1 are the specific volume and pressure ahead of the shock wave and v_2 and p_2 refer to the same properties behind the shock front. These equations apply to the propagation both of a pure shock wave (case 1) and of a detonation wave (case 2). A third hydrodynamic equation expressing the condition of conservation of energy is most generally written as

$$E_2 - E_1 = \Delta Q + \tfrac{1}{2}(p_2 + p_1)(v_1 - v_2) \tag{6-3}$$

Case 1. For a pure impact wave ΔQ is zero, and the change in internal energy $E_2 - E_1$ when no chemical reaction results from passage of the shock front is determined merely by differences in temperature and pressure. It is interesting to note that even acoustic waves travel with a modified velocity if the system is capable of a chemical transformation [*39a*] such as

$$N_2O_4\ (g) \rightleftarrows 2NO_2\ (g)$$

Shock waves propagated through such media should likewise exhibit a modified behavior and in particular should be strongly absorbed [*65*].

Case 2. When the passage of a shock wave front leads to chemical rearrangements at such a rate that the chemical energy released can contribute to the propagation of the wave front, then a stable detonation wave can result. In this case, ΔQ in Art. 8 is the heat liberated by the reaction of 1g of the explosive at constant volume.

These three equations do not uniquely define the velocity of propagation of the wave. An additional assumption due to Chapman and Jouguet is made which can be expressed in the form that the front velocity has the sonic value a with respect to the moving material behind the front (see III,D and G), i.e.

$$D_0 - U = a \tag{6-4}$$

N,7. Equations of State for Detonation. For the propagation of a detonation wave these four equations can be solved if an equation of state be assumed for the detonation products. The conditions of temperature, pressure, and specific volume in the unchanged material can be readily defined. The thermodynamic properties of the detonation products can often be approximately calculated by treating them as a gas. At a temperature of several thousand degrees such as prevails in the detonation products, Van der Waals attractions can be ignored, but the high compressions give to the molecular volumes of the product molecules a controlling influence on the detonation velocity.

The simplest practicable equation of state for the product gases is $p_2(v_2 - b) = RT$. For such a gas

$$C_p/C_v = \gamma \quad \text{and} \quad C_p = C_v + R \tag{7-1}$$

The adiabatic equation which then applies to detonation is

$$p_2(v_2 - b)^{\gamma} = \text{const} \tag{7-2}$$

Schmidt [75] adopts the above equation of state for the molecules produced in the detonation of condensed explosives, and derives a general expression for the detonation velocity

$$D_0 = \frac{v_1}{(v_1 - b)} \frac{v_1}{v_2} \sqrt{8.31\gamma n_2 T_2} \tag{7-3}$$

where n_2 is the number of gas moles formed from one kilogram of a solid or liquid explosive.

More complicated equations of state have been assumed by other workers to calculate the characteristic parameters of detonation in condensed systems of [18,32a,33,44,68]. A test of the validity of the hydrodynamic equations is provided by predictions of detonation velocities for gaseous systems in which the covolume term b is unimportant [3, 48]. In solid and liquid systems the calculation of detonation velocities does not permit a complete test of the theory. All that can be done is to calculate covolumes from observed detonation velocities and to compare these with physical theories of molecular compressibility at high pressures. Cook has pointed out that of all the parameters characteristic of detonation in condensed systems only the detonation temperature may be used to check the accuracy of the various equations of state employed. All other parameters are insensitive to the particular equation of state adopted. The results of spectroscopic estimates of detonation temperatures are too uncertain to test any particular equation of state at the present time [18].

The development of hydrodynamic and thermodynamic equations to characterize a detonation wave does not give any clue to the physicochemical processes which determine the rate of liberation of the energy of chemical decomposition. Under marginal conditions the rate of energy release is a controlling factor in the propagation of detonation. Systematic studies of the factors that influence the "marginal" propagation of detonation have provided the necessary experimental counterpart to theoretical speculation on the physical chemistry of detonation. Such experiments are of considerable practical importance in addition to their theoretical significance.

N,8. A Finite Time of Energy Release in Detonation. In the detonation of a medium with comparatively simple molecules such as ozone, it is just conceivable that the molecular rearrangement of chemical bonds which releases the energy of detonation can occur in single collisions such as $2O_3 \rightarrow 3O_2$ though the bond structure of ozone makes this somewhat unlikely. In the passage of a detonation wave through a homogeneous explosive with complex explosive molecules such as pentaerythritol tetranitrate, even mere spatial requirements would demand a number of steps to permit the atoms to take up their new molecular patterns.

$$C(CH_2NO_3)_4 \rightarrow 2N_2 + 4H_2O + 2CO + 3CO_2$$

The over-all reaction that maintains the detonation wave by its energy release probably occurs with the intervention of molecular fragments and atoms. Free radicals such as OH and CH almost certainly intervene but even larger fragments would be sufficiently long-lived to play some part in the reaction mechanism. A finite number of collisions between these intermediate fragments implies a finite time τ before the irreversible chemical changes that maintain the detonation wave can be completed. Even longer times for the completion of irreversible chemical changes are to be expected, and are found in the case of heterogeneous explosives such as amatol (trinitrotoluene + ammonium nitrate) where the molecules of the two components have to mix before chemical reaction can be completed.

Over and above these irreversible chemical changes, any mixture of molecules, especially when CO_2 or H_2O are present, will undergo a shift in equilibrium composition following the pressure and temperature changes in the detonation wave. In theoretical calculations it is generally assumed that these equilibria are established in a very short time compared with the time of transit of the detonation wave. Provided this holds, the energy changes that accompany shifts in equilibrium composition can be allowed for in the calculation of parameters of the ideal detonation wave [*18,33*].

However, as has been stated above, the calculated detonation velocity is not very sensitive to the precise equilibria assumed. More refined methods of assessing the conditions in a detonation wave may eventually show that chemical equilibrium, and even the partition of thermal energy between the vibrational, rotational, and translational degrees of freedom of the gas molecules in a detonation wave, lag behind the value for infinitely slow changes of pressure and temperature (cf. I,H). Results on the detonation of H_2-O_2 with admixed argon or helium suggest that significant delays in partition do in fact occur [*3*].

N,9. The Finite Time of Energy Release in Relation to Lateral Venting. According to elementary hydrodynamic theory, detonation velocity depends on conditions of T and p in the region immediately behind the zone in which energy release is complete. In the earlier formulations of the theory no account was taken of the effect of lateral expansion over this finite reaction zone, so that the detonation velocities were calculated as being independent of the dimensions of the charge. Experimentally it has long been known that measured detonation velocities decrease as the charge radius decreases and that detonation fades out if the charge diameter is sufficiently small. A theory of the effect of charge dimensions on stable detonation was developed in 1942 by Jones and published in part in 1947 and 1950 [*32a*,*34*].

It is convenient to refer to the thickness X of the reaction zone in which irreversible energy release is completed and to define a reaction time τ by the equation

$$X = D\tau \tag{9-1}$$

To relate τ defined in this way with the progress of chemical reaction in the reaction zone, a correction factor of about $\frac{4}{3}$ must be introduced [*32a*]. Then

$$\tau_1 = \tfrac{4}{3}\tau = \frac{1}{K_\tau} \tag{9-2}$$

where K_τ is an over-all specific velocity constant for the irreversible chemical reactions, whose temperature dependence can be formally represented by a standard activation equation

$$K_\tau = \frac{kT}{h} e^{\Delta S^*/R} e^{-\Delta H^*/RT} \tag{9-3}$$

In what follows it will be seen that the effects of lateral venting can be expressed in terms of the dimensionless parameter X/R where R denotes the charge radius. This implies that two detonative reactions with very different reaction times should exhibit similar behavior at radii where X/R has the same value. Some consequences of this for weak detonations, e.g. of large masses of ammonium nitrate, have been discussed by Ubbelohde [*78*; *79*; *84*, p. 464].

This theory of the relation between charge dimensions and detonation velocities has been tested in a number of experimental studies. Because it can be treated both experimentally and theoretically with some success, the case which has received most consideration is that of a detonation wave traveling with its front normal to the axis of a cylindrical charge of radius R. The full lines in Fig. N,9 illustrate the pressure and temperature distribution behind such a detonation front when it is assumed that over the length of the finite reaction zone no lateral expansion occurs. If appreciable lateral expansion occurs before the reaction is complete there will be a drop in the numerical values of p and T and a consequent drop in the observed detonation velocity. This state of affairs is illustrated by the dashed lines in Fig. N,9. In this figure the ordinate is the ratio of the pressure (or temperature) at any point behind the detonation front to the pressure (or temperature) established behind the reaction zone of an ideal nonexpanding charge.

Important consequences of lateral venting are:

1. The observed detonation velocity (and brisance) are less than the ideal hydrodynamic values calculated for no venting.
2. The extent of venting, i.e. of lateral expansion of the reaction zone, is determined by the hydrodynamic properties of the medium confining the charge.
3. The detonation front is curved.

Quantitative expressions for the effect of lateral expansion on detonation parameters have been developed by two rather different methods

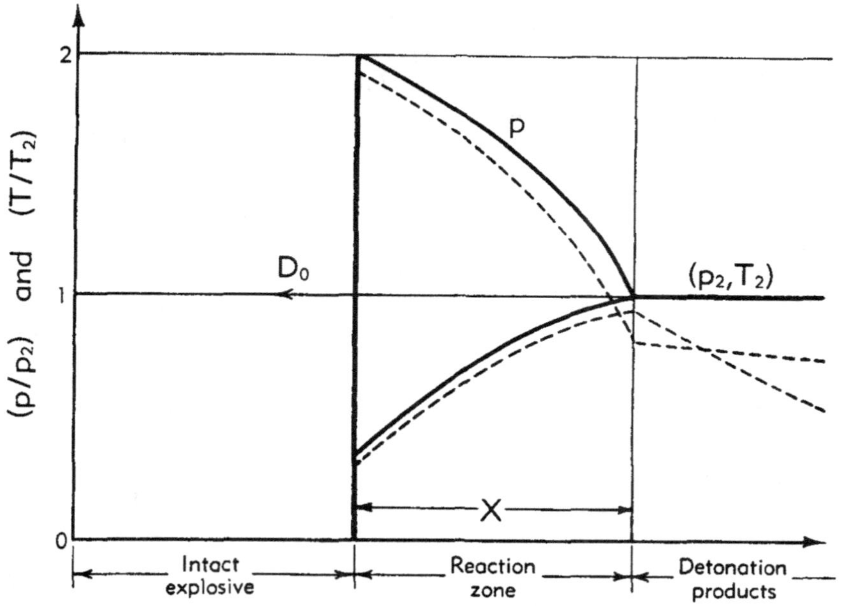

Fig. N,9. The form of the distribution of temperature and pressure behind a detonation point. (1) Full lines: no lateral expansion of reaction zone. (2) Dashed lines: allowing for lateral expansion.

[*32a,26*]. Numerically similar results are obtained by the two methods for the special conditions that can be attained experimentally. Both consider that lateral expansion of the reaction zone modifies only the hydrodynamic equation for continuity of mass. Jones summarizes the general effect of lateral expansion in the equation

$$\left(\frac{D_0}{D}\right)^2 \cong 1 + 2.25(r^4 - 1) \tag{9-4}$$

where r is the relative expansion of a central stream tube at the end of the reaction zone. This quantity r is then determined for three conditions of peripheral confinement. D is the observed velocity, D_0 the limiting velocity at infinite charge radius.

1. For bare charges, values of r for various values of X/R have been tabulated. From measurements of the ratio D_0/D, the value of r can be derived from Eq. 9-4 and hence values of X/R interpolated from the published tables.
2. For cased charges, if the casing is treated as thin compared with X, laminar conditions apply, and the relation between boundary conditions and detonation velocities follows approximately the equation

$$\left(\frac{D_0}{D}\right)^2 = 1 + \frac{9X^2\rho}{8RS} \tag{9-5}$$

ρ is the density of the charge confined in a laminar tube with the value S for the mass/unit of area of the charge surface.

3. For the infinitely thick case, if the casing is so thick that the shock wave set up in it by the head of the detonation wave does not reach the exterior of the case before energy release is completed, the conditions cannot be distinguished from those for an infinitely thick casing. The relationship for the relative expansion at the end of the reaction zone is

$$r_1 = 1 + \frac{X}{R} \tan B \tag{9-6}$$

where the values of the expansion angle B are calculated from the compressibility of the material of the casing, which determines the angle between the expanded inner wall and its original direction. Substituting Eq. 9-6 in Eq. 9-4 an approximate relationship is

$$\left(\frac{D_0}{D}\right)^2 = 1 + 9\frac{X}{R} \tan B \tag{9-7}$$

When D_0/D is not much greater than unity, all three conditions (1, 2, and 3) give a linear plot of $1/D^2$ against $1/R$ which can be extrapolated to $1/R \to 0$ to yield a value of D_0, the limiting detonation velocity of the charge for no venting. The applicability of these equations to the evaluation of reaction times from the approximate relationship of Eq. 9-1 has been tested experimentally by detonating the same explosive under different conditions of confinement. Very fair agreement is generally obtained, as is verified from the linear plots of $1/D^2$ against $1/R$ recorded by Copp and Ubbelohde [*16*] for various types of confinement. These give the same value of D_0 irrespective of the individual detonation velocities, and consistent reaction times calculated from detonation velocities which may differ by 1000 m/sec (see Art. 10). Analysis of a wide selection of experimental results has permitted further verification [*26*].

Effects of finite reaction times on the Chapman-Jouguet condition in detonation waves, and effects of curved detonation fronts have recently been reexamined [*38a*].

N,10. Data on Energy Release in Stable Regimes. Studies on times of energy release up to the present have been largely restricted to solid high explosives, though they should in fact give valuable information in gaseous detonation as well.

The use of a graphical method, with the appropriate value of D_0/D to give the corresponding value of X/R, is illustrated in Fig. N,10a. For

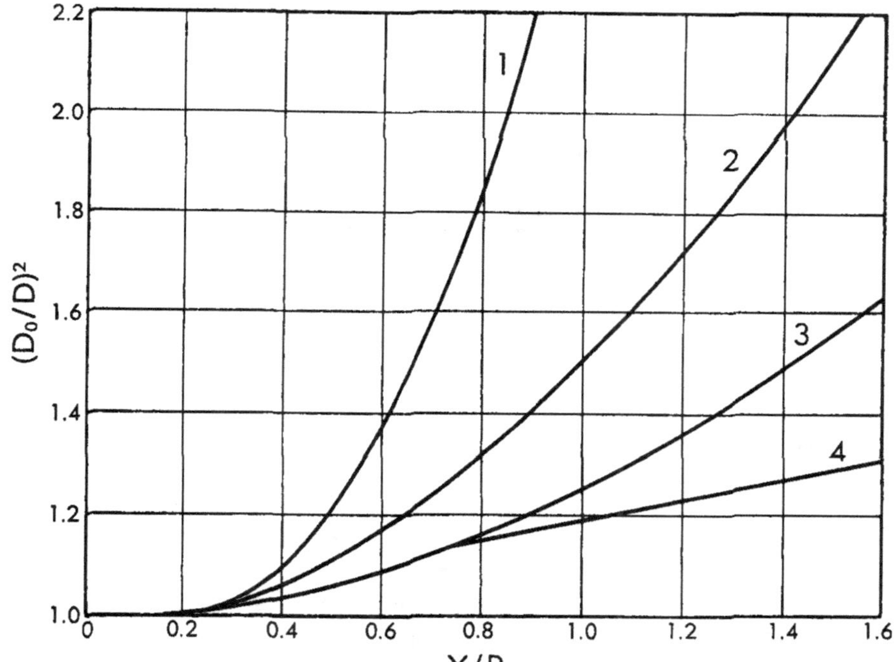

Fig. N,10a. The effect of confinement on detonation velocity. (1) Bare charge. (2) Laminar confinement $2S/R\rho = 5$. (3) Laminar confinement $2S/R\rho = 10$. (4) Infinite confinement, TNT in steel.

heavily cased charges a plot of $1/D^2$ against $1/R$ with extrapolation to give D_0 is illustrated in Fig. N,10b which records data for the homogeneous explosive TNT and the heterogeneous 60/40 amatol. Observed detonation velocities were independent of the wall thickness for each charge diameter so that the conditions approximate those for infinite confinement (see Eq. 9-6). Reaction zone thicknesses X are obtained from the slopes of the lines drawn in Fig. N,10b using calculated expansion angles B. (For cast TNT in iron, B is 1°18′ and in lead it is 3°3′.) The precise value of B depends on the detonation pressure (cf. Jones, in [*64*]).

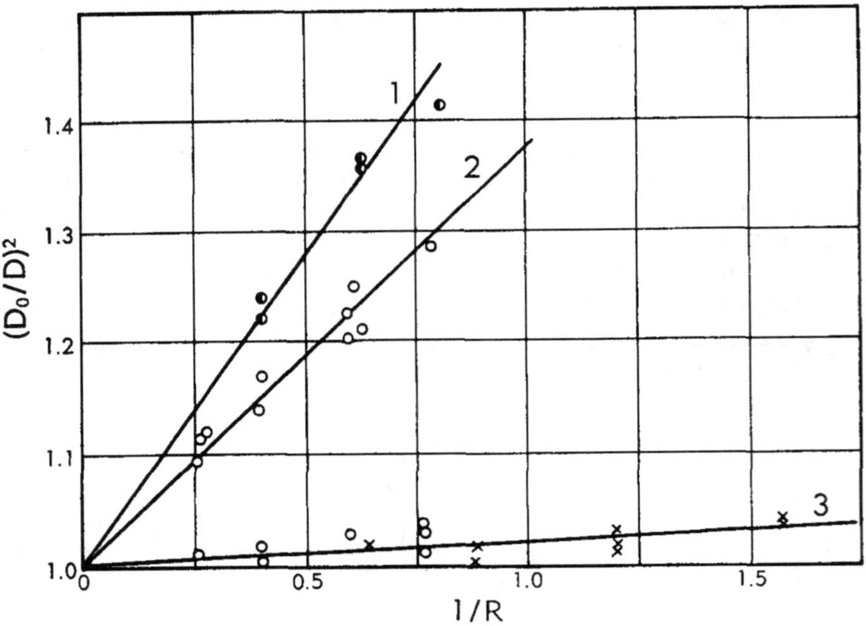

Fig. N,10b. The effect of charge radius for heavily confined charges.
1. 60/40 amatol in lead casing, D_0 = 6500 m/sec.
2. 60/40 amatol in steel tubing, D_0 = 6500 m/sec.
3. TNT cast in steel tubing, D_0 = 6900 m/sec.

Reaction zone thicknesses and the associated reaction times (Eq. 9-1 and 9-2) give important information about the physical chemistry of the mechanism of energy release in a detonation wave. Typical results for the reaction zone thickness of a variety of high explosives are collected in Table N,10a[*26*]. As will be seen, in cast TNT the thickness X is of the order of 0.2 cm. At the estimated detonation pressure and temperature the number of binary molecular collisions would be of the order of 10^{34}/sec if ordinary gas-kinetic processes could be used to describe

Table N,10a. Some reaction times (τ) in granular explosives.

Explosive	*Density, g/cm³*	*Confinement*	*D_0, m/sec*	*X, cm*	*Reaction time τ, microsec*
Picric acid	0.90	glass	5300	0.22	0.40
RDX	0.90	glass	5900	0.08	0.10
TNT	1.61	steel (thick)	7000	0.21	0.3
TNT	1.60	carton	6950	0.21	0.3
Amatol coarse	1.50	carton	6500	1.95	3.0
Amatol coarse	1.50	steel (thick)	6500	2.14	3.3
Amatol coarse	1.50	lead (thick)	6500	1.43	2.2
Amatol fine	1.50	steel (thick)	6500	1.43	2.2
Amatol fine	1.50	carton	6500	1.36	2.1

molecular encounters at such high pressures. This is certainly not permissible. All that can safely be said in the present state of the kinetic theory as applied to high temperatures and pressures is that for a complex molecule such as TNT many molecular collisions are required to complete the rearrangement of bonding that is involved in the release of chemical energy. Corresponding data are unfortunately not available for the detonation of simple molecules such as ozone or acetylene where a much thinner reaction zone would be anticipated (cf.[*83*, p. 566]).

When various inert solids are mixed with TNT, the detonation velocity can be greatly reduced without modifying its dependence on charge radius; for example, the magnitude of the reaction time is sensibly unaltered by the incorporation of up to 50 per cent by weight of inert fillers such as common salt. At first sight it might be thought that the explosive merely detonates in the interstices between the grains of inert material with its own characteristic reaction time. But calculation shows that even during the time for energy release the inert material abstracts energy from the detonation zone both by thermal conduction and by its contraction, and thus cools the product gases. An obvious indication of this is the decrease in D_0 when inert fillers are present.

Table N,10b. Effects of inert fillers on detonation velocity.

Composition of mixture	*Density, g/cm³*	*Nominal density of explosive in the interstices, g/cm³*	*Nominal value of D_0, m/sec*	*Experimental value of D_0, m/sec*
NaCl/TNT 60/40	1.85	1.50	6610	5520
NaCl/TNT 50/50	1.84	1.62	7030	6100
BaSO/TNT 25/75	1.94	1.63	7060	6640
BaSO/TNT 15/85	1.80	1.62	7030	6740
TNT pure	1.60	1.60	6950	6950
PETN pure	1.000	1.000	5500	5500
NaCl/PETN 10/90	1.000	0.938	5280	4950
NaCl/PETN 40/60	1.000	0.716	4520	3830
NaCl/PETN 70/30	1.000	0.309	3550	2350

The data for TNT-inert filler mixtures were obtained by Copp and Ubbelohde [*16,17*]; those for PETN-salt mixtures have been reported by Patterson [*58*] and analyzed, without reference to reaction times, in terms of entrainment of the inert particles with the gaseous explosion products. Further examples, including some mixtures of practical importance, are quoted by Taylor [*77*].

The nominal density of explosive in the interstices can be calculated by allowing for the mass and volume of the inert filler and the related nominal detonation velocity from an empirical equation which for TNT gives

$$D_0 = 4850 + 3500(\rho - 1)$$

(Marshall quoted by Copp and Ubbelohde [*16,17*]). From the dependence of detonation velocity on charge radius (for thick steel cases), the conclusion is that the reaction times are 0.3 to 0.5 microseconds for all mixtures. Detailed calculations involving the compressibility and heat conductivity of the filler and neglecting entrainment [*16,17*] show that the detonation temperature may be decreased as much as 500°C by these additions. Since this has no appreciable effect on the reaction time τ, the physicochemical processes in the detonation wave cannot depend on a supply of activation energy mainly controlled by a factor $e^{-\Delta H^*/RT}$ (Eq. 9-3). The time required for energy release refers instead to temperature-insensitive processes such as the gasification of the explosive, and diffusion of the molecular fragments to their final reaction partners. This view, that the reaction time of detonative energy release in granular explosives refers primarily to transport processes, is borne out by quantitative correlations between reaction time and average grain dimensions of explosives such as TNT, the amatols, and ammonium picrate [*16,17,22,26*]. For example, in the case of the amatols both the lengthened reaction time compared with TNT and the magnitude of D (see Table N,10a) agree with the notion that the major part of the reaction time is used up in erosion of the ammonium nitrate granules, and that the rate processes controlling the subsequent reaction of the gasified ammonium nitrate (or molecular fragments) cannot have a large temperature coefficient [*16,17*]. The important role of granule size in determining the characteristics of energy release in the detonation of solids is further emphasized by its influence over such properties as shock sensitivity and fading. It is not unlikely that large single crystals of an explosive such as TNT would prove to be nondetonating. An experimental complication in cast charges is that the crystal size will in general increase toward the center of the charge. This factor can complicate the analysis of venting effects and modify the shape of the wave front [*77b*].

The experimental observation (Table N,2a) that the propagation of detonation is not only dependent on particle size but on the distribution of sizes has led to the suggestion [*26*] that chemical reaction is initiated by the impact wave at the highly stressed load-bearing points and spreads into the explosive granules. Such a view has analogies with ultrashearing mechanisms of microinitiation of reaction (Art. 15).

N,11. Mechanisms of Activation in the Propagation of Detonation. As explained above, a study of venting phenomena shows that, with the condensed explosives studied to date, the process of energy release in successive layers of undetonated material does not involve a large temperature coefficient. The reasons for this will be apparent when the characteristics of the detonation zone are considered. In what follows, the kinetic energy and photochemical factors 1, 2, and 3 refer to physicochemical activation processes which are common to all explosives, whether solid, liquid, or gaseous. In solid granular explosives a fourth factor—the process of transition from solid to gas—calls for special consideration, and may in fact control the onset of detonative activation.

1. The passage of a shock wave through various media can cause very considerable rises in temperature, particularly in gases (cf. [*65*]). In detonation, the energy to produce these high temperatures is provided by the chemical energy liberated. Thus in TNT of density 1.6 g/cm^3, detonation leads to a maximum temperature of about 4000°K and a maximum pressure of about 2×10^5 atm. At the front of the detonation wave the undetonated material is subject to the impact of molecules (including active fragments) with translational energy corresponding to the detonation temperature.

2. Superimposed on this random thermal motion there is the hydrodynamic mass motion of the products. The magnitude of this mass flow velocity U can be assessed from Eq. 6-1. U is nearer to D_0 the larger the specific volume of the undetonated material, i.e. for gaseous explosives. For condensed explosives, U is almost independent of density and for TNT at a density of 1.6 g/cm^3 is about 1300 m/sec. For reasons detailed more fully elsewhere [*83*, p. 567] it seems likely that translational energy is of prime importance for bond activation in detonation. The mass flow can greatly reduce the requirements for thermal kinetic energy. In the detonation wave in TNT, the proportion of molecules and molecular fragments with sufficient kinetic energy (due to 1 and 2 above) to promote bond rearrangements on collision must be of the order of unity. A calculation for gaseous detonation, using a different approach [*47*], likewise indicates that, in many known stable detonation waves, chemical activation involves collisions with molecular fragments for which adequate activation energy will always be available.

This statement does not necessarily apply for marginal or "weak" detonations (cf. [*78*]), where the supply of activation energy may be only just adequate and may control detonative activation. More experimental evidence is required to elucidate such limiting cases.

3. The undetonated material is exposed to radiation from the detonation zone. Photochemical activation of successive layers of explosive is a possibility that must be kept in view, though no definite evidence for such a mode of activation appears to have been obtained.

4. In solid granular explosives, the granules must undergo processes of liquefaction and gasification. From the behavior of granules of inert materials mixed with explosives [*16*] it seems likely that, when a detonation wave (cf. Fig. N,9) traverses a granular explosive, the particles require time for attrition and erosion which, in the first instance, are surface processes. The high shearing stresses between the particles, and the mass flow of gaseous detonation products, probably both contribute to the par-

ticle breakup and gasification. It must be clearly realized, however, that in the fragmentary state of present experimental knowledge about reaction times no final statement can be made about the relative importance of the above factors. Two experimental investigations which might help to clear up the problem include:

a. A determination of how the reaction time for a granular explosive at various densities compares with that for the liquid or vapor of the same explosive maintained at corresponding densities by suitable control of the temperature and pressure of the undetonated charge.

b. A systematic evaluation of the reaction time in detonating gaseous systems with a range of molecular complexities, to determine whether τ is much smaller in systems at low densities than for granular explosives; and how it depends on the number of bond rearrangements required for energy release.

CHAPTER 3. LIMITING CONDITIONS FOR STABLE DETONATION

N,12. The Decay and Failure of Detonation. The conditions that are essential for detonation to be set up in an explosive charge and for its propagation with a stable velocity are determined largely by the efficiency of the activation processes, whereby detonative energy release is brought about in the undecomposed explosive at the head of the detonation wave. If this efficiency is too small, then the detonation front moves forward with a decreasing velocity and finally gives way to a sound wave. The phenomenon of fading and the conditions under which it occurs can be studied in condensed explosives in a variety of simple ways (see the methods of Ubbelohde, Tranter, and Cybulski described by Eyring, et al [*26,82*]. A photographic record of a fading detonation is reproduced in Plate N,2a.

For any given solid explosive in which detonation is initiated in a cylindrical charge from a standard impulse, it is found that for every condition of confinement there is a critical radius of charge below which stable detonation cannot be achieved. This critical radius increases (1) as lateral venting, conditioned by peripheral confinement, increases; (2) as the average grain size of the explosive is increased; (3) as the proportion and size of inert or slow-reacting fillers are increased; and (4) as the charge density is increased. All these factors affect p, T, and U and consequently influence the physicochemical processes of energy release.

A tentative but suggestive discussion of the phenomenon of fading and its dependence on the rate of chemical change in the reaction zone has been given by Eyring, et al [*26*]. The basis of this treatment is the assumption that the reaction zone thickness X, and therefore the over-all specific velocity constant for the chemical reaction are functions of the detonation velocity of bare charges (see Eq. 9-2 and 9-3). The final expression for the dependence of X on D is quoted for reference:

$$\frac{X}{X_0} = \frac{D_0}{D} e^{\left(\frac{\Delta H^*}{RT_0}\right)\left(\frac{D_0^2}{D^2}-1\right)} \tag{12-1}$$

X_0 is the reaction zone thickness in an ideal nonventing charge, and T_0 the corresponding detonation temperature. This treatment predicts: (1) There is a critical radius determined by $\Delta H^*/RT_0$ below which stable detonation cannot be established. (2) For any charge radius greater than the critical radius for stable detonation there can be two steady velocities. The high velocity corresponds to the normally observed detonation velocity, and it is tempting to correlate the lower velocity with the "low velocity detonation" observed in the case of liquid explosives. (3) An analysis of the experimental results (see also above) suggests that $\Delta H^*/RT_0$ is close to zero; that is, the physicochemical processes occurring in the reaction zone require only a very small over-all activation energy.

Though this discussion is suggestive, experimental evidence is as yet quite inadequate to establish which of a number of possible rate processes controls the reaction time τ. Presumably owing to the effects on the transmission of activation, a *sudden* change of diameter or of direction of a detonation wave can choke down propagation over some distance, or stop it suddenly [*5*, pp. 186, 517]. (In solids and liquids the effects may differ from those in gases owing to the very different density of the explosive system.) Detonation can also be choked down by incorporating in the charge a gap of some inert material, which may be air.

When fading occurs due to excessive venting, it normally begins at the boundary of a charge and gradually eats inward radially into the detonation wave as it dies out. Even in stable detonation the wave front in a cylindrical charge is curved outward. Herzberg and Walker, and Hurwitz and Strecker (quoted by Eyring, et. al [*26*]), have used rotating-mirror cameras to demonstrate that the wave front in a detonating column of high explosive is often meniscus-shaped. By way of illustration, the bending-back near the edge of the detonation front is of the order of 0.4 mm for TNT of density 1.55 g/cm^3 in a charge 16 mm in diameter, and as much as 3.7 mm for a 55/45 amatol charge 31 mm in diameter. The wave front curvature depends on crystal size, charge density, and peripheral confinement.

Fading effects are of considerable practical importance both in the designing of high explosive charges to detonate efficiently and in designing safe combustion trains so as to choke down the possibility of a build-up of detonation.

N,13. Macroinitiation of Detonation. Graded Detonative Impulses. Detonation decay, as the wave travels down a charge, is due to inadequate activation of the successive layers of explosive. An alternanative procedure for investigating the marginal activation which will just promote detonation is to expose test charges to "graded" detonative impulses. To distinguish this procedure from other forms of sensitiveness it is convenient to term it the macroinitiation of detonation. Published work refers almost wholly to detonative impulses from a standard charge. These are progressively degraded by interposing layers of inert material. Various devices have been used to degrade a detonative impulse. Though these are almost wholly empirical, they can give useful relative information about the marginal pickup of detonation of the test explosives.

Fig. N,13 represents the most widely used procedure, the "double cartridge" or "air-gap" test for pickup. An initiator explosive charge *A* is carefully standardized as to composition, density of packing, diameter, length, and confinement. It is initiated by a standard detonator at one end, and a receiver cartridge *B* of test explosive is placed at various

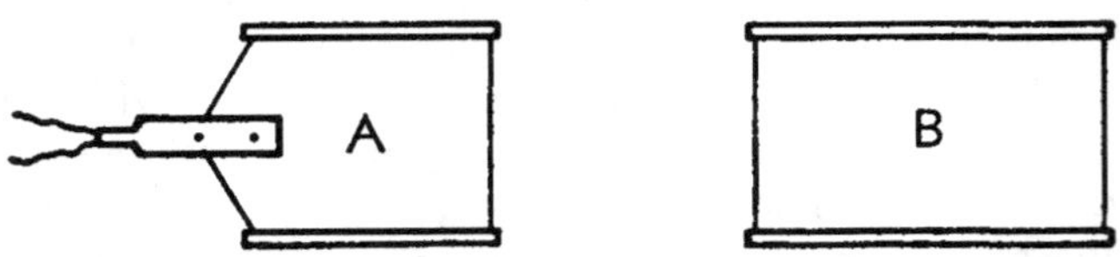

Fig. N,13. Graded detonative impulses gap test. *A* = initiation charge, *B* = acceptor charge. The material of the gap and its length determine the nature of the initiating impulse impressed on the acceptor charge.

distances from the initiator charge, coaxially. As the air gap is progressively increased the percentage of acceptor cartridges which attain detonation passes statistically from 100 per cent "go" to 100 per cent "no." The "gap" which measures ease of initiation of detonation may be defined with respect to this statistical curve, e.g. as the distance for 50 per cent probability of pickup. Various acceptor explosives can be calibrated fairly readily by this means according to their relative ease of pickup. Other facts about the activation of detonation can also be established. For example, increasing the confinement or the diameter of the initiator charge *A* facilitates activation, since it increases the 50 per cent gap for a given acceptor explosive *B*. This is in general agreement with the conclusions in Art. 12 about the effects on fading of diameter and confinement.

With regard to a complete theoretical interpretation of the results, the air-gap test suffers not only from the empirical nature of the calibration, but also because several forms of disturbance travel across the gap from the initiating charge. Photographic methods show the projection of luminescent molecules (flame), of particles (if the detonation products include solids), and of a shock wave; over short distances it is not easy to segregate the effects of these different forms of energy transfer. To disentangle the initiating effect of the shock wave from that of the hot product gases, Laffitte and Patry [*42,43*] and Payman and Woodhead [*62*] have determined the axial velocities of the shock wave and of the luminous products at various distances from their origin. A typical set of results (Table N,13) obtained for a cylindrical charge of tetryl (100 by 13 mm, density 0.9 g/cm^3) shows that initially the two phenomena move together, but that as the distance from the charge increases, the axial velocity of the luminous products decays more rapidly than that of the shock wave. More than one mechanism of initiation of detonation can be effective in the acceptor cartridge according to the distance of separation from the initiator cartridge [*43*]. Photographic methods of distinguishing between the various effects are open to criticism on account of

Table N,13. Axial velocities of luminous products (gas flow) and shock waves projected across an air gap [*42*]. Charge of tetryl [*43*].

Distance, cm	u_1 *velocity of shock wave, m/sec*	u_2 *velocity of gases (flame), m/sec*
1–3	6250	6250
30	4000	4000
50	3500	3500
60	2700	2000

this inadequate separation effect. For the theoretical interpretation of the macroinitiation of detonation a fundamentally sounder procedure would be to use pure shock waves of increasing intensity. Such waves can be produced, for example, by sudden rupture of a diaphragm confining a gas under pressure. Unfortunately the highest intensities of shock wave so far attained are not sufficient to initiate any but the most sensitive condensed explosives. However, by detonating mixtures of H_2 with O_2 in the gas chamber, very strong shocks can readily be produced. These have been used to initiate decomposition of a wide array of solids [*18a*].

A modification of the gap test in which different thicknesses of an inert medium such as water, wax, or paper sheets are used instead of air, has the advantage that the principal effect transmitted is the shock wave; although since the thicknesses of degrading mediums which can be used are much smaller than in the air-gap test, it is doubtful whether there is much increase in the accuracy with which changes of pickup sensitiveness can be detected. A form of graded impulse which is more suitable for easily detonated systems, e.g. for gases, involves the use of gradually increasing amounts of lead azide conveniently pressed to standard density in a pellet [*81*, p. 223] and initiated by a hot wire; the critical quantity can be related, for example, to the amount corresponding to

50 per cent initiation of detonation. In studies with detonable gas mixtures [3] it is found that, as the intensity of an impacting shock wave increases, the response of the detonable mixture accords with expectations. The response is indistinguishable from nonreacting mixtures for weak shock waves. Slow combustion turning to detonation is set up by shock waves whose peak temperature exceeds the flame ignition temperature of the mixture. Immediate detonation is induced by shock waves sufficiently strong to produce detonative activation. Though none of these methods of grading the activation of detonation is fully quantitative, owing to the absence of any accurate means of measuring the high detonation pressures set up, they show that detonation is activated more easily the lower the density of the acceptor charge. This is because the pressure corresponding to stable detonation in the acceptor charge is smaller, the lower the charge density ρ, in accordance with the relationship (6-2) which can be written as ($p_2 \gg p_1$ in detonation)

$$p_2 = D_0 U \rho \tag{13-1}$$

where $\rho = 1/v_1$. An impact pressure comparable with p_2 probably has to be exerted across the gap before stable detonation can be set up. Again, if the acceptor explosive is heavily confined it picks up more easily because the impact pressure set up across the gap cannot be vented as readily laterally.

With gases the relative ease of setting up detonation varies with composition, though systematic work on this topic appears to be scanty (cf. [*39*; *53a*; *83*, p. 566]).

Information obtained from graded impulses is complementary with the information in Art. 12 on critical venting of stable detonation, which will cause fade-out. These two phenomena merely present different aspects of the physicochemical activation of detonation in successive layers of explosive. There is, however, a difference in that flame particles decay across an air gap in a rather different way from the decrease in detonative pressure and temperature, on increasing the venting. Careful comparison of the two procedures does not appear to have been made, but would probably repay further study.

N,14. The Transition from Macrocombustion to Detonation. All systems which can release chemical energy in a detonation wave by rearrangement of atomic linkages can obviously also rearrange these linkages under the more normal conditions of temperature and pressure in a process conveniently, though not very accurately, termed combustion. In both detonation and combustion a chemical reaction with decrease in free energy takes place. Detonation products as collected, for example, in a steel bomb, usually have a somewhat different chemical composition from both the products of ordinary combustion and from the immediate products of detonation, but this is merely due to the freezing in of chemical equilibrium at some stage as the detonation products expand [*30,33*]. The problem to be discussed in this article is: What determines whether a potential explosive will burn quietly or detonate? In all cases where significant information has been obtained, photographic methods have given important information.

Gaseous combustion. Gaseous combustion and its transformation to detonation has been studied extensively [*6,7,25*]. When a combustible mixture is ignited at the open end of a long glass tube which is closed at the other end, *uniform flame* may be propagated down the tube at velocities which may reach the order of 40 m/sec. In this propagation of flame, activation of combustion is transferred from the burning gases to the layer of the mixture ahead of the flame at a rate controlled by factors such as flame temperature, thermal conductivity, and diffusion of radicals and hot molecules.

In other cases the flame may be *nonuniform.* Nonuniform flame may decay in velocity in a vibratory manner until it is extinguished. It may also accelerate to an enhanced velocity, resulting from the build-up of pressure behind the flame front, which is maintained by adiabatic combustion of the explosive medium. This intermediate region of acceleration is apparently unstable since the flame front vibrates, and changes beyond a certain point to detonation with a much higher velocity. Wave speed records have been obtained of the transition from flame combustion to detonation and, in particular, of the role of successive compression (or shock waves) moving in the direction of the flame front (see, e.g. [*7*]).

Combustion of a solid or liquid fuel. In the combustion of a solid or liquid fuel such as a propellant in a closed tube, generally speaking, the rate of combustion increases as a function of pressure (see, e.g. [*28a*]). Typical rates of burning r may be illustrated from data (see below) for colloidal propellants [*56*] for which $r = a + bp$ where a and b are con-

Pressure p of nitrogen applied, kg/cm^2	50	100	400	1000	5000	10000
Burning rate r, mm/sec	13	19.1	66.5	119	490	950

stants. When the tube walls are chosen of sufficient strength, the combustion products build up greater and greater pressures due to inadequate venting at the orifice. Eventually a sudden transition to a markedly higher velocity takes place, and the tube is shattered beyond this point. Wave speed records can be used to show that a uniform velocity of detonation is attained in a fairly short distance from this transition. Corresponding phenomena have also been observed on the semimicroscale in thin films of nitroglycerin [*54*]. When an open unconfined combustible mixture is ignited, the venting of the product gases is usually sufficient to prevent acceleration to detonation, unless the masses concerned are so large that the substance itself confines the combustion process, and permits the build-up of pressure and the setting-up of a shock wave. The conditions discussed above are of great practical importance in devising safety rules in handling potentially explosive systems, which, when the mass is large, may include chemicals not normally classed as dangerous.

Although no complete theory of alternation from combustion to detonation has been proposed, it is possible to interpret some of the physicochemical phenomena described in the preceding articles (cf. also [*83*], p. 566). A system capable of sustaining detonation will always detonate if given a sufficiently intense impulse. The provision of this necessary impulse from the chemical reactions accompanying a burning process is the chief problem in the transition from combustion to detonation (cf. also [*36,72,73,74,75*]). Whereas ordinary combustion or explosion involves the propagation of activation energy through molecular collisions at moderate pressures, detonation has, as its essential characteristic, activation by passage of a shock wave through the medium. The change-over from combustive activation to detonative activation must involve the build-up of a shock wave of sufficient intensity to ensure that chemical decomposition occurs at the rate necessary for detonation. All the evidence suggests that the change-over is abrupt. The shock wave which is built up has two characteristic hydrodynamic features:

1. The pressure and temperature exhibit a discontinuity (or very steep rise) at the wave front. A physicochemical mechanism for the build-up of this pressure must be feasible if combustion is to change over to detonation.
2. The material behind a shock front has a high velocity in the direction of propagation. This mass flow must likewise be built up to permit the transition from combustion to detonative activation.

It has been suggested that a shock wave develops in the combustion region behind a flame front by the combination of streams of gas moving outward from reaction centers, rather like the development of a single wave front by combination of Huygens wavelets. This theory, though it has not been put on a fully quantitative footing, indicates certain controlling factors in the change from combustion to detonation:

1. The possibility of building up an intense shock wave is essential if combustion is to be replaced by detonation. This cannot occur unless the confining walls are sufficiently strong or the mass of explosive is sufficiently large to permit a build-up of pressure. On the other hand, preventing reflection of shock waves from the walls of the confining vessel by roughening or lining with felt, or by modifying the shape of the vessel, should choke down detonation.

2. It should be possible to hinder the build-up required before detonation is fully established by adding negative catalysts to slow down the development of reaction behind the flame front. Negative catalysts appear to have little or no influence on *stable* detonation once it is established, but should have controlling effects on the critical transition phenomena. By way of illustration, experimental evidence has been obtained of the choking down of detonation in lead azide by changing the medium surrounding the crystals [*81*].

N,15. Microinitiation of Detonation. The Sensitiveness of Explosives. In the preceding articles, various aspects of the macroinitiation of detonation are discussed. The important feature common to these examples is the activation of detonative decomposition as a result of the passage through the system of a shock wave of considerable extension. In the practical handling of explosive or combustible systems, explosion and detonation can also develop as a consequence of much smaller initial energy transfers. In view of the need to minimize handling risks, considerable empirical and theoretical research has been carried out on the "sensitiveness" of explosives, i.e. their relative liability to develop explosion and detonation when subjected to a scale of increasing localized disturbances.

A wide variety of mechanical and other means of supplying localized activation energy has been studied. Before devising safety precautions the explosive is subjected in turn to mock setups of all the effects it may encounter during fabrication or in use. But from the basic scientific standpoint, only a comparatively small number of processes of activation have been sufficiently clarified to warrant detailed physicochemical study.

Explosions due to out-of-balance self-heating. All known explosive systems liberate heat when undergoing spontaneous chemical change. If the change takes place sufficiently slowly, the heat evolved is conducted away as fast as it is liberated and only a small rise in temperature of the mass occurs. But when a sufficient mass of the explosive is warmed, out-of-balance self-heating may occur due to the inadequate loss of heat to the surroundings. This leads to progressive rise of temperature and explosion. The process has been investigated in some detail [*81*]. It leads only to explosive combustion and not to detonation in the first instance and is not further discussed here.

Impact tests. Taylor and Weale [*77a*] found that the *kinetic energy* of a falling hammer at the moment of impact on a heap of solid explosive was critical for determining the statistical probability of ignition. In other cases the dynamic variables are more complex [*66*]. The energy of the blow for, say 50 per cent ignitions, is proportional to the mass of the explosive for a large number of impact machines [*26*]. For high explosives, this energy is of the order of 25 cal/g. If this energy were uniformly dispersed throughout the solid as heat it would produce a uniform temperature rise of about 70°C, which would have no effect on the stability of most explosives. Some form of concentration of the stresses imposed on the explosive must occur leading to local instability, which is then spread through the whole mass. Localized high intensity stresses include at least three kinds:

1. Localized "hot spots." Hot spots can be created whose maximum temperature is limited by the melting point of the impacting or sliding surfaces. The creation of such hot spots has been demonstrated visually [*9a,10*], independent of the presence of explosive. When the size, maximum temperature, and persistence of such hot spots are sufficient, then initiation of explosive reaction by purely thermal mechanisms is probable. In principle this mechanism of microinitiation is no different from the spark ignition of gaseous explosive mixtures [*49*]. The importance of hot spots created in this way was demonstrated by the results of Copp and Ubbelohde [*17*] (cf. theory by Robertson and Rideal [*69*]). The sensitiveness of high explosives to impact when confined between a steel anvil and a brass cover is markedly increased by the presence of even one or two hard grit particles. When the explosive is protected from the brass and steel surfaces by thin skins of metals of lower melting points this enhancement of sensitivity by grit particles is eliminated (Table

Table N,15a. The effect of low melting skins on the sensitization of cyclonite by 0.2 per cent of carborundum grit.

Confining surfaces	*Lowest melting point of impacting surfaces, °K*	*Per cent increase in sensitiveness produced by grit*
brass-cyclonite-steel	1700	51
brass-Al-cyclonite-Sn-steel	930	34
brass-Sn-cyclonite-Sn-steel	500	0

N,15a). Such effects can be explained if the carborundum grit produces, on impact, hot-spot temperatures from which explosive reaction develops above 500°K. In these experiments maximum hot-spot temperatures are limited by the melting point of the metal skins.

Similar conclusions emerge from the results of Bowden and Gurton [*8*] on impact and friction sensitivity of PETN in which the conditions are such that the melting point of the grit controls the maximum hot-spot temperature (Table N,15b).

2. "Ultrashearing" of explosives. Although "thermal" hot spots account for the effects of foreign "grit" on impact and friction sensitiveness, the possibility of a transfer of mechanical energy to direct rupture of chemical bonds, and subsequent reaction, has been suggested

especially for solid high explosives starting from various theoretical and experimental considerations [*13,57,76,79*].[1]

Initiating explosives which are salts of comparatively high lattice energy can, in principle, store bond energy at the crystal surfaces on breakup of the crystal. Even for soft explosives of low melting point, certain types of ultrashear in impact tests may involve such high flow rates that the stress can lead to bond rupture. Effects of this kind may explain why, in the absence of all grit, disks of thin tin foil increase the impact sensitiveness of cyclonite in the Rotter test by at least 100 per cent [*17*].

3. Adiabatic compression effects. If the explosive contains cavities of gas (particularly when the substance is liquid or jellified) and these cavities are suddenly compressed, the adiabatic rise in temperature for quite moderate impacts can exceed 1000°C. Consequent self-heating,

Table N,15b. Effect of grit on the friction and impact sensitiveness of PETN.

Grit added	*Hardness of grit, Moh's scale*	*Melting point of grit °C*	*Explosion efficiency*	
			Friction	*Impact*
Nil	—	—	0	2
$K_2Cr_2O_7$	2–3	398	0	0
AgBr	2–3	434	50	6
Glass	7	800	100	100
Calcite	3	1339	100	43

leading to burning and subsequent detonation, is greater when the bubbles contain oxygen rather than nitrogen. Various techniques [*8,9,10,11,20,21, 29*] show that, in the case of high explosives, adiabatic compression leads first to microcombustion which then builds up into detonation.

The problem common to all cases of microinitiation is the way in which explosion *develops* from the region of local disturbance. One mechanism of development is the process of self-heating referred to above. In the case of initiating explosives it seems likely that there is a second process of development which involves more directly the build-up of a detonation wave [*21,78,79,80*]. The process of build-up of detonation in a condensed system has quite close physicochemical analogies with the development of a visible image in a photosensitive grain from small initial disturbances of various kinds. In the sensitiveness of explosives it is a basic physicochemical problem to determine whether the energy in the initial disturbance is first converted locally into heat, which then leads to the growth of thermal decomposition and to detonation, or whether the energy in the initial disturbance can be directly transmitted into activation energy, i.e. much more economically.

[1] This view has been disputed by several authors in private communications.

CHAPTER 4. THE DEVELOPMENT OF RESEARCH IN THE PHYSICAL CHEMISTRY OF DETONATIVE PROCESSES

N,16. Basic Detonation Parameters. From the preceding articles it can be stated quite generally that in spite of the interest and technological importance of detonative processes, much of the physical chemistry is still descriptive rather than metrical. The domains of time, pressure, and temperature involved in detonation do not greatly overlap other fields of inquiry, which has hindered the development of suitable instrumentation.

Difficulties in the measurement of the time intervals seem likely to yield to procedures such as those described in Art. 2. Measurements of pressure are still very indirect. Information about the equations of state of matter at high pressures and temperatures is at present rudimentary [*33*]. Data of this kind are also of interest in geophysical problems. Important extensions of available equations of state for inert molecules may eventually emerge from precision measurements on detonation as affected by inert additions. These can be investigated as a heterogeneous mixture in solid explosives [*16*] and in homogeneous admixture in gaseous or liquid explosives. Some of the theoretical problems connected with such inert additions are still unsolved. A direct method of measuring detonation pressures is highly desirable. Pressure bar techniques [*64*, p. 19] do not permit the use of the high pressures characteristic of condensed explosives. An alternative route might be the measurement of mass flow velocities behind detonation waves. With the development of X-ray flash radiography, precision values of U may eventually become more readily accessible than precision values of p_2 in Eq. 6-1 and 6-2. Initial rates of expansion of various charge casing materials could also yield values of p_2 when sufficiently accurate measurements can be made.

With regard to the measurement of detonation temperatures, this has not been satisfactorily achieved. This problem is to some extent academic. It may not prove possible to define a detonation "temperature" as a unique parameter defining the statistical distribution of energy among the various degrees of freedom in the system, since these distributions may show relaxation times comparable with τ [*64,3*]. "Temperature" probably gives an accurate description of the distribution of translational energy at all stages.

N,17. Studies on the Marginal Propagation of Detonation. From what has been said above, it seems likely that the most promising opportunities for developing the physical chemistry of detonation lie in this field at the present time. Properly designed researches can be of practical importance as well as of basic interest. Extensions of the techniques for evaluating τ to examples of stable detonation of homogeneous explosives should be particularly useful in disentangling the effects of a granular structure of the explosive. The nature of the erosive processes in granular explosives warrants further physicochemical study. A range of inert granular additions with various melting and boiling points could be useful in such investigations. Gas-kinetic processes such as "collisions" also call for more refined theoretical treatment at the high temperatures and pressures in detonation. The possibility of weak geological detonations [*78*; *84*, p. 464] also merits careful study.

N,18. Direct Microinitiation. The possibility of direct bond rupture by ultrashearing stresses is not finally elucidated and may be of fundamental importance for detonative activation. A cognate physicochemical problem on which some current progress is being achieved is in the disintegration of high polymers by ultrasonic waves [*51*,*53*].

N,19. Cited References.

1. Alentsev, M., and Sobolev, N. *Compt. rend. acad. sci. U.R.S.S. 51,* 691 (1946).
2. Becker, R. *Z. Physik 4,* 393 (1921).
3. Berets, D., Greene, E., and Kistiakowsky, G. *J. Am. Chem. Soc. 72,* 1080, 1086 (1950).
4. Berthelot, M., and Vieille, P. *Compt. rend. 94,* 101, 149, 822 (1882).
5. Bone, W., and Townend, D. T. A. *Flame and Combustion in Gases.* Longmans, Green & Co., 1927.
6. Bone, W., and Frazer, R. *Trans. Roy. Soc. London A228,* 197 (1929).
7. Bone, W., Frazer, R., and Wheeler, W. *Trans. Roy. Soc. London A235,* 29 (1935).
8. Bowden, F., and Gurton, O. *Proc. Roy. Soc. London A198,* 337, 350 (1949).
9. Bowden, F., Mulcahy, M., Vines, R., and Fyffe, A. *Proc. Roy. Soc. London A188,* 291 (1947).

9a. Bowden, F., and Singh, K. *Nature 172,* 378 (1953).

10. Bowden, F., Stone, M., and Tudor, G. *Proc. Roy. Soc. London A188,* 329 (1947).
11. Bowden, F., Yoffe, A., and Vines, R. *Proc. Roy. Soc. London A188,* 311 (1947).
12. Breidenbach, H. *Rev. Sci. Instr. 20,* 899 (1949).
13. Carl, R. *J. Franklin Inst. 230,* 75 (1940).
14. Cairns, R. *Ind. Eng. Chem. 36,* 79 (1944).
15. Chapman, D. *Phil. Mag. 47,* 90 (1899).
16. Copp, J., and Ubbelohde, A. *Trans. Faraday Soc. 44,* 646 (1948).
17. Copp, J., and Ubbelohde, A. *Trans. Roy. Soc. London A241,* 248 (1948).
18. Cook, M. *J. Chem. Phys. 15,* 518; *16,* 554 (1947).

18a. Cook, M., and Olson, F. *Inst. for Study of Rate Processes, Univ. Utah, Tech. Rept. 4,* Sept. 1954.

19. Cottrell, T. L. *Trans. Faraday Soc. 47,* 337 (1951).
20. Courtney-Pratt, J. *Research London 2,* 287 (1949).
21. Courtney-Pratt, J. *Proc. Roy. Soc. London A204,* 27 (1950).
22. Cybulski, W., Payman, W., and Woodhead, D. *Proc. Roy. Soc. London A197,* 51 (1949).
23. Dautriche, L. *Compt. rend. 143,* 641 (1906).
24. Dixon, H. *Trans. Roy. Soc. London A184,* 97 (1893).
25. Dixon, H. *Trans. Roy. Soc. London A200,* 315 (1903).
26. Eyring, H., Powell, R., Duff, G., and Parlin, R. *Chem. Revs. 45,* 69 (1949).

27. Friederich, W. *Z. ges. Schiess-u. Sprengstoffw. Nitrocellulose 28*, 2 (1933).
28. Gibson, F. *Rev. Sci. Instr. 25*, 226 (1954).
28a. Gibson, R. *J. Phys. Chem. 54*, 847 (1950).
29. Gray, P. *Trans. Faraday Soc. 46*, 848 (1950).
30. Haid, A., and Schmidt, A. *Z. ges. Schiess-u. Sprengstoffw. Nitrocellulose 34*, 38 (1939).
31. Henriot, E., and Huguenard, E. *J. phys. radium 8 (6)*, 433 (1927).
32. Jones, E. *Proc. Roy. Soc. London A120*, 603 (1928).
32a. Jones, H. *Proc. Roy. Soc. London A189*, 415 (1947).
33. Jones, H., and Miller, A. R. *Proc. Roy. Soc. London A194*, 480 (1948).
34. Jones, H. *Proc. Roy. Soc. London A204*, 9 (1950).
35. Jost, W. *Z. physik. Chem. B42*, 136 (1939).
36. Jost, W. *Explosion and Combustion Processes in Gases*. McGraw-Hill, 1946.
37. Jouguet, E. *Mécanique des Explosifs*. Doin, Paris, 1917.
38. Khariton, Y., and Ratner, S. *Compt. rend. acad. sci. U.R.S.S. 41*, 293 (1943).
38a. Kirkwood, J., and Wood, W. J. *J. Chem. Phys. 22*, 1915, 1921 (1954).
39. Kistiakowsky, G., Knight, H., and Malin, M. *J. Chem. Phys. 20*, 876, 884, 994 (1952).
39a. Kistiakowsky, G., and Richards, W. *J. Am. Chem. Soc. 52*, 4661 (1930).
40. Laffitte, P., and Parisot, A. *Congr. chim. ind., Compt. rend. 18., Nancy*, 930 (1938).
41. Laffitte, P. *Ann. physique 4*, 587 (1925).
42. Laffitte, P., and Patry, M. *Compt. rend. 192*, 744 (1931).
43. Laffitte, P., and Patry, M. *Compt. rend. 192*, 948 (1931).
44. Landau, L., and Stanyukovich, K. *Compt. rend. acad. sci. U.R.S.S. 46*, 362 (1945).
45. Lawrence, R. *Geophysics 9*, 1 (1944).
46. Lehalleur, J. P. *Traite des Poudres, Explosifs et Artifices*. J. B. Baillière et fils, Paris, 1935.
47. Lewis, B. *J. Am. Chem. Soc. 52*, 3122 (1930).
48. Lewis, B. and Friauf, J. *J. Am. Chem. Soc. 52*, 3905 (1930).
49. Lewis, B., and von Elbe, G. *Combustion, Flame and Explosion in Gases*. Academic Press, 1951.
50. Mallard, E., and Le Chatelier, H. *Ann. mines 4 (8)*, 274, 335 (1883).
51. Marinesco, N. *Les propriétés piezo chimiques des ultra-sons*. Hermann, Paris, 1937.
52. Marshall, A. *Explosives*, 2nd ed. J. & A. Churchill, London, 1932.
53. Melville, H. W., and Murray, A. R. *Trans. Faraday Soc. 46*, 996 (1950).
53a. Mooradian, A., and Gordon, W. *J. Chem. Phys. 19*, 1116 (1951).
54. Mulcahy, M., and Vines, R. *Proc. Roy. Soc. London A210*, 191, 226 (1947).
55. Muraour, H., and Michel-Levy. *Mem. artillerie franç. 21*, 263 (1947).
56. Muraour, H., and Fauveau, J. *Compt. rend. 231*, 1043 (1950).
57. Nash, T., Powell, W., and Ubbelohde, A. *Trans. Roy. Soc. London A241*, 272 (1948).
58. Patterson, S. *Research London 1*, 221 (1947–1948).
59. Payman, W., and Walls, N. *J. Chem. Soc. 123*, 420 (1923).
60. Payman, W., and Woodhead, D. *Proc. Roy. Soc. London A132*, 200 (1931).
61. Payman, W., Shepherd, W., and Woodhead, D. *Safety in Mines Research Board London Papers 99*, 1 (1937).
62. Payman, W., and Woodhead, D. *Proc. Roy. Soc. London A163*, 575 (1937).
63. Payman, W., and Shepherd, W. *Proc. Roy. Soc. London A186*, 294 (1946).
64. Penney, W. G., et al *Proc. Roy. Soc. London A204*, 1 (1950).
65. Penney, W. G., and Pike, H. H. M. *Repts. Progr. in Phys. 13*, 46 (1950).
66. Powell, W., Skelly, H., and Ubbelohde, A. *Trans. Roy. Soc. London A241*, 287 (1948).
67. Ratner, S. *Compt. rend. acad. sci. U.R.S.S. 42*, 265 (1944).
68. Ratner, S. *Acta Physicochim. U.R.S.S. 42*, 357 (1947).
69. Robertson, A., and Rideal, E. *Proc. Roy. Soc. London A195*, 135 (1948).

70. Robertson, Sir R., and Garner, W. E. *Proc. Roy. Soc. London A103*, 539 (1923).
71. Schmidt, A. *Z. ges. Schiess-u. Sprengstoffw. Nitrocellulose 27*, 1, 45, 82 (1932).
72. Schmidt, A. *Z. ges. Schiess-u. Sprengstoffw. Nitrocellulose 30*, 364 (1935).
73. Schmidt, A. *Z. ges. Schiess-u. Sprengstoffw. Nitrocellulose 31*, 8, 37, 80, 114 (1936).
74. Schmidt, A. *Z. ges. Schiess-u. Sprengstoffw. Nitrocellulose 33*, 121, 280, 312 (1938).
75. Schmidt, A. *Z. physik. Chem. A189*, 88 (1941).
76. Tamman, G., and Kroger, C. *Z. anorg. Chem. 169*, 1 (1928).
77. Taylor, J. *Detonation in Condensed Explosives.* Clarendon Press, Oxford, 1952.
77a. Taylor, W., and Weale, A. *Proc. Roy. Soc. London A138*, 92 (1932).
77b. Tranter, T. *Nature 174*, 81 (1954).
78. Ubbelohde, A. *Proc. Roy. Soc. London A204*, 25 (1950).
79. Ubbelohde, A. *Research London 3*, 207 (1950).
80. Ubbelohde, A. *Trans. Roy. Soc. London A241*, 198 (1948).
81. Ubbelohde, A., and Woodward, P. *Trans. Roy. Soc. London A241*, 222, 238 (1948).
82. Woodhead, D., and Wilson, R. *Nature 167*, 565 (1951).
83. *Third Symposium on Combustion, Flame and Explosion Phenomena.* Williams & Wilkins, 1949.
84. *Fourth Symposium on Combustion.* Williams & Wilkins, 1953.

PART SIX

ENERGY PRODUCTION BY NUCLEAR REACTIONS

SECTION O

ENERGY PRODUCTION BY NUCLEAR REACTIONS

H. SOODAK

O,1. Introduction. From one point of view our present civilization can be said to depend upon an accident of nature. The comparatively rapid burial of fallen carboniferous trees deep beneath the atmospheric oxygen is the circumstance which led to the formation of the coal and oil deposits by which our civilization is now energized. Except for these deposits the entire crust of the earth is in chemical equilibrium and consequently useless as a chemical energy source. Another noteworthy exception is living matter which not only is an energy source in itself, but was the source of our coal and oil. On second thought, therefore, the real accident might seem to be the existence of life, that specialized kind of matter which can manufacture energy sources from stable compounds. This life process, according to the first law, requires energy. The ultimate source of this energy is the sun which, as we now know, is powered by the nuclear burning of hydrogen into helium.

With the ever-increasing power demands and the constantly decreasing supply of chemical fuels, the importance of nuclear energy is clearly evident. It is the purpose of this article to discuss the various nuclear species regarded as possible energy sources. The important considerations will include, aside from some general background material, the types of exothermic processes, the magnitude and form of the energy release per process, the rate of the process as a function of conditions, and finally the practical realization. It will be seen as a result of this discussion that there is an enormous amount of energy stored up in nuclear matter, an amount which can satisfy a power-hungry civilization for ages. It will also be seen that to tap this energy store generally requires temperatures much too high to be contained at present, except by large massive bodies such as the stars. Here again, however, there occurs an accident of nature which allows the realization of a controlled energy release by means of a neutron-induced fission chain reaction in uranium-235. This chain reaction is seen to produce directly usable power and to allow the production of other chain-reacting materials and possibly useful nonfissionable fuel materials

O,2. Nuclei.

Nuclear forces. The physical universe is composed of matter and radiation. The properties of matter are determined by the three known kinds of forces, gravitational, electromagnetic, and nuclear, among which there is no clearly apparent relationship. Gravitational forces are important only in the cosmic domain and will therefore not be considered in this discussion. In the cases of electromagnetic and nuclear forces, the most highly developed theories are the so-called field theories which resolve the situation into matter and radiation components. A charged particle (matter) is said to be a source of field (radiation), and this field exerts a force on a second charged particle. In this way the difficult idea of action at a distance is avoided.

The electromagnetic forces are extremely well understood. The fundamental charge is the usual electric charge and is quantized in units of the electronic charge. The field is the usual electromagnetic field and, according to quantum mechanics, is quantized into units called photons. The fundamental force law is that of the inverse square. Starting from the orbital picture of the atom whereby the electrons run around the central positively charged nucleus, the electromagnetic theory is sufficient to explain all non-nuclear properties of matter.

Nuclear forces, on the other hand, are not well understood. Although field theories exist, they have not yet clarified the nuclear situation to any great extent. In these theories, the carriers of nuclear charge are the nucleons (the proton and the neutron) and the field is quantized into mesons. The mesons are particles of mass greater than that of the electrons but less than that of the nucleons. A fundamental complication arises from the fact that the proton carries not only nuclear charge, but electric charge as well. Another complicating fact is that several kinds of mesons have been observed. In view of these considerations, it is not surprising that most of our present knowledge of nuclear matter consists of purely experimental facts, and that most of our understanding is based on semi-empirical theories.

The most important property of nuclear forces is that on the whole they are attractive forces. This, of course, is necessary to explain the existence of nuclear matter. Next in importance is the short range nature of nuclear forces. Within the characteristic nuclear dimensions, of the order of 10^{-13} cm, the forces are large, being $\sim$50 times as large as electrostatic forces, and drop rapidly to zero as distances are increased, in sharp contrast to the inverse square law which extends out to large distances. It is precisely this difference in behavior that explains our exact knowledge of electromagnetic forces (which were discovered and measured macroscopically) and our comparative ignorance of nuclear forces as well. Since their short range nature makes direct measurement of nuclear forces impossible, further details concerning their nature can be only inferred from results of collision experiments and of energy measurements. It is therefore appropriate at this point to halt the discussion of nuclear forces as such, and to proceed along the main line to nuclei, their energies and reactions.

The natural nuclei. According to present ideas, the nuclei consist only of protons and neutrons. As a consequence, the nuclei form a two-parameter family and can be represented by points in a two-dimensional plot. The coordinates most often used are Z, the atomic number or number of protons, and A, the mass number or total number of nucleons. In representing a nucleus by its chemical symbol, Z is used as a left subscript and A as a right superscript. Thus a general nucleus is represented

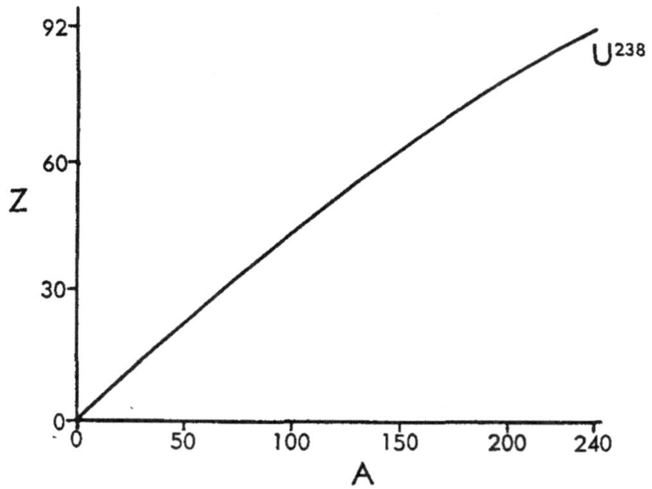

Fig. O,2a. The stability line.

by $_Z(\ \ )^A$ or $_ZX^A$; and some particular nuclei are $_0n^1$ or n for the neutron, $_1H^1$ or p for the proton, $_1H^2$ or d for the deuteron (the deuterium nucleus), $_2He^4$ or α for the helium nucleus, and so on. The subscript giving the value of Z is not necessary of course, since the chemical symbol determines the atomic number, and is often omitted. In this discussion the terms nucleus, isotope, and element are used in their broadest sense and are interchangeable.

The natural nuclei, those found in nature, do not spread out over an area in a Z, A plot. They fall, rather, with a very slight scatter along the so-called stability line (Fig. O,2a). Some aspects of the situation that are of interest are as follows:

1. For small A, up to about 40, the number of protons and the number of neutrons are very nearly equal. That is $Z \cong \frac{1}{2}A$. As A is in-

creased, the relative number of protons drops, falling to $Z \cong 0.4A$ at the high end of the stability line.

2. For each value of A up through 30, there is just one natural nucleus, except for the values 5 and 8 for which there is no natural nucleus.
3. For even values of A, there is in general either one or two nuclei, and in some cases there are three. All but six of the natural nuclei with even A also have even Z. The exceptions are ${}_1H^2$, ${}_3Li^6$, ${}_5B^{10}$, ${}_7N^{14}$, ${}_{19}K^{40}$, and ${}_{71}Lu^{176}$.
4. For odd values of A, there is usually only one nucleus. In several cases there are two. The odd A natural nuclei are evenly distributed between odd Z and even Z.

The relative abundance of the existing nuclei is, according to geochemical and astronomical evidence, generally constant throughout the

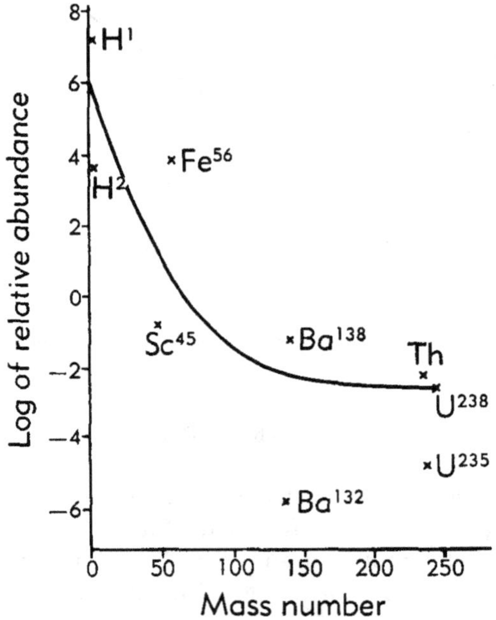

Fig. O,2b. Relative abundance of the elements in the universe (after [1] by permission).

universe. One important exception is the extreme lack on the earth of hydrogen and helium, which are the most abundant elements in the universe as a whole. During the early period of the earth's history these gases, being so light, must have escaped from the earth's gravitational field. Another important exception is the tendency of terrestrial uranium to concentrate in the crust of the earth. A graph showing the relative abundance of the elements in the universe would be a very jagged line with sharp corners at each element [*1*, p. 305]. Fig. O,2b is a plot giving a smooth line as the average tendency of abundance versus mass number. Also shown are points off the line representing some of the most and least abundant of the elements, and also some points of special interest.

A number of the naturally-occurring nuclei are unstable and are consequently radioactive. For example, ${}_{92}U,^{235}$ ${}_{92}U^{238}$, and ${}_{90}Th^{232}$ are unstable and are the parents of the three natural radioactive families comprising most of the nuclei with mass numbers larger than 208. There exist also natural radioactive nuclei that are not members of families, ${}_{19}K^{40}$, ${}_{37}Rb^{87}$, ${}_{62}Sm^{148}$, ${}_{71}Lu^{176}$, and ${}_{75}Re^{187}$. Since these radioactive nuclei are continually decaying, it is reasonable to assume that they have not always been in existence, and that they as well as the stable nuclei were all formed in some original cosmic era. In fact, both radioactivity and astronomical evidence indicate the age of formation of the elements to be about five billion years ago. The radioactive nuclei which still exist are either constantly being formed as is the case with the daughters of the radioactive families, or else have a sufficiently slow rate of decay as is the case with the other nuclei mentioned above. In the original process of the creation of the elements, many nuclei must have been formed which no longer exist. These nuclei are off the stability line, either above, below, or to the right, and have completely decayed into the stable natural nuclei. Such nuclei can and have been artificially produced, and they do, in fact, decay into natural nuclei.

O,3. Nuclear Energies.

Mass-energy equivalence. Questions of stability lead directly to considerations of energy. Nuclear energy differences are measured by determining the energy released in a nuclear reaction or the energy required to initiate the reaction. Energy ratios are measurable by means of the mass spectrograph, in which ions of the atoms are subjected to electric and magnetic fields, and move differently depending on their mass. Such spectroscopic analyses yield directly the values of mass ratios. It is found that the mass of an atom is always less than the sum of the masses of its constituent protons, neutrons, and orbital electrons. This apparent discrepancy is explained by the presence of a binding energy, together with Einstein's equivalence relationship

$$E = mc^2 \tag{3-1}$$

which says that mass and energy are essentially the same quantity, but measured in units differing by the square of the velocity of light, c. Thus a particle of mass m at rest contains an energy (the rest energy) of $E = mc^2$. If the particle is in motion with a kinetic energy E_k, its mass is no longer m but is rather $m + E_k/c^2$. Similarly a compressed spring has a larger mass than the uncompressed spring by an amount equal to the potential energy divided by c^2; and the mass of CO is smaller than the sum of the masses of C and O by the binding energy divided by c^2. In the case of chemistry, the binding energies are extremely small compared to the rest energies, by a factor of about 10^{-9}, and as a result they cannot be obtained from measurements of mass. Nuclear binding energies, however, are sufficiently large, $\sim$1 per cent of the rest energy, to appreciably affect the mass. The measured mass ratios, according to Eq. 3-1, directly yield the values of the energy ratios.

The results of such measurements are usually expressed in a table of relative atomic masses (energies). The unit, called an atomic mass unit, or amu, is such that the mass of the O^{16} atom is exactly 16. The chemical atomic weight unit, on the other hand, is such that the average mass of the naturally-occurring isotopes $O^{16,17,18}$ weighted by their relative abundances is exactly 16. The relative abundance of O^{16} being 99.76 per cent, the two units are practically equal, differing by only 0.03 per cent. Since Avogadro's number N of oxygen atoms have a mass of 16g, and one atom a mass of 16 amu, it follows that

$$1 \text{ amu} = \frac{1}{N} \text{ g}$$

Combining this with the result of Eq. 3-1, that

$$1 \text{ g (energy)} = c^2 \text{ erg}$$

yields

$$1 \text{ amu} = \frac{c^2}{N} \text{ erg}$$

Changing now into the practical energy unit of nuclear physics, the million electron volt, mev, by means of the conversion

$$1 \text{ mev} = 1.601 \times 10^{-6} \text{ erg}$$

and inserting the values $c = 2.998 \times 10^{10}$ cm/sec and $N = 6.025 \times 10^{23}$ gives

$$1 \text{ amu} = 931 \text{ mev} \tag{3-2}$$

Another point worth mentioning is that the mass tables give the values of the atomic mass rather than the nuclear mass. The mass of the Z atomic electrons is included. Since the number of electrons is usually conserved in nuclear reactions, and since the electronic energies are small compared to the nuclear energies, the energy differences computed from such tables are generally equal to nuclear energy differences.

Binding energy. For the purposes of this discussion it is more convenient to express the data by means of graphs and semiempirical formulas, rather than in tabular form. The important quantities are the total atomic mass M, the total binding energy E_b, and the binding energy per particle, E_b/A. The relationships between these quantities are[1]

$$E_b = [Zm(\mathrm{H}) + (A - Z)m(\mathrm{n})] - M \tag{3-3}$$

where M is the mass of an atom containing Z protons and $A - Z$ neutrons in the nucleus, and Z orbital electrons; m(H) is the mass of the hydrogen atom, one proton and one electron; and m(n) is the mass of the neutron. The binding energy is thus the energy equivalent of the amount by which the mass of the aggregate is smaller than the mass of its component parts.

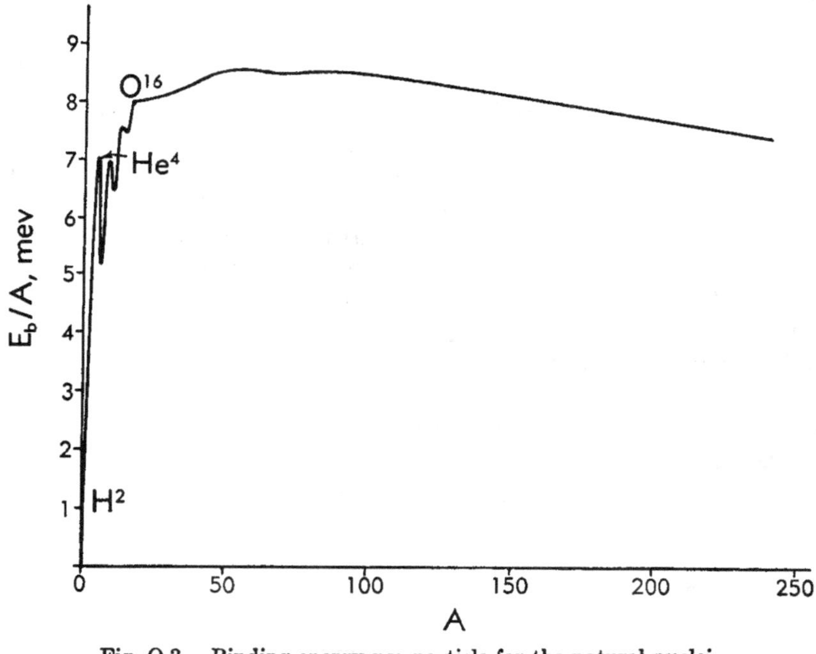

Fig. O,3. Binding energy per particle for the natural nuclei.

It is equal to the energy required to separate the nucleus into its component protons and neutrons, and to the energy released in the formation of the nucleus from free protons and neutrons. The binding energy per particle is simply the average energy by which a nucleon is bound to the nucleus, and is a direct measure of the relative stability of the nucleus. The larger the binding energy per particle, the greater is the stability.

A slightly smoothed-out plot of the measured binding energy per particle for the natural nuclei is presented in Fig. O,3. It is seen that except for some periodicity in the case of the light elements and some over-all smooth variation, the binding energy per particle is quite con-

[1] In Eq. 3-3, as in succeeding equations of a similar nature, mass and energy are treated as quantities having the same units. The factor of c^2 and the conversion relation (3-2) are tactily implied.

stant and is about 8 mev. The curve shows a maximum value of 8.7 mev at $A \cong 60$ and drops to 7.6 mev at the heavy end of the plot.

The over-all constancy of E_b/A, plus the observed fact of quite constant density of nuclear matter, form the basis of the *liquid drop* model. Collision experiments and radioactivity data both indicate that all nuclei have closely the same density, and together with data on the electric charge distribution within the nucleus show that the nuclei deviate from the spherical shape by only a few per cent. The nuclei are thus considered as spherical drops of constant density; and according to the experiments just mentioned, the radius of a nucleus containing A nucleons is given by

$$R = 1.5A^{\frac{1}{3}} \times 10^{-13} \text{ cm} \tag{3-4}$$

which leads to a value of $\sim 2 \times 10^{14}$ g/cm³ for the density of nuclear matter. As regards energy, the nuclei behave differently from actual liquid drops in several important respects. First of all, the nucleus is electrically charged and thus contains an energy of repulsion. This electrostatic effect causes the E_b/A to drop off at the higher values of A, rather than rise slowly to an asymptotic value as is the case with liquid drops. There is thus a most stable size of nuclear matter corresponding to A about 60. Also, nuclei containing too many protons are not sufficiently stable to have withstood the few billion years test of time (as evidenced by the fact of no natural nuclei with $Z > 92$) and have split into the smaller, more stable nuclei. In spite of this electrostatic repulsion of the protons, the natural nuclei still consist of not far from equal numbers of protons and neutrons rather than predominantly of neutrons. This behavior is understandable in terms of a symmetry energy proportional to the square of the number of unpaired nucleons, a pair meaning a neutron and a proton.

The first four terms of the following semiempirical formula for the binding energy are respectively the volume energy, the surface energy, the electrostatic energy, and the symmetry or pairing energy.

$$E_b \text{ (mev)} = 14.6A - 14.6A^{\frac{2}{3}} - 0.60\frac{Z^2}{A^{\frac{1}{3}}} - 77\frac{T^2}{A} + \delta \tag{3-5}$$

where

$$T = \tfrac{1}{2}(A - 2Z) = \tfrac{1}{2} \text{ number of unpaired nucleons}$$

and

$$\begin{aligned} \delta &= 0 \quad \text{for } A \text{ odd} \\ &= 3.4A^{-\frac{3}{4}} \quad \text{for } A \text{ even, } Z \text{ even} \\ &= -3.4A^{-\frac{3}{4}} \quad \text{for } A \text{ even, } Z \text{ odd} \end{aligned}$$

The last term is introduced to account for the difference in behavior between odd A and even A nuclei, as previously mentioned. The coefficients have been arranged so as to give the minimum mass to the nuclei on the stability line, and so that these nuclei give a good fit with their measured binding energies. Eq. 3-3 may be written as

$$M = \frac{m(\mathrm{n}) + m(\mathrm{H})}{2} A + [m(\mathrm{n}) - m(\mathrm{H})]T - E_b \tag{3-6}$$

which shows the dependence of the atomic mass on the n, H mass difference.[2] The quantity T is, as before, the number of unpaired nucleons. Minimizing the mass with respect to Z (for constant A) by setting $dM/dZ = 0$, using Eq. 3-5 for E_b without the δ term, and using the following values

$$\begin{aligned} m(\mathrm{n}) &= 1.008945 \text{ amu} \\ m(\mathrm{H}) &= 1.008131 \text{ amu} \\ m(\mathrm{n}) - m(\mathrm{H}) &= 0.000814 \text{ amu} = 0.76 \text{ mev} \end{aligned} \tag{3-7}$$

gives then for the equation of the stability line

$$\frac{Z}{A} = \frac{0.505}{1 + 0.0078A^{\frac{2}{3}}} \tag{3-8}$$

Eq. 3-5 gives a reasonably good fit to the over-all behavior of the actual binding energies of nuclei, but does not account for the structural details which give rise to the periodicity of four in the case of the light elements, and to the larger evidenced periodicity of the so-called magic numbers. The reasons for introducing the semiempirical formula in addition to the data of Fig. O,3 are, first, that it can be used for nuclei off the stability line as well as for those on, and second, that it can be used for general analytic considerations.[3]

Excited states. As in the atom and molecule, the nucleus is a quantized system and can exist in states of various energy levels. The discussion up to this point has been concerned only with the ground or normal states of the nuclei. The presence of the higher excited states causes resonance effects to take place in nuclear collisions. Knowledge of the excited states can therefore be obtained by observing such resonances, as well as by studying the radiations emitted by an excited nucleus as it makes transitions down to its ground state. The data on the subject indicate that the spacing of the excited energy levels depends on the structural details of the nuclei, but nevertheless allow some general conclusions to be drawn. For excitation energies up to a few mev, the level-spacing averages in the neighborhood of 1 mev for the light nuclei, and decreases down to about $\frac{1}{2}$ mev for the heavy nuclei. As the excitation energy is increased, the level-spacing drops, the drop being much sharper

[2] The fact that $m(\mathrm{H})$ is less than $m(\mathrm{n})$ favors the presence of protons rather than neutrons. In light nuclei this effect is important. Thus ${}_2\mathrm{He}^3$ is more stable than ${}_1\mathrm{H}^3$ in spite of the electrostatic repulsion between the two protons in the ${}_2\mathrm{He}^3$ nucleus.

[3] Although the coefficients of Eq. 3-4 and 3-5 have recently been amended, the older values used above are entirely adequate for the purposes of this chapter.

in the case of the heavy nuclei. At an excitation energy of about 10 mev, for example, the average level-spacing is ~0.4 mev for A near 10, ~0.03 mev for A near 30, and drops down to the order of only 30 ev for A near 100, holding at this low value for the heavier elements.

Energy comparison. Before going on to the subject of nuclear reactions, it is interesting first to remark on the different orders of magnitude of the energies involved in physical, chemical, and nuclear rearrangements of matter. Due to the short range nature of nuclear forces and the electrostatic repulsion of the protons, the stable nuclei exist as essentially spherical drops of the order of 10^{-12} cm in radius. These positively charged drops attract electrons to form neutral atoms which, although not sharply defined in extent, can be described as fuzzy balls with radii of about 10^{-8} cm. The atoms combine into molecules by means of ionic and covalent forces, and the molecules in turn combine into the orderly crystalline and the less orderly liquid and amorphous structures. Table O,3 presents an order-of-magnitude comparison of the energies involved in these three types of processes by giving values in some special cases.

Table O,3. Comparison of energies.

Type of process	*Effect of process*	*Energy per process*	*Calories per mole*	*Case*
Nuclear	Rearrangement of nucleons to form different nuclei	2 mev	$46{,}000 \times 10^6$	Fusion of a neutron and a proton to form a deuteron
		200 mev	$46{,}000 \times 10^8$	Fission of a heavy nucleus
Chemical	Rearrangement of atoms to form different molecules	1 ev	26,000	$C + O \rightleftarrows CO$
Physical	Rearrangement of molecules to form different structures	0.4 ev	10,000	Vaporization of H_2O at ~100°C

Some understanding of the energy values can be gained from the following simple but rough analysis. It would be expected that the total energies should not be far from the potential energies. For the chemical process, it is reasonable to approximate the potential energy as that of two electrons separated by 10^{-8} cm. This energy, which should be an overestimate, is 15 ev. For the physical process, the energy should be smaller by a factor of perhaps 5 since the molecules interact over somewhat larger distances, and since the forces involved are somewhat smaller, being of the polarization kind. In the nuclear case, the energy is approximated as the potential energy of two nucleons separated by a distance equal to the range of the nuclear forces which is $\sim 2.8 \times 10^{-13}$ cm. Since this is smaller by a factor of 4×10^4 than the distances involved in a chemical process, and since the nuclear forces are, within their range, about 50 times as great as electric forces, the nuclear energy should be larger than the chemical energy by a factor of $50 \times 4 \times 10^4$ or 2×10^6.

O,4. Nuclear Reactions.

CONSERVATION LAWS. It is a general principle of nature that any process can occur as long as the required conservation laws are satisfied. The important laws for this discussion[4] are the conservation of energy and of electric charge. The rate at which it goes depends, of course, on the details of the process. As in chemistry, a nuclear reaction may be expressed by means of an equation, a skeleton form of which is

$$_{Z_1}(\ )^{A_1} + {_{Z_2}}(\ )^{A_2} + \cdots + x_1 + x_2 + \cdots \rightarrow {_{Z'_1}}(\ )^{A'_1} + {^*_{Z'_2}}(\ )^{A'_2} + \cdots + x'_1 + x'_2 + \cdots + E \quad (4\text{-}1)$$

In Eq. 4-1, each symbol of the form $_Z(\ )^A$ represents a nucleus of proton number Z and mass number A in its ground state. An excited state is symbolized by an upper left asterisk. Non-nuclear entities which take part in the process are represented above as the x's. The important ones are the electron e^- or β^-, the positron or positive electron e^+ or β^+, the photon or gamma ray γ, and the neutrino ν, which will be described subsequently. The letter E stands for the energy release and is positive for an exothermic, and negative for an endothermic reaction. The arrow points only toward the right, in contrast with the chemistry case, because nuclear reactions are not usually described as going to equilibrium. They are practically always one-way reactions in the exothermic direction, although the endothermic reactions can be realized by the use of accelerating machines. The reason for this is that the energies involved, of the order of mev, correspond to temperatures of the order of 10^{10} °K. The exothermic reactions proceeding in the stars do release energies with such high potentials, but the dissipation is sufficiently rapid to lower the temperatures of the stars to the order of 10^7 °K. Thus, except for those reactions which have abnormally low E, the only period in the history of the universe, in which reactions could have gone both ways, is the era of the original creation of the elements.

The law of conservation of charge imposes on Eq. 4-1 the strict condition

$$Z_1 + Z_2 + \cdots + \text{charge on } x\text{'s} = Z'_1 + Z'_2 + \cdots + \text{charge on } x'\text{'s} \quad (4\text{-}2)$$

[4] Also conserved are linear momentum, angular momentum, and some symmetry characteristics.

The balance equation that corresponds to the chemical conservation of atoms is the conservation of the number of nucleons,

$$A_1 + A_2 + \cdots = A_1' + A_2' + \cdots \tag{4-3}$$

This balance relation is not a general law of nature but rather a restriction on the types of process considered. Neutrons and protons can, in theory, be created or destroyed in pairs, but the energy required to initiate such processes is measured in billions of electron volts. The restriction therefore does not appear to have any practical significance for this discussion.

The meaning of the energy release E is such that the kinetic energy, including the photon and neutrino energies if present, of the product entities E_k' is greater by E than the kinetic energy E_k of the reacting entities. That is,

$$E_k' = E_k + E \tag{4-4}$$

This extra energy must, according to the mass-energy equivalence (3-1), arise from an equivalent decrease of mass. Thus

$$M_0 = M_0' + E \tag{4-5}$$

where M_0 and M_0' are respectively the sum of the masses of the reactants (nuclei and others) and of the products. For all reactions except those in which positrons take part, it will be seen that Eq. 4-5 can be rewritten as

$$E = M_1 + M_2 + \cdots - [M_1' + M_2' + \cdots] \tag{4-6}$$

where the masses appearing are the atomic rather than nuclear masses, and the masses of the non-nuclear entities are not counted. By combining this result with Eq. 3-3 which defines the binding energy, using the conservation relation (4-3), and inserting the numbers of Eq. 3-7, the energy release can be written as

$$E = \Delta E_b + 0.76 \text{ mev } \Delta Z$$

where

$$\Delta E_b = [(E_b)_1' + (E_b)_2' + \cdots] - [(E_b)_1 + (E_b)_2 + \cdots] \tag{4-7}$$

and

$$\Delta Z = [Z_1' + Z_2' + \cdots] - [Z_1 + Z_2 + \cdots]$$

As examples of the use of Eq. 4-7, consider the two processes mentioned in Table O,3. In the first, the energy release is in the form of a gamma ray, and the reaction is

$${}_0n^1 + {}_1H^1 \rightarrow {}_1H^2 + \gamma + E$$

The binding energy is zero for the neutron and the proton since they are single nucleons. According to Fig. O,3 the binding energy of the deuteron is ~1 mev per particle, or ~2 mev. Using these values in Eq. 4-7 gives the value 2 mev for E. Thus if the reactants, the neutron and proton, have no kinetic energy, the gamma ray has an energy of $\sim$2 mev.[5] For the second case consider the fission of ${}_{92}U^{238}$ into two fragments of equal size. This process takes place in two distinct stages. The first stage is the fission into ${}_{46}Pd^{119}$ plus ${}_{46}Pd^{119}$ if it is assumed that the charge also divides in half. These fragment nuclei, however, are below the stability line and by a process to be described subsequently (beta decay), transform into the stable nucleus of mass number 119 which is ${}_{50}Sn^{119}$. The total process is then

$$ {}_{92}U^{238} \rightarrow 2\,{}_{50}Sn^{119} + x\text{'s} + E $$

Inserting the values of the E_b/A as given by Fig. O,3 into Eq. 4-7 results in

$$ E \cong 2(8.4)(119) - (7.6)(238) + (0.76)(8) \cong 200 \text{ mev} $$

Most of this $\sim$200-mev energy release, as will be seen, is in the form of kinetic energy of the fragments. The remainder is carried by the x's.

SPONTANEOUS REACTIONS. Reactions can be grouped into two general classes, the spontaneous type and the collision type. A spontaneous reaction involves only one reactant entity and is by its nature exothermic. A collision process is one involving more than one reactant and may be either exothermic or endothermic. It will be seen that for reasons of rate, the only important collision processes are those involving two reactants.

For a spontaneous reaction the rate is described by the decay rate λ, or the mean life τ, or the half life $\tau_{\frac{1}{2}}$, which are related by

$$ \begin{aligned} \lambda\tau &= 1 \\ \tau_{\frac{1}{2}} &= \tau \ln 2 = 0.693\tau \end{aligned} \tag{4-8} $$

Observation of the rate at which the products are produced shows in all cases that the process is statistical in nature, and that the average rate of loss of reactant nuclei is proportional to the amount present. That is

$$ \frac{dn}{dt} = -\lambda n \tag{4-9} $$

where n is the amount (number of atoms or moles, etc.) of reactant nuclei and λ is the decay rate. The solution of Eq. 4-9 is

$$ \frac{n}{n_0} = e^{-\lambda t} = e^{-t/\tau} = 2^{-t/\tau_{\frac{1}{2}}} \tag{4-10} $$

where n is the amount present at time t and n_0 the amount at time zero. The process goes exponentially with time, the amount of reactant material dropping by a factor of e every mean life, and by a factor of 2 every half life. If the energy release of the reaction is E (mev per nucleus or calories

[5] The fact that almost all the energy is carried off by the gamma ray, and very little by the deuteron, is a consequence of the law of conservation of linear momentum

per mole, etc.), then the rate of energy production $P = -E(dn/dt)$ is

$$P = n_0 \frac{E}{\tau} e^{-t/\tau} \tag{4-11}$$

Thus for a process with $E = 10^{10}$ cal/mole and $\tau = 1$ day $= 24$ hr

$$\frac{P}{n_0} = 0.5e^{-t/24\,\text{hr}} \text{ megawatts per initial mole}$$

It should be emphasized that the energy released in such a process can in no way be controlled except by varying the amount of reactant material.

Strictly speaking, the mean life can vary anywhere from infinity down to $\tau_0 \cong 10^{-21}$ to 10^{-22} sec which is the characteristic nuclear time.[6] If $\tau \ll 10^{-10}$ sec, the process is too rapid for the exponential behavior to be observable by present electronic techniques, and the existence of the reactant nucleus can only be inferred. If $\tau \gg 10^{16}$ yr, the products are formed too rarely to be counted, and the reactant nucleus can be described as stable.

The statistical and exponential behavior of a spontaneous process finds a natural explanation in terms of quantum theory. According to quantum theory it may be said that an unstable reactant nucleus does not decay continuously into its products but retains its identity until a definite and unpredictable instant is reached, at which time it transforms rapidly into the products. What can be predicted is the probability dp that the decay takes place during the time interval dt. Since the reactant nucleus does not change until the decay occurs, this probability cannot be a function of time but must simply be proportional to dt. The constant of proportionality is the decay rate. Thus

$$dp = \lambda dt \tag{4-12}$$

Since the total number of processes, $-dn$, in the time dt is equal on the average to the probability of decay per nucleus, dp, multiplied by the number of nuclei, n, Eq. 4-12 leads directly to Eq. 4-9.

Spontaneous reactions fall into three main groups which are:

1. Gamma decay, or emission of a photon by a nucleus in an excited state on its way down to the ground state. In this process the nuclear species is not changed.
2. Beta decay, or emission of an electron and a neutrino by a nucleus off the stability line. In this process the mass number A is unchanged, but Z is either increased or decreased by unity depending on whether the emitted electron is e^- or e^+. The product nucleus is always closer to the stability line than the reactant nucleus. The

[6] $\tau_0 \cong R/v$ where R is the nuclear radius and v is the velocity of a nucleon having a kinetic energy of several mev.

reaction is effectively a transformation between a neutron and a proton within the nucleus.

3. Fragmentation, or the splitting of the nucleus into fragments. The main types of splitting are alpha decay and fission. In the alpha decay process, there are only two fragments and one of them is an alpha particle (helium nucleus). The term fission is usually reserved for the case in which the nucleus splits into two large fragments of comparatively equal size, and perhaps some smaller fragments as well, such as neutrons.

These three types of reactions are now considered in turn, the emphasis being placed on the general behavior of energies and rates.

Gamma decay. The process of gamma decay can be represented by

$$\overset{*}{_Z}(\quad)^A \rightarrow {_Z}(\quad)^A + \gamma + E \tag{4-13}$$

where the final state of the nucleus may or may not be an excited state. The energy release is carried off by the gamma ray and is

$$E = U - U' \tag{4-14}$$

where U and U' are respectively the excitation energies of the initial and final states. A closely related process to that of Eq. 4-13 is the process of internal conversion, whereby the energy is released as kinetic energy of an orbital electron rather than in the form of a gamma ray. Although not a true representation, this process can be thought of as a gamma decay followed by a photoelectric effect on an orbital electron in the same atom. The processes of gamma decay and internal conversion are in direct competition for the energy release. For large E, the gamma is dominant, while the internal conversion wins out for small E. The dividing energy depends on the atomic number of the decaying nucleus, increasing with Z, and depends also on considerations of angular momentum. The situation is represented in Fig. O,4a [*2*, p. 281].

The radiation, as indicated in the figure, falls fairly cleanly into discrete groups depending on considerations of angular momentum. For each group there is a simple functional relationship between the decay rate for gamma decay λ_γ and the energy release E. There is also for each group a relationship for the decay rate of internal conversion, λ_e, as a function of E and Z. The total decay rate λ of the reactant nucleus is given by the sum

$$\lambda = \lambda_\gamma + \lambda_e$$

and the half life is therefore

$$\tau_{\frac{1}{2}} = \frac{0.693}{\lambda_\gamma + \lambda_e}$$

The straight lines in Fig. O,4a give the dependence of the gamma decay

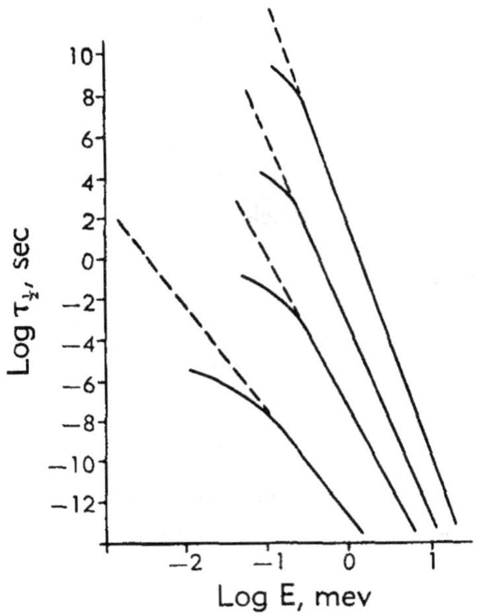

Fig. O,4a. Half life dependence on energy for gamma decay (after [2]).

rates on the energy, being plots of $0.693/\lambda_\gamma$ versus E. The solid lines are plots of $\tau_{\frac{1}{2}}$ against E for the special case of $Z = 35$. For higher (lower) values of Z, the curved portions would branch off the straight lines at larger (smaller) values of E.

Of possible importance as fuel materials are the excited states with mean or half lives measured in hours, days, or longer.[7] Such nuclei, if

Table O,4a. Gamma decay processes.

Nucleus	Energy release, mev	Half life
$^{*}_{30}Zn^{69}$	0.4	14 hr
$^{*}_{32}Ge^{72}$	0.7	3×10^{-7} sec
$^{*}_{48}Cd^{115}$	0.5	2.3 day
$^{*}_{50}Sn^{119}$	0.2	13 day
$^{*}_{52}Te^{129}$	0.1	32 day

producible in quantity, can be transferred from the production site and serve as fuel material for some reasonable period of time. Questions of producibility are discussed briefly in the next article. Table O,4a presents some typical cases of gamma decay and/or internal conversion, paying attention to the long-lived nuclei.

[7] Excited nuclei with half or mean lives sufficiently long to be observable ($\tau > 10^{-10}$ sec) are said to be isomeric to their ground states and are given the name "isomer."

Beta decay. The beta decay process can be expressed as

$$_{Z}(\quad)^{A} \rightarrow {}_{Z+1}(\quad)^{A} + e^{-} + \nu + E \tag{4-15}$$

for an electron process, and

$$_{Z}(\quad)^{A} \rightarrow {}_{Z-1}(\quad)^{A} + e^{+} + \nu + E \tag{4-16}$$

for a positron process. The symbol ν stands for the unobservable neutrino, an entity conceived by Pauli for reasons given in the following paragraph. The reactant nucleus may or may not be in an excited state,[8] while the product nucleus is often formed in an excited state. If the reactant nucleus is far from the stability line, the product nucleus will be off the stability line, and further beta decays will follow until the final product is stable. The energy release E is given by Eq. 4-5 in which the masses are those of the bare nuclei, rather than the masses of the neutral atoms used in Eq. 4-6. To get from Eq. 4-5 to Eq. 4-6 in the case of the e^{-} process, it is only necessary to add to each side of Eq. 4-5 the mass of Z electrons. The same procedure applied to the e^{+} process of Eq. 4-16 results in two extra electrons on the right side of the equation. The energy release for the e^{+} process is therefore less than that given by Eq. 4-6 and 4-7 by two electron masses or $\sim$1 mev.

As already indicated, the neutrino has never been observed. What is observable is the production of the product nucleus and of the electron. The product nucleus is formed with a negligible kinetic energy, while the electron is formed with a kinetic energy which varies from process to process (for the same reaction) between the values zero and E, the energy release. An averaging of the kinetic energy from process to process gives about $\frac{1}{2}E$ as the result, and calorimetric measurements show that the energy absorbed is just this amount. In order to account for the missing half of the energy release (and also some missing angular momentum), Pauli invented the neutrino. The neutrino is thus supposed to carry off the missing energy and is supposed to be practically unabsorbable by matter. Experiments designed to detect the neutrino indicate that it can penetrate millions of miles of solid matter without losing its energy.[9] To explain all this it is necessary to consider the neutrino as an entity having no electric charge (this is also required by the conservation of charge) and practically no mass. If ever a method is discovered whereby neutrinos can be absorbed, a useful supply of energy would be opened up since a significant fraction of the sun's energy is carried by neutrinos.

A process closely related to and in direct competition with the positron decay is that called K capture. This process proceeds by the nucleus cap-

[8] When the reactant nucleus is in a long-lived isomeric state, the process of beta decay can compete successfully with that of γ decay.

[9] Recently, nuclear reactions have been detected that were induced by neutrinos coming from a reactor.

turing a K-shell orbital electron and simultaneously emitting a neutrino and can be expressed by

$$_{Z}(\ \ )^{A} \rightarrow {}_{Z-1}(\ \ )^{A} - e^{-} + \nu + E \qquad (4\text{-}17)$$

The transformation $_{Z}(\ \ )^{A} \rightarrow {}_{Z-1}(\ \ )^{A}$ can occur either by K capture or by e^{+}. According to the argument given previously, the energy release is larger for K capture than for e^{+} by $\cong$1 mev. Thus if the mass of $_{Z-1}(\ \ )^{A}$ is less than the mass of $_{Z}(\ \ )^{A}$ by an amount smaller than 1 mev, the only allowed process is that of K capture.

The energy release in beta decay reactions can be estimated by inserting the semiempirical formula (3-5) for the binding energy into the beta decay form of Eq. 4-7. The resulting order of magnitude is $E \sim 1$ mev. There are cases where, because of particulars, the energy is many times smaller than this, but no cases in which it is much larger. Although as in the case of gamma decay, the beta processes can with some reason be grouped into classes, the overlap between classes is sufficiently large to obscure the situation, and no such grouping will be given. In general, the decay rate tends as the $\frac{7}{2}$ power of the energy release. Table O,4b presents data on some typical and some especially interesting cases, attention again being paid to the long-lived nuclei.

Table O,4b. **Beta decay processes.**

Nucleus	Activity	Energy release, mev	Half life
$_{0}n^{1}$	β^{-}	0.76	$\sim\frac{1}{2}$ hr
$_{1}H^{3}$	β^{-}	0.015	11 yr
$_{4}Be^{7}$	K	0.86	53 day
$_{6}C^{14}$	β^{-}	0.16	5.7×10^{3} yr
$_{11}Na^{24}$	β^{-}	1.4	14 hr
$_{15}P^{32}$	β^{-}	1.7	14 day
$_{18}A^{35}$	β^{+}	4.4	1.8 sec
$_{27}Co^{61}$	β^{-}	1.3	1.8 hr
$_{37}Rb^{87}$	β^{-}	0.13	6.3×10^{10} yr
$_{91}Pa^{233}$	β^{-}	0.4	27 day
$_{93}Np^{239}$	β^{-}	0.7	2.3 day

Fragmentation and the potential barrier. Spontaneous fragmentation of a natural nucleus in the ground state is possible if the total mass of the fragments is less than the mass of the original nucleus, or in other words if the energy release is positive. Since this criterion is, according to Eq. 4-7, generally equivalent to an increase in binding energy, a look at Fig. O,3 shows that spontaneous splitting is possible for the heavier elements, the fragments being closer to the maximum of the curve than the reactant nucleus. If it is assumed that the nucleus splits into two fragments of the same composition (ratio of Z to A), the reaction can be written as

$$_{Z}(\ )^{A} \rightarrow {}_{xZ}(\ )^{xA} + {}_{(1-x)Z}(\ )^{(1-x)A} + E \tag{4-18}$$

where x can vary between 0 and 0.5. The product nuclei in general lie below the stability line and will subsequently rise to stability by a series of negative beta decays. The total energy release E_{tot} is given by

$$E_{\text{tot}} = E + E_{\beta} \tag{4-19}$$

where E_{β} is the energy involved in the beta decays, and E is the energy release of the fragmentation process (4-18) and is carried by the fragments as kinetic energy.

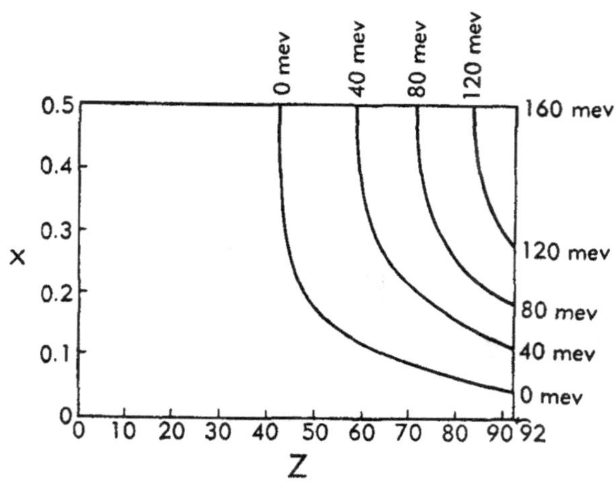

Fig. O,4b. Fission energy contour plot (after [1] by permission).

Combining Eq. 3-5, 4-7, and 4-18, it is seen that

$$E = 14.6A^{2/3}[1 - x^{2/3} - (1 - x)^{2/3}] + 0.60\,\frac{Z^2}{A^{1/3}}\,[1 - x^{5/3} - (1 - x)^{5/3}] \tag{4-20}$$

Maximizing this expression with respect to x shows that the energy release is a maximum for splitting into equal fragments and that this maximum is given by

$$E_{\max} = E_{1/2} = 0.22\,\frac{Z^2}{A^{1/3}} - 3.8A^{2/3} \tag{4-21}$$

Results of computations of E as a function of x for the natural nuclei are presented in Fig. O,4b which is a plot of isoenergetic contour lines. It is seen that all natural nuclei with Z larger than about 40 are in principle unstable. The fact of their existence must therefore prove that the half life for the fragmentation process must be at least $\sim 10^9$ yr in order of magnitude. Some of these nuclei have half lives that are easily observable.

A listing of these is given in Table O,4c along with some interesting cases of non-natural or artificial nuclei. The following analysis of the fragmentation process explains the slow rates involved, and also points out the reasons why the most probable modes of splitting are those of alpha decay, and of fission.

Table O,4c. Fragmentation processes.

Nucleus	Mode of decay	Energy release, mev	Half life
$_{62}Sm^{148}$	α	2.0	1.6×10^{11} yr
$_{88}Ra^{226}$	α	4.8	1.6×10^{2} yr
$_{90}Th^{232}$	α	4.2	1.3×10^{10} yr
$_{92}U^{233}$	α	4 8	1.6×10^{5} yr
$_{92}U^{235}$	α	4.5	8.8×10^{8} yr
$_{92}U^{238}$	α	4 2	4.5×10^{9} yr
$_{92}U^{238}$	fission	200	3×10^{15} yr
$_{94}Pu^{239}$	α	5.2	2.4×10^{4} yr

If, during a process, a system must pass through configurations of energy greater than the amount available, there is said to be an activation energy E_a, defined simply as the maximum energy deficit. According to classical mechanics, such a process cannot proceed unless this energy

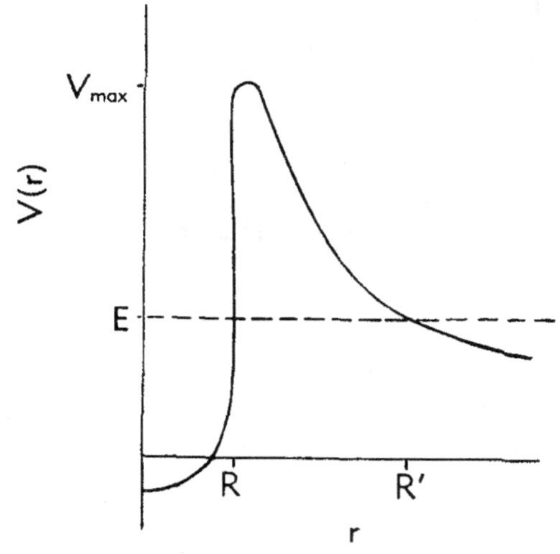

Fig. O,4c. The potential barrier.

deficit is externally supplied. Quantum mechanics, however, allows the system to pass through the so-called potential barrier in a manner similar to the case of electromagnetic radiation penetrating through a sufficiently thin sheet of opaque material. A picture of the potential barrier for the fragmentation process of Eq. 4-18 is given in Fig. O,4c. The solid line is a plot of the potential energy V as a function of the separation r between the fragments.[10] For large distances, V is simply the Coulomb potential of electrostatic repulsion,

$$V = x(1 - x)\,\frac{Z^2e^2}{r} \tag{4-22}$$

where e is the electronic charge. As the separation approaches the nuclear radius R the nuclear forces take over and cause the pictured potential well. The horizontal dotted line refers to the available energy, or energy release, E. Within the triangular barrier, extending from R to R', exist the energy deficit configurations. The activation energy is simply

$$E_a = V_{max} - E \tag{4-23}$$

where V_{max} is the maximum value of the potential energy curve, and according to the above remarks, should be somewhat less than

$$V_{max} \cong x(1 - x)\,\frac{Z^2e^2}{R} \tag{4-24}$$

In Eq. 4-24 the distance R may be considered as the sum of the radii of the separate fragments. For alpha decay ($xZ = 2$), V_{max} varies from $\sim$12 mev for $Z = 40$ to $\sim$22 mev for $Z = 92$. For fission, V_{max} is in the neighborhood of 180 mev for nuclei near uranium.

Quantum mechanics predicts that the probability p_p, for an incident particle of energy E to penetrate the potential barrier, is given by

$$p_p \cong e^{-\frac{2}{\hbar}(2m)^{\frac{1}{2}}\overline{(V-E)^{\frac{1}{2}}}(R'-R)} \tag{4-25}$$

where m is the mass of the incident particle, $\overline{(V - E)^{\frac{1}{2}}}$ is the average square root energy deficit, and $\hbar$ is Planck's constant divided by 2π. Since the small fragment inside the nucleus can be imagined to hit against the barrier $1/\tau_0 \sim 10^{-21}$ to 10^{-22} times per second, the probability per second (which equals λ, Eq. 4-12) for fragmentation[11] can be approximated by

$$\lambda \cong \frac{1}{\tau_0}\, e^{-\frac{2}{\hbar}(2xM)^{\frac{1}{2}}\overline{(V-E)^{\frac{1}{2}}}(R'-R)} \tag{4-26}$$

where M is the mass of the reactant nucleus. It is seen that λ is large if the product

$$x^{\frac{1}{2}}[\overline{(V - E)^{\frac{1}{2}}}(R' - R)]$$

is small. The first factor favors small fragmentation, since it is easier for

[10] This picture is reasonable if one of the fragments is very small as in alpha decay. If both fragments are large, as in fission, the situation is more complicated, involving the exact deformations of the shape of the reactant nucleus during the process of splitting.

[11] See footnote 10.

a light particle to penetrate the barrier. Small fragmentation occurs via alpha decay because He^4 is the lightest nucleus for which spontaneous fragmentation is possible (positive energy release). The second factor in the product favors large fission since the barrier region shrinks as x is increased toward 0.5. For nuclei near uranium, for example, the activation energy drops from ~29 mev for alpha decay, down to ~6 mev for fission. A glance at Fig. O,4c shows that the width $R' - R$ of the barrier narrows rapidly as the activation energy is decreased. When applied to alpha decay, Eq. 4-26 predicts the observed rates to within a few orders of magnitude, and explains the extremely wide range of rates, the nonexistence of nuclei heavier than uranium, and the apparent stability of the heavier natural nuclei. For example when applied to the alpha decay of $_{80}Hg^{200}$, the result is $\tau = 1/\lambda \cong 10^{35}$ yr.

Since the penetration probability p_p approaches the order of unity as E_a approaches zero, the rate of fragmentation can be tremendously increased by supplying the nucleus with some extra energy. Thus the process

$$ {}^{*}_{Z}(\ \ )^{A} \rightarrow \text{fission} \tag{4-27}$$

whereby an excited nucleus undergoes fission can be made to go as rapidly as $\tau \cong \tau_0$, if the excitation energy is sufficiently high. The same remark applies to alpha decay as well. Also competing for the energy are gamma decay and neutron emission. The latter can occur if the excitation energy is greater than the energy with which the neutron is bound to the nucleus, and can be expressed as

$$ {}^{*}_{Z}(\ \ )^{A} \rightarrow {}_{Z}(\ \ )^{A-1} + \text{n} \quad \text{or} \quad {}^{*}_{Z}(\ \ )^{A} \rightarrow {}^{*}_{Z}(\ \ )^{A-1} + \text{n}' \tag{4-28}$$

depending on whether the product nucleus is left in the ground or an excited state. The total decay rate λ is given by the sum of the rates of the individual reactions. Thus

$$\lambda = \lambda_f + \lambda_\alpha + \lambda_n + \lambda_{n'} + \lambda_\gamma \tag{4-29}$$

where the symbols are self-explanatory. The probability that the nucleus decays by process k (any one of f, α, n, n$'$, γ) is

$$p_k = \frac{\lambda_k}{\lambda} \tag{4-30}$$

Bohr and Wheeler [*3*] have estimated these various decay rates for the nucleus U^{236}. Fig. O,4d presents their results as plots of $1/\lambda_k$ as functions of the excitation energy, and indicates how the different decay possibilities gain and lose importance as the excitation energy is varied. Nuclei near U^{236} would be expected to behave in a generally similar fashion. At excitation energies much higher than those shown, other reactions become energetically possible, such as the star reaction whereby the nucleus breaks up into many fragments.

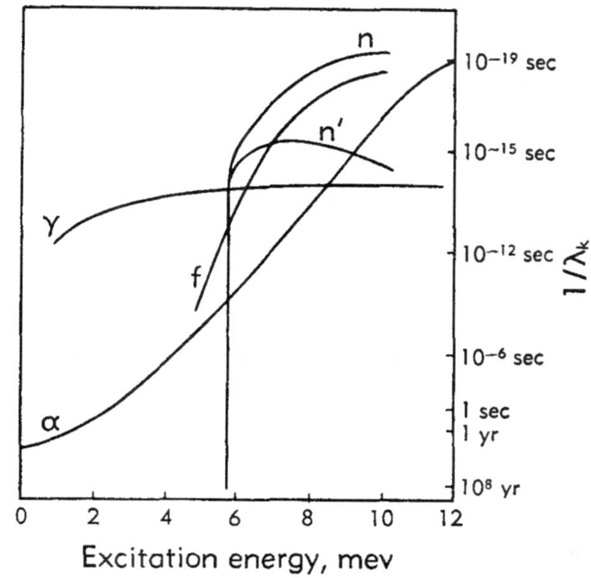

Fig. O,4d. Rates of decay of excited U^{236} nucleus (after [3]).

COLLISION REACTIONS.

The cross section. The main features of collision reactions can be understood in terms of the compound nucleus, a concept introduced by Bohr. Thus the reactions

$$\begin{array}{ll} n + U^{235} \rightarrow \text{fission} & n + U^{235} \rightarrow U^{235} + n \\ n + U^{235} \rightarrow Th^{232} + \alpha & n + U^{235} \rightarrow {}^*U^{235} + n' \\ & n + U^{235} \rightarrow U^{236} + \gamma \end{array} \tag{4-31}$$

are viewed as a succession of two independent reactions. The first is the formation of the compound nucleus

$$n + U^{235} \rightarrow {}^*U^{236}$$

where the excitation energy is the sum of the kinetic energy of the incident neutron and the energy by which the neutron is bound to the nucleus. The formation of the compound nucleus is then followed in short order by one of the f, α, n, n$'$, γ processes previously described, with relative probabilities as given qualitatively in Fig. O,4d. The rate of any of the reactions of Eq. 4-31 is, according to this view, given by the rate of formation of the compound nucleus multiplied by the probability (4-30) that the compound nucleus decays into the reactants in question.

For a two-reactant process the rate is proportional to the concentrations of the reactants, to their relative velocity, and to a quantity having units of area, called the cross section and denoted by σ. If two reactants are mixed with concentrations, C_1 and C_2, measured in nuclei per cm^3, and if their velocity distributions are such that $F(v)dv$ is the probability that their relative velocity lies between v and $v + dv$, then the number of successful reactions $\mathcal{R}$ taking place per cm^3 per sec is given by

$$\mathcal{R} = C_1C_2 \int_0^\infty v\sigma(v)F(v)dv \tag{4-32}$$

where $\sigma(v)$ is the cross section for the reaction in question and is generally a function of the relative velocity. When several reactions can occur with the same reactants, the cross section for each can be written as

$$\sigma_k = p_k\sigma_c \tag{4-33}$$

where σ_c is the cross section for formation of the compound nucleus and p_k is the probability for reaction k. The cross section σ_c is then the sum of the σ's of the individual reactions and is called the total cross section. The rate of energy production P/V per cm^3 per sec is then

$$P/V = E\mathcal{R} \tag{4-34}$$

where E is the energy release per reaction. When Eq. 4-32 is applied to a mixture of reactants in thermal equilibrium at some high temperature T (thermonuclear reaction), $F(v)$ is the relative velocity distribution corresponding to Maxwellian velocities of the individual reactants. A simple interpretation of the cross section as a target area can be made by applying Eq. 4-32 to a case for which the relative velocity has a definite value. Such a situation exists in a beam experiment whereby reactant 2 is bombarded by a beam composed of reactant 1, the beam having a velocity v. The reaction rate is

$$\mathcal{R} = C_1C_2v\sigma(v) = C_1v\Sigma \tag{4-35}$$

which can be read as the current density of the beam C_1v multiplied by

$$\Sigma = C_2\sigma(v) \tag{4-36}$$

The quantity Σ can thus be interpreted as the target area of a cubic centimeter of reactant 2, and σ can therefore be considered as the target area per nucleus. The reciprocal of Σ,

$$l = \frac{1}{\Sigma} = \frac{1}{C_2\sigma(v)} \tag{4-37}$$

has units of length and is equal to the average distance that a reactant 1 nucleus must travel (at velocity v) through reactant 2 before making a successful reaction. The reaction cross section σ is sometimes of the order of magnitude of the actual nuclear cross section which according to Eq. 3-4 is about

$$\pi R^2 \sim 10^{-24} \text{ cm}^2 = 1 \text{ barn} \tag{4-38}$$

Since atomic concentrations in solids and liquids are of the order of magnitude of 10^{23} per cm^3, the mean path corresponding to $\sigma \cong 1$ barn is $l \cong 10$ cm.

It was previously mentioned that the processes of three or more reactants are too slow to be of consequence. A rough but simple argument suffices to prove this. For a three-reactant process to take place, reactants 1 and 2 must meet successfully within the effective volume τ_{eff} of a reactant 3 nucleus. The rate of this reaction is therefore of the order of a two-reactant rate multiplied by the effective volume of the third reactant per cm^3 of mixture $C_3\tau_{\text{eff}}$. Since C_3 is at most $\sim 10^{23}$ per cm^3 and τ_{eff} is generally not much larger than the actual nuclear volume, $\sim 10^{-36}$ cm^3, the multiplication factor is $\sim 10^{-13}$.

If Eq. 4-32 is applied to the case of two liquid reactants at ordinary concentrations, $C_1 \cong C_2 \cong 10^{23}$, with a cross section of ~ 1 barn, and an average relative velocity of $v \sim 10^5$ cm/sec corresponding to thermal motion at room temperature, the result is $\mathcal{R} = 10^{27}/\text{cm}^3$ sec. Assuming an energy release of $E \cong 1$ mev, the energy production is $P/V \cong 10^{14}$ w/cm^3. These cosmic results imply that either there are no exothermic reactions between natural nuclei, or that the cross sections at room-temperature velocities are extremely far below their naturally expected values, the actual nuclear areas. Since there is a wide variety of known exothermic reactions, the conclusion is that the cross sections are low. The reason for this is clear from Fig. O,4c. Because of the necessity of penetrating the potential barrier, the cross section is lowered below the natural value by a factor equal to the probability of penetration (4-25). As seen from the discussion relating to Eq. 4-26, this factor can be of tremendous importance. The activation energy which is a measure of the penetration probability is

$$E_a = V_{\max} - E_k \tag{4-39}$$

where E_k is the incident kinetic energy of the reactant nuclei[12] and

$$V_{\max} = \frac{Z_1 Z_2 e^2}{R_1 + R_2} \tag{4-40}$$

as in Eq. 4-24, where Z_1, Z_2, R_1, R_2 are the atomic numbers and nuclear radii of the reactants. The height $V_{\max}$ of the potential barrier facing the proton and the alpha particle is plotted in Fig. O,4e as a function of Z_2. The curves are based on values of the nuclear radii as given by Eq. 3-4.

Types of reactions. It is clear from these considerations that the important reactions by virtue of rate are those involving only light elements (low $V_{\max}$), and those involving the neutron as one of the reactants.

[12] The difference between the fragmentation process and the process considered here of fusion into the compound nucleus lies only in the direction in which the barrier is penetrated. In Eq. 4-23, the energy release is precisely equal to the kinetic energy of the separated fragments.

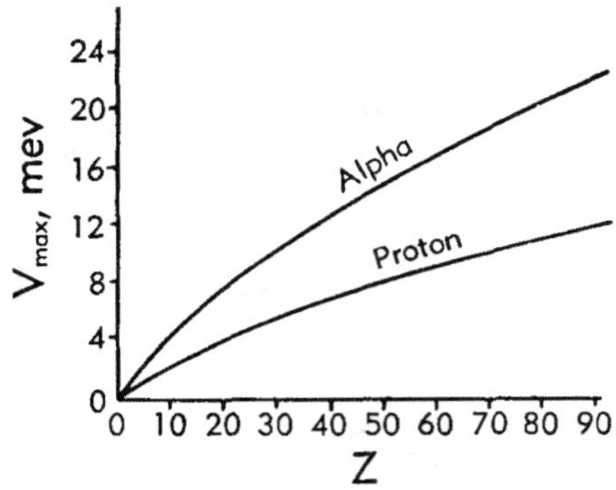

Fig. O,4e. Heights of potential barrier for proton and alpha particle as functions of Z.

Since the neutron is electrically neutral, there is no potential barrier preventing its entry into any nucleus, light or heavy.

Table O,4d presents a list of some exothermic light element reactions along with the values of the energy release E, the height V_{max} of the potential barrier, and the temperature T_{max} at which the average thermal energy is equal to V_{max}. This temperature is given by

Table O,4d. Exothermic light element reactions.

Reaction	E, mev	V_{max}, mev	T_{max}, 10^{10}°K
$H^1 + H^1 \rightarrow H^2 + e^+ + \nu$	0.41	0.48	0.6
$H^1 + H^2 \rightarrow H^3 + \gamma$	6.2	0.42	0.5
$H^1 + B^{10} \rightarrow C^{11} + \gamma$	8.6	1.5	2
$H^1 + C^{12} \rightarrow N^{13} + \gamma$	2.0	1.8	2
$H^1 + Li^6 \rightarrow He^3 + He^4$	3.9	1.0	1
$H^1 + Li^7 \rightarrow 2He^4$	17.3	1.0	1
$H^1 + N^{15} \rightarrow C^{12} + He^4$	4.9	1.9	2
$H^2 + H^2 \rightarrow He^3 + n$	3.3	0.38	0.4
$H^2 + H^2 \rightarrow H^3 + H^1$	4.0	0.38	0.4
$H^2 + H^3 \rightarrow He^4 + n$	17 6	0 36	0.4
$H^2 + Li^6 \rightarrow 2He^4$	22.1	0.93	1
$H^2 + C^{12} \rightarrow C^{13} + H^1$	2.7	1.6	2
$H^3 + H^3 \rightarrow He^4 + 2n$	11.4	0.33	0.4
$He^4 + Li^7 \rightarrow B^{11} + \gamma$	8.5	1.6	2
$He^4 + Be^9 \rightarrow C^{13} + \gamma$	10.5	2.1	2
$He^4 + B^{10} \rightarrow C^{13} + H^1$	4.0	2.6	3

$$kT_{max} \cong V_{max} \tag{4-41}$$

where k is the Boltzmann constant. The reactions of the form

$$X + Y \rightarrow Z + \text{non-nuclear entities}$$

in which the reactants simply fuse into a larger nucleus are called pure fusion reactions, and are expected to be exothermic for light nuclei. (The net effect is a motion to the right in the direction of the maximum of the curve in Fig. O,3.) The other listed reactions may be described as "essentially" fusion reactions as in

$$H^2 + C^{12} \rightarrow C^{13} + H^1$$

which is equivalent to the absorption of a neutron, or as periodicity reactions as in

$$H^1 + Li^7 \rightarrow 2He^4$$

which takes specific advantage of the periodicity of four.

Table O,4e presents the important types of neutron reactions along with some illustrative reactions and some brief remarks. Using the idea

Table O,4e. Neutron-induced reactions.

Type	Case	E, mev	Remarks
n, γ	$n + {}_ZX^A \rightarrow {}_ZX^{A+1} + \gamma$	~6–8	X ~ any nucleus
n, n	$n + {}_ZX^A \rightarrow {}_ZX^A + n$	0	X ~ any nucleus
n, n′	$n + {}_ZX^A \rightarrow {}^*_ZX^A + n'$	$-U$	U = excitation energy X ~ heavier nucleus
n, 2n	$n + Be^9 \rightarrow Be^8 + 2n$ $n + C^{12} \rightarrow C^{11} + 2n$	−1.7 −17.5	Always endothermic
n, p	$n + N^{14} \rightarrow C^{14} + H^1$	0.56	Can be exothermic or endothermic
n, α	$n + Li^6 \rightarrow H^3 + He^4$	4.7	Can be exothermic or endothermic
n, f	$n + U^{233} \rightarrow$ fission $n + U^{235} \rightarrow$ fission $n + Pu^{239} \rightarrow$ fission	200	Reaction proceeds with slow neutrons
	$n + U^{238} \rightarrow$ fission	200	Requires neutrons of kinetic energy ~ 1 mev.

of the compound nucleus, the first step in any neutron reaction can be written as

$$n + {}_ZX^A \rightarrow {}^*_ZX^{A+1} \tag{4-42}$$

The excitation energy U_c of the compound nucleus is equal to

$$U_c = B + E_k \tag{4-43}$$

where E_k is the incident energy of the reactants and B is the energy with which the neutron is bound to the compound nucleus. The energy B is not identical with the binding energy per particle given in Fig. O,3, since the E_b/A is averaged over all the neutrons and protons.[13] Thus in the flat region of the curve, $A \sim 60$, it can be said that $B = E_b/A$. For the lighter elements the average binding increases as particles are added to the nucleus, and it would be expected that B would be somewhat larger than E_b/A except for the very light elements for which the shell structure is important. For the heavier elements the situation is reversed, and the binding energy B of the added neutron is less than E_b/A. At the heavy end of the elements, $A \cong 238$, B has dropped to about 6 mev. The next step in the reaction is the decay of the compound nucleus resulting in one of the listed reaction types with probabilities depending on the particular nucleus and on the value of the excitation energy U_c.

For low incident energies, $E_k \ll 1$ mev, the prominent reactions for most elements are the n, γ and the n, n, the latter being dominant in the case of light nuclei. In the elastic scattering process, the product kinetic energy is equal to the reactant kinetic energy and the nucleus X is left in the same state. The only effect of this process is thus a transfer of kinetic energy from the incident neutron to the reactant nucleus. It is by this process that neutrons come into thermal equilibrium with the reactant matter. In the neutron-capture reaction, the excited compound nucleus decays to its ground state by gamma emission. This decay may occur in one step, by a direct transition to the ground state, or in several steps (a cascade). The total energy of the emitted radiations is simply U_c which equals B for low values of E_k. The n, γ process can be a source of potentially useful fuel materials. Since the product nucleus tends to have an excess of neutrons it is often radioactive, decaying by β^-. Another possibility is that during the cascade process, the product nucleus may be trapped in a long-lived state.

Also for low incident energy occur the less prominent n, p and n, α processes in the case of the lighter elements. The restriction of the n, p and n, α to the lighter elements is due to the potential barrier hindering the emission of charged particles from the compound nucleus. These reactions occur only when the energy release E is not too far below V_{max}. The n, p reaction usually results in a β^- active product nucleus.

The fission reaction can take place if the excitation energy U_c is close to or greater than the activation energy. This occurs only in the case of the heaviest elements, for which the fission activation energy has dropped to ~6 mev. Fission with slow neutrons, low E_k, is therefore possible if $B \geqq 6$ mev, which holds for the nuclei U^{233}, U^{235}, ${}_{94}Pu^{239}$ but not for

[13] The stability line represents the balance relationship between the Z protons and the A-Z neutrons, the Coulomb repulsion of the protons being balanced by the apparent repulsion due to the unpaired neutrons. It would therefore be expected for the natural nuclei that the average binding energy per proton be the same as that per neutron, both then being equal to the grand average of E_b/A.

U^{238}. For the reaction to succeed with U^{238}, the incident kinetic energy[14] must be equal to or greater than 6 mev $- B$. This process has been called fast fission. The specific importance of the fission process lies in the emission of $\sim$2.5 neutrons, on the average, from the excited fragments, more than enough to allow a chain reaction. The fission fragments, having an excess of neutrons and being formed in highly excited states, undergo the process of neutron emission (4-28) thus accounting for two neutrons, one from each fragment. In 50 per cent of the cases the product nucleus of such a neutron emission is still sufficiently excited to emit yet another neutron. In 0.75 per cent of the cases a neutron is emitted following a β^- decay of one of the fragments on its way up to the stability line. Such neutrons are termed *delayed.*

As the incident energy is increased up to the order of 1 mev and higher, the n, n′ and n, 2n reactions become possible along with the above-mentioned fast fission. In the case of inelastic scattering, the emitted neutron has less kinetic energy than the incident neutron, and the nucleus is left in an excited state. This process competes successfully with the n, n process only for the heavier nuclei and is always endothermic by an amount equal to the excitation energy of the final state. Inelastic scattering is an important method of slowing fast neutrons with kinetic energy of several mev down to below $\sim$1 mev. If the incident kinetic energy is sufficiently large, the excitation of the product nucleus may be high enough to emit yet another neutron, and the result is an n, 2n process. This reaction is thus more endothermic than the n, n′. The n, 2n reaction occurs also for some of the lighter nuclei, but as a single process rather than the described two-step process. Here again the reaction must be endothermic. The n, 2n process does not lead to a chain because it is not the dominant reaction, and because it is endothermic. The product neutrons do not have sufficient energy to maintain the chain.

Behavior of cross sections. As already mentioned, the cross section σ for any reaction is a function of the relative velocity of the reactants or, in other words, a function of the incident kinetic energy. The exact form of this function depends on the type of reaction, the particular nuclei involved, the location of the excited energy levels of the compound nucleus, and so on. The general rules which follow, pictured in Fig. O,4f, are theoretically reasonable and are borne out by experimental observation. Detailed effects such as resonances are not considered, although they can be of extreme importance in some particular reactions.

1. For sufficiently large incident energy E_k (from one to several mev depending on the process) the total cross section approaches the order of $\sigma_c \cong \pi R^2 \cong 1$ barn.

[14] This difference in the behavior of B is due to the odd-even effects. Since the product nucleus in Eq. 4-42 has even A and even Z in the case of U^{233}, U^{235}, Pu^{239}, the binding energy (3-5) is expected to be larger than it is for the U^{238} case.

2. For lower E_k the cross section can rise above this value in the case of an exothermic neutron reaction. The reason for this rise is that the neutron can hit the reactant nucleus even when not aimed directly at it, because of the attractive nuclear force. This effect grows larger as the incident energy approaches zero. The n, n reaction is the exception to this rule since the same force which attracts the neutron also holds it to the nucleus.

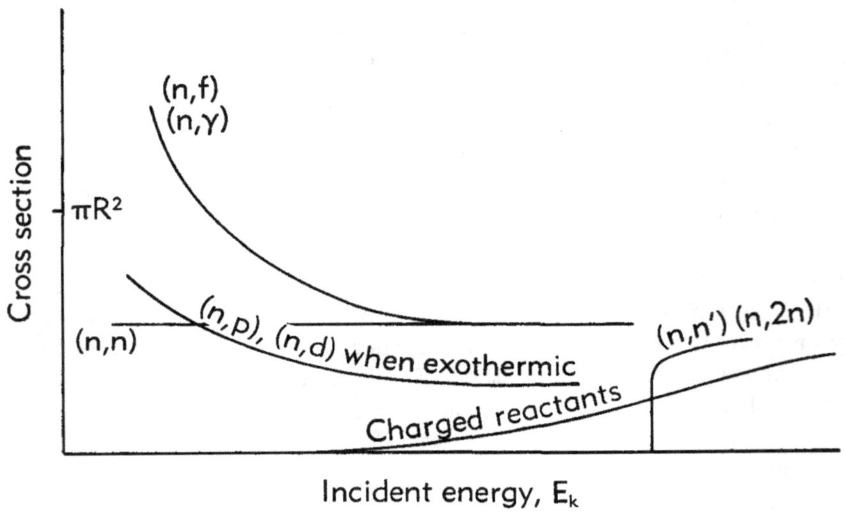

Fig. O,4f. Behavior of cross sections.

3. For the charged-particle reactions of Table O,4d, the short range nuclear attraction is overshadowed by the electrostatic potential barrier, and the cross section is reduced by p_p, the probability that the incident reactants penetrate the barrier and form the compound nucleus. According to Eq. 4-25 this probability approaches the order of unity as the incident energy approaches V_{max}. Only for such energies is the cross section of a charged particle reaction of the order of πR^2.
4. An effect encountered in both neutron and charged-particle reactions is the potential barrier hindering the exit of the product nuclei from the compound nucleus, in case both products are electrically charged. This reduces the cross section by a factor of $p_{p'}$, the probability of penetrating the exit barrier. For the slow fissionable elements this effect is not critical because the binding energy of the neutron is sufficient to overcome the potential barrier. The probability $p_{p'}$ is negligible, however, for the case of slow neutrons on U^{238}, thus preventing slow fission. This effect also cuts down the n, p and n, α cross sections, allowing these reactions to occur

prominently only for some select light nuclei. For the light-element reactions of Table O,4d, the energy release is larger than V_{max}, thus causing $p_{p'}$ to be of the order of unity. The cross sections for these reactions are thus regulated by p_p.

5. For endothermic reactions the cross section is zero until the incident energy is sufficiently large ($E_k = -E$) to make the reaction energetically possible. As E_k rises above this value, the cross section rises and proceeds according to the above rules.

The important conclusions are that the cross sections for the neutron reactions are of the order of barns and higher, especially for low incident energies, while the cross sections for the light-element reactions are of the order of barns only for high incident energies, and drop to extremely low values for incident energies corresponding to ordinary temperatures. As computed on page 637, the reaction rate of a process of cross section $\cong 1$ barn involving reactants at ordinary solid densities, $C \cong 10^{23}$ per cm^3, and at ordinary temperatures, has the tremendous value of $\mathcal{R} \cong 10^{27}$ reactions per cm^3 per sec. To obtain a neutron reaction rate of 10^{13} per cm^3 per sec, leading to a power density of $P/V \cong 1$ w/cm^3 for a 1-mev energy release, requires a neutron current density of $C_n v \cong 10^{14}$ neutrons per cm^2 per sec. The requirement in the case of a light-element reaction is an incident energy sufficiently high to result in an effective cross section of $\sigma \cong 10^{-14}$ barns.[15] The chances of meeting these requirements by means of chain reactions, thermonuclear reactions, and accelerating machines are discussed in the next article.

Other reactions. Reactions not yet discussed are those involving gamma rays as reactants, such as

$$\gamma + H^2 \rightarrow H^1 + n \tag{4-44}$$
$$\gamma + U^{238} \rightarrow \text{fission}$$

which are called photodisintegration and photofission respectively; very high energy reactions such as meson production and the star reaction (Eq. 4-27) which occur in the cosmic rays and can be produced by the high energy accelerating machines; and the non-nuclear interactions between nuclei and the orbital electrons of the medium through which the nuclei move.

The gamma ray reactions have no significance as energy sources. They have comparatively small cross sections, $\sigma \cong 10^{-3}$ barns, and usually require energetic photons, $\sim$2 mev, for the above endothermic photodisintegration (the binding energy of H^2), and $\sim$6 mev for the photofission (the required activation energy). To obtain useful energies from such photoreactions would require unattainable radiation densities.

[15] At such energies the incident velocity (4-35) increases by a couple of factors of 10 and thus it is not necessary to attain a cross section of $\sim 10^{-14}$ barns.

The very high energy reactions, although of basic importance and spectacular as well, also have no apparent significance as energy sources. The limited information on the subject holds no clue to the possibility of a chain reaction. The natural course of events whereby energetic particles are slowed down by inelastic processes would appear probable to break a possible chain almost as soon as it started.

The non-nuclear interactions are of primary importance to this discussion because they compete directly with the energy source reactions, and because they determine the form of the final energy output of such reactions. A neutron passing through matter is not impeded by the atomic electrons[16] and moves in a straight line at constant velocity until it encounters a nucleus. Protons, alpha particles, or any charged nuclei, however, are strongly influenced by the atomic electrons of the medium through which they move, losing energy at an extremely rapid rate as they ionize the atoms along their path. The distance they traverse, called the range X, before coming into thermal equilibrium with the medium is given roughly by the following formula

$$X(E_i) = \frac{3.2 \times 10^{-3}}{\rho} \frac{E_i^{\frac{3}{2}}}{Z^2 A^{\frac{1}{2}}} \text{ cm} \tag{4-45}$$

where E_i is the kinetic energy of the moving particle in units of mev, Z is its proton number, A its mass number, and ρ is the density in g/cm^3 of the medium material. A proton of initial energy 1 mev, for example, has a range of 10^{-3} cm in a medium of density 3.2 g/cm^3. In the same medium a 1-mev alpha particle has a range of $\sim 10^{-4}$ cm. The time taken for these slowing-down processes is extremely short, of the order of 10^{-12} sec.

Since a successful nuclear reaction involving charged particles requires an incident energy sufficiently high to overcome the potential barrier, it can be concluded that the slowing-down process reduces the possibility of such a reaction by a large factor. If $l(E_i)$ is the mean path (Eq. 4-37) for the reaction with incident energy E_i, the probability s that a successful reaction take place before the incident reactant is thermalized must satisfy the inequality

$$s < \frac{X(E_i)}{l(E_i)} \tag{4-46}$$

since the cross section decreases as the incident reactant slows down. Applying Eq. 4-46 to the case of a 5-mev alpha particle results in $s < 10^{-3}/10 = 10^{-4}$, using the value $l \cong 10$ cm corresponding to a cross section of ~ 1 barn. Thus, less than one in 10^4 of such incident alpha particles results in a nuclear reaction while slowing down. The remainder

[16] There is some slight interaction due to the fact that both the neutron and electron have a magnetic moment.

become thermalized and may then lead to a nuclear reaction if the temperature is sufficiently high. A secondary conclusion is that energy released in a nuclear reaction in the form of kinetic energy of nuclei is rapidly converted into heat at the position of release. If the reaction occurs in a gaseous medium, the conversion into heat takes place along the greater range since the density is smaller. The conversion into heat is a complicated process involving further ionizations by the knocked-on electrons, recombinations, reionization, and so on.

Energy release in the form of energetic electrons is also converted into heat fairly rapidly, but over longer distances. A very rough formula for the range of electrons is

$$X(E_i) = \frac{E_i}{2\rho}\ \text{cm} \tag{4-47}$$

where E_i is the kinetic energy in mev and ρ is the density in g/cm^3. A 1-mev electron has a range of about 0.2 cm in aluminum. Electrons, therefore, do not present any shielding difficulties.

Serious shielding difficulties occur for energetic neutrons and for energetic gamma rays since these can penetrate many centimeters of solid matter. The only solution of the shielding problem is sufficient mass of shield although special materials can be of help. In shielding against neutrons, for example, a good inelastic scatterer can be used to great advantage by slowing the faster neutrons down to more manageable speeds.

O,5. Summary and Conclusions. The limited extent of the naturally existing nuclei, $Z \leqq 92$, and their confinement along the stability line (Fig. O,2a) are understood as direct consequences of energetic considerations. Experimental and theoretical observations show that the stability of the nuclei as measured by the binding energy per nucleon, Fig. O,3, increases to a maximum as the mass number increases to $A \cong 60$, and then decreases for heavier nuclei. It follows (Fig. O,4b) that all nuclei with $Z > \cong 40$ are in principle unstable against splitting into smaller and lighter nuclei, and that the very light nuclei are in principle unstable against fusing together (Table O,4d) to form heavier nuclei. The explanation of their general apparent stability is the unobservably slow rate of transformation due to the activation energy required to overcome the electrostatic potential barrier (Fig. O,4c and Eq. 4-25). For the heaviest natural nuclei, which are the least stable, the rate of fragmentation is easily observable, having half lives comparable to the age of the universe, $\sim 10^9$ yr (Table O,4c). For nuclei with $Z > 92$, the half life against fragmentation is small compared to 10^9 yr, thus explaining their nonexistence. All natural nuclei therefore, except those in the neighborhood of $A \cong 60$, can be described as being in a state of metastable equilibrium and represent a tremendous reservoir of energy. The energies released in these splittings and fusions are larger than chemical energies by factors of $\sim 10^6$ to 10^8 and are generally converted into heat very rapidly.

The possibilities of avoiding or overcoming the potential barriers, and thus tapping the energy reservoir, are discussed below. Also discussed are the possibilities of production of natural nuclei in excited states, and of artificial nuclei since these may be useful as fuel materials if their half lives are comparatively long. The excited state nuclei release energy in the form of gamma rays. The artificial nuclei are off the stability line and transform into the natural nucleus of the same mass number by beta decay.

Light-element power. Thermonuclear reactions. The fusion reactions having the greatest chance of realization are those among the lightest elements because it is for these that the potential barrier is lowest. The problem of obtaining energy from these reactions is not one of density since the reactants are natural nuclei, but is a straightforward problem of cross section (4-32). As estimated on page 643, it is necessary to have an effective cross section of only $\sigma_{\text{eff}} \cong 10^{-14}$ barns to have a respectable power density of ~ 1 w/cm^3 in a mixture of reactants at ordinary solid or liquid densities.

Such low values of σ_{eff} are easily attainable by use of accelerating machines since they correspond to incident energies of the order of ~ 0.1 mev. In fact by accelerating the incident nuclei up to the order of $V_{\max}$ (Table O,4d and Fig. O,4e), the cross section may approach the order of barns. There are two drawbacks, however, to the use of accelerators as energy tappers. The first, and the killing drawback, is the fact of Eq. 4-46. Although accelerators can be built to have a beam energy efficiency of ~ 30 per cent and higher, only one in thousands of the accelerated particles yields an energy profit. The remainder simply lose their energy by heating up the target and become thermalized. The second drawback concerns the density limitations. The beam currents obtained in present accelerators are not higher than the order of milliamperes which corresponds to beam energies of the order of kilowatts and energy profits of the same order, assuming that every particle results in the desired nuclear reaction.

The remaining possibility is to retain the ordinary reactant densities, but at a sufficiently high temperature to result in an effective cross section of $\sim 10^{-14}$ barns or higher. Since the required temperatures, as will be seen shortly, may be described as unearthly, the problems involved here concern the kindling of the reaction, and the control and containment of the resulting high temperatures. As an illustration the reaction

$$\mathrm{H}^2 + \mathrm{H}^2 \rightarrow \mathrm{He}^3 + \mathrm{n} + 3.3 \text{ mev} \tag{5-1}$$

will be considered. This is one of the fastest of the light-element reactions since its barrier height is one of the lowest (Table O,4d) and the resulting required temperatures are lower than those of the other reactions. The results of computations[17] on this reaction are presented in Table O,5. The first column gives the temperature of the mass of deuterium. The

Table O,5. Thermonuclear fusion of deuterium.

T, °K	σ_{eff}, barns	$\mathcal{R}$, per cm^3 per sec	P/V, w per cm^3
10^5	2×10^{-36}	6×10^{-8}	3×10^{-20}
5×10^5	10^{-23}	6×10^5	3×10^{-7}
10^6	2×10^{-14}	2×10^{15}	10^3
3×10^6	10^{-9}	2×10^{20}	10^8
10^7	6×10^{-5}	2×10^{25}	10^{13}

second column gives the effective cross section σ_{eff} defined according to Eq. 4-32 as

$$\mathcal{R} = C_1 C_2 \bar{v} \sigma_{eff} \tag{5-2}$$

where

$$\bar{v} \cong \left(\frac{kT}{m}\right)^{\frac{1}{2}} \tag{5-3}$$

is the average thermal velocity of the deuterium nuclei of mass m at T°K. Combining these equations and inserting the values $C_1 = C_2 = 10^{23}/\text{cm}^3$ corresponding to a deuterium density of $\rho \cong 1$ g/cm^3 results in

$$\mathcal{R} = 10^{49} T^{\frac{1}{2}} \sigma_{eff} \tag{5-4}$$

The third column gives the values of $\mathcal{R}$ in reactions per cm^3 per sec, and the fourth column gives the resulting power density, $P/V = E\mathcal{R}$, where $E = 3.3$ mev is the energy release.

It is seen therefore that temperatures close to 10^6°K are required in order to obtain useful power. The problems of control and containment of such very high temperatures have taxed the powers of man. The stars, however, with their huge mass and gravity thrive on even higher temperatures and do, in fact, feed on thermonuclear reactions. At the conditions $T \cong 10^7$°K and $\rho \cong 10^2$ g/cm^3 which obtain inside the core of stars like the sun, the reaction of Eq. 5-1 would proceed at a rate so rapid that the deuterium would be burned up within a matter of minutes. The deuterium reaction therefore cannot be the source of the sun's energy, although it may have been in the early history of the sun. The reaction that is thought to feed stars hotter than the sun was discovered by Weizsacker and Bethe and is called the CN cycle. It runs as follows:

[17] These computations were carried out by R. A. Alpher and are reported in [*1*, p. 267].

$$\begin{aligned} H^1 + C^{12} &\rightarrow N^{13} + \gamma \\ N^{13} &\rightarrow C^{13} + e^+ + \nu \\ H^1 + C^{13} &\rightarrow N^{14} + \gamma \\ H^1 + N^{14} &\rightarrow O^{15} + \gamma \\ O^{15} &\rightarrow N^{15} + e^+ + \nu \\ H^1 + N^{15} &\rightarrow C^{12} + He^4 \end{aligned} \tag{5-5}$$

The net effect of this chain of reactions is the conversion of hydrogen into helium and can be written as

$$4H_1 \rightarrow He^4 + 3\gamma + 2e^+ + 2\nu + 27 \text{ mev} \tag{5-6}$$

Of the total energy release, ~2 mev or 8 per cent is carried by the neutrinos. For the sun and similar or cooler stars, the energy source is thought to be a noncatalytic fusion of four nuclei into a helium nucleus. The initial process of this direct fusion is the reaction

$$H^1 + H^1 \rightarrow H^2 + e^+ + \nu \tag{5-7}$$

listed in Table 0,4d.

Neutron power. The fission chain. The failure of charged nuclei reactions even in the most favorable cases of light nuclei leaves the field open only to the neutron-induced reactions (Table O,4e). Since there is no potential barrier for the neutron, the cross sections for the reactions can be of the order of barns and higher. The problem is, therefore, to maintain a high density of neutrons which can induce the desired reaction. The number of reactions occurring per cm³ per sec is given by

$$\mathcal{R} = C_n C \bar{v} \bar{\sigma} \tag{5-8}$$

where C_n is the number of neutrons per cm³, C is the number of reactant nuclei per cm³, $\bar{v}$ is the average speed of the neutrons, and $\bar{\sigma}$ is the result of averaging the cross section over the neutron velocity spectrum. Inserting the values $C \cong 10^{23}$, $\sigma \cong 10^{-24}$ cm² and assuming a ~1-mev energy release, the power density (Eq. 4-34) is

$$\frac{P}{V} = 1.6 \times 10^{-14} C_n \bar{v} \text{ w/cm}^3 \tag{5-9}$$

and $\mathcal{R}$ is

$$R = 0.1 C_n \bar{v} \tag{5-10}$$

where the product $C_n\bar{v}$ is the neutron current density, or flux, and is measured in neutrons per cm² per sec. A power density of ~1 w/cm³ thus requires a neutron flux of ~10^{14}. For ordinary thermal velocities of $\bar{v} \cong 10^5$ cm/sec this flux corresponds to a neutron density of $C_n \sim 10^9$/cm³. To maintain this neutron density it is necessary to produce the neutrons at the same rate $\mathcal{R}$ at which they disappear, namely ~10^{13} per cm³ per sec. The decay of neutrons into protons by beta decay has no significant effect on the neutron density, because the half life $\tau_{\frac{1}{2}} \sim \frac{1}{2}$ hr is much longer than the time interval between the birth of the neutron and its death via the nuclear reaction. This time interval should be of the order of the mean path (Eq. 4-37) of the neutron, $l \cong 10$ cm, divided by the velocity, $\sim 10^5$ cm/sec, and is then $\sim 10^{-4}$ sec. Thus only a negligible fraction, 10^{-4} sec/$\frac{1}{2}$ hr, of the neutrons decay.

As indicated, a reaction must result in more than one product neutron in order to be successful as an energy source. The exothermic n, γ reaction is excluded, since the resulting gamma ray has only a small chance of producing even one neutron, (see page 643). The n, n reaction is zerothermic and serves only to slow down the neutrons. If the substance in which the neutron moves has a low average atomic weight and absorbs neutrons comparatively slowly, the n, n reaction causes the neutrons to become thermalized. The inelastic n, n′ reaction is endothermic and neither multiplies neutrons nor supplies energy. The n, 2n reaction, although multiplying the neutrons, is endothermic and cannot then support a chain. The product nuclei do not have sufficient energy to continue the reaction. The only remaining reaction is fission, which as an accomplished fact leads to a successful chain reaction.

Of all the nuclei with $Z > \sim 40$ which are fissionable in principle, only the heaviest have sufficiently low activation energies to allow fission to be induced by neutrons of reasonable energies. Of these, the nuclei U^{233}, U^{235}, and Pu^{239} are fissionable by slow neutrons, while the fission of U^{238} and other nuclei in that neighborhood requires neutrons with energy of ~ 1 mev. The opening wedge into the energy reservoir of the heavy nuclei is the fission of U^{235}, the only naturally-occurring slow-neutron fissionable element. The existence of sufficient amounts of U^{235} is the first important accident concerning fission. If the alpha decay half life of U^{235} were smaller by a factor of only 3 than the actual value of $\sim 9 \times 10^8$ yr, the present concentration of U^{235} as compared with U^{238} would be the futile amount of 1 part in 10^6 rather than the actual 1 part in 140. The second important accident concerning fission is the emission by the excited fragments of ~ 2.5 neutrons per fission, more than enough to maintain the chain. Even with the multiplication of 2.5, however, the production of a successful chain is not a direct consequence. Natural uranium does not support a chain because of excessive neutron capture in U^{238}. Only by purifying the U^{235} by separation of the isotopes or by mixing the natural uranium with a proper low atomic weight moderating material can a chain be realized. In the first method the unwanted neutron capture is reduced and, in the second method, advantage is taken of the variation of the cross sections as functions of the neutron energy.

There are two compensating factors which more than make up for the low concentration of U^{235}. The first is a combination of a large fission cross section, ~ 500 barns for slow neutrons, and a large energy release of ~200 mev per fission[18] corresponding to the equivalence of 3×10^{10} fissions per sec = 1 watt. Eq. 5-9 and 5-10, derived on the basis of $C \cong 10^{23}$, $\sigma \cong 1$ barn, and $E \cong 1$ mev, are still correct for the actual values of σ and E for concentrations $C \cong 10^{18}$ per $cm^3 = (1/200)(1/500) \times 10^{23}$. Such a concentration of U^{235} atoms corresponds to a dilution of natural uranium by a factor of $\sim 10^3$. The second, and in the long view, more important compensation is that the chain reaction of U^{235} allows the tapping of the energy stored up in the more abundant U^{238} and the still more abundant Th^{232}. If these elements are placed in a U^{235} reactor, the following reactions occur

$$\begin{aligned} n + {}_{90}Th^{232} &\rightarrow {}_{90}Th^{233} + \gamma & \quad n + {}_{92}U^{238} &\rightarrow {}_{92}U^{239} + \gamma \\ {}_{90}Th^{233} &\rightarrow {}_{91}Pa^{233} + e^- + \nu & {}_{92}U^{239} &\rightarrow {}_{93}Np^{239} + e^- + \nu \\ {}_{91}Pa^{233} &\rightarrow {}_{92}U^{233} + e^- + \nu & {}_{93}Np^{239} &\rightarrow {}_{94}Pn^{239} + e^- + \nu \end{aligned} \tag{5-11}$$

thus converting U^{238} and Th^{232} into the easily fissionable Pu^{239} and U^{233}. Although these produced nuclei decay by alpha emission, their half lives (Table O,4c) are many civilizations long.

The major characteristic of the operation of these chain reactions is that the neutron flux and consequent power density can be maintained at any desired level. Power-density limitations are not inherent to the operation, but are due rather to difficulties of strength of materials, heat removal and related problems, and to difficulties of control. The question of control introduces the third accident concerning fission, the emission by the fission products of the delayed neutrons. If it were not for the delayed neutrons, the problems of start-up and maintenance of power level would be too sensitive to be feasible.

Artificial nuclei and power. The use of artificial and excited nuclei as energy sources has two disadvantages. The first arises from the necessity of their production. Since accelerating machines are grossly inefficient in terms of successful nuclear reactions, these artificial nuclei can be produced in quantity only by means of neutron reactions inside a chain reactor. The production of one such nucleus must therefore cost at least one neutron—a neutron which could be used to more advantage by causing a fission and thus yielding a greater energy profit, or by producing a more valuable, easily fissionable nucleus. This disadvantage applies to all artificial nuclei except those which are produced in the process of fission. The fission products are therefore the most economical of artificial nuclei, since they are the least wasteful and can be produced in the largest amounts, two product nuclei being formed for each fission. Also their

[18] Of this ~200 mev, ~160 is carried as kinetic energy by the fragments (Fig. O,4b) several mev by the n's and γ's emitted by the excited fragments immediately following the splitting, and some 20-odd mev by β's and γ's emitted as the fragments rise to stability.

extraction from the fissionable material is possible. Other artificial nuclei can also be produced in quantity, but in only fractional amounts by comparison.

The second disadvantage is the uncontrollability of such energy sources. According to Eq. 4-11, the rate of energy production P is determined completely by the mean life τ, the energy release E, and the amount of material. Some degree of control is possible by mixing various amounts of sources having different values of τ and E. In this way the dependence of P as a function of time can be regulated to suit the desired performance. For such mixtures, Eq. 4-11 should be replaced by the corresponding sum of terms. The fission products are a case in point, since fission occurs in a variety of different splittings, and since each fragment emits several betas and gammas[19] during its rise to stability. The power, P_{fp}, of such a fuel mixture is proportional to the power P_r of the reactor in which the products were formed, and is given by Way and Wigner [*4*] as[20]

$$P_{fp} = 0.07P_r[t^{-0.2} - (t + T)^{-0.2}] \tag{5-12}$$

where the times t and T are in seconds. The meaning of Eq. 5-12 is best illustrated by an example. A reactor of power $P_r = 1$ w/cm^3 is started up with fresh uranium and runs at power for $T = 10$ days. At this time the uranium from, say, 1 cm^3 of the reactor is removed and the fission products separated out. Then at a time, say, $t = 1$ day after the removal of the uranium, the power from the fission products is

$$P_{fp} = 0.07(1 \text{ w/cm}^3)[(\sim 10^5)^{-0.2} - (\sim 10^6)^{-0.2}]$$
$$= 3 \times 10^{-3} \text{ watts per cm}^3 \text{ of reactor}$$

The reaction best suited to the production of other artificial nuclei is that of neutron capture n, γ, because it is a prominent reaction and often leads to an e^- emitter or to an excited product. For example, the radioactive nuclei Na^{24}, P^{32}, and $^*Zn^{69}$, of Table O,4b and O,4a, are the result of neutron capture in Na^{23}, P^{31}, and Zn^{68}, respectively. There is a serious problem, however, in connection with the separation of the radioactive products since they have the same chemistry as the reactant mass. Chemical separation methods are nevertheless possible [*2*] depending on the breakage of the chemical bond in the molecule containing the activated atom. No such separation problem is met in the case of artificial nuclei produced by the n, p reaction. This reaction is, however, rarely

[19] The process of the beta decay often results in an excited product nucleus and thus leads to the emission of gammas.

[20] Way and Wigner derive this formula by considering the fission products as a statistical assembly. The half life of each product is taken as depending on the binding energy via an approximate empirical correlation. The binding energy is determined by a semiempirical formula in terms of the charge. The charge distribution used is based on empirical results.

prominent. An exception is the reaction

$$n + N^{14} \rightarrow C^{14} + H^{1} \tag{5-13}$$

which results in the production of perhaps the most important of radioactive nuclei, C^{14}. The activity of C^{14} (Table O,4b) allows a detailed study of basic biological processes by means of tracer techniques.

O,6. Cited References and Bibliography.

Cited references

1. Gamow, G., and Critchfield, C. L. *Theory of Atomic Nucleus and Nuclear Energy-Sources*. Oxford Univ. Press, 1949.
2. Segrè, E., and Helmholz, A. C. *Revs. Mod. Phys. 21*, 271 (1949).
3. Bohr, N., and Wheeler, J. A. *Phys. Rev. 56*, 426 (1939).
4. Way, K., and Wigner, E. P. *Phys. Rev. 73*, 1318 (1948).

Bibliography

Gamow, G. *Atomic Energy in Cosmic and Human Life*. Macmillan, 1946.
Goodman, C. D., et al. *The Science and Engineering of Nuclear Power*, Vol. I and II. Addison-Wesley, 1947, 1948.
Lapp, R. E., and Andrews, H. L. *Nuclear Radiation Physics*. Prentice-Hall, 1948.
Murray, R. L. *Introduction to Nuclear Engineering*. Prentice-Hall, 1954.
Ridenour, L. N. *Modern Physics for the Engineer*. McGraw-Hill, 1954.
Stephens, W. E. *Nuclear Fission and Atomic Energy*. The Science Press, Lancaster, Pa., 1948.
Stephenson, R. *Introduction to Nuclear Engineering*. McGraw-Hill, 1954.
Stranathan, J. D. *The "Particles" of Modern Physics*. Blakiston, 1942.

INDEX

GPSR Authorized Representative: Easy Access System Europe - Mustamäe tee 50, 10621 Tallinn, Estonia, gpsr.requests@easproject.com

www.ingramcontent.com/pod-product-compliance
Lightning Source LLC
LaVergne TN
LVHW052206200926
848481LV00001B/40